APPLIED MECHATRONICS

A. SMAILI

Mechanical Engineering and Mechatronics Department
Hariri Canadian University—Meshref

F. MRAD

Electrical and Computer Engineering Department
American University of Beirut—Beirut, Lebanon

New York   Oxford
OXFORD UNIVERSITY PRESS
2008

Oxford University Press, Inc., publishes works that further Oxford University's
objective of excellence in research, scholarship, and education.

Oxford New York
Auckland   Cape Town   Dar es Salaam   Hong Kong   Karachi
Kuala Lumpur   Madrid   Melbourne   Mexico City   Nairobi
New Delhi   Shanghai   Taipei   Toronto

With offices in
Argentina   Austria   Brazil   Chile   Czech Republic   France   Greece
Guatemala   Hungary   Italy   Japan   Poland   Portugal   Singapore
South Korea   Switzerland   Thailand   Turkey   Ukraine   Vietnam

Published by Oxford University Press, Inc.
198 Madison Avenue, New York, New York 10016
http://www.oup.com

Oxford is a registered trademark of Oxford University Press

Library of Congress Cataloging-in-Publication Data

Smaili, A. (Ahmad)
     Applied mechatronics / A. Smaili, F. Mrad.
       p. cm.
     Includes bibliographical references and index.
     ISBN-13: 978-0-19-530702-3
      1. Mechatronics—Textbooks. I. Mrad, F. (Fouad) II. Title.

TJ163.12.S63 2006
621—dc22   2006049473

Printing number: 9 8 7 6 5 4 3 2 1

Printed in the United States of America
on acid-free paper

To our children and their generation
They bear the responsibility to fashion certainty for an uncertain world

To the Hariri Foundation
For opening the global gates of knowledge for scores of underprivileged

Contents

Preface   xx

Acknowledgments   xxi

1. Mechatronics: An Introduction   1

Objectives   1

1.1   What is Mechatronics?   1

1.2   Essential Skills for Mechatronics   2

1.3   Why is Mechatronics Important?   2

1.4   Components of a Mechatronic System   2

1.5   Brain for Mechatronics   4

Related Reading   5

Questions   5

Problems   5

Project   6

2. Elements and Analysis of Electric Circuits   7

Objectives   7

2.1   Introduction   7

*2.2   Electric Field (EE and ME Basic)   7

*2.3   Current and Voltage (EE and ME Basic)   8

*2.4   Elements of an Electric Circuit (EE and ME Basic)   10

    2.4.1   Circuit Conditions   10

    2.4.2   Electric Circuit Sources   10

    2.4.3   Circuit Load   11

    2.4.4   Circuit Ground   11

2.5   Circuit Analysis (EE and ME Basic)   12

    2.5.1   Circuit Elements in Series and in Parallel   12

Sections summarizing information from courses prerequisite to Mechatronics are marked in the Table of Contents. Readers should focus more on the sections that are new to them.

* = **EE and ME Basic:** from general engineering sciences

† = **EE Basic:** mechanical engineers should spend more time studying these sections

‡ = **ME Basic:** electrical engineers should spend more time with these sections

2.5.2 *Kirchhoff's Laws* 13

2.5.3 *Equivalent Circuit Representation* 14

*2.6 Resistor (EE and ME Basic) 15

2.6.1 *Voltage Divider* 16

2.6.2 *Bridge Circuit* 18

2.6.3 *Small-Signal Resistance* 19

2.6.4 *Resistance-Based Sensors* 19

2.6.5 *Measuring Electrical Resistance* 19

*2.7 Capacitor (EE and ME Basic) 20

2.7.1 *Capacitor Applications* 25

*2.8 Inductor (EE and ME Basic) 25

2.8.1 *Magnetic Effect of an Electric Current* 25

2.8.2 *Electromagnetic Force* 28

2.8.3 *Self-Inductance* 28

2.8.4 *Inductor-Based Devices* 29

*2.9 Alternating Current (EE and ME Basic) 33

2.9.1 *Steady-State and Frequency Response* 34

2.9.2 *Complex Number Representation of Voltage and Current* 34

*2.10 Impedance (EE and ME Basic) 35

2.10.1 *Generalized Voltage Divider* 36

2.10.2 *Circuit Loading* 37

2.10.3 *Impedance Matching* 38

†2.11 Power (EE Basic) 39

2.11.1 *Average Power* 40

2.11.2 *Reactive Power* 42

2.11.3 *Power Factor* 42

†2.12 Signals and Signal Sources (EE Basic) 42

2.12.1 *Signal Sources* 43

†2.13 Time Domain Analysis (EE Basic) 44

2.13.1 *Differentiators* 46

2.13.2 *Integrators* 47

2.14 Passive Filters 47

2.14.1 *Low-Pass Filters (LPFs)* 49

2.14.2 *High-Pass Filters (HPFs)* 51

2.14.3 *Band-Pass and Band-Reject Filters* 53

2.14.4 *Notch and Trap Filters* 55

2.15 Noise and Interference in Circuits 55

2.15.1 *Guarding Against Electromagnetic Interference (EMI)* 55

2.15.2 *Bypass Capacitors* 56

2.16 Grounding 56

2.16.1 *Ground Loops* 56

2.16.2 Grounding Techniques   57

2.16.3 Galvanic Isolation   58

2.17 Summary   59

Related Reading   59

Questions   60

Problems   60

Laboratory Projects   63

3.   **Diode, Transistor, and Thyristor Circuits   64**

Objectives   64

†3.1   Semiconductors (EE Basic)   64

†3.2   Diodes (EE Basic)   65

3.3   Diode Applications   67

3.3.1 Rectification   67

3.3.2 Diode Limiter, or Clipper   70

3.3.3 Diode Clamp   70

3.3.4 Inductive Load and Diode Protection   71

3.3.5 Temperature Sensor   72

3.3.6 Varactor   72

3.4   Zener Diodes   72

3.5   Light-Emitting Diode (LED)   74

3.6   Photodiode   75

†3.7   Transistors (EE Basic)   76

†3.8   Bipolar Junction Transistor (BJT) (EE Basic)   76

3.8.1 Transistor Characteristics   78

3.8.2 Transistor States   79

3.8.3 DC Biasing of the BJTs   81

3.8.4 Basic BJT Circuits   83

3.9   Phototransistor   86

†3.10   Field-Effect Transistor (FET) (EE Basic)   87

3.10.1 JFETs   87

3.10.2 MOSFETs   88

3.11   Main Features of FETs and BJTs   91

3.12   Power Transistors   91

3.12.1 Packages   92

3.12.2 Power Bipolar Transistors   92

3.12.3 Darlingtons   93

3.12.4 Power MOSFETs   94

3.12.5 Insulated Gate Bipolar Transistors (IGBTs)   96

3.13   Thyristors   97

3.13.1 Silicon-Controlled Rectifiers (SCRs)   97

3.13.2 Gate Turn-Off (GTO)   100

3.13.3 TRIAC   100

3.14 Optocouplers   101

3.15 Summary   102

Related Reading   102
Questions   103
Problems   103
Laboratory Projects   105

4. **Operational Amplifier (Op-Amp) Circuits   106**
    Objectives   106
    4.1 Introduction   106
    *4.2 Op-Amp Basic Symbol (EE and ME Basic)   108
    *4.3 Circuit Model (EE and ME Basic)   108
    4.4 Ideal Op-Amp Behavior   110
    4.5 Common Op-Amp ICs   110
    4.6 Basic Op-Amp Circuits   112

    4.6.1 Inverting Amplifier   112

    4.6.2 Noninverting Amplifier   113

    4.6.3 Follower   114

    4.6.4 Differential Amplifier   114

    4.6.5 Instrumentation Amplifier   116

    4.7 Linear Circuit Applications   118

    4.7.1 Summing Amplifier (Adders)   118

    4.7.2 Integrators   119

    4.7.3 Differentiators   120

    4.8 Nonlinear Op-Amp Circuits   121

    4.8.1 Comparators   121

    4.8.2 Schmitt Triggers   122

    4.8.3 Rectifiers   123

    4.8.4 Limiters   124

    4.9 Nonideal Op-Amp Behavior   125

    4.9.1 Feedback with Finite-Gain Amplifiers   125

    4.9.2 Offset Voltage and Bias Currents   126

    4.10 Active Filters   128

    4.10.1 Filter Circuits and Frequency Characteristics   130

    4.10.2 Filter Types   130

    4.11 Power Op-Amps   133

    4.12 Summary   133

    Related Reading   133
    Questions   134
    Problems   134
    Laboratory Projects   137

5. Digital Logic and Logic Families   138
Objectives   138
[†]5.1 Digital Signals (EE Basic)   138
[†]5.2 Combinational and Sequential Logic Circuits (EE Basic)   139
5.3 Clock Signals   140
[†]5.4 Boolean Algebra and Logic Gates (EE Basic)   141
    5.4.1 Basic Functions and Gates   141
    5.4.2 Boolean Laws and Theorems   143
    5.4.3 Karnaugh Maps   146
    5.4.4 Design of Combinational Logic Circuits   147
5.5 Integrated Circuits and Logic Families   148
    5.5.1 Logic Levels   149
    5.5.2 Noise Immunity   149
    5.5.3 Fan-Out   149
    5.5.4 Power Dissipation   150
    5.5.5 Propagation Delay   150
5.6 TTL Logic Family   151
    5.6.1 TTL Designations   151
    5.6.2 TTL Versions   151
    5.6.3 Output Configurations   151
    5.6.4 TTL Characteristics   154
5.7 The CMOS Family   155
5.8 Interfacing CMOS and TTL   156
    5.8.1 Interfacing TTL to CMOS   157
    5.8.2 Interfacing CMOS to TTL   158
5.9 Flip-Flops   158
    5.9.1 Set–Reset (SR) Flip-Flop   158
    5.9.2 Trigger (or T) Flip-Flop   159
    5.9.3 Clocked D Flip-Flop   160
    5.9.4 J-K Flip-Flop   161
5.10 Buffers and Drivers   162
    5.10.1 Bus Drive and Termination   163
5.11 Counters and Registers   164
5.12 Decoders and Encoders   165
5.13 Multiplexers and Demultiplexers   166
5.14 The 555 Timer   167
    5.14.1 Operating Modes   168
5.15 Phase-Locked Loop (PLL)   170
5.16 Glossary of Logic Terms   171
5.17 Summary   173

Related Reading   173
Questions   174
Problems   174
Laboratory Projects   175

6. Microcontrollers and Programming   177
   Objectives   177
   †6.1 Computers and Computer Programs (EE Basic)   177
       6.1.1 Microprocessor or Microcontroller   178
   6.2 Overview of the 9S12C MCUs   180
       6.2.1 Central Processing Unit (CPU12)   182
       6.2.2 System Bus   183
       6.2.3 System Clocks   185
       6.2.4 Operating Modes   185
       6.2.5 Memory Map   187
       6.2.6 Programming Basics   188
       6.2.7 CPU12 Programming Registers   192
       6.2.8 Instruction Queue   197
   6.3 Addressing Modes   198
       6.3.1 Inherent (INH) Mode   199
       6.3.2 Immediate (IMM) Mode   199
       6.3.3 Extended (EXT) Mode   200
       6.3.4 Direct (DIR) Mode   200
       6.3.5 Relative (REL) Mode   200
       6.3.6 Indexed Mode   201
   6.4 Instruction Set of the CPU12   207
       6.4.1 Data-Handling Instructions   207
       6.4.2 Arithmetic Instructions   213
       6.4.3 Special Math Instructions   218
       6.4.4 Logic Instructions   220
       6.4.5 Data-Compare and -Testing Instructions   220
       6.4.6 Condition Code Register Instructions   221
       6.4.7 Program-Control Instructions   222
       6.4.8 Miscellaneous Instructions   230
   6.5 Assembler Directives   230
       6.5.1 Section Definition Directives   231
       6.5.2 Constant Definition Directives   231
       6.5.3 Data Allocation Directives   232
       6.5.4 Assembly Control Directives   232
       6.5.5 Listing File Control   232
       6.5.6 Conditional Assembly   233
       6.5.7 Macro Control   233

6.6 Development of an Assembly Language Program 233

    6.6.1 Program Strategies 234

    6.6.2 Source Code Structure 235

    6.6.3 Conversion from Assembly Code to Machine Code 236

    6.6.4 Debugging Tools 237

6.7 High-Level Language 237

    6.7.1 C-Programming for the 9S12C MCUs 238

6.8 Development Tools for the MC9S12C 242

    6.8.1 Background Debug Mode (BDM) 243

6.9 16-Kbyte Flash Module 244

    6.9.1 Security 244

    6.9.2 Flash Protection 245

    6.9.3 Flask Clock 246

    6.9.4 Flash Configuration (FCNFG) 246

    6.9.5 Flash Operations 246

6.10 Microchip PIC Microcontrollers 248

    6.10.1 Architectural Overview of the Microchip PIC 18F452 MCU 249

    6.10.2 Instruction Set 249

    6.10.3 Pipelining 250

    6.10.4 Clocking Scheme 250

    6.10.5 Memory Organization 250

    6.10.6 Addressing Modes 250

    6.10.7 I/O Ports 250

    6.10.8 Timers 251

    6.10.9 Compare/Capture/PWM (CCP) Module 251

    6.10.10 Analog-to-Digital Conversion (ADC) Module 251

    6.10.11 Interrupt Structure 251

    6.10.12 PIC Development Suite 252

6.11 Summary 252

Related Reading 252

Questions 253

Problems 253

Laboratory Projects 255

7. Parallel I/O and Interrupt Mechanism 256

Objectives 256

7.1 Introduction 256

7.2 Parallel Input/Output (I/O) 257

    7.2.1 Common Port Features 258

    7.2.2 Specific Port Features 259

7.3 Mechanical Switches   262

    7.3.1 Interfacing Binary Switches   263

    7.3.2 Switch Debounce   264

7.4 Interfacing Keyboards   265

    7.4.1 Hardware Decoding   266

7.5 Displays   267

    7.5.1 Light-Emitting Diodes (LEDs)   268

7.6 Interfacing LED Displays   271

    7.6.1 Software Decoding   272

    7.6.2 Multiplexed Displays   277

    7.6.3 Hardware Decoding   279

7.7 LCD Displays   279

7.8 Interrupt Mechanism   280

    7.8.1 Maskable and Nonmaskable Interrupts   281

    7.8.2 Interrupt Process   281

    7.8.3 Vectored Priority Interrupt   282

    7.8.4 Interrupt and Reset Vectors   284

    7.8.5 Stacking the Registers   284

7.9 Resets   285

    7.9.1 External Pin RESET   285

    7.9.2 Power-On Reset   286

    7.9.3 COP Failure Reset   286

    7.9.4 Clock Monitor Reset (CMR)   287

    7.9.5 Reset Sequence   287

7.10 Nonmaskable Interrupt (XIRQ)   288

7.11 Maskable Interrupts   289

7.12 Summary   292

Related Reading   292
Questions   292
Problems   293
Laboratory Projects   294

8. Serial Interface Facility   295

Objectives   295

8.1 Introduction   295

8.2 Serial Communication Interface (SCI)   296

    8.2.1 Communications Protocol (Framing)   296

    8.2.2 Data Transfer (Baud) Rate   297

8.3 SCI Registers   298

    8.3.1 Data Register   298

    8.3.2 SCI Control Registers   299

    8.3.3 SCI Status Registers   301

8.4 SCI Operation   302

    8.4.1 SCI Configuration   302

    8.4.2 Transmit Operation   303

    8.4.3 Receive Operation   304

8.5 Interfacing the 9S12C with the RS232 Port   306

8.6 Serial Peripheral Interface (SPI)   307

    8.6.1 Port M Data Direction Register (DDRM)   308

    8.6.2 SPI Baud Rate Register (SPIBR)   308

8.7 SPI Registers   309

    8.7.1 SPI Data Register (SPIDR)   309

    8.7.2 SPI Control Registers   309

    8.7.3 SPI Status Register (SPISR)   311

8.8 SPI Topologies   311

8.9 SPI Operation   313

8.10 I/O Expansion of the 9S12C   315

    8.10.1 Output Port Expansion   315

    8.10.2 Input Port Expansion   317

8.11 Summary   319

Related Reading   319
Questions   319
Problems   320
Laboratory Projects   320

9. **Programmable Timer Facility   321**
  Objectives   321

9.1 Introduction   321

9.2 Timer Module in the 9S12C MCU   322

    9.2.1 Free-Running Counter (TCNT)   323

    9.2.2 Timer Overflow   324

    9.2.3 Clearing the Timer Flag   324

9.3 Output Compare   324

    9.3.1 Output Compare Registers   325

    9.3.2 General Setup for the Output Compare Operation   326

    9.3.3 Operation of the OC7   330

    9.3.4 Forced Output Compare   332

9.4 Input Capture Facility   333

    9.4.1 Input Capture Pins and Registers   333

9.5 Pulse Accumulator   337

    9.5.1 Pulse Accumulator Count Register (PACNT)   337

    9.5.2 PA Enable and Active Edge Detection   338

    9.5.3 PA Operating Modes   338

9.6 Real-Time Clock   341

9.7 Pulse-Width Modulation (PWM)   342

9.8 Summary   348

Related Reading   349
Questions   349
Problems   349
Laboratory Projects   350

10.   **Analog-to-Digital (A/D) and Digital-to-Analog (D/A) Conversion   351**
       Objectives   351

10.1 Introduction   351

10.2 Fundamentals of A/D Conversion   352

   10.2.1 Resolution   353

   10.2.2 I/O Mapping   353

   10.2.3 Aliasing   356

   10.2.4 Amplitude Uncertainty   357

   10.2.5 Sample and Hold (S/H)   358

   10.2.6 Multiple Sensor Inputs   360

10.3 A/D Conversion Techniques   360

   10.3.1 Integrating ADCs   360

   10.3.2 Successive-Approximation ADC   363

   10.3.3 Flash ADC   364

10.4 ADC Facility of the 9S12C MCU   365

   10.4.1 Voltage References   365

   10.4.2 ATD Registers   366

   10.4.3 ATD Setup   366

   10.4.4 Conversion Time   368

   10.4.5 Channel Selection   368

   10.4.6 Channel Sampling and Conversion Results   369

   10.4.7 Input Signal Range   373

10.5 Digital-To-Analog Conversion (DAC)   376

   10.5.1 Components of a D/A Converter (DAC)   377

   10.5.2 Output Voltage   377

   10.5.3 Range   378

   10.5.4 Resolution   379

   10.5.5 Accuracy   379

   10.5.6 Bipolar DACs   380

   10.5.7 DAC ICs   381

10.6 Summary   382

Related Reading   382
Questions   382
Problems   383
Laboratory projects   384

11. Sensors and Their Interface   385

Objectives   385

11.1 Introduction   385

11.2 Classification of Sensors   386

11.3 Smart Sensors   387

11.4 Sensor Models and Response Characteristics   388

11.5 Sensor Characteristics   390

11.6 Signal Conditioning   392

11.6.1 Amplification   393

11.6.2 Conversion   393

11.6.3 Filtering   393

11.6.4 Impedance Buffering   394

11.6.5 Modulation/Demodulation   394

11.6.6 Linearization   394

11.6.7 Grounding and Isolation   395

11.7 Potentiometer Sensors (Pot)   396

11.8 Light Detectors   398

11.8.1 Materials for Light Detectors   400

11.8.2 Types and Modes of Operation of Light Detectors   400

11.8.3 Applications of Light Detectors   400

11.9 Photoresistor (Photocell)   400

11.9.1 Materials for Photocells   401

11.9.2 Interfacing a Photocell to the 9S12C   401

11.10 Photodiode   402

11.10.1 Photodiode Types   402

11.10.2 Photodiode Characteristics   403

11.10.3 Operating Modes   403

11.10.4 Applications   404

11.11 Phototransistor   405

11.11.1 Phototransistor Characteristics   405

11.11.2 Applications   406

11.12 IR Emitter/Detector Packages   407

11.12.1 Optical Interrupter   407

11.12.2 Optical Coupler (Optical Isolator)   408

11.12.3 Optical Reflectors   408

11.12.4 NIR Receiver/Demodulator Sensors   409

11.13 Optical Encoder   410

11.14 Pyroelectric Sensor   412

11.14.1 Signal Conditioning   414

11.15 Thermal Detectors   414

11.15.1 Thermocouple   415

11.15.2 Thermopiles   418

11.15.3 Theremoresisitive Devices 418

11.15.4 Thermodiode 423

11.15.5 Thermotransistor 424

11.16 Heat Flux Sensor 425

11.17 Magnetic Sensors 426

11.17.1 Magnetic Reed Switch 426

11.17.2 Hall-Effect Device 427

11.18 Strain Gauges 430

11.18.1 Bridge Circuit 431

11.18.2 Strain-Gauge Measurement 433

11.19 Acoustic Measurement 434

11.19.1 Properties of Wave Propagation 434

11.19.2 Acoustic Sensors 436

11.19.3 Types of Transducer Element 437

11.19.4 Types of Measurements 440

11.20 Piezoelectricity 443

11.20.1 Piezoelectric Effect 443

11.20.2 Piezoelectric Use in MEMS 444

11.20.3 Constitutive Relations in One Dimension 444

11.20.4 Piezoelectric Sensor 445

11.20.5 Piezoelectric Mass-Sensitive Chemical Sensor 447

11.21 Resolver 448

11.22 Tachometer 449

11.23 Capacitive Sensors 449

11.24 Inductive Sensors 451

11.24.1 Motion-Detection Sensor 451

11.24.2 Linear Variable Differential Transformer (LVDT) 451

11.25 Four- to 20-mA Transmitters 453

11.25.1 Voltage-to-Current Converter 454

11.26 Summary 454

Related Reading 455
Questions 455
Problems 456
Laboratory Projects 459

12. Electric Actuators 460

Objectives 460

12.1 Actuators 460

12.2 DC Motors 461

12.2.1 Principles of Operation of a DC Motor 461

12.2.2 Modeling of DC Motor Behavior 465

12.2.3 Heat Dissipation in DC Motors 472

       12.2.4 Velocity Profile Optimization   *473*

       12.2.5 Inertia Matching   *474*

       12.2.6 Motor Selection   *476*

       12.2.7 Servo Amplifiers   *479*

       12.2.8 DC Motor Servo Drive   *482*

       12.2.9 Interfacing DC Motors to the 9S12C   *485*

       12.2.10 DC Servos   *490*

    12.3   Stepper Motors   491

       12.3.1 Characteristics of a Stepper Motor   *491*

       12.3.2 Classification of Stepper Motors   *491*

       12.3.3 Principle of Operation   *494*

       12.3.4 Step Angle   *498*

       12.3.5 Electrical Model of an Energized Coil   *499*

       12.3.6 Drive Methods   *501*

       12.3.7 Stepper Motor Performance   *503*

       12.3.8 Interfacing Stepper Motors to the 9S12C MCU   *509*

    12.4   AC Induction Motors   515

       12.4.1 Three-Phase Motors   *516*

       12.4.2 Speed Control of the Induction Motor   *519*

    12.5   Summary   524

    Related Reading   524

    Questions   524

    Problems   525

    Laboratory Projects   526

13.   **Control Schemes   527**

    Objectives   527

    13.1   Introduction   527

       13.1.1 History of Control   *527*

       13.1.2 Open-Loop Control   *529*

       13.1.3 Closed-Loop Control   *529*

    13.2   Classical Control   530

       13.2.1 Mathematical Modeling   *530*

       13.2.2 Transfer Function   *532*

       13.2.3 Transient and Steady-State Analyses   *533*

       13.2.4 Root Locus   *537*

       13.2.5 Frequency Response   *543*

       13.2.6 Lag-Lead Compensator   *549*

       13.2.7 Proportional-Integral-Derivative (PID) Controller Design   *557*

    13.3   State-Space-Based Control Strategies   565

13.4 Adaptive Control   571
    13.4.1 Gain Scheduling   571
    13.4.2 Model-Reference Adaptive Control (MRAC)   572
    13.4.3 Self-Tuning Regulators   573
13.5 Digital Control   574
    13.5.1 Discretization Techniques   574
    13.5.2 Emulation   575
    13.5.3 Direct Digital Control   575
13.6 Intelligent Control   576
    13.6.1 Fuzzy Logic Control Design   576
13.7 Adaptive Fuzzy Logic Controllers   583
    13.7.1 Introduction   583
    13.7.2 Fuzzy Model-Reference Adaptive Controller   583
    13.7.3 Membership-Tuning Adaptive Controller   586
13.8 Experimental Comparative Analysis   591
    13.8.1 Hardware Platform
    13.8.2 Digital Control Workstation   591
13.9 Conclusion   599
Related Reading   599
Questions   600
Problems   601

14. Case Studies   603
Objectives   603
14.1 Introduction   603
14.2 Case Study 1: Autonomous Mobile Robot   604
    14.2.1 Introduction   604
    14.2.2 Mechanical Design Alternatives   605
    14.2.3 Design Specifications   606
    14.2.4 Electronic Circuits and Interfacing   612
    14.2.5 Software Design   618
    14.2.6 Case Outcomes   620
    References   621
14.3 Case Study 2: Wireless Surveillance Balloon   621
    14.3.1 Problem Definition   621
    14.3.2 Design   621
    14.3.3 Parts   626
    14.3.4 Case Outcomes   635
    References   636
14.4 Case Study 3: Firefighting Robot   636
    14.4.1 Problem Statement   636
    14.4.2 Design Alternatives   638

14.4.3 Implementation   639

14.4.4 Case Outcomes   647

References   649

14.5 Case Study 4: Piezo Sensors and Actuators in Cantilever Beam Vibration Control   649

14.5.1 Introduction   649

14.5.2 Modeling of the Cantilever Beam and PZT Actuator   650

14.5.3 Beam Experimental Setup   652

14.5.4 Instrumentation Setup   654

14.5.5 Controller and Software   658

14.5.6 Simulation and Experimental PID Results   661

14.5.7 Simulation and Experimental Fuzzy Results   665

14.5.8 Conclusions   668

14.5.9 Case Outcomes   668

[†]Appendix A: DC Power Supply (EE Basic)   670

Appendix B: Pinout of Selected ICs   672

Appendix C: Instruction Set, Addressing Modes, and Execution Times for the MC9S12C   674

Appendix D: MC9S12C Registers and Control Bit Assignments   676

Appendix E: Using the CodeWarrior Integrated Development Environment (IDE)   678

Appendix F: ASCII Code Table   680

[†]Appendix G: Number Systems (EE Basic)   681

[‡]Appendix H: Mechanisms For Mechatronics (ME Basic)   691

Index   706

Preface

The ease and affordability of integrating and packaging microelectronics into practical products for sensing, conditioning, processing, and controlling signals and systems have driven the design and implementation of electromechanical systems to the multidisciplinary approach of mechatronics. It is essential that twenty-first-century engineers acquire strong skills in the technologies behind these products. The aim of this book is to bring together in one text the many components of mechatronics.

Applied Mechatronics is intended for junior-/senior-level engineering students in mechanical, electrical, computer, or civil engineering programs. In addition, this book can be a very useful all-in-one reference book for practicing engineers in the fields of instrumentation, industrial automation, and product development.

The authors assume that most of the students studying mechatronics have a basic background in electrical circuits, dynamic systems modeling, and introductory computer language. Icons in the Table of Contents highlight sections that cover prerequisite topics electrical engineers would have already seen († = **EE Basic**), or that mechanical engineers would have already seen (‡ = **ME Basic**). Prerequisites from basic engineering sciences are marked (* = **EE and ME Basic**). Students can concentrate more attention on the basics that are new or unfamiliar to them. Instructors can omit these sections in lectures and focus on the core ideas of mechatronics integration. Thus, students from all engineering backgrounds can be equally prepared for the course according to their own needs.

Applied Mechatronics is the culmination of many years of teaching and lab development in the areas of electronics, instrumentation, microprocessors, and control systems. The strength of the text is derived from its comprehensive nature in realizing embedded engineering systems. Specifically, this book covers the required information to measure events with sensing devices, condition sensed signals through electronics interface circuits, process the conditioned data by a microprocessor with necessary memory and ports (specifically the Motorola 9S12C microcontroller), develop the adequate control command for the process, and realize the driving circuitry for the actuators to achieve the desired behavior. Embedding these skills is not a cascading task, rather a global design effort distributing functions among software and hardware agents.

The book's material is complemented with many practical examples using MATLAB, SIMULINK, and LabVIEW simulations and experimentation.

The authors believe that the most effective way to use the material in this book is through collaborative project-based learning. This format allows students to practice integrating technologies as interdisciplinary teams and to work on open-ended problems.

Ahmad Smaili
Fouad Mrad

Acknowledgments

Meaningful things in life are the result of the efforts of so many. This book would not have been possible were it not for the help of so many who crossed the authors' paths and contributed, directly or indirectly to its completion. The first author is greatly indebted to Prof. Ed Griggs, Chair of the Mechanical Engineering Department, for having the vision to support the author's efforts to establish a mechatronics laboratory at Tennessee Technological University.

All the students who took the mechatronics, instrumentation, and control courses taught by the authors at Tennessee Technological University and the American University of Beirut are greatly acknowledged; they are too numerous to name here. Special appreciation goes to the students of the senior projects that were selected and included in Chapter 14, "Case Studies." We particularly want to thank A. Atallah, N. Atallah, B. Chaaya, J. Chamoun, G. Dib, H. El-Sayed, A. Farchoukh, M. Hasanieh, K. Joujou, S. Maakaron, M. Sidani, and F. Zein eddine. We thank the reviewers who commented on this book throughout its development and contributed enormously to its ideas and its accuracy.

Chaouki Abdallah	University of New Mexico
Lawrence Agbezuge	Rochester Institute of Technology
David Alciatore	Colorado State University
Ron Averill	Michigan State University
Jay D. Bernheisel	Union University
Sridhar S. Condoor	Saint Louis University
Mo-Yeun Chow	North Carolina State University
Marcelo Dapino	Ohio State University
Gary Fedder	Carnegie Mellon University
Burford J. Furman	San Jose State University
John Gardner	Boise State University
Stephen Heinrich	Marquette University
Kam K. Leang	Virginia Commonwealth University
Kevin Lynch	Northwestern University
Mohammed Mahinfalah	North Dakota State University
Sanford Meek	University of Utah
Ibrahim Miskioglu	Michigan Technological University
William Murray	California Polytechnic University, San Luis Obispo
R. Ben Mrad	University of Toronto
Mark Nagurka	Marquette University
Satish Nair	University of Missouri—Columbia
Jim Ostrowski	University of Pennsylvania
Stewart Prince	California State University, Northridge

Karl Seeler   Lafayette College
Ramavarapu S. Sreenivas   University of Illinois—Urbana Champaign
Naiqian Zhang   Kansas State University

And last but not least, sincere thanks are due to the Oxford University Press team for encouraging us to stay the course and working very hard to secure adequate review and professional production of the book. We especially thank Danielle Christensen, Karen Shapiro, Dawn Stapleton, Adriana Hurtado, Annika Sarin, Liz Cosgrove, and their group in New York.

Mechatronics: An Introduction

Thoughtful engagement with the material presented in this chapter will enable the student to:

1 Realize the importance of mechatronics in entrepreneurial ventures

2 Understand the integrative nature of mechatronics

3 Appreciate the role of mechatronics in the modern era

4 Describe the main components of a mechatronic system

5 Appreciate the role of microcontroller technology for embedded applications

1.1 WHAT IS MECHATRONICS?

In today's competitive markets, engineers face an ongoing challenge to produce complex engineering systems with a high level of performance, reliability, and value at a low price. This demand is brought about by advances in microprocessor technologies (small and cheap) that make what was state of the art yesterday fade in comparison with what might be possible tomorrow. To persevere in this highly competitive atmosphere, engineers must be able to integrate a number of relevant technologies. Mechatronics provides a platform to best accomplish this integration from the earliest stages of the design process.

Mechatronics is an interdisciplinary engineering field comprising the design and development of smart electromechanical systems. Its aim is to integrate various technologies, including electronics, mechanical devices, real-time control, microprocessors, materials, and human–computer interaction, from the very early stages of the conceptual design process and throughout the embodiment phases to introduce to the market simple, smart, high-quality, and competitive products in a short time. Mechatronics is key to a wide range of products, such as disk drives (video and CD), cameras and camcorders, flexible autonomous systems, process controllers, avionics, appliances, smart weapons, and power tools. Modern automobiles, the so-called *electronic vehicles*, involve a large number of mechatronics subsystems to manage engine controls, comfort control, antilock brakes, active suspension, collision avoidance, drive by wire, safety features, and so on.

1.2   ESSENTIAL SKILLS FOR MECHATRONICS

A mechatronics engineer is one who views a system as a whole and offers optimum solutions to a multivariable problem. To perform correctly, contemporary systems and products rely on harmonious interactions between mechanical systems, sensors, actuators, and computers to realize multifunctional, flexible, smart, and precision machines. Therefore, a mechatronics engineer must be able to transcend barriers that existed in the past between various engineering disciplines and acquire the necessary skills and expertise that enable her/him to:

- Select, design, and integrate mechanical components and drives,
- Select sensors, design, and implement appropriate signal-conditioning (SC) circuits,
- Select and drive appropriate actuators,
- Develop mathematical models of the processes involved,
- Design and implement appropriate control schemes, and
- Use microprocessor software and hardware to build target systems and interfaces.

1.3   WHY IS MECHATRONICS IMPORTANT?

The breathtaking speed of technology advances is influencing to a large extent the future and spirit of the world in which we live. Technology, properly harnessed and liberally distributed, has the potential to break not just geographical barriers but human ones as well. Economies in the era of globalization are becoming dependent on knowledge and technology. As microprocessor technologies continue to advance, becoming ever smaller, cheaper, and more powerful, successful products of yesterday fade in comparison with tomorrow's possibilities. Competing in a global market requires the commercialization of knowledge and technology to yield flexible multifunctional products that are better, cheaper, and more intelligent quickly. To this end, engineers involved in the product realization process must master technology as it develops and quickly integrate it into products well ahead of the competition. Mechatronics, as an interdisciplinary engineering field, plays a key role in achieving this goal. The importance of mechatronics is evidenced by the myriad smart products that we take for granted in our daily lives, from the little robotic toy that could climb walls to all the stuff that constitutes a modern electronic vehicle, including engine controls, antilock braking systems, active suspension systems, collision avoidance, drive by wire, electronic muffler, comfort, and so on.

The new economic reality represents a new frontier that is being challenged, and leaving it unguarded will cause dire results. Commercializing technology presents a challenge to society and an urgency to meet the competition head on. In the era of globalization, entrepreneurial ventures will play a major role in economic growth, and mechatronics engineers are very likely to embark on such ventures.

1.4   COMPONENTS OF A MECHATRONIC SYSTEM

Figure 1.1 shows a typical microprocessor-based mechatronic application. It consists of many subsystems, including the following.

- *Mechanical subsystem.* This constitutes the machine or device being controlled. The device itself may involve a great number of internal individual or interrelated mechanisms with their individual subsystems.

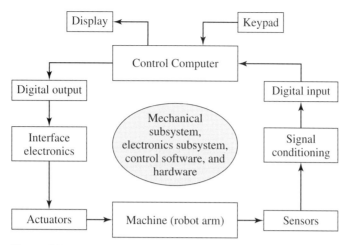

Figure 1.1 Components of a mechatronic system.

- *Sensors* to measure the controlled variables. Sensors of all types provide the computer with information it needs to perform monitoring and control tasks.
- *Actuators* convert digital signals into physical or chemical quantities as required to correct errors in meeting the required performance.
- *Instrumentation* subsystem. This involves all necessary signal-conditioning (SC) circuits to interface actuators, sensors, and other devices to the control computer.
- *Control computer*. Proper control of a process requires intense processing of information representing a complex interplay between sensing the environmental conditions that affect the outcome of the process, interpretation of the sensed values, recognizing the effect of these values on the integrity of the operation, and initiating a response by signaling the actuators to take the necessary corrective actions. Coordinating all these activities is the responsibility of the control computer, which involves embedded microprocessor systems. It handles the flow of information between the various components by integrating the software and hardware dedicated to controlling the process in question.

The *hardware*, also referred to as the *target system*, consists of all components needed to manage the inputs and outputs of the mechatronic device. This includes sensors, actuators, mechanical components, power supplies, control computer, and the SC circuits interfacing the computer with the sensors and actuators. As a first step, a prototype of the hardware may be built using wire-wrapped boards to test the application software before the final product is realized.

The *software* manages all activities of the target system. It consists of the program (or code), which is developed to meet the application requirement. Development of the application code can begin concurrently with hardware prototype development once the system specifications have been completed. The code could be modified as it is developed to reflect design changes and/or new requirements as they emerge.

The application code may be developed using machine language, assembly language, or a high-level language such as C. The *machine code*, or object code, contains binary codes that represent the instructions to the computer. It is the only language the microprocessor can understand and execute. High-level languages such as C use words and statements that are easily understood. A program written in a high-level language must be translated into executable machine language using a program called a *compiler*. Assembly language represents a middle ground between the extremes of machine language and the high-level language. An assembly language program is written using *mnemonics* in which each statement corresponds to a machine instruction. The assembly program must be translated into machine language by a special program called an *assembler.*

1.5 BRAIN FOR MECHATRONICS

Real-time control actions and information processing are essential features of a mechatronic system that are typically performed by an embedded microprocessor system. The microprocessor provides the target system with flexibility and intelligence. An even higher form of flexibility can be realized using microcontroller systems. A microcontroller unit (MCU) is often called an *embedded controller* because it is used as one component in a larger system. It is a highly integrated, programmable single-chip integrated circuit (IC) that includes all resources required for real-time control. It integrates on a single chip the microprocessor, memory, input/output (I/O) interface circuits, clock, analog-to-digital converter, programmable timer, address and data busses, and many other features. Many MCU families are available from various sources. Different MCU families have comparable features, offering a wide range of options to suit application cost and performance requirements. Chapters 6–10 of this text focus on programming various features of the MC9S12C MCUs from Freescale Semiconductors. Figure 1.2 shows a detailed block diagram of an MCU-based mechatronics system and the text chapters that relate to the different blocks.

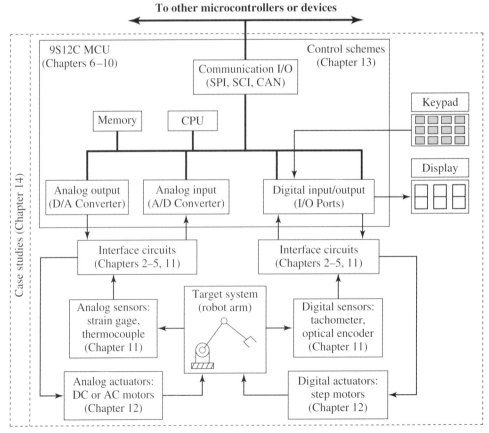

Figure 1.2 Roadmap to mechatronics as presented in the text.

RELATED READING

D. G. Alciatore and M. B. Histand, *Introduction to Mechatronics and Measurement Systems*, 2nd ed. New York: McGraw-Hill, 2002.

D. Auslander, "What Is Mechatronics?" *IEEE/ASME Transactions on Mechatronics*, vol. 1, no. 1, 1996, pp. 5–9.

C. F. Bergh, A. Kita, and C. I. Ume, "Development of a Mechatronics Course in the School of Mechanical Engineering at Georgia Tech," *Proc. of the 1999 IEEE/ASME International Conference on Advanced Intelligent Mechatronics* (AIM '99).

J. E. Carryer, "The Design of Laboratory Experiments and Projects for Mechatronics Course," *Mechatronics*, vol. 5, no. 7, 1995, pp. 787–797.

W. K. Durfee, "Designing Smart Machines: Teaching Mechatronics to Mechanical Engineers Through a Project-Based, Creative Design Course," *Mechatronics*, vol. 5, no. 7, 1995, pp. 775–785.

N. Kyura and H. Oho, "Mechatronics: An industrial Perspective," *IEEE/ASME Transactions on Mechatronics*, vol. 1, no. 1, 1996, pp. 10–15.

Making Sense, *ME Magazine*, Jan. 2001, pp. 44–46.

S. Meek, S. Field, and S. Devasia, "Mechatronics Education in the Department of Mechanical Engineering at the University of Utah," *Mechatronics*, vol. 13, no. 1, 2003, pp. 1–11.

W. R. Murray and J. L. Garbini, "Embedded Computing in the Mechanical Engineering Curriculum: A Course Featuring Structured Laboratory Exercises," *Journal of Engineering Education*, vol. 86, no. 3, 1997, pp. 285–290.

S. Shooter and M. McNeil, "Interdisciplinary Collaborative Learning in Mechatronics at Bucknell University," *Journal of Engineering Education*, vol. 91, no. 3, 2002, pp. 339–344.

M. Wald (Editor-in-Chief) and T. Kurfess (Guest Editor), Special issue on Mechatronics Education. *International Journal of Engineering Education*, vol. 19, no. 4, 2003.

QUESTIONS

1.1 What is mechatronics?

1.2 Identify the elements of a mechatronic system.

1.3 Explain the term *embedded*.

1.4 What is meant by *target system*?

1.5 What is the difference between a microprocessor and a microcontroller?

1.6 Explain the difference between machine language, assembly language, and high-level language.

1.7 What is an assembler?

1.8 What is a compiler?

1.9 What constitutes an entrepreneur?

PROBLEMS

1.1 Write a short essay about an entrepreneurial venture that culminated in a start-up company created to produce and market a mechatronics product the idea of which was born in college. Suggested sources for your search are: *Technology Review* (published by MIT), *ME Magazine* (published by the American Society of Mechanical Engineers, ASME), and PRISM (Published by the American Society of Engineering Education, ASEE).

1.2 Do the following: for any one of these mechatronic systems: CNC machine, copy machine, robot, elevator, camera, traffic management system, vacuum cleaner, or any other machine.

a. Describe briefly (in one paragraph) its operation.

b. Identify the sensors used, their type, and the signals they measure.

c. Identify the actuators used.

d. Provide a block diagram of the system indicating the major components and the interconnection between them.

PROJECT

Investigate a pressing problem in your community, and envision a mechatronic system that would, if realized, contribute to the solution of that problem. Such problems include water resources management, sewage treatment, renewable energy, traffic safety, emergency rescue operation, security system, entertainment, library management, planting and sustaining trees, etc.

   a. Describe the problem, and establish the need to create a mechatronics device that would offer a solution to the problem; then draft a mission statement.
   b. Guided by the mission statement, establish a set of specifications for the device.
   c. Generate a viable concept for the device.
   d. Divide up the device into subsystems (energy subsystem, mechanical subsystem, signal flow, etc.), and explain the role of each subsystem and possible ways to implement it.
   e. Identify the sections in the text that include the required information that helps you in realizing each subsystem.

Consider this to be the project you will realize as you engage the material in the text.

2

Elements and Analysis of Electric Circuits

OBJECTIVES

Thoughtful engagement with the material presented in this chapter will enable the student to:

1 Understand the basic principles of electricity
2 Perform analysis of passive *RLC* circuits in time and frequency domains
3 Develop understanding of loading errors and ways to reduce their effects
4 Acquire skills on passive filter design
5 Recognize basic interference issues and grounding problems

2.1 INTRODUCTION

This chapter presents a review of the basic principles of electricity, passive circuit elements, and passive circuits, which form the necessary building blocks of signal-conditioning circuits and mechatronics devices. The goal is to provide students with a good intuitive understanding of circuit design, implementation, and behavior.

2.2 ELECTRIC FIELD

The most elemental electrical quantity is the *electric charge q* in units of coulombs (C) or ampere-seconds (A-s). Electric charges exist in one of two kinds, arbitrarily called *positive* and *negative*. The charge of the electron is negative and equals -1.602×10^{-19} C, meaning a total of 6.22×10^{18} electrons equal to 1 C. Electric charge can produce a force f in units of newtons (N) that repels other charges of the same sign and attracts other charges of opposite sign. The magnitude of the force that is felt equally by each of the charges is given by *Coulomb's law*:

$$f = \frac{k|q_1 q_2|}{d^2} \tag{2.1}$$

where q_1 and q_2 are the charges of the two charged bodies, k is a constant depending on the units used and the properties of the medium surrounding the charge, and d is the distance separating the charges. Equation (2.1) implies that the presence of a charge creates a region of influence called a *field* in its vicinity, which gives rise to a force when another charge is introduced into the field. This force will progressively weaken as the new charge is placed in more remote positions. If the field is set up by electric charges, it is called *electric field* and $k = 1/4\pi\epsilon$, where ϵ depends on the electrical properties of the medium, for vacuum $\epsilon = \epsilon_0 = 8.854 \times 10^{-12}$ (C^2/N.m^2).

Coulomb's law is similar to the law of gravitational attraction, except that the electric force may be attractive or repulsive according to the sign of charges involved. Usually the electrical force between charged particles is much stronger than their gravitational attraction.

Charges can move from one place to another due to a process called the *triboelectric effect* (prefix *tribo* means "pertinent to friction"). In spite of this effect, electric charge is conserved. Charge separation and redistribution is due to the friction (*tribo*) that results from rubbing two objects against each other, a person's walking on a carpet, atmospheric electricity, and so on.

2.3 CURRENT AND VOLTAGE

When an electric field is imposed on a conducting circuit, the work performed on the electrons sets them in motion, resulting in energy transfer to other parts in the circuit. The flow of electrons results in an *electric current i* in units of amperes (A). The current past a point or *through* a device is the time rate of flow of electric charge

$$i(t) = \frac{dq(t)}{dt} \tag{2.2}$$

and is positive in the direction opposite to the flow of electrons.

The *voltage*, or *potential difference*, v in units of volts (V) is the cost in energy, or amount of work, required to move a unit positive electric charge *across* two points in a field. It is expressed as

$$v = \frac{dw}{dq} \tag{2.3}$$

where w is the work (or energy) in units of joules (J). If the voltage is that of a source of electrical energy it is called *electromotive force* (emf). Equation (2.3) implies that if the work done in moving 1 C of charge between two points is 1 J, then the potential difference between the two points is 1 V. This is analogous to moving a mass in a gravitational field.

The total work, or energy, w associated with the movement of q coulombs between two points is

$$w = \int_0^q v\,dq \tag{2.4}$$

If the work is being done at a constant rate and the total charge q is moved through a voltage v volts in t seconds, then the power p [W or J/s] is

$$p = \frac{vq}{t} = vi \tag{2.5}$$

The total electric flux ϕ_E (C) through a closed surface is equal to the total charge q_S enclosed by the surface and is given by *Gauss's law*:

$$\Phi_E = \oint_s \mathbf{D}\cdot d\mathbf{s} = q_S \tag{2.6}$$

where $\mathbf{D}$ (F/m) is the electric flux density vector, which is related to the electric field vector $\mathbf{E}$ (V/m) by the constitutive relation

$$\mathbf{D} = \varepsilon\mathbf{E} \tag{2.7}$$

Note that a boldface roman letter indicates a vector quantity.

Two types of electric circuits are available, *direct current*, or DC, circuits and *alternating current*, or AC, circuits. DC circuits involve constant (time-independent) voltage and current signals (Fig. 2.1a), whereas voltage and current signals involved in AC circuits vary with time, usually sinusoidally (Fig. 2.1b). The two circuit types are depicted in Fig 2.1. Power in AC circuits varies with time and is called *instantaneous* power.

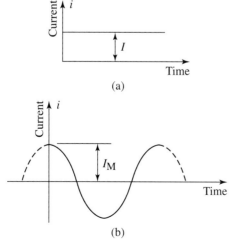

Figure 2.1 Steady direct current (a) and alternating current (b).

2.4 ELEMENTS OF AN ELECTRIC CIRCUIT

An electric circuit consists of three basic components, as shown in Fig. 2.2: the *source*, the *load*, and the *ground* reference. By convention, the current flows through the circuit from the anode to the cathode, opposite to the direction of the actual electron flow.

2.4.1 Circuit Conditions

The simple circuit shown in Fig. 2.3 depicts three distinct circuit conditions. When the switch is closed, the two wires conduct the current from the battery through the lamp and back, forming a *complete*, or *closed*, *circuit* (Fig. 2.3a). When the switch opens (Fig. 2.3b), an *open-circuit* condition occurs, where no current flows ($i = 0$) and no energy transfer would take place. If the insulation in the base of the lamp breaks down and becomes conducting or if a wire is accidentally connected between points c and d around the lamp or between points a and b around the battery, a *short-circuit* condition (Fig. 2.3c) develops, where the source outputs (destructively) high currents, with a negligible amount flowing through the lamp. In practice, *fuses* or *circuit breakers* are inserted in a circuit to automatically open the circuit if short-circuit faults occur.

2.4.2 Electric Circuit Sources

Two types of ideal electrical energy sources are recognized, an *ideal voltage source* and an *ideal current source*. Common symbols used to indicate a voltage source and a current source are shown in Figs. 2.4a and b, respectively. Ideal sources are available in DC or AC with internal resistance, inductance, or capacitance.

An ideal voltage source is a two-terminal circuit element that provides a voltage across its terminals that is an independently specified function of time $V_s(t)$ regardless of load resistance. However, a *real voltage source* is power limited and can only supply a limited amount of current. It is usually modeled as an ideal voltage source in series with a small resistance. Voltages

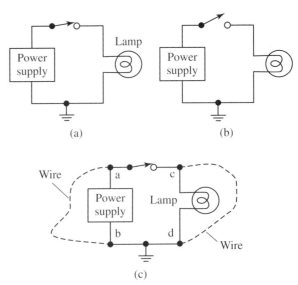

(a)   (b)

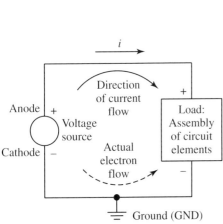

Figure 2.2 Components of an electric circuit.

(c)

Figure 2.3 Circuit conditions: closed, or complete, circuit (a), open circuit (b), and short circuit (c).

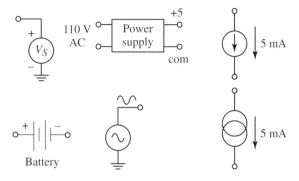

Figure 2.4 Schematic representations of ideal voltage sources (a) and current sources (b).

are generated by doing work on charges in many ways: by electrochemical means in batteries, by magnetic forces in generators, by photovoltaic conversion of photons energy in solar cells, and so on. A voltage source provides energy to the circuit across its two sides, known as the *anode* and the *cathode*. The anode is the positive side, where electrons are attracted, and the cathode is the negative side, where electrons are released.

Current flows through a circuit element when a voltage is placed across its terminals. An ideal current source is a two-terminal circuit element that maintains a constant current through the circuit connected to it that is independently specified function of time $I_s(t)$ regardless of load resistance or applied voltage. Thus, it is capable of supplying across its terminals any voltage needed by the connected circuit. However, a *physical* current source can only provide a limited amount of power, and its current output deviates from the ideal value $I_s(t)$. In most cases, the ideal representation is sufficient for engineering purposes. In other cases, the source is represented by an ideal current source in parallel with a small resistance.

Another class of sources is recognized, where the voltage or current they source depends on the voltage or current values in another part of the circuit. Important physical devices involving such controlled sources include the electric generator and the transistor.

2.4.3 Circuit Load

The load in a circuit is an assembly of *circuit elements* designed to perform a specific function. Three types of *passive* circuit elements have been recognized; each exhibits specific current–voltage relationships that have been established on the basis of experiments. *Passive* refers to any element that does not require an external power supply for its operation. The three basic elements are the *resistor*, characterized by its resistance R in units of ohms (Ω); the *capacitor*, characterized by its *capacitance* C in units of farads (F); and the *inductor*, characterized by its *inductance* L in units of henrys (H). The electrical symbols of these elements are shown in Fig. 2.5. The resistor dissipates energy into heat, the capacitor stores electrostatic energy, and the inductor stores electromagnetic energy. Electric circuits built from a combination of R, L, C elements are collectively referred to as *passive circuits*. These elements are discussed in detail in the following sections.

2.4.4 Circuit Ground

Voltage is not an absolute quantity, but rather a measure of the potential difference between two points. The circuit includes a central reference connection to provide a common ground potential

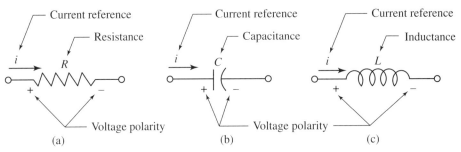

Figure 2.5 Schematic representation of resistance (a), capacitance (b), and inductance (c).

among all instruments and power sources used in the circuit or system. The voltage potential is assumed to be zero. Two types of references are recognized, earth ground and reference ground. *Earth ground* refers to a connection to the potential of the massive earth. Many instruments connect to the earth grounded via the third prong found in most electrical outlets wired to the building electrical system for safety. *Reference ground*, also called return path or *signal common*, is the reference potential of interest, which may or may not be wired to earth ground. Everywhere on a connecting wire in a circuit is assumed to be at the same voltage with respect to ground. This assumption may not be strictly true when high-frequency signals or low impedances are involved. Impedance describes algebraically the dynamic relation between the voltage and current in an electric circuit. Further distinction is made between *signal ground* and *chassis ground*. The latter is attached to the metal case enclosing an instrument to provide user safety if there are internal faults in the instrument. It is internally connected to the ground wire on the power cord and may not be connected to the signal ground. Circuit grounding issues are discussed in Section 2.16.

2.5 CIRCUIT ANALYSIS

The goal of circuit analysis is to determine voltages across—and currents through—elements in a circuit. Voltages and currents in a purely resistive circuit are real quantities, whereas in circuits containing inductors and capacitors in addition to resistors, voltages and currents are represented as complex quantities **V** and **I**. The opposition to current flow through circuit elements is also in general a complex quantity called *impedance* **Z** in units of ohms (Ω). Complex representations of currents, voltages, and impedances are treated later in the chapter. In the following section the general rules that facilitate analysis of *RLC* circuits are introduced, namely, equivalent series and parallel connections of circuit elements, Kirchhoff's voltage and current laws, and equivalent circuit representations.

2.5.1 Circuit Elements in Series and in Parallel

Series Connection

In a series connection, current flows through a single path. If N circuit elements with impedances $\mathbf{Z}_1, \mathbf{Z}_2, \ldots, \mathbf{Z}_N$ are connected in series, they can be replaced by a single equivalent impedance **Z** given by

$$\mathbf{Z}_e = \sum_{i=1}^{N} \mathbf{Z}_i \tag{2.8}$$

A circuit containing two impedances in series is called a *voltage divider*, because the voltage is divided among the two impedances. The dominant portion of the voltage will be across the larger impedance.

Parallel Connection

In a parallel connection, current flows through several paths because the voltage across the common terminals is the same. If N circuit elements with impedances $\mathbf{Z}_1, \mathbf{Z}_2, \ldots, \mathbf{Z}_N$ are connected in parallel, they can be replaced by a single equivalent impedance given by

$$\frac{1}{\mathbf{Z}_e} = \sum_{i=1}^{N} \frac{1}{\mathbf{Z}_i} \tag{2.9}$$

A circuit containing two impedances in parallel is called a *current divider*, because the current is divided between the two impedances. The dominant portion of the current flows through the smaller impedance.

2.5.2 Kirchhoff's Laws

Kirchhoff's laws relate voltages and currents in a circuit, with one law relating current flow into a contour and a second relating voltages around a loop. The two laws form the basis for circuit analysis.

Kirchhoff's Voltage Law (KVL)

This law states that *the sum of voltage drops on the circuit elements around a closed current path in a circuit is zero*,

$$\sum_{i=1}^{N} \mathbf{V}_i = 0 \tag{2.10}$$

Though the circuit must be closed, the conductors need not be. A consequence of the Kirchhoff voltage law (KVL) is that circuit elements connected in parallel have the same voltage across them. Applying KVL to the circuit in Fig. 2.6a gives

$$\mathbf{V}_1 - \mathbf{V}_2 - \mathbf{V}_3 - \mathbf{V}_4 = 0 \tag{2.11}$$

Kirchhoff's Current Law (KCL)

This law states that *the algebraic sum of all currents flowing into a closed contour or node is zero*, or *the sum of currents into a node equals the sum of the currents out*,

$$\sum_{i=1}^{N} \mathbf{I}_i = 0 \tag{2.12}$$

Applying KCL to the circuit in Fig. 2.6b gives

$$\mathbf{I}_1 - \mathbf{I}_2 + \mathbf{I}_3 - \mathbf{I}_4 = 0 \tag{2.13}$$

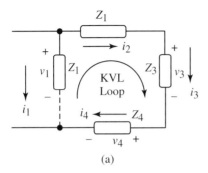

(a)

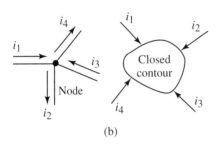

(b)

Figure 2.6 Kirchhoff's voltage law (a) and current law (b).

2.5.3 Equivalent Circuit Representation

Analysis of complex circuits is greatly simplified by applying Thevenins's and Norton's theorems. *Thevenin's theorem* states that any two-terminal linear network of impedances and voltage sources can be replaced by a *Thevenin equivalent circuit* consisting of a single impedance $\mathbf{Z}_{TH}$ in series with an ideal single voltage source $\mathbf{V}_{TH}$. As demonstrated in Fig. 2.7,

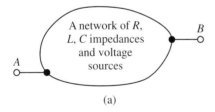

(a)

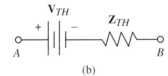

(b)

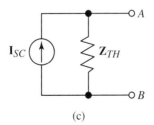

(c)

Figure 2.7 A network of R, L, C impedances and sources (a) and its equivalent Thevenin (b) and Norton (c) circuits.

any number of sources and impedances can be reduced to an equivalent circuit with one source and one impedance. $\mathbf{V}_{TH}$ is defined as the open-circuit voltage ($\mathbf{V}_{OC}$) of the original circuit, which is the output voltage measured between the output terminals when no load is attached on the output. It may be calculated if the circuit is known, or it may be measured if it is not. $\mathbf{Z}_{TH}$ is defined as the ratio of the open-circuit voltage $\mathbf{V}_{OC}$ to the short-circuit current $\mathbf{I}_{SC}$:

$$\mathbf{Z}_{TH} = \frac{\mathbf{V}_{TH}}{\mathbf{I}_{SC}} = \frac{\mathbf{V}_{OC}}{\mathbf{I}_{SC}} \tag{2.14}$$

$\mathbf{I}_{SC}$ is the current that flows from the circuit output to ground if the output is short-circuited to ground.

Another equivalent circuit representation is the *Norton equivalent circuit*. According to Norton's theorem, any two-terminal linear network of resistors and voltage sources can be replaced by an ideal single current source $\mathbf{I}_{SC}$ in parallel with $\mathbf{Z}_{TH}$.

2.6 RESISTOR

A resistor is a two-terminal passive circuit element used to relate voltage to a current, and vice versa. The voltage v across the terminals is proportional to the current i through the resistor. The proportionality constant is called *resistance R* in ohms (Ω). The resistor dissipates energy into heat. *Ohm's law* relates the volume charge density $\mathbf{J}$ (A/m^2) and electric field intensity $\mathbf{E}$ (V/m) by

$$\mathbf{J} = \sigma \mathbf{E} \qquad \text{or} \qquad \mathbf{E} = \rho \mathbf{J} \tag{2.15}$$

where σ and ρ are, respectively, the electric conductivity (A/(V.m)) and resistivity of the material, with $\rho = 1/\sigma$ ((V.m)/A). For a long bar of length L and cross section A, $E = V/L$ and $J = I/A$, and Eq. (2.15) gives

$$R = \frac{v}{i} = \frac{\rho L}{A} \tag{2.16}$$

Electrical resistance impedes current flow through it, which is analogous to resistance to fluid flow through a pipe, heat flow through a wall, and damping in a mechanical system. The reciprocal of the resistance is called the *conductance G = 1/R* in units of mhos.

A resistor is made out of a conducting material, such as thin metal or carbon film, or from a wire of poor conductivity. Figure 2.8 shows various forms of resistors used in circuits: the

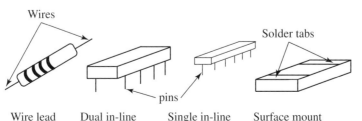

Figure 2.8 Available resistor packages.

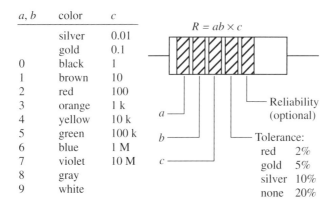

a, b	color	c
	silver	0.01
	gold	0.1
0	black	1
1	brown	10
2	red	100
3	orange	1 k
4	yellow	10 k
5	green	100 k
6	blue	1 M
7	violet	10 M
8	gray	
9	white	

$R = ab \times c$

Reliability (optional)

Tolerance:
red   2%
gold   5%
silver   10%
none   20%

Figure 2.9 Wire resistance code.

wire lead, surface mount, dual in-line package (DIP), and the single in-line package (SIP). The SIP contains multiple resistors and is made to fit conveniently into printed circuit boards.

Resistors are easy to use, but their behavior is not ideal. Resistors are characterized by the power they can safely dissipate, temperature coefficient of resistance, thermal noise, voltage coefficient—the extent to which R depends on applied v, stability with time, humidity, inductance, and other parameters. Allowable resistor power dissipation depends on air flow, thermal conduction via the resistor leads, and circuit density. The power rating of a resistor refers to the average power dissipation and may be substantially exceeded for a few seconds or more. The rated power and tolerance of the resistors used depend on the application; $\frac{1}{4}$-watt resistors with a 5% tolerance are common.

Wire-lead resistors of standard values are coded by four colored bands, a, b, c, and tol, listed with their respective values in Fig. 2.9. The corresponding resistance is determined by

$$R = ab \times c \pm \text{tol}\,(\%) \tag{2.17}$$

The $\pm$ indicates that a value could vary within $+tol$ to $-tol$ from the nominal. Resistors with carbon composition are the most frequently used type, with typical values from 1Ω to about $22\ \text{M}\Omega$. Nonstandard resistor values may be obtained by connecting standard resistors in series and/or parallel. Combining series and parallel resistors is based on the rules of Eqs. (2.8) and (2.9), with resistances used instead of impedances. It is not necessary to compute resistor values or other circuit component values to many significant places because the components themselves have a certain tolerance, and a properly designed circuit is not as sensitive to precise components' values.

A parallel combination of two resistors forms a *resistive current divider* circuit Fig. 2.10a, whereas a series combination of two resistors forms a *resistive voltage divider* circuit Fig. 2.10b.

2.6.1 Voltage Divider

The voltage divider is used ubiquitously in electronic circuits. It serves to attenuate a signal to avoid saturating the input stage of a processing circuit such as an analog-to-digital converter. The divider attenuates the signal linearly as long as the current demand of the connected load is much smaller than the current drawn from the source. For a reference input voltage v_R, the simple voltage divider shown in Fig. 2.10b, produces a predictable output voltage $v_O \leq v_R$

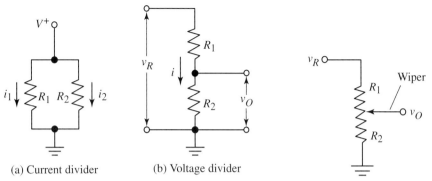

(a) Current divider   (b) Voltage divider

Figure 2.10 Two resistors connected in parallel (a), and in series as voltage divider circuit (b).

Figure 2.11 A potentiometer circuit.

always. With no load on the output (open circuit), KVL and Ohm's laws (Eq. (2.16)) gives the current i and the output voltage in resistor R_2 as

$$i = \frac{v_R}{R_1 + R_2}$$

$$v_O = iR_2 = \frac{R_2}{R_1 + R_2} v_R$$

(2.18)

If a load is connected between the output terminals, it may drain significant current and reduce the voltage references produced by the divider. The reduction in the output voltage is called *loading error*.

The voltage divider circuit is often used to provide a constant voltage source of a few millivolts from a readily available source of a few volts. It also plays a major role in processing signals from various sensors, as will be seen in Chapter 11.

When the combination of R_1 and R_2 in Fig. 2.10b is made from a single resistor, R_1 and R_2 could vary as shown in Fig. 2.11; the voltage divider is called a *potentiometer*, or *pot*. The pot provides a range of resistance values controlled by a mechanical screw, knob, or linear slide. A *trim pot* is used to adjust (trim) the resistance in a circuit. The pot is also used to measure the position of a moving object

The Thevenin and Norton equivalents of the voltage divider shown in Fig. 2.12a are shown in Figs. 2.12b and 2.12c, respectively. The Thevenin equivalent circuit consists of a voltage

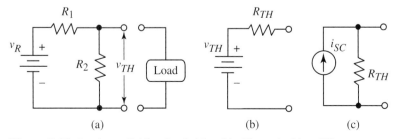

(a)   (b)   (c)

Figure 2.12 A voltage divider circuit (a) and its Thevenin (b) and Norton equivalent circuits (c).

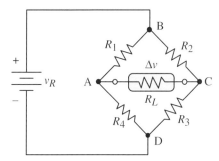

Figure 2.13 DC Wheatstone bridge circuit with load resistance.

source $v_{TH} < v_R$ in series with a resistor R_{TH} given by

$$v_{TH} = v_R \frac{R_2}{R_1 + R_2} \quad \text{and} \quad R_{TH} = \frac{v_{TH}}{i_{SC}} = \frac{R_1 R_2}{R_1 + R_2} \tag{2.19}$$

For example, a voltage divider with resistors $R_1 = R_2 = 10\ \text{k}\Omega$ and driven by a perfect 30-volt battery is equivalent to a perfect 15-volt battery in series with a 5-kΩ resistor.

2.6.2 Bridge Circuit

Bridge circuits are commonly used in instrumentation to convert a change in the impedance of a sensor to a change in voltage. The DC Wheatstone bridge shown in Fig. 2.13 is the simplest and most commonly used bridge circuit. It consists of four resistive arms with resistances R_1, R_2, R_3, and R_4. The load resistor R_L represents the input impedance of a connected device, such as a voltmeter to measure the voltage difference Δv between points A and C. In instrumentation, the connected device is usually a high-input impedance instrumentation amplifier. In the following analysis R_L is assumed to be infinite and the circuit is an open circuit. Using the voltage divider rule and with some algebra, the voltage difference Δv, also called *offset voltage*, is

$$\Delta v = \frac{R_1 R_3 - R_2 R_4}{(R_1 + R_4)(R_2 + R_3)} v_R \tag{2.20}$$

The offset voltage Δv will be zero if the resistors are chosen to satisfy

$$R_1 R_3 = R_2 R_4 \tag{2.21}$$

This is known as the *null condition*. Equations (2.20) and (2.21) are the basis of bridge circuit applications.

Thevenin Equivalent

When the null condition is not met, and since the load resistance R_L is finite, current i_L will flow through the load. One way to determine i_L is to find the Thevenin equivalent of the bridge circuit between terminals A and C. The Thevenin voltage is simply the open-circuit voltage given by Eq. (2.20). The Thevenin resistance R_{TH} is the resistance seen between terminals

A and C when the supply is replaced by its internal resistance. The internal resistance of the supply is insignificant compared with the arm resistances, and the Thevenin resistance is simply the bridge resistance R_B, which is found to be

$$R_B = \frac{R_1 R_4}{R_1 + R_4} + \frac{R_2 R_3}{R_2 + R_3} \tag{2.22}$$

The current through the load resistance is simply

$$i_L = \frac{V_{TH}}{R_{TH} + R_L} \tag{2.23}$$

and the bridge offset with a finite load resistance becomes $\Delta v_L = i_L R_L$. The ratio of this offset voltage to the open-circuit offset voltage is

$$\frac{\Delta v_L}{\Delta v} = \frac{1}{1 + R_B/R_L} \tag{2.24}$$

Equation (2.24) is used to assess loading errors associated with finite load resistance. Refer to Section 2.10.2 for more details on loading errors.

2.6.3 Small-Signal Resistance

In electronic devices for which i is not proportional to v, it is useful to know the slope dv/di of the v-i curve. This is called the *small-signal resistance*, incremental resistance, or dynamic resistance. Dynamic resistance is an important specification of the *Zener diode*, discussed in Chapter 3.

2.6.4 Resistance-Based Sensors

Many forms of resistors are manufactured to be used as sensors to measure various physical quantities. The potentiometer is a variable resistance sensor used to measure the position of a moving object. Thermoresitive devices such as the *resistive temperature detector* (RTD) and the *thermistor* are temperature-dependent resistors and are used to measure temperature and other, related variables, such as heat flow rate and thermal conductivity. The *strain gauge* is a resistive sensor where the resistance changes with the applied stain. It is used to measure strain and other quantities derived from it, such as stress, force, and pressure. The *photoresistor* is a light-dependent resistor used to measure light intensity by measuring the resistance change caused by imposed light. These sensors are discussed in detail in Chapter 11.

The signal from a resistive-based sensor may be processed by replacing one of the resistive elements in the voltage divider circuit or the bridge circuit by the sensor element. The voltage measured at the output of the circuit is related to the physical quantity being measured.

2.6.5 Measuring Electrical Resistance

Electrical resistance may be measured by one of two techniques: constant voltage and constant current. In the constant-voltage method, a known voltage is applied across the unknown resistance and the resulting current through it is measured. This method is used to measure

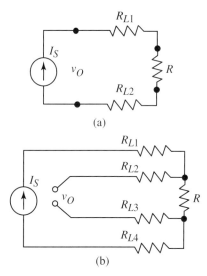

(a)

(b)

Figure 2.14 Measuring resistance: 2-wire (a) and 4-wire (b)

ultrahigh resistance in the 10^8–10^{16}-Ω range typical of leakage resistance of a capacitor, insulator resistance, or relay contact isolation. The *voltage coefficient* of the resistance, a measure of how the resistance reacts to various magnitudes of test voltages, can be ascertained.

A constant-current method is used to measure resistances below 200 MΩ and is the basis of digital multimeters (DMMs) measurements. A known current is sourced through the unknown resistance, and the voltage developed across the resistance is measured and Ohm's law applied to determine the resistance. This method may be implemented by either the two-wire measurement or the four-wire measurement, as shown in Fig. 2.14. The two-wire method sources and senses on the same two leads, whereas in the four-wire method the current is sourced through two wires and the voltage drop is measured across two other wires. The four-wire method eliminates the effect of the small lead resistance of the test wires (hundreds of milliohms), which may be relevant in low-resistance measurements.

2.7 CAPACITOR

A capacitor is a two-terminal passive element that stores energy in the form of an electric field. Capacitors are essential in nearly every circuit of practical importance, such as waveform generators, oscillators, filters, integrators, differentiators, and sample-and-hold circuits, among many more. Figure 2.15 shows two basic capacitor forms. The capacitor in Fig. 2.15a is formed from two parallel conducting plates separated by an insulator material called a *dielectric*. Applying a voltage v across the terminals of a capacitor results in positive charges $+q$ and negative charges $-q$ to be stored on the opposite plates. For an ideal capacitor, the constitutive relation relating the charge to the applied voltage is

$$q = Cv \tag{2.25}$$

The constant C is the *capacitance* of the capacitor, in units of farads (F or C/V). The capacitance is a property of the dielectric material, the plate geometry, and the distance between the plates. C for the two-parallel-plate capacitor with plates of area A (m^2) separated a

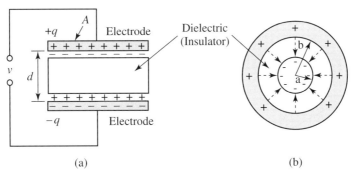

Figure 2.15 Two capacitor types: parallel plates (a) and cylindrical (b).

distance d (m) that is much smaller than the plate area by a dielectric material of dielectric constant κ is

$$C = \frac{\kappa \varepsilon_0 A}{d} \tag{2.26}$$

where the permittivity constant $\epsilon_0 = 8.8542 \times 10^{-12}$ (C^2/N-m^2). Table 2.1 gives the dielectric constant for various dielectric materials. The dielectric increases the capacitance as a result of permanent or induced electric dipoles in the material. The capacitance of a capacitor with n dielctric layers is

$$C = \frac{\varepsilon_0 A}{d_1/\kappa_1 + d_2/\kappa_2 + \cdots + d_n/\kappa_n} \tag{2.27}$$

where κ_i and d_i are, respectively, the dielectric constant and the gap of the ith dielectric layer.

Figure 2.15b is a capacitor consisting of two concentric cylinders of radii a and b. If the conductors overlap for a length l and if $l \gg b$, then the capacitance of this arrangement may be determined by

$$C = \frac{2\kappa \varepsilon_0 \pi l}{\ln(b/a)} \tag{2.28}$$

TABLE 2.1 Dielectric Constant of Selected Gap Materials at 25°C

Material	κ	Frequency (Hz)	Material	κ	Frequency (Hz)
Air	1.00054	0	Porcelain	6.5	0
Ceramic (alumina)	4.5–4.8	10^6	Pyrex glass (7070)	4.0	10^6
Epoxy resins	3.65	10^3	Rubber (neoprene)	6.6	10^3
Ferrous oxide	14.2	10^8	Rubber (silicone)	3.2	10^3
Lead nitrate	37.7	6×10^7	Silicon resins	3.85	10^3
Nylon	3.5	10^3	Teflon	2.04	10^3–10^8
Paper	3.5	0	Vacuum	1	—
Plexiglass	3.12	10^3	Water	78.5	0

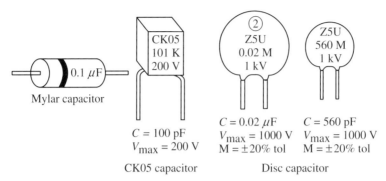

Figure 2.16 Various capacitor packages and associated markings.

A farad is very large quantity, and capacitances on the order of microfarads (μF) or pico-farads (pF) appear in circuits. Typical C values range between 1 pF and 1000 μF. A larger capacitance value requires a larger overlapping area and a closer gap. This is usually accomplished by placing a conductor onto a thin insulating dielectric, such as aluminized Mylar film rolled up into a small cylindrical configuration.

The primary types of commercial capacitors are the ceramic disc capacitors, tantalum capacitors, Mylar capacitors, metallized mica, and metal foils with oxide insulators known as electrolytics. Electrolytic capacitors are normally used when high voltage and large capacitances are needed. Figure 2.16 shows some capacitor packages and associated markings, and Table 2.2 gives the characteristics of some commercial capacitors. The leakage resistance R_{Leakage} is the effective resistance of the discharge path through the dielectric, which measures the capacitor's ability to retain its charge. This is an important parameter in *timing* circuits where the capacitor is connected in series with a resistor to produce a specific time delay in the circuit. The R_{Leakage} through the capacitor's dielectric adds to the series resistance and changes the time constant of the circuit given by $\tau = RC$. For a small τ, capacitors with high-leakage resistance are recommended. For large τ, capacitors with low-leakage resistance, such as tantalum capacitors, are generally used.

Since DC current does not flow through a capacitor, an electric field is established by the displacement of charges from one side of the capacitor through the conducting circuit to the

TABLE 2.2 Comparison of Selected Capacitors

	Type	C	V_{MAX}	Tolerance (%)	R_{Leakage} (MΩ)
Electrostatic	Mica	1 pF to 0.1 μF	500–75,000	± 1 to ± 20	1,000
	Ceramic (low loss)	1 pF to 0.001 μF	6,000	± 1 to ± 20	1,000
	Ceramic (high κ)	100 pF to 0.1 μF	up to 100	± 1 to ± 20	30–100
	Mylar	5,000 pF to 20 μF	100–600	± 20	10,000
	Teflon	1,000 pF to 2 μF	50–200	Excellent	Best
	Polystyrene	500 pF to 20 μF	up to 1,000	± 5	10,000
	Paper (oil soaked)	1,000 pF to 50 μF	100–100,000	± 10 to ± 20	100
Electrolytic	Tantalum	0.1 μF to 500 μF	6–10	Poor	100
	Aluminum	0.1 μF to 0.5 F	up to 500	Very poor	1

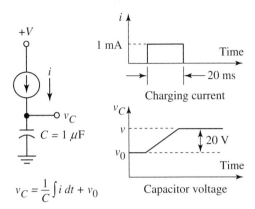

$$v_C = \frac{1}{C} \int i\, dt + v_0$$

Figure 2.17 A constant current charges the capacitor at a constant rate dv/dt.

other side. The displacement of charge is called a *displacement current*, since current appears to flow momentarily through the device. From Eqs. (2.2) and (2.25), the integral form of the voltage–current relationship in a capacitor is

$$v = \frac{1}{C} \int_0^t i\, dt \tag{2.29}$$

The voltage rises as a ramp when a constant current is supplied to the capacitor. Refer to Fig. 2.17, where a 1-mA current supplied to a 1-μF capacitor for 20 ms increases the voltage across the capacitor by 10 V. Equation (2.29) implies that the voltage across a capacitor cannot change instantaneously unless an infinite current is applied. Differentiating Eq. (2.29) gives the derivative form of the current–voltage relation in a capacitor as

$$i = C \frac{dv}{dt} \tag{2.30}$$

which implies that the current through a capacitor is proportional to the rate of change of voltage. Equations (2.29) and (2.30), respectively, define the capacitor as having *integral causality* and *derivative causality*. Referring to Eq. (2.30), causality means that one variable (say v) is considered to be defined by the system or an external input and that the other variable (i) is defined by the equation that relates the two variables.

The rules stated in Eqs. (2.8) and (2.9) are applied to find the combination of parallel and series capacitances, respectively, where capacitances are used instead of impedances.

Electrostatic Energy

Assuming an isotropic medium with continuous charge distribution, the electrostatic (potential) energy stored in a capacitor is determined by Eqs. (2.3) and (2.29) as

$$w_e(q) = \int_0^t vi\, dt = \int_0^q v\, dq \tag{2.31}$$
$$= \frac{1}{2} C v^2$$

The positive and negative charges on the opposite electrodes of the capacitor set up an electrostatic force of attraction F_e between them. If one of the capacitor's plates is allowed to move, the force would cause plate displacement. Assuming a lossless conservative system, and applying the *principle of virtual work*, the differential change in the electrostatic energy dw_e stored on the capacitor is equal to the differential change in the mechanical work dw_{mec} done by the force as it displaces the movable plate by $d\mathbf{x}$:

$$dw_e = dw_{mec} = \mathbf{F}_e \cdot d\mathbf{x} \tag{2.32}$$

The electrostatic force on the movable plate is given by

$$\mathbf{F}_e = \nabla w_e = \frac{\partial w_e}{\partial x}\mathbf{i} + \frac{\partial w_e}{\partial y}\mathbf{j} + \frac{\partial w_e}{\partial z}\mathbf{k} \tag{2.33}$$

Equations (2.31) and (2.33) form the basis for the analysis and design of electrostatic actuators, an important actuator in the design of microelectromechanical systems (MEMSs).

Example 2.1: Electrostatic Actuator—Derive the equation of motion of the parallel-plate electrostatic actuator shown in Fig. 2.18. m is the mass of the moving upper plate, k is the stiffness of the spring connected between the upper plate and the fixed frame, and c represents the equivalent viscous damping induced due to air flow resulting from the up/down motion of the upper plate. The actuator is controlled by a voltage source that is modeled as an ideal source v_{in} in series with a resistor R. Initially, when the applied voltage and spring force are both zero, the gap between the plates is x_0. Neglect the fringing fields at the edges of the capacitor plates.

Solution: The force of attraction between the oppositely charged plates is F_z. Substituting Eqs. (2.25) and (2.26) into Eq. (2.31) and applying the third term of Eq. (2.33) to the result yields the actuator force,

$$F_z = \frac{1}{2}\frac{q^2}{\varepsilon A} \tag{2.34}$$

Applying Newton's second law on the m, k, c mechanical subsystem results in the differential

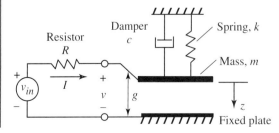

Figure 2.18 Electrostatic actuator driving a mass, damper, and spring system.

equation that defines the behavior of the actuator as

$$m\ddot{z} + b\dot{z} + kz + \frac{1}{2}\frac{q^2}{\varepsilon A} = 0 \tag{2.35}$$

If at $t = 0$ the position of the mass is at x_0, then $z = x - x_0$, and Eq. (2.35) becomes

$$m\ddot{x} + b\dot{x} + kx = kx_0 - \frac{1}{2}\frac{q^2}{\varepsilon A} \tag{2.36}$$

2.7.1 Capacitor Applications

Capacitors have a wide range of electronic applications, which include timing, filtering, trimming and tuning, bypassing, coupling, and sensing. *Trimming* and *tuning* capacitors are variable capacitors used in circuits (commonly radio frequency circuits) designed to be sensitive to only one frequency. The desired frequency is set by adjusting the variable capacitor. In *filtering*, capacitors are required to remove AC voltages superimposed on the DC output of a DC power supply. Capacitors are part of many filter designs to remove noise from a pure signal. The impedance of the capacitor decreases with frequency. Therefore, placing a capacitor between any point in the circuit and ground will allow DC and slowly varying signals to pass but will short higher-frequency signals to ground. The capacitor used for this purpose is called a *bypass* capacitor. *Coupling* serves the opposite function to bypassing. A coupling capacitor is used to pass or couple a particular AC signal from one stage in a circuit to the next. Depending on the frequency involved, either electrolytic or electrostatic units may be used in bypassing and coupling. Capacitive-based sensors used to measure a variety of physical quantities are discussed in Chapter 11.

2.8 INDUCTOR

The *inductor* is a two-terminal passive element that stores energy in the form of a magnetic field. The most common form of an inductor is the coil, a wire wound with N turns. The inductor requires a voltage directly proportional to the time rate of change of the current. The constant of proportionality, called *inductance L* in units of henrys (H), is related to the magnetic field in the circuit. For a better understanding of inductance and inductive-based devices such as motors and solenoids, a brief review of the basics of electromagnetic theory and circuits is presented next.

2.8.1 Magnetic Effect of an Electric Current

When a current flows through a conductor, the space around the wire develops into an energy storage region called *magnetic field*, existing simultaneously with the electric field. Like a permanent magnet, the magnetic field exerts forces and does work on any magnetic object that exists in the vicinity of the field.

A magnetic field is mapped as continuous and closed magnetic lines called *flux lines*. The lines surrounding a straight conductor are circular, as shown in Fig. 2.19. The direction of the

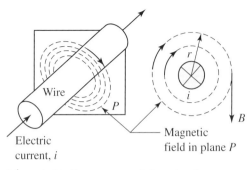

X symbol indicates current is into the plane

Figure 2.19 Circular magnetic field generated around a current-carrying conductor.

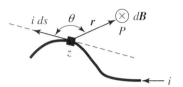

Figure 2.20 Dependence of magnetic field on the direction and magnitude of electric current.

lines depends on the direction of the electric current. It is determined by the right-hand rule, as follows: Grasp the conductor by the right hand while the thumb points in the direction of current flow; the flux direction will be that in which the fingers wrap around the conductor. The number of flux lines passing through a given surface of area A is a measure of the magnetic field ϕ, in units of lines in US units and webers in SI units, for that surface.

The magnetic flux density **B** at a point in space is represented by a vector tangent to a magnetic field line at that point. The SI units of **B** are webers/m², or teslas (T). **B** is large if the flux lines are close together, small otherwise. The density of the lines is greatest near the conductor. At a point P in space located by a radial distance r from a current-carrying conductor as shown in Fig. 2.20, **B** is defined by the *Biot–Savart law* as

$$d\mathbf{B} = \frac{\mu}{4\pi} \times \frac{i d\mathbf{s} \times \mathbf{r}}{r^3} \tag{2.37}$$

where $\mu = \mu_0 \mu_r$ is the permeability of the medium in which the field exists, $\mu_0 = 4\pi \times 10^{-7}$ (Tm/A) is the permeability of the vacuum, or *permeability constant*, and μ_r is the *relative permeability* of the medium. $d\mathbf{s}$ is a differential element of the conductor, a vector quantity tangent to the wire in the direction of the current. $\mathbf{r}$ is the position vector from element $d\mathbf{s}$ to the point in space.

Magnetic flux density is related to the magnetic field intensity **H** [A/m] by

$$\mathbf{B} = \mu \mathbf{H} \tag{2.38}$$

The magnetic flux ϕ [Wb = (T-m²)] through a closed surface is defined as

$$\phi = \oint_s \mathbf{B} \cdot d\mathbf{s} \tag{2.39}$$

When the flux is distributed uniformly over an area, the density at any point in that area is

$$B = \frac{\phi}{A} \tag{2.40}$$

The ability of a magnetic circuit to produce a magnetic flux in a coil is known as the *magnetomotive force* F_{MM} in units of ampere-turns and is defined as the line integral of the magnetic field intensity

$$F_{MM} = \int_l \mathbf{H} \cdot d\mathbf{l} \tag{2.41}$$

$$= \phi \mathfrak{R}$$

where $\mathfrak{R}$ is the *reluctance* of the magnetic circuit given by

$$\mathfrak{R} = \frac{l}{\mu A} \tag{2.42}$$

where l is the length, A is the cross-sectional area, and μ is the permeability of the magnetic material. A current i flowing through a coil with N turns results in F_{MM} given by

$$F_{MM} = Ni \tag{2.43}$$

If a coil has widely spaced turns, the magnetic field tends to cancel between the wires, as illustrated in Fig. 2.21a. The magnetic flux sustained by a coil with closely packed helix is depicted in Fig. 2.21b. The arrowheads indicate the north-to-south direction of the magnetic field lines.

The energy stored in a magnetic field is expressed as

$$w_m = \frac{1}{2} \int_v \mathbf{B} \cdot \mathbf{H} \, dv \tag{2.44}$$

For a linear isotropic medium, Eq. (2.44) becomes

$$w_m = \frac{1}{2} \int_v \mu |\mathbf{H}|^2 dv = \frac{1}{2} \int_v \frac{1}{\mu} |\mathbf{B}|^2 dv \tag{2.45}$$

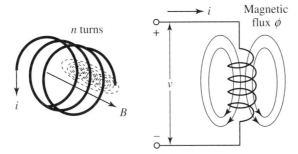

(a)   (b)

Figure 2.21 Formation of a magnetic flux in a coil (a); flux linkage in a conductor (b).

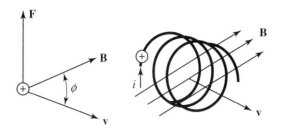

Figure 2.22 Spiral movement of an electric charge in a magnetic field caused by the induced force.

The magnetic field on a conductor exerts a magnetostatic force on any permeable object in the field. Using the principle of virtual work, the differential change in magnetic energy dw_m, assuming a lossless conservative system, is equal to the differential change in the mechanical energy, dw_{mec}:

$$dw_m \equiv dw_{mec} = \mathbf{F}_m \cdot d\mathbf{s} \tag{2.46}$$

Thus, the magnetostatic force is given by

$$\mathbf{F}_m = \nabla w_m = \frac{\partial w_m}{\partial x}\mathbf{i} + \frac{\partial w_m}{\partial y}\mathbf{j} + \frac{\partial w_m}{\partial z}\mathbf{k} \tag{2.47}$$

2.8.2 Electromagnetic Force

If an electric charge q is moved through a fixed magnetic field $\mathbf{B}$ at a velocity $\mathbf{v}$, a deflecting force $\mathbf{F}$ called *Lorentz force* will act on it, causing its velocity to change and making it experience a spiral motion, as shown in Fig. 2.22. The deflection force is always at right angles to the plane formed by $\mathbf{v}$ and $\mathbf{B}$ and is given by the cross product

$$\mathbf{F} = q\mathbf{v} \times \mathbf{B} \tag{2.48}$$

This equation indicates that the units of $\mathbf{B}$, teslas, is equivalent to $(N/C)/(m/s)$.

The deflection force acting on the charge results in an electric field that leads to a voltage difference in a conducting material, producing an electric current. The Lorentz force is the basis of the *Hall effect* and the *magnetoresistive effect* exploited in sensors.

If mobile charges within a conductor of length L traverse a magnetic field, the charges and the conductor experience a force according to

$$\int d\mathbf{F} = \int_0^L dq\, \mathbf{v} \times \mathbf{B} = \int_0^L i\, d\mathbf{s} \times \mathbf{B} \tag{2.49}$$

This relation is the basis for the operation of DC motors.

2.8.3 Self-Inductance

When the magnetic flux ϕ generated in a coil of wire carrying a current varies, a voltage is induced in any circuit linked to the varying magnetic flux, including the circuit that produced the flux. *Faraday's law* relates the induced voltage to the changing flux as

$$v = \frac{d\phi}{dt} \tag{2.50}$$

Lenz's law defines the direction of ϕ so as to produce a current opposing the flux changes. If all N turns in a coiled wire have the same cross-sectional area, the flux in each turn will be the same, and the induced voltage is

$$v = \frac{d(N\phi)}{dt} = N\frac{d(BA)}{dt} = \frac{d\lambda}{dt} \tag{2.51}$$

where $\lambda = N\phi$ is the total magnetic flux linking the coil and is called *flux linkage* (V-s or weber).

The magnetic flux generated in a coil of wire carrying a current produces a magnetic field that links the coil and permeates the medium around it. If the current in the coil changes with time, the magnetic flux will also change and a voltage is induced in the circuit. This *self-induced* voltage is found by Eq. (2.50) as

$$v = N\frac{d\phi}{dt} = N\frac{d\phi}{di}\frac{di}{dt} = L\frac{di}{dt} \tag{2.52}$$

where

$$L = N\frac{d\phi}{di} \tag{2.53}$$

is the *self-inductance* of the circuit, in units of henrys (H) when ϕ is in webers (weber). Inductors used in practical circuits are specified in millihenrys (mH). Inductance is analogous to a moving mass in a mechanical system.

Equation (2.52) indicates that, in the absence of infinite voltage, the current in an inductor cannot change instantaneously. For instance, putting a constant voltage across an inductor causes the current to rise as a ramp; 1 volt across a coil of 1 henry produces a current that increases at a rate of 1 A/s. For an ideal coil, the flux is proportional to the current and Eq. (2.52) becomes

$$L = \frac{N\phi}{i} \tag{2.54}$$

The value of the inductance L depends on the coil geometry and the properties of the core material on which it is wound. High-permeability ferromagnetic materials, such as iron (or iron alloys such as permalloy and AlNiCo, laminations, or powder) and ferrite, a black nonconductive brittle magnetic material, are far more receptive to magnetic flux than is air or free space. Such core materials can concentrate and confine the predominant portion of the flux within itself and multiply the inductance of a given coil by its permeability. These materials, however, are not suitable for all applications, because they are prone to magnetic saturation and losses. The core may be formed in a variety of shapes; the rod and the toroid are the most common. Inductors are heavily used in radio frequency (RF) circuits. They also form the basis of a variety of practical devices.

2.8.4 Inductor-Based Devices

Relays

A relay is a device that makes or breaks an electrical current, much like an electrically controlled switch. When its coil is energized, a normally open (NO) contact will close and a normally

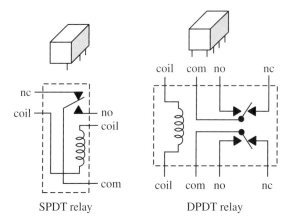

Figure 2.23 Electromechanical relays.

SPDT relay   DPDT relay

closed (NC) contact will open. There are two types of relays: Electromechanical relays use moving mechanical contacts, and solid-state relays are all electronic, with no moving parts.

Electromechanical relays are available in all shapes and sizes and with different contact configurations. They range from small reed relays for low-current applications to heavy-duty power relays that can switch load voltages up to 200 volts DC and 140 volts AC (RMS, or root mean square). Reed relays are usually rated for less than 1 A of load currents and are controlled by voltage levels of 5, 12, and 24 volts. Consequently, a reed relay may be controlled from an open-collector transistor-transistor logic, or TTL, output (discussed in Chapter 4). Reed relays are available in DIPs for mounting in standard IC sockets.

Figure 2.23 shows the internal structure of a single-pole double-throw (SPDT) and a double-pole double-throw (DPDT) electromechanical relay. The input side consists of a small coil. When different voltages are applied to the two lines marked *coil*, the resulting current creates a magnetic field inside the device. This field attracts a metal lever to which the internal switch contacts are attached. Activation of the lever in turn disconnects one circuit and connects the other. When no voltage is applied, the line marked *com*, for *common*, is connected to the *normally closed* pin *nc*. When voltage is applied across the coil, com is disconnected from nc and connected to the *normally open* line *no*.

High-power relays are often placed in series with high-power loads to control load currents above 1 A. They are available in a variety of power ratings and contact configurations to control both DC and AC current levels. Power relays cannot be directly controlled by digital signals; an intermediate solid-state relay must be used in this case.

Solenoids

A solenoid is an electromechanical actuator that converts magnetic field energy into a linear mechanical motion, applying in the process a sudden force on a connected object. It is formed by winding a coil with a closely packed helix of N turns around a rod made of a ferrous material such as hard steel. The rod, also called the plunger (Fig. 2.24a), may be free or spring loaded. Exciting the coil with a current creates a magnetic field that exerts a force on the plunger and causes it to move, or actuate. If the length of the coil helix l is much greater than its radius r, the flux density inside the solenoid is

$$B = \frac{\mu N i}{l} \qquad (2.55)$$

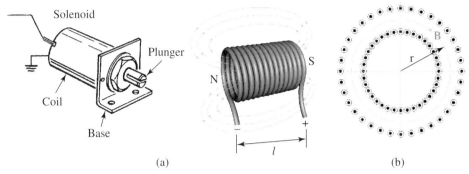

(a)   (b)

Figure 2.24 A solenoid (a) and a toroid (b).

Solenoids are available as DC or AC devices. The solenoids must be energized by an inter-mediate solid-state relay if they are used in a digital control system. Solenoids are rated for either intermittent duty or continuous duty. Intermittent-duty solenoids can be energized for short periods of time. Continuous-duty solenoids can be energized for longer periods of time without damage.

The *toroid* is the rotary counterpart to a solenoid. The closely packed helix coil is wound around a core material in the shape of a toroid (Fig. 2.24b). The flux density inside the toroid is

$$B = \frac{\mu}{2\pi} \frac{iN}{r} \tag{2.56}$$

where r is the radius within the circular core where magnetic field is calculated.

Mutual Inductance

Two circuits linked by the same magnetic field are said to be *coupled*. The circuit element used to represent magnetic coupling is shown in Fig. 2.25 and is called *mutual inductance M* in units of henrys (H). The voltage v_2 (or v_1) induced in one of two coupled circuits due to cur-rent i_1 (or i_2) flowing through the other circuit is expressed as

$$v_2 = M \frac{di_1}{dt} \qquad \text{and} \qquad v_1 = M \frac{di_2}{dt} \tag{2.57}$$

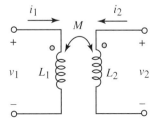

Figure 2.25 Mutual inductance circuit.

If currents flow through the two coupled circuits simultaneously, the voltage–current relations become

$$v_1 = L_1 \frac{di_1}{dt} + M \frac{di_2}{dt}$$

$$v_2 = L_2 \frac{di_2}{dt} + M \frac{di_1}{dt}$$

(2.58)

Mutual inductance is the basic circuit of many devices, including the *resolver* and the *linear variable differential transformer* (LVDT) displacement sensors, discussed in Chapter 11.

Transformers

A transformer is a device for transferring the energy of AC signals across two closely coupled coils called the *primary coil* with N_P turns and the *secondary coil* with N_S turns, as shown in Fig. 2.26. The two coils are electrically insulated from each other. The transfer of energy involves a change in voltage levels, currents, and impedances. The transformation ratio is the direct ratio of turns $n = N_S/N_P$. If $N_S < N_P$, the transformer is called a *step-up* transformer; otherwise it is called a *step-down* transformer. The coupling coefficient of a transformer is given by

$$K = \frac{M}{\sqrt{L_P L_S}}$$

(2.59)

For an ideal transformer, $K = 1$; but for a practical transformer, $K < 1$ because of anomalies such as hysteresis, leakage, and heating of the core. In ideal transformers, voltages, currents, and impedances are transformed as

$$v_S = \frac{N_S}{N_P} v_P$$

$$i_S = \frac{N_P}{N_S} i_P$$

(2.60)

$$Z_S = \left(\frac{N_S}{N_P} \right)^2 Z_P$$

whereas power remains unchanged and a step-up transformer gives higher voltage at lower current.

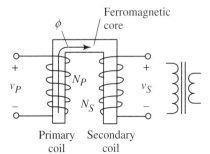

Figure 2.26 Ideal transformer and its electrical symbol.

Transformers are specified by the secondary voltage in RMS volts at full transformer rated-power load, the power rating (V-A or W), and the regulation factor, which is the percentage increase in the secondary voltage when the load is removed. For example, a 15-V-A transformer delivers 1.5 A when its secondary voltage is 10 V. Transformers that are typically used in instruments may have secondary voltages from 10 to 50 volts, with current ratings ranging from 0.1 to 5 A.

Transformers are very useful in a variety of applications. In signal conditioning, transformers are used to change the level of the AC line signal voltage to a lower value, isolate electrical signals between different parts of a circuit to eliminate ground loop problems, isolate an instrument from actual connection to the power line, convert impedance levels or match impedances for maximum power transfer, and transfer radio frequency signals between various parts of a circuit. Transformers also make it possible to generate electricity at the most economical generator voltage, transfer power at the most economical transmission voltage, and utilize power in a particular device at the most economical voltage.

2.9 ALTERNATING CURRENT

Figure 2.27 shows cosine waveforms for a current and a voltage, where T is the *period* of one *cycle*. The number of cycles per second is the waveform *frequency* $f = 1/T$, in units of cycles/s, or hertz (Hz). The equations of the current and voltage as functions of time are

$$v = V_M \sin(\omega t + \alpha)$$
$$i = I_M \sin(\omega t + \beta)$$

(2.61)

where i and v are instantaneous values and I_M and V_M are the corresponding current and voltage amplitudes, respectively. The *angular frequency* ω in units of radians/s is

$$\omega = \frac{2\pi}{T} = 2\pi f$$

(2.62)

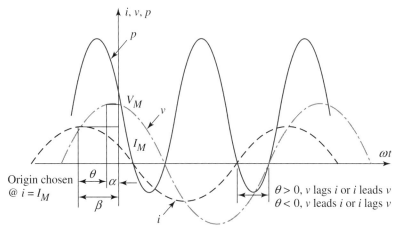

Figure 2.27 Instantaneous voltage, current, and power in a circuit excited by a sine wave.

While the origin of the waveform in Fig. 2.27 is chosen where i has a maximum positive value, it may be chosen at any other convenient point. The current and voltage are displaced from one another by a phase angle θ. If θ is positive (negative), the voltage *lags* (*leads*) the current, or the current *leads* (*lags*) the voltage by the angle θ.

The amplitude is one of many ways to characterize the magnitude of a sine wave or any other signal. Other specifications include *peak-to-peak* (p-p) amplitude and the *effective* amplitude, also known as the *root mean square*, or RMS, value. The RMS letters, read in reverse, indicate the process of evaluating the RMS value: Square (S) the ordinate of the wave, find the mean (M) of the squared wave, and take the square root (R) of the result. It follows that the RMS values for a time-varying current and voltage waveforms are

$$I_{RMS} = \sqrt{\frac{1}{T}\int_0^T i^2 dt}$$

$$V_{RMS} = \sqrt{\frac{1}{T}\int_0^T v^2 dt}$$

(2.63)

For a sinusoidal signal, $I_{RMS} = I_M/\sqrt{2}$ and $V_{RMS} = V_M\sqrt{2}$.

2.9.1 Steady-State and Frequency Response

Analysis of circuits with time-varying voltages and currents is performed in the *time domain* and the *frequency domain*. The time domain approach deals with the variations in time of v and I, whereas frequency domain deals with the variation in frequency of the amplitude and phase. Both methods provide important information about the behavior of the circuit.

A circuit that contains linear circuit components (resistors, capacitors, and inductors) and linear amplifiers is called a *linear circuit*. The forced response of a linear circuit driven by periodic functions, such as a sine wave, or time-invariant function (DC) is often referred to as the *steady-state response*. A linear circuit has the property that its output, when driven by two input signals, equals the sum of its individual outputs when driven by each signal in turn. When linear circuits are excited by AC signals of a given frequency ω, the current through and voltage across every element in the circuit at steady state will be AC signals of the same frequency, with, at most, changed amplitude and phase. The magnitudes of the currents everywhere in the circuit will be proportional to the magnitude of the driving voltage.

If a linear circuit is excited by a sine wave and the frequency f is varied over the bandwidth (frequency range) of interest, the output to the input ratio is referred to as a *frequency response*. This ratio has a magnitude and a phase angle, and both are functions of frequency f. The manner in which the amplitude ratio varies with frequency defines the behavior of linear circuits.

2.9.2 Complex Number Representation of Voltage and Current

The steady-state analysis of AC circuits is simplified by using the algebra of complex numbers. Sinusoidal voltages and currents are represented by the complex quantities

$$\mathbf{V} = V_M e^{j\alpha} = V_M\angle\alpha$$

$$\mathbf{I} = I_M e^{j\beta} = I_M\angle\beta$$

(2.64)

where **V** and **I** are called *phasors* and

$$e^{j\theta} = \cos\theta + j\sin\theta \qquad \text{where} \qquad j = \sqrt{-1} \tag{2.65}$$

is *Euler's identity*. Actual voltages and currents are obtained by multiplying their complex numbers by $e^{j\omega t}$ and then taking the real part as

$$\begin{aligned}
v &= \text{Re}(\mathbf{V}e^{j\omega t}) = \text{Re}(\mathbf{V})\cos\omega t - \text{Im}(\mathbf{V})\sin\omega t \\
i &= \text{Re}(\mathbf{I}e^{j\omega t}) = \text{Re}(\mathbf{I})\cos\omega t - \text{Im}(\mathbf{I})\sin\omega t
\end{aligned} \tag{2.66}$$

For example, a complex voltage $\mathbf{V} = 10j$ corresponds to a real voltage that changes in time as $v = -10 \sin \omega t$ volts.

2.10   IMPEDANCE

The ratio of the voltage across a two-terminal element to the current through it is in general a complex quantity called *impedance* **Z** in units of ohms (Ω). Any linear circuit may be described by the generalized Ohm's law, where the word *resistance* is replaced with *impedance*:

$$\mathbf{V} = \mathbf{IZ} \qquad \text{or} \qquad \mathbf{I} = \frac{\mathbf{V}}{\mathbf{Z}} \tag{2.67}$$

The ratio of the current to the voltage is also a complex quantity, called *admittance* **Y** in units of mhos. Impedance and admittance may be expressed by their real and imaginary parts as

$$\begin{aligned}
\mathbf{Z} &= R + jX \\
\mathbf{Y} &= G + jB
\end{aligned} \tag{2.68}$$

The real parts of **Z** and **Y** are the resistance R and the conductance G, and the imaginary parts are the *reactance X* and the *susceptance B*. The reactance of a capacitor is found by substituting $v = \Re(V_M e^{j\omega t})$ into Eq. (2.30) to give

$$i = -V_M C\omega \sin\omega t = \Re\left(\frac{V_M e^{j\omega t}}{-j/\omega C}\right) = \Re\left(\frac{V_M e^{j\omega t}}{X_C}\right) \tag{2.69}$$

The symbol $\Re$ signifies real part of a complex quantity. Thus, the reactance of a capacitor is

$$X_C = \frac{-1}{\omega C} \tag{2.70}$$

A similar analysis for an inductor gives

$$X_L = \omega L \tag{2.71}$$

The reactances of the capacitor and the inductor are also the impedances of these elements. In summary, the formulas for the impedance of a resistor, a capacitor, and an inductor are

$$\mathbf{Z}_R = R \quad \text{Resistor}$$

$$\mathbf{Z}_C = \frac{-j}{\omega C} \quad \text{Capacitor} \tag{2.72}$$

$$\mathbf{Z}_L = j\omega L \quad \text{Inductor}$$

The impedance of a circuit of only inductors and capacitors is purely imaginary. The v and i waveforms are always 90° out of phase, and the circuit is said to be purely *reactive*. Many AC circuits are analyzed by the rules of Section 2.5.

Equations (2.72) indicate that the impedances of the capacitor and the inductor depend on the frequency of the signal. The voltage across an inductor leads the current through by 90°. For a DC signal (an AC signal with a zero frequency), the impedance of an inductor is zero and the inductor acts as a short circuit, whereas at very high frequencies ($f \to \infty$), the inductor has infinite impedance and behaves as an open circuit.

For a capacitance, the voltage lags the current by 90°. For a DC signal, the impedance of a capacitor is infinite and the capacitor acts as an open circuit; but at a very high frequency ($f \to \infty$), the capacitor has zero impedance and behaves as a short circuit.

2.10.1   Generalized Voltage Divider

The simple resistive voltage divider can be generalized by replacing either or both resistors in Fig. 2.10 by a capacitor or an inductor or by a more complicated network of R, L, and C, as shown in Fig. 2.28. In general, the division ratio $\mathbf{V}_{IN}/\mathbf{V}_{OUT}$ is not constant but depends on frequency. A straightforward analysis leads to the generalized voltage divider relations

$$\mathbf{I} = \frac{\mathbf{V}_{IN}}{\mathbf{Z}_{TOTAL}} = \frac{\mathbf{V}_{IN}}{\mathbf{Z}_1 + \mathbf{Z}_2}$$

$$\tag{2.73}$$

$$\mathbf{V}_{OUT} = \mathbf{I}\mathbf{Z}_2 = \mathbf{V}_{IN}\frac{\mathbf{Z}_2}{\mathbf{Z}_1 + \mathbf{Z}_2}$$

These results are applied in the analysis of many important signal-conditioning circuits, including integrators, differentiators, and filters.

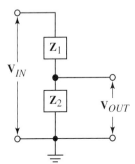

Figure 2.28 Generalized voltage divider.

2.10.2 Circuit Loading

Figure 2.29a shows two connected devices representing a common fragment of an electronic circuit. For the discussion to follow, device A represents a signal *source*, such as a sensor, whose output drives device B, which represents the *load*. The resistive circuit of Fig. 2.29b consists of a load R_{LOAD} and a voltage divider source. The voltage divider is replaced by its Thevenin equivalent, source voltage V_{TH} in series with the divider's *equivalent internal resistance* R_{TH}. The concept of equivalent internal resistance applies to all physical sources, including batteries, oscillators, amplifiers, and sensing devices.

When the load resistor is connected, the voltage divider's output v_O drops in comparison to the open-circuit voltage V_{TH}. The level of attenuation v_O/V_{TH} depends on the relative values of R_{LOAD} and R_{TH}. If R_{LOAD} is smaller than or comparable to R_{TH}, the load draws current from the source, causing V_{OUT} to attenuate. This undesirable attenuation is called *circuit loading*. If R_{LOAD} is made much greater than R_{OUT}, the load will not draw significant current from the source and loading error is insignificant. In general, loading effects can be reduced by using a voltage source (power supply) constructed from active components, e.g., operational amplifiers, whose internal resistance R_{OUT} is typically in milliohms (see Chapter 4), and v_O can be made adjustable.

The input impedance $\mathbf{Z}_{IN}$ of a device is determined as the ratio of the rated input voltage to the corresponding current through the input, while the output terminal is maintained as an open circuit. Meanwhile, the output impedance $\mathbf{Z}_{OUT}$ is the ratio of the open-circuit (no-load) voltage to the short-circuit current, both at the output port.

To generalize the concept of loading effects, Fig. 2.30 shows a sensor-amplifier arrangement very common in instrumentation and the equivalent frequency-dependent voltage divider. The single-output sensor is represented by its Thevenin equivalent circuit, a voltage source v_S in series with the sensor's internal impedance $\mathbf{Z}_{OUT}$. The sensor signal drives the amplifier, whose input impedance is $\mathbf{Z}_{IN}$. An important design objective is to reduce loading errors. This can be accomplished by ensuring that $\mathbf{Z}_{OUT}$ is much smaller than $\mathbf{Z}_{IN}$ so that the source is lightly loaded by the receiving circuit, resulting in only insignificant attenuation of

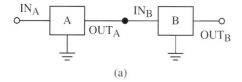

(a)

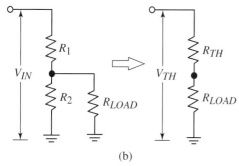

(b)

Figure 2.29 A circuit fragment (a) and a voltage divider source driving a load (b).

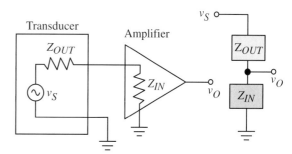

Figure 2.30 Transducer-amplifier circuit depicting circuit loading as a voltage divider.

the signal. This goal is articulated in this rule of thumb: $\mathbf{Z}_{OUT}$ of the source must be less than or equal to $1/10$ of $\mathbf{Z}_{IN}$ of the load.

Relaxing the general goal of making $\mathbf{Z}_{OUT} \ll \mathbf{Z}_{IN}$ may be possible in some situations. For example, loading the source may be tolerated if the signal levels remain constant after a load with a known and constant $\mathbf{Z}_{IN}$ is permanently connected. In applications where $\mathbf{Z}_{IN}$ varies with signal level, having a low $\mathbf{Z}_{OUT}$ ensures linearity; otherwise, the level-dependent voltage divider would cause distortion.

If the coupled signal is a current rather than a voltage, the situation is reversed and $\mathbf{Z}_{IN} \ll \mathbf{Z}_{OUT}$. Meanwhile, if the source is a current source, then $\mathbf{Z}_{OUT} = \infty$.

Circuit loading can be prevented by inserting a buffer between the source and the load. The buffer is a unity gain op-amp that has a very large input impedance and a very small output impedance. Op-amp buffers are discussed in Chapter 4.

Example 2.2: Effect of loading–A temperature sensor has an output impedance of 4 kΩ and converts the temperature to a voltage signal with a sensitivity of 10 mV/°C. The sensor signal is connected to the input terminals of an amplifier. The input impedance of the amplifier is 10 kΩ, and its output voltage is 20 times that of the input. Determine the amplifier's output when the temperature is 50°C, and compare it with the ideal value.

Solution: Figure 2.30 is a representation of the sensor amplifier connection for this example. From the available data, the unloaded output of the sensor when the temperature is 50°C is $v_S = (10 \text{ mV/°C})(50°\text{C}) = 0.5$ V. For an amplifier gain of 20, the corresponding output of the amplifier will be 10 V. The actual amplifier input by Eq. (2.73) is

$$V_{IN} = v_S \left[\frac{R_{IN}}{R_{IN} + R_{OUT}} \right] = 0.5 \times \left[\frac{10}{10 + 4} \right] = 0.357 \text{ V}$$

The corresponding amplifier output will be $0.357 \times 20 = 7.14$ V. Clearly, the loading had caused a major reduction of $10 - 7.14 = 2.86$ V at he amplifier's output.

2.10.3 Impedance Matching

Impedance matching refers to making the input impedance of the load equal the output impedance of the source; that is, $\mathbf{Z}_{IN} = \mathbf{Z}_{OUT}$. Impedance matching is important in many applications involving radio frequency (RF), video, and audio signals and microprocessor-based applications where transmission lines carry high-frequency signals. It is also important in communication devices that use low power levels. The consequence of not properly matching the impedances of the source and the receiver is that the receiver will reflect the high-frequency components of the signal back toward the source, resulting in oscillations known as *ringing*. These oscillations take time to subside unless the signal is properly terminated at the receiving end. The characteristic impedance $\mathbf{Z}_O$ of long transmission lines could also cause ringing if the cable carrying the signal is not properly terminated. The characteristic impedance of conductors in printed circuit boards or twisted pairs made from ordinary insulated wire can be as high as 100 Ω. Impedance matching is illustrated in Fig. 2.31. The 50-Ω resistor placed in parallel with the higher-impedance load matches the input impedance of the receiving network to the output impedance of the function generator.

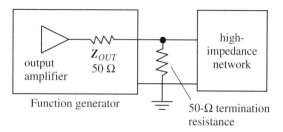

Figure 2.31 An example of signal termination.

Impedance matching is also important in applications where it is necessary to draw the largest possible amount of power from a source. One can show that for a given source resistance R_{OUT} the maximum power transferred to the load in Fig. 2.29 occurs when $R_{LOAD} = R_{OUT}$. When $R_{LOAD} = 0$ or $R_{LOAD} = \infty$, the power being transferred is zero. In some applications, impedances are not exactly matched to reduce power consumption. In most low-frequency measurements ($<$1 kHz), matching the output impedance of the transducer with the input impedance of the amplifier provides no net benefit and may not be desirable.

2.11 POWER

The *instantaneous power* in any circuit element is the time rate of work done in moving electrons in an electric field. Mathematically it is expressed as

$$p = \frac{dw}{dt} \tag{2.74}$$

Substituting Eq. (2.3) into Eq. (2.74) gives

$$p = v\frac{dq}{dt} = vi \tag{2.75}$$

The voltages and currents in reactive circuits (circuits containing capacitors and inductors) are frequency dependent and cannot simply be multiplied with each other as Eq. (2.75) suggests. Figure 2.32a shows a simple circuit in which the alternating voltage

$$v = V_M\cos\omega t \tag{2.76}$$

is applied to a network whose equivalent impedance is

$$\mathbf{Z} = Z\angle\theta = R + jX \tag{2.77}$$

The current through the network is

$$i = I_M\cos(\omega t + \theta) \tag{2.78}$$

The instantaneous power supplied to the network by the source is, from Eq. (2.75),

$$p = vi = V_M\cos\omega t I_M\cos(\omega t + \theta) \tag{2.79}$$

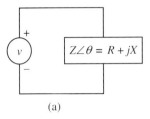

(a)

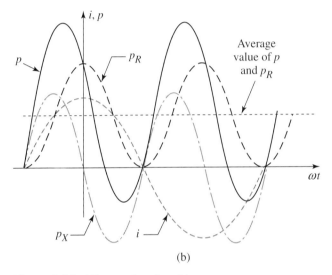

(b)

Figure 2.32 AC network and total instantaneous power p, power taken by resistance p_R and by reactance p_X.

Using trigonometric identities, Eq. (2.79) may be written as

$$p = vi = \frac{V_M}{\sqrt{2}} \frac{I_M}{\sqrt{2}}[\cos\theta(1 + \cos 2\omega t) - \sin\theta\sin 2\omega t] \qquad (2.80)$$

The voltage, current, and power waves are shown in Fig. 2.32b.

2.11.1　Average Power

Since the current and voltage in AC circuits reverse twice each cycle, the resulting power is time varying. Alternating currents and voltages are useful because the time-average value of this varying power is not zero. The *average power* consumed by an arbitrary circuit is found by

$$p_{avg} = \frac{1}{T}\int_0^T vi\, dt \qquad (2.81)$$

where T is the period for one complete cycle. Instead of using Eq. (2.81), the average power over a complete cycle can be easily computed by

$$p_{avg} = \text{Re}(\mathbf{VI}^*) = \text{Re}(\mathbf{V}^*\mathbf{I}) \qquad (2.82)$$

the symbol * indicates complex conjugate. The average power consumed by an AC network with inductors, capacitors, and resistors is expressed as

$$p = |\mathbf{V}_{RMS}||\mathbf{I}_{RMS}|\cos\theta = V_{RMS}I_{RMS}\cos\theta \tag{2.83}$$

where $|\mathbf{V}_{RMS}|$ and $|\mathbf{I}_{RMS}|$ are the complex rms amplitudes and their product is the RMS power p_{RMS}. θ is the phase angle between the current and the voltage. This relation can also be concluded from Eq. (2.80) because the time-average values of the terms $\cos 2\omega t$ and $\sin 2\omega t$ are zero.

The resistance is the only circuit element that dissipates a net amount of energy, or a definite average power, in form of heat. The corresponding instantaneous power dissipated by the resistor is

$$\begin{aligned}
p_R &= iv_R = i(iR) \\
&= (I_M\cos\omega t)(I_M R\cos\omega t) \\
&= I_{RMS}^2 R(1 + \cos 2\omega t)
\end{aligned} \tag{2.84}$$

The average time value of $\cos 2\omega t$ is zero, and the average power consumed by a resistor can be expressed as

$$p_{avg} = V_{RMS}I_{RMS} = I_{RMS}^2 R = \frac{V_{RMS}^2}{R} \tag{2.85}$$

While inductance and capacitance affect the instantaneous power, they do not contribute to the average power. The reason is that the energy stored in the inductor's magnetic field when the current applied through it increases is returned when the applied current decreases. Similarly, the energy stored in the capacitor's internal electric field when the voltage applied across its terminals increases is recovered when the applied voltage decreases. Therefore, capacitors and inductors do not contribute to a net energy transfer in a circuit. Referring to the circuit in Fig. 2.32a, the instantaneous power in the inductor is

$$\begin{aligned}
P_{X_L} &= iv_L = iL\frac{di}{dt} \\
&= -I_M\cos\omega t L I_M\omega R\sin\omega t \\
&= -I_{RMS}^2 X_L\sin 2\omega t
\end{aligned} \tag{2.86}$$

The instantaneous power taken by the capacitor is

$$\begin{aligned}
P_{X_C} &= iv_C = i\frac{1}{C}\int i\,dt \\
&= I_M\cos\omega t\left(\frac{1}{\omega C}\right)I_M\sin\omega t \\
&= -I_{RMS}^2 X_C\sin(2\omega t)
\end{aligned} \tag{2.87}$$

Hence, the average reactive power is zero because the time-average values of $\sin(2\omega t)$ in Eqs. (2.86) and (2.87) are zero.

2.11.2 Reactive Power

The cyclic variation of the instantaneous power associated with a reactive circuit is undesirable because it constitutes loading on the equipment without contributing to energy transfer. This type of power is called *reactive power* or *wattless power*, measured in reactive volt-amperes, or var. To emphasize the distinction between reactive power and average power, the latter is often referred to as the *active power* or *real power*.

Inductive reactive power is, by convention, defined as positive reactive power. Since the instantaneous power oscillations in a capacitance are of opposite sign to those of an inductance, capacitive reactive power is negative. Thus an inductor draws positive reactive power from the system, and a capacitor draws negative reactive power. A capacitor may then be viewed as a source of positive reactive power.

2.11.3 Power Factor

The two expressions in Eqs. (2.83) and (2.85) are equivalent. The reduction factor $\cos\theta$ in Eq. (2.83) appears because, in the general case, only a portion of the voltage in the circuit appears across the equivalent resistance. The $\cos\theta$ term is called the *power factor* since the average power dissipated by the network depends on it. The power factor ranges from 0 for a purely reactive circuit to 1 for a purely resistive circuit. A power factor of less than 1 indicates the existence of some component of reactive current. A circuit in which the current lags the voltage (i.e., inductance circuit) is said to have a *lagging* power factor. A circuit in which the current leads the voltage (i.e., capacitance circuit) is said to have *a leading* power factor.

Reactive currents do not result in the delivery of useful power to the load, but the cost to the power company in terms of heat dissipated in the resistance of generators, transformers, and wiring is high. Residential users are billed for only real power, but the power company charges industrial users according to the power factor. Therefore, power factor is significant in large-scale electrical power distribution systems. Capacitor yards, which are usually built behind large industrial facilities with heavy machinery such as motors, prevent the reactance of the inductive machines from returning through the distribution lines by cancelling them, thereby reducing the power factor.

2.12 SIGNALS AND SIGNAL SOURCES

In addition to the sine wave, certain patterns of time variation of waveforms, voltages that change in time in a particular way, are also of special significance in electronics. Some of the frequently encountered signals, shown in Fig. 2.33, are the ramp, sawtooth wave, triangle wave, square wave, pulse, step, spike, and noise.

Noise signals can be considered a signal that varies randomly in both frequency and amplitude. Noise may be inherently generated within the circuit components or it may be the result of external factors. Noise voltages can be specified by their frequency spectrum (power spectral density, or power per hertz) or by their amplitude distribution. A commonly existing noise is *band-limited white noise*. This kind of noise exhibits a constant power spectral density over a wide frequency range. If the noise amplitudes are normally distributed, the noise is referred to as *Gaussian white noise*. Noise generated by resistors is of this type and is detrimental to sensitive measurements of any kind.

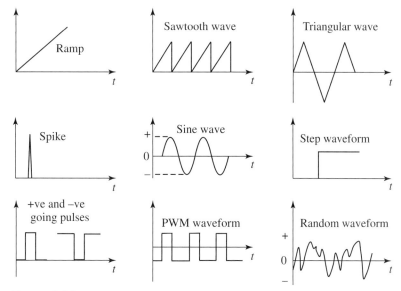

Figure 2.33 Various forms of signals.

Two signals are usually compared by the ratio of their amplitudes. Because this ratio can be very high, a logarithmic measure called the *decibel* (dB) is usually used. For example, the amplitudes and power ratios of the output signal to the input signal are

$$dB = 20 \log_{10}(A_{OUT}/A_{IN}) \quad \text{or} \quad dB = 10 \log_{10}(p_{OUT}/p_{IN}) \tag{2.88}$$

where A_{IN} and A_{OUT} represent the RMS amplitudes of any physical quantity, such as voltage or current, and p_{IN} and p_{OUT} are the power levels of the input and output signals, respectively.

2.12.1 Signal Sources

Signal sources generate a variety of signal types and provide analysis capabilities required for instrumentation and control systems. The three most common types of signal sources are signal generators, pulse generators, and function generators.

Signal generators (SGs) are sine wave oscillators capable of generating a wide range of frequency coverage (typically 50 kHz to 50 MHz) with accurate control of amplitude. Special types of signal generators are the sweep generator and the frequency synthesizer. The sweep generator can sweep its output frequency repeatedly over some range. A frequency synthesizer can generate sine waveforms with precise frequencies that could be digitally set to many significant figures, eight or more.

Pulse generators can generate pulses with the capability to control and adjust important pulse features, such as pulse width, output rate, amplitude, polarity, and rise time. Some pulse generators provide for two waveforms to be simultaneously generated, with adjustable spacing and output rate.

Function generators can generate sine, square, and triangle output waveforms with the desired amplitude and frequency (typically 0.01 Hz to 100 MHz). In addition, the devices offer DC offset setting (a constant DC voltage added to the signal), frequency sweep capabilities (often in several modes, such as linear and logarithmic), and modulation (the product of two analog waveforms).

2.13 TIME DOMAIN ANALYSIS

Figure 2.34a shows a simple RC circuit, in which a charged capacitor is placed across a resistor. Using Eq. (2.30) and Kirchhoff's voltage law, we see that the behavior of the circuit is governed by the first-order differential equation

$$RC\frac{dv}{dt} + v = 0 \tag{2.89}$$

If switch S_2 is closed while switch S_1 is open, the capacitor charges to an initial voltage v_0. If S_2 is opened while S_1 is closed, the capacitor discharges according to

$$v = v_0 e^{-t/\tau} \tag{2.90}$$

where $\tau = RC$ is the *time constant* of the circuit in seconds with R in ohms and C in farads. Equation (2.90) is shown in Fig. 2.34b.

Figure 2.35a shows another RC circuit, in which the battery source is connected at $t = 0$ and provides a step input voltage v_{IN}. The equation for this circuit is simply

$$RC\frac{dv}{dt} + v = v_{IN} \tag{2.91}$$

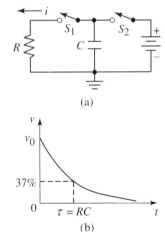

(a)

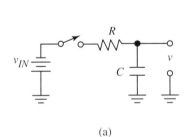

(a)

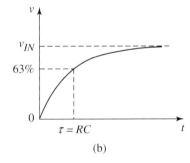

(b)

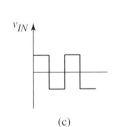

(b)

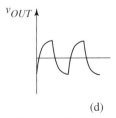

(c)

(d)

Figure 2.34 RC circuit (a) and its discharge characteristics (b).

Figure 2.35 RC circuit (a), capacitor charge characteristics (b), square wave input with two different frequencies (c) and corresponding capacitor output (d).

Solving Eq. (2.91) yields

$$v = v_{IN} + Ae^{-t/\tau} \tag{2.92}$$

Applying the initial condition $v(t = 0) = 0$ gives $A = -v_{IN}$ and Eq. (2.91) becomes

$$v = v_{IN}(1 - e^{-t/\tau}) \tag{2.93}$$

The response of the circuit is shown in Fig. 2.35b. The voltage v_C across the capacitor will asymptotically approach the steady-state value $v_{ss} = v_{IN}$. At $t = 0.7\tau$, $v_C = 0.5\ v_{IN}$; at $t = \tau$, $v_C = 0.63\ v_{IN}$; at $t = 4\tau$, $v_C = 0.9817v_{IN}$; and at $t = 5\tau$, $v_C = 0.99v_{IN}$. Changing v_{IN} to a different value, say to 0, means the capacitor will exponentially decay toward that new value at the rate $e^{-t/\tau}$. If the capacitor is driven by the square wave of Fig. 2.35c, it will charge/discharge in the manner depicted in Fig. 2.35d.

Example 2.3: Time-delay circuit—Figure 2.36a shows an *RC* circuit hooked to two non-inverting CMOS (complimentary metal oxide-semiconductor) buffers. Buffers are integrated circuits (ICs) that play many useful roles in digital circuits (discussed in Chapter 5). For now, the buffer in Fig. 2.36 will generate an output voltage equal to its supply voltage V_S if v_{IN} is more than $0.5V_S$. The overall circuit provides a time delay between the input pulse and the output pulse. Develop the timing diagram for the voltages across the capacitor and the output terminals.

Solution: The first buffer outputs a logic HIGH when the voltage at its input reaches a value that is one-half the DC supply voltage used to power it. The buffer's low output impedance prevents the *RC* circuit from loading it, and its output is a replica of the input signal, with 10.5-μs delay. The output buffer switches to a logic LOW as the *RC* circuit discharges to 50% output in 0.7τ, 10.5 μs after the input pulse becomes LOW. The resulting timing diagrams are shown in Fig. 2.36b.

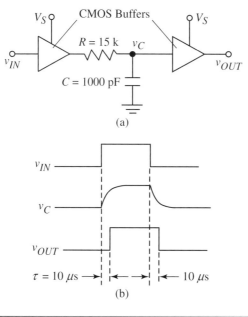

(a)

(b)

Figure 2.36 *RC* circuit hooked up to two buffers to generate a delayed digital signal.

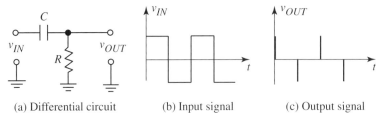

(a) Differential circuit   (b) Input signal   (c) Output signal

Figure 2.37 A differentiator circuit and its output when driven by a square wave.

2.13.1 Differentiators

The differentiator generates an output signal proportional to the time rate of change of the input signal. The RC circuit shown in Fig. 2.37a performs this function. The current through the capacitor is given by

$$i = C \frac{d}{dt}(v_{IN} - v) = \frac{v}{R} \tag{2.94}$$

If R and C, and thus the product RC, are small enough so that $dv/dt \ll dv_{IN}/dt$, the output voltage v from Eq. (2.94) will be proportional to the rate of change of the input waveform:

$$v = RC \frac{dv_{IN}}{dt} \tag{2.95}$$

The change in voltage across the capacitor at the transition is zero, and the load seen by the input is R. The load resistance may therefore load the input if R is chosen to be too small when attempting to satisfy small-time-constant requirements. The output generated for a square wave input shown in Fig. 2.37b is depicted in Fig. 2.37c.

Differentiator circuits are useful for detecting leading edges and trailing edges of pulse signals in digital circuits. Figure 2.38 shows a RC differentiator circuit connected to two buffers. The differentiator generates spikes at the transitions of the input signal v_{IN}, and the output buffer converts the spikes to a square pulses with small amplitude. In practical circuits, the buffer would have a built-in diode and the output corresponding to a negative spike will be small.

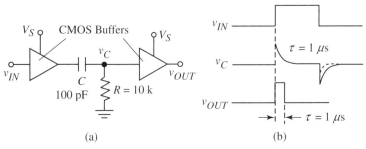

(a)   (b)

Figure 2.38 RC-based leading edge detection circuit.

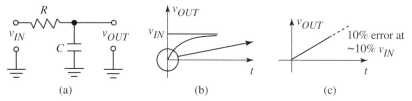

Figure 2.39 A passive integrator RC circuit.

2.13.2 Integrators

An integrator circuit produces an output proportional to the integral of the input signal. A simple RC integrator is shown in Fig. 2.39a. If the time constant $\tau = RC$ is large to maintain $v \ll v_{IN}$, that is, if i is proportional to v_{IN}, then

$$C\frac{dv}{dt} = \frac{v_{IN}}{R} \quad \text{or} \quad v = \frac{1}{RC}\int_0^{t_1} v_{IN}(t)\,dt + v(0) \tag{2.96}$$

where t_1 is the integration time and $v(0)$ is the initial voltage across the capacitor. Equation (2.96) indicates that the output voltage is the integral of the input voltage over the integration time from 0 to t_1 scaled by a gain factor of $1/\tau$. If the input to the integrator is a square wave, the capacitor charges exponentially, as shown in Fig. 2.39b. A ramp approximates the first part of the exponential response. If τ is large compared to the period of the square wave, only the ramp portion representing the integral of a constant will appear, and the output is a sawtooth waveform with a voltage directly proportional to time.

An exact integrator is realized if the input is a current i instead of a voltage. A practical current source is actually approximated as a large voltage across a large resistance.

The integrator is used in many practical applications, including feedback control systems, analog-to-digital converters, and function generators.

2.14 PASSIVE FILTERS

Filtering is one of the most important signal-conditioning techniques in instrumentation and control. Filters are designed to attenuate or, ideally, to inhibit a band of frequencies of an input signal while transmitting or passing the frequency band of interest unaltered.

Two main types of filters are defined, analog filters and digital filters. Analog filters are further classified as passive filters and active filters. Passive filters are built from a combination of passive R, L, and C elements, while active filters include op-amps instead of inductors in their design to eliminate losses of the pure signal. For high-frequency applications, passive filters are favored because op-amps have limited bandwidth. This section briefly introduces useful passive filters. Active filters are discussed in Section 4.10.

Because the impedance of capacitors and inductors depends on frequency, passive filters may be designed to allow passage of the frequency band of interest while rejecting or attenuating frequencies associated with noise. The number of independent energy storage inductors and capacitors in the circuit determines the order of the filter, and we speak of first-order filters, second-order filters, and so on. Filtering usually leads to significant improvement in measurement and signal processing. The transfer function of the filter is the ratio of the output voltage $\mathbf{V}_{OUT}$ to the input voltage $\mathbf{V}_{IN}$ and has a gain magnitude and a phase. Because the amplitude and phase response of realizable analog filters are related, it is not possible to design a filter that exhibits specific amplitude response and phase response simultaneously. The magnitude may be arbitrarily specified, but the phase follows a corresponding causal relationship. In general, the interest in filter design lies in the magnitude of the transfer function and not in the phase angle.

Depending on the frequency band to attenuate and/or transmit, filter circuits can be arranged in a variety of ways, but all exhibit one of the following response types: low-pass, high-pass, band-pass, band-reject, notch, and trap filters. Figure 2.40 shows the ideal and actual gain amplitude response of the first three types, with key response features indicated. The response behavior of the other types may be easily inferred. The *passband* (PB) is the range of frequencies that are allowed to pass through the filter relatively unattenuated, that is, with a gain amplitude of 1 (0 dB). The passband ends at the *cutoff frequency* f_C, where the gain magnitude attenuates to 0.707 (-3 dB). Beyond this point the response of the filter drops off through a *transition region* to the *stopband* (SB) region, where the range of frequencies is rejected or significantly attenuated. The stopband region begins at a point of some minimum attenuation, such as -40 dB.

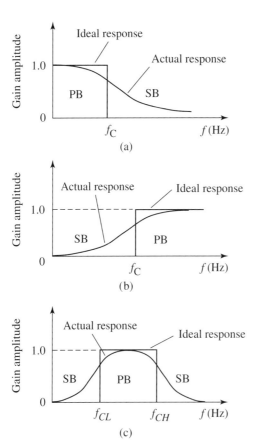

Figure 2.40 Typical amplitude response characteristics of a low-pass filter (a), a high-pass filter (b), and a band-pass filter (c).

2.14.1 Low-Pass Filters (LPFs)

First-Order LPF

Figure 2.41a shows an RC circuit that exhibits a first-order low-pass filter response. The capacitor blocks (short-circuits) the high frequencies and allows only low frequencies to pass. The output v_O is approximately equal to the input at low frequencies ($f > 1/RC$) and goes to zero at high frequencies. Using the voltage divider rule, the transfer function of this LPF is

$$\frac{\mathbf{V}_{OUT}}{\mathbf{V}_{IN}} = \frac{1}{1 + j\omega\tau} \tag{2.97}$$

where $\tau = RC$ is the time constant of the filter. The magnitude is

$$\left|\frac{\mathbf{V}_{OUT}}{\mathbf{V}_{IN}}\right| = \frac{1}{\sqrt{1 + (\omega RC)^2}} = \frac{1}{\sqrt{1 + (f/f_C)^2}} \tag{2.98}$$

and the phase angle is $\phi = \tan^{-1}(-\omega\tau)$. The frequency response of the filter is shown in Fig. 2.41b on a linear scale and in Fig. 2.41c on a log scale. The corner frequency corresponding to the -3-dB breakpoint is defined as $f_C = f_{(-3\text{dB})} = 1/2\pi\tau$.

Inductors could be used instead of capacitors in combination with resistors to make a low-pass filter. However, RL filters are rarely employed because the inductor is a large and more expensive element and deviates further from the ideal performance as compared with capacitors.

The RC low-pass filter and the RC integrator have identical circuits. The key to achieving good integration is that the signal frequency must be well above the -3-dB point.

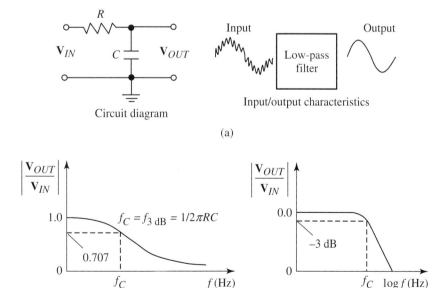

Figure 2.41 RC first-order low-pass filter and its response characteristics.

Example 2.4: Design of an LPF—A noise signal of frequency 1 MHz is imposed on a measurement signal whose frequency is less than 1 kHz. Design an LPF to attenuate the noise to 1%. Assess the effect of the design on the measurement signal at its maximum frequency of 1 kHz.

Solution: The filter's corner frequency f_C to provide the desired 1% attenuation at 1 MHz is determined from Eq. (2.98), after recognizing that $\omega = 2\pi f$ and $f_C = 1/2\pi RC$:

$$\left|\frac{V_{OUT}}{V_{IN}}\right| = \frac{1}{\sqrt{1 + (f/f_C)^2}} = \frac{1}{\sqrt{1 + (1\,\text{MHz}/f_C)^2}} = 0.01$$

$$\Rightarrow f_C = 10\,\text{kHz}$$

Selecting $C = 0.1\,\mu\text{F}$, R is then found to be

$$R = \frac{1}{2\pi C f_C} = \frac{1}{2\pi \times 0.1\,\mu F \times 10\text{kHz}} = 1.59\,\text{k}\Omega$$

To assess the effect of the filter on the measured signal at maximum frequency, use Eq. (2.98) again to determine the level of attenuation:

$$\left|\frac{V_{OUT}}{V_{IN}}\right| = \frac{1}{\sqrt{1 + (f/f_C)^2}} = \frac{1}{\sqrt{1 + (1/10)^2}} = 0.995$$

At max frequency, the signal attenuates by only 0.5%.

Second-Order LPF

Adding an inductor to the circuit of the first-order LPF in series with the voltage source results in the second-order LPF shown in Figure 2.42a. The result is an increase in the attenuation of the transfer function. The transfer function is

$$\frac{\mathbf{V}_{OUT}}{\mathbf{V}_{IN}} = \frac{1}{\left[1 - (\omega/\omega_n)^2\right] + j(2\zeta\omega/\omega_n)} \tag{2.99}$$

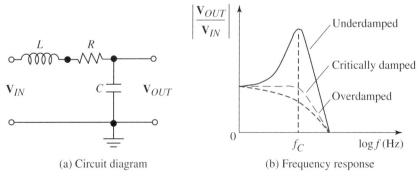

(a) Circuit diagram   (b) Frequency response

Figure 2.42 RLC second-order passive low-pass filter and typical response characteristics.

where $\omega_n = 1/(LC)^{1/2}$ is the *undamped natural frequency* of the filter, also its corner frequency, and $\zeta = (R/2)(C/L)^{1/2}$ is the *damping ratio*. The *quality factor* of the filter is given by $Q = 1/2\zeta = \omega_C/\Delta\omega$, where $\Delta\omega$ is the 3-dB bandwidth. The frequency response magnitude and phase functions of the filter are

$$\left|\frac{\mathbf{V}_{OUT}}{\mathbf{V}_{IN}}\right| = \frac{1}{\sqrt{[1 - (\omega/\omega_n)^2]^2 + [2\zeta\omega/\omega_n]^2}}$$

$$\phi = \tan^{-1}\left[\frac{-2\zeta(\omega/\omega_n)}{1 - (\omega/\omega_n)^2}\right]$$

(2.100)

The filter response depends directly on the values of the damping ratio ζ and the undamped natural frequency ω_n. Three response forms of interest are defined, depending on the value of the damping ratio. When the damping ratio ζ is less than 1, the filter amplifies the input in the region of ω_n and the gain function assumes a resonant peak before decaying to zero at high frequencies. This response form is the *underdamped response*, shown in Fig. 2.42b. As ζ approaches zero, the resonant peaks becomes narrower and the resonant peak increases. When ζ is greater than 1, the gain decreases slowly with frequency due to the dissipation in the larger resistor and the response form is known as the *overdamped response*. If $\zeta = 1$, the response form is known as a *critically damped* response and represents a transition between the underdamped and the overdamped response forms. The gain magnitude is flat at low frequencies and rolls off toward zero at higher frequencies.

It is possible to achieve sharper roll-off characteristics by cascading multiple filter sections in series. Filters containing resistive elements are less efficient because the heat dissipated by the resistors is energy drawn from the signal.

2.14.2 High-Pass Filters (HPFs)

First-Order HPF

Interchanging the R and the C of the LPF results in the first-order high-pass filter (HPF) circuit shown in Fig. 2.43a. The capacitor blocks (short-circuits) low frequencies and allows only high frequencies to pass. The output V_{OUT} is approximately equal to the input at high frequencies ($>1/RC$) and goes to zero at low frequencies. The transfer function of the HPF is

$$\frac{\mathbf{V}_{OUT}}{\mathbf{V}_{IN}} = \frac{j\omega\tau}{1 + j\omega\tau}$$

(2.101)

where $\tau = RC$ is the time constant of the filter. The phase angle is $\phi = \tan^{-1}(-\omega\tau)$. The capacitor blocks the passage of DC currents. The DC-blocking ability of the capacitor is one of its most frequently exploited attributes. The magnitude of the transfer function is

$$\left|\frac{V_{OUT}}{V_{IN}}\right| = \frac{\omega RC}{\sqrt{1 + (\omega RC)^2}} = \frac{f/f_C}{\sqrt{1 + (f/f_C)^2}}$$

(2.102)

where f_C is the corner frequency of the filter at the -3-dB breakpoint and is given by $f_C = f_{(-3\text{-dB})} = 1/2\pi\tau$. The frequency response of the filter is shown in Fig. 2.43b in linear scale and Fig. 2.43c in log scale.

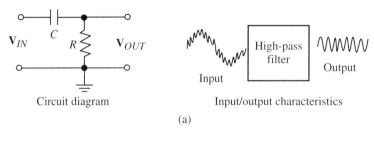

Circuit diagram   Input/output characteristics

(a)

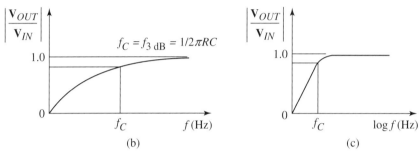

(b)   (c)

Figure 2.43 *RC* first-order passive high-pass filter and its frequency response characteristics.

The *RC* high-pass filter and the *RC* differentiator have identical circuits. The key to achieving good differentiation is that the signal frequency be well below the −3-dB point.

Example 2.5: Design of a High-Pass Filter−An MCU generates a pulse train to drive a stepper motor at 2000 Hz (pulses per second). A 50-Hz noise is superimposed on the pulse train (clean signal). Design a filter to reduce the effect of the 50-Hz noise while maintaining the attenuation of the clean signal to less than 3 dB.

Solution: The 3-dB attenuation in the clean signal is equivalent to

$$3 \text{ dB} = 20\log\left|\frac{V_{OUT}}{V_{IN}}\right| \Rightarrow \left|\frac{V_{OUT}}{V_{IN}}\right| = 10^{-3/20} = 0.707$$

We apply Eq. (2.102) as follows:

$$\left|\frac{V_{OUT}}{V_{IN}}\right| = \frac{f/f_C}{\sqrt{1 + (f/f_C)^2}} = \frac{1200/f_C}{\sqrt{1 + (1200/f_C)^2}} = 0.707$$

Solving this equation renders a corner frequency at $f_C = 1200$ Hz. Incidentally, the corner frequency has the same value as the pulse frequency. To assess the level of attenuation in the 50-Hz noise, Eq. (2.102) is applied again, with $f = 50$ Hz and $f_C = 1200$ Hz. This gives

$$\left|\frac{V_{OUT}}{V_{IN}}\right| = \frac{f/f_C}{\sqrt{1 + (f/f_C)^2}} = \frac{50/1200}{\sqrt{1 + (50/1200)^2}} = 0.042$$

The filter attenuates the 50-Hz noise by 96%.

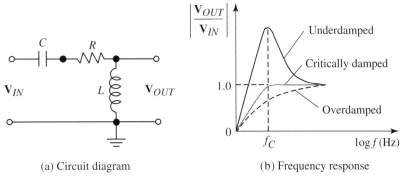

(a) Circuit diagram   (b) Frequency response

Figure 2.44 *RLC* passive second-order high-pass filter and its frequency response characteristics.

Second-Order HPF

If we interchange the capacitor and the inductor of the second-order LPF, the second-order HPF shown in Fig. 2.44a emerges. The corner frequency is the same as for the LPF. The transfer function becomes

$$\frac{\mathbf{V}_{OUT}}{\mathbf{V}_{IN}} = \frac{\omega^2}{\left[1 - (\omega/\omega_n)^2\right] + j(2\zeta\omega/\omega_n)} \tag{2.103}$$

The frequency response magnitude and phase functions of the filter are

$$\left|\frac{\mathbf{V}_{OUT}}{\mathbf{V}_{IN}}\right| = \frac{\omega^2}{\sqrt{\left[1 - (\omega/\omega_n)^2\right]^2 + \left[2\zeta\omega/\omega_n\right]^2}}$$

$$\phi = tan^{-1}\left[\frac{2\zeta(\omega/\omega_n)}{1 - (\omega/\omega_n)^2}\right] \tag{2.104}$$

The response of the filter is shown in Fig. 2.44b.

2.14.3 Band-Pass and Band-Reject Filters

If an *RC* low-pass filter is cascaded in series with an *RC* high-pass filter, the magnitude ratio of the resulting circuit becomes

$$\left|\frac{V_{OUT}}{V_{IN}}\right| = \frac{1}{\sqrt{1 + (f/f_{LP})^2}} \frac{1}{\sqrt{1 + (f_{HP}/f)^2}} \tag{2.105}$$

where f_{LP} and f_{HP} are the corner frequencies (Hz) of the low-pass and the high-pass filters, respectively. If the value of f_{LP} is greater than that of f_{HP}, the circuit is a band-pass filter with a bandwidth of $\Delta f_{BP} = f_{LP} - f_{HP}$. Typical gain amplitude response of a BPF is shown in Fig. 2.40. If, on the other hand, f_{HP} is greater than f_{LP}, the circuit is a band-reject filter. Usually an op-amp follower or buffer (see Section 4.6.3) is placed between the low-pass and the high-pass circuits to prevent loading.

Example 2.6: Wien's network—The circuit shown in Fig. 2.45 is known as Wien's network.

a. Determine the ratio of the output voltage v_o to the input voltage v_i.

b. Simplify the relation obtained in (a) if $C_1 = C_2 = C$ and $R_1 = R_2 = R$.

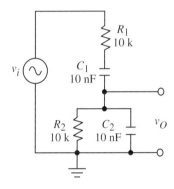

Figure 2.45 Wien's network.

c. Establish the condition that makes the phase difference between v_o and v_i vanish, and determine the attenuation that results from satisfying the condition.

d. What are the benefits of Wien's network based on the results found in (b)–(d).

Solution:

a. The network consists of two impedances, Z_1 and Z_2, forming a voltage divider circuit. The impedances and the resulting input–output relation are:

$$Z_1 = R_1 + \frac{1}{j\omega C_1}, \qquad Z_2 = \frac{1}{(1/R_2) + j\omega C_2}$$

$$\frac{v_O}{v_i} = \left(\frac{Z_2}{Z_1 + Z_2} \right)$$

b. If $C_1 = C_2 = C$ and $R_1 = R_2 = R$, then

$$\frac{v_O}{v_i} = \frac{1}{3 - j\sqrt{(1 - \omega^2 R^2 C^2)}/\omega RC}$$

c. The condition for the phase difference to vanish is to make the imaginary part in the denominator of the preceding equation zero. That is,

$$\omega^2 R^2 C^2 = 1 \qquad \text{which yields} \qquad \omega = \frac{1}{RC}$$

$$\frac{v_O}{v_i} = \frac{1}{3}$$

d. The results show that if Wien's network can identify an input signal vi that has a fixed by producing an output signal v_O in phase with the input signal but and has a magnitude equal to one-third that of v_i.

2.14.4 Notch and Trap Filters

An ideal notch filter allows only one frequency to pass while all other frequencies are rejected. Refer to Problems 2.18 and 2.19 for examples of the notch filter. The trap filter has opposite characteristics, it rejects only one frequency.

2.15 NOISE AND INTERFERENCE IN CIRCUITS

As mentioned earlier, noise is any electrical or magnetic signal that is unintentionally imposed on the signal of interest. Noise may cause serious accuracy problems, especially in circuits involving small voltage signals, such as those produced by sensors. Noise may exist in one of two forms, interfering noise and random noise.

Interfering noise may come from external sources, or it may develop on ground lines in the form of ground loops. External sources impose noise in the form of electromagnetic interference (EMI) comprising electric fields, or *E*-fields (V/m) and magnetic fields, or *H*-fields (A/m). Any circuit is susceptible to EMI and is also a potential source of electromagnetic (EM) emissions. An electric field will always exist between any two conductors separated by a distance *d* and at different potentials; a current flowing through a conductor will always form a magnetic field around it; and alternating voltages and currents in a network of conductors generate EM waves comprising *E*- and *H*-fields that propagate in free space at right angles and at the speed of light (3×10^8 m/s). The radio frequency (RF) energy (3 kHz to 3000 MHz) inherent in those fields tends to follow conductors and may cause interference in the operation of other equipment.

Random noise is inherent in every component of which a circuit is made. Assessing the influence of random noise on system performance usually requires statistical analysis. Examples of random noise include thermal noise, shot noise, and flicker noise. *Thermal noise* exists in any dissipative element that can be modeled as a resistor. It is called *Johnson noise* if the element is an electric resistor (See Problem 11.1). Johnson noise is a *white noise* because its power spectral density (PSD) function is constant over all frequencies. The PSD provides information concerning the dependence of the value of a random variable at one frequency on the value of the variable at another frequency. *Shot noise* represents fluctuations associated with DC currents created by independent charge carriers (electrons) that must cross a potential barrier such as a *pn* junction. *Flicker noise* is a noise current associated with diodes and field-effect transistors (FETs). It arises due to the capture and release of charge carriers in localized trap states in the semiconductor. This type of noise dominates at low frequencies because it depends on $1/f$ and varies with the magnitude of the DC current.

2.15.1 Guarding Against Electromagnetic Interference (EMI)

Different kinds of protection are needed to guard against EMI, depending on the offending field. There is no single method that would solve all EMI problems, but the following recommendations provide guidelines that would help reduce their effect.

In many systems, such as vehicles, interference levels can be attenuated by simply moving the conductors from air to the proximity of the vehicle body (ground). However, at microwave frequencies, the level of attenuation diminishes and the amplitude of the signal may increase. A properly packaged multilayer printed circuit board (PCB) layout is crucial to reducing the effect of EMI.

Protection from E-fields

Electric shielding provides immunity against static and time-varying electric-field interference. The shield is an electrically conductive (nonmagnetic) material, usually aluminum or copper, that encloses a conductor or a circuit. The main purpose of shielding is to cancel mutual capacitances between neighboring conductors. In most practical cases, the signal conductors are shielded from the capacitive pick up of the 60-Hz (or 50-Hz) signal surrounding power lines. To be effective, electric shielding should conform to the following rules.

1. The electric shield must be connected to a constant voltage, such as the zero-signal reference potential of the circuit it encloses.
2. If the signal is earthed, the shield should be connected to the point where the signal is also earthed.
3. If the circuit within the shield is driven by a power transformer, the secondary of the transformer may be placed inside the shield while the primary coil is placed on the outside.

Protection from *H*-fields

Electric shielding does not provide attenuation of the *H*-field interference. In *RF* applications, electric shields become waveguides. While heavy and expensive, the use of tightly twisted-wire pairs similar to those used in telephone lines provides an efficient way to attenuate *H*-field interference. Industrial facilities are required to provide immunity against *H*-fields up to 20 V/m, with a maximum allowed interference of 0.4 V. This level of immunity can be achieved by using single-shielded interconnect cables and/or providing low-pass filtering of input and output lines. Using plastic cabinets to enclose circuits provides immunity up to 100 V/m. Embedded applications involving fast microprocessors, switching power supplies, and communication lines are usually prone to magnetic emissions. In those cases, the effects of unwanted emissions can be reduced by (1) applying conductive coating to the plastic enclosure, (2) enclosing offending components in local shields, (3) attaching ferrite beads on the cable, or (4) using a metal cabinet.

2.15.2 Bypass Capacitors

Bypass capacitors can prevent noise and oscillations in analog circuits. They can also prevent false triggering and memory loss in digital circuits. In practical applications, it is common to place a 0.1-μF capacitor across the leads of a DC motor to help reduce the effect of voltage spikes caused by commutation. Power supply lines may carry high-frequency noise. A 0.1-μF capacitor connected between the pins of the power supply and ground creates a short-circuit path and helps eliminate the effect of the high-frequency noise. Fast-changing waveforms in digital circuits produce AC voltage spikes. To stop their propagation to other circuits, a 0.1-μF bypass capacitor is usually placed across the power supply pins of each IC in the circuit.

2.16 GROUNDING

2.16.1 Ground Loops

A ground is a point or plane in a circuit that serves as a reference potential for signal and power supply lines. A circuit may have several grounding points. The wires connecting the different ground points provide a path for currents to flow between them, such as currents returning from a circuit to the power supply. Current flow through the low but finite resistance of the ground wires causes the voltage levels at the different ground points to be different. This results in the formation of the potentially dangerous *ground loops*. For example, it is not uncommon to have ground points in the same building to differ by a few tens of millivolts AC at 60 Hz. While such levels do not pose a safety threat, they make remote measurements difficult. Therefore, proper circuit grounding is essential to insure the integrity of circuit operation.

Additionally, ground loops are susceptible to EMI through capacitive and inductive coupling, which will always exist between two circuits even if the circuits are not directly connected. A changing voltage in a conductor creates an electric field that couples with a nearby conductor due to the finite capacitance between the two conductors. For example, the coupling capacitance between two parallel insulated wires 2.5 mm apart is about 50 pF/m, and the primary-to-secondary capacitance in an unscreened medium power line-voltage transformer is 100–1000 pF. On the other hand, a changing current in a conductor creates a magnetic field that couples with a nearby conductor inductively due to the mutual inductance between the two conductors. Magnetic field interference is also a manifestation of Faraday's law of induction. It couples with a conductor if the conductor carries a current while moving through the magnetic field or if the conductor is stationary in a changing magnetic field. For these reasons, ground conductors must be designed to be short and to have low resistance and low inductance.

2.16.2 Grounding Techniques

The method used to avoid interference voltage in ground loops largely depends on the connected circuits. Having a keen knowledge of the characteristics of ground wiring and the return current they carry between circuits is essential to minimizing ground loop problems. Three grounding methods are depicted in Fig. 2.46. The single-point series ground connection shown in Fig. 2.46a results in interference voltage that may be significant. If this connection is unavoidable, the most offending circuit must be placed close to the common reference point. Referring to Fig. 2.46a, if $Z_1 = Z_2 = Z_3 = 4\ \Omega$ and $i_1 = 50$ mA, $i_2 = 1$ mA, and $i_3 = 1\ \mu$A, the return current of circuit 1 alone causes a voltage drop of 400 μV at the low side of circuit 2 and 200 μV at the "ground" of circuit 3. Using single-point parallel grounding at a central point, shown in Fig. 2.46b, overcomes this problem, at the cost of requiring a more elaborate circuit layout. The parallel grounding method is also preferred for small systems involving low-frequency signals ($<0.03 \times \lambda$; 9m at 1 MHz). Persisting ground loop problems in low-frequency circuits may be eliminated through galvanic isolation.

Multipoint grounding in a large system is usually unavoidable. The multipoint parallel grounding shown in Fig. 2.46 may be preferred for high-frequency circuits ($>0.15\lambda$; 4.5 m at 10 MHz) over single-point grounding because it results in lower ground impedance. The distance between the connected points should be kept as short as possible. Plating the surface of the ground conductors helps reduce ground loop impedance.

Mechatronic systems usually involve three subsystems: analog circuits, digital circuits, and motor drives. Applying the following guidelines helps reduce interference voltages associated with ground loops (refer to Fig. 2.47).

1. Provide a separate power path and a separate ground path for each subsystem.
2. Whenever possible, power each subsystem with a separate power supply.
3. Use one common ground point for the analog circuits and one for the digital circuits, and then tie both ground points to a single point to reduce ground impedances.

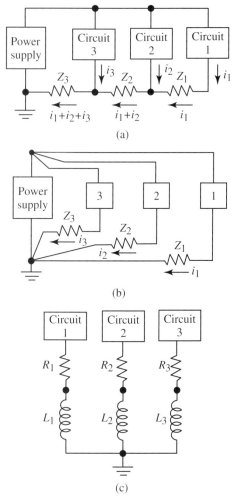

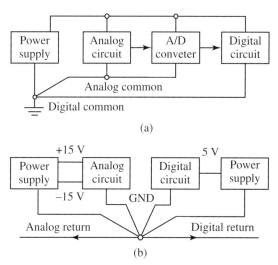

Figure 2.46 Series ground connection (a), parallel ground connection (b), and multipoint parallel grounding (c).

Figure 2.47 Using separate power and ground paths (a) and separate power supplies (b) for analog and digital circuits.

2.16.3 Galvanic Isolation

Galvanic, or ohmic, *isolation* allows the transmission of signal levels through a gap along the signal path. The device that provides isolation is called the *isolator*. Isolators break ground loops, provide floating grounds between circuits with high voltage potentials, and provide isolation between high-power circuits and signal circuits. In instrumentation, isolation is a form of signal conditioning used to isolate the sensor signal from the computer to prevent damage to the computer. Isolators provide protection for workers in high-voltage environments and for patients from malfunctioning medical instruments.

Galvanic isolation may be achieved by capacitive, optical, or magnetic means. *Capacitor couplers* are applied in both digital and analog circuits to transmit small signals across the isolation barrier with minimal power. Capacitor couplers are usually limited to single-channel coupling due to a size limit imposed on capacitors, and then are not suitable for power signal transmission. The *optical isolator*, or optocoupler (see Section 3.14 and Chapter 11), combines an LED made of gallium-arsenide-phosphate (GaAsP) or aluminum-gallium-arsenide (AlGaAs) with a photodiode or a phototransistor, or fiber optics, to transmit signals by light. Optical isolators are suitable for digital coupling but not appropriate in high-power applications. They can be built into a circuit that transmits analog signals over single or multiple channels. *Magnetic coupling* using transformers with modulated carrier and coil type are used mostly in DC/DC converters. However, they are also being used in coupling digital circuits and in sophisticated analog isolators to move analog signals across the barrier. Transformers are difficult to use and generally impossible to produce as an integrated circuit. An isolation transformer has equal secondary and primary voltages. Hall-effect devices may also be used for magnetic coupling. If a conductor is placed in a magnetic field and a current perpendicular to the direction of the magnetic field flows through it, a voltage is generated across the terminals perpendicular to the directions of the current and magnetic field, a consequence of the Lorentz force. This is the Hall effect. Hall-effect sensors are discussed in Chapter 11.

2.17 SUMMARY

This chapter introduced basic electrical components and principles. The material started with basic electrical principles and components and progressed further to analyze and design basic conditioning circuits using passive components (RLC). This chapter laid the foundation for electrical signal conditioning and excitation circuits. In addition, various popular passive filter designs were presented and used in practical examples. The mechatronics products integrate many of the basic functions, circuits, and best practices presented in this chapter.

RELATED READING

D. G. Alciatore and M. B. Histand, *Introduction to Mechatronics and Measurement Systems*, 2nd ed. New York: McGraw-Hill, 2002.

I. Cochin and W. Cadwallender, *Analysis and Design of Dynamic Systems*, 3rd ed. Reading, MA: Addison-Wesley, 1997.

D. Curington, "Isolation—Is Your Measurement System Safe?" *Sensors*, vol. 17, no. 9, September 2000.

P. Horowitz and W. Hill, *The Art of Electronics*. Cambridge, UK: Cambridge University Press, 1989.

C. D. Johnson, *Process Control Instrumentation Technology*, 7th ed. New York: Prentice Hall, 2002.

S. Kamichik, *IC Design Projects*. Indianapolis: Prompt Publications, 1998.

S. E. Lyshevski, *Nano- and Microelectromechanical Systems*. Boca Raton, FL: CRC Press, 2001.

F. Mims, *Engineer's Mini-Notebook: Schematic Symbols, Device Packages, Design and Testing*. Radio Shack Archer Catalogue No. 276-5017, 1988.

G. Novacek, "The Shocking Truth about EMC, Part 1: Design for Compatibility," *Circuit Cellar*, no. 117, 2000.

G. Novacek, "The Shocking Truth about EMC, Part 2: Practical Application," *Circuit Cellar*, no. 118, 2000.

R. Pallas-Arney and J. G. Webster, *Sensors and Signal Conditioning*, 2nd ed. New York: Wiley Interscience, 2001.

P. Pickering, "A System Designer's Guide to Isolation Devices," *Sensors*, vol. 16, no. 1, January 1999.

E. Ramsden, "Interfacing Sensors and Signal Processing Components," *Sensors*, May 1998.

G. Rizzoni, *Principles and Applications of Electrical Engineering*, 5th ed. New York: Mc-Graw Hill, 2006.

S. D. Senturia, *Microsystem Design*. Dordrecht, Netherlands: Kluwer Academic, 2001.

QUESTIONS

2.1   Describe electric field.

2.2   Explain triboelectric effect.

2.3   Describe the relation between electric charge, current, voltage, power, and work.

2.4   What is an open circuit and a closed circuit?

2.5   What is the difference between ground and earth?

2.6   What is the difference between signal ground and chassis ground?

2.7   Describe various types of electric sources.

2.8   What is the role of a circuit breaker?

2.9   How can you determine the Thevenin equivalent of a resistive network?

2.10 How can you determine the Norton equivalent of a resistive network?

2.11 Explain the null condition of a Wheatstone bridge.

2.12 Describe the role of five resistive-based sensors.

2.13 What is the difference between regular and electrolytic capacitors?

2.14 List six roles of capacitors, and explain each.

2.15 Explain magnetic field.

2.16 Explain the Hall effect and the magnetoresistive effect.

2.17 Describe three important specifications of a transformer.

2.18 Explain what a solenoid is

2.19 How does a relay operate?

2.20 What is meant by *steady-state response*?

2.21 What is impedance?

2.22 What is the decibel?

2.23 Explain loading effect and its ramification on measurements.

2.24 What should the relation be between the impedances of a sensor and the following measuring device when measuring voltage?

2.25 What should the relation be between the impedances of a sensor and the following measuring device when measuring current?

2.26 Why is AC power used in virtually all commercial and public utility systems?

2.27 Explain power factor and its effect on domestic and industrial electrical bills.

2.28 Why are capacitor yards built behind large factories?

2.29 Explain the role of filters in mechatronics.

2.30 Explain the role of Wien's network.

2.31 Describe the response characteristics of low-pass, high-pass, band-pass, notch, and trap filters.

2.32 Explain electromagnetic interference, and describe ways to reduce its effects.

2.33 Why is isolation important?

2.34 Explain the various isolation methods.

2.35 What constitutes an isolation transformer?

PROBLEMS

2.1   Using Thevenin's theorem, determine the effective output resistance and voltage v_O between A and B of the circuit shown in Fig. 2.48.

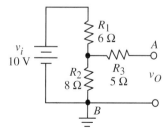

Figure 2.48

2.2   Show that when R_{LOAD} in the circuit shown in Fig. 2.29b is equal to the internal resistance of the source,

the power in the load is maximum for a given source resistance.

2.3   A current of 1 mA charges a 1-μF capacitor. How long does it take the ramp to reach 10 volts?

2.4   What is the current flow through the secondary winding of a 50-VA transformer if the secondary voltage is 25 volts?

2.5   Determine the voltage and power ratios for a pair of signals with the following decibels ratios: (a) 3 dB; (b) 6 dB; (c) 10 dB; (d) 20 dB.

2.6   Show that for a step waveform signal v_{IN}, the rise time for the capacitor in the circuit of Fig. 2.35 is 2.2 time constants.

2.7   Determine the energy required to build a charge q in a capacitor through a resistor and the energy stored in the capacitor at the end of transient. Explain the difference.

2.8 A 200-turn iron-core solenoid with an air gap ($\mu = 2000\ \mu_0$) has a length of 0.3 m and a uniform cross section of 0.01 m^2. Determine its self-inductance.

2.9 Find an expression for the self-inductance of a toroid that has a mean radius r and a rectangular cross section $2a \times b$. Determine also the magnetic energy of the solenoid.

2.10 If in the circuit shown in Fig. 2.49 $R_1 = R_2 = 10$k and $C = 0.1\ \mu$F, find $v_{OUT}(t)$ and sketch it. Indicate $v_{OUT}(t)$ at τ and 5τ.

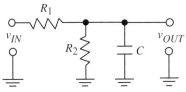

Figure 2.49

2.11 Show that adding a series capacitor of value $C = 1/\omega^2 L$ makes the power factor in a *series RL* circuit equal 1.0.

2.12 Determine the power factor for the circuit shown in Fig. 2.50.

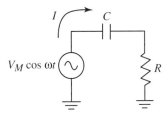

Figure 2.50

2.13 At what frequency does an RC low-pass filter attenuate by 6 dB (output voltage equal to half the input voltage)? What is the phase shift at that frequency?

2.14 Determine the voltage across C_2 after the circuit shown in Fig. 2.51 reaches steady state.

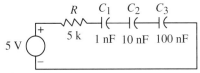

Figure 2.51

2.15 The circuit shown in Fig. 2.52 is an inductor, with a DC resistance of 30 ohms, connected in parallel with a 60-ohm resistor. The circuit is powered by a 24-V battery through a switch. When the switch is opened, what is the peak power dissipated by the 60-ohm resistor?

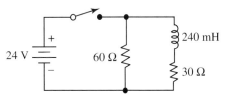

Figure 2.52

2.16 In Figure 2.53, a loudspeaker is connected between terminals A and B. V is a voltage source. What should the impedance of the loudspeaker be so that it receives maximum power?

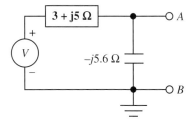

Figure 2.53

2.17 The circuit in Fig. 2.54 is a cascade of two identical RC-RC filters with the same transfer function. The transfer function of the composite system will not be simply the product of the transfer functions of the individual circuits, due to loading problems. Determine the condition that would eliminate loading problems.

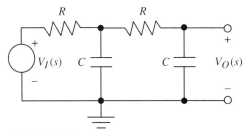

Figure 2.54

2.18 Find the transfer function for the filter circuit shown in Fig. 2.55. Plot the gain amplitude v_{OUT}/v_{IN} versus frequency, and describe its filtering behavior.

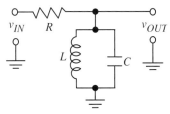

Figure 2.55

2.19   Find the transfer function for the LC filter circuit shown in Fig. 2.56. Plot the gain amplitude v_{OUT}/v_{IN} versus frequency, and describe its filtering behavior.

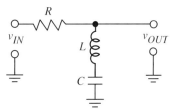

Figure 2.56

2.20   It is desired to design a band-reject filter by cascading a low- and a high-pass filter with critical frequencies 100 Hz and 10 kHz, respectively. Ensure that the input impedance of the second stage is 10 times the output impedance of the first stage at 5 kHz. Use LabVIEW to show the variation of amplitude ratio of the response with frequency within the range of 10 Hz to 100 kHz.

2.21   The circuit shown in Fig. 2.57 is the bridge form of Wien's network. This circuit may be used as an error signal detector because the magnitude and phase of the output voltage will be zero for a specific input signal frequency f_0. If the input signal frequency deviates from f_0, then the magnitude and the phase of v_0 change. Verify the operation of the network, and explain how it will detect whether the input signal frequency has decreased or increased.

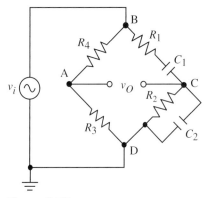

Figure 2.57

2.22   The RLC resonator network shown in Fig. 2.58 behaves as a band-pass filter.

   a. Determine the transfer function $T(s)$ of the network, and express it in the following form:

$$T(s) = \frac{V_O(s)}{V_I(s)} = \frac{a_1 s}{s^2 + (\omega_0/Q)s + \omega_0^2}$$

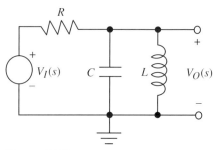

Figure 2.58

   b. Give expressions for the pole frequency ω_0 and the pole quality factor Q in terms of R, L, and C.

   c. Determine expressions for the frequencies ω_1 and ω_2 at which the magnitude response is 3 dB below its maximum value at ω_0. The bandwidth BW of the filter is defined as $\omega_2 - \omega_1$. Use $R = 10$ kW and design the filter to obtain $\omega_0 = 104$ rad/sec and a bandwidth of 100 Hz.

   d. Obtain Bode plots for the filter, and label on the magnitude plot ω_0, ω_1, and ω_2.

2.23   Use Matlab/Simulink or LabVIEW to determine the response of the mass in the system shown in Fig. 2.18 to a step input using the given parameters' values.

Parameter	Symbol	Value
Area	A	100
Permittivity	ϵ	1
Initial gap	g_0	1
Minimum gap	g_{min}	0.01
Mass	m	1
Damping constant	b	0.5
Spring constant	k	1
Resistance	R	0.001

2.24   Derive expressions for the magnetic energy and the mechanical force for the microelectromagnet shown in Fig. 2.59.

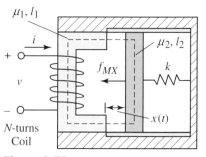

Figure 2.59

LABORATORY PROJECTS

2.1 Select a resistor and a capacitor, measure their values with a DMM, and use a function generator and an oscilloscope to build all *RC* circuits discussed in the chapter. For each circuit, verify the results acquired from the experiments with the values calculated using corresponding relations.

2.2 Find manufacturer's specifications for each of the following devices: potentiometer, solenoid, DPDT relay, transformer, LVDT, capacitive proximity sensor, and optoisolator. Summarize the specs of each device in one page. Imagine three possible practical circuits that might be built using these components, sketch them, and build one of them.

2.3 Build the Wien's network of Fig. 2.45 and verify the results of Example 2.6.

3

Diode, Transistor, and Thyristor Circuits

OBJECTIVES

Thoughtful engagement with the material presented in this chapter will enable the student to:

1 Understand the basic principles of electronics

2 Explain and use various diode applications

3 Understand the fundamentals of BJTs and FETs

4 Analyze electronics circuits

5 Utilize diodes, transistors, and thyristors in designing and building mechatronics projects

3.1 SEMICONDUCTORS

The degree to which current can flow in a material depends on how tight the valence electrons are bound to their atoms. In this regard materials are classified as conductors, insulators, and semiconductors. The valence electrons in a metallic *conductor* are loosely bound to their atoms and can migrate freely as the metal is subjected to an electric field. The electrons in an *insulator* material are tightly bound to their atoms, impeding the flow of electric current. The conduction properties of *semiconductor* materials, such as silicon, germanium, and cadmium, are in between. For example, the properties of a semiconductor can be altered, depending on the amount of incident light or the operating temperature, to cause its electrons to move into the electric field when a voltage is applied.

The most widely used semiconductor material is silicon (Si). Silicon is the only semiconductor that, when exposed to air, grows a native oxide layer at the surface. Essentially glass, the oxide layer adds the convenience of creating both insulators and conductors in silicon processing.

Each silicon atom has four valence electrons, which join with neighboring atoms to form a complete bond, that is, an insulator. If one of the Si atoms is replaced (or doped) with a donor dopant that has five valence electrons (group V atoms), such as arsenic (Ar), phosphorous (P), or antimony (Sb), one electron will remain unpaired. This extra electron can easily be freed by an electric field to carry electric charge and provide electricity. This donor electron is called an *n-type*, n standing for a *negative charge carrier*. If, on the other hand, the Si atom is doped with an atom that has three outer electrons (group III atoms), such as gallium (Ga) or boron (B), an electron vacancy called a *hole* will be formed. Any free electron can then wander in and fill this vacancy, causing the hole to jump from atom to atom, conducting an electric current in the process. This *positive charged carrier* is called a *p-type*. The two types of charge carriers are not completely symmetric. Holes are not as mobile as electrons.

3.2 DIODES

If a *p*-slice (majority carriers holes) and an *n*-slice (majority carriers electrons) semiconductors are joined together, the *p-n* junction device shown in Fig. 3.1 is formed. The extra electrons from the donor atoms on the *n*-side diffuse across the junction and attach themselves to the acceptor atoms on the *p*-side. The result is the formation of a *depletion region* that ties up the majority charge carriers and inhibits current flow. If a voltage is applied to the *p-n* junction with the anode connected to the *p*-side and the cathode connected to the *n*-side, the majority carriers get attracted to the far-side terminals, cross the *p-n* junction, mix and recombine, and become neutral. New charge carriers supplied by the terminals result in a continuous current flow, and the junction is said to be *forward biased*. If the voltage is connected in reverse, the majority carriers move away from the junction as they get attracted to the nearby terminals where electron–hole pairs recombine on either side, forming a region with no charge carriers, known as the depletion region, of about 10^{-6} m. The *p-n* junction is said to be *reverse biased*, with negligible current flow through the device.

The *p-n* junction is known as the junction *diode*, with the *p-n* slices sandwiched between two metallic terminals. The symbol and schematic representation of a diode device (e.g., 1N314) are

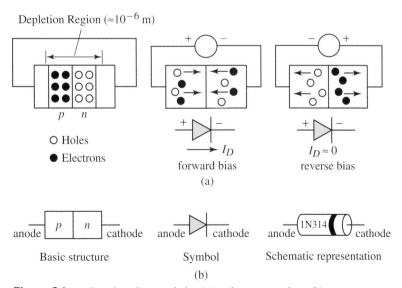

Figure 3.1 *p-n* junction characteristics (a) and representations (b).

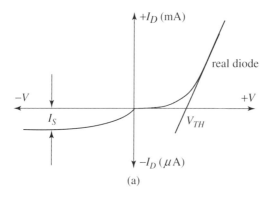

(a)

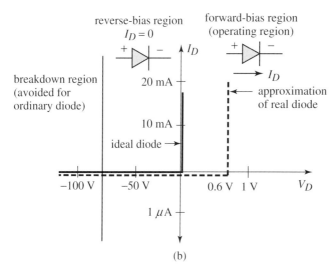

(b)

Figure 3.2 V-I characteristics of a *p-n* junction: real (a) and ideal (b).

shown in Fig. 3.1b. The direction of forward current flow is from the anode to the cathode, as pointed by the diode's arrow at the anode's terminal. The diode is a passive device that exhibits the nonlinear *V-I* characteristics shown in Fig. 3.2 and described by the *diode equation* given by

$$I = I_S\left[\exp\left(\frac{q_e v}{\eta k_b T}\right) - 1\right] \tag{3.1}$$

where q_e is electronic charge (1.602×10^{-19} C), T is the absolute temperature (K), k_b is Boltzmann's constant (1.38×10^{-23} J/K), and η is a nonideality multiplicative factor. For an ideal diode, $\eta = 1$; for a nonideal behavior, η is typically 1.5. At room temperature (293 K), $q_e/k_b T = 40$. The constant reverse saturation current I_S depends on many factors, including the junction area A, the doping profile, and temperature. At room temperature (25°C), $I_S \approx 25$ nA; it may become 7 mA at 150°C. As a rule of thumb, the I_S doubles its initial value at room temperature for every 10°C. Equation (3.1) may be rearranged as

$$V = (\eta kT/q_e)\ln(I/I_S + 1) \tag{3.2}$$

This relation is the basis for *proportional-to-absolute-temperature* (PTAT) temperature measurement sensors.

TABLE 3.1 Specifications of Selected Diodes

Part Number	$V_{R(max)}$	$I_{R(max)}$	$V_{F(continuous)}$	$I_{F(continuous)}$	$V_{F(peak)}$	$I_{F(peak)}$	C (10 V)
1N914	75 V	5 µA	0.75 V	10 mA	1.1 V	0.1 A	1.3 pF
FJT1100	30 V	0.001 µA	—	—	1.1 V	0.05 A	1.2 pF
1N4002	100 V	50 µA	0.9 V	1000 mA	2.3 V	25 A	15 pF
ID101	30 V	10 pA at 10 V	0.8 V	1 mA	1.1 V	0.03 A	0.8 pF

Equation (3.1) indicates that the forward current flow through the diode remains insignificant until the applied voltage reaches the threshold voltage V_{TH} of the diode. V_{TH} depends on the semiconductor material used; it is about 0.75 V for a Si diode and about 0.25 V for a Ge diode.

As soon as the applied voltage surpasses V_{TH}, current flow through the diode increases with v according to the slope $\Delta I/\Delta V$, which is equal to $1/R_D$, the reciprocal of the diode dynamic resistance R_D. For a Si diode, $R_D \approx 25\ \Omega$; for a Ge diode, $R_D \approx 65\ \Omega$. The resistance to current flow in the reverse direction is very high.

The ideal diode behavior depicted in Fig. 3.2 is usually assumed in circuit analysis and design, where the diode's V_{TH} value is used in the calculations. Current flow in the reverse direction for a typical diode, such as the 1N914, is in the nanoampere range and is almost never of any consequence until the *reverse breakdown voltage*, typically 75 V, is reached. The diode acts as a one-way switch, permitting current to flow in one direction but not in reverse. This asymmetric conduction of an ideal diode is analogous to the function of a fluid check valve. The diode switches on and off in the nanosecond range.

A variety of diode types with different *V-I* characteristics are available. The *hot carrier diode* (Schottky diode) has a threshold voltage of about 0.25 V. A device known as the *back diode* has nearly zero forward voltage drop; however, its usefulness is limited due to its very low reverse breakdown voltage. The *tunnel diode* has a negative resistance within a region of the *V-I* curve. Diodes are further classified as *signal diodes*, used mainly in signal-conditioning circuits, and *power diodes*, used in power electronics applications.

Table 3.1 gives the characteristics of selected diodes. Important specifications include the repetitive peak reverse voltage $V_{R(max)}$ and current $I_{R(max)}$ before breakdown occurs, continuous and peak forward current I_F through the diode, and voltage V_F across the diode.

3.3 DIODE APPLICATIONS

Diodes are very important devices in electronic circuits. Some of their applications are given next.

3.3.1 Rectification

The asymmetric behavior of a real diode is used to convert an alternating power into unidirectional, but pulsating, DC power. In Fig. 3.3 a sinusoidal input waveform of voltage V_S is applied to a load R_L through a diode. Since the resistance of a reverse-biased diode is very large, the diode will inhibit current flow through it during the negative half of the sine input. If $V_S > V_{TH}$, the output will be nonzero only during the positive half cycle of the input sinusoid, as shown in Fig. 3.3b. This process is called *rectification*, and the associated circuit is called a *half-wave rectifier* because it conducts only one-half of the input waveform. The diode must be able to sustain a reverse voltage swing known as *peak inverse voltage*, or PIV.

The pulsations in the output waveform are usually smoothed out by connecting the output of the rectifier to an *RC* low-pass filter. However, the resistor is generally omitted or a small R

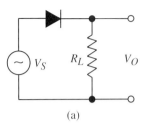

(a)

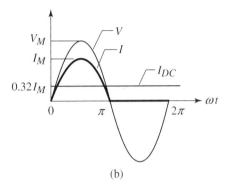

(b)

Figure 3.3 A half-wave rectifier: wiring diagram (a) and output waveform (b).

is used, because it is the energy storage capability of the capacitor that is utilized to smooth out the pulsations and to provide an output current with a small ripple. The capacitance used must be chosen to satisfy the relation

$$R_{Load} C \gg 1/f \qquad (3.3)$$

where f is the ripple frequency, which is twice the frequency of the AC supply ($f = 120$ Hz for a 60-Hz AC supply). The output ripples may also be reduced by using LC filters, but the large values needed for L makes this option a more expensive one.

Example 3.1: Reducing pulsations of a rectified waveform–Figure 3.4 shows a half-wave rectifier circuit. The transformer T with a power rating of 3 V A isolates the mains and steps down the input voltage to produce 0—6 V RMS at 0.5 A. The voltage is applied to a load resistance $R_L = 100$ ohms through a diode that can sustain the 0.5 A at ambient temperature without overheating. If an electrolytic capacitor of value $C = 10,000$ μF is added to smooth out the pulsations in the output waveform, determine the resulting peak-to-peak ripple. Neglect the forward voltage drop across the diode.

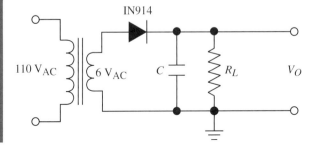

Figure 3.4 Half-wave rectifier circuit for Example 3.1.

Solution: The diode should have a PIV value of $6 \times \sqrt{2} = 8.4$ V. The capacitor will charge during the positive half cycle to this voltage in a very short time. During the negative half cycle, the capacitor discharges through R_L with a current $I = 8.4/100 = 84$ mA for a period of $T = 1/50 = 20$ ms. Using Eq. (2.30), the change in the output voltage, known as the *ripple*, will be

$$\Delta V_o = \frac{I \times T}{C} = \frac{(84 \times 10^{-3}) \times (20 \times 10^{-3})}{10,000 \times 10^{-6}} = 168 \text{ mV}$$

Applying the AC mains power to four diodes connected in a bridge circuit results in a *full-wave rectifier* (e.g., DF01, DF04 ICs), shown in Fig. 3.5a. The whole input sinusoidal waveform (50 Hz) will be transformed into a unidirectional output waveform of a frequency twice the mains (100 Hz), as shown in Fig. 3.5b. The inclusion of the RC filter reduces the ripple voltage, as shown in Fig. 3.5c.

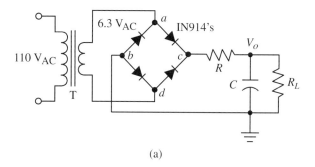

(a)

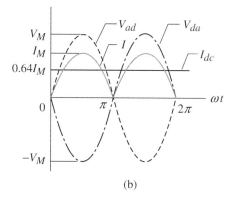

(b)

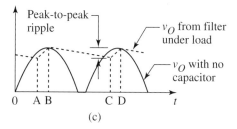

(c)

Figure 3.5 A full-wave bridge rectifier: wiring diagram (a), output waveform (b), with RC filter (c).

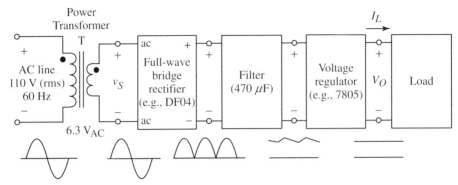

Figure 3.6 A complete power supply system.

The diode rectifier is an essential building block of the DC power supply, shown as a block diagram in Fig. 3.6. It consists of a transformer, a rectifier, a filter, and a voltage regulator. The voltage regulator reduces the ripples and stabilizes the magnitude of the DC output voltage of the supply against variations caused by changes in load current. A Zener shunt regulator or a dedicated IC regulator (e.g., MC78xx series, MC79xx series, LM340, LM723) may be used as the voltage regulator stage. Refer to Appendix A for a 5-V_{DC} power supply circuit and the part's list used to construct it.

In addition to their use as sine wave rectifiers, diodes are used to rectify other signals by generating a unipolar waveform. For example, the diode in the circuit shown in Fig. 3.7 is used to rectify a differentiated square wave. The result is a pulse corresponding to the rising edges of the square wave.

3.3.2 Diode Limiter, or Clipper

In this application, the diode is used to limit the amplitude range of a time-varying input signal. The diode circuit shown in Fig. 3.8 prevents the output from exceeding approximately +4.6 V. Diode clamps are always used on all inputs of CMOS integrated circuits to protect them from static electricity during handling. A diode limiter is placed on the input of an op-amp to prevent the op-amp output from being saturated.

3.3.3 Diode Clamp

A diode clamp is used to change the DC level of the baseline of a waveform. If an input sinu-soid that swings about the reference is applied to the capacitor in Fig. 3.9, the output voltage will depend on the initial charge of C. This may be undesirable. By adding a diode as shown,

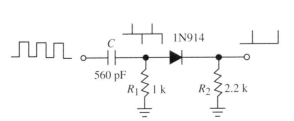

Figure 3.7 Rectified differentiator.

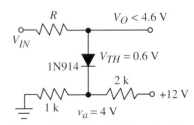

Figure 3.8 A diode limiter circuit with a voltage reference.

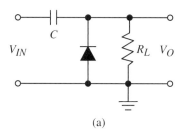

(a)

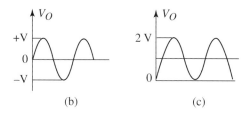

(b)   (c)

Figure 3.9 Diode clamp (a) and a clamp with voltage reference (b).

the dependency of the baseline value of V_o on the initial capacitor charge will be eliminated because the diode will conduct if the peak voltage V_o becomes more negative than the reference level. Consequently, the output voltage swing will always be the peak-to-peak value of 2 V with respect to the reference, and the DC level has been restored, which is why the diode clamp is also referred to as a DC restorer.

3.3.4 Inductive Load and Diode Protection

The current in an inductor cannot be turned off immediately. So when the current supplied to an inductive load is interrupted, the voltage across the inductor suddenly rises, and it keeps rising until it forces current flow. This phenomenon, *inductive kick*, is illustrated in Fig. 3.10a. Inductive kick could damage the electronic devices dedicated to controlling the inductive load. For inductors driven from DC, this problem can be solved by placing a diode across the inductor, as shown in Fig. 3.10b.

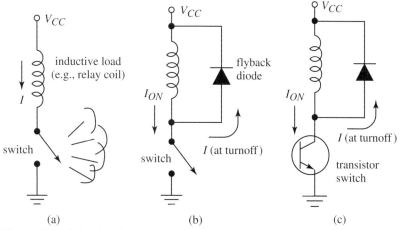

(a)   (b)   (c)

Figure 3.10 Inductive kick (a) and using a flyback diode to block the inductive kick and protect a switch (b) and a transistor (c).

Example 3.2: Effect of the flyback diode—A supply voltage $V_s = 12$ V is applied to a solenoid with a coil inductance $L = 1$ mH and a current rating of $I = 1$ A. Determine the voltage across the inductor when the current is switched off. What would be the effect of adding a diode across the inductor? Assume the current is switched off the solenoid in 0.2 μs.

Solution: The voltage spike generated by switching the current off is

$$V = V_s - L_a \frac{dI}{dt} = 12 - \frac{1(\text{mH}) \times 1(\text{A})}{0.2(\mu\text{s})} = 512 \text{ V}$$

If a diode is placed across the inductor, the voltage excursions will then be

$$V = V_s + 0.7 \text{ V} = 12.7 \text{ volts}$$

which is much smaller than 512 V. A fast-recovery diode is required (0.2 μs) for this application. Also, the diode must have a high current rating.

3.3.5 Temperature Sensor

The *p-n* junction diode is used to measure temperature by exploiting the limited linear range of its output voltage variation with temperature. A stable constant-current source is essential for this application because the diode voltage is a function of current. In most of their usable range, diodes are more sensitive and exhibit higher linearity but less repeatability than a thermocouple, or a resistance thermometer. Measurement accuracy within $\pm 1°$C can be achieved. A high-precision gallium arsenide (GaAS) diode thermometer may have an accuracy of ± 0.002 K for a temperature range from 14 to 300 K. Diode thermometers should not be used in the presence of a magnetic field greater than 1 tesla.

3.3.6 Varactor

A varactor is a diode in which the *p*- and *n*-regions are doped such that the capacitance that normally forms near the *p-n* junction can be precisely controlled by a reverse-biased voltage. The capacitance C is inversely related to the applied voltage V by

$$C(V) = \frac{C_0}{(1 + V/V_0)^n} \tag{3.4}$$

where V_0 is the junction potential when no voltage is applied. It is measured experimentally and ranges between 0.5 and 0.7 V. C_0 is the capacitance with zero-bias voltage, also measured experimentally; n depends on the doping profile (0.33 for graded junction, 0.5 for abrupt junction, and 1 or 2 for a hyperabrupt junction).

3.4 ZENER DIODES

The Zener diode, also called an *avalanche*, is a special class of diodes that can sustain reverse voltages greater than the PIV of a regular diode. Zeners may be made with low reverse-breakdown voltage, called *Zener voltage* V_Z. The symbol and *V-I* characteristics of

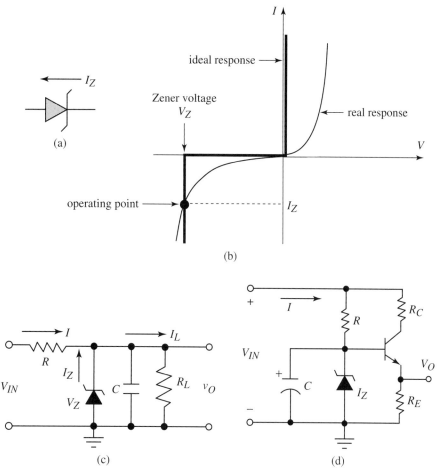

Figure 3.11 Zener diode: symbol (a), *V-I* characteristics (b), a simple voltage Zener regulator (c), and a Zener regulator with follower for increased output current (d).

a Zener are shown in Figs. 3.11a and 3.11b. If the applied voltage becomes equal to V_Z, the output remains roughly constant over a wide range of load currents larger than I_Z, the Zener current, because the Zener's dynamic resistance R_Z is small, on the order of 5–50 Ω. This property gives the Zener its main use as a *voltage regulator* to provide a constant-voltage source at some point inside a circuit from a higher, unregulated voltage some-where else within the two voltage regulation circuits, as shown in Figs. 3.11c and 3.11d. Table 3.2 gives the characteristics of selected Zener diodes. Referring to Fig. 3.11d, the following relations apply:

$$V_O - V_Z = IR$$
$$I = I_L + I_Z \tag{3.5}$$

The load current is simply the transistor base current I_B. The power dissipated in the Zener diode is simply $P_Z = I_Z V_Z$. The Zener must be able to sustain the total current I without over-heating when the load current is zero (open circuit).

TABLE 3.2 Specifications of Selected Zeners

Part Number	Zener Voltage V_Z	Test Current I_{ZT}	Tolerance (±%)	Temp. Coeff. Max (ppm/°C)	P_{DISS} Max (W)	Description
1N821A	6.2 V	7.5 mA	5	±100	0.4	Reference Zener
1N829A	6.2 V	7.5 mA	5	±5	0.4	Reference Zener
1N5221A	2.4 V	20 mA	10	−850	0.5	Regulator Zener
1N5231A	5.1 V	20 mA	10	±300	0.5	Regulator Zener

3.5 LIGHT-EMITTING DIODE (LED)

This is a special type of diode that emits light when forward biased. As Fig. 3.12 indicates, a typical LED has two leads, with the anode usually the longer lead. The LED is usually encapsulated in red, green, yellow, or blue plastic material. Typical specifications of different-color LEDs are given in Table 3.3. The LED may be connected in one of two configurations, *common anode* and *common cathode*. In the common-cathode connection, the cathode is tied to ground and the input voltage V_{IN} is applied to the anode. The LED turns on if V_{IN} is greater than the voltage drop across the LED required to turn it on, or the turn-on voltage, $V_{Turn-ON}$ which is typically between 1.5 and 2.5 V, depending on the LED material. The intensity of emitted light depends on the current flowing through the LED; typically, between 5 and 25 mA of current is sufficient to light the LED. A *current-limiting resistor* is placed in series with the LED to prevent excess current flow that could destroy the diode and possibly the circuit controlling it. A 330-Ω resistor is typically used with LEDs involved in 5-volt digital circuit designs. In general, the resistor is determined to satisfy (see Fig. 3.12c)

$$V_{IN} = V_{Turn-on} + I(R + R_D) \tag{3.6}$$

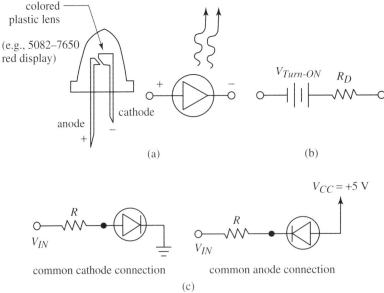

Figure 3.12 Light-emitting diode and symbol (a), equivalent circuit parameters (b), and connection procedure (c).

TABLE 3.3 Typical Characteristics of Various LEDs

Color	R_D (Min)	R_D (Typical)	R_D (Max)	$V_{\text{Turn-on}}$
Standard red	—	3 Ω	7 Ω	1.55 V
High-efficiency red	17 Ω	21	33	1.55
Yellow	15	25	37	1.6
Green	12	19	29	1.65

Example 3.3: Current-limiting resistor–Determine the current-limiting resistor to use with a yellow LED connected in the common-cathode configuration. The LED is driven from a +5-V source and requires 15 mA to turn on.

Solution: Using Eq. (3.6) and Table 3.3, R is determined as

$$R = \frac{V_{IN} - V_{Turn\text{-}ON} - IR_D}{I} = \frac{5 - 1.6 - (15 \times 10^{-3}) \times 25}{(15 \times 10^{-3})} = 201.6\ \Omega$$

3.6   PHOTODIODE

A photodiode converts photons (light energy) into charge carriers (electric energy), specifically one electron and one hole per photon. A small voltage is generated across the photodiode's semiconductor junction when exposed to light, which increases as the light intensity increases. This mode of photodiode operation is referred to as the *photovoltaic* mode, which is the basis of fabrication of *solar cells*. The open-circuit voltage produced by the incident light is determined by $V_O = (kT/q)/\ln(I_L/I_S)$, where I_L is the photogenerated current. The efficiency of solar energy conversion associated with CdS or GaAs materials is about 15%.

A standard photodiode is constructed from a *p-n* junction with a transparent, nonreflective silicon dioxide window placed on the *p*-layer, to allow light to enter and strike the diode, and a metal backplate attached to the *n*-layer, to allow electrical contact to this region. The *p-i-n* photodiode includes a high-resistance intrinsic *i*-layer between the *p*- and *n*- regions of a standard *p-n* diode. *p-n* and *p-i-n* diodes are often mounted on an insulative substrate and sealed within a metal case, as shown in Fig. 3.13.

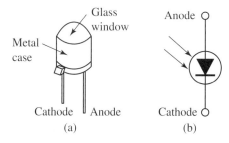

Figure 3.13 Photodiode's appearance (a) and symbol (b).

The photodiode can also operate in the photoconductive mode, where the resistance changes with light intensity is used in a variety of sensing schemes, such as the measurement of light intensity, and fire detection, among many more. Section 11.10 discusses photodiode circuits and interfacing in more detail.

3.7   TRANSISTORS

A transistor is a *three-terminal* device made from semiconductors, usually silicon or germanium. The transistor is the principle *active* device of every electronic circuit, from the simplest amplifier to the most elaborate digital computer. Depending on the manufacturing technology, transistors are broadly classified into two types, *bipolar junction transistors* (BJTs) and *field-effect transistors* (FETs). BJTs form the essential ingredients of the *transistor-transistor logic* (TTL) family of *integrated circuits* (ICs) and are widely used in analog signal-conditioning and drive circuits. FETs are used primarily in the *complimentary metal oxide semiconductor* (or CMOS) family of ICs. FETs are favored in power-switching and digital devices.

Transistors are also classified as either signal-level devices or power devices, depending on their power-handling capabilities. Signal-level transistors are typically used in signal processing, such as amplification of sensor signals, and in logical manipulations of bits. Power devices, on the other hand, are employed in areas in which high currents and voltages are involved, such as in audio systems and in motor drives. A brief overview on transistors is provided in the following sections.

3.8   BIPOLAR JUNCTION TRANSISTOR (BJT)

The BJT consists of three layers of *p*-type and *n*-type semiconductors arranged to form either an *npn* or a *pnp* transistor. In the *npn* transistor, a thin *p*-layer is sandwiched between two *n*-layers; in the *pnp* transistor, a thin *n*-layer is sandwiched by two *p*-layers. The basic structure, electronic symbol, associated voltages and currents, and packages of small-signal *npn* and *pnp* transistors are shown in Fig. 3.14. The terminals of a BJT are known as the base (B), the collector (C), and the emitter (E). The base is the middle and controlling lead. The B, C, and E lie along the top surface of the silicon wafer. The direction of internal current flow and the polarities of the applied voltages for the *npn* transistor are opposite to those in the *pnp* transistor. In addition, the magnitudes of currents and voltages for the two types differ.

In general, *npn* transistors perform better than their *pnp* counterparts because electrons are more mobile than holes. *npn* transistors cost less to produce and are used primarily in low-power, high-frequency applications, whereas *pnp* transistors are used in high-power applications where high voltages and currents are involved. Using complementary pairs (*npn-pnp*) in instrumentation circuit designs simplifies the design considerably.

Depending on the circuit application, each of the three terminals of a transistor may be used as an input terminal, an output terminal, or a common terminal, allowing for three possible configurations: the common emitter (CE), the common collector (CC), and the common base (CB). The common-base configuration is less commonly used. The most frequently used connection of a *npn* transistor is the common-emitter configuration, shown in Fig. 3.15. It will form the basis of the discussion hereafter. In this configuration, an input voltage applied to the base causes an output collector current (I_C) to flow and a collector-to-emitter voltage drop (V_{CE}) to occur. The voltage between the base and the emitter controls the flow of current from

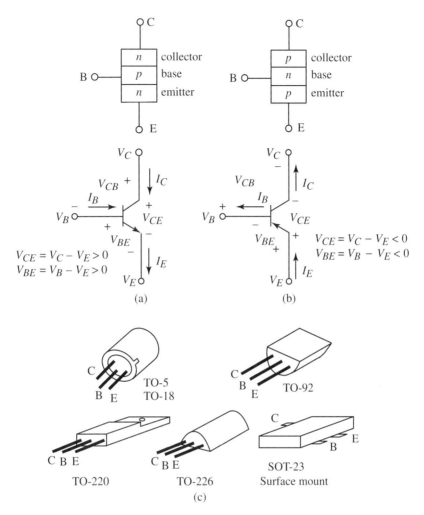

Figure 3.14 Bipolar junction transistors: *npn* transistor (a), *pnp* transistor (b), and transistor packages (c).

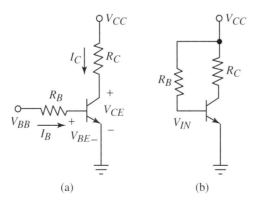

Figure 3.15 Biasing a BJT with two supplies (a) and a single supply (b).

TABLE 3.4 Characteristics of Selected BJTs

Part Number	Type	Case	V_{CE}	$I_{C(MAX)}$	Power	$f_{T(MAX)}$ (MHz)	β	Substitute
2N3904	*npn*	TO-92	40 V	200 mA	350 mW	300	30–100	2N2222
2N3906	*pnp*	TO-92	40 V	200 mA	350 mW	250	30–100	2N2907
TIP29	*npn*	TO-220	40 V	1A	30 W	3	15–75	TIP29/A/B/C
TIP30	*pnp*	TO-220	40 V	1A	30 W	3	15–75	TIP30/A/B/C
2N3055	*npn*	TO-3	60 V	15A	350 mW	2.5	20–70	
MJ2955	*pnp*	TO-3	60 V	15A	350 mW	2.5	20–70	

the collector to the emitter, I_C, which is determined by the value of I_B. The common emitter provides for current amplification and has the highest power gain of other configurations.

BJTs are *current-controlled* devices. A bias current applied through the base terminal initiates the flow of a much larger collector current. Figure 3.15 shows two biasing schemes. Two DC sources are used in Fig. 3.15a, collector supply V_{CC} and the base supply V_{BB}. A more practical biasing circuit is the simple circuit in Fig. 3.15b, which uses only one supply connected to the positive terminal V_{CC}. DC biasing of a BJT is discussed further in Section 3.8.3.

The internal construction of the transistor determines its performance characteristics and the maximum ratings that must not be exceeded. Specific ratings of a transistor include maximum allowable values for collector current I_C, base current I_B, collector-to-emitter voltage V_{CE}, power dissipation $(I_C V_{CE})$, base-to-emitter voltage V_{BE}, and temperature, among other specifications. Table 3.4 gives the maximum ratings of selected BJTs.

3.8.1 Transistor Characteristics

The base–emitter and base–collector circuits behave as two diodes connected back to back, as illustrated in Fig. 3.16. Normally the base–emitter diode is a conducting *p-n* junction and must be forward biased with a base-to-emitter voltage $V_{BE} = V_B - V_E$. For a Si transistor, $V_{BE} \approx 0.7$ V; for a Ge transistor, $V_{BE} \approx 0.3$ V. The base–emitter reverse breakdown voltage for a Si transistor is small, about 6 V. The base–collector diode meanwhile is a reverse-biased *n-p* junction.

The behavior of a transistor is described by its input-to-output characteristics, as shown in Fig. 3.17 for a common-emitter *npn* configuration. The input characteristics shown in Fig. 3.17a are similar to those for a forward-biased diode. The collector current curves are plotted against the collector–emitter voltage V_{CE} for different values of the base current I_B.

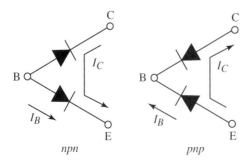

Figure 3.16 Two back-to-back *pn*-junction diodes represent the behavior of a BJT.

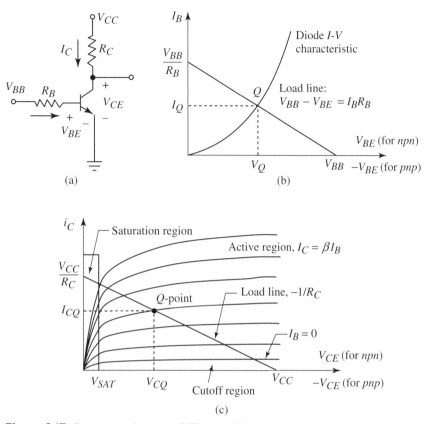

Figure 3.17 Common-emitter *npn* BJT (a) and its input characteristics (b) and output characteristics (c).

3.8.2 Transistor States

The characteristic curves of Fig. 3.17 indicate that the transistor can be made to operate in one of the following three distinct regions: the cutoff region, the active region, and the saturation region.

Cutoff state. When V_{BE} is less than V_{TH}, one diode drop (0.6 V for Si), the transistor is in the *cutoff* (or *open*) *state* and $I_C = I_B = 0$. A typical value of 5 μA with a 5-V input voltage corresponds to a DC *cutoff resistance* of 1 MΩ.

Active state. When $V_{BE} > V_{TH}$, the base-to-emitter diode becomes forward biased and puts the transistor in its *active state*. At this point, a small current flowing into the base I_B controls a much larger current flowing into the collector I_C. This is the property from which the transistor derives its usefulness as a current amplifier. The collector current I_C is roughly proportional to the base current I_B:

$$I_C = h_{FE}I_B = \beta I_B \tag{3.7}$$

where h_{FE}, or β, is the forward *current amplification factor*, also known as the DC current gain. Although the typical value of β is 100, it varies between 50 and 250 for different transistor specimens of the same type. β also depends upon I_C, V_{CE}, and temperature and should be used in design with care. The current flowing through the emitter is simply

$$I_E = I_C + I_B = (1 + \beta)I_B \cong I_C \tag{3.8}$$

The supply voltage V_{CC} is divided between the voltage drop across the load R_C and the voltage drop across the transistor V_{CE}:

$$V_{CC} = I_C R_C + V_{CE} \tag{3.9}$$

which indicates the dependency of the collector current I_C on the load resistance R_C. Rearranging Eq. (3.9) gives the *load-line equation*, expressed as

$$I_C = \frac{V_{CC}}{R_C} - \frac{V_{CE}}{R_C} \tag{3.10}$$

From Eq. (3.9), $V_{CE} = 0$ at $I_C = V_{CC}/R_C$, and $V_{CE} = V_{CC}$ at $I_C = 0$. The intersection of the load line with the corresponding base current curve defines the operating point, Q-point in Fig. 3.17 c. The Q-point is defined by three parameters: I_C, I_B, and V_{CE}. Thus, for a given value of I_B, I_C is calculated by Eq. (3.7) and the load line relation then gives V_{CE}.

As I_B rises and V_{BE} slowly increases to 0.7 V, I_C will rise exponentially, as shown in Fig. 3.17b. The load voltage $I_C R_C$ also increases and V_{CE} continues to drop toward ground until it reaches approximately 0.2 V. At this point, the transistor becomes *saturated* with a corresponding collector current I_{CS}. Equation (3.7) no longer applies and I_C is found by Eq. (3.10) instead. At saturation, $I_B \geq I_{CS}/\beta$, where I_{CS} is the transistor's saturation current. V_{CE} will be a few hundreds of millivolts, 0.2 V for most transistors, and does not vary with R_C because of the very high slope of the transistor's *saturation line*. Thus, the collector cannot drop below ground; otherwise the collector to base diode would be forward biased.

Example 3.4: Analysis of a BJT circuit—For the basic circuit and transistor characteristics shown in Fig. 3.18, find the pertinent transistor voltage and current if V_{IN} is: (a) 0.1 V, (b) 1.0 V, and (c) 5 V. Assume $V_{TH} = 0.7$ V.

Solution:

   a. Since $V_{BE} < 0.7$ V, the transistor is cut off, $I_B = I_C = 0$, and $V_{CE} = V_{CC} = 30$ V. The load line shown in Fig. 3.18 is drawn from the $V_{CE} = 30$ V point to the maximum possible collector current, found by Eq. (3.10) to be $I_{CS} = V_{CC}/R_C = 30/1.2$ k$\Omega = 25$ mA.

   b. Since $V_{IN} > 0.7$ V, the current flowing through the base is

$$I_B = \frac{V_{IN} - V_{BE}}{R_B} = \frac{1 - 0.7 \text{ V}}{10 \text{ k}\Omega} = 30 \ \mu A$$

The operating point P is where the load line and the $I_B = 30 \ \mu A$ base current curve intersects. The values of V_{CE} and I_C corresponding to this point are $V_{CE} \cong 16$ V and $I_C \cong 11$ mA.

    Usually, the characteristic curves are not readily available in graphical form, and the key device specifications must be read from the manufacturer's tables. For this example, using a

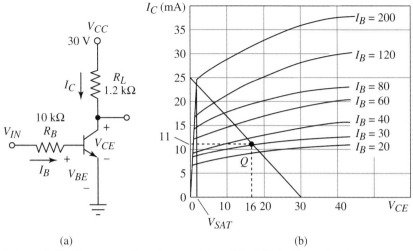

Figure 3.18 Common-emitter characteristics of the BJT for Example 3.4.

current gain of $\beta = 250$ yields $I_C = I_B\beta = 7.5$ mA, compared to 11 mA. Also $V_{CE} = V_{CC} - I_C R_L = 30$ V $- (7.5$ mA $\times 1.2$ k$\Omega) = 21$ V, compared to 16 V.

c. The base current when $V_{IN} = 5$ V is

$$I_B = \frac{V_{IN} - V_{BE}}{R_B} = \frac{5 - 0.7 \text{ V}}{10 \text{ k}\Omega} = 430 \text{ }\mu\text{A}$$

Meanwhile, the base current at saturation is where a vertical line corresponding to V_{SAT} intersects the load line. Here it is about $I_{BS} = 190$ μA. Since $I_B > I_{BS}$, the transistor is saturated. Then the collector's saturation current is $I_{CS} = 24$ mA. As an alternative approach to using the graph, we have

$$I_{CS} = \frac{V_{CC} - V_{CE}}{R_C} = \frac{30 - 0.2 \text{ V}}{1.2 \text{ k}\Omega} \cong 25 \text{ mA}$$

$$I_{BS} = \frac{I_{CS}}{\beta} = \frac{25 \text{ mA}}{250} \cong 100 \text{ }\mu\text{A}$$

Again $I_B > I_{BS}$ and the transistor is saturated and the preceding I_{CS} holds. Since the current gain is nonlinear, if a more representative gain near saturation is 100, then $I_{BS} = 250$ μA instead of the 190 μA found earlier.

3.8.3 DC Biasing of the BJTs

BJT transistors involved in the design of discrete devices such as voltage or current amplifiers require biasing (1) to establish a *constant* emitter DC current to reduce the effects of temperature and β variations and (2) to locate the bias point, also called the *quiescent point* or *Q*-point, to allow maximum swing in the output signal as the input is applied at the base.

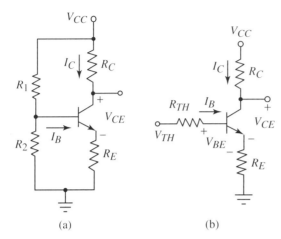

Figure 3.19 DC biasing circuit (a) and its Thevenin equivalent (b).

(a)   (b)

For amplifier operation, the Q-point located in the active region of the $I_C - V_{CE}$ plane describes the DC condition of the amplifier. Since an AC signal is normally superimposed on the operating base-to-emitter voltage V_{BE} in the amplification region, the Q-point is subject to a swing in either direction. The Q-point is set by V_{BE} and V_{CC} at the intersection of the load line and the $I_C - V_{CE}$ curve corresponding to the associated base current I_B. In most cases the quiescent point is located to provide enough range to allow the output waveform to have a maximum symmetrical signal swing without *clipping* (flattening of the top or bottom of the waveform); that is, the swing should be approximately equal in each direction so that the positive peak of V_{CE} is less than V_{CC} and the negative peak of V_{CE} is not below a few tenths of a volt. In the region around the quiescent point, any small variation in the input signal causes a proportional change in the output signal, and the gain will remain approximately constant.

Figure 3.19 shows the most commonly used biasing method with one power supply only. In practice, the quiescent point is set by using a fixed DC supply (no input signal present), and the circuit elements are selected so as to bias the collector–base and emitter–base in appropriate magnitude and polarity. To make I_E insensitive to temperature and β variations, the design should satisfy these two conditions: $V_{BB} \gg V_{BE}$ and $R_E \gg R_B/(\beta + 1)$.

If the values of the circuit elements and the transistor gain β are known, the biasing point can be determined. Applying KVL around the base loop, the base current is found by

$$I_B = \frac{V_{TH} - V_{BE}}{R_{TH} + \beta R_E} \tag{3.11}$$

The collector current is then calculated by Eq. (3.10). Once I_B and I_C are known, V_{CE} is determined by applying KVL around the loop formed by the collector, emitter, and DC supply V_{CC} to give

$$V_{CE} = V_{CC} - R_C I_C - R_E I_E = V_{CC} - \beta(R_C + R_E)I_B \tag{3.12}$$

The resistance R_E at the emitter increases the input resistance by $(1 + \beta)R_E$. However, it also reduces the effective voltage gain. Thus, R_E is used if a maximum input resistance is desired; otherwise it is kept at zero for a maximum voltage gain.

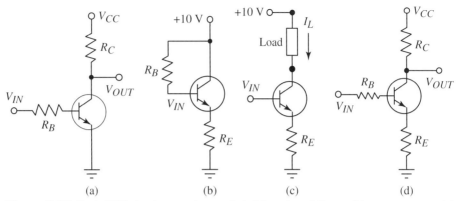

Figure 3.20 Basic BJT circuits: transistor switch (a), emitter-follower (b), current source (c), and common-emitter follower (d).

3.8.4 Basic BJT Circuits

The most important BJT circuits are the transistor switch, the emitter follower, the current source, and the amplifier. Schematics of these basic circuits are shown in Fig. 3.20. The main difference among these circuits is the output terminal used.

Transistor Switch

The term *transistor switch* applies when a small control current results in a larger current flow in another circuit. The transistor acts as a switch when driven into saturation. Four typical transistor switch applications are shown in Fig. 3.21: a solenoid driver (Fig. 3.21a), an LED driver (Fig. 3.21b), a relay driver (Fig. 3.21c), and a lamp driver (Fig. 3.21d). When the mechanical switch in Fig. 3.21d is open, the input base voltage is zero and the transistor is in the cutoff state, $I_C = I_B = 0$, and the lamp is off. The resistor connected between the base and ground guarantees that the base is at ground when the switch is open. When the switch is closed, V_{BE} rises to 0.6 V and the transistor becomes active. The drop across the base resistor is then 9.4 V and the base current is $I_B = 9.4$ mA. Then $I_C = 100$ mA because the transistor is saturated and V_{CE} cannot go below ground, which is required if more than 100 mA is drawn ($V_{SAT} = 0.2$ V). In a transistor switch circuit design, it is important to choose R_B to get excess I_B, to use a diode in series with the collector to prevent collector–base conduction on negative swings, and to protect the transistor with a diode across inductive loads, as in Fig. 3.21c.

Emitter-Follower

The *common-collector amplifier* shown in Fig. 3.20b has no collector resistance, and the output terminal is the emitter, with R_E considered as the load impedance. The circuit is known as an emitter-follower because the output (emitter voltage) is the same as (follows) the input (base voltage) but less by one diode drop; that is, $V_E = V_B - 0.6$ V and the voltage gain is slightly less than 1. The input impedance (impedance looking into the base) of the emitter-follower is $R_{IN} \cong (1 + \beta)R_E$, and the output impedance (impedance looking into the emitter) is $R_{OUT} \cong R_S / (1 + \beta)$, R_S is the impedance of the source connected to the circuit. Having a low output impedance and a high input impedance gives the emitter-follower circuit its main use, as a buffer between a source and a load. Although the emitter-follower has practically no voltage gain, it has current gain, resulting in a net power gain. An *npn* emitter-follower can *source* large current but can only *sink* limited current through its emitter.

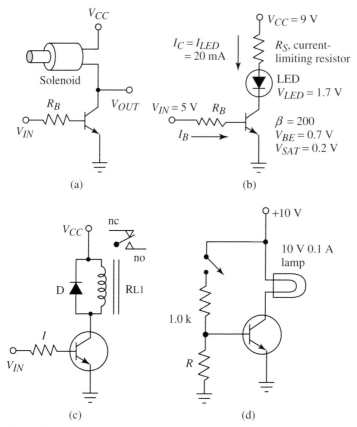

Figure 3.21 Transistor switch applications: basic switch (a), LED driver (b), relay driver (c), and lamp driver (d).

Figure 3.22 shows an AC-coupled emitter-follower biased by a voltage divider. Resistors R_1 and R_2 of the voltage divider are selected such that: (1) $R_1 \approx R_2$, to put the base voltage midway between V_{CC} and ground with no input signal, and (2) $R_1/R_2 \ll \beta R_E$, to make the DC impedance of the voltage divider small compared with the input impedance of the follower;

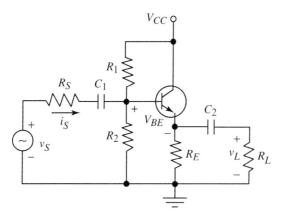

Figure 3.22 An AC-coupled emitter-follower circuit.

that is, current flowing through the voltage divider should be large compared with the current drawn by the base.

C_1 is chosen to form a high-pass filter with the impedance it sees as a load, which consists of the parallel combination of the input impedance of the follower with the Thevenin equivalent of the voltage divider. C_2 forms a high-pass filter with the load impedance R_L (unknown), to put the corner frequency (the 3-dB point) below the lowest frequency of interest.

Current Source

The basic current source circuit is shown in Fig. 3.20c. The collector current is simply $I_C \approx (V_B - 0.6 \text{ V})/R_E$, which is independent of V_{CC} as long as the transistor is not saturated $(V_C > V_E + 0.2 \text{ V})$. The base voltage can be provided from a low-impedance voltage divider, shown in Fig. 3.23, or from a Zener diode biased from V_{CC}. A current source can provide a load with a constant current only over a finite range of load voltage. Furthermore, varying the base voltage results in a voltage-programmable current source. The impedance looking in the collector (or collector impedance) of a current source is very large (measured in megaohms). Current sources provide inputs to integrators, sawtooth generators, and ramp generators. They are also employed as the active loads for a high-gain amplifier stage and as emitter sources for differential amplifiers.

Common-Emitter Amplifier

If the load in a current source is replaced with the resistor R_C and the collector is used as the output terminal, the result is the common-emitter amplifier shown in Fig. 3.20d. Once the Q-point has been established by a biasing circuit, an input voltage can be applied through *coupling capacitors* C_1 and C_2, as shown in Fig. 3.24. C_1 and C_2 isolate the DC signals of the biasing circuit from the input signal V_S and load resistance R_L, respectively. If C_1 is removed, R_S and R_2 become parallel and V_B is disturbed. If C_2 is removed, V_{CC} would depend on R_L.

Capacitor C_1 connects, or couples, the source to the transistor, causing the collector voltage to vary and to block the flow of a direct current back to the source. The high-pass filter that is formed by the capacitor in combination with the parallel resistance of the base biasing

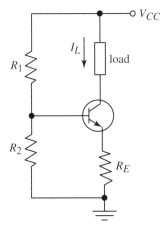

Figure 3.23 A current source circuit.

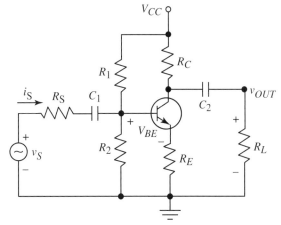

Figure 3.24 Common emitter-follower amplifier circuit.

resistors R_1 and R_2 passes all frequencies of interest, provided that C_1 satisfies the relation $C_1 \geq 1/[2\pi f(R_1 \| R_2)]$. This circuit is a voltage amplifier, with a gain

$$\text{gain} = \frac{v_{OUT}}{v_s} = -\frac{R_C}{R_E} \tag{3.13}$$

which is negative, indicating a phase shift of $180°$. The input impedance (as seen by the input signal) of this circuit is the Thevenin equivalent of the voltage divider ($R_B = R_1 \| R_2$) in parallel with $(1 + \beta)R_E$. The output impedance of the amplifier is R_L, or, more precisely, the parallel combination of R_L with the very large collector impedance.

3.9 PHOTOTRANSISTOR

A phototransistor is a photojunction device with construction and appearance similar to those of an ordinary BJT transistor, but it is often packaged in a metal case with a glass window, much like a photodiode. Although the phototransistor has a base lead as well as an emitter and a collector lead, the base lead is seldom used. Current flow through a phototransistor is directly proportional to the amount of incident light. Any increase in light intensity results in an increase in the base current, which causes a large increase in the collector current flowing through the transistor due to the transistor's amplifying ability. The phototransistor is used in the same way as a photodiode in its photoconductive mode. A phototransistor is more sensitive to light variations than is a photodiode, making it more suitable for a wider variety of applications. However, a phototransistor does not respond as quickly to changes in light intensity as a photodiode, and, therefore, it is not suitable for applications requiring extremely fast response. The phototransistor is characterized by an input impedance that varies from $1000\ \Omega$ to $1\ M\Omega$ in most low-level DC circuits, a response time that is inversely proportional to incident light, moderately high intensity levels, a yield response much less than 1 millisecond, and poor response to green and blue colors but acceptable to reds and near-infrared.

The phototransistor is usually combined with a spectrally matched infrared light-emitting diode, both shown in Fig. 3.25a, to form an IR emitter/detector pair that is the basis of many commercially available optical sensors. The circuit connection in Fig. 3.25b gives the IR pair its usefulness as an optical sensor in a variety of applications. The resistor R_2 can be adjusted

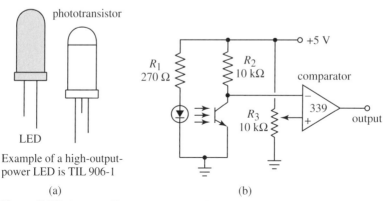

Figure 3.25 A spectrally matched IR LED/phototransistor pair: packages (a) and typical circuit connection (b).

to vary the sensitivity of the phototransistor. The output of the transistor can be connected to a control circuit such as a comparator. This circuit may be used as a counter or in control applications requiring specific actions whenever the light level reaches a certain level. It may also be used to measure position and speed of rotating ot translating objects. Refer to Chapter 11 for several applications involving IR emitter/detector.

3.10 FIELD-EFFECT TRANSISTOR (FET)

A field-effect transistor, or FET, has three terminals, called the *gate* G, the *drain* D, and the *source* S, that correspond in many ways to the base, collector, and emitter of BJTs. The FET is a voltage-controlled device; the load current is controlled by a voltage applied at the gate.

An FET has no *p-n* junction in the current path. The current is conducted across a narrow channel between the drain and the source by only one type of carrier, electrons in an *n*-channel FET and holes in the *p*-channel FET. For this reason the FET is called a *monopolar* device.

FETs can be made with two different kinds of gate construction: the *junction* FET, or JFET, and the *metal-oxide-semiconductor* FET, or MOSFET. Furthermore, FETs are available with two types of channel doping. If the device has a physically implanted channel, it is said to be a *depletion* type, since the applied gate voltage depletes original free carrier electrons in the channel. On the other hand, if no channel is provided, the device is an *enhancement* type. Figure 3.26 shows the different types of FETs.

3.10.1 JFETs

JFETs are available with either an *n*-channel or a *p*-channel. The structure, circuit symbols, and characteristics of *n*-channel JFETs are shown in Fig. 3.27, and a selected list of JFETs is given in Table 3.5. As Figure 3.27 shows, the gate in a JFET forms a semiconductor junction with the underlying channels. The gate channel diode in an *n*-channel JFET will conduct when V_{GS} is more than 0.5 V. A V_{GS} value greater than 0.5 destroys the operation of the device by permitting gate current; thus a JFET should always be reverse biased ($V_{GS} < 0$) with respect to the channel, to prevent gate current. Consequently, JFETs are *always depletion-mode devices*.

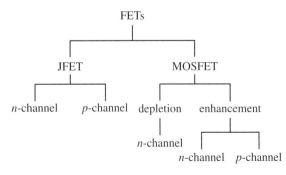

Figure 3.26 A family tree of field-effect transistors.

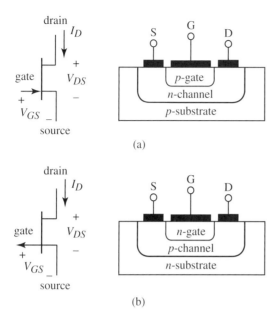

Figure 3.27 Single-level JFETs: n-channel (a) and p-channel.

3.10.2 MOSFETs

The MOSFET is made with the gate region separated from the conducting channel by a thin layer of SiO_2 (glass) grown onto the channel. The gate oxide creates a capacitor between the gate and the source, so the steady-state gate current is zero, which makes it easier for MOSFETs to be driven from a microprocessor or a CMOS logic gate. MOSFETs can be a *depletion* type or an *enhancement* type, the enhancement type being the more important. The enhancement MOSFET can be either an n-channel with a p-substrate, referred to as NMOS, or a p-channel with an n-substrate, referred to as PMOS. The structure and symbols of MOSFETs are shown in Fig. 3.28, with a selected list given in Table 3.6. The junctions are fabricated to maintain separate regions of like charge carriers when the device is off. Applying an electric field causes a channel to form if a sufficient number of charge carriers accumulate near the surface of the substrate under the gate to create, in effect, a region that connects the like-charge carriers of the source and drain regions. The movement of the charge carriers through the channel is held back by the effective resistance from the drain to the source. This resistance depends on many factors, such as the mobility of the carriers and the width of the channel.

TABLE 3.5 Characteristics of Selected JFETs

Part Number	Type	V_{GSS}	$I_{DSS(MAX)}$	$V_{GS(OFF)},$ $V_{P(MAX)}$	$C_{iss(MAX)}$ (pF)	$C_{rss(MAX)}$ (pF)
2N4338	n-channel	50 V	0.6 mA	1 V	6	2
2N4416	n-channel	30	15	6	4	0.8
2N5457	n-channel	25	5	6	7	3
2N5484	n-channel	25	5	3	5	1
2N5114	p-channel	30	90	10	25	7
2N5460	p-channel	40	5	6	7	2

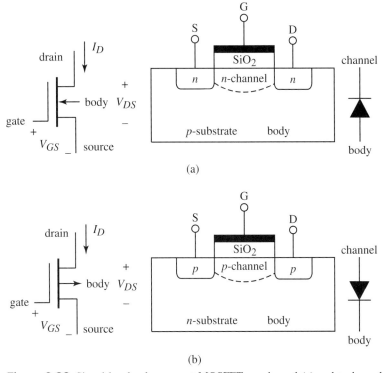

Figure 3.28 Signal-level enhancement MOSFETs: *n*-channel (a) and *p*-channel (b).

The output characteristics of an *n*-channel MOSFET is shown in Fig. 3.29. With zero or negative gate bias ($V_{GS} = 0$), the device is off (nonconducting). The device is driven into conduction by bringing the gate positive with respect to the source. The minimum value of gate–source voltage V_{GS} that is required to establish a channel is called the *threshold voltage* $V_{GS(TH)}$, which varies between 0.5 and 5 volts, below which the MOSFET is put in the cutoff state. At $V_{GS} = V_{GS(TH)}$, the drain current is very small, almost 0 ($I_D \approx 0$). When $V_{GS} > V_{GS(TH)}$, the drain current I_D increases almost linearly with V_{DS} for small values of V_{DS}. When V_{DS} becomes sufficiently large and $V_{GS} > V_{GS(TH)}$, the channel becomes maximally open and the current I_D becomes saturated and remains constant even as V_{DS} increases. This is the saturated

TABLE 3.6 Characteristics of Selected MOSFETs

Part Number (Channel)	$R_{DS(ON), Max}$ @ V_{GS}	$V_{GSS(TH)}$ MAX	$I_{D(ON)}$ (min) ($V_{DS} = 10$ V)	$V_{DS}/$ V_{GS}	I_{GSS}	$C_{rss(MAX)}$
SD210 (*n*)	45 Ω @ 10 V	2 V	—	30 V/40 V	0.1 nA	0.5 pf
SD211 (*n*)	45 Ω @ 10 V	2 V	—	30 V/15 V	10 nA	0.5 pf
VN2222L (*n*)	8 Ω @ 5 V	2.5 V	750 mA	60 V/40 V	0.1 nA	5 pf
2N4351 (*n*)	300 Ω @ 10 V	5 V	3 mA	25 V/35 V	0.01 nA	2.5 pf
2N4352 (*p*)	600 Ω @ 10 V	6 V	2 mA	25 V/35 V	0.01 nA	2.5 pf
3N172 (*p*)	250 Ω @ 20 V	5 V	5 mA	40 V/40 V	0.2 nA	1 pf

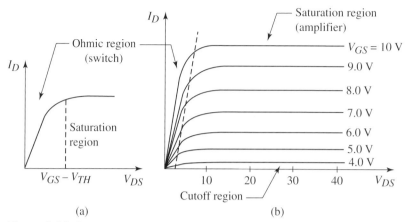

Figure 3.29 Output characteristics of an enhancement-mode signal-level MOSFET.

state of the transistor. If the gate voltage is high enough, usually 10 V, the drain current stays in the constant-resistance region and V_{DS} is minimal, less than 1 V. The inverse of the slope of the I_D–V_{DS} curve in the linear region is the on-resistance of the MOSFET given by

$$R_{DS} = \frac{V_{DS}}{I_D} \tag{3.14}$$

The performance p-channel MOSFET is characterized by a higher gate threshold voltage, higher R_{ON}, and lower saturation current than the n-channel MOSFET.

In a depletion-mode MOSFET, the p-type substrate is doped with n-type atoms to increase channel conduction, even with zero gate bias. The gate must be reverse biased a few volts to cut off the drain current.

The distinction between the depletion and enhancement types is illustrated in Figure 3.30, in which a graph of the drain current I_D is plotted against V_{GS} for a fixed value of drain voltage V_D. The enhancement n-channel draws no drain current until $V_{GS} > 0$, whereas the depletion device operates at nearly its maximum value of drain current I_D when $V_{GS} = 0$.

Depletion-mode MOSFET is ideal for signal amplification because of its wider operating range. However, it is rarely used in practice except in radio frequency applications. The

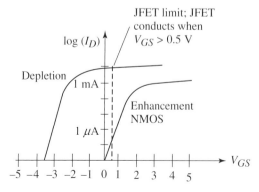

Figure 3.30 Comparison between the depletion-mode and enhancement-mode FETs; drain current I_D vs. V_{GS} at a fixed value of drain voltage V_D.

enhancement-mode MOSFET is ideal for switching operations with open drain–source. This type is analogous to the *npn* BJT. In normal operation the drain is more positive than the source, $V_{DS} > 0$. Current will flow through the drain–source terminals when the gate is "forward biased" by making $V_{GS} > 0$. I_D is determined by the value of V_{GS}, as depicted in Fig. 3.29.

3.11 MAIN FEATURES OF FETs AND BJTs

Here is a summary of the main attributes to remember about FETs and BJTs.

1. FETs are voltage-controlled devices, whereas BJTs are current-controlled devices that incur power dissipation in the bias resistor.
2. The effective input impedance of an FET is very high, greater than 10^8 ohms. This important attribute allows the FET to be used as an amplifier of small signals in high-impedance circuits, such as pyroelectric sensors, discussed in Section 11.14.
3. A consequence of attribute 1 is that the FET's *gate draws no current*. This makes the FET easy to drive by a microprocessor or a CMOS gate.
4. There is no offset voltage associated with the activation of the FET as a switch, whereas a BJT requires a minimum base-to-emitter voltage V_{BE} to operate.
5. FETs are less noisy than BJTs, which makes them more suitable to act as the input stage for low-level-signal amplifiers, used extensively in FM receivers.
6. FETs provide better thermal stability than BJTs because their parameters are less sensitive to temperature variations.
7. FETs are cheaper to manufacture and can be packed with higher densities. MOSFETs are used in very large-scale integrated (VLSI) circuits, such as microprocessors and memory chips.
8. The metal oxide insulation of a MOSFET is susceptible to breakdown from static electricity.

3.12 POWER TRANSISTORS

Small-signal transistors are limited to low-power applications involving small voltages and currents. Power transistors, on the other hand, are made with more semiconductor material, to provide the higher drive currents and voltages required in medium-to-high DC power applications. The package of a power transistor is made large to handle the high heat dissipation more efficiently.

Power transistors are generally used to switch power to a load ON and OFF, the H-bridge arrangement used in DC and stepper motor drives is one example. A transistor switch remains ON as long as the base (or gate) is active. Ideally a switch has infinite impedance in the OFF state and no voltage drop (no power dissipation) in the ON-state. However, a real transistor switch experiences a leakage current on the order of microamps in the OFF state, a voltage drop across its terminals on the order of 1–3 V in the ON state, and a switching time on the order of microseconds. A power device consumes the largest amount of power as it switches through the linear range and is most efficient when it is fully ON.

Power transistors are available in three types: bipolar transistors, Darlingtons, and MOSFETs. Bipolar transistors can handle lower currents than Darlingtons or MOSFETs. A

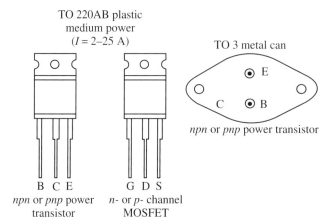

Figure 3.31 Power transistor packages.

Darlington is simply two bipolar devices connected together to increase the current gain. In general, MOSFETs are preferred over bipolar transistors, mainly because FETs are controlled by a gate voltage, whereas bipolar transistors are controlled by a base current. Darlingtons can handle currents up to 400 A and voltages up to 1000 V and can dissipate power up to 20 kW. Corresponding limits for MOSFETs are 100 A, 1000 V, and 5 kW. The limits imposed on the voltage is dictated by the *p-n*-junction breakdown in the OFF state, whereas current limits depend on the wire capacity of the device. The power that can be switched to a load is limited by the small power dissipation of the chip.

3.12.1 Packages

Power semiconductors are available in various packages designed to dissipate heat. Figure 3.31 shows many of the available packages. Devices in the TO-3 metal have higher power limits because they can dissipate heat more efficiently. However, heat sinks are always required for all power transistor applications, except for the low-power ones.

3.12.2 Power Bipolar Transistors

A power bipolar transistor is manufactured with the back side of the Si wafer used as the collector terminal, and the current flows vertically through the device. Bipolar transistors are typically larger, often use back-side connections, and seldom integrate a large number of different devices. Current gains are typically on the order of 20–50 A. Table 3.7 gives the

TABLE 3.7 Power Transistors

Part Number		Package	$I_{C(MAX)}$	$V_{CEO(MAX)}$	$\beta @ I_C$	P_{DISS} ($T_C = 25°C$)
npn	pnp					
2N5191	2N5194	TO-126	4 A	60 V	100 @ 0.2A	40 W
2N5979	2N5976	TO-127	5 A	80 V	50 @ 0.5A	70 W
MJE3055	MJE2955	TO-127	10 A	60 V	50 @ 2A	90 W
2N3055	MJ2955	TO-3	15 A	60 V	50 @ 2A	115 W
2N5886	2N5884	TO-3	25 A	80 V	50 @ 10A	200 W
TIP73		TO-228AB	15 A	40 V	20–150 @ 5A	

$V_{CE(SAT)} = 0.4$ V (typ), V_{BE} (on) $= 0.8$ V (typ)

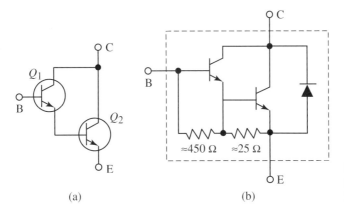

Figure 3.32 Darlington power transistor: general circuit (a) and TIP 663 Darlington circuit (b).

(a)   (b)

characteristics of some available bipolar transistors. Other bipolar transistors commonly used include the *npn* TIP 31 and TIP 41 and the *pnp* TIP 32 and TIP 42.

Bipolar transistors require substantial base currents to control large collector currents. Since microprocessors and logic gates can only source or sink small amounts of currents, an interface circuit is often needed to connect the controlling device to the load. The added complexity involved in the bias network and the power dissipation through the base resistor precludes the use of bipolar transistors in applications requiring high current gains.

3.12.3 Darlingtons

A bipolar power transistor is usually driven by another lower-rated power transistor to deliver high base currents and boost the overall current gain. If the two BJTs are hooked together as shown in Fig. 3.32, a Darlington power transistor is formed with three external connections B, C, and E and that has an overall gain β equal to the product of the two transistor gains, $\beta = \beta_1 \times \beta_2$. The emitter current of Q_1 is the base current of Q_2. Current gains of the two transistors are controlled during their manufacture so that the overall gain varies linearly over a range of collector current when the device is used as an amplifier. As a result, Darlingtons have high values of $V_{CE(SAT)}$, resulting in larger switching losses than for the discrete design. The resistor between the emitter and the base is employed to speed up switching and to prevent leakage current through Q_1 from biasing Q_2 into conduction. A Darlington provides high currents in a small space and makes an excellent input stage for amplifiers where very high input impedance is necessary. Table 3.8 gives the specifications of selected Darlingtons.

TABLE 3.8 Selected Darlingtons

Part Number		Package	$I_{C(MAX)}$	$V_{CEO(MAX)}$	β @ I_C	P_{DISS} ($T_C = 25°C$)
npn	*pnp*					
2N6038	2N6035	TO-126	4 A	60 V	2,000 @ 2A	40 W
2N6044	2N6041	TO-127	8 A	80 V	2,500 @ 4A	75 W
2N6059	2N6052	TO-3	12 A	100 V	3,500 @ 5 A	150 W
2N6284	2N6287	TO-3	20 A	100 V	3,000 @ 10A	160 W
TIP663		TO-3	30 A	300 V	500–10,000 @ 5A	

$V_{CE(SAT)} = 0.8$ V (typ), V_{BE} (on) $= 1.4$ V (typ)

Figure 3.32(b) shows the *npn* TIP663 Darlington. The TO-3 package has a built-in fly-back diode for protection from an inductive kick. The 2N6282 *npn* Darlington provides a current gain of 2400 (typically) at a collector current of 10 A.

Example 3.5: Darlington motor drive–A TIP663 Darlington rated at 20 A continuous I_C is used to drive a DC motor connected as a T bridge configuration for ON–OFF control. The motor is rated at $V_M = 10$ V, and its current demand is 2 A in continuous drive and 10 A in the start/stall mode. The supply voltage is 12 V. Assuming that the installed heat sink maintains the temperature of the motor at 25°C, determine the efficiency of the drive in the continuous and start/stall modes.

Solution: The manufacturer's charts indicate that the TIP663 has the following characteristics :

at $I_C = 2$ A: $V_{CE(SAT)} = 0.8$ V, $\beta = 1500$, $I_B = 1.3$ mA (4-mA overdrive)

at $I_C = 10$ A: $V_{CE(SAT)} = 1.2$ V, $\beta = 1000$, $I_B = 10$ mA (30 mA overdrive)

The total power dissipation in the Darlington is the sum of the power dissipation in the base and the collector. For the continuous-drive mode, this amounts to

$$P_{diss} = (V \times I_B) + (V_{CE(SAT)} \times I_C) = (12 \times 0.004) + (2 \times 0.8) = 1.6 \text{ W}$$

The resulting efficiency is

$$[(2 \times 12) - 1.6]/(2 \times 12) = 93\%$$

The motor drive voltage is the difference between the supply voltage and the voltage drop across the transistor; that is,

$$V_M = 12 - 0.8 = 11.2 \text{ V}$$

Similarly, for start/stall drive, the power dissipation amounts to

$$P_{diss} = (V \times I_B) + (V_{CE(SAT)} \times I_C) = (12 \times 0.01) + (10 \times 1.2) = 12 \text{ W}$$

The resulting efficiency is

$$[(10 \times 12) - 12]/(10 \times 12) = 90\%$$

The motor drive voltage in this case is

$$V_M = 12 - 1.2 = 10.8 \text{ V}$$

3.12.4 Power MOSFETs

The circuit symbol of an *n*-channel power MOSFET is shown in Fig. 3.33. The MOSFET requires an input voltage of at least 10 V to switch power to a load when used as a power-switching device, although low-threshold MOSFETs are also available. The device that controls a power MOSFET must be able to sustain the small leakage current, on the order of

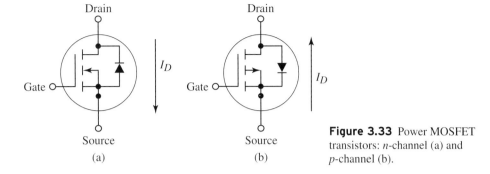

Figure 3.33 Power MOSFET transistors: *n*-channel (a) and *p*-channel (b).

microamps, typical of a CMOS input. Power MOSFETs are also used as linear amplifiers. For example, a pair of MOSFETs form the output stage of a high-fidelity audio system to drive speakers.

Referring to Fig. 3.29, the power MOSFET can operate in one of three states: a cutoff state, where $V_{GS} \leq V_{GS(TH)}$; an active or saturation state, where $V_{DS} \geq V_{GS} - V_{GS(TH)}$; and a linear or constant-resistance state, where $V_{DS} \leq V_{GS} - V_{GS(TH)}$. When a voltage is applied to the gate, the ON-state resistance R_{ON} to current flow through the power-handling part of the device between the drain and the source is low. High-current-power MOSFETs are designed to have extremely low ON-state resistance, to minimize power dissipation. If V_{GS} is high, usually about 10 V, I_D stays in the linear region, where the power MOSFET operates in the switched ON-state where V_{DS} is minimal.

The back-side connection of a MOSFET is used as the drain. The central region is connected to the source, creating a source–drain diode. The TO 220 is a popular MOSFET package because it is easy to mount on a printed circuit board and attaching a heat sink to it is simple. Using an oversized power MOSFET in a typical design simplifies the cooling requirements and may eliminate the need for a fan. Table 3.9 gives specifications of six 50-V TO-220 package power MOSFETS, known as HEXFETs, by International Rectifier (www.irf.com). Table 3.10 gives other MOSFETs manufactured by International Rectifier along with comparable devices.

Several power MOSFETS are conveniently packages on a single DIP IC suitable for relatively low-power applications. The IRFG113 is a 14-DIP IC with four MOSFETS. Each

TABLE 3.9 International Rectifier 50-V Power MOSFET Specifications

Part TO-220	R_{ON} (Ω)	I_D (cont) 100°C case	I_D (cont) 25°C case	P_{MAX} (W) 100°C case	P_{MAX} (W) 25°Ccase
IRFZ40	0.028	32	51	50	125
IRFZ42	0.035	29	46	50	125
IRFZ30	0.05	19	30	30	75
IRFZ32	0.07	16	25	30	75
IRFZ20	0.10	10	15	16	40
IRFZ22	0.12	9	14	16	40

voltage drop = $I_{DS} \times R_{DS(ON)}$; power dissipation = $V_{DS} \times I_{DS}$

TABLE 3.10 *n*-Channel Power MOSFETs

Part Number	R_{ON} (Ω)	I_{DS} (cont)	V_{DS}	Similar Packages
IRF510[1]	0.6	4 A	100 V	MTP4N10, VN1110N5
IRF520[1]	0.25	8 A	100 V	BUZ72A, VN1210N5
IRF540[1]	0.08	25 A	100 V	MTP25N10
IRF610[1]	1.5	3 A	200 V	VN1220N5
IRF620[1]	0.8	5 A	200 V	MTP5N20
IRF630[1]	0.4	9 A	200 V	MTP8N20, BUZ232
IRF640[1]	0.18	18 A	200 V	—
IRF150[2]	0.06	40 A	100 V	2N6764
IRF250[2]	0.09	30 A	200 V	2N6766
IRF450[2]	0.4	12 A	500 V	2N6770

V_{GS} = 10 V; $V_{GS(TH)}$ = 4 V
[1]TO-220
[2]TO-3

MOSFET has a 60-V breakdown voltage, and a 1.0-Ω ON-resistance, can handle 0.85-A maximum currents, and can dissipate 1.4 W at 25°C and 0.6 W at 100°C.

Running large currents through a transistor causes it to heat up, and its response to temperature increase becomes an issue to contend with. Higher current flows through a BJT as its temperature increases. If a hot BJT is placed in parallel with another BJT, it *hogs* all the current, resulting in a condition called *thermal runaway*. High temperatures cause the ON-resistance of a MOSFET to increase, until a stable operating point is reached. Thus, MOSFETs do not experience thermal runaway and can be arranged in parallel.

3.12.5 Insulated Gate Bipolar Transistors (IGBTs)

As stated earlier, the power BJT requires substantial base currents to control high collector currents. On the other hand, power MOSFETs provide extremely high current gains, but the value of the V_{DS} in a saturated-mode MOSFET is larger than the V_{CE} of a comparable saturated BJT. The result in both cases is high power dissipation. A good alternative to power MOSFETs and BJTs is the insulated gate bipolar transistor IGBT. This device combines in a single package the high current gain of a MOSFET with the low V_{CE} of a power BJT. As shown in Fig. 3.34, the IGBT is equivalent to a Darlington pair consisting of a MOSFET and a BJT. The MOSFET controls the base current of the BJT, whereas the BJT handles the main load current between the collector and emitter. One disadvantage of an IGBT over a standard BJT is that it

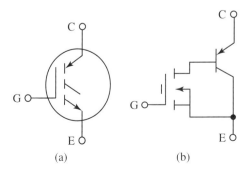

Figure 3.34 IGBT: schematic symbol (a) and equivalent circuit (b).

TABLE 3.11 Characteristics of Selected IGBTs

Part Number	Package	BV_{CES} (Min)	I_C (@110°C)	$V_{CES (SAT)}$(Typ)	Application
FGA15N120AN	TO-3P	1200 V	15 A	2.4 V	Induction heating
FGH30N6S2	TO-247	600	20	2	Motor, powerconversion
FGL60N100D	TO-264	1000	42	2.5	Induction heating
FGP40N6S2	TO-220	600	35	1.9	Motor, power conversion
HGTG30N60C3D	TO-247	600	30	1.5	Motor

has slower switching times. The characteristics of several IGBTs manufactured by Fairchild Semiconductors (www.fairchildsemi.com) are given in Table 3.11. Gate driver ICs for IGBTs are also available. The MC33153 IC single IGBT gate driver by ON Semiconductor (www.onsemi.com) provides for AC induction motor control, brushless DC motor control, and uninterruptable power supplies. It can also be used to drive power MOSFETs and BGTs. The IC provides a high current output stage of 1-A source and 2-A sink.

3.13 THYRISTORS

Thyristors are semiconductor devices intended for switching and control applications involving AC or DC high power loads associated with motors, lamps, heaters, solenoids, and others. They may be used for ON–OFF control or for phase control. Table 3.12 gives a list of various thyristor devices. The following section provides an overview of their operation.

3.13.1 Silicon-Controlled Rectifiers (SCRs)

The silicon-controlled rectifier is a *pnpn* device with three terminals: the gate G, or trigger, the anode A, and the cathode K. Its symbol, internal structure, and current voltage characteristics are shown in Fig. 3.35. When the voltage applied at the gate V_G is more positive than the cathode voltage V_K by the *gate threshold voltage* V_{GT}, a small *threshold current* I_{GT} (as low as hundreds of microamps) flows through the gate. If I_{GT} is sustained for a very short time known as the *gate-controller-turn-on time,* a unidirectional current up to 100 A is triggered to flow from the anode to the cathode. The SCR is very sensitive; even current flow through a human body could trigger the device. The impedance of the load connected to the anode dictates the limit of the current conducted through the SCR. Hundreds of load amps can be switched on with little power loss and small voltage drop (typically 1–2 V) across the SCR.

TABLE 3.12 Characteristics of Selected SCRs and Triacs

Part Number	Voltage (Vdrm)	Current (RMS, DC)	I_H	I_{GT}	V_{TH}
TIC 106Y/F/A/B	30/60/100/200	3.2 A	5 mA	60 μA	0.6 V
MCR 106-1/-2/-3/-4	30/60/100/200	4.0 A	5 mA	200 μA	1.0 V
TIC 126-B/-D	200/400	8.0 A	70 mA	5 mA	0.8 V
TIC 226-B/-D	200/400	8.0 A	60 mA	50 mA	2.5 V

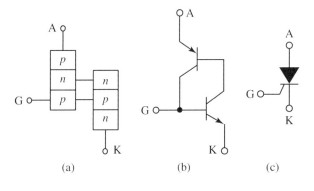

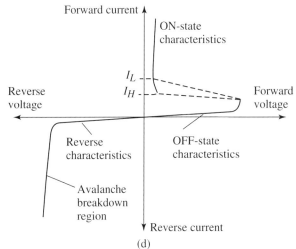

Figure 3.35 Thyristor: construction (a), equivalent circuit (b), schematic symbol (c), and *V-I* characteristics (d).

In DC applications, once the anode current reaches the SCR's *latching current* I_L, the device remains ON and the current continues to flow even if the trigger signal is removed. The SCR can only be turned OFF by commutation, a process that reduces the load current below its holding current I_H (1–50 mA; more for larger units) for a sufficient period of time to allow the SCR to return to its blocking state. Forced commutation is accomplished by using a combination of capacitors, inductors, and switching devices that would reduce the load current to zero. In AC applications, commutation occurs naturally when the voltage crosses the zero point at the end of the conducting half cycle. If the gate is active, the anode current of the SCR connected to an AC source will be half-rectified. A diode bridge rectifier produces a full-rectified wave.

Figure 3.36 shows external components used to protect the SCR from load transients and to ensure trouble-free operation. The series resistor R_S is introduced to prevent gate burnout that may occur if the drive source at the gate has low impedance. The resistor R_G (1 kΩ typical) between the gate and the cathode (or MT1 in a TRIAC circuit) serves two purposes: to drain off any current leaking from the anode to the gate, and to reduce the likelihood of a low-trigger-current SCR to be false triggered by spurious voltage spikes, circuit noise, or static electric discharge. Some large SCRs integrate such a resistor internally. Noisy currents may cause the cathode voltage to exceed the gate voltage by more than 10 V. The diode D1 inserted as shown limits this voltage to <1 V and eliminates this potential threat.

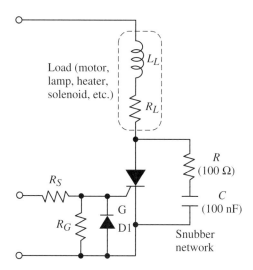

Figure 3.36 A thyristor with external components used to ensure proper operation.

The SCR is characterized by the maximum rate of change of commutating voltage dV/dt and the maximum rate of change of commutating current dI/dt. Introducing an inductor of a few millihenrys in series with the load would limit the level of dI/dt that is likely to be generated when the load power is switched ON. If the load is a DC motor, enough inductance will be readily available; but if the load is, for example, a resistive heater, the external inductance is likely to be needed.

The transients generated when power to the anode is switched ON may cause false triggering, even if the gate is off. To prevent this condition, an *RC snubber circuit* across the anode–cathode (across the MT1-MT2 in a TRIAC) is used to keep dV/dt within limits. The snubber resistance and the capacitor are selected by treating the snubber/load combination as a critically damped series *RCL* network, yielding

$$C = [(V^2)_{MAX}]/[L_L(dV/dt)^2_{MAX}] \quad \text{and} \quad (R + R_L) = 2\sqrt{L_L/C}$$

SCRs are used in many DC applications, such as motor drives. The SCR is also used as *crowbar* device, shown in Fig. 3.37, which is used to speed up fuse burnout when the current rises above a predetermined level. The SCR is connected in parallel with the output of a DC power supply, with a fuse or a large series resistance placed ahead of it to limit short-circuit

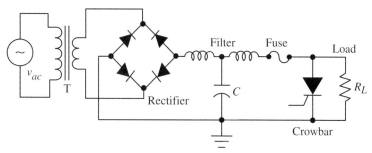

Figure 3.37 A power supply using an SCR as a crowbar for overvoltage protection.

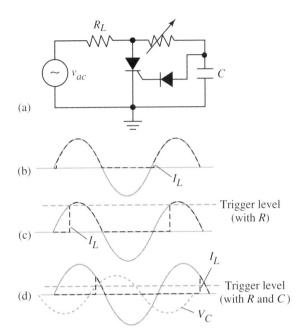

Figure 3.38 Timed-proportional control using a thyristor (a) and response to various triggering levels (b)–(d).

currents and to protect the SCR and the power supply from possible damage. A sensing device connected to the gate detects the overvoltage condition when it occurs and applies a voltage between the gate and the cathode to trigger the SCR and to force the fuse to blow.

In AC power control applications, the SCR can be latched by triggering the gate at any point in time along the waveform, thus providing time-proportional control. Various possibilities are shown in Fig. 3.38. The SCR can be triggered almost immediately at the beginning of every positive half-cycle by simply connecting the SCRs gate to the anode through a normal rectifying diode. Inserting a variable resistor as shown causes a delayed triggering signal to occur anywhere during the first half of the waveform's positive cycle. If it is desired to trigger the SCR at a threshold level beyond the waveform's peak, a phase-shifting capacitor is added.

3.13.2 Gate Turn-OFF (GTO)

A gate turn-OFF thyristor is a variant of the SCR constructed to reverse-trigger the gate with a negative signal and to turn off the device with DC sources instead of forced commutation. GTOs accommodate higher switching frequencies than SCRs, but they require more complicated gate-triggering circuitry, have a higher ON-state voltage drop (typically 3.5 V), and can control a lower load power.

3.13.3 TRIAC

A TRIAC is a three-terminal thyristor constructed from one piece of a semiconductor material, appropriately doped and layered to allow current conduction in both directions. As shown in Fig. 3.39, the device is equivalent to two antiparallel SCRs with one common gate electrode. The other two terminals, designated *main terminal* 1 (MT1) and *main terminal* 2 (MT2), are not interchangeable. The TRIAC is triggered by applying a gate current from the MT1 side of the circuit. All characteristics and ratings of SCRs apply equally to TRIACs.

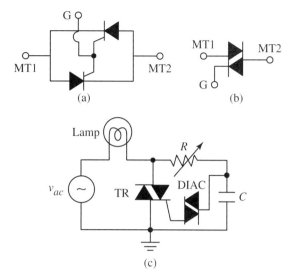

Figure 3.39 A TRIAC: equivalent circuit (a), schematic symbol (b), and lamp dimmer circuit (c).

A trigger voltage in the range of 1–2 V and currents between 10 and 50 mA trigger main currents from 1 A to more than 100 A to flow. Being bidirectional, TRIACs are favored in AC applications, whereas SCRs are preferred in DC applications.

TRIACs provide simple and economical means for AC power control. They are limited to low-power applications in which load currents are less than 100 A and line frequencies are less than 60 Hz. Phillips Semiconductors (www.semiconductors.phillips.com) and other manufacturers offer TRIACs with a wide range of operational characteristics. Typical applications of TRIACs include household appliances such as vacuum cleaners, motor speed controls, and light dimmer switches. In those applications, the polarities of the MT2 and gate terminals are always the same. TRIACs are usually clamped to a heatsink by clip mounting, screw mounting, or riveting.

The positive and negative triggering voltages of a TRIAC are usually not the same, resulting in nonsymmetrical current waveforms. Symmetry can be improved by triggering pulse timing via an external device. For example, placing a DIAC in series with the gate usually suffices. A DIAC is used as a trigger device for a TRIAC. The DIAC triggers on positive and negative voltages. When the voltage across a DIAC rises to a certain value (commonly 27 V), the device switches from the OFF to the ON state in a very short time. When a DIAC and a TRIAC are included in a single package, the device is called a QUADRAC.

Other members of the thyristor family include the silicon unilateral switch (SUS), the unijunction transistor (UJT), and the silicon bilateral switch (SBS). The SUS shows negative resistance and is used as a triggering device for SCRs. The Neon lamp also has a negative resistance and may be used for triggering many applications. The SBS is another thyristor triggering device with a trigger voltage that can be programmed by the control electrode. The UJT has a negative input resistance, which makes it useful as an oscillator in low-frequency applications.

3.14 OPTOCOUPLERS

Optocouplers, also known as optoisolators, integrate in a single package an IR LED source with a photosemiconductor such as phototransistor, photo-Darlington, or photothyristor. The package provides an optical means for information transfer across an insulating air gap. The result is complete isolation between the output on the detector side and the input on the source side. Isolation is required in embedded applications to prevent high voltages from being applied to the microprocessor if it is connected directly to the outside world. The LED is usually driven by a low-voltage circuit, such as microprocessor, or logic gates. The detector side is often a part of high-voltage DC or AC circuit. Optocouplers offer the following benefits: Then protect drive circuitry from transients occurring on the high-voltage side, eliminate ground loop problems, and protect workers from high voltages. Isolator ICs provide an internal light path instead. Figures 11.19a and b represent typical optocoupler packages with a reflective light path and an external air gap light path, respectively. Isolator ICs provide an internal light path instead. Fairchild Semiconductor (www.fairchildsemi.com) offers the 4N2x (x = 5, 6, 7, 8), 4N3x (x = 5, 6, 7), and the H11Ax (x = 1–5) series of general-purpose optocouplers with phototransistor output and the MOC30xyM (x = 1, 2 and y = 0, 1, 2) series with TRIAC driver output.

3.15 SUMMARY

Electronic components and fundamentals of semiconductors were covered in this chapter. Diodes were presented as an electronic circuit-conditioning component and a sensor/source of light. BJT (current control) and FET (voltage control) transistors were presented based on their principles of operation and the useful configurations they compose in switching and amplifying. These electronic components and associated popular circuits form the heart of analog signal conditioning of most interfacing circuits. In addition, the chapter presented many off-the-shelf commercial diodes and transistors, along with suppliers and technical specifications. The chapter concluded with power electronic components and application circuits that are typically needed in high-power applications (e.g., actuator driving). The material in this chapter is essential in mechatronics products for interfacing the microprocessor with the actuated processes.

RELATED READING

D. G. Alciatore and M. B. Histand, *Introduction to Mechatronics and Measurement Systems*, 2nd ed. New York: McGraw-Hill, 2002.

P. Horowitz and W. Hill, *The Art of Electronics*. Cambridge, UK: Cambridge University Press, 1989.

J. L. Jones, B. A. Seiger, and A. M. Flynn. *Mobile Robots: Inspiration to Implementation*, 2nd ed. A. K. Peters, 1999.

S. Kamichik, *IC Design Projects*. Indianapolis, Prompt Publications, 1998.

G. McComb, *The Robot Builder's Bonanza: 99 Inexpensive Robotics Projects*. Blue Ridge Summit: TAB Books, 1987.

F. Mims, *Engineer's Mini-Notebook: Basic Semiconductor Circuits*. Radio Shack Archer Catalogue No. 276-5013, 1986.

F. Mims, *Engineer's Mini-Notebook: Optoelectronics Circuits*, Radio Shack Archer Catalogue No. 276-5012A, 1986.

G. Rizzoni, *Principles and Applications of Electrical Engineering*, 4th ed. New York: McGraw-Hill, 2002.

A. Sedra and K. Smith, *Microelectronic Circuits*. 5th ed. New York: Oxford University Press, 2003.

A. K. Stiffler, *Design with Microprocessors for Mechanical Engineers*. New York: McGraw-Hill, 1992.

Website: www.allaboutcircuits.com

QUESTIONS

3.1 Explain the *n*-type and *p*-type semiconductor.

3.2 What is the difference between the behavior of an ideal diode and a real diode?

3.3 Describe possible applications of diodes.

3.4 Explain how a diode is used as a variable capacitor.

3.5 Describe the main role of the Zener diode.

3.6 List the main sections of a power supply, and explain the role of each.

3.7 Why do electronic circuits require a stable power supply?

3.8 Explain the transistor states.

3.9 Describe the basic transistor circuits.

3.10 Explain the difference between *npn* and *pnp* BJTs.

3.11 Describe members of the FET family.

3.12 List the differences between a BJT and a MOSFET.

3.13 Describe the thermal runaway condition.

3.14 What is a Darlington circuit?

3.15 Explain the difference between small-signal transistors and power transistors.

3.16 Describe the operation of the thyristor, the GTO, and the TRIAC.

3.17 How is the SCR used as a crowbar device?

3.18 What does a snubber network consists of? What is its role?

3.19 Explain the various types of optocouplers and the role they play in circuit design.

PROBLEMS

3.1 Assuming ideal diode behavior, sketch the output waveform for v_O in the circuit shown in Fig. 3.40 if the input signal is $v_i = 6\sin(2\pi t)$.

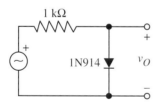

Figure 3.40

3.2 Assuming ideal diodes, sketch the output waveform for v_O in the *full-wave rectifier* circuit shown in Fig. 3.41 if $v_i = 1.5 \sin(4\pi t)$.

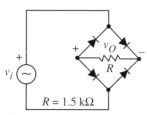

Figure 3.41

3.3 Assuming ideal diodes, sketch the output voltage V_O for the clipping circuit shown in Fig. 3.42 over three cycles of the input $v_i = 3 \sin(2\pi t)$.

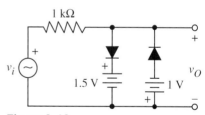

Figure 3.42

3.4 Determine the currents and voltages in the photo-interrupt package shown in Fig. 3.43.

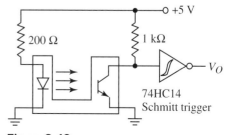

Figure 3.43

3.5   Plot the output voltage v_O for the circuit shown in Fig. 3.44, assuming ideal diodes, if $R = 2\ k\Omega$ and $v_i = 5\sin(\pi t)$ volts.

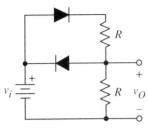

Figure 3.44

3.6   Assuming that the diode and transistor are ideal, sketch for the circuit shown in Fig. 3.45 the ON/OFF curve for the LED on a graph for the following v_i values:

$0 < t < 1$ s, v_i ramps from 4 V to 0 V
$1 < t < 2$ s, $v_i = 5$ V
$2 < t < 4$ s, $v_i = -2$ V
$4 < t < 8$ s, $v_i = 2$ V

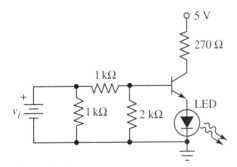

Figure 3.45

3.7   Design a driver for the LED indicator shown in Fig. 3.46.

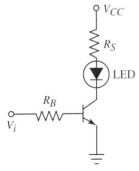

Figure 3.46

3.8   Design and build the inverter circuit shown in Fig. 3.47 to provide $+6$ V across the 600-Ω load using a 15-V supply voltage.

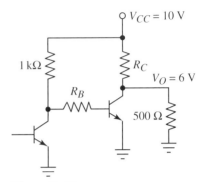

Figure 3.47

3.9   A coil is characterized by a series combination of a 24-mH inductor and an 18-Ω resistor (Fig. 3.48).

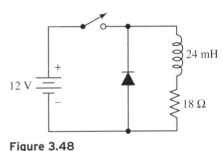

Figure 3.48

When the switch is opened after the circuit reaches steady state and the current through the diode is zero, the flyback voltage generated across the coil forces the diode into conduction. The flyback diode is selected such that its continuous forward current rating I_F is equal to or greater than the maximum peak current it has to sustain. Determine:

a.   The maximum current through the diode.

b.   The diode's minimum reverse breakdown voltage.

c.   The flyback voltage developed across the coil, ignoring switching times.

3.10   Determine the required minimum power ratings for the Zener diode and the series resistor R_S in the circuit shown in Fig. 3.49 if the maximum current drawn by the load is 80 mA.

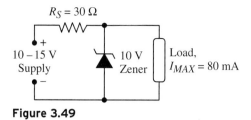

Figure 3.49

3.11 Design a 5-volt Zener diode regulator that can deliver 50 mA of current to a load. The input voiltage to the regulator is 9 volts.

3.12 The infrared emitter circuit shown in Fig. 3.50 is designed to drive four separate IR LEDs. When installed in a weather station, the four IR emitters provide a wider "data beam" than a single emitter and the alignment tolerance between the weather station and the receiver can be relaxed. The IR receiver is isolated from the temperature extremes experienced by the weather station. What would be the problems associated with the design if the nominal forward voltage drop across the IR LEDs is 2 V and the intended forward current is ~20 mA?

3.13 Explain the function of the circuit shown in Fig. 3.51 and the purpose of R_{B1} and R_{B2}; then determine v_O if $v_i = 1.08$ V_{DC} and the transistor $\beta = 120$.

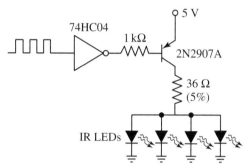

Figure 3.50

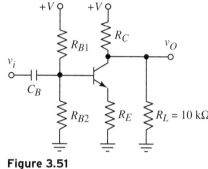

Figure 3.51

3.14 Obtain the specs for the 4N25 and the MOC3011 optocouplers from the Fairchild Semiconductor website and summarize its characteristics.

LABORATORY PROJECTS

3.1 Obtain the specs of the MC7805 and the LM732 voltage regulators, and use one of them to build a 5 V_{DC} power supply.

3.2 Combine the IR emitter/detector pair circuit of Fig. 3.25 with the solenoid drive of Fig. 3.21a to form a circuit so that if the light path between the emitter/detector pair breaks, the solenoid is activated to, say, open a door.

3.3 Build a lamp dimmer using a TRIAC.

4

Operational Amplifier (Op-Amp) Circuits

OBJECTIVES

Thoughtful engagement with the material presented in this chapter will enable the student to:

1　Understand the role of op-amps in signal conditioning
2　Acquire knowledge of various linear and nonlinear op-amp circuits
3　Apply op-amp rules to analyze op-amp circuits
4　Understand nonideal op-amp behavior and ways to minimize their effects
5　Design and analyze op-amp filters

4.1　INTRODUCTION

There are two basic types of electronics signals and techniques, analog and digital. An *analog signal* is an AC or DC voltage or current that varies continuously (smoothly) over a specified range. Analog signals exist in a variety of forms, as illustrated in Fig. 4.1, the sine wave being the most common. Radio and audio signals are two examples of analog signals. Electronic circuits that process analog signals are called *analog*, or *linear*, circuits.

Mechatronic systems involve measuring relevant physical quantities using appropriate sensors. The basic components of a measurement system, shown in Fig. 4.2, include a sensor subject to a stimulus and a signal-conditioning circuit that connects the signal to the MCU. In general, the sensor converts the stimulus, such as temperature, into a continuous analog electrical signal, usually voltage. The raw sensor signal may not be useful because it may be too small or too noisy, may contain wrong information due to poor transducer design or installation, may have a DC offset due to the transducer and instrumentation design or calibration, or may not be compatible with the input requirements of the processing device. Therefore, the raw analog signal must be conditioned before it can serve a useful function. The circuit used for this purpose is called *signal-conditioning* (SC) circuitry, or interface electronics. A

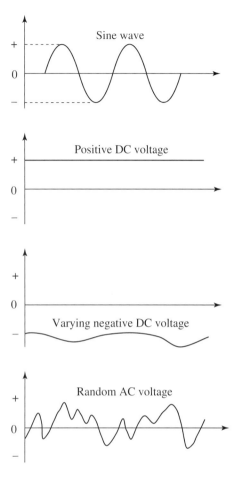

Figure 4.1 Analog signal waveforms.

detailed knowledge of the characteristics of the input and output parts of the SC is essential because they affect the performance of externally connected components. Depending on the characteristics of the signal being measured, the SC circuit performs various functions, which include *amplification, filtering, modulation and demodulation, linearization, impedance buffering, analog-to-digital* (A/D) and *digital-to-analog* (D/A) *conversions, isolation,* and other important functions.

The operational amplifier (op-amp) is one of the fundamental building blocks of SC circuits for mechatronic systems. The op-amp is a linear, DC-coupled, differential amplifier IC that has a very high gain. It plays a very important role in almost all of the aforementioned SC functions. This chapter provides an overview of the various op-amp circuits used as the building blocks in the design of SC circuits.

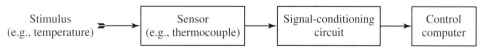

Figure 4.2 Interface electronics matches the signal-conditioning formats from a sensor to the control computer.

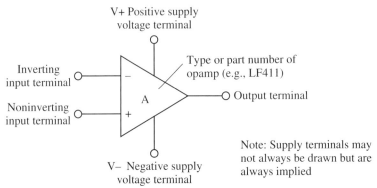

Figure 4.3 The standard op-amp schematic symbol.

4.2 OP-AMP BASIC SYMBOL

The universal symbol for the op-amp is shown in Fig. 4.3. It has a single output terminal and two input terminals, the inverting (−) and the noninverting (+). The (−) terminal is always drawn as the upper of the two inputs. The output terminal is always drawn at the apex of the triangle. The output goes positive when the (+) input goes more positive than the (−) input, and vice versa. The (+) and (−) symbols do not mean that one input is kept positive with respect to the other; each simply implies the relative phase of the output that is important to maintain the feedback negative. The op-amp is connected to the power supply via the positive $V+$ and the negative $V-$ terminals. The $V+$, $V-$, and output terminals are all referenced to a common ground, although the op-amp has no ground terminal.

4.3 CIRCUIT MODEL

A simplified op-amp equivalent circuit is shown in Fig. 4.4. The input circuit is equivalent to a high-input-impedance R_D connecting the two input terminals. The output circuit is equivalent to a controlled-voltage source in series with an output resistance R_O connected to the output terminal. The average of the two input voltages is called the *common-mode voltage* (CMV) v_C:

$$v_C = \frac{v_2 + v_1}{2} \tag{4.1}$$

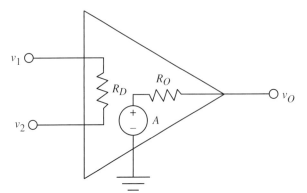

Figure 4.4 Op-amp equivalent circuit diagram.

and the difference between the two input voltages is called the *differential voltage v_D*:

$$v_D = v_2 - v_1 \tag{4.2}$$

The op-amp responds only to the differential signal and ignores any signal common to both inputs. This property is called *common-mode rejection*. Any voltage difference between the two input terminals causes a current flow through R_D. The output voltage of the op-amp is obtained by multiplying the differential voltage by the op-amp *differential gain A*; that is, the op-amp differential gain is given by

$$A = \frac{v_o}{v_D} \tag{4.3}$$

Operational amplifiers are never used without a negative feedback because of their enormous open-loop gain. The feedback provides for constant closed-loop gain and reduces nonlinearity and distortion. Because the closed-loop gain is much smaller than the open-loop gain, the op-amp characteristics depend more on the feedback network and less on the amplifier's open-loop characteristics.

The linear range of the op-amp circuit is determined by the power supply voltages. The op-amp saturates and is said to be out of the amplifier's linear range when v_o exceeds the saturation voltage V_{SAT}. Any increase in v_i results in no change in v_o. Thus, the op-amp output cannot swing beyond the supply voltages. Actually, for most op-amps, the output voltage can swing only to within 2 V of the supplies, that is, ∓ 13 V if the supply is ± 15 V. There must always be a feedback at DC in the op-amp circuit. Otherwise the op-amp is guaranteed to go into saturation.

Although op-amps are usually powered from a ± 15 V power supply, they can also operate from split supplies of lower voltages or from unsymmetrical supply voltages (e.g., $+12$ and -3 V), provided that the total supply voltage (V^+-V^-) is within specifications. Generating a reference voltage above ground enables the op-amp to operate from a single supply (e.g., 12 V), provided that minimum supply voltage, output swing limitations, and maximum common-mode input range are maintained.

The output transistors of an op-amp limits its maximum output current to about 20–25 mA for most op-amps. At 12 V, the load connected to the output must be greater than 480 Ω. If the op-amp is driving a low-impedance load, the output current may be large enough to heat the device. This causes input voltage drift as a result of offset voltage change with temperature. High-gain amplifiers require output loads of 10 kΩ or more to minimize the effect of drift problems.

Many op-amps have a relatively small differential input voltage limit. The maximum voltage difference between the inverting and noninverting inputs might be limited to as little as 5 volts in either polarity. Breaking this rule will cause large currents to flow, which degrades or destroys the op-amp.

Precision and accuracy are important performance criteria of op-amps in DC applications. Thus, specifications such as bias currents and offset voltages are important, whereas frequency response is not. The opposite is true in AC applications, where slew rate and frequency response of the op-amp are far more important. *Slew rate* (SR) refers to the maximum rate at which the op-amp can change its output. For example, if an op-amp has SR $= 10$ V/μs, then its output cannot change from, say, -5 to $+5$ V in less than 1 V/μs. Offset voltage, bias currents, and other nonideal op-amp characteristics are discussed in Section 4.9.

4.4 IDEAL OP-AMP BEHAVIOR

A circuit design task usually begins by assuming ideal op-amp behavior. This is a reasonable assumption because the characteristics of IC op-amps approach the ideal ones. After the initial design is accomplished, the effect of the nonideal op-amp characteristics are investigated to assess their influence on circuit performance. If the effects are not important, the design is complete; otherwise, additional design is needed. Three "*golden rules*" define the behavior of an ideal op-amp. Although these rules apply to op-amp circuits with negative feedback, they can be applied to almost all op-amp circuit designs.

Rule 1. The output attempts to bring about *the voltage difference between the two input terminals to zero*; that is,

$$v^+ = v^- \tag{4.4}$$

Because the op-amp voltage gain is so high (e.g., the DC open-loop gain for the LF 411 is approximately 200,000), a fraction of a millivolt between the input terminals will swing the output over its full linear range (± 13 V), and consequently the voltage difference is ignored. The op-amp does not actually change the voltages at the inputs, for this is not possible, but it checks the input terminals and, if possible, swings the output voltage around, which in turn helps the external feedback network to bring the input differential to zero. This will not be possible in the case of a poor design.

Rule 2. The op-amp draws very little input current because its input impedance is infinite (30 fA to 250 nA; 0.2nA for the LF 411). Neglecting this current gives credence to the claim that *no current flows into either input terminal*; i.e.,

$$i^+ = i^- = 0 \tag{4.5}$$

Rule 3. With the loop closed, the $(-)$ input will be driven to the potential of the $(+)$, or reference, input.

In all op-amp circuits to follow, Rules 1 and 2 will be applied if the op-amp is in the active region, that is, if the inputs and outputs are not saturated at one of the supply voltages. From the three op-amp rules evolve five fundamental assumptions for ideal op-amp circuits analysis and design (values in parentheses indicate real op-amp behavior):

1. Infinite voltage gain, $A = \infty$ (80–170 dB).
2. Infinite bandwidth, BW $= \infty$ (1 kHz to 1 GHz).
3. Infinite input impedance, $R_D = \infty$ (10^6 Ω, 1–20 pF).
4. Zero output impedance, $R_O = 0$ (1–1000 Ω). This implies that the output voltage does not depend on the output current.
5. Zero input offset voltage; that is, $v_O = 0$ when $v_i = 0$.

4.5 COMMON OP-AMP ICs

The op-amp is the basic component on which most SC circuits are based. Table 4.1 gives the characteristics of IC op-amps commonly used in the design of SC circuits. An example of a versatile op-amp that performs well in many design situations is the LF411. This op-amp is available as a single 8-pin DIP package, shown in Fig. 4.5, or in a dual version IC

TABLE 4.1 Selected General-Purpose Op-Amps

Part Number (Manuf.)	V_{CC} (V)	$V_{OS\,(MAX)}$ (mV)	$I_{OS\,(MAX)}$ (nA)	$I_{B\,(MAX)}$ (nA)	SR (V/μs)	Description
LF411 (NS)	36 max 10 min	2	0.1	0.2	15	General purpose
OP-27E (PM)	44 max 8 min	0.025	35	40	2.8	Precision, low noise
TL-081B (TI)	36 max 7 min	3	0.01	0.2	13	All-purpose bi-FET, low power, low noise, many package styles
741C (FA)	36 max 10 min	6	200	500	0.5	General purpose; dual = 1458, quad = 348
LM358N (NS)	±16 or +32	7	50	250	0.6	8-pin DIP dual, low power, bipolar or single power supply
TL082CN (MOT)	±18	15	0.2	0.4	13	8-pin DIP dual FET, high speed, high input impedance
MC33171 (MOT)	3–44 ±1.5–±22	6.5	40	200	2.1	bipolar, low power, split or single supply

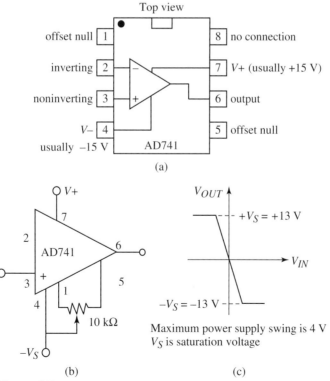

Figure 4.5 Pin connection of a general-purpose DIP package op-amp (a), connection for offset null (b), and output voltage (c).

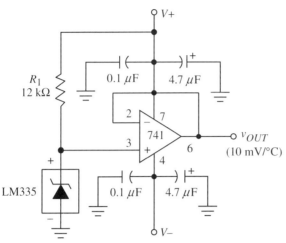

(a) Measuring temperature using the LM335 temperature sensor IC

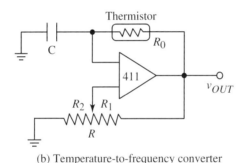

(b) Temperature-to-frequency converter

Figure 4.6 Applications using the 741 op-amp.

designated LF412. Pin 1 is identified by a dot at the corner (or by a notch at the end) of the package. The other pin numbers are counted counterclockwise from the top. The LF 411 is easy to use and costs less than $1. The internal structure of the LF411 contains 24 transistors, 11 resistors, and 1 capacitor. The "offset null" pins, also known as "balance" or "trim," are provided to externally correct for the small asymmetries that are unavoidable during op-amp manufacture.

Another commonly used op-amp is the LM741. This op-amp can be used for signal amplification, differentiation, integration, sample-and-hold, and many other useful applications. The TL071 op-amp, manufactured by Texas Instruments, has the same pin configuration as the 741, but because it has FET inputs, it has a larger input impedance (10 MΩ) and a wider BW. Figure 4.6 shows several applications using the 741 and 411 op-amps.

4.6 BASIC OP-AMP CIRCUITS

4.6.1 Inverting Amplifier

Figure 4.7 shows the basic circuit of the inverting op-amp. Point *A* is called the *summing point* because the input and output signals sum at this junction. Since the noninverting input at *B* is at ground, by Rule 1 point *A* is also at a 0-V potential and is called a *virtual ground*. This

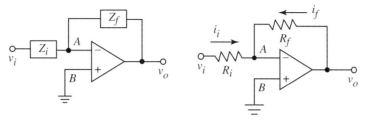

Figure 4.7 The basic inverting op-amp circuit.

implies two things: (a) The voltage across R_f is v_o and (b) the voltage across R_i is v_i. By Rule 2, no current flows into the input of the op-amp; that is, $i_i + i_f = 0$ and $i_i = -i_f$. By Ohm's law, $i_i = v_i/R_i$ and $i_f = v_o/R_f$, we have $v_o/R_2 = -v_i/R_1$. In other words, the *voltage gain* is

$$A = \frac{v_o}{v_i} = -\frac{R_f}{R_i} \tag{4.6}$$

The inverting op-amp provides sign inversion and for desired gain changes with the proper choice of R_f/R_i. It is important to note that even though the basic op-amp has high Z_{in} and a low Z_{out}, the input and output impedances of the op-amp used in circuit design are not the same as those of the op-amp. The output impedance of the inverting op-amp is a fraction of an ohm, and its input impedance is $Z_{in} = R_i$ because point A is always a virtual ground. The low input impedance may not be an issue if the inverting op-amp is driven from another op-amp of very low output impedance. However, the low input impedance is an undesirable feature in many applications. For example, amplifiers with large closed-loop voltage gains require relatively small R_i values. In this case, increasing the input impedance comes at the cost of decreasing available gain. Alternatively, R_f could be increased, but there is a practical limit to the maximal value of R_f.

The inverting amplifier is less demanding of the op-amp than is the noninverting amplifier and thus may yield a better performance. Additionally, the virtual ground in the inverting amplifier provides for combining several signals without interaction.

The foregoing analysis applies for DC as well, and the op-amp is referred to as a DC amplifier. In the inevitable presence of input offset voltage, a zero (grounded) input produces an output. For example, for a 411-based inverting op-amp with a gain of $A = 100$, the output could be as large as ±0.2 V when the input is grounded. If the signal source is offset from ground, a coupling (sometimes called *blocking*) capacitor is used to block DC offset while providing signal coupling. A blocking capacitor is also used if the input signal is AC.

It is often better not to ground B directly, but rather through a resistor. This would reduce the errors due to bias currents. The value of the resistor should be equivalent to the parallel combination of R_i and R_f. If the amplifier is AC-coupled, then $R = R_f$ is used. Other solutions to bias current problems are also available.

4.6.2 Noninverting Amplifier

The noninverting op-amp circuit is shown in Fig. 4.8a. By Rule 1, v_i also exists at the inverting input. By Rule 2, i_f must flow through R_i to ground, since no current can flow through the

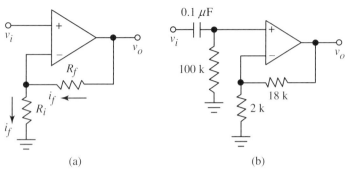

(a)

(b)

Figure 4.8 The basic noninverting op-amp circuit.

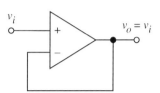

Figure 4.9 The unity-gain amplifier-follower.

op-amp. By the voltage divider rule,

$$v_i = \frac{R_i}{R_i + R_f} v_o$$

$$\frac{v_o}{v_i} = \frac{R_i + R_f}{R_i} = 1 + \frac{R_f}{R_i} \geq 1 \tag{4.7}$$

This gain is an approximation since the input impedance of the op-amp is assumed infinite (with the LF411 op-amp it is 10^{12} Ω or more). The output impedance is still a fraction of an ohm.

The noninverting op-amp is also a DC amplifier. For a capacitively coupled AC signal source, provision for a DC return path to the ground for the very small input bias current is required, as illustrated in Fig. 4.8b. The indicated component values provide a gain of 10 and a low 3-dB point frequency of 16 Hz. A main attribute of the noninverting amplifier is its high input impedance. However, its operation is more demanding of the op-amp.

4.6.3 Follower

If R_i in the noninverting op-amp is set equal to infinity and R_f is set to zero, the circuit reduces to that shown in Fig. 4.9. By Rule 1, v_i must also exist at the inverting input, which is connected directly to v_o. Therefore $v_o = v_i$ and the output follows the input voltage. This *unity-gain amplifier* is used as a *buffer* to isolate one circuit from the loading effects of a following stage. It is also useful as an *impedance converter*, from a high input impedance to a low output impedance, and in A/D and D/A converters (refer to Chapter 10 for more details). In A/D converters, it is used to provide a constant input impedance. It is used in D/A converters to provide a high-impedance load, required for correct operation. It is also used following the capacitor in a sample-and-hold circuit to prevent capacitor discharge during conversion time.

4.6.4 Differential Amplifier

A combination of the inverting and the noninverting amplifier results in a circuit with unique characteristics of its own called the *differential amplifier*, shown in Fig. 4.10, which is used to

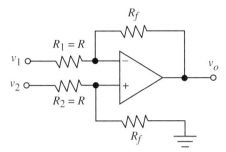

Figure 4.10 The classic differential amplifier circuit.

subtract two analog signals. The differential op-amp can reject an interference noise signal common to the two inputs, such as the 60-Hz AC line noise, thermal noise, and magnetic noise. This circuit requires precise resistor matching. By the voltage divider rule, the voltage at the noninverting input is

$$v_+ = \frac{R_f}{R_f + R_2} v_2 \tag{4.8}$$

By Rule 1, the voltages at the inverting and the noninverting inputs must be equal. Therefore, the top half of the circuit acts as an inverter. The current in the top half of the circuit is

$$i = \frac{v_1 - v_+}{R_1} = \frac{v_+ - v_o}{R_f} \tag{4.9}$$

Substituting Eq. (4.8) into Eq. (4.9) yields

$$v_o = -\left(\frac{R_f}{R_1}\right)v_1 + \left(1 + \frac{R_f}{R_1}\right)\left(\frac{R_f}{R_2 + R_f}\right)v_2 \tag{4.10}$$

If $R_1 = R_2 = R$, then Eq. (4.10) reduces to

$$v_o = \frac{R_f}{R}(v_2 - v_1) \tag{4.11}$$

Eq. (4.11) indicates that if the two inputs are tied together and driven by the same voltage source so that the common-mode voltage v_C is $v_1 = v_2$, then $v_o = 0$. The term *common mode* is used because both inputs have a common driving voltage. The common-mode gain A_C of the differential amplifier circuit is zero.

If $v_1 \neq v_2$, the differential voltage gain A_D is equal to R_f/R. In practice the differential amplifier cannot perfectly reject the common-mode voltage, and the output voltage is

$$v_o = A_D v_D + A_C v_C \tag{4.12}$$

The ability of a differential amplifier to reject CMV is measured quantitatively by the common-mode rejection ratio (CMRR), expressed as

$$\text{CMRR} = \frac{A_D}{A_C} \tag{4.13}$$

Substituting A_C from Eq. (4.13) into Eq. (4.12) yields

$$v_o = A_D v_D \left(1 + \frac{v_C}{\text{CMMR} \times v_D} \right) \tag{4.14}$$

The CMRR may range from 100 in some applications to greater than 10,000 for high-quality biopotential amplifiers. The CMRR of an op-amp is reduced with increase in frequency.

4.6.5 Instrumentation Amplifier

A single differential amplifier op-amp has low input impedance. This may be satisfactory for low-input-impedance sources, such as strain-gauge bridge circuits, but it is not satisfactory for high-impedance sources. A solution to this problem is the *instrumentation amplifier* (IA), shown in Fig. 4.11. It consists of two stages. The input stage consists of two noninverting amplifiers, A1 and A2, and three resistors, $2R_3$ and R_4. This stage forms a differential-in/differential-out amplifier whose outputs are connected to the inputs of the differential amplifier stage. Connecting the R_4 resistors of the noninverting amplifiers by R_3 avoids connection to ground. The common-mode gain of this stage is $A_C = 1$. The combination A3, $2R_1$, and $2R_2$ forms the differential amplifier stage that passes the differential signal while rejecting the common-mode signal of the input stage. The LM102 or OP07 may be used for A1 and A2, whereas LM107 is suitable for A3.

The common-mode gain A_C of the first stage is determined by considering the two noninverting amplifiers separately. If $v_3 = v_4$, then by Rule 1, v_3 appears at the inverting inputs of both op-amps, and no current flows through R_3, because the same voltage appears at both ends of R_3. By Rule 2, current flowing out of the input terminals of the op-amp is zero, so no current flows through the R_4's. Therefore, voltage v_3 appears at both op-amp outputs and $A_C = 1$.

The differential gain A_G of the first stage is determined when $v_3 \neq v_4$. The voltage drop across resistor R_3 is then $v_3 - v_4$. This causes a current to flow through R_3, and by Rule 2 the same current also flows through both R_4's. The output voltage is

$$v_1 - v_2 = i(R_4 + R_3 + R_4) \tag{4.15}$$

while the input voltage is

$$v_3 - v_4 = iR_3 \tag{4.16}$$

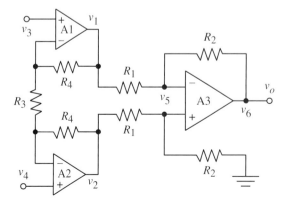

Figure 4.11 The instrumentation amplifier.

The differential gain is then

$$A_D = \frac{v_1 - v_2}{v_3 - v_4} = \frac{2R_4 + R_3}{R_3} \tag{4.17}$$

Since $A_C = 1$, Eq. (4.13) indicates that the CMRR of the first stage is determined by the A_G. Since $A_C = 1$ for the first stage, the overall A_C of the IA is zero because the inputs to the second stage are equal. For the differential amplifier stage, Eq. (4.11) gives

$$v_6 = \frac{R_2}{R_1}(v_2 - v_1) \tag{4.18}$$

The overall gain of the IA is obtained by multiplying the gains of each stage. That is, substituting Eq. (4.17) into Eq. (4.18) gives

$$v_6 = (v_4 - v_3)\frac{R_2}{R_1}\frac{2R_4 + R_3}{R_3} \tag{4.19}$$

The IA has high input impedance, high CMRR, and a gain that is adjustable by changing R_3. It also rejects the 60- or 50-Hz interference that is common to both inputs of the IA. IAs are used at the front end of a signal-data-aquisition system in order to amplify the output of low-level sensors, including thermocouples, RTDs, and strain-gauge bridges. They are also used to amplify bioelectric signals such as those from electrocardiograms (ECGs) and electroencephalograms (EEGs). A normally harmless 60- or 50-Hz current can cause cardiac arrest under certain circumstances. As a result, manufacturers of bioelectric amplifiers, especially EEG and ECG equipment, use isolation amplifiers that provide as much as 10^{12} Ω of isolation between the patient and the AC power line.

The Burr Brown INA117 is a precision IA that can withstand 200 V of common-mode voltage by dividing the input resistively inside the device. However, the input and output sides are not electrically separate and share a common return path; i.e. it does not provide for galvanic (ohmic) isolation. It is a good candidate for high-side current-sensing applications.

Analog Devices manufactures a large array of IAs. The AD524 has a high CMRR and can extract differential signals at the millivolt level from the 5 V of common-mode signals present in the circuit. Its gain can be selected by external jumpers. At a gain of 100, CMRR is specified as 100 dB minimum up to 60 Hz, which helps reduce the maximum output error due to common-mode input.

Example 4.1: The electrocardiogram amplifier—The multiple op-amp circuit shown in Fig. 4.12 is the ECG amplifier. The IA is used as a preamplifier stage since (1) it is able to reject interference, (2) it has a high noise immunity, allowing only the desired signal to be amplified through the remaining stages of the amplifier, (3) it has high input impedance and therefore represents the load seen by the ECG electrodes, and (4) it provides high CMRR. The 120-kΩ potentiometer permits maximization of the CMMR. Explain the role of the circuit components.

Solution: Since electrodes may produce an offset potential of up to 0.2 V, the gain of the instrumentation amplifier is intentionally low (from Eq. (4.19), the gain is 40) to prevent saturation.

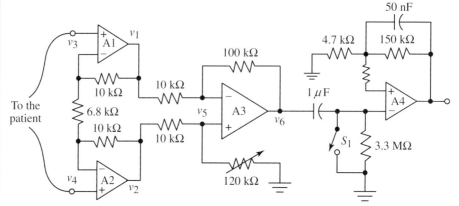

Figure 4.12 The electrocardiogram amplifier (adapted from Tompkins and Webster, 1988).

No coupling capacitors are used at the input because they would block the op-amp bias current. The 1-μF coupling capacitor and the 3.3-MΩ resistor form the high-pass filter, allowing frequencies above 0.05 Hz to pass, as calculated from the transfer function of the simple high-pass filter. The output stage of the ECG amplifier is a noninverting amplifier with a gain of approximately 32 (Eq. (4.7)). The 50-nF capacitor and the 150-kΩ resistor form a low-pass filter that passes frequencies up to 100 Hz. The 3.3-MΩ resistor at the inverting input balances bias-current source impedances. When the output saturates, S_1 may be momentarily closed in order to charge the 1-μF capacitor. This is done after lead switching or defibrillation to return the output to the linear region. Popular 741 op-amps work well in this circuit, but op-amps with low bias current may be desired to keep DC current through the electrodes small.

4.7 LINEAR CIRCUIT APPLICATIONS

4.7.1 Summing Amplifier (Adders)

Op-amp circuits can perform analog computations and other linear operations. The circuit in Fig. 4.13 is a variation of the inverting op-amp that can sum several input voltages. Each input connects to the inverting input through a weighting resistor. Point A, called the summing junction because it sums all inputs and feedback currents, is a virtual ground, so the input current is

$$i_f = i_1 + i_2 + \cdots + i_n$$

$$= \frac{v_1}{R_1} + \frac{v_2}{R_2} + \cdots + \frac{v_n}{R_n} \tag{4.20}$$

Since the inverting input is at zero voltage, then

$$v_o = -i_f R_f$$

$$= -R_f \left(\frac{v_1}{R_1} + \frac{v_2}{R_2} + \cdots + \frac{v_n}{R_n} \right) \tag{4.21}$$

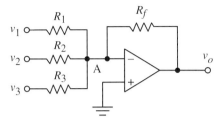

Figure 4.13 Basic summing op-amp.

Resistor R_f controls the overall gain of the circuit. The inputs can be positive or negative. Each resistor in the $R_1 - R_n$ network determines the weighting factor and input impedance of the corresponding channel. If they are unequal, a weighted sum will be obtained. Consider four inputs, each of 1 volt or zero, representing the binary values 1, 2, 4, and 8. Using input resistors of 10K, 5K, 2.5K, and 1.25K results in an output (in volts) that is equal to the binary count input. This scheme can easily be expanded to several digits. It forms the basis for the construction of digital-to-analog converters (D/A) discussed in Chapter 10.

4.7.2 Integrators

Op-amps provide for the making of very accurate integrator circuits. The voltage signal generated at the output of the op-amp circuit shown in Fig. 4.14 is proportional to the integral of the input signal. The output voltage is equal to the voltage across the feedback capacitor. From circuit analysis, $i_f = v_i/R$ because the inverting input is a virtual ground; and by the voltage–current relation of a capacitor, the output voltage over time t_1 is

$$v_o(t) = -\frac{1}{RC} \int_0^{t_1} v_i(t)\, dt + v_{ic} \tag{4.22}$$

where v_{ic} is the initial-condition voltage of the capacitor. The input can be a current instead of a voltage, in which case R will be omitted.

The integrator operates as an open-loop amplifier at DC. Consequently, a feedback must always be present at DC, otherwise, if v_i is applied indefinitely or if the input is grounded, op-amp offsets and bias currents will cause v_o to decrease until it is equal to the op-amp saturation voltage. The integrator circuit shown in Fig. 4.15 helps mitigate this issue. The integration is accomplished in a three-cycle sequence: reset cycle (S_1 closed and S_2 open),

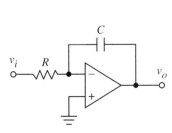

Figure 4.14 Simple integrating op-amp.

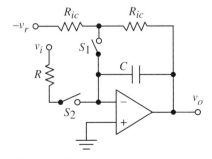

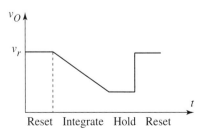

Figure 4.15 Practical integrating op-amp.

integrate cycle (S_1 open and S_2 closed), and hold cycle (both S_1 and S_2 open). During the reset cycle, the op-amp functions as an inverter with a gain of -1 ($R_i = R_f = R_{ic}$). Accordingly, the output v_o reaches the initial capacitor charge, which is $v_{ic} = -v_r$. The output voltage v_o is held constant during the hold cycle for subsequent processing.

4.7.3 Differentiators

The voltage signal generated at the output of the op-amp circuit shown in Fig. 4.16a is proportional to the time rate of change of the input signal, dv_i/dt. Assuming $dv_i/dt > 0$, the output voltage v_o is

$$v_o = RC\frac{dv_i}{dt} \tag{4.23}$$

The differentiator circuit in Fig.4.16a is susceptible to noise and stability problems at high frequencies. This occurs because the ideal gain curve of the differentiator intersects the open-loop gain curve at the frequency $f_i = 1/(2\pi R_i C)$, as indicated in Fig. 4.16b. Adding a resistor R_i as shown in Fig. 4.16c forces the gain curve of the differentiator to bend over at f_i. It can be shown that a stable circuit is realized if R_i is selected according to

$$R_i = [R_f/A_0\omega_b C]^{1/2} \tag{4.24}$$

where $A_0\omega_b$ is the gain-bandwidth product of the op-amp. Op-amps with high slew rate and gain-bandwidth product (such as the LU356) are well suited for the construction of differentiator circuits.

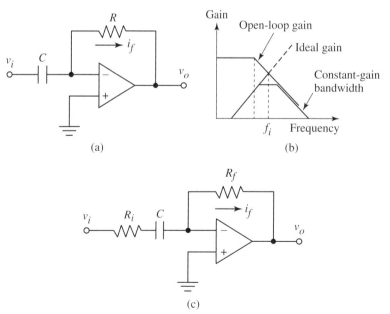

Figure 4.16 Differentiator op-amp: basic circuit (a), gain-frequency behavior (b), and practical differentiator (c).

Example 4.2: Design of signal-conditioning circuit—Op-amps represent the backbone of signal-conditioning circuits for instrumentation and control applications. In this example, an op-amp circuit is required to transform the range of the output voltage of a sensor V_S to within the voltage range of the analog-to-digital converter V_X (see Chapter 10). Assuming that a linear relationship exists between V_S and V_X,

$$V_X = AV_S + V_R \qquad (4.25)$$

Figure 4.17 shows an op-amp circuit that implements this relation. The circuit consists of an inverting op-map A1 and an op-amp summing amplifier A2. The output of the inverting op-amp is $V_A = V_S(-R_2/R_1)$. The voltage divider provides a constant voltage $V_B = [R_6/(R_5 + R_6)] \times V_R$, which is summed with V_A via the summing amplifier. To implement the relation in Eq. (4.25), the resistors in the circuit are selected such that

$$A = \frac{R_2}{R_1}\frac{R_4}{R_3} \qquad \text{and} \qquad \frac{R_4}{R_7} = 1 \qquad (4.26)$$

Note that the voltage divider's supply V_R in this case should be negative, or V_B should be inverted by an inverting op-amp with a gain of 1.

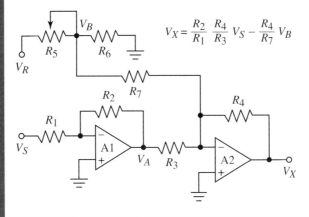

$$V_X = \frac{R_2}{R_1}\frac{R_4}{R_3} V_S - \frac{R_4}{R_7} V_B$$

Figure 4.17 Op-amp circuit for Example 4.2.

4.8 NONLINEAR OP-AMP CIRCUITS

4.8.1 Comparators

A comparator is a circuit that compares two analog voltages at its two input terminals and generates a signal at its output terminal relative to the state of the two inputs. The simplest comparator circuit is the high-gain differential amplifier shown in Fig. 4.18a, with its transfer characteristics shown in Fig. 4.18b. Because the voltage gain typically exceeds 100,000, the inputs will have to be equal to within a fraction of a millivolt to prevent the output from saturation. A reference (trip or trigger) voltage V_R is applied to the noninverting terminal, and the variable signal v_i is fed to the inverting terminal. When v_i becomes greater than V_R, the output v_o

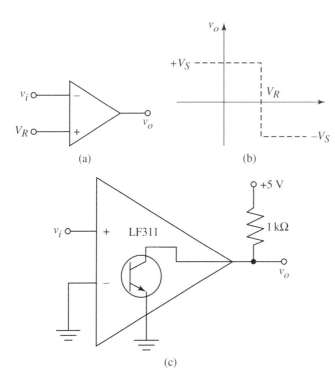

Figure 4.18 Op-amp comparator: circuit diagram (a), transfer characteristics (b), and open-collector output (c).

swings to the negative saturation voltage $-V_S$. For $v_i < V_R$, the output $v_o = +V_S$. With its output being either HIGH or LOW, the comparator is used to communicate between the analog and digital domains. It is also used to generate alarm signals in control applications. Comparators are also used as integral parts in constructing A/D and D/A converters (see Chapter 10).

Comparator IC circuits offer a more flexible solution than just an op-amp. A comparator IC usually has an open-collector output with a grounded emitter, as shown in Fig. 4.18c. It also has a high slew rate to change the output states as quickly as possible with minimum response time. Good comparator ICs include the LM101A, LM306, LM311, LM393, NE527, and TL372 op-amps. The MC3405 from Motorola integrates two differential-input op-amps and two comparators on a single IC.

4.8.2 Schmitt Triggers

The simple comparator circuit in Fig. 4.18c may not handle noisy or slowly varying signals properly. If the input signal variation is very slow, the output may "jiggle," causing a digital signal to slowly ramp from LOW to HIGH over a finite period of time. Furthermore, if the input is noisy, the output may jiggle between LOW and HIGH several times as the input passes through the trigger point, as shown in Fig. 4.19. This may wreak havoc in circuits used to interpret the comparator's output. A cure to these two problems is the *Schmitt trigger* circuit shown in Fig. 4.20a. The circuit has two threshold voltage levels, depending on the output state, the HIGH V_H and the LOW V_L. Referring to Fig. 4.20c, as the input signal rises from a low value, the output remains HIGH until the varying signal reaches the positive-going threshold V_L ($+4.76$ V) and the output swings to a LOW. It remains LOW until the input signal falls to the negative-going threshold V_H ($+5.0$ V). A noisy input is less likely to produce multiple triggering. The difference between the two threshold levels is called *hysteresis*.

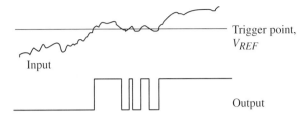

Figure 4.19 Slowly varying noisy signal and output from the circuit shown in Fig. 4.18c.

The 74HC132 and the 4093 ICs are two examples of quad, 2-input NAND Schmitt triggers. The 74HC7014 Schmitt trigger (see pinout in Appendix B) from Phillips Semiconductors (www.semiconductors.phillips.com) contains six precision noninverting Schmitt triggers. This IC triggers at $V_L = 3.1$ V and $V_R = 2.9$ V.

4.8.3 Rectifiers

To properly function, resistor-diode rectifiers are required to overcome the forward voltage drop of the diode. Thus, they are not suitable for precision applications involving voltages below 0.7 V (SI diodes). An op-amp circuit with a diode placed within the feedback loop cures

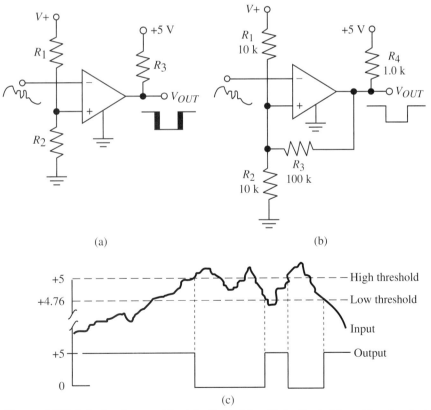

Figure 4.20 Schmitt trigger: without feedback (a), with feedback (b), and input–output characteristics (c).

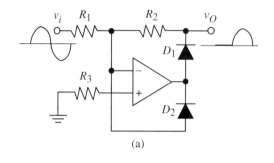

(a)

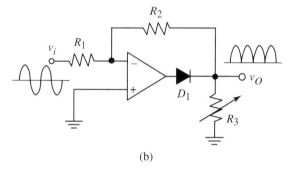

(b)

Figure 4.21 Precision half-wave rectifier (a) and full-wave rectifier (b).

this problem because the gain of the op-amp reduces the voltage limitation. Figures 4.21a and 4.21b show precision half-wave and full-wave rectifiers, respectively.

4.8.4 Limiters

A limiter circuit is used to impose a limit on the output. Figure 4.22 shows a feedback limiter in which a Zener diode is used as a feedback impedance of an inverting op-amp. The values of the Zener's forward and reverse voltages establish the limits of the output voltage. If v_i is negative, the output of the op-amp is positive and the Zener diode is reverse biased. This drives the output to the fixed value of the Zener's breakdown voltage V_z. If, on the other hand, v_i is positive, the output of the op-amp becomes negative, which forward-biases the Zener diode and limits the output to the forward voltage drop (typically 0.7 V). Reversing the Zener connection would change the polarity of the output voltage.

The output of the op-amp should not be allowed to reach the saturation voltage, because a saturated op-amp needs a long recovery time before it can operate linearly again. Operation

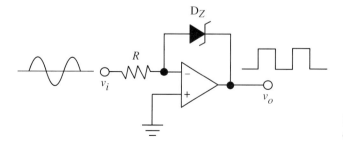

Figure 4.22 Active limiter circuit.

of digital logic circuits requires much lower positive or negative limits (0–5 V for TTL). Accordingly, a limiter circuit is usually used to reduce the typical op-amp voltage swing (± 13 V).

4.9 NONIDEAL OP–AMP BEHAVIOR

The performance of a real op-amp departs from the assumed ideal behavior. Nonideal op-amp properties limit the operating range of the op-amp circuits. The circuit designer should have intimate knowledge of the practical op-amp characteristics and their effects on the performance of the associated circuits. The most serious of the nonideal op-amp behaviors are the finite gain, finite bandwidth, bias currents, and offset voltages.

4.9.1 Feedback with Finite-Gain Amplifiers

Figure 4.23 shows the variation of open- loop and closed-loop gain versus frequency f. The numbers are typical of most general-purpose op-amps, such as the 741. Although the open-loop gain A_{OL} is high at DC and low frequencies, it begins to fall off at a rather low frequency (10 Hz in this case). To ensure stability and nonoscillatory behavior, an op-amp IC includes a *frequency compensation* network, usually a single capacitor, which causes the op-amp to exhibit the single-time constant, low-pass response, indicated by the -20-dB "roll-off." The gain magnitude of an op-amp at any frequency is determined by

$$|A(j\omega)| = \frac{\omega_t}{\omega} = \frac{f_t}{f} \tag{4.27}$$

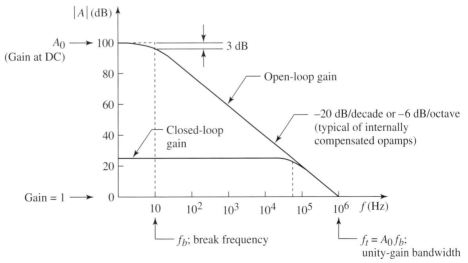

Figure 4.23 Open- and closed-loop gains of a general-purpose (e.g., 741) internally compensated op-amp.

TABLE 4.2 Comparison of Op-Amp Technologies (Full Temperature Range, 0–70°C)

	Bipolar (TLE2037C)	BiFET (TLE2071C)	CMOS (TLC2201C)
I_B (max)	150 nA	5 nA	100 pA
$I_{OS(MAX)}$	150 nA	1.4 nA	100 pA
dI_B/dT	Minimal	Doubles every 10°C	Doubles every 10°C
V_{OS} (max)	145 μV	6 mV	600 μV
dV_{OS}/dT (typical)	0.4 μV/°C	3.2 μV/°C	0.5 μV/°C
CMRR (min)	98 dB	68 dB	85 dB
SR @ unity gain (min)	5 V/μs	23 V/μs	1.5 V/μs

where f_t is the unity-gain bandwidth, or *gain-bandwidth product* (GBWP), given by $f_t = A_0\omega_b$, where A_0 is the DC gain and f_b is the break or corner frequency that corresponds to a gain drop of 3 dB from A_0. $f_t = \omega_t/2\pi$ is usually specified on the op-amp data sheet. Bandwidth is the band of frequencies over which the gain of the amplifier is almost constant, to within a specified level of decibels, usually 3 dB.

4.9.2 Offset Voltage and Bias Currents

The inevitable imbalance in the op-amp inverting and noninverting input stages (differential pair) provides a source of op-amp errors, due to input offset voltage and input bias currents. Refer to Table 4.2 for important characteristics of three op-amp technologies.

Offset Voltages

If both inputs of an ideal op-amp are tied together, the output will be zero. With a real op-amp, the output will tend toward one of the supply rails. The voltage required to bring the output to zero is the input-offset voltage, V_{OS}. It is a voltage that is equal in magnitude and of opposite polarity to a voltage that should be applied externally between the two input terminals on the op-amp to bring the op-amp output to its ideal value of 0 V. Figure 4.24a shows an op-amp circuit model with input-offset voltage. The value of V_{OS} exhibited by a general-purpose

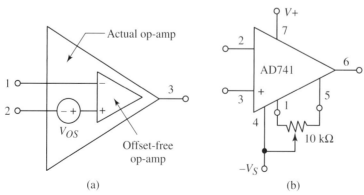

Figure 4.24 Circuit model for an op-amp with input-offset voltage V_{OS} (a) and means of trimming the output due to V_{OS} to zero (b).

op-amp ranges between 1 and 5 mV. The effect of V_{OS} on the output of inverting and noninverting op-amps is given by

$$v_O = V_{OS}(1 + R_f/R_i) \tag{4.28}$$

Offset voltages are nullified by connecting the wiper of a pot inserted between the offset null terminals of the op-amp to the negative supply voltage, as shown in Fig. 4.24b. Since V_{OS} varies with temperature, nulling will only be valid at that nulling temperature. Additionally, the V_{OS} increases approximately 3 µV per year. Another factor that causes an input-offset voltage is the change in the bias point caused by variations in the supply voltages. The measure that defines the ability of an op-amp to reject this effect is the *power supply rejection ratio* (PSRR), defined as

$$\text{PSRR} = \frac{dV_{CC}}{dV_{OS}} \tag{4.29}$$

PSSR is measured in decibels, and it decreases as frequency increases.

Bias Current

Operation of the op-amp requires biasing its two input terminals with DC currents called *input bias currents*, indicated in Fig. 4.25. The average (magnitude) of the two bias currents, I_{B1} and I_{B2}, is called the input bias current,

$$I_B = \frac{I_{B1} + I_{B2}}{2} \tag{4.30}$$

Bias currents I_{B1} and I_{B2} would be equal and their effects would cancel out if the op-amp's differential pair are perfectly matched. However, mismatch in the differential pair is inevitable, which gives rise to an input-offset current, defined as the difference between the two bias currents,

$$I_{OS} = |I_{B2} - I_{B1}| \tag{4.31}$$

I_{OS} is typically 10–25% of I_B. Typical values for general-purpose bipolar op-amps are $I_B = 100$ nA and $I_{OS} = 10$ nA. FET op-amps have a much smaller I_B, on the order of picoamps. If $I_B^+ = I_B^-$, the bias currents cancel out. Since this is not generally possible, $V_{OUT} = R_f \times I_{OS}$.

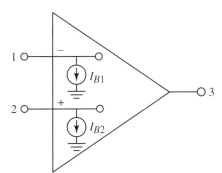

Figure 4.25 The op-amp input bias currents represented by two current sources I_{B1} and I_{B2}.

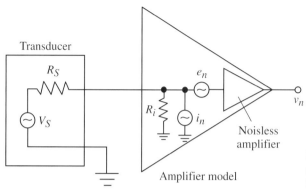

Figure 4.26 Typical frequency response characteristics of a low-pass filter.

The effect of I_B on the output DC voltage of a buffer circuit may be reduced by introducing in the feedback path a resistor whose value is equal to the source resistance. For the inverting op-amp circuit shown in Fig. 4.7, inserting the resistance R_3, which is equal to the parallel equivalent of R_i and R_f, in series with the noninverting input lead reduces the effect of bias currents. If the op-amp is AC coupled, R_3 should be equal to R_f. For more complex circuits, the effects of I_{OS} is minimized if the DC resistance seen by the noninverting terminal and that seen by the inverting input, both with respect to ground, are equal. The effect of the source resistance should be included, whereas the output resistance is negligible and may be neglected.

Table 4.2 indicates that bipolar op-amps are better than CMOS when considering V_{OS} but worse when considering I_B. Using smaller source and feedback resistors reduces bias-current effects in bipolar op-amps. The effect of bias currents in FET op-amps may be neglected, except for the temperature effects.

Amplifier Noise Model

Any sensor measurement will likely require amplification before it reaches the computer. Depending on the signal level generated by the sensor, amplifier noise may be significant and must be considered for a more accurate measurement. Figure 4.26 shows the Thevenin equivalent of a sensor hooked to an amplifier with inherent noise. The amplifier noise is characterized by two independent sources, an equivalent current noise i_n and voltage noise e_n, both of which vary with frequency. Voltage noise is modeled as a voltage source in series with the input of a noise-free amplifier. *Current noise* is modeled as a source in parallel with the amplifier input. The noise current flows through a parallel combination of the output impedance R_S of the sensor and the input impedance R_i of the of the amplifier and converts to a voltage noise. The combined noise effect of i_n and e_n results in a corresponding output voltage v_n estimated by

$$v_n = \sqrt{e_n^2 + \left[i_n \frac{R_S R_i}{R_S + R_i} \right]^2}$$

(4.32)

Chapter 11 presents a detailed treatment of sensors and measurements.

4.10 ACTIVE FILTERS

The discussion of passive filters and response characteristics in Section 2.14 applies to active filters, which are the focus of this section. Figure 4.27 shows typical gain and phase plots of a low-pass filter frequency response. The *passband* is the range of frequencies that are allowed

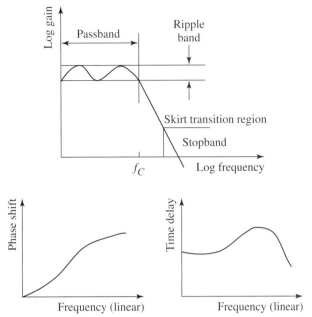

Figure 4.27 First-order low-pass filter: frequency response.

to pass through the filter relatively unattenuated. The response within the passband exhibits *ripples* that are confined within a band called the *ripple band*. The passband ends at the *cutoff frequency* f_c, where the gain magnitude attenuates by -3 dB from the maximum gain. Beyond this point the response of the filter drops off through a *transition region* referred to as the *skirt* to the *stopband* region, where the range of frequencies is rejected or significantly attenuated. The stopband region begins when the level of attenuation reaches some minimum, such as -40 dB. The *phase shift* of the output signal relative to the input signal is important because a signal entirely within the passband will emerge with its waveform distorted if the time delay of the different frequencies going through the filter is not constant.

Including inductors in the design helps passive filters fashion a response with desired passband flatness, sharp transition, and steep falloff outside the band. However, inductors are bulky and expensive, and their behavior suffers from inherent anomalies, which include significant series resistance, nonlinearity, distributed winding capacitance, and susceptibility to magnetic pickup or interference. Moreover, realizing maximal passband flatness and sharp falloff outside the band leads to a more complex design and a large number of components. Also, improved amplitude response comes at the expense of worse transient response and phase-shift characteristics.

By using op-amps instead of inductors in a filter design, it is possible to fashion ideal *RLC* filter characteristics without using inductors. Since the op-amp is an active element, the inductorless filters are known as active filters. Like passive filters, active filters can be used to make low-pass, high-pass, band-pass, and band-reject filters, with a choice of filter type according to the desired maximal flatness of passband, steepness of skirts, or uniformity of time delay versus frequency. Also, "all-pass filters" with flat amplitude response but with custom phase versus frequency variation may be realized, as well as the opposite—a filter with constant phase shift but tailored amplitude response.

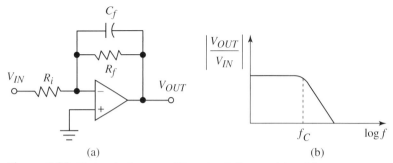

Figure 4.28 First-order low-pass filter: circuit diagram (a) and frequency response (b).

4.10.1 Filter Circuits and Frequency Characteristics

Depending on the operating frequency range, a filter either passes or attenuates input signal components. Figure 4.28 shows a first-order low-pass filter obtained by adding a feedback capacitor C_f to the basic inverting op-amp circuit. The frequency response of this circuit is almost identical to that of the first-order passive low-pass filter shown in Fig. 2.41. Meanwhile, if an input capacitor C_i is added to the basic inverting op-amp circuit, the result is the first-order high-pass filter circuit shown in Fig. 4.29. The frequency response of this filter is almost identical to that of the passive high-pass filter shown in Fig. 2.43. Op-amp filters have a very low output impedance, rendering loading effects in cascaded circuits insignificant.

Figures 4.30 and 4.31 show second- and third-order active filters. The second-order filter is obtained by cascading two RC filters and applying positive feedback to increase the gain at the corner frequency. A third-order filter has an additional RC filter. Higher-order filters may be realized by cascading second-order and third-order filters (see Example 4.3). Also, by cascading low- and high-pass filters, broadband band-pass filters can be achieved. Other configurations yield narrowband and notch filters.

4.10.2 Filter Types

Many methods can be deployed to design filter circuits with desired response characteristics. The *Butterworth filter* provides for a smooth response at all frequencies. The response magnitude is maximally flat in the passband and monotonic attenuation at high rate beyond the

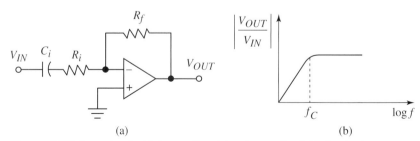

Figure 4.29 Second-order high-pass filter (a) and normalized third-order low-pass filter (b).

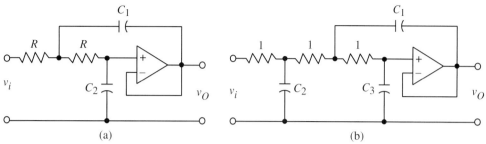

Figure 4.30 Second-order low-pass filter (a) and normalized third-order low-pass filter (b).

cutoff frequencies. However, the steepness of the transition at the corner frequency is proportional to the order of the filter; the higher the order is, the closer the response comes to the ideal LPF. These features make voltage control and wide-range tuning easier because they set all cascaded sections to the same frequency.

The *Chebyshev filter* provides for a sharper roll-off than the Butterworth filter. It has the highest attenuation rate, at the cost of a ripple in the response within the passband. The higher the attenuation rate, the higher the ripple. Different Chebyshev tables are utilized, depending on the amount of ripple that can be tolerated in the passband. This filter optimizes the steepness of the transition from passband to stopband.

Bessel filters have the ability to pass signals within the passband without distortion of their waveforms caused by phase shifts. It has a maximally flat time delay in response to a step input. However, the attenuation rate is very gradual.

The response of each of these filters can be produced with a variety of different filter circuits. They are all available in low-pass, high-pass, and band-pass versions. There is no all-around best circuit; various circuits excel in one or another desirable property.

The values given in Table 4.3 simplify filter design. The tabulated values correspond to a corner frequency $\omega_0 = 1$ rad/sec and filter resistors $R_0 = 1\ \Omega$. The values listed for all filter capacitors C_0 are in units of farads. Since a farad is a very large quantity, these values are not practical and require scaling, which can be accomplished via the relation

$$\omega_0 R_0 C_0 = \omega RC \qquad (4.33)$$

The designer chooses the values for ω and R, and then Eq. (4.33) is solved for C.

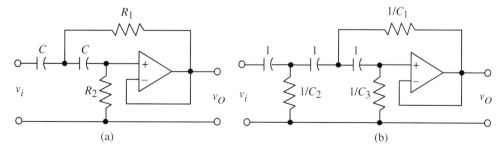

Figure 4.31 Second-order high-pass filter (a) and normalized third-order high-pass filter (b).

TABLE 4.3 Tables for Cascaded Filter Design

Poles	C_1	C_2	C_3	C_1	C_2	C_3
		Bessel			Butterworth	
2	9.066 E − 1	6.799 E − 1		1.414 E + 0	7.071 E − 1	
3	1.423 E + 0	9.880 E − 1	2.538 E − 1	3.546 E + 0	1.392 E + 0	2.024 E − 1
4	7.351 E−1	6.746 E − 1		1.082 E + 0	9.241 E − 1	
	1.012 E + 0	3.900 E − 1		2.613 E + 0	3.825 E − 1	
5	1.009 E + 0	8.712 E − 1	3.095 E − 1	1.753 E + 0	1.354 E + 0	4.214 E − 1
	1.041 E + 0	3.098 E −1		3.235 E + 0	3.089 E − 1	
6	6.352 E − 1	6.098 E − 1		1.035 E + 0	9.660 E − 1	
	7.225 E − 1	4.835 E − 1		1.414 E + 0	7.071 E − 1	
	1.073 E + 0	2.561 E − 1		3.863 E + 0	2.588 E − 1	
		2-dB Chebyshev			0.25-dB Chebyshev	
2	2.672 E + 0	5.246 E − 1		1.778 E + 0	6.789 E−1	
3	2.782 E + 1	3.113 E + 0	3.892 E − 2	8.551 E + 0	2.018 E + 0	1.109 E − 1
4	4.021 E + 0	1.163 E + 0		2.221 E + 0	1.285 E + 0	
	9.707 E + 0	1.150 E − 1		5.363 E + 0	2.084 E − 1	
5	1.240 E + 1	4.953 E + 0	1.963 E − 1	5.543 E + 0	2.898 E + 0	3.425 E − 1
	1.499 E + 1	7.169 E − 2		8.061 E + 0	1.341 E − 1	
6	5.750 E + 0	1.769 E + 0		3.044 E + 0	1.875 E + 0	
	7.853 E + 0	2.426 E − 1		4.159 E + 0	4.296 E − 1	
	2.146 E + 1	4.902 E − 2		1.136 E + 1	9.323 E − 1	

Example 4.3: Filter design–An application requires a five-pole low-pass Bessel filter with a corner frequency of $f_C = 300$ Hz and an input resistance of 40 kΩ. The five-pole design may be obtained by cascading third- and second-order filters in series. Determine the required values for the resistors and capacitors.

Solution: The desired five-pole Bessel filter is shown in Fig. 4.32. Substituting the corresponding parameters from Table 4.3 into Eq. (4.33) gives:

$$C_{1A} = \frac{\omega_0 R_0 C_0}{\omega R} = \frac{(1)(1)(1.009)}{2\pi(300)(50,000)} = 10.7 \text{ nF}$$

$$C_{2A} = \frac{0.8712}{2\pi(300)(50,000)} = 9.24 \text{ nF}$$

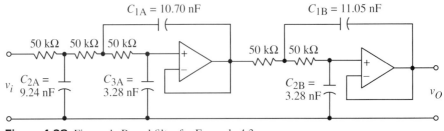

Figure 4.32 Five-pole Bessel filter for Example 4.3.

$$C_{3A} = \frac{0.3095}{2\pi(300)(50{,}000)} = 3.28 \text{ nF}$$

$$C_{1B} = \frac{1.041}{2\pi(300)(50{,}000)} = 11.05 \text{ nF}$$

$$C_{2B} = \frac{0.3089}{2\pi(300)(50{,}000)} = 3.28 \text{ nF}$$

Figure 4.32 shows the complete design.

4.11 POWER OP-AMPS

The output of low-power op-amp ICs is limited to about 20–25 mA. Only in special designs can the output current reach 100 mA. Obviously, these op-amps are not suitable to control high-current devices such as DC motors and audio equipment, among others. Amplifiers for such applications may be constructed from discrete components, including transistors operating in the linear range. However, power op-amps offer a more efficient alternative because they are easy to use, obey all op-amp rules, are built with matched components to provide a more linear output, have built-in protection circuits, and produce less electrical noise.

A typical power op-amp essentially combines a power transistor with an op-amp. The result is a device that can handle output voltages above 300 V. APEX Microtechnology (www.apexmicrotech.com) manufactures power op-amps offering a wide range of drive capabilities. The PA01, for example, is a low-cost, highly linear power op-amp capable of providing high output currents up to ± 5 A and high supply voltages up to ± 30 V. The μA759 power op-amp from Fairchild Semiconductors (www.fairchildsemi.com) has characteristics similar to the 741 op-amp, but with an output power booster that can deliver more than 300 mA through a 50-Ω load. Power op-amps are available with more power stage, such as the L272 dual version from SGS Semiconductors (www.thomson.com)

4.12 SUMMARY

Analog signal conditioning is based on improving signal properties. The operational amplifier is the most popular and useful component found in conditioning mechanisms' input and output signals for the majority of mechatronics systems. This chapter presented the op-amp as a commercial component with technical specifications and then available packaged circuits and functions. The chapter covered the theory, model, design, and practical aspects of popular op-amp circuits (linear, nonlinear, nonideal, and filters). Practical examples that deploy these op-amp circuits were also presented, with full discussion of practical implementation issues, such as noise and loading effects.

RELATED READING

D. G. Alciatore and M. B. Histand, *Introduction to Mechatronics and Measurement Systems*, 2nd ed. New York: McGraw-Hill, 2002.

J. DiBartolomeo, "Op-Amp Specifications—Parts I–IV," *Circuit Cellar*, nos. 117–120, April–July 2000.

P. Horowitz and W. Hill, *The Art of Electronics*. Cambridge, UK: Cambridge University Press, 1989.

J. Fraden, *AIP Handbook of Modern Sensors: Physics Design and Application*. New York: AIP, 1993.

S. Kamichik, *IC Design Projects.* Indianapolis: Prompt Publications, 1998.

G. Rizzoni, *Principles and Applications of Electrical Engineering*, 4th ed. New York: Mc-Graw Hill, 2002.

W. J. Tompkins and J. G. Webster, *Interfacing Sensors to the IBM PC.* New York: Prentice Hall, 1988.

John G. Webster (ed.), *The Measurement, Instrumentation, and Sensors Handbook.* Boca Raton, FL: CRC Press, 1998.

QUESTIONS

4.1 Explain the schematic symbol of an op-amp.

4.2 Describe the various stages of an op-amp.

4.3 What are the ideal characteristics of an op-amp?

4.4 Explain how and why the actual behavior of an op-amp differs from its ideal characteristics.

4.5 Define *open-loop gain*.

4.6 Why is a feedback loop almost always needed in practical op-amp circuits?

4.7 What op-amp specifications are important in DC applications? AC applications?

4.8 What is common-mode voltage?

4.9 What is differential voltage, and what is the maximum differential input voltage?

4.10 What happens to the op-amp if the differential input voltage exceeds its maximum limit?

4.11 What type of power supply do most op-amps require?

4.12 When does an op-amp operate in its linear range?

4.13 Define *CMRR*.

4.14 Define *slew rate of an-amp*.

4.15 Explain how a Schmitt trigger can introduce hysteresis in a circuit.

4.16 Explain offset voltages, and discuss ways to reduce their effects.

4.17 Explain bias currents and offset current, and discuss ways to reduce their effects on op-amp performance.

4.18 Describe the noise model of an amplifier and its ramification on measurements.

4.19 Define *PSRR*, and discuss how it affects op-amp behavior.

4.20 State the advantages and disadvantages of passive and active filters.

4.21 Explain the various filter parameters.

4.22 Compare the performances of Butterworth, Chebyshev, and Bessel filters.

PROBLEMS

4.1 What is the output current in the circuit shown in Fig. 4.33 for a given input voltage V_{IN}?

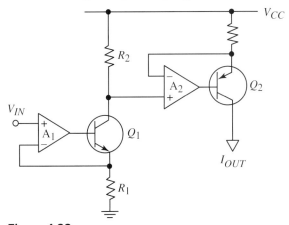

Figure 4.33

4.2 The circuit shown in Fig. 4.34 is the "Howland" current source. Show that if the resistors are chosen such that $R_3/R_2 = R_4/R_1$, then $I_{LOAD} = -V_{IN}/R_2$.

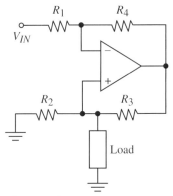

Figure 4.34

4.3 A thermistor measures temperature over the range 0–100°C. Its resistance over the range decreases non-linearly from 12,432 to 1,384 Ω. The circuit shown in Fig. 4.35 will generate a corresponding voltage of 0–10 V, which will interface an ADC (AD converter). Determine appropriate values for the resistors.

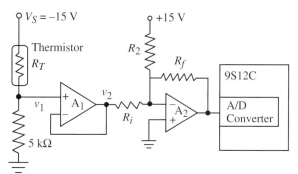

Figure 4.35

4.4 An op-amp integrator is constructed with 1-MΩ input resistor and a 0.1-μF feedback capacitor so that $\tau = RC = 1/10$ s. If it is desirable to integrate for 1 second, how reliable is a general-purpose op-amp for this application?

4.5 By flipping a switch, the circuits shown in Fig. 4.36 can be used to invert or amplify without inversion. Show that, depending on the switch position, the voltage gain is either +1 or −1.

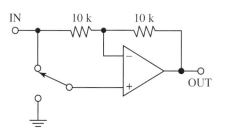

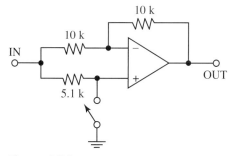

Figure 4.36

4.6 If in the circuit shown in Fig. 4.37 the 1-kΩ resistor is disconnected from ground and connected to a third signal source v_3, use superposition to determine v_o in terms of v_1, v_2, and v_3.

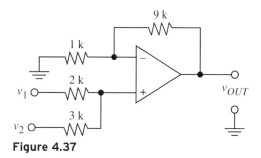

Figure 4.37

4.7 The circuit shown in Fig. 4.38 functions as a phase shifter. It is also known as a *first-order all-pass filter*. Derive an expression for the transfer function V_O/V_I. Find expressions for the magnitude and phase of the response.

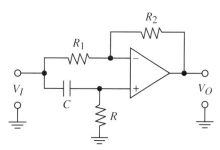

Figure 4.38

4.8 The circuit shown in Fig. 4.39 is an adjustable output voltage regulator. Assume that the basic op-amp is ideal. Find the regulated output voltage V_O.

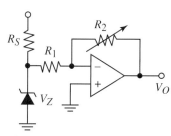

Figure 4.39

4.9 For $V_{IN} = 1.0$ V in the circuit shown in Fig. 4.40, what is the output voltage, V_{OUT}?

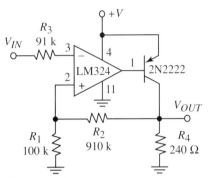

Figure 4.40

4.10 For the circuit shown in Fig. 4.41, select standard 1% resistors for R_F and R_B to allow an input range of 0–5 V. Assume an ideal op-amp and an input range of 0–2.5 V for the ADC. *Hint*: The amplifier is inverting, so full scale (5 V) at the input must map to 0 V at the A/D converter input, and 0 V at the amp input must map to 2.5 V at the ADC input.

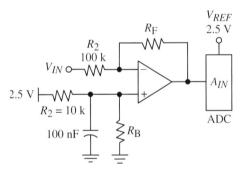

Figure 4.41

4.11 The noninverting op-amp circuit shown in Fig. 4.42 is used in a circuit to measure capacitance change in a capacitive-based sensor. Resistance R_b may be required to be on the order of tens, even hundreds, of gigaohms when converting currents from a piezo or pyroelectric sensors (see Chapter 11). Determine the response of this circuit to a step function in the time domain.

4.12 The op-amp circuit shown in Fig. 4.43 provides a constant-current source, for instance, to a resistive temperature detector (RTD). The circuit consists of an op-amp, an AD589 with 1.23-V bandgap, a trim pot, and a current-sense resistor R_1. Capacitor C_1 bypasses the AD589 at high frequencies to maintain stability. Determine the value of the stable current the circuit provides. What is the role of the trim pot?

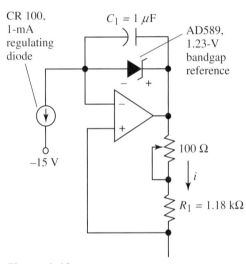

Figure 4.43

4.13 Figure 4.44 shows an op-amp used to form an oscillator circuit. Develop an expression for the frequency of oscillations in terms of the parameters shown.

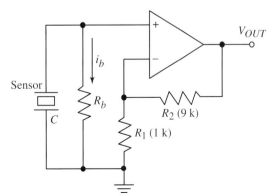

Figure 4.42

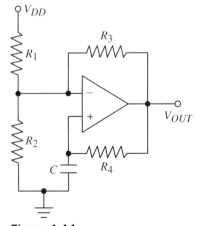

Figure 4.44

4.14 Consider an ideal op-amp in proper (nonsaturated) operation. In this case both of the input currents are approximately zero: $i^- \approx 0$ and $i^+ \approx 0$. The current at the output may flow either into or out of the op-amp, depending on the additional feedback circuitry and loads. Explain how this is possible in light of Kirchhoff's current law, which must always be satisfied.

4.15 Consider the inverting summer shown in Fig. 4.13 with input voltages v_1 and v_2 only. Suppose that the power supply voltages to the op-amp are symmetric at ± 15 V. Obtain a constraint on the input voltages that will ensure proper (nonsaturated) operation of the circuit.

4.16 An op-amp has a CMRR of 80 dB. Two sets of signals are applied to its inputs: one is 0.2 V and 0.1 V and the other is 4.5 V and 4.7 V. Determine the common-mode voltages, differential voltages, and output voltages for both input sets. Discuss the results.

4.17 A differential amplifier is constructed with $R_f = 470$ kΩ and $R = 2.7$ kΩ. The amplifier generates an output of 807 mV when the inputs are $v_1 = v_2 = 2.5$ V. Determine the Common Mode Rejection and ratio.

4.18 A battery-powered system causes the rails of an op-amp to change by $\pm 10\%$ from the nominal 3 V. Determine the output error caused by this change if the op-amp has a PSRR of 80 dB and 100 dB.

4.19 A sensor outputs 0–1 volts. Develop a voltage-to-current converter so that this becomes 0–10 mA. Specify the maximum load resistance if the op-amp saturates at ± 10 V.

4.20 A sensor generates an output voltage in the range 0.1–0.5 V. Connect the sensor to a signal-conditioning circuit (refer to Example 4.2) to generate a voltage for the full range of the analog-to-digital converter, which is 0–5 V. Sketch the circuit, label all parameters, and give values for the associated components.

4.21 The Wien's network of Example 2.7 is connected in the feedback loop of an op-amp as shown in Fig. 4.45. Prove that sinusoidal oscillation will occur when the op-amp gain is greater than 3 by adjusting R.

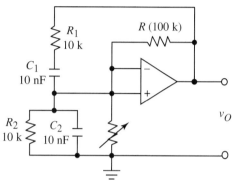

Figure 4.45

4.22 Figure 4.46 shows a basic logarithmic op-amp circuit. Determine the ratio of the output voltage to the input voltage. (*Hint*: Use the diode relation.)

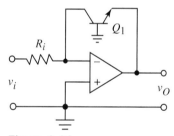

Figure 4.46

LABORATORY PROJECTS

4.1 Select an op-amp, resistors, and capacitors and build the circuits in Problems 4.7, 4.10, and 4.12. Experiment with their responses using a function generator and an oscilloscope. Measure the values, determine the corner frequency for the filter circuit, and compare that with what you see on the oscilloscope.

4.2 Build a temperature-activated projectile that will shoot a small ball when the temperature reaches a threshold value. The ball should hit a lever that will cause the water in a tank to flow. Use an IC temperature sensor, a comparator, a transistor, and a solenoid to accomplish the task. Suggested parts: LM324 comparator, 12 V $\times$ 1 A solenoid, TIP31 transistor, and LM335 temperature sensor IC.

5

Digital Logic and Logic Families

OBJECTIVES

Thoughtful engagement with the material presented in this chapter will enable the student to:

1 Understand the functions of basic logic circuits

2 Describe the characteristics of various logic families

3 Understand the role of buffers and drivers

4 Know how to interface between TTL and CMOS ICs

5 Analyze and design logic networks

5.1 DIGITAL SIGNALS

Digital signals are essentially a series of pulses or rapidly changing voltage levels that vary in discrete steps or increments, as illustrated in Fig. 5.1. At any time, a digital signal can have one of two possible values, called *logic levels* and assigned the binary value 0 or 1; 0 represents logic LOW and 1 represents logic HIGH. Depending on the logic family used, 0 and 1 also represent a range of voltage levels. In the transistor-to-transistor logic (TTL) integrated circuit family, logic 0 represents any voltage between 0 V and 0.8 V, and logic 1 represents the range from $+2$ to $+5$ V. Electronic circuits that process digital signals are called *digital logic circuits*. A digital device in which the binary state 1 is the more positive of the two voltage levels is said to use *positive logic*; and if the binary state 1 is the more negative of the two voltage levels, the device is said to use *negative logic*. A general block diagram of a digital system is shown in Fig. 5.2. Each input represents the voltage level (a number) of a specific quantity, such as 1 bit of a binary number, 1 bit of some binary code such as BCD or ASCII, a control level, a sensor signal, and so on. Appendix G provides an overview of different number systems, their representations, conversion processes from one to another, and arithmetic operations. Fundamental to understanding digital circuits is Boolean algebra, which is discussed next.

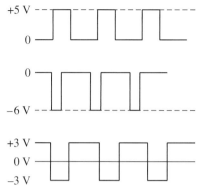

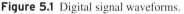

Figure 5.1 Digital signal waveforms.

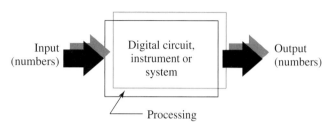

Figure 5.2 Block diagram of a digital system.

5.2 COMBINATIONAL AND SEQUENTIAL LOGIC CIRCUITS

Logic devices are constructed from a few basic circuits called *logic elements*. There are two types of basic logic elements: decision making and memory. The basic *decision-making* element is called a *gate*, shown schematically in Fig. 5.3. The gate takes two or more binary inputs and generates an appropriate output signal based on the logical operation of the gate and the *current states* of its inputs. Gates are generally combined to form sophisticated and complex decision-making networks called *combinational circuits*.

A general block diagram of a combinational logic circuit is shown in Fig. 5.4. The circuit converts binary inputs into binary outputs based on the rules of mathematical logic and independent of a specific input timing. Decoders, encoders, multiplexers, demultiplexers, and binary comparators are examples of combinational logic.

The basic *memory* element is a bistable storage device known as a *flip-flop*, or FF. Flip-flops are also known as *latches* or *bistable multivibrators*. A general block diagram of a flip-flop is shown in Fig. 5.5. The flip-flop has two or more inputs, used to cause the flip-flop to switch between two output stable states that depend on the *previous state* of its inputs. The outputs are the complement of each other and are usually assigned the symbols Q and Q'. The "prime" symbol indicates complement operation. The Q output defines the state of the flip-flop: LOW (bit 0) or HIGH (bit 1). Once an input signal causes the flip-flop to go to a given state, it retains that state, or "remembers" it, even after the input signal is removed. This is where the flip-flop derives its memory characteristics.

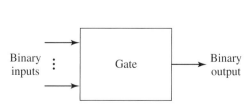

Figure 5.3 Block diagram of a digital logic gate.

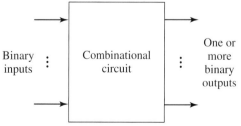

Figure 5.4 Block diagram of a combinational logic circuit.

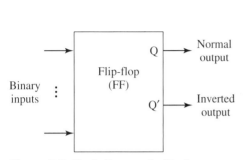

Figure 5.5 Block diagram of a flip-flop.

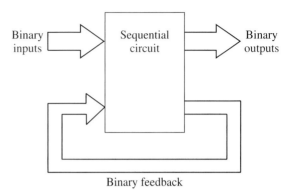

Figure 5.6 Block diagram of a sequential logic circuit.

Many of the basic memory elements can be combined to store complete binary numbers or words. Most memory elements are interconnected with combinational circuits to form another, more sophisticated form of memory elements known as *sequential circuits*. A general block diagram of a sequential logic circuit is shown in Fig. 5.6. Counters, shift registers, timers, sequencers, clocks, and microprocessors are examples of sequential logic circuits.

Most digital logic circuits are in the form of integrated packages commonly referred to as integrated circuits (IC) chips. Logic gates and their different types, such as AND, OR, NAND, and DECODER, are ICs constructed from a combination of transistors and resistors. Integrated circuits are usually packaged in metal TO5 cans, flat pack, and dual-in-line (DIP) packages. The DIP packages are most common, the TO5 cans are the most expensive, and the flat pack are designed for applications where circuit board space is limited.

5.3 CLOCK SIGNALS

Most digital systems operate as *synchronous sequential systems*, where the sequence of operations takes place at regularly spaced intervals synchronized by a master trigger signal called the *clock*, or CLK. The clock generates rectangular pulses at a fixed frequency and distributed to all parts of the system. The pulses are usually one of the forms shown in Fig. 5.7. A system is usually made to perform operations at times when the clock signal is making a transition

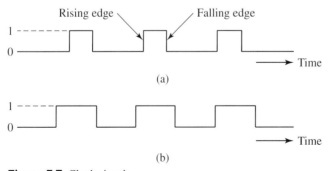

Figure 5.7 Clock signals.

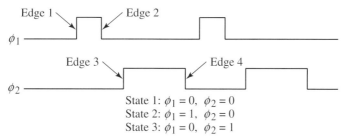

Figure 5.8 Two-phase clock signals.

from 1 to 0 or from 0 to 1. The 0-to-1 transition is called the *rising* (or *positive-going*) *edge*, and the 1-to-0 transition is called the *falling* (or *negative-going*) *edge* of the clock signal. If the device is made to respond to a rising edge, it is called a *positive edge-triggered* device. If it is made to respond to a falling edge, it is called a *negative edge-triggered* device.

The timing of many digital circuits and computers is controlled by two or more related clock signals. A common combination, shown in Fig. 5.8, utilizes two clock signals, identified by the symbols ϕ_1 and ϕ_2 and referred to as phase 1 and phase 2, respectively. The two signals provide four different edges and three different states per period, compared to only two edges and two states per period for a single clock signal.

5.4 BOOLEAN ALGEBRA AND LOGIC GATES

The basic mathematics needed for the study of two-state logic designs of digital systems is the two-state Boolean algebra. Boolean algebra is a simplified mathematical system for binary functions used to express all the various logic functions in a convenient mathematical format.

5.4.1 Basic Functions and Gates

The three basic Boolean operations on which all other operations are based are the inversion operation, interpreted as Complement ($'$); the multiplication operation, interpreted as AND (.); and the addition operation interpreted as OR ($+$). These basic functions are implemented by three corresponding logic gates: the inverter (or NOT) gate, the AND gate, and the OR gate.

The logical value of any Boolean function represents the output of the associated logic gate and can be represented on a truth table. A *truth table* is a plot that specifies the value of a logic expression for all possible combination values of the variables in that expression. In general, the truth table for an n-variable logic expression will have 2^n combinations, or rows. The symbol, truth table, logic expression, timing, and circuit applications of the NOT, AND, and OR gates are given in Figs. 5.9–5.11, respectively.

Boolean algebra provides a convenient means for analyzing and expressing operations in digital circuits. By applying the rules and laws of Boolean algebra, logic equations that characterize the functions of a given network can be reduced to the minimum number of components to accomplish the same functions. When properly applied in the design of logic circuits, Boolean algebra usually results in the simplest, least expensive, and most efficient logic circuit design. The basic Boolean laws (X is a state value of 0 or 1) for the OR function are:

$$X + 0 = X \qquad X + 1 = 1$$
$$X + X = X \qquad X + X' = 1$$

(5.1)

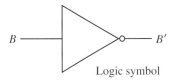

Input	Output
B	B'
0	1
1	0

Truth table

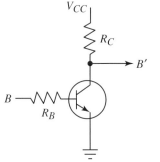

A transistor logic inverter

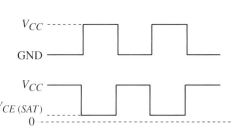

Input and output waveforms of
a transistor logic inverter

Figure 5.9 The NOT (or inverter) gate.

for the AND function are

$$X \cdot 1 = X \qquad X \cdot 0 = 0$$
$$X \cdot X = X \qquad X \cdot X' = 0$$

(5.2)

The basic Boolean laws for double inversion is

$$(X')' = X$$

(5.3)

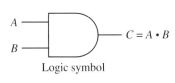

Logic symbol

Inputs		Output
A	B	C
0	0	0
0	1	0
1	0	0
1	1	1

Truth table

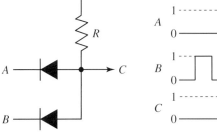

Two-input diode AND gate

Input and output waveforms of the diode AND gate

Figure 5.10 The AND gate (e.g., 7408, 4081).

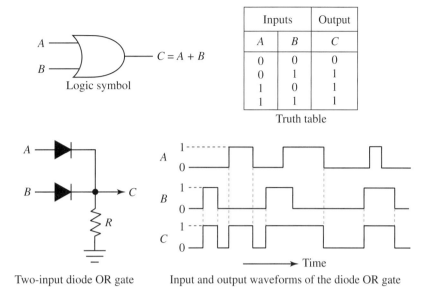

Inputs		Output
A	B	C
0	0	0
0	1	1
1	0	1
1	1	1

Truth table

Two-input diode OR gate   Input and output waveforms of the diode OR gate

Figure 5.11 The OR gate (e.g., 7432, 4071).

The three basic gates can be used in combination to form a variety of other decision-making gates, with associated Boolean functions. The NAND, NOR, EX-OR, and EX-NOR gates are shown in Figs. 5.12–5.15, respectively. Figure 5.12 indicates that the operation of the NAND gate can be implemented by an AND gate and a NOT gate. Similarly, Fig. 5.13 shows the operation of a NOR gate, which can also be implemented by combining an OR and a NOT gate. In general, the term *logic equivalence* is used when the same logic operation can be implemented in more than one way. Appendix B provides the pinout of several ICs, including the NOT, AND, NAND, and OR gates.

5.4.2 Boolean Laws and Theorems

The function of a logic circuit is completely defined by a logic (or Boolean) equation. The Boolean equation is formed by applying the basic operations NOT, AND, and OR to one or more input variables or constants. The Boolean equation identifies each term that will provide an output of 1. The terms are then ORed or ANDed together to provide all the logic terms that provide a 1 in the output. From the Boolean equation, we can determine the number of AND and OR gates, how many inputs each gate contains, and how many inverters are required to realize the logic circuit. The Boolean expression can be written directly from the truth table. In general, the function of a logic circuit can be simplified and converted to a

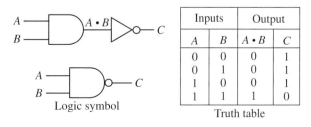

Inputs		Output	
A	B	A • B	C
0	0	0	1
0	1	0	1
1	0	0	1
1	1	1	0

Truth table

Figure 5.12 The NAND gate (e.g., 7400, 4011).

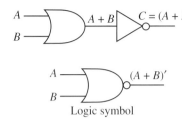

Inputs		Output	
A	B	A + B	C
0	0	0	1
0	1	1	0
1	0	1	0
1	1	1	0

Logic symbol   Truth table

Figure 5.13 The NOR gate (e.g., 7402, 4001).

simpler network that can be physically realized using logic gates. This can be done by applying the following laws and theorems of Boolean algebra.

Commutation Laws:

$$X + Y = Y + X$$
$$X \cdot Y = Y \cdot X$$

(5.4)

Association Laws:

$$(X + Y) + Z = X + (Y + Z)$$
$$(X \cdot Y) \cdot Z = X \cdot (Y \cdot Z)$$

(5.5)

Distribution Laws:

$$X \cdot (Y + Z) = (X \cdot Y) + (X \cdot Z)$$
$$X + (Y \cdot Z) = (X + Y) \cdot (X + Z)$$

(5.6)

Simplification Theorem:

$$(X \cdot Y) + (X \cdot Y') = X$$
$$X + (X \cdot Y) = X$$
$$(X + Y') \cdot Y = X \cdot Y$$
$$(X + Y) \cdot (X + Y') = X$$
$$X \cdot (X + Y) = X$$
$$(X \cdot Y') + Y = X + Y$$

(5.7)

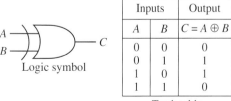

Inputs		Output
A	B	C = A ⊕ B
0	0	0
0	1	1
1	0	1
1	1	0

Logic symbol   Truth table

Figure 5.14 The EX-OR gate (e.g., 7486, 4070).

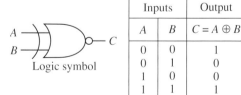

Inputs		Output
A	B	C = A ⊕ B
0	0	1
0	1	0
1	0	0
1	1	1

Logic symbol   Truth table

Figure 5.15 The EX-NOR gate (e.g., 74266, 4077).

DeMorgan's Theorem:

$$(X + Y + Z + \cdots)' = X' \cdot Y' \cdot Z' \cdots$$
$$(X \cdot Y \cdot Z \cdots)' = X' + Y' + Z' + \cdots$$

(5.8)

Duality:

$$(X + Y + Z + \cdots)^D = X \cdot Y \cdot Z \cdots$$
$$(X \cdot Y \cdot Z \cdots)^D = X + Y + Z + \cdots$$

(5.9)

Multiplying Out and Factoring Theorems:

$$(X + Y) \cdot (X' + Z) = X \cdot Z + X' \cdot Y$$
$$(X \cdot Y) + (X' \cdot Z) = (X + Z) \cdot (X' + Y)$$

(5.10)

Consensus Theorem:

$$(X \cdot Y) + (Y \cdot Z) + (X' \cdot Z) = (X \cdot Y) + (X' \cdot Z)$$
$$(X + Y) \cdot (Y + Z) \cdot (X' + Z) = (X + Y) \cdot (X' + Z)$$

(5.11)

Example 5.1: Boolean expression–The logic network shown in Fig. 5.16 has three inputs, A, B, and C, and one output, Z. From the corresponding truth table, determine the Boolean expression and the gate realization of the network.

Solution: The truth table indicates that four terms provide an output of 1: $AB'C'$, ABC', $AB'C$, and $A'BC'$. The Boolean equation, formed by ORing those terms, is

$$Z = AB'C' + ABC' + AB'C + A'BC'$$

(5.12)

This function can be realized by a network of four 3-input AND gates, three inverters, and one 4-input OR gate, as shown in Fig. 5.16. The logic expression in Eq. (5.12) is in the sum-of-products form.

Inputs			Output
A	B	C	Z
0	0	0	0
0	0	1	0
0	1	0	1
0	1	1	0
1	0	0	1
1	0	1	1
1	1	0	1
1	1	1	0

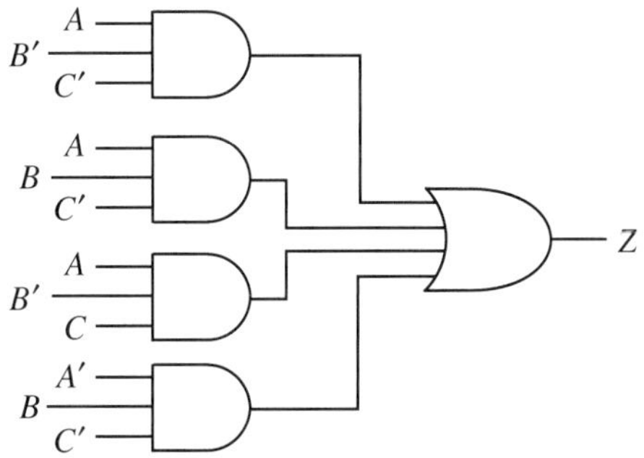

Figure 5.16 Logic network for Example 5.1.

Example 5.2: Applying Boolean rules—Simplify the logic function of Eq. (5.12) using Boolean rules and laws.

Solution: Applying the rules and laws of Boolean algebra, the Boolean expression is simplified successively as follows:

$$
\begin{aligned}
Z &= AB'C' + ABC' + AB'C + A'BC' \\
&= AB'C' + AB'C + BC'A + BC'A' \\
&= AB'(C' + C) + BC'(A + A') \\
&= AB'(1) + BC'(1) \\
&= AB' + BC'
\end{aligned}
\tag{5.13}
$$

This equation corresponds to the network shown in Fig. 5.17.

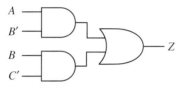

Figure 5.17 Logic circuit implementation of Eq. (5.13).

5.4.3 Karnaugh Maps

Two problems arise when using Boolean algebraic procedures: (1) The procedures are difficult to apply in a systematic way, and (2) it is difficult to tell when a minimum is reached. Using Karnaugh maps (or K-maps) as an alternative to Boolean algebra overcomes the aforementioned difficulties and helps reduce logic equations to their minimum form in a systematic way.

A Karnaugh map provides a method of plotting the logic circuit's outputs. It also provides a convenient format for combining like terms and equivalent terms. A K-map requires a square, or cell, for each term that represents a possibility of 1. Thus, if the equation has 4 variables, the truth table will require 2^4, or 16, input combinations (16 rows), and the K-map will require a grid of 16 cells to plot all of the output possibilities. DeMorgan's theorems are extremely important when designing discrete digital circuits using K-Maps. Unless the NOTs are distributed, a Boolean equation cannot be plotted in a K-Map. Once the grid of cells is set up, the process of implementing K-maps involves three steps: plotting, grouping, and reading. The grouping operation requires combining the cell plots marked on the K-map grid, if possible. The following rules apply to the grouping process.

1. Only adjacent vertical and horizontal cells can be grouped.
2. Diagonal groups are illegal.
3. Make groups as large as possible. However, group sizes must be in powers of 2, i.e., 2 cells, 4 cells, 8 cells, etc.
4. All plots must be grouped, if possible, using the preceding rules.
5. A given plot may be used to form more than one group, thereby creating overlapping groups.
6. If no grouping is possible, the equation is in its simplest form and cannot be reduced further.

To read the groups and generate a simplified equation, each group within the K-map must be read and a fundamental product terms generated for each. The fundamental product terms are then ORed together to produce the final sum-of-products equation.

Example 5.3: Applying Karnaugh maps–Use the K-map to reduce the truth table for Example 5.1.

Solution: The truth table is plotted on an 8-cell grid by transferring the logic 1 outputs to the grid cells shown in Fig. 5.18. Reading the two groups yields the same result as obtained in Example 5.1.

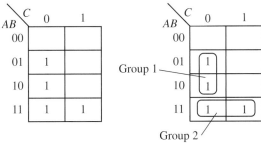

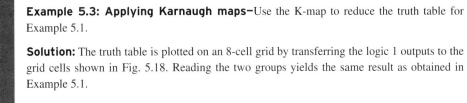

Figure 5.18 Karnaugh map for Example 5.3.

5.4.4 Design of Combinational Logic Circuits

The goal of a logic design is to realize a function using the least number of gates. The number of gates through which a signal must pass between the input and output terminals defines the number of levels in a gate network. The minimum sum of products (or product of sums) yields a minimum two-level gate network. However, the minimum number of gate levels may not be what gives the minimum number of gates or gate inputs.

A set of gates capable of realizing all switching functions is said to be functionally complete. A set of gates is functionally complete if it can realize AND, OR, or NOT. Minimum two-level AND-OR, NAND-NAND, OR-AND, and NOR-OR networks can be realized using the minimum sum of products as a starting point. Minimum two-level OR-AND, NOR-NOR, AND-NOR, and NAND-AND networks can be realized using the minimum product of sums as a starting point. Design of multilevel, multioutput NAND-gate or NOR-gate networks is most easily accomplished by first designing a network of AND and OR gates. The following guidelines help in designing a combinational switching network.

1. Define the problem in words, and write "quasi-logical" statements that can be translated into Boolean expressions.
2. Set up a truth table that specifies the outputs as a function of the input variables. A "don't care" or "X" is used as an output for a given combination of values for the input variables that can never occur at the network inputs.
3. Derive simplified Boolean expressions for the output functions using Boolean algebra or K-maps.
4. Manipulate the simplified algebraic expressions into the proper form, depending on the type of gates to be used in realizing the network.

Example 5.4: Logic system design–An alarm system design requires the alarm to operate as follows:

> The <u>alarm</u> will ring **IFF** the <u>alarm switch</u> is turned on and the <u>door</u> is not closed,
> **OR** it is <u>after 6 p.m.</u> **AND** the <u>window</u> is not closed.

Write an algebraic expression corresponding to this sentence, and give a network that realizes it.

Solution: The first step is to associate a Boolean variable with each phrase in the given sentence. The following assignments for variables are used:

> The alarm will ring (Z) **IFF** The alarm switch is on (A) **AND** The door is not closed (B') **OR** It is after 6 p.m. (C) **AND** The window is not closed (D')

This set of assignments implies that if $Z = 1$, the alarm will ring. If the alarm switch is closed, $A = 1$; and if it is after 6 p.m., $C = 1$. If the variable B is used to represent the phrase "the door is closed," then B' represents "the door is *not* closed." Thus $B = 1$ if the door is closed, and $B' = 1$ ($B = 0$) if the door is not closed. Similarly, $D = 1$ if the window is closed, and $D' = 1$ if the window is *not* closed. Using this assignment of variables, the given sentence may be translated into the following Boolean expression:

$$Z = AB' + CD' \tag{5.14}$$

Equation (5.14) may be implemented by the network shown in Fig. 5.19. The output Z is connected to the alarm so that it will ring when $Z = 1$.

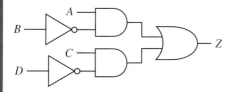

Figure 5.19 Network realization for Eq. (5.14).

5.5 INTEGRATED CIRCUITS AND LOGIC FAMILIES

A collection of different types of ICs made with the same technology and having the same electrical characteristics is called a *logic family*. Popular logic families include transistor-transistor logic TTL (T^2L), emitter-coupled logic (or ECL), integrated injection logic (or I^2L), and complementary metal oxide semiconductor (or CMOS). The TTL and the CMOS families are the most widely used, especially in sensor technologies. The TTL logic family is based on the bipolar junction transistor, and the CMOS family is based on the metal oxide field-effect transistor, or MOSFET (or MOS for short). Table 5.1 shows the basic characteristics of the four aforementioned logic families. A brief overview of the important characteristics of logic

TABLE 5.1 Primary Characteristics of TTL, CMOS, ECL, and IIL Logic Families

Characteristic	TTL	CMOS	ECL	IIL
Fan-out	10	50+	25	2
Power dissipation per gate (mW)	2–22	0.01 static 1.2 at 1 MHz	25–60	<0.07
Avg. propagation delay (ns)	2–40	5.1–90	0.75–2.9	20–50
Noise immunity	Good	Very good	Good	Fair to good
Noise generation	High	Low to medium	Low to medium	Low
Typical supply voltage (V)	5	+1.5–18	−5.2	+1–10.15
Logic levels	Binary 0: +0.4 V Binary 1: +3.6 V	Binary 0: 0 V Binary 1: $+V_{DD}$	Binary 0: −1.75 V Binary 1: −0.9 V	Binary 0: <+0.1 V Binary 1: +0.7 V
Type of logic	Current sinking	Current sinking, current sourcing	Unsaturated, current sourcing	Current sinking
Temperature range (°C)	−55 to +125 0–70	−55 to +125 −40– +85	−55–+125	0–70
Avg. clock rate (MHz)	15–120	5–10	200–1000	1–10
Basic gate form	Positive NAND Negative NOR	Positive NOR Negative NAND	OR/NOR	Positive NAND or NOR

families to be considered when designing digital circuits is given next, namely, logic level, noise immunity, fan-out, power dissipation, and speed.

5.5.1 Logic Levels

Logic levels refer to the voltage values and bands corresponding to the binary 1 and 0 states of a given type of logic family. Examples are included in Table 5.1.

5.5.2 Noise Immunity

Noise immunity is the measure of susceptibility of a logic circuit to noise pulses on the inputs and outputs of the logic circuit. Noise could be any extraneous and undesired signal generated within the device itself during high-speed switching or externally from noisy industrial environments. All digital circuits have built-in noise immunity to reject noise spikes of amplitudes from 10 to 50% of supply voltage.

5.5.3 Fan-Out

Fan-out is generally expressed in terms of the number of logic loads or gates a logic gate can *drive* while maintaining the desired LOW or HIGH levels. The output drive capability of a driving gate is expressed in terms of its current-*sourcing* or current-*sinking* capabilities. When the output is HIGH, the gate can source a current I_{OH} to n gates, each demanding an input current of I_{IH}; thus, $n = I_{OH}/I_{IH}$. Meanwhile, when the output is LOW, a gate can sink a current I_{OL} from n gates, each supplying a current of I_{IL}; thus, $n = I_{OL}/I_{IL}$. Since the input impedance of a gate is large (input capacitance is typically 5–10 pF), fan-out for a HIGH is never important. A logic gate may have a fan-out of 10, indicating that 10 gate inputs can be attached to the output of this logic circuit and still maintain proper operation according to the manufacturer's specifications. When heavy loads are involved, noninverting gates or buffers are usually used to provide additional drive.

5.5.4 Power Dissipation

Power dissipation is the amount of power (mW) the components of a typical logic gate or other circuit drain from the power supply. In addition to increasing the heat within the device, power dissipation drives up the scale of the power supply. It is an important consideration if the device is used in a remote location (space) or if it is battery powered. TTL consumes 10 mW/gate, whereas CMOS draws no power unless it is in the act of switching the logic. In this mode, power dissipation in CMOS increases almost linearly up to 10 mW at 1-MHZ switching frequency.

5.5.5 Propagation Delay

Propagation delay is a measure of the *speed* of operation of a logic circuit. It defines how fast a digital circuit responds to a change in the input level. Propagation delay of a signal through a basic inverter of a gate is illustrated in Fig. 5.20. Propagation delay t_p is the average of two time delays: t_{pHL}, the time delay when the output changes from HIGH to LOW; and t_{pLH}, the time delay when the output changes from LOW to HIGH. On average, TTL is about 5–10 times faster than CMOS.

The two main criteria for selecting a logic family in general are power dissipation and speed. In all logic circuits, speed and power dissipation are directly dependent on one another. For most digital applications, higher speed and low-power dissipations are desirable. The two most widely used logic families are the TTL and the CMOS. A brief overview of the two families and their interface is given next.

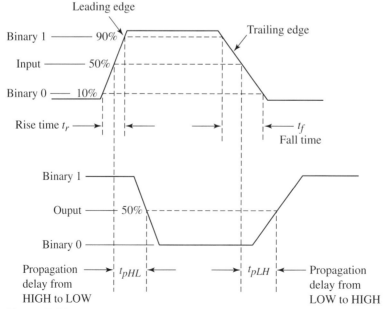

Figure 5.20 Propagation delay.

5.6 TTL LOGIC FAMILY

The TTL family is the most popular and most widely used type of digital bipolar ICs. Its popularity is attributed to the following: It has low cost; it is available in a wide variety of small-scale integrated (SSI) logic gates and medium-scale integrated (MSI) functional circuits; it is easy to use; and it has high performance characteristics and interfacing capabilities. SSI is usually limited to circuits that perform only one function, such as inverters, flip-flops, and NAND, NOR, AND, and OR gates. SSI packages usually contain one to four gates, six inverters, and one or two flip-flops. Medium-scale integration usually contains a variety of gates to form a complete functional circuit that performs more complex functions, usually on the order of 12–100 gates. Examples of MSI packages are address decoders, multiplexers, decoders, registers, and counters. Large-scale integrated (LSI) circuits, such as microprocessors, contain on the order of a hundred to a few thousand gates.

5.6.1 TTL Designations

TTL gates are designated as 54xx and the 74xx. Both numbering schemes are identical except for their operating temperature range. The 54xx series is guaranteed for use from $-55°C$ to $+125°C$, whereas the 74xx series is rated for service from $0°C$ to $70°C$. Each IC has a number stamped on the top. The *prefix* identifies the manufacturer, followed by the gate number, then a *midfix* indicating the version, and a *suffix* indicating the package type (refer to Table 5.2 for details).

5.6.2 TTL Versions

The TTL family is available in the following versions (series): the low-power (L), the high-speed (H), the Schottky (S), the low-power Schottky (LS), the advanced Schottky (AS), the advanced low-power Schottky (ALS), and the Fairchild advanced Schottky (F, for fast). All series operate with a $+5$ V ± 0.5 V power supply. A comparison of the speed and power dissipation of important TTL series is given in Table 5.3. The advanced LS series has the best speed–power dissipation product, an important design criterion. This series is often used with microprocessor-based applications.

5.6.3 Output Configurations

All versions of TTL gates come in three different types of output configurations: totem-pole output, open-collector output, and three-state output. Totem-pole and open-collector NAND gate circuits are shown in Figs. 5.21a and b. In the *totem-pole* arrangement, the output circuit

TABLE 5.2 Designations of TTL Logic Family ICs

Prefix–Manufacturer	Midfix–Version	Suffix–Package
TI—Texas Instruments	S—Shottky	N—plastic DIP
AM—Advanced Micro	LS—low-power S	J—ceramic DIP
AP—Apex	AS—advanced S	W—flat
LM—National Semiconductors	ALS—advanced LS	
AD—Analog Devices	F—fast	

TABLE 5.3 Typical Speed and Power Dissipation of TTL Series

TTL Name	Propagation Delay (ns)	Power Dissipation (mW)	Speed-Power Product (pJ)
Schottky (S)	3	19	57
Low-power Schottky (LS)	2	9.5	19
Advanced Schottky (AS)	2	8	16
Advanced LS	4	1.3	5.2
Fast (F)	3.5	5.4	18.9

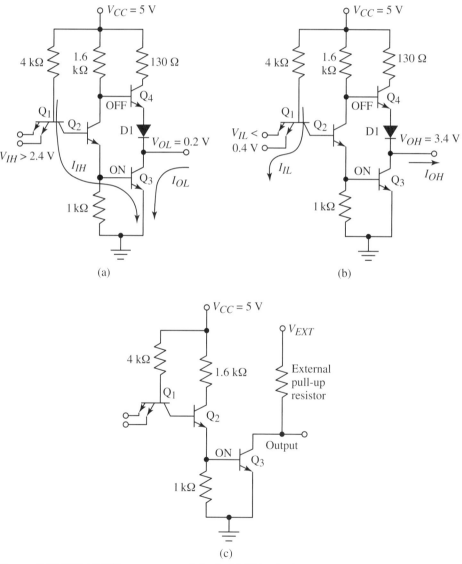

Figure 5.21 TTL NAND gate: totem-pole logic LOW (a), logic HIGH (b) outputs and open-collector output (c).

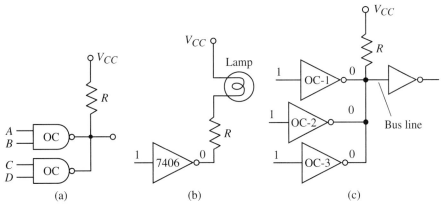

Figure 5.22 Wire ANDing two NAND gates (a), using open-collector inverter for lamp ground (b) and open-collector forming a common bus line (c).

consists of the two transistors Q_3 and Q_4 stacked one on the other. The totem-pole construction is the type used most frequently in gate design, especially in functional complex packages such as decoders and microprocessors. Two outputs of a totem-pole gate should never be wired together, for this connection cannot yield a valid output LOW, and, if excessive current is drawn, the gate will be damaged. Thus the output of a totem-pole gate should drive the input of another.

The *open-collector* (or OC) arrangement shown in Fig. 5.21c has no output drive capability, and one is provided in the form of an external *pull-up resistor* of several thousands ohms. When the output from an OC gate is high, the output transistor of the gate is cut off. The OC output voltage takes on the value of the external voltage V_{EXT} tied to the pull-up resistor, provided that the OC drives a high-impedance load, such as another gate. Figure 5.22 shows three major applications of OC gates: Wire-ANDing in Fig. 5.22a, driving a lamp or a relay in Fig.5.22b, and forming a common data bus line in Fig.5. 22c. *Wire ANDing* refers to several OC gates tied together with a single external pull-up resistor. Devices connected to a one-wire bus are logically ANDed onto the bus to make the idle state HIGH, allowing any device to pull the bus LOW. The PIC16F84 microcontroller provides one output pin, with an open-collector configuration, used for a transmission bus.

The *three-state output* TTL circuit is a special version of a totem-pole construction that has two inputs and one output. The three-state noninverting buffer shown in Fig. 5.23 has one data

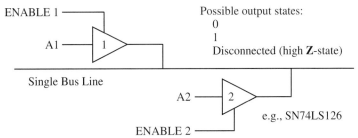

Figure 5.23 Noninverting three-state buffer (special version of totem-pole construction).

input and one enable/disable input called ENABLE, or simply E. The two inputs allow the device to take on one of three distinct output states: 1, 0, or disabled, analogous to the three possibilities for an electric light: ON, OFF, and unplugged. The third state represents a very high impedance (OFF) and is called the *high-impedance* or *high-Z* state (also referred to as open, disconnected, disabled, or floating). This state is essentially equivalent to disconnecting the TTL totem-pole output circuit from the output pin on the IC. Three-state output circuits are useful in digital systems when data from different devices are transferred over a common line called a *bus*. All microprocessors and their peripherals, including memory, use three-state outputs, which allow the microprocessor to control the timing of each device to access the bus. Examples of logic devices with three-state logic are buffers, inverters, flip-flops, memories, and almost all microprocessors and their interface chips.

5.6.4 TTL Characteristics

All TTL series have almost the same electrical characteristics and operate from a $+5V \pm 0.5$ V power supply. The primary variables of interest for TTL devices are indicated in Fig. 5.23. They are:

V_{IH} (V_{IL}) is the voltage when the input is HIGH (LOW).

V_{OH} (V_{OL}) is the voltage when the output is HIGH (LOW).

I_{IH} (I_{IL}) is the current when the input is HIGH (LOW).

I_{OH} (I_{OL}) is the current when the output is HIGH (LOW).

A summary of typical TTL characteristics is given in Table 5.4 for the totem-pole and open-collector output NAND and/or inverter gates, although V_{OH} has no meaning for open collectors. The logic characteristics for three-state output gates are given in Table 5.5. The minus sign in Tables 5.4 and 5.5 is the convention adopted for current sourced by the gate, regardless of input or output. Practically, a gate can source currents higher than the I_{OH} (max) values given in Table 5.4 but at a lower gate voltage, an important factor when driving transistors. Meanwhile, a gate can accept sink currents higher than I_{OL} (max) but at higher gate voltages, a factor to consider when driving a lamp.

TABLE 5.4 Logic Characteristics for TTL Totem-Pole NAND and Inverter Gates

	Low Power	High Speed	Schottky	Schottky (3-State)	Low-power Schottky	Low-Power Schottky (3-State)	Units
V_{IH} (min)	2	2	2	2	2	2	V
V_{IL} (max)	0.7	0.8	0.8	0.8	0.8	0.8	V
V_{OH} (min)	2.4	2.4	2.7	2.4	2.7	2.4	V
V_{OL} (max)	0.4	0.4	0.5	0.5	0.5	0.5	V
I_{IH} (max)	10	50	20	20	20	20	μA
I_{IL} (max)	−0.18	−2	−2	−2	−0.4	−0.4	mA
I_{OH} (max)	−0.2	−0.5	−1	6.5	−0.4	−2.6	mA
I_{OL} (max)	3.6	20	20	20	8	24	mA
I_{OZ} (off state)	—	—	—	−50	—	−20	μA

TABLE 5.5 Logic Characteristics of the 400B and 74C CMOS

	5 V	10 V	15 V	Units
V_{IH} (min)	3.5	7.0	11.0	V
V_{IL} (max)	1.5	3.0	4.0	V
V_{OH} (min at $I_{OH} = 0$)	4.95	9.95	14.95	V
V_{OL} (max)	0.5	0.5	0.5	V
I_{OH} (min)	$-0.44V_{OH}$ = 4.6 V	$-1.1V_{OH}$ = 9.5 V	$-0.3V_{OH}$ = 13.5 V	mA
I_{OH} (typical)	-0.88	-2.25	-8.8	mA
I_{OL} (min)	$0.44V_{OL}$= 0.4 V	$1.1V_{OL}$ = 0.5 V	$3.0V_{OL}$ = 1.5 V	mA
I_{OL} (typical)	0.88	2.5	8.8	mA
Propagation delay	160	65	50	ns

5.7 THE CMOS FAMILY

The CMOS logic family is based on the enhancement type of MOSFETs. The gate region in a MOSFET is blocked from the conducting channel by an insulating layer of silicon dioxide, making the input impedance very high. Thus, gate input currents are practically nonexistent. A CMOS NAND gate and its equivalent circuit are shown in Figs. 5.24a and 5.24b. In the OFF state, the transistor offers a very high-impedance path for current. When the transistor is switched ON, it conducts as a simple resistor according to Ohm's law. Unlike TTL gates, the output HIGH will be the same as the supply voltage and the output LOW will be the same as ground for an open-circuit load. This can be easily verified by applying the voltage divider rule in Fig. 5.24b.

CMOS families include the following: The 400B and 74C are equivalent and compatible, but the 74C is made to be pin compatible with TTL. The 74HC high-speed CMOS has the same speed as the 74LS; the 74AC advanced CMOS has the same speed as the 74F or 74AS and the 74ACT high-speed CMOS with TTL threshold. Supply voltage is not bound to the 5 V of TTL. V_{DD} can range from +3 to +18 V for the 400B and 74C families. The characteristics of the 400B and 74C CMOS are given in Table 5.5. These variables form the basis for logic connections.

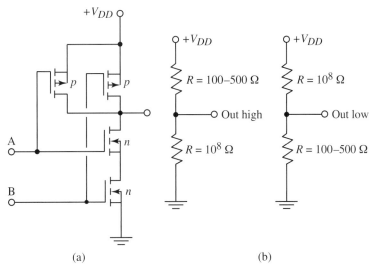

(a)   (b)

Figure 5.24 CMOS NAND gate (a) and its equivalent output circuit (b).

5.8 INTERFACING CMOS AND TTL

CMOS and bipolar TTL ICs are the two process technologies most widely used to develop signal-conditioning electronics for sensors. Because of the large input impedance of a gate, attention is usually focused on the output when interfacing TTL and CMOS devices. A comparative summary of the main characteristics of the LS Schottky and the 5-V CMOS is given in Table 5.6.

Power Consumption CMOS operates on less power than TTL. CMOS consumes power only when switching between logic levels. In low-switching applications, less than 1 μW is consumed per gate, compared to 1 mW per gate for low-power TTL. Lower power consumption makes CMOS devices ideal for battery-driven applications.

Speed CMOS is five to ten times slower than TTL. CMOS transistors have large resistances, and driving only a little capacitance results in large *RC* delay times. Just a few picofarads from other components or wiring in proximity to outputs can make the system unreliable.

Supply Voltage CMOS offers a wider supply voltage range than TTL. The voltage range for CMOS is from 3 V to 18 V, whereas TTL is tightly regulated at 5 V. Analog components (e.g., op-amps) need a 12- to 15-V power source. Using TTL logic with analog devices would require another +5 V of power supply. That can be eliminated if CMOS devices are used instead.

Noise Immunity CMOS has a higher tolerance for noise. For logic LOW, CMOS has a range of 0–1.5 V, with even a better margin at a higher supply voltage, compared with 0–0.4 V for TTL. A TTL HIGH must fall in the range of 2.4–5 V, while a 15-V CMOS will function over the range of 11–15 V. CMOS also has a low susceptibility to EMI.

Electrostatic Charge CMOS has infinite input impedance. This allows the buildup of an electrostatic charge to where a 100-V potential can damage the thin oxide layer insulating the input from the substrate. (A person walking on a carpet can easily generate several thousand volts.) The internal diodes connected between input and ground pins do not always provide the necessary protection. It is a standard practice to store all CMOS ICs on conducting foam or foil. It is also important to ensure that all pins of a CMOS device are connected to supply or ground and not left loose.

Drive Capability While CMOS has negligible input current demand, it cannot supply very much source or sink current as compared to TTL. CMOS has large fan-out when driving CMOS, but it can only drive one or two other TTLs.

TABLE 5.6 Comparison of TTL and CMOS

	LS TTL	CMOS (5 V)
Supply voltage	+5 V	3–18 V
Speed	9.5 ns	160 ns (5 V)
Input impedance	High	Infinite
Input current	0.4 mA	0
Power consumption	2 mW	0 unless in switching
Sink current I_{OL}	8 mA (max)	0.88 mA (typical)
Source current I_{OH}	−0.4 mA (max)	−0.88 mA (typical)
I_{IL}	0.4 mA (max)	Draws no current
I_{IH}	20 μA (max)	Draws no current
Logic LOW	0–0.4 V	0–1.5 V (<0.2 of power supply)
Logic HIGH	2.4–5 V	≥4 V (0.8 of power supply)

Availability of Circuits   TTL provides the designer with a wider variety of basic gate circuits than CMOS. Most CMOS gates have a TTL equivalent, although not necessarily pin compatible. Certain TTL circuits, such as the open-collector logic, have no CMOS equivalent. On the other hand, the pure resistive nature of the switched-ON CMOS lends itself to functions not possible with TTL.

Applicability for Sensor Technologies   CMOS is well suited for high-volume commercial production of sensors because it lends itself to automated electronic programming of sensor and batch calibration. On the other hand, manual thick-film or thin-film laser trimming is required to calibrate TTL-fabricated sensors. Batch calibration using laser trimming is virtually impossible and is unlikely to be accurate.

5.8.1   Interfacing TTL to CMOS

Guidelines for interfacing between TTL and CMOS families are depicted in Fig. 5.25. Since CMOS inputs do not draw any current, only the voltage must be matched when TTL drives CMOS. TTL LOW output ranges from 0 to 0.4 V and is not a problem because the low input of a 5-V CMOS ranges from 0 to 1.5 V. TTL HIGH, however, can be as low as 2.4 V, which is too low for a CMOS HIGH input. This mismatch is resolved by adding a pull-up resistor to the TTL output that drives CMOS to raise the input voltage to 5 V. The minimum pull-up resistor (2 kΩ is typical) is sized so that the TTL LOW sink current is within I_{OL}(max). To interface TTL to a higher-voltage CMOS, a buffer/driver IC such as the TTL 7406 or 7407 is used. The

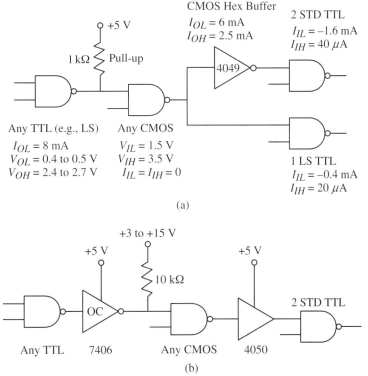

Figure 5.25 Interfacing CMOS to TTL: 5-V CMOS (a) and 3- to 15-V CMOS (b).

buffer/driver provides input protection against the higher voltage. Refer to pages 574–575 in Horowitz and Hill (1989) for more details on interfacing various logic families.

5.8.2 Interfacing CMOS to TTL

When a 5-V CMOS drives TTL, CMOS LOW is of concern. As noted in Table 5.6, a 5-V CMOS can sink 0.44 mA or source 0.88 mA. Therefore, a CMOS driving TTL LOW is capable of driving two low-power TTL loads or one low-power Schottky TTL load. If driving larger TTL loads is needed, a CMOS 4049 hex buffer can be used, because it can sink 6 mA or source 2.5 mA.

5.9 FLIP-FLOPS

The basic types of flip-flops that are commonly used in digital circuits are the set/reset (or SR), trigger (or T), D, and J-K flip-flops. All flip-flops are capable of set and reset. Normally the *set* condition means the flip-flop is storing a 1, and the *reset* means the flip-flop is storing a 0. Most flip-flops are edge triggered, meaning that a flip-flop changes states on either the leading or the trailing edge of the input pulse. As stated earlier, the fundamental operating rule of a flip-flop is that the flip-flop outputs be complements.

5.9.1 Set-Reset (*SR*) Flip-Flop

The symbol and truth table of the *SR* flip-flop are shown in Fig. 5.26. This network contains two logic gates and is commonly referred to as *set/reset*. An input $S = 1$ sets the output to $Q = 1$, and an input R resets the output to $Q = 0$. The *SR* flip-flop can have ambiguous states,

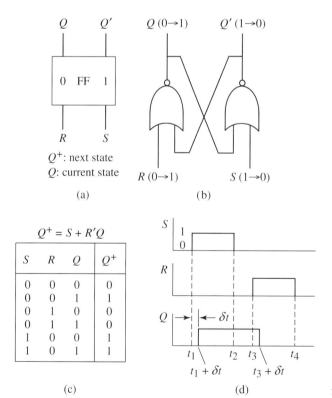

$$Q^+ = S + R'Q$$

S	R	Q	Q^+
0	0	0	0
0	0	1	1
0	1	0	0
0	1	1	0
1	0	0	1
1	0	1	1

(c)

(d)

Figure 5.26 *SR* (set/reset) flip-flop.

occurring when both outputs are high or low at the same time. Therefore, R and S cannot simultaneously be 1 because both outputs will be 1, violating the requirement that outputs be complements. The characteristic equation of the SR flip-flop is $Q^+ = S + R'Q$, with $SR = 0$. The $(+)$ symbol indicates the state of the flip-flop after the clock pulse. The duration of S and R must normally be at least as great as the propagation delay (δt) of the gate to cause a change in the state of Q. The timing diagram for the RS flip-flop is shown in Fig. 5.26d. The SR flip-flop is usually used as a *switch debouncer* (refer to Section 7.3.2 for this application).

5.9.2 Trigger (or *T*) Flip-Flop

The symbol, construction, truth table, and timing diagram of the trigger (or T) flip-flop is shown in Fig. 5.27. A T flip-flop can be constructed from an RS flip-flop and two AND gates, as shown. The T flip-flop has a single input. Applying a pulse to this input causes the output state to change. The characteristic equation is $Q^+ = T \oplus Q$, or $Q^+ = TQ' + T'Q$. The timing diagram indicates that if Q is 0, pulsing T also pulses S and sets the flip-flop. On the other hand, if Q is 1, a pulse on T pulses R and causes the state Q to change to 0. Proper operation of this flip-flop requires a finite time delay Δt between the time the S (or R) is pulsed and the time the output state of the flip-flop changes. Adding a capacitance or external logic circuitry within the RS flip-flop would serve to create the necessary time delay.

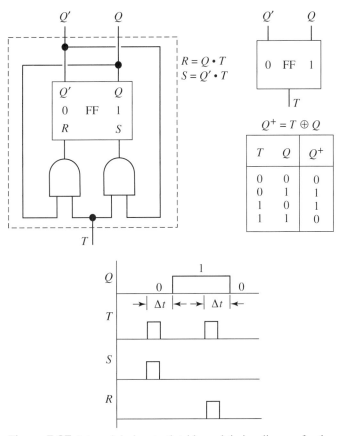

Figure 5.27 Internal design, truth table, and timing diagram for the T (trigger) flip-flop.

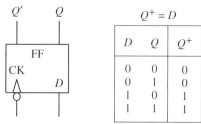

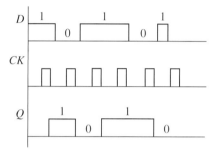

∧: indicates edge triggered
○: indicates falling edge

Figure 5.28 Clocked D flip flop forms the basis of a latch (e.g., 4013).

5.9.3 Clocked *D* Flip-Flop

The symbol and truth table of the clocked D flip-flop is shown in Fig. 5.28. It contains four logic gates and an inverter. The state of the D flip-flop after the clock pulse is equal to the input before the clock pulse. Therefore the characteristic equation is $Q^+ = D$. The arrowhead on the flip-flop symbol identifies the clock input, and the small circle (inversion symbol) indicates that the state occurs on the trailing, or negative-going, edge of the clock pulse. The D flip-flop does not have the ambiguity problem of the RS flip-flop. D flip-flops are particularly useful in computer hardware because registers, buffers, and other groups of memory functions can be readily constructed. Connecting the Q' output to the D input results in a divide-by-2 circuit whereby the output at the Q terminal has half the frequency of the signal at the CLK input (see Problem 5.10).

The D flip-flop is the basis of what is called a *latch*. The D latch is similar to the clocked D flip-flop, except that it can change states during the HIGH portion of an ENABLE signal rather than on the appropriate edge of the CLK.

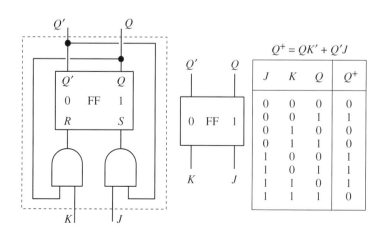

Figure 5.29 The *J-K* flip-flop.

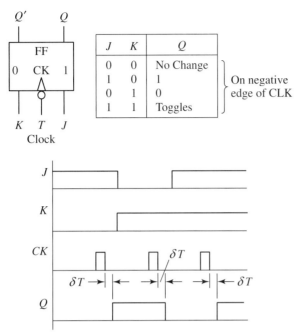

Figure 5.30 The edge-triggered (clocked) *J-K* flip-flop.

5.9.4 *J-K* Flip-Flop

The symbol, truth table, and timing diagram of the *J-K* flip-flop are shown in Fig. 5.29. The *J-K* flip-flop combines the features of the *SR* and *D* flip-flops, plus many other functions. A 1 input applied to *J* or *K* alone acts exactly like an *S* or *R* input, respectively. Unlike the *SR* flip-flop, it is permissible to simultaneously apply an input of 1 to both *J* and *K*, in which case the flip-flop changes state just like a *T* flip-flop. When *J* and *K* are pulsed at the same time, timing problems may arise if the pulses are too long or if they do not arrive at exactly the same time. To overcome such timing problems, the clocked *J-K* flip-flop shown in Fig. 5.30 is used, in which the flip-flop changes state in response to the negative edge of the clock pulse rather than in response to a change in *J* or *K*. The characteristic equation for clocked and unclocked *J-K* flip-flops is $Q^+ = JQ' + K'Q$.

One way to realize the clocked *J-K* flip-flop is with two *SR* flip-flops connected in a *master–slave* arrangement, as shown in Fig. 5.31. The state change of the master takes place

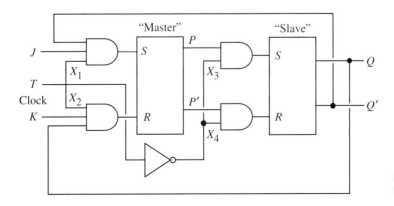

Figure 5.31 The master–slave *J-K* flip-flop.

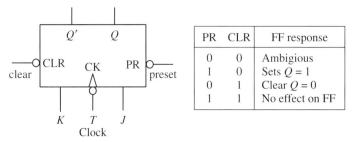

Figure 5.32 CLR and PR inputs override the clock and *J-K* inputs.

on the rising edge of the clock pulse, and the state change of the slave follows on the falling edge. To ensure proper operation, any changes in the *J* and *K* inputs must occur between clock pulses.

Clocked IC flip-flops often have one or more additional inputs that can be used to set the flip-flop to an initial state independent of the clock. The inputs are called *asynchronous inputs*. Figure 5.32 shows a clocked *J-K* flip-flop with two asynchronous inputs, clear (CLR) and preset (PR) inputs. These inputs override the clock and the *J-K* inputs. The small circles on these inputs indicate that a logic 0 is required to clear or set the flip-flop.

5.10 BUFFERS AND DRIVERS

Buffer ICs play important roles in logic circuits. (1) They provide interface compatibility between ICs of different logic families; (2) they provide temporary storage areas for peripheral devices such as registers of I/O devices, and (3) they provide the means by which several devices connected to a common bus are allowed orderly access to the bus. Two types of buffer ICs are available to perform these functions, two-state buffers and tri-state buffers, with both types available in inverting and noninverting varieties. A two-state buffer, such as the noninverting buffer shown in Fig. 5.33, provides interface compatibility between ICs of different logic families. The AM2812 buffer IC provides temporary data storage (up to 32 bytes) when data is to be sent to a device at a rate faster than the rate at which the device can process the data.

In addition to the input and output lines, a tri-state buffer contains a third line, called the ENABLE, or E line. Various arrangements of tri-state buffers are available, depending on whether the input to the E line and the output state of the device are active LOW or active HIGH, as indicated in Fig. 5.34. Active LOW, indicated by the circle at the end of the E line, means that an input of 0 applied to the ENABLE line would enable the device, whereas an input of 1 disables it. IC devices such as random access memory (RAM) and read-only memory (ROM) have their own built-in three-state buffers. The buffers on a chip are enabled or disabled by one or more pins on each chip, called *chip selects* (CS).

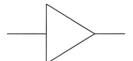

Figure 5.33 A two-state noninverting buffer.

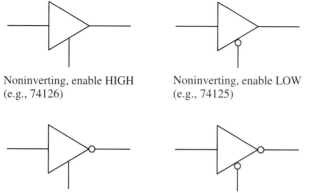

Noninverting, enable HIGH
(e.g., 74126)

Noninverting, enable LOW
(e.g., 74125)

Inverting, enable HIGH

Inverting, enable LOW

Figure 5.34 Types of three-state buffers.

5.10.1 Bus Drive and Termination

Information between interconnected logic devices and circuit boards flows in the form of signals along a bus of coaxial cables, twisted pairs, flat ribbon cables, or fiber-optic cables. When the bus is long and feeds into several logic devices (high fan-out), two important issues must be considered: bus driving and signal termination. Special logic function ICs called buffers/line drivers can provide increased current capabilities to drive bus lines. The driver boosts the logic signal to a level required for the operation of the logic devices in interconnected logic circuits and peripherals. High-speed TTL can sink up to 20 mA, whereas buffers/drivers can sink up to 64 mA (e.g., the advanced Schottky series). Beyond this level, transistors must be used. Selected buffer/driver ICs are given in Table 5.7. The 7447 IC driver decodes BCD code and drives a seven-segment display. The MC1489 interfaces TTL to transmission lines.

TABLE 5.7 Selected Buffer/Driver ICs

IC number	I_{OH}/I_{OL} (mA)	Description
74LS06	$-/40$	Hex inverting buffer/driver with OC high-voltage output. Ideal for indicator lamps and relays. Eliminates mismatch associated with interfacing different logic families. The 74LS06 has inverting inputs
74LS33	$-/24$	Quad 2-input positive NOR buffer with OC outputs
74LS37	$-1.2/24$	Quad 2-input NAND buffers. The 74LS38 has OC outputs
74ABT240	$32/64$	Inverting octal buffer/line driver with three-state outputs
74AS241	$-15/64$	Octal noninverting buffer
74LS241	$-15/24$	Octal buffer and line driver with three-state outputs
74ABT245	$-24/24$	Noninverting buffer, octal bidirectional transceiver with three-state inputs/outputs
74LS365	$-2.6/24$	Hex bus driver with three-state outputs
74LS47		Common anode (sink current); it decodes a 4-bit (*nibble*) binary-coded decimal (BCD) input into a decimal number from 0 through 9 in the standard 7-segment display format and directly drives a 7-segment display. Uses for this circuit include counters, timers, multimeters, and other numeric displays

Long lines, and short lines carrying high-frequency signals, behave as *transmission lines* and must be properly terminated, otherwise the high-frequency parts of the signal traveling on the line will be reflected back at the receiving end, causing oscillations known as *ringing* between the driver and receiver. This is due to the inherent capacitance and inductance of the transmission line (wiring inductance is roughly 5 nH/cm), known as characteristic impedance $\mathbf{Z}_O$. For a ribbon cable, $\mathbf{Z}_O \approx 200\ \Omega$; for conductors in a typical printed circuit board and for twisted pairs made from ordinary insulated wire, $\mathbf{Z}_O \approx 100\ \Omega$. *Signal termination* is realized if the value of the termination impedance (usually the resistor) is equal to the value of the characteristic impedance. A method to terminate a TTL bus is to include at the receiving end a voltage divider from +5-V supply to ground. The parallel combination of the resistors is roughly equal to $\mathbf{Z}_O$. For example, if $\mathbf{Z}_O \approx 100$-Ω, connecting a 180-Ω resistor to +5-V side and a 390-Ω resistor to ground suffices. Another method applicable to a TTL and CMOS bus is to insert a series resistor/capacitor from the data line to ground. The resistor value should be close to $\mathbf{Z}_O$, and a capacitance value of 100 pF is generally adequate.

5.11 COUNTERS AND REGISTERS

Many devices, such as registers, counters, and clocks, are essential to many computer functions, such as timing, sequencing, and storage operations. These devices are built from a combination of flip-flops. The counting ability of a binary counter depends on the number of FFs it contains. A counter with N FFs can count up to $2^N - 1$, for a total of 2^N different states, called MOD numbers, of the counter. For example, a 4-bit counter contains 4 FFs and can count up to %1111 (binary), for a possible 16 state changes. Applying one extra pulse to a fully loaded counter will cause all stages to reset to zero. The basic binary counter uses the 8, 4, 2, 1 weighted code. It can be further divided into segments to represent any number system. In addition to counting pulses, all counters can perform frequency division. The signal at the least-significant-bit (LSB) FF changes at a rate exactly one-half that of the input pulse, and the state of the most-significant-bit (MSB) FF divides the input frequency by the MOD number of the counter. A D FF such as the 4013 or a *J-K* FF such as the 112 can be used to divide the input signal by 2. Counters that can count in either direction are called UP and DOWN counters. Examples of counter ICs are the TTL 74LS90 and the CMOS CD4017 decade counters, the 74LS193 binary Up/Down counter, and the CD4553 BCD counter.

Figure 5.35 shows the pin assignment of the CD4017 decade counter. The CD4017 consists of a five-stage decade counter and an output decoder. The decade counter advances one count at the leading edge of the clock pulse when the clock enable and reset signals are low. The output decoder converts the binary code to a decimal number. The carryout signal completes one cycle for every 10 clock pulses, as shown in the timing diagram.

Most of the mathematical and I/O operations a computer performs involve registers. While the counter is a special case of an FF register, other types of FF registers, such as buffer registers used for storage and shift registers, are also available. An example of an IC register is the data-latch register. This type of register uses a D-type latch, in which each Q output latches (holds) the logic level presented on its corresponding D input when the ENABLE input is HIGH. When the ENABLE is LOW, the output cannot change even with changing corresponding input. The edge-triggered data register uses the edge-triggered D FF. Its operation is similar to a data-latch register, with the exception that the data inputs affect only the data outputs at the instant when CLK makes low-to-high (or high-to-low) transition. The tri-state registers contain tri-state buffers at their outputs to allow data bussing. This is a very important feature for devices, such as microprocessors, memory chips, and A/D and D/A converters, that

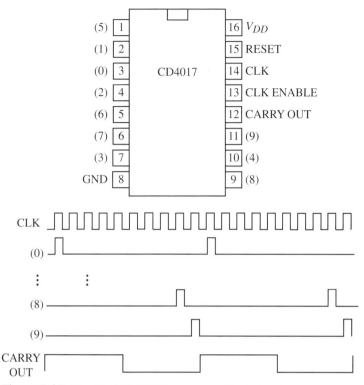

Figure 5.35 The CD4017 CMOS counter and its timing diagram.

are commonly connected to a data bus. The TTL 74173 and the CMOS 4076B and 74C173 are examples of tri-state registers. Tri-state registers eliminate *bus contention* problems from occurring. Bus contention may arise when two or more sets of outputs are simultaneously connected to a common data bus. Bus contention may damage the IC.

5.12 DECODERS AND ENCODERS

An *n*-bit binary code can have 2^n different combinations; each may represent a different piece of information. A decoder is a logic circuit that takes the *n*-bit logic code input and then, out of the 2^n possible output combinations, activates the corresponding output signal. The encoder performs the opposite operation to that of a decoder. It generates a binary output code corresponding to the input that has been activated. An encoder is commonly used in applications involving array switches. The encoder generates a binary number corresponding to the particular switch pressed. A priority encoder produces a binary code for the *highest*-numbered input that is activated if more than one input is activated.

The 2^n outputs generated by an *n*-to-2^n line decoder are defined by:

$$y_i = m_i, \qquad i = 0 \text{ to } 2^n - 1 \quad \text{(noninverted outputs)}$$
$$y_i = m_i' = M_i, \quad i = 0 \text{ to } 2^n - 1 \quad \text{(inverted outputs)} \tag{5.15}$$

where m_i and M_i are the *minterm* and *maxterm* (see glossary of terms in Section 5.16) of the *n*-input variables. Figure 5.36a shows a 3- to 8-line decoder along with its truth table. Exactly

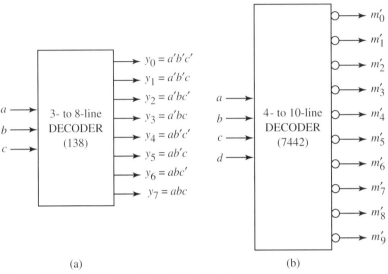

(a)   (b)

Figure 5.36 3- to 8-line and 4- to 10-line decoders.

one of the output lines will be 1 for each combination of values of the input variables. Figure 5.36b illustrates a 4- to 10-line decoder. This decoder has inverted outputs. For each combination of values of the inputs, exactly one of the output lines will be 0. When a binary-coded decimal digit is used as an input to this decoder, one of the output lines will go low to indicate which of the 10 decimal digits is present.

The 74LS47 decoder/driver decodes a 4-bit binary-coded decimal (BCD) input into a decimal number from 0 to 9 in the standard seven-segment display format and directly drives a seven-segment common-anode (sink current) display. The 74LS48 is similar to the 74LS47; it drives common-cathode (source current) displays. See Section 7.5 for details on seven-segment displays.

5.13 MULTIPLEXERS AND DEMULTIPLEXERS

A multiplexer (or MUX), also called a *data selector*, has two groups of inputs: the data inputs and the control inputs. The control inputs are used to select one of the data inputs and connect it to the output terminal. Figure 5.37 shows a 4-to-1, an 8-to-1, and a 2^n-to-1 multiplexer. In

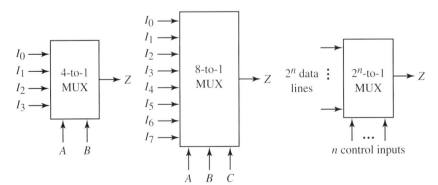

Figure 5.37 4-to-1, 8-to-1, and 2^n-to-1 multiplexers.

general, a multiplexer with n control inputs can be used to select any one of 2^n data inputs. The general equation for the output of a MUX with n control inputs and 2^n outputs is

$$Z = \sum_{k=0}^{2^n - 1} m_k I_k \tag{5.16}$$

where m_k is a minterm of the n control variables and I_k is a data input. For a 4-to-1 MUX, the output using Eq. (5.16) will be

$$Z = A'B'I_0 + A'BI_1 + AB'I_2 + ABI_3 \tag{5.17}$$

Multiplexers are frequently used in digital system design to select the data to be processed or stored, e.g., in A/D converters. They can also be used to realize combinational logic functions. A 4-to-1, an 8-to-1, and a 16-to-1 multiplexer can be used to realize 3-, 4-, and 5-variable functions, respectively, without added logic gates. Multiplexers are commonly available in integrated circuit packages in the following configurations: quadruple 2-to-1, dual 4-to-1, 8-to-1, and 16-to-1.

A demultiplexer has a single input port and many output ports. It transmits data from the input port to one of the output ports as determined by the SELECT inputs.

5.14 THE 555 TIMER

The 555 timer is the most popular IC timer used for applications requiring fixed timing or clock signals. Many versions of the 555 timer are manufactured by various vendors: NE555 by Signetics and Fairchild, LM555 by National Semiconductor, the MC1555 by Motorola, etc. The timer usually comes in 8-pin T packages or in 8-pin DIP packages; both have the same pin assignments. The 556 IC timer is a 14-pin DIP that combines two 555 timers on a single IC. A typical 555 timer contains 23 transistors, 2 diodes, and 16 internal resistors. These components are arranged in five basic components, as shown in Fig. 5.38: two comparators,

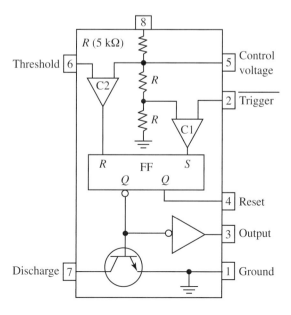

Figure 5.38 The internal structure of the 555 timer.

C1 and C2, one resistor ladder network, one flip-flop (FF), a discharge transistor Q_1, and an output driver.

Typical 555 characteristics (e.g., NE555) include a supply voltage V_{CC} with any value between 5.5 and 16 V. For $V_{CC} = +5$ V ($+15$ V), supply current is 3–6 mA (10–15 mA). The timer is capable of sourcing or sinking 200 mA, enough to drive small motors or operate even large relays and other low-power devices. Power dissipation in the chip is 600 mW. The timer also generates a large (≈ 150 mA) supply-current glitch during each output transition, requiring the use of a large bypass capacitor near the IC.

The internal 5k-resistor network divides the supply voltage applied at pin 8 so that $(1/3)\,V_{CC}$ is applied to the noninverting input of comparator C1 and $(2/3)\,V_{CC}$ to the inverting input of comparator C2. A voltage applied at pin 5 overrides the ladder voltage. However, pin 5 is not used and is usually tied to ground via an 0.01-μF capacitor to improve noise immunity.

5.14.1 Operating Modes

With the appropriate addition of resistors and capacitors to control timing, the timer can be configured to operate in one of two modes: the *monostable mode* and the *astable* mode.

Monostable Mode

Figure 5.39a shows the monostable connection of the 555 timer. A negative pulse on the trigger input (pin 2) generates a one-shot pulse at the timer output, pin 3. The operating sequence is as follows. When the voltage at the trigger input is less than $(1/3)\,V_{CC}$, the output of comparator C1 is LOW, which clears the FF. A flip-flop LOW produces a HIGH driver output and turns OFF the discharge transistor. The external capacitor C_1 will begin charging toward 5 volts through resistor R_1. When the voltage at threshold input (pin 6) has reached the control voltage, $(2/3)\,V_{CC}$, the threshold input is triggered and comparator C2 sets the FF. A HIGH flip-flop causes the discharge transistor Q_1 to turn on, connecting capacitor C_1 to ground. The timer output then goes LOW. The voltage at pin 6 at this quiescent state is effectively $V_{CE\,(SAT)} \approx 0.2$ V. The width of the generated pulse is given by

$$\frac{2}{3}V_{CC} = V_{CC}\left[1 - \exp\left(\frac{-T}{R_1 C_1}\right)\right] \qquad \text{or} \qquad T = 1.1 R_1 C_1 \qquad (5.18)$$

Practical limits are 100 pF $\leq C_1 \leq$ (any value up to leakage) and 1 k$\Omega \leq R_1 \leq$ 10 MΩ. Monostable-mode applications include timing, detection of missing pulse, switch debounce, touch switches, and signal conditioning.

Connecting pin 5 via an 0.1-μF capacitor C_2 to ground obviates the problem of false triggering that might exist in some applications. Additionally, if power leads are long or if a circuit seems to malfunction, an 0.1-μF capacitor is placed across pins 8 and 1.

Astable Mode

Figure 5.39b shows the 555 timer connected to operate in the astable mode. Pins 6 and 2 are tied together to provide self-triggering and cause the circuit to function as an oscillator. When power is applied, capacitor C_1 discharges and triggers the timer. This will cause the output to go HIGH, the discharge transistor Q_1 to turn off, and the capacitor to begin charging toward 10 volts through $R_1 + R_2$. When it has reached the control voltage, $(2/3)\,V_{CC}$, the threshold input is triggered and comparator C2 sets the FF. A HIGH flip-flop causes discharge transistor Q_1 to turn on, connecting capacitor C_1 to ground. This cyclic operation continues, with the

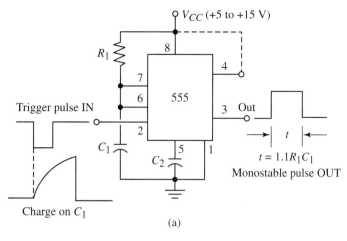

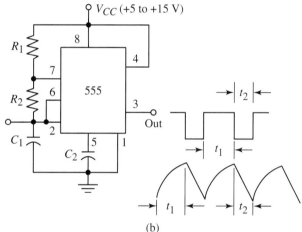

Figure 5.39 Basic operating modes of the 555 timer: monostable mode (a) and astable mode (b).

voltage across C_1 varying between $(1/3)\, V_{CC}$ and $(2/3)\, V_{CC}$. The timer cannot produce a true square wave. The period of a cycle is determined by

$$T = t_1 + t_2 = [0.693(R_1 + R_2)C_1] + [0.693R_2C_1]$$
$$T = 0.693(R_1 + 2R_2)C_1 \tag{5.19}$$

The duty cycle is defined as

$$\text{Duty cycle} = \frac{t_1}{T} = \frac{1 + R_2/R_1}{1 + 2R_2/R_1} \tag{5.20}$$

Adjusting the resistance ratio controls the duty cycle. The frequency of the output signal is

$$f = \frac{1}{T} = \frac{1.44}{(R_1 + 2R_2)C_1} \tag{5.21}$$

Applications of the timer in the astable mode include oscillator circuits, LED and lamp flashers, pulse generation, logic clocks, tone generation, and security alarms.

Example 5.5: Interface senors to the 555 timer–Explain how the 555 timer can be used in the measurement of various physical parameters.

Solution: The externally added capacitor and resistors determine the timing behavior of the 555 timer. This suggests that instead of connecting resistors and capacitors with fixed values, resistive and capacitive sensors are used instead to generate a variety of effects based on changes in physical quantities such as temperature, light intensity, and humidity. The timer circuit of Fig. 5.40 is used to generate a tone or to flash an LED at frequencies dependent on the resistance R_1 or the capacitance C_1. Resistor R_1 may be a potentiometer for position measurement, a photocell for light intensity measurement, a thermistor for temperature measurement, a strain sensor for strain measurement, or a chemoresistor for chemical measurement. Capacitance C_1 may be a humidity sensor, a level sensor, a proximity sensor, an occupancy sensor, etc.

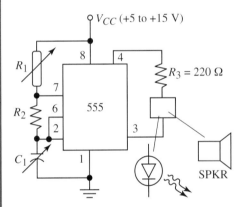

Figure 5.40 Interfacing resistive and capacitive sensors to the 555 timer IC.

5.15 PHASE-LOCKED LOOP (PLL)

The PLL is a circuit used to synchronize two sinusoidal waveforms. A typical PLL consists of three main parts: a voltage-controlled oscillator (VCO), a phase comparator, and a low-pass filter. Three components are involved in the frequency-capturing process, as follows. The VCO internally generates a reference signal at a frequency f_r. The phase comparator compares the reference frequency to the frequency of the input signal f and generates a voltage error proportional to the difference between the two frequencies, $\Delta f = f_r - f$. The pulsating error voltage is then smoothed by the LPF, amplified, and fed back to the control input of the VCO as a DC error signal, which is a measure of the input frequency. The VCO changes the frequency of the reference signal in the direction of f. This process continues until a frequency match occurs and the PLL is phase-locked to the input signal. The VCO locked frequency is a clean replica of the input frequency, which is brought out for external use. The PLL is characterized by its capture range, the frequency band it is capable of tracking. The PLL tracks changes in the locked input frequency that fall within its *lock range*. While the LPF limits the bandwidth of the PLL and reduces the speed with which it can track changes in the locked frequency, it reduces the effect of high frequencies associated with voltage spikes on the ability of the PLL to operate properly.

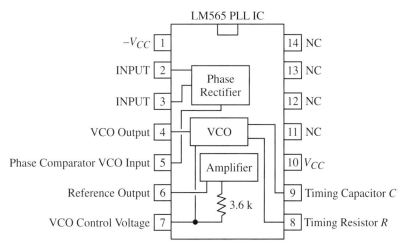

Figure 5.41 Pinout of the LM565 PLL IC.

Many ICs that implement the PLL function are available. The LM 565 (Figure 5.41) and the CD 4046 CMOS are two examples. Both ICs find use in a wide range of applications, including frequency-shift keying (FSK), tone decoding, FM discrimination, telemetry receivers, function generation, frequency multiplication, clean signal regeneration. The LM565 may be powered from a unipolar supply of at least $+12$ V but not more than $+24$ V. A bipolar supply with a negative rail of at least -6 V but not more than -12 V may also be used. An externally connected timing capacitor C and resistor R (to pins 9 and 8, respectively) set the frequency of the VCO at $f = 0.3/RC$ to a range between 0.1 Hz and 500 kHz. Refer also to the tone decoder IC, the LM567, introduced in Section 11.19. The CMOS 4046 is suited for battery-powered applications.

5.16 GLOSSARY OF LOGIC TERMS

This section summarizes the definitions of logic terms used in the design of logic circuits.

Complementation The complement of an expression is formed by replacing each variable with its complement: 0 with 1, 1 with 0, AND with OR, and OR with AND. The complement of any Boolean expression can easily be found by successively applying DeMorgan's theorems.

DeMorgan's Theorems DeMorgan's theorems are used to distribute the NOTs over individual variables or terms, allowing for the ones that appear in the output column of a circuit truth table to be separated. These theorems are:

$$(A + B)' = A'B'; \text{ the complement of the sum is the product of the complements}$$
$$(AB)' = A' + B'; \text{ the complement of the product is the sum of the complements}$$

Literal Each appearance of a variable or its complement in an expression is called a literal. When an expression is realized using logic gates, each literal in the expression corresponds to a gate input. The following expression, which has three variables, has 10 literals

$$ab'c + a'b + a'bc' + b'c'$$

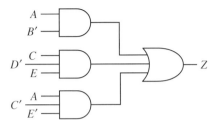

Figure 5.42 Realizing a sum-of-products expression.

Sum of Products An expression is said to be in sum-of-products form when all products are the products of single variables only. It is the end result when an expression is fully multiplied. For example:

$$AB' + CD'E + AC'E' = Z$$

A sum-of-products expression can always be realized directly by one or more AND gates feeding a single OR gate at network output, as shown in Fig. 5.42. Inverters required to generate the complemented variables have been omitted.

Product of Sums An expression is in product-of-sums form when all sums are the sums of single variables. For example:

$$(A + B')(C + D' + E)(A + C' + E')$$

A product-of-sums expression can always be realized directly by one or more OR gates feeding a single AND gate at network output, as shown in Fig. 5.43. Inverters required to generate the complemented variables have been omitted.

Duality Given a Boolean expression F, the dual F^D is formed by replacing AND with OR, OR with AND, 0 with 1, and 1 with 0 while variables and complements are left unchanged. This rule for forming the dual can be summarized as follows:

$$[f(X_1, X_2, \ldots, X_n, 0, 1, +, \cdot)]^D = f(X_1, X_2, \ldots, X_n, 1, 0, \cdot, +)$$

If a given combinational network realizes a function F when input and output variables are defined according to positive logic, the same network will realize the dual function F^D when the input and output variables are defined according to negative logic.

Minterm A minterm of n variables is a product of n literals in which each variable appears exactly once in either true or complemented form but not both. A minterm expansion is a

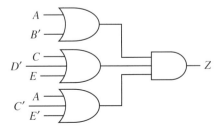

Figure 5.43 Realizing a products-of-sum expression.

TABLE 5.8 Minterms and Maxterms for Three Variables

Row Number	A B C	Minterms	Maxterms
1	0 0 0	$A'B'C' = m_0$	$A + B + C = M_0$
2	0 0 1	$A'B'C = m_1$	$A + B + C' = M_1$
3	0 1 0	$A'B\,C' = m_2$	$A + B' + C = M_2$
4	0 1 1	$A'B\,C = m_3$	$A + B' + C' = M_3$
5	1 0 0	$A\,B'C' = m_4$	$A' + B + C = M_4$
6	1 0 1	$A\,B'C = m_5$	$A' + B + C' = M_5$
7	1 1 0	$A\,B\,C' = m_6$	$A' + B' + C = M_6$
8	1 1 1	$A\,B\,C = m_7$	$A' + B' + C' = M_7$

standard sum of products. Table 5.8 lists all minterms of the three variables A, B, and C. The complement of a minterm is the corresponding maxterm. For example:

$$
\begin{aligned}
f(A, B, C) &= (A'BC) + (AB'C') + (AB'C) + (ABC') + (ABC) \quad &\text{(sum of products)}\\
&= m_3 + m_4 + m_5 + m_6 + m_7 \quad &\text{(minterm expansion)}\\
&= m_0 + m_1 + m_2 = M_0 M_1 M_2 \quad &\text{(maxterm expansion)}
\end{aligned}
$$

Maxterm A maxterm of n variables is the sum of n literals in which each variable appears exactly once in either true or complemented form but not both. A maxterm expansion is a standard product of sums. Table 5.8 lists all maxterms of the three variables A, B, and C. The complement of a maxterm is the corresponding minterm.

$$
\begin{aligned}
f(A, B, C) &= (A + B + C)(A + B + C')(A + B' + C) \quad &\text{(product of sums)}\\
&= M_0 M_1 M_2 \quad &\text{(maxterm expansion)}\\
&= M_3 M_4 M_5 M_6 M_7 = m_3' + m_4' + m_5' + m_6' + m_7'
\end{aligned}
$$

5.17 SUMMARY

This chapter covered the digital logic components and their interfacing, design, and applications. Most of the mechatronics products cannot be described as analog or digital; rather, they are a mixture of both types of components and signals. This chapter stressed the interfacing between the analog and digital parts of the system. Basic digital logic was presented as it is needed in the design of the digital networks composed of gates that materialize a logical equation. The interface of many digital networks was discussed as serving useful functions to realize desired discrete logic performance of the system.

RELATED READING

D. G. Alciatore and M. B. Histand, *Introduction to Mechatronics and Measurement Systems*, 2nd ed. New York: McGraw-Hill, 2002.

P. Horowitz and W. Hill, *The Art of Electronics.* Cambridge, UK: Cambridge University Press, 1989.

S. Kamichik, *IC Design Projects.* Indianapolis: Prompt Publications, 1998.

S. E. Lyshevski, *Nano- and Microelectromechanical Systems*. Boca Raton, FL: CRC Press, 2001.

F. Mims, *Engineer's Mini-Notebook: Digital Logic Circuits*. Radio Shack Archer Catalogue No. 62-5014, 1986.

F. Mims, *Engineer's Mini-Notebook: 555 Timer IC Circuits*. Radio Shack Archer Catalogue No. 62-5010, 1984.

G. Rizzoni, *Principles and Applications of Electrical Engineering*, 4th ed. New York: Mc-Graw Hill, 2002.

C. H. Roth Jr., *Fundamentals of Logic Design*, 3rd ed. St. Paul: West, 1997.

QUESTIONS

5.1 Define *fan-out*.

5.2 Define *propagation delay*.

5.3 Name four logic families.

5.4 What do TTL and CMOS stand for?

5.5 Explain the major differences between TTL and CMOS logic families.

5.6 What are some precautions necessary when handling CMOS ICs?

5.7 What are the LOW and HIGH voltage levels for TTL and CMOS?

5.8 What is the major concern in interfacing TTL to CMOS?

5.9 What is the major concern in interfacing CMOS to TTL?

5.10 Why is CMOS the most widely used sensor and signal-conditioning technology?

5.11 Explain the difference between the clocked *D* flip-flop and the *D*-type latch.

5.12 Explain the difference between flip-flop synchronous and asynchronous inputs.

5.13 What is meant by a *data bus*? *Bus contention*?

5.14 What is the function of a decoder, an encoder, a multiplexer, and a demultiplexer?

5.15 What is the role of a tri-state buffer?

5.16 What is the internal structure of a basic binary counter?

5.17 Explain how a binary counter performs frequency division.

5.18 Explain the two operating modes of the 555 timer.

5.19 Explain how to avoid false triggering in 555 timer circuits.

5.20 Explain the role of the major components of a PLL IC and the overall operation of the device.

PROBLEMS

5.1 Implement the functions of basic logic gates with switches and LEDs.

5.2 Simplify the following Boolean expression:

$$\overline{\overline{\overline{A \cdot B} + C}} + \overline{\overline{A} + \overline{\overline{B} \cdot C}}$$

Sketch the original and the simplified function.

5.3 Is it possible to use the TTL 74LS04 inverter gate to drive the transistor circuit of Example 3.4?

5.4 Verify that a CMOS gate can drive the transistor circuit in Problem 5.3.

5.5 A load resistance is to be powered by a 10-V source and switched ON by a 2N3904 general-purpose transistor. What is the maximum load that can be controlled?

5.6 Select a TTL gate to drive a DC lamp connected to a 30-V source that operates at 1 W of power.

5.7 Determine the number of (a) standard TTL inputs and (b) low-power Schottky TTL inputs that a TTL can drive.

5.8 Determine the number of Schottky TTL inputs that a low-power Schottky TTL will drive.

5.9 What is the equivalent logic function that results when two open-collector TTL inverter outputs are tied together?

5.10 Explain how a single *D* FF or a *J-K* FF can be connected to divide the input clock frequency by 2.

5.11 What is the frequency of oscillation of the circuit shown in Fig. 5.44? Assume each inverter delay time is T.

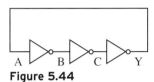

Figure 5.44

5.12 The oscillator circuit shown in Fig. 5.45 includes two 74FC04 CMOS inverters powered from a +5-V source (not shown). Determine R_2 if the desired frequency of the output signal is 40 kHz and $R_1 \approx 100$ k, and $C = 0.001$ μF.

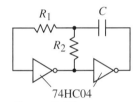

Figure 5.45

5.13 Design a 4-bit binary counter from 4 D FFs.

5.14 Design a simple circuit based on combinational logic that generates pulses at twice the frequency of an input signal.

5.15 Design a combinational logic system for a vending machine that dispenses either tea or coffee. The machine should dispense tea when the money button and tea button are high. It should dispense coffee when the money and coffee buttons are both high. If both the coffee and tea buttons are pressed after the money criterion is satisfied, the machine will dispense tea. Let C, T, and M represent the coffee button, tea button, and money criterion, respectively. Construct a "quasi-logical" statement and the truth table, use a Karnaugh map to simplify the Boolean expression, and then construct an appropriate network.

5.16 The speed (S), load weight (W), and rate of loading (R) in a conveyor system must be kept below threshold levels for safe operation. The signal-conditioning circuit generates a logic HIGH (1) when the sensor output exceeds the corresponding threshold value; otherwise the output is logic LOW (0). Two conditions are deemed unsafe and warrant the initiation of an alarm:

a. $S = 0$, $W = R = 1$

b. $S = 1$, $R = 0$

Determine the Boolean expression that describes the required alarm output.

5.17 A car alarm system is designed to operate under the following conditions: when the alarm system switch A is activated, when the ignition switch I is deactivated, and when one door or all four doors D_1–D_4 are open. The alarm features a 15-second delay T before the alarm goes off, allowing the rightful owner to enter the car and deactivate the alarm switch A. All switching timing signals assume a logic HIGH when activated and logic LOW when deactivated.

a. Select appropriate values for the circuit elements of a 555 timer operating in a monostable mode and triggered by the opening of any door, to generate the 15-s delay.

b. Derive appropriate Boolean expressions to implement the desired control functions, and construct a suitable logic circuit.

LABORATORY PROJECTS

5.1 Design and build a circuit that will indicate the number of persons who enter through a door. At your disposal you have, in addition to resistors and capacitors, an infrared emitter/detector pair, a counter, a seven-segment display, and its buffer/driver IC.

5.2 Use the 555-timer and the CD4017 decade counters to design an automatic light timer to turn several houselights on at night when occupants are away to deter would-be thieves from breaking in.

5.3 Design and build a bidirectional motor drive from the following components: a small DC motor (probably equipped with a gear reduction unit), a 555 timer, one red and one green LED, two toggle switches, a small propeller, resistors, capacitors, potentiometer, and transistors. The motor rotates in the clockwise (CW) direction and the green LED turns ON when a switch is toggled; the motor rotates in the reverse direction and the red LED lights up when the other switch is toggled. The motor speed should also be adjusted by turning a potentiometer wheel.

5.4 Design and build a home (or a museum, car, bank, power plant, etc.) security alarm system to provide safety features against burglary, fire, and flooding. It should include, at the least, the following features.

a. The system should automatically be activated at night (when it is dark) and deactivated during the day (when it is light) if desired.

b. The circuit should light up a lamp at night and turn it OFF during the day.

c. When active, the system should turn ON an alarm (e.g., a buzzer) and actuate a solenoid or a motor that activates a sprinkler system and opens an emergency exit when a fire is detected.

d. When active, the system should activate a solenoid or a DC motor that would shut the door to, say, the sleeping quarters in a home, if an intruder is detected through a window or a door.

e. If water is detected on the floor, the system should provide an alert signal.

f. Provide a panel of LEDs that would give indication to where the intrusion, fire, etc. is taking place.

You may use a thermistor, a smoke detector, or a phototransistor to detect fire, a photocell as a light detector, and an IR emitter-detector pair as a break-in sensor, etc. Include other features from your imagination. Use your architectural skills to build a nice model of the house from recyclable items.

5.5 Design and build a circuit to drive a small DC motor. Use the 555 timer with a variable resistor R_1 to generate a pulse-width-modulating signal to vary the speed of the motor. Experiment with the circuit, and determine the lowest and highest possible duty cycles that can be generated by the timer.

5.6 Design and build a 4-channel transmitter using the 555 timer to generate signals at the following frequencies: 1 kHz, 1.2 kHz, 1.4 kHz, and 2.1 kHz. A 567 (PLL) IC should decode the signal, and, when a desired frequency is detected, the system should generate the following effects: light a lamp, run a fan, sound a buzzer, and activate a solenoid, one effect for each frequency.

6

Microcontrollers and Programming

OBJECTIVES

Thoughtful engagement with the material presented in this chapter will enable the student to:

1 Understand the basic operation of a CPU and program execution concepts

2 Explain the difference between machine, assembly, and high-level languages

3 Understand the operation of the condition codes

4 Learn and apply assembly instructions and assembler directives

5 Write assembly language programs with proper hierarchy

6.1 COMPUTERS AND COMPUTER PROGRAMS

A *computer* is a machine that processes *numerical* data. Actions taken by the computer occur under program control. The program consists of a series of instructions that the computer *reads*, *interprets*, and then *executes* in an orderly manner. The term *fetch* describes a read action, and the term *decode* describes an interpret action. To decode an instruction means to translate it into an action that the computer must take.

Instructions that make up the computer program are binary numbers arranged in a sequential order and stored in a set of physical locations in the computer's *memory*. A number called an *address* identifies a single memory location. A computer program can start at any address in memory so long as the computer is told of the address. *Transfer-of-control instructions* can direct the computer to any out-of-sequence address. A group of binary digits (bits of 0s and 1s) that occupies a memory location is called a *word*. The computer handles each word as a single unit. Thus, the word is the fundamental unit of information used in the computer. A word may be (1) a piece of data to be handled, (2) an instruction that tells the computer which operation to perform, (3) an ASCII character representing a letter or a symbol, or (4) an address that tells the computer where data is located. The length, in bits, of the word that can be handled by a computer is called the *word length*.

A computer in general has an *input* medium to accept data; a *memory* where data is stored; a *control* section to decode instructions; a *calculating* section to process data; *decision*-making circuitry to enable alternate courses of action based on data; an *output* medium to provide results to the outside world; and a system *bus*, the set of conductors to transmit information among various facilities. The term *I/O* refers to input or output operations. The *hardware* refers to physical parts of a computer, and the *software* refers to the supporting programs. The term *computer architecture* is used to describe a computer's hardware and the way various components are integrated.

6.1.1 Microprocessor or Microcontroller

Real-time control action and information processing are essential features of a mechatronics system, and they are typically performed by an *embedded* (hidden) microprocessor. Components of an embedded application and the chapters in which they are discussed are shown in Fig. 6.1. A microcontroller unit (MCU) provides a higher form of flexibility and embedded intelligence than a microprocessor used alone. The MCU is a highly integrated, programmable single-chip IC that includes the microprocessor, memory, I/O interface circuits, clock, analog-to-digital

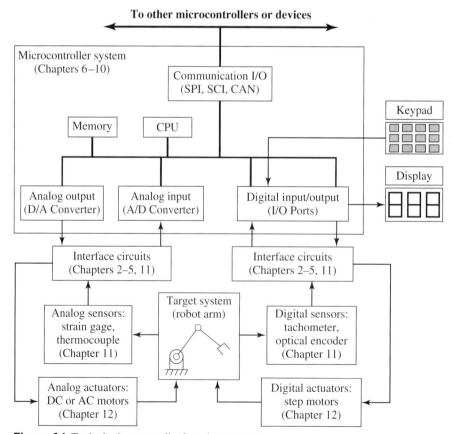

Figure 6.1 Typical microcontroller-based target system.

TABLE 6.1 Selected Freescale Semiconductor's 16-Bit Flash-Based MCUs (www.freescale.com)

Product Number	RAM/ EEPROM	I/O Pins	ATD (10 Bits)	Serial Interface	Timer Channels	PWM (Channels/Bits)	Bus Frequency (MHz)
MC9S12C32	2048/32000	60	8	CAN, SPI, SCI	8	6/(8,10)	16, 25
MC9S12C64	4096/65536	60	8	SPI/SCI	8	6/8	25
MC9S12C96	4096/98304	60	8	SPI/SCI	8	6/8	25
MC9S12C128	4096/131072	60	8	SPI/SCI	8	6/8	25
MC9S12H128	6000/131072	99	8	CAN 2.0 A/B, IIC, SCI, SPI	8	3,6/8,16	16
MC9S12DJ128	8192/131072	59, 91	8	CAN 2.0 A/B, IIC, J1850, SCI, SPI	7, 8	4,8/8,16	25
MC9S12D32	4096/32000	91	8	CAN 2.0 A/B, IIC, SCI, SPI	8	4,8/8,16	25

RAM: random access memory; EEPROM: electrically erasable programmable read-only memory; ATD: analog-to-digital converter; PWM: pulse-width modulation; CAN: controller area network; SCI: serial communication interface (UART); SPI: serial peripheral interface; IIC: inter-integrated circuit bus.

converter, programmable timer, data and address busses, and many other features. A large number of MCU families is available with comparable features, offering a wide range of options to suit application cost and performance requirements. MCU families are classified according to the word size, that is, the number of data bits, the core CPU can handle at one time. Common word lengths are 8 and 16 bits. Table 6.1 shows a sample of 16-bit MC9S12C MCUs of the HCS12 family from Freescale Semiconductor (www.freescale.com); Table 6.2 gives a sample of the 8-bit PIC family from Microchip Technology (www.microchip.com).

To provide an integrated treatment of embedded systems throughout, this text focuses on the 52-pin version of the MC9S12C32 MCU. All members of the 9S12C family are software compatible and use the same 16-bit *central processor unit* core (CPU12), but they differ in the I/O resources and the amount of memory available onboard to suit a wide range of an application's cost and performance requirements.

TABLE 6.2 Selected Microchip's 8-Bit PIC MCUs (www.microchip.com)

Product Number	Words/ Length	SRAM/ EEPROM	I/O Pins	ATD Channels (10 Bits)	Serial Interface	Timers (8/16 Bits)	CCP (10 Bits)	Max Speed (MHz)
PIC12F675	1024/14	64/128	6	4	—	1/1	—	20
PIC16F684	2048/14	128/256	12	8	—	2/1	—	20
PIC16F874	4096/14	192/128	33	8	USART/ IIC/SSP	2/1	2	20
PIC16C72	2048/14	128/0	22	5	IIC/SSP	2/1	1	20
PIC17C44	8192/16	454/0	33	0	USART	2/2	2	33
PIC18F452	32768/16	1536/0	34	8	2 USART/ IIC/SSP	2/3	2	40
PIC18F6520	32768/16	2048/1024	52	12	2 USART/ IIC/SSP	3/2	5	25

PSP: parallel slave port; SSP: synchronous serial port; USART: universal synchronous/asynchronous receiver/transmitter; CCP: capture/compare/PWM.

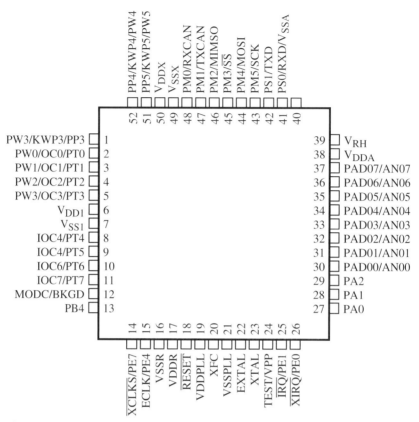

Figure 6.2 The 52-pin LQFP version of the Motorola MC9S12C32 microcontroller.

The 9S12C is a flash-based MCU that combines small size, high speed, low power, and high noise immunity for general-purpose and automotive applications. Its fully static design allows operations at frequencies down to DC, allowing for further reduction in power consumption. In addition to the onboard resources, the MCU provides a mechanism that allows the CPU to use external resources not available on the chip. The 9S12C is available in three versions: 80-pin quad flat pack (QFP), 52-pin LQFP, and 48-pin LQFP. Figure 6.2 shows the 52-pin LQFP version.

6.2   OVERVIEW OF THE 9S12C MCUs

Figure 6.3 is a *block diagram* of the 80-pin version of the 9S12C MCU showing the major onboard facilities and how they relate to the pins on the unit. The pin designations shown in bold are not available on the 52-pin chip. The MCU communicates with the *outside world* through the pins by which actuators, sensors, displays, keyboards, and peripherals can be controlled. The MC9S12C MCU possesses the following modules from which it derives its usefulness.

- The central processing unit (CPU12) controls the operation of the MCU with an extensive instruction set and an *arithmetic logic unit* (ALU) to perform arithmetic calculations and decision-making functions on data as wide as 20 bits.

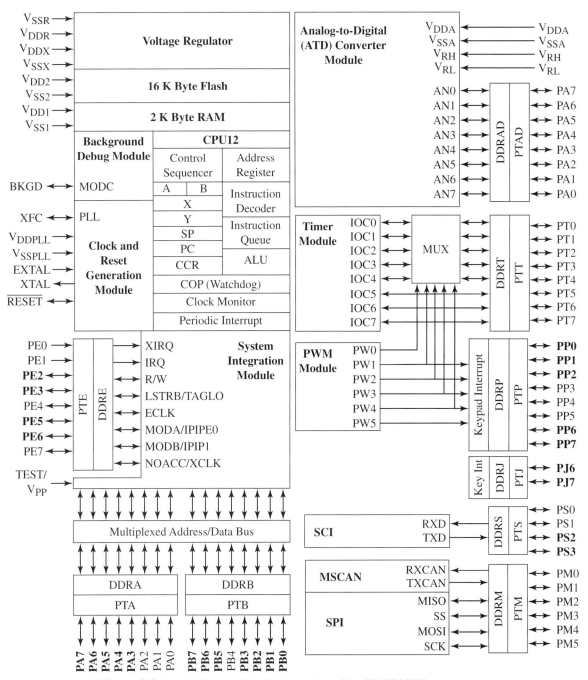

Figure 6.3 Block diagram of the 80-pin version of the 9S12C MCU.

- Flash-based electrically erasable programmable read-only memory (EEPROM) and random access memory (RAM) store application programs and associated information.
- Eight ports can be used for general-purpose *input/output* (I/O) or can be configured to control and access available subsystems on the chip. The ports include interface adapters that

connect the MCU with the outside world for control of various parallel and serial input/output devices. The maximum drive current of an output pin is 25 mA when the output is either high or low.

- A timer module (TIM) can be configured to function as input capture (IC) for frequency and period measurements, output compare (OC) to generate fixed- or variable-pulse waveforms to drive actuators, and pulse accumulator (PA) to count input pulses.
- A pulse-width modulator subsystem generates PWM signals with programmable period and duty cycle at a wide range of frequencies.
- An analog-to-digital (ATD) converter module provides a 10-bit ATD conversion for 8 channels of analog input signals from sensors.
- A serial interface module manages serial data transfer through one asynchronous serial communication interface (SCI), one synchronous serial peripheral interface (SPI), and a controller area network (CAN).
- A phase-locked loop (PLL) module runs the MCU from a time base that is different from that provided by the incoming oscillator clock.
- An interrupt controller unit manages many independent interrupt sources, from internal peripherals, such as the timer, ATD, and PWM, and from I/O devices, such as printers and general-purpose interrupt inputs.
- A bus system carries binary information between the CPU, memory, I/O devices, and registers.
- A 512-byte register stores specialized information and configures the operation of chip modules.
- A clock and reset generator (CRG) module generates a 16-MHz or 25-MHz clock signal that acts as a master timer for all modules.
- Power supply connections accept voltages from 2.97 to 5.5 V, and an internal voltage regulation subsystem produces 2.5 V.
- Single-wire background debug mode (BDM) supports program development.

The impetus here is to harness students' understanding of how to develop application programs to use the various modules on the the MCU rather than how they are designed or manufactured.

6.2.1 Central Processing Unit (CPU12)

A block diagram of the 16-bit CPU12 is shown in Fig. 6.4. The CPU manages the operation of the 9S12C using seven programming registers to store information temporarily: two 8-bit *accumulators* A, B that could be combined as the 16-bit accumulator D, two 16-bit *index registers* X and Y, a 16-bit *stack pointer* SP, a 16-bit *program counter* PC, and a 8-bit *condition code register* CCR. The built-in *arithmetic logic unit* (ALU) receives input data from two sources—One is the CPU register, and the other is either memory or other data register—and performs arithmetic or logic operations on the data that could be as wide as 20 bits. Depending on the internally generated control signals, the ALU can then add the data, subtract one from the other, multiply one by the other, divide one by the other, perform some logic operations, such as logical AND, OR, XOR, and complementation, and arithmetic and logic shift left or right in addition to computationally involved operations such as interpolation of tabular data entries and fuzzy logic operations. The CPU12 also contains the necessary

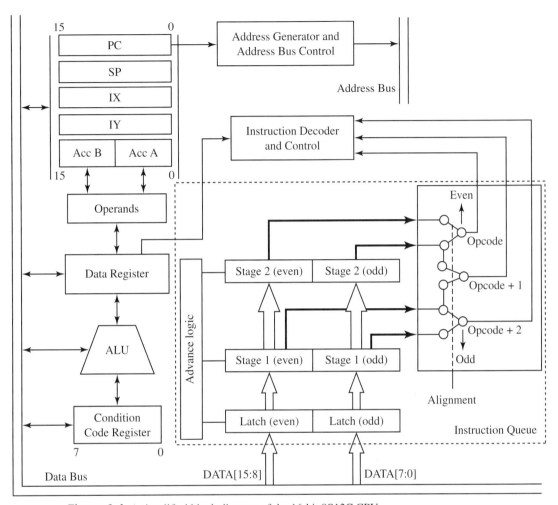

Figure 6.4 A simplified block diagram of the 16-bit 9S12C CPU.

circuits to execute the application program stored in memory and to control and provide timing reference for memory and all subsystems. The CPU12 manages the transfer of data between the MCU and external devices through the pins of eight input/output (I/O) ports: ports A, B, E, M, P, S, T, and AD. The 9S12C MCUs contain 31, 35, and 60 I/O pins on the 48-pin, 52-pin, and 80-pin chips, respectively. The instantaneous maximum drive limit on each I/O pin is -25-mA current sourcing and $+25$-mA current sinking. Note that all V_{SS} pins on the chip must be tied. Bypass capacitors with high-frequency characteristics should also be placed as close as possible to the MCU to protect the ports from voltage and current spikes that may result, depending on load demand.

6.2.2 System Bus

Binary information flows between the CPU and the various modules, memory, I/O, and registers over a group of parallel conductors known as the system bus that carry three types of logic signals: data signals, address signals, and control signals. The CPU12 has a full internal

16-bit *data bus* (DATA [15:0]) that carry 16-bit-(2 bytes)-wide data signals at one time, although a narrow 8-bit option may be selected to interface with 8-bit external memory devices. The bracketed numbers refer to the bit locations in the associated mnemonic. The CPU12 supports several data types, including single-bit and 8-bit data, 16-bit signed and unsigned integers, and 16-bit unsigned fractions. *Data signals* represent instruction opcodes and values of different variables and can be either input or output, depending on whether the CPU is reading or writing data. Data going to or coming from the data bus are temporarily stored in the data register while the CPU completes the current operation.

The *address signals* travel on a 16-bit-wide *address bus* (ADDR [15:0]). The address bus is an output bus from the CPU and an input bus to other devices. It enables the CPU12 to access $2^{16} = 65,536$ (64 Kbytes) memory locations. The 64K address space is equivalent to 256 ($FF) pages, each holding 256 address locations. It includes the memory onboard the chip and the external memory used for data storage, internal registers, I/O devices, and the registers associated with the I/O ports. Each segment occupies a specific block of the memory space. The manner in which memory blocks are arranged is called the *memory map*. The application program may occupy any memory block within the address space. Each memory location has its own address, and the CPU references them as normal memory locations. The address of the memory location or I/O device involved in a current operation is temporarily held in the *address register*, which sets the condition of the address bus. A list of all registers and associated addresses and bits is provided in Appendix D.

Figure 6.5 shows how the CPU interacts with a memory address. When the CPU signals memory, data transfer between memory and the data bus occurs. Address $1C43 in Fig. 6.5 points to the memory location where the data byte $5C is stored. The 16-bit hexadecimal (hex) address consists of two bytes, the most significant byte (MSB), or *high byte* ($1C), refers to the memory *page*, and the least significant byte (LSB), or *low byte* ($43), indicates the location of data in the page.

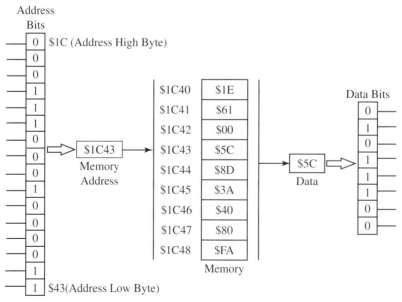

Figure 6.5 Representation of the addressing memory and data storage.

Control signals coordinate the operation of the CPU with other circuits on the MCU. These signals are not the same as the signals sent out to control external devices, such as a motor. The *control bus* transmits control signals along many lines. Each line may be either a control output from the CPU or a control input to the CPU.

6.2.3 System Clocks

All operations performed within the MCU, including execution of instructions, transfer of signals along busses, and register updates, occur in a particular order and specific timing pattern under the control of various signals generated internally by the *clock and reset generator* (CRG) module. The 9S12C can operate with either an internal or an external clock source at a nominal frequency of 16 MHz or 25 MHz, depending on the speed option. A 32- (or 50-) MHz external *quartz crystal* connects to the crystal output pin XTAL and the crystal driver pin EXTAL on the MCU. The internal oscillator generates the system clock (SYSCLK), which provides the CPU (core) clock reference, as shown in Fig. 6.6. Alternatively, a CMOS-compatible clock connects to the EXTAL pin, while the XTAL pin is left unterminated. The CRG module divides the 32- (or 50-)-MHz SYSCLK by 2 to produce the required 16- (or 25-)-MHz bus clock frequency from which two clock signals are further derived and used as basic timing references; the clock available on the ECLK pin and the PCLK provides reference to various modules on the chip. Hereafter, all timing calculations will assume a 16-MHz bus frequency for a clock period of $1/(16 \times 10^6)$ cycles/s, or 62.5 ns/cycle.

6.2.4 Operating Modes

The signals on the BKGD, MODA, and MODB external pins are latched into the MODC, MODB, and MODA bits of the MODE register at address $000B on the rising edge of the

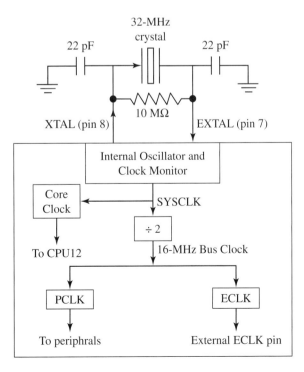

Figure 6.6 Driving the 9S12C by a 32-MHz external crystal.

TABLE 6.3 Hardware Mode Select Options

BKGD	MODA	MODB	Selected Mode	
1	0	0	Normal single chip	No external address or data busses needed
1	0	1	Normal expanded narrow	Allows access to 8-bit external memory
1	1	0	Normal peripheral	Reserved
1	1	1	Normal expanded wide	Ports A and B carry ADDR and DATA
0	0	0	Special single chip	MCU is in active BDM mode
0	0	1	Special expanded narrow	For emulation of normal expanded-narrow mode
0	1	0	Special peripheral	External master controls CPU for testing on-chip peripherals
0	1	1	Special expanded wide	For emulation of normal expanded-wide and normal single-chip modes

RESET signal and become available thereafter. The value settings of these bits configure the CPU to operate in one of the four normal modes or in one of the four special modes, namely, single-chip, expanded-narrow, expanded-wide, and peripheral mode. The eight possible operating modes and corresponding pin codes are given in Table 6.3. Each mode has an associated default memory map and external bus configuration.

With an external clock source and a reset circuit, the MCU contains all the resources it needs to operate and manage all I/O functions. In the *normal single-chip mode*, which is ideal for applications that require no external resources. In this mode, ports A, B, and E are available for general-purpose I/O. In the special single-chip mode, the background debug mode controls the operation of the CPU by the serial commands through the BKGD pin. Some registers and bits that are protected in the normal modes from inadvertent changes can be accessed in the special modes for special purposes, such as emulation and testing.

The *normal expanded-wide mode* provides for external resources to be accessed by the address, data, and control busses. In this mode, ports A and B carry the address and data; address bits ADDR[15:8] and data bits DATA [15:8] are multiplexed onto port A, whereas address bits ADDR[7:0] and data bits DATA [7:0] are multiplexed onto port B. The *special expanded-wide mode* is used for emulation of normal expanded-wide and normal single-chip modes. Port E bits PE[1:0] are dedicated to bus control output functions.

The *normal expanded-narrow mode* allows access to 8-bit external RAMs or EEPROMs. Address bits ADDR[15:8] and data bits DATA [7:0] are multiplexed onto port A, whereas port B carries address bits ADDR[7:0]. External 16-bit data is handled in two consecutive bus cycles. The *special expanded-narrow mode* can be used for emulation of normal expanded-narrow mode.

The *special peripheral mode* renders the CPU inactive, to allow an external master to control on-chip peripherals for testing purposes. The *normal peripheral mode* is dedicated to testing the MCU during manufacture.

Low-Power Operating Modes

The 9S12C MCU can be made to operate in one of three main low-power modes: the stop mode, the pseudo-stop mode, and the wait mode. The stop and the pseudo-stop mode are activated by executing the STOP instruction. In the stop mode, the oscillator and all peripheral clocks stop. The MCU comes out of this if a reset is initiated or an external interrupt is generated. In the pseudo-stop mode, the oscillator continues to run and the real-time interrupt and the watchdog timer continue to operate, but all other peripherals are turned OFF. In the pseudo-stop mode, the chip consumes more current but requires shorter wake-up times.

The MCU enters the wait mode when the WAI instruction is executed. In this mode the MCU stops executing program instructions, but all peripherals remain active unless they are individually turned off by the associated bit in a local register.

It is recommended that all unused peripherals in the normal run mode be disabled to save power consumption.

6.2.5 Memory Map

The 16-bit program counter (PC) allows the CPU12 to access a total of 64K memory locations within an address space of $0000–$FFFF. The memory space on the 9S12C chip encompasses three blocks: Internal RAM and EEPROM, external memory, and 512 Kbytes of I/O registers. The *memory map* describes the organization and specific area within the 64K address space allocated for each block. The operating mode, the application requirements, and the nature of the CPU operation affect memory organization. Figure 6.7 shows the memory map for the 9S12C32 MCU.

The user assigns the placement of RAM, EEPROM, and I/O registers within the standard 64K space by programming associated registers. RAM can be assigned to any 2-Kbyte boundary within the standard 64K space. The RAM[15:11] bits in the INITRM, *initialization of internal RAM position register*, at address $0010 defines the upper five bits of the base address for the internal RAM array. The RAMHAL bit in the INITRM register causes the RAM to be aligned to either $0000 if it is clear or to $FFFF if it is set. Flash EEPROM can be mapped to either the upper or the lower one-half of the 64K space by the value settings of the EE[15:11] bits in the INITEE, *initialization of internal EEPROM position*, *register* at address $0012. The EEON bit in the INITEE register must be set to allow read access to the EEPROM block. Its

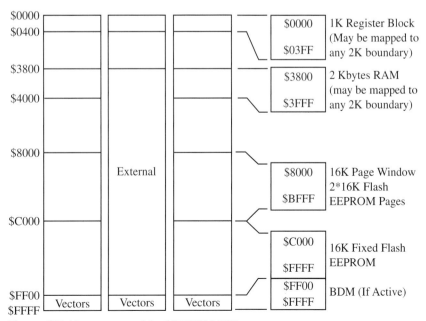

Figure 6.7 Memory map of the 9S12C32 MCU.

default value is 1 in single-chip mode. The I/O register block may be mapped to occupy the first 512 Kbytes of any 2K boundary within the 64K map space. The upper five bits of the base address of the register block are a zero followed by the values programmed into the REG[14:11] bits in the INITRG, *initialization of internal registers position register*, at address $0011. The default base address is $0000, which will be assumed hereafter.

The all-important bits in the mapping registers are protected from being mistakenly written to by writing to them during the initialization phase of program execution, whether the default values are chosen or not. A write to a mapping register should be followed immediately by a NOP (no operation) instruction because the write goes into effect between the two cycle after the write. Mapping registers can be written once in a normal mode or anytime in a special mode, but they can be read anytime. In case of a mapping conflict, priority is given to the BDM ROM (if BDM is active), register space, RAM, EEPROM, and external memory, in that order.

The 9S12C provides a mechanism that allows the CPU to address external flash EEPROM or ROM beyond the 64K standard memory space allowed by the core CPU12 architecture. The added memory is accessed through a 16K window through the $8000–$BFFF address block. The value settings of the PXI[5:0] bits of the PPAG, *page index register*, at address $0300 defines the address for the 16K flash block to be accessed through the window. For the 9S12C MCU, flash page $3F is visible through $C000–$FFFF if the ROMON bit in the MISC, *miscellaneous system control register*, at address $0013 is set, whereas flash page $3E is visible through $4000–$7FFF if the ROMHM bit in the MISC register is clear while the ROMON bit is set.

6.2.6 Programming Basics

The application program consists of a series of lines or statements that provide the CPU access to all MCU resources and allow memory locations and I/O devices to be examined or modified. Each statement includes the instruction that the CPU executes. The CPU12, like any other CPU, can recognize and execute a program called *machine language* only if its instructions are coded in binary (0s and 1s). By convention, a machine code uses hexadecimal (hex) to make it simple to read and debug. However, machine coding is impractical and time consuming, especially for large application programs. Alternatively, the application program is written in *assembly language* or in a *high-level language* such as C. In either case, the program must eventually be converted to the equivalent machine code for execution. Although programming details are discussed in Section 6.6, a brief overview of assembly programming basics is included here to facilitate understanding of CPU12 infrastructure and operation. Section 6.7 provides a brief overview on C programming for the CPU12.

Assembly Language

An *assembly language* program is developed to provide direct control of the MCU hardware using specific instructions for the target CPU. The instruction set that the CPU12 can recognize and execute is given in Appendix C. Although the level of assembly language is higher than that of machine coding, it is still regarded as a low-level programming language because it is strictly based on the low-level operation of the device. Programming with assembly language requires detailed knowledge of the instruction set and all registers and bits of the MCU. Assembly language is convenient for small programs that do not exceed a few hundred lines for intensive-hardware applications, such as I/O control and signal processing. The program that converts the assembly *source code* into its equivalent machine code for the target processor is called the *assembler*.

Description of an Assembly Language Instruction

Each assembly code statement in the source code consists of up to four fields: the label field, the instruction field, the operand field, and the comment field, as demonstrated by the following statements. The meanings of the symbols will be explained shortly.

Label	Mnemonic	Operand	Comments
temp	EQU	$7100	; Address where "temp" is stored
tmax	EQU	70	; EQU directive assigns 70 to "tmax"
	ORG	$C000	; ORG directive defines start address of user code at $C000
loop	LDAA	temp	; Load temp into A. "loop" = $C000 points to the opcode of LDAA
	ADDA	#$18	; Add offset to A
	CMPA	#tmax	; Compare the value in A to tmax, that is check temp – tmax
	BGT	hot	; Branch to "hot" if temp > tmax
	BRA	loop	; Branch always to "loop"
hot	JSR	htroff	; Jump to subroutine "htroff" to turn a heater off.

The suffix "%" indicates a binary number, the suffix "$" represents a hexadecimal number, and no suffix indicates a decimal number. Refer to Appendix G for an overview of numbering systems and arithmetic.

The *instruction field* contains either an assembler directive or an instruction. An assembler directive is not executable; it is simply a guide to the assembler on how to translate the instructions into machine code. The foregoing code fragment begin with two assembler directives. The first EQU, *equate* directive, defines the address in which the current value of the label "temp" is stored. The second EQU assigns a value of 70 to the name "tmax". The ORG directive instructs the assembler to hold the following code starting at address $C000. Because assembler directives are not executable instructions, the actual value assigned by the assembler to the label "loop" will be $C000. Assembler directives are further discussed in Section 6.5.

The function of an assembly language instruction is described by an easy-to-remember three- or four-letter code known as a *mnemonic*, or aid to programming, concocted from a description of the operation itself. For example, LDAA is the mnemonic for a *load accumulator A* operation, CMPA is a mnemonic that signifies a *compare* operation, BGT signifies a *branch if greater than* operation, and so on.

Each instruction contains bytes of information consisting of two parts, the opcode (or operation code) and the operand. The *opcode* identifies the operation to be performed and whether the CPU needs to fetch more operand bytes. The instruction opcode may be one or two bytes long, depending on the instruction. The opcodes and operand for the LDAA, ADDA, CMPA, BGT, BRA, and JSR instructions are, respectively, $86, $8B, $A1, $2E, $20, and $16.

The *operand field* contains either the immediate data or the address location of the data for the corresponding instruction. The instruction may require one or more operands or no operand at all. The operands associated with the LDAA, ADDA, CMPA, BGT, BRA, and JSR instructions are, respectively, temp, $18, #tmax, hot, loop, and htroff.

The different ways the operand arguments are written enable instructions to reference memory and to locate its operand in several ways, referred to as *addressing modes*. The same instruction may have many opcodes, one for each possible address mode. The LDAA instruction has eight possible opcodes, allowing for eight different operand argument constructs, whereas the CMPA instruction has only one opcode. The 174 different instructions in the instruction set accommodate 566 possible opcodes, 90 of which are two bytes each and the rest one byte each.

Example 6.1: Reading information using Appendix C–Referring to Appendix C, indicate for each instruction the associated opcode, operands, number of bytes, number of clock cycles required to execute it, and the addressing mode. Assume X and Y are pointing to $7400 and $7420, respectively.

Solution

INST	Opcode	Operands	Number of Bytes	Number of Cycles	Addressing Mode
INCB	5C	—	1	2	Inherent (INH)
LDAA #$5C	86	5C	2	2	Immediate (IMM)
ADDA $02	9B	$02	2	3	Direct (DIR)
SUBB $503E	F0	$50 3E	3	4	Extended (EXT)
LDAA $56, X	A6	$56 74 00	4		Indexed (IDX)
STAA 4, Y	18 A7	4 74 20	3	5	Indexed (IDX)

An instruction may begin with a label, which allows the program to refer to a line by a name instead of more cryptic numbers, to make the code easier to read. The *label field* may be a label that identifies a line in a program or a variable name. A *label* marks the address of the opcode byte in the object code generated for that line. For example, the labels "loop" and "brake" are references to the associated line numbers that will be assigned by the assembler. A *name* in the label field represents a variable to which the assembler assigns the value defined by the *assembler directive* EQU. For example, using the instruction "LDAA #tmax" instead of "LDAA #70" allows one to refer to the number 70 by the name "tmax", which is easier to remember. Anywhere "tmax" is used in the program it is assigned the value $20. The label must begin with a letter in the first column and may consist of any number of alphanumeric characters in addition to the underscore ("_") character.

The text on a line following a semicolon ";" represents the *comment field*, which will be ignored by the assembler. The user adds comments to describe the program and make it easy to read and understand. Blank lines or lines that begin with a semicolon in column 1 are also treated as comment lines.

When any data is retrieved into the CPU from memory, it is placed in the instruction register, which is a part of the instruction decoder and control circuit. The instruction decoder examines the opcode bytes of the instruction, decodes them, and informs the CPU of the required sequence of events to execute the instruction. The control circuit produces different combination of control signals for each instruction. These signals control the flow of data between registers and the bus interface, generate external control signals for the control bus, and control the operation of the ALU.

Example 6.2: Basic format of an assembly language program–An important feature of a mechatronics system is to read sensor signals and take appropriate actions. A thermistor is used as a sensor to measure the temperature "`temp`" of fluid tank. Develop an assembly code to continuously monitor the sensor output, compare it with a maximum temperature "`tmax`", and turn on an emergency light (red LED display) if `temp` ≥ `tmax`.

Solution: Figure 6.8 shows a schematic of a possible setup. The output from the temperature sensor is fed through a Schmitt trigger to a comparator, whose output is connected to pin PT[0] of port T at address $0240. The emergency light is connected to pin PB[4] of port B, configured as an output pin. The following code fragment manages the desired operation.

```
        XDEF  Entry, main              ; export symbols
        INCLUDE 'mc9s12c32.inc'        ; include derivative specific
                                         macros
MY_Variables: SECTION                  ; variable/data section
temp      EQU       $0240              ; Address of Port T where
                                         temperature is stored
tmax      EQU       20                 ; Maximum temperature value
display   EQU       $0002              ; Address of Port B for output
                                         display

MyCode:   SECTION                      ; code section
main:
Entry:    MOVB      #$FF, DDRB         ; Configure Port B for output
                                         ($0002 is DDRB address)
          MOVB      #$00, DDRT         ; Configure Port T as input
                                         ($0243 is DDRT address)
loop:     LDAA      temp               ; Get temperature
          CMPA      tmax
          BLT       alert
          BRA       loop
alert     LDAA      #$10               ; Loads %00010000 into A
          STAA      display            ; Turns LED connected to pin
                                         PB[4] on
```

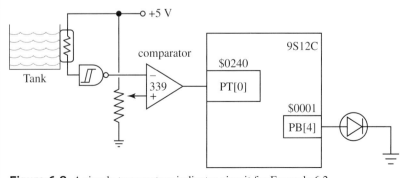

Figure 6.8 A simple temperature indicator circuit for Example 6.2.

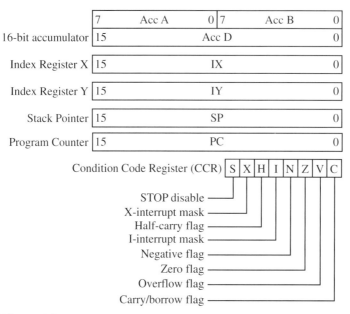

Figure 6.9 CPU registers of the 9S12C32 MCU (programmer's model).

6.2.7 CPU12 Programming Registers

Figure 6.9 shows the seven registers that are an integral part of the CPU12. These registers are available to the programmer but cannot be addressed as other memory locations.

Accumulators A, B, and D

The two 8-bit *accumulators* A and B (hereafter referred to as A and B) are general-purpose accumulators used by many instructions as a source of operands and the destination for the result of arithmetic and logic operations and data manipulations. Instructions involving two-operand bytes can identify either A or B as the source of one (8-bit) operand and the destination of the result. For example, the instruction

```
SUBA    #bias
```

subtracts the content of the 1-byte variable called "bias" from the value in A and puts the result back into A.

The accumulators also play an intermediary role in constructing superinstructions from simpler ones. For example, to read input port B at address $0001 and save its contents to the RAM location "level" for later use, one executes the following sequence:

```
LDAA  $0001   ;Move data from Port B into A
STAA  level   ;Move the contents of A into
              location "level"
```

Accumulators A and B may be used interchangeably with most operations; however, the ABX, ABY, TAP, TPA, DAA, ABA, SBA, and CBA instructions involve accumulator A only. Instructions involving two 16-bit operands may use the combination of the two 8-bit accumulators A and B as a 16-bit double accumulator D to hold one of the operands and to store the 16-bit result.

Index Registers (X and Y)

The 16-bit index registers X and Y are used with the index addressing mode for indexing operations. The combination of two index registers facilitates operations such as moves and manipulation of data stored in two separate tables. The operand argument consists of a post-byte followed by 1, 2, or 3 extension bytes. A typical instruction may add a 5-bit, a 9-bit, or a 16-bit constant offset, or the contents of an accumulator, to the contents of the X or Y register to form the effective address of the instruction's operand. The instruction

```
LDAB 4, X
```

adds (temporarily) the offset 4 to the value in X to determine the operand address. The operand is read from this address and then loaded into B.

Program Counter (PC)

The *program counter* (PC) is a 16-bit register that holds the address of the opcode of the next instruction to be executed. At the start of an instruction cycle, the CPU knows the number of operations required to execute the current instruction and automatically increments the PC to point to the memory address where the opcode of the following instruction is stored. If the application program starts at address $C000, the PC counts sequentially from $C000, and, at any given instant, the count indicates the memory address of the opcode byte of the following instruction. The PC can be used in all indexed addressing modes except auto increment/decrement operations.

Stack Pointer (SP)

The 16-bit *stack pointer* register holds the address of the *last-used* location on the stack. The *stack* is any memory block within the 64K address space reserved to hold the current state of the system during subroutine calls and interrupts. The stack may also be used for temporary storage of data and, less commonly, to pass parameters to subroutines. The stack must be initialized at the start of the program to point to the next-higher address of the memory block because the SP decrements before the first data is pushed onto the stack. If RAM occupies the $3800–$3FFF space in the memory map and the highest RAM bytes are set aside for the stack, the SP is then initialized at the beginning of the program to $4001, the next-to-highest RAM address, by executing LDS $4001. The stack grows down in memory and the SP decrements by 1 every time a data byte is pushed onto the stack. All indexed mode instructions can use the SP as a base register. The stack structure is shown in Fig. 6.10. Space may be removed or added to the stack, as illustrated by the following instructions:

```
LEAS  -3, SP  ; Allocates 3 bytes to the stack
LEAS  4, SP   ; Deallocates 4 bytes off the stack
```

If the stack pointer decrements enough to cause a push operation to interfere with other data in memory, a *stack overflow* error occurs, causing the MCU to behave erratically. The only way to recover from this situation is to RESET the MCU.

Operation of the Stack The SP is used by the CPU to store the return address temporarily during subroutine call. When a subroutine call is initiated, by a *jump-to-subroutine instruction* (JSR)

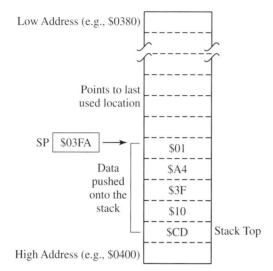

Figure 6.10 Stack structure.

or by a *branch-to-subroutine instruction* (BSR), the CPU automatically loads the SP (lower byte first) with the address of the instruction following the one that called the subroutine. Then it loads the PC with the start address of the subroutine. When the CPU executes the last statement in a subroutine, a *return from subroutine* (RTS) or a *return from call* (RTC), it pulls the return address off the stack top and returns it to the PC. This fully automatic mechanism also handles nested subroutine calls, where one subroutine is called from within another one.

The SP always points to the address of the last-used location on the stack. Push instructions are used to load data onto the stack, and pull instructions retrieve data from the stack. The CPU decrements the SP following a push operation and increments the SP after a pull operation. Pull operations are executed in reverse order to push operations, to restore the registers to their original values. This is especially important if the data involves register information. Figure 6.11 demonstrates the effect of executing the following push/pull instructions on the stack:

```
LDAA        #$5A
PSHA
LDX         #$2D51
PSHX
PULX
PULA
```

Condition Code Register (CCR)

The 8-bit CCR contains five status indicators, or flag bits (C, V, Z, N, H), two interrupt mask bits (X, I), and one STOP disable bit (S). The CCR bits occupy the bits indicated in Fig. 6.9. A status bit is *set* if its corresponding value is 1 and *clear* if its value is zero. Bit manipulation instructions can be used to set or clear any flag bit. The value of a status bit is updated according to the result of the last instruction that has effect on that bit. Not all instructions affect the condition code bits. The condition code column in Appendix C indicates the effect of each instruction on the status bits using the following symbols:"–" means no effect; "0" means the

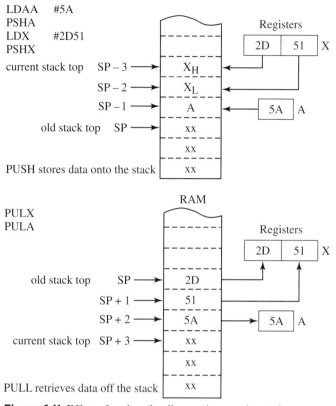

```
LDAA    #5A
PSHA
LDX     #2D51
PSHX
```

Figure 6.11 Effect of push and pull operations on the stack.

bit is cleared; "1" means the bit is set; "Δ" means the bit is affected by operation; "fl" means the bit may be cleared or remain set but is not set by operation; "⇑"means the bit may be set or remain cleared but is not cleared by operation; "?" means the bit may be changed by operation but the final state is not affected; "!" means the bit is used for a special purpose.

The C, or *carry/borrow*, flag will generally be set under the following conditions: (1) an addition operation that produces a carry, (2) a subtraction operation that produces a borrow, and (3) a shift or rotate operation that shifts out a 1 into C. The converse will clear the bit. Bit C also acts as an error flag for multiply and divide operations. The carry bit is always ignored when performing 2s-complement arithmetic. All subtract or compare instructions set C if the absolute value of the subtrahend is greater than the absolute value of the minuend.

The V, or *overflow*, flag will set when the result of a 2s-complement arithmetic exceeds the 2s-complement range, $-128 \leq R \leq 127$ for 8-bit and $-32768 \leq R \leq 32767$, signaling an incorrect answer for signed numbers. This will occur when, after an arithmetic operation, the carry from the most significant bit (MSB) differs from the carry from the second MSB. All subtract or compare instructions set V if the absolute value of the subtrahend is greater than the absolute value of the minuend.

The Z, or *zero*, flag will set if the result of the last ALU operation is zero. Compare instructions inherently perform subtraction, and the condition codes, including Z, reflect the result. INX, INY, DEX, and DEY instructions affect the Z flag only. All subtract or compare instructions clear Z if the absolute value of the subtrahend is greater than the absolute value of the minuend.

The N, or *negative*, flag will set when the last ALU operation produces a negative result. In a 2s-complement system, a result with 1 in the MSB is interpreted as negative and sets the N flag, and a 0 in the MSB indicates a positive value and clears the N flag. All subtract or compare instructions set N if the absolute value of the subtrahend is greater than the absolute value of the minuend.

The I, or *mask*, flag is the mask bit for all maskable interrupt sources, including the interrupt associated with the IRQ interrupt pin on the chip (see Fig. 6.3). If the I bit is set, all maskable interrupt sources are masked and will be ignored by the CPU. Interrupts are discussed in Section 7.8.

The H, or *half-carry*, flag will set if a carry from the lower four bits occurred during an 8-bit addition. The H bit is used only with binary-coded decimal (BCD) addition (refer to Appendix G). The H flag is used as a signal to add 6 to the binary sum to convert it the correct BCD result. A DAA instruction inserted immediately after an addition instruction involving accumulator A (ADDA, ABA, and ADCA) instructs the CPU to use the values in A, C, and H to determine the correct BCD result and stores it into A.

The S, or *stop disable*, flag causes the program to treat the STOP instruction as a NOP (no operation) if it had been set.

The X bit is the *mask* flag for the XIRQ interrupt pin on the chip. Interrupts from the XIRQ pin are disabled (masked) when the X bit is set and are enabled when the X bit is clear. The X bit is automatically set after a reset and while an XIRQ interrupt routine is being serviced, disabling XIRQ interrupts. Once enabled, XIRQ interrupts cannot be disabled under program control. Upon completing the XIRQ interrupt service routine, XIRQ interrupts are reenabled by restoring the state of the X bit from the stack.

Example 6.3: Effect of instructions on condition codes–Verify that the condition code status bits C, V, Z, N, and H are affected as indicated when the last instruction in each group is executed.

Solution: Refer to the last column in Appendix C to verify the status bits indicated here.

```
LDAA    #%01001001
SUBA    #%10000000   ; A = %11001001; N = 1, Z = 0, V = 1,
                       C = 1
LDAA    #%01111111
ADDA    #%00000001   ;A = %10000000; H = 1, N = 1, Z = 0, V=1,
                       C = 0.
LDAB    #$C6
SUBB    #%11000110   ;B = %00000000; N = 0, Z = 1, V = 0,
                       C = 0.
LDAA    #%00001000
ADDA    #%11110100   ;A = %11111100; H = 0, N = 1, Z = 0,
                       V = 0, C = 0.
LDAA    #%01111001
LDAB    #%00011000
ABA                  ;A = %10010001; H = 1, N = 1, Z = 0,
                       V = 1, C = 0.
```

6.2.8 Instruction Queue

To process information faster, the CPU12 has a built-in instruction queue mechanism, similar to the concept of pipelining implemented by other MCUs, such as the PIC family. At least three bytes of object code are cached in the instruction queue at the start of every instruction, providing the CPU with immediate access to the 8-bit object code plus two more data bytes. The instruction queue is part of the block diagram in Fig. 6.4, which shows how data flows through the instruction pipe in two 16-bit stages. A 16-bit latch is used as a buffer to hold a data word of object code that arrives on the data bus of the CPU12 before the first byte of the next instruction occupies the lower half of the leading pipe stage and the instruction pipe is ready to advance. The pipe logic and not the CPU controls proper alignment of the queue. The queue mechanism provides for many instructions to be executed in one clock cycle, with no delays for fetching. If the three byte is all that is needed to execute the instruction, then the CPU does not need to fetch for any further information. If more bytes are needed, they are always fetched as aligned 16-bit words.

Example 6.4: Object code and program execution history–Consider this instruction list.

```
ORG     $C000   ; Start address of the code is the start of the
                  fixed 16k byte Flash EEPROM
CLRB            ; Clear B
LDAB    $3800   ; Loads with $80 from (M) = $3800
ADCB    $3801   ; Add byte at (M) = $3801 and C to B and store
                  result into B
ADCB    $3802   ; Add byte at (M) = $3802 and C to B and store
                  result into B
```

Write the corresponding object code and the execution history if the initial conditions are A = $B1; addresses $3800–3802 contain $80, $A1, and $2F; and flags bits HNZVC are 11011.

Solution: The *object code* indicates the address, opcode, and corresponding operands of each instruction in the code.

	PC	Opcode	Operand	
CLRB	C000	5F		
LDAB	C001	D6	38	00
ADCB	C003	D9	38	01
ADCB	C005	D9	38	02

Execution history refers to the status of the registers, memory, and CCR flags following the execution of each instruction.

B	(3800)	(3801)	(3802)	HNZVC	Operations	Notes
B1	80	A1	2F	11011	Initial conditions	
00	80	A1	2F	10100	0 → B	H not affected
80	80	A1	2F	11000	M → B	H not affected
21	80	A1	2F	00011	B+M+C→ B	+ve sum, V = 1
51	80	A1	2F	10000	B+M+C→ B	+ve sum, V = 0

6.3   ADDRESSING MODES

An instruction can access a memory location in many ways, referred to as the addressing modes, depending on how its argument is written. Each instruction is provided with all addressing modes relevant to its operation. Addressing modes allowed by the CPU12 and the abbreviations used hereafter are: *inherent* (INH) mode, *immediate* (IMM) mode, *extended* (EXT) mode, *direct* (DIR) mode, *relative* (REL) mode, and *indexed mode*. With a postbyte mechanism, the indexed mode accommodates many forms: *indexed* 5-bit offset (IDX), *indexed auto postdecrement/-increment* mode (IDX), *indexed* 9-bit offset mode (IDX1), *indexed* 16-bit offset mode (IDX2), *indexed-indirect* 16-bit offset mode ([IDX2]), and *indexed-indirect* with accumulator D offset ([D,IDX]).

Table 6.4 gives examples of various program instructions involving all possible addressing modes with various forms of operand arguments. The following symbols are used to indicate different operands:

abd	Accumulator may be A, B, or D
xysp	Register may be index X or Y, SP, or PC
opr8i	8-bit immediate data
opr16i	16-bit immediate data
opr8a	8-bit address
opr16a	6-bit address
oprk	*k*-bit offset; *k* = 3-, 5-, 9-,16-
rel8	8-bit relative offset
rel16	16-bit relative offset

TABLE 6.4. Instruction Formats and Examples in All Addressing Modes

Mode	General Format	Example	Operation
INH	**INST**	**INCB**	Increment B, No operand
REL	**INST** *rel8*	**BNE** there	Branch if not equal to "there"
IMM	**INST** *opr8i*	**LDAA** #$0A	load immediate byte #$0A into A
IMM	**INST** *opr16i*	**LDY** #$039A	Load immediate word #$039A into Y
DIR	**INST** *opr8a*	**ADDA** $0A	Add data at $0A to A, store result in A
EXT	**INST** *opr16a*	**SUBA** $C10A	Subtract data at $C10A from A, store result in A
IDX	**INST** *opr3,* −*xys*	**LDAA** 2, −X	Predecrement X by 2, load data at X into A
IDX	**INST** *opr3,* +*xys*	**STAA** 2, +X	Preincrement X by 2, store data in A into X
IDX	**INST** *opr3, xys*−	**LDAA** 2, X−	Load data at X into A, postdecrement X by 2
IDX	**INST** opr3, xys+	**LDAA** 2, X+	Load data at X into A, postincrement X by 2
IDX	**INST** *abd, xysp*	**LDAA** B, X	Load data at (B + X) into A
IDX	**INST** *opr5, xysp*	**LDAA** −12, X	Load data at (X − 12) into A
IDX1	**INST** *opr9, xysp*	**LDAA** 150, X	Load data at (X + 150) into A
IDX2	**INST** *opr16, xysp*	**LDAA** 2100, X	Load data at (2100 + X) into A
[D,IDX]	**INST** [*d, xysp*]	**LDAA** [D,X]	Load data at (D + X) into A
[IDX2]	**INST** [*opr16, xysp*]	**STAA** [8000, X]	Store data in A at (8000 + X)

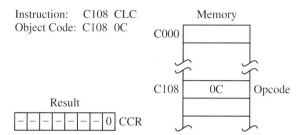

Figure 6.12 Format of an INH mode instruction and its execution process.

A quick look at Appendix C reveals that a large number instructions may be used in many addressing modes and that a large number of instructions use similar formats.

6.3.1 Inherent (INH) Mode

Inherent addressing deals with registers instead of memory and is usually used for internal operations, such as shifting and clearing the contents of a register. The opcode for the INH mode instruction has no operands, or the operands are available in the CPU registers. The CPU has all the information it needs to execute the instruction. Opcodes of INH mode instructions are usually one or two bytes and may require 1–12 CPU clock cycles to execute. The execution of the following INH mode instruction is shown in Fig 6.12:

```
$C1008   CLC   ; Clears carry bit
```

6.3.2 Immediate (IMM) Mode

Immediate addressing mode is used to load a register with an immediate constant value. The operand is the immediate data prefixed with the "#" symbol. The data may be 8 bits or 16 bits long, depending on the base register used in the instruction, and requires one or two cycles to execute. The execution of the following IMM mode instructions is shown in Fig. 6.13:

```
$EA10   LDX   #$70A2   ; Load immediate word #$70A2 into X
```

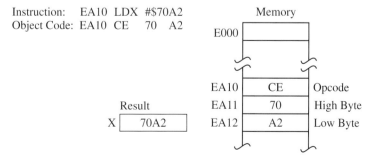

Figure 6.13 Format of an IMM mode instruction and its execution process.

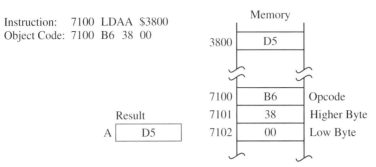

Figure 6.14 Format of an EXT mode instruction and its execution process.

6.3.3 Extended (EXT) Mode

The operand of an EXT mode instruction is the effective address of data, which can be anywhere within the 64K address space. Most EXT mode instructions are three bytes long; the first byte is the opcode, and the second and third bytes are the absolute address of the operand; then take three or four cycles to execute. The execution of the EXT mode instruction is shown in Fig. 6.14:

```
$7100   LDAA   $3800   ; Load data at $3800 into A
```

6.3.4 Direct (DIR) Mode

Direct addressing is similar to the extended addressing, except its use is limited to address operands in page 00 (addresses $0000–$00FF) of the memory space. A DIR mode instruction usually involves an opcode byte and an address byte and requires two to three cycles to execute. The execution of the DIR mode instruction is shown in Fig. 6.15:

```
$C100   LDAA   $20   ; Load data at address $20 into A
```

6.3.5 Relative (REL) Mode

Relative addressing mode is only used with branch instructions. A branch instruction directs the program execution elsewhere in the program. The operand of a branch instruction is a

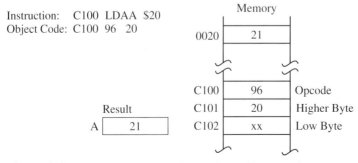

Figure 6.15 Format of a DIR mode instruction and its execution process.

signed 2s-complement offset, to allow forward and backward branching relative to the current address held in the PC. Adding the relative offset to the current PC value specifies the address of the instruction to be executed next. Short branch instructions are two bytes long; one byte is for the opcode and the other is the 8-bit relative offset for branching within the range −128–+127. The corresponding long branch instructions are four bytes long: two bytes for the opcode and a 16-bit relative offset allowing a branch range from −7001 to +7000. Branch instructions are introduced in Section 6.4.7. The following instructions demonstrate the use of short and long branch instructions.

```
BLE    fire   ; Short branch to "fire" if result of
               previous instruction is Less or Equal
LBNE   $E010  ; Long Branch to $E010 if result of
               previous instruction is Not Equal
```

If the branch condition is true, the relative offset is added to the current content of the PC to form the effective branch address; otherwise, control proceeds to the instruction immediately following the current branch instruction. The relative offset is determined by the assembler, provided the current and destination instructions are within the allowed range. If they are not, the assembler generates an error. If the branch condition is true, the destination address PC_{des} is calculated by

$$PC_{des} = PC_{cur} + rr \tag{6.1}$$

where PC_{cur} is the address of the instruction following the current branch instruction and rr is the relative offset. For short branch instructions, the ranges of rr are $\$00 \leq rr \leq \$7F$ (remaining bits are 0s) for *forward branch* and $\$80 < rr < \$FF$ (remaining bits are 1s) for *backward branch*. The following examples demonstrate the use of Eq. (6.1):

```
$6120 BRA    $7A  ; forward branch (PCdes = $6122 +
                    $007A = $619C)
$6120 BRA    $8A  ; backward branch (PCdes = $6122 +
                    $FF8A = $60AC)
                  ; Verification: $60AC − $6122 =
                    $FF8A (ignore FF)
```

Note that the absolute start address does not affect relative addresses.

6.3.6 Indexed Mode

The CPU12 implements a postbyte mechanism followed by no, one, or two extension bytes after the 8-bit or the 16-bit opcode to construct five possible forms of indexed addressing modes, further extending the operational capabilities of that mode.

Indexed Signed Offset Mode

In this mode, a constant signed offset is added to a base register to create the address of the memory location affected by the instruction. The base register is X, Y, SP, or PC, and the offset may be 5-bit, for an offset range from −16 to +15 from base register, 9-bit for an offset range from −256 to +255 from base register, and 16-bit offset for an offset range from

Instruction: C100 LDAA $65, Y
Object Code: C100 96   20

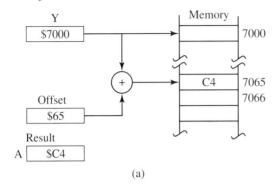

(a)

Instruction: C100 LDAA –$65, Y
Object Code: C100 E6   65

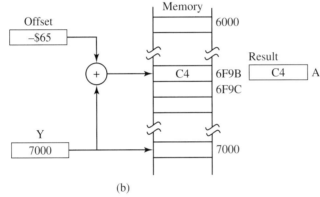

(b)

Figure 6.16 Format of indexed signed offset mode instructions.

−32K to +32K from base register. The following instructions demonstrates the use of the three possible offsets:

```
$C100    LDAA    $65, X     ; Load data at M = 6+X into A
$C100    LDAA    -$65, Y    ; Store data in B at M = Y-22
$C100    CLR     $40C0, X   ; Clear data at M = X + $40C0
```

Figure 6.16 shows the execution of the first two instructions.

Indexed Accumulator Offset Mode

The offset in this mode is the content of the 8-bit A or B or the 16-bit D, for a possible offset range from −32K to +32K. The offset is added to X, Y, SP, or PC to determine the effective address (EA) of the memory location to be operated on by the instruction. The execution history of the following instruction is shown in Fig. 6.17:

```
LEAX    B, Y   ; Load effective address (B + Y) into X
```

Instruction: LEAX B, Y
Object Code: 1A xb

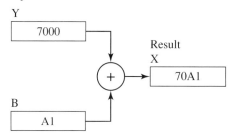

Figure 6.17 Format of indexed accumulator offset mode instructions.

Indexed Auto Pre-/Postincrement/Decrement Mode

In this mode, the value in the base register X, Y, or SP can be automatically incremented by 1 through 8 or decremented by −8 through −1 during execution of the instruction before or after indexing operation. The result of execution is affected by a preindexing operation but not by a postindexing operation. In either case the affected register retains the new value. Here are examples of instructions with the four possible formats:

```
INC    4, −Y  ; Predecrement Y by 4, then increment data
              pointed by Y
DEC    7, Y−  ; Decrement data pointed to by Y, then postdecre-
              ment Y by 7
ORAB   2, Y+  ; Inclusive-OR A with data pointed at by Y, then
              postincrement Y by 2
ORAB   2, +Y  ; Preincrement Y by 2, then Inclusively OR A with
              data pointed at by Y
```

Figure 6.18 shows the execution of these instructions.

Indexed Indirect Offset Mode [IDX2]

In this mode, the 16-bit offset included in the instruction is added to the base index register X, Y, SP, or PC to define the extended address <EA> of the memory location to be operated on by the instruction. The following two instructions demonstrate the use of this mode:

```
CPX [$7200, Y]   ; Compare X to data at EA = $7200 + Y
EORA [$0410, X]  ; Exclusive-OR A with data at EA = $04100 + X
```

Figure 6.19 shows the execution of the EORA instruction.

Indexed Indirect Accumulator D Mode [D, IDX]

This mode is similar to [IDX2] mode, but the contents of accumulator D is the offset. In this case D is added to the base register X, Y, SP, or PC to form the extended memory address targeted by the instruction. Figure 6.20 shows the execution of the instruction

```
COM   [D, Y]   ; 1s complement dat at EA = D + Y
```

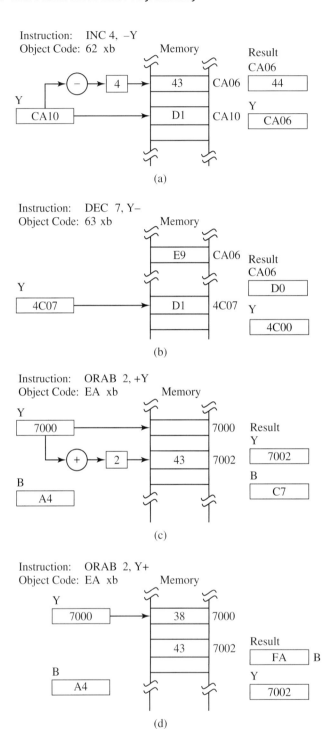

Figure 6.18 Format of indexed auto pre-/postincrement/decrement mode instructions.

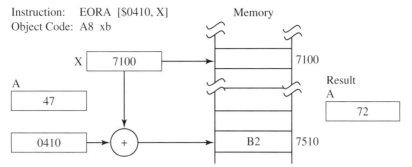

Figure 6.19 Format of indexed indirect offset mode instruction.

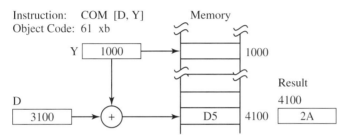

Figure 6.20 Format of indexed indirect accumulator D mode instruction.

Example 6.5: Array manipulation using IDX mode–Many applications require access to blocks of data stored in the memory space. For instance, the elements of the rows and columns of a 2D matrix are stored as a linear array of numbers. A convenient way to address individual elements within the matrix is to have a pointer pointing to the start of the row or column and an offset to address the element within the row or the column. The required memory address is the sum of the start or base address and the offset. Develop an assembly code to add the two linear arrays shown in Fig. 6.21. Each array consists of 16 elements, and the corresponding pairs of elements are to be added to one another and stored back in the first array. Refer to Example 6.19 for a C code version for this example.

Solution

```
; Define Variables

        ORG     $0380
count   DS.B    1                    ; Allocate one space at address
                                     $0380 for the variable count

; Define Equates

array1  EQU  $00                     ; Define offset address of array1
array2  EQU  $10                     ; Define offset address of array2
```

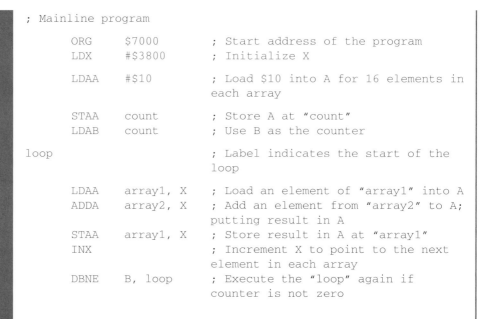

```
; Mainline program
        ORG     $7000       ; Start address of the program
        LDX     #$3800      ; Initialize X

        LDAA    #$10        ; Load $10 into A for 16 elements in
                            each array

        STAA    count       ; Store A at "count"
        LDAB    count       ; Use B as the counter

loop                        ; Label indicates the start of the
                            loop

        LDAA    array1, X   ; Load an element of "array1" into A
        ADDA    array2, X   ; Add an element from "array2" to A;
                            putting result in A
        STAA    array1, X   ; Store result in A at "array1"
        INX                 ; Increment X to point to the next
                            element in each array
        DBNE    B, loop     ; Execute the "loop" again if
                            counter is not zero
```

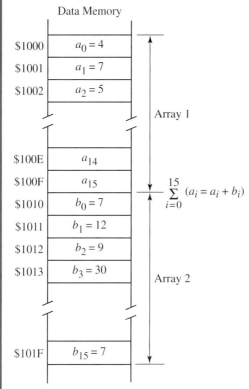

Data Memory

$1000 $a_0 = 4$
$1001 $a_1 = 7$
$1002 $a_2 = 5$

Array 1

$100E a_{14}
$100F a_{15}

$$\sum_{i=0}^{15} (a_i = a_i + b_i)$$

$1010 $b_0 = 7$
$1011 $b_1 = 12$
$1012 $b_2 = 9$
$1013 $b_3 = 30$

Array 2

$101F $b_{15} = 7$

Figure 6.21 Storage location of array 1 and array 2 to be added.

6.4 INSTRUCTION SET OF THE CPU12

From a functional point of view, the instruction set of the CPU12 is classified into the following seven functional groups: (1) data-handling instructions, (2) arithmetic instructions, (3) special math instructions, (4) logic instructions, (5) data compare and test instructions, (6) condition code instructions, (7) program control instructions, and (8) miscellaneous instructions. A brief and general overview of the instructions in each group is presented in the following sections. The mnemonics in each group will be presented in tabular form with examples and description of their operation. Appendix C provides a list of all instructions in alphabetical order, and for each instruction it describes its operation and provides the corresponding Boolean expression, opcode, and operands for each addressing mode, execution cycles, and the effect of the instruction on the condition code bits. Hereafter, the symbols used and their meaning are as follows: (M) indicates an address of a memory location or immediate byte; (M:M+1) represents two consecutive bytes of a 16-bit operand, (M) is the high byte and (M+1) is the low byte; $(X_H:X_L)$ refers to the high and low bytes of the 16-bit register X, and similar designation applies to Y, SP, and PC. Since instructions that use X and Y as base registers perform identically, X will be used in related discussions.

6.4.1 Data-Handling Instructions

The mnemonics belonging to this group are given in Table 6.5 in the form of examples. The associated instructions manage the transfer of data between memory, peripherals, and I/O devices on the one hand and the CPU on the other. They can be further divided into five subgroups: (i) data-movement instructions, (ii) data-transfer and -exchange instructions, (iii) data-modify instructions, (iv) shift and rotate instructions, and (v) min/max instructions.

Data-Movement Instructions

This group of instructions is used to move data between the CPU and any appropriate destination. Load and store accumulators or 16-bit registers may use all addressing modes except the INH and REL modes. Instructions that handle data pushes onto the stack and data pulls off the stack are available in INH mode only. Load effective address <EA> instructions may be used in the IDX, IDX1, and IDX2 modes. They load the base register (X, Y, or S) with the <EA> calculated by adding the postbyte and the extension bytes. Figure 6.22 shows the operation of the LEAX instruction given in Table 6.5.

The MOVB and MOVW instructions involve mixed addressing modes to allow movement of a byte or a word from a source to a destination. The source and destination of a move may be a memory location or a register. In the instruction

```
MOVW    2, X+, 2,-Y
```

X points to the source data and Y points to the destination area. The instruction reads a word at X, postincrements X by 2, then preincrements Y by 2, and finally deposits the data at Y − 2. Move instructions do not modify the CPU registers other than the one used as a destination. Figure 6.23 shows the operation of the MOVB instruction in Table 6.5.

Data-Transfer and -Exchange Instructions

The mnemonics in this group and their operations are given in Table 6.6. All transfer and exchange instructions involve CPU registers and are INH Mode instructions, except the TFR

TABLE 6.5 Data-Movement Instructions

Instruction		Operation Performed
Load and Store Instructions		
LDAA	$3800	Load data byte at (M) = $3800 into A
LDAB	-20, X	Load data byte at (M) = X-20 into B
LDD	[500,X]	Load data word at (M:M+1) = 500 + X into D
LDX	$9100	Load data word at (M:M+1) = $9100 into X
LDY	[D, X]	Load data word at (M:M+1) = D + X into Y
LDS	4, X+	Load data word at (M:M+1) = B + SP into SP
LEAX	B, Y	Load <EA> = B + Y into X
LEAY	-10, SP	Load <EA> = SP − 10 into Y
LEAS	200, Y	Load <EA> = Y + 200 into Y
STAA	$0001	Store dat byte in A into (M) = $0001
STAB	-$A2, Y	Store (M) into B
STD	-$8601, X	Store (D) at M:M+1
STX	$3900	Store (X) at M:M+1
STY	-2000, X	Store (Y) at M:M+1
STS	12, Y	Store (S) at M:M+1
Push and Pull Instructions (All INH mode)		
PSHA		Push A onto Stack
PSHB		Push B onto Stack
PSHC		Push CCR onto Stack
PULA		Pull A from Stack
PULB		Pull B from Stack
PULC		Pull CCR from Stack
PULD		Pull D from Stack
PULX		Pull X from Stack
PULY		Pull Y from Stack
Byte and Word Move Instructions		
MOVB	2,X+,4,−Y	Move a data byte from (M_1) = X to M_2 = Y − 4
MOVW	8,+X,8,Y−	Move a data word from (M_1) = X + 8 to M_2 = Y

and EXG instructions. Transfer operations do not affect the source register. The TFR instruction allows transfers from any source register R_1 to any destination register R_2, where R_1 and R_2 may be A, B, C, D, X, Y, or SP. The EXG instruction performs bidirectional exchange between any two registers R_1 and R_2. Exchange D with X (XGDX) or exchange D with Y (XGDY) instructions provide the means for arithmetic manipulations to be performed on the data in X or Y. Once the required manipulations are done, applying the exchange instruction again updates the index register and restores D to its original value.

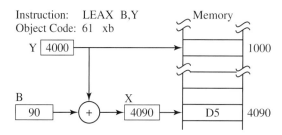

Figure 6.22 Format of the load effective address instruction.

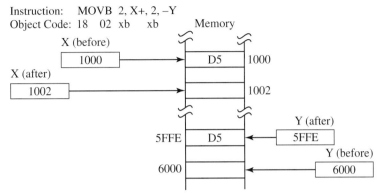

Instruction: MOVB 2, X+, 2, −Y
Object Code: 18   02   xb   xb

Figure 6.23 Format of the MOVB instruction.

Data-Modify Instructions

Table 6.7 gives the instructions in this group in the form of examples. The mnemonic that uses accumulator A ends with the letter A; replacing it with the letter B makes the instruction use accumulator B instead. Similarly a mnemonic that uses register X ends with the letter X; replacing it with the letter Y or S makes the instruction use register Y or SP, respectively. Most instructions in this group are INH mode instructions. This group includes instructions that decrement or increment memory or a register, take 2s- or 1s-complement of memory or accumulator, and clear the contents of a memory location or accumulator. The argument for the bit set BSET and bit clear BCLR instructions includes an operand address and a special byte called the *mask* (mm) to set or clear operand bits. The bit clear BCLR instruction

```
BCLR   -20, X, #$F0
```

logically ANDs the data byte at address (−20, X) with the complement of the mask byte $F0. If at address (X − 20) the byte $FF is stored, it will be replaced with $0F after the instruction executes. The bit set instruction BSET

```
BSET $4800, #$C2
```

TABLE 6.6 Data-Transfer and -Exchange Instructions

Instruction	Operation Performed
TAB	Transfer the data in A to B
TBA	Transfer the data in B to A
TXS	Transfer the data in X to SP
TSX	Transfer the data in SP to X
TYS	Transfer the data in Y to SP
TSY	Transfer the data in SP to Y
TFR X, Y	Transfer the data from register R_1 to R_2
EXG Y, SP	Exchange the data between the two registers R_1 to R_2
XGDX	Exchange the data between D and X
XGDY	Exchange the data between D and Y

TABLE 6.7 Data-Modify Instructions

Instruction		Operation Performed
DEC	$7000, Y	Decrement (M) = Y + $7000, putting result in M
DECA		Decrement A, putting result in A
DEX		Decrement X, putting result in X
INC	4, −Y	Increment (M) = Y−4, putting result in M
INCA		Increment A, putting result in A
INX		Increment X, putting result in X
NEG	2, X+	Take 2s-complement of (M) = X, putting result in M, and postincrement X by 2
NEGA		2s-complement A
COM	$4600	1s-complement M
COMA		1s-complement A
CLR	[D, X]	Clear M
CLRA		Clear A
BCLR	−20, X, $F0	AND bits in (M) = X − $20 with 0s to clear them
BSET	$4800, $C2	OR bits in (M) = $4800 with 1s to set them

logically ORs the memory byte at address $4800 with the mask byte $C2 to set bits. If the original byte at $4800 is $F5, it will be replaced with $F7. COM and COMA (B) instructions take the 1s-complement of the number in memory and A (or B), respectively.

Shift and Rotate Instructions

This group of instructions, listed in Table 6.8, performs arithmetic and logical shift and rotate operations on accumulators A or B or on memory. The mnemonic for an arithmetic instruction begins with the letter A; replacing it with the letter L changes the instruction to a logical shift

TABLE 6.8 Shift and Rotate Instructions

Mnemonic		Operation Performed	Remarks
Shift A and B Instructions			
ASL	$C400	Arithmetic shift left data at (M) = $C400	Multiply by 2
ASLA		Arithmetic shift left A	Fig. 6.24c
ASR	2, -X	Arithmetic shift right data at (M) = X − 2	Divide by 2
ASRA		Arithmetic shift right A	Fig. 6.24d, e, and f
Rotate A and B Instructions			
ROL	-$65, Y	Rotate left dat at (M) = Y −$65	
ROLA		Rotate left A	Fig. 6.24a
ROR	$2100, X	Rotate right data at (M) = X + $2100	
RORA		Rotate right A	Fig. 6.24b
Shift and Rotate D Instructions			
ASLD		Arithmetic shift left D: $0 \rightarrow D[0]$; $D[i] \rightarrow D[i+1]$; $D[15] \rightarrow C$	
LSRD		Logical shift right $D[0] \rightarrow D[15]$; $D[i+1] \rightarrow D[i]$; $D[0] \rightarrow C$	

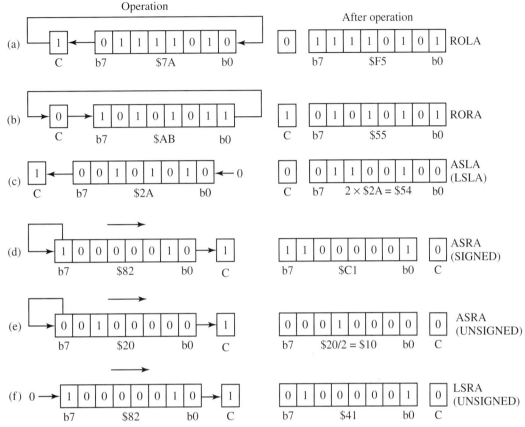

Figure 6.24 Examples of shift/rotate instructions.

instead. Also, the mnemonic that uses accumulator A ends with the letter A; replacing it with the letter B makes the instruction use accumulator B instead. Shift instructions LSLD, LSRD, and ASLD involve accumulator D to operate on 16-bit operands. All shift and rotate functions involve the carry bit C.

Figure 6.24 shows examples of shift/rotate operations on A. The rotate instructions in Figs. 6.24a and b rotate the contents of the operand without altering its bits. The RORA instruction rotates the bits in A right; the C bit value is placed in the MSB, and the LSB becomes the new C value. If C is set or cleared prior to a rotate instruction, the user can control the bit shifted at the open end of the operand. Rotate instructions may be used to examine bits from sensor outputs or switches. Executing the same rotate instructions eight consecutive times restores the original number.

Execution of an arithmetic shift instruction, as illustrated in Figs. 6.24c–e, places a zero into the LSB of the operand. This is equivalent to multiplying the operand by 2. Applying ASL to a byte eight consecutive times annuls all its bits. Note that logical shift left LSLA and arithmetic shift left ASLA instructions are identical. Arithmetic shift right instructions assume 2s-complement signed numbers and preserves the sign (MSB) bit of the operand. Thus, shifting a byte to the right fills it up with 1s or 0s, depending on the value of the sign bit. A shift-to-right operation divides the operand by 2, as demonstrated in Fig. 6.24d, where the ASRA operation divides the negative number $82 by 2 to yield $C1.

Logical shift operations are illustrated in Figs. 6.24c and f. Executing a logical shift left (right) instruction places "0" in the LSB (MSB) of the operand. Logical shifts may be used to clear all bits after each one is examined using the C bit. Logical shift right instructions LSR do not retain the sign bit as arithmetic shift right instructions ASR do. Therefore, they can be used for unsigned divide by 2, as shown in Fig. 6.24f.

Example 6.6: Use of shift and rotate operations–Determine the values in A, B, and the condition flag C as the instructions that follow the LDAB instruction in the following sequence are executed.

```
CLC             ;clear C the bit
LDAA    #$C1    ;initial A data
LDAB    #$AA    ;initial B data
RORA
ASRA
LSLD
ASRB
```

Solution: The contents of A and B and the value of C will be:

	A	B	C
RORA	$60	$AA	1
ASRA	$30	$AA	0
LSLD	$51	$54	0
ASRB	$51	$2A	0

Min and Max Instructions

This group of instructions is listed in Table 6.9. The MINA instructions compares two 8-bit unsigned values, one in A and the other in memory, and stores their minimum in A. MAXA does the same, but the maximum of the two bytes is stored in A. MINM and MAXM instructions perform similar operations, but the result of comparison is stored in memory instead of

TABLE 6.9 Min and Max Instructions

Mnemonic	Operands	Operation Performed
MAXA	pmax, X	Maximum of (A) and (M) = pmax + X is put in A
MAXM	[$C700, X]	Maximum of (A) and (M) = X + $C700 is put in M
MINA	1, X+	Minimum of (A) and (M) = X is put in A, X is postincremented by 1
MINM	−650, X	Minimum of (A) and (M) = X − 650 is put in M
EMAXD	[D,X]	Maximum of (D) and (M) = D + X is put in D
EMAXM	[$8600, X]	Maximum of (D) and (M) = X + $8600 is put in M
EMIND	1,−Y	Minimum of (D) and (M) = Y - 1 is put in D
EMINM	3,SP+	Minimum of (D) and (M) = SP is put in M, SP is postincremented by 3

A. The same instructions starting with the letter E perform the same operation, with two exceptions: the operands are two 16-bit unsigned numbers and D is used instead of A. Min and Max instructions facilitate the implementation of linear programming algorithms, such as the Simplex algorithm and in the fuzzification stage of a fuzzy logic system.

Example 6.7: Determine the minimum byte in an array—Write a code to determine the minimum unsigned byte in a block of data arranged in a table of 20 entries.

Solution: The following code finds the minimum unsigned byte in the block stored at addresses beginning at `table`. The number of bytes is stored at address `count`.

```
        INCLUDE  "display.asm" ; display is defined in another
                                 source code
; declare constants used in the program
stack   EQU      $4001
; define data used in the program
        ORG      $3800          ; place data in ram to
                                 enable the use DIRECT
                                 addressing mode
minbyte      DS.B  1            ; reserve one byte at address
                                 $3800 to minbyte
count   DC.B  $20               ; assign $20 for count
table   DC.B  $3A,$2C,$01,$10,$42,$A2,$7D,$2F,$6E ; assigns
                 address $3803 to first and
        DC.B  $81,$29,$B2,$4D,$0A,$2B,$CE,$54,$E6 ; place one
                 byte in each address

;main program begins here
        ORG      $C000
        LDS      #stack
        LDAB     #count-1       ; Load count into B
        LDX      #table         ; Point to first element in data
                                 block
        LDAA     0, X           ; Load first byte into A
loop    MINA     1, +X          ; Minimum of two consecutive
                                 bytes
        STAA     minbyte        ; Store current minimum into
                                 minbyte
        DECB                    ; Decrement counter B
        BNE  loop               ; Repeat loop until B = 0
        JSR  display            ; Display minbyte
here    BRA  here
        END
```

6.4.2 Arithmetic Instructions

The group of instructions listed in Table 6.10 handles arithmetic operations on a variety of operands. The mnemonic that uses accumulator A ends with the letter A. It is replaced with the letter B to form an accumulator B–based instruction that functions identically. Similarly a mnemonic

TABLE 6.10 Arithmetic Instructions

Mnemonic Operand	Operation Performed
Add and Subtract Instructions	
ABA	Add (A) to (B), putting result in A
ABX	Add (B) to (X), putting result in X
ADCA $8F00	Add (C) and (M) = $8700 to A, putting result in A
ADDA #$7A	Add (M) = #$7A to (A), putting result in M
ADDD $00	Add(M:M+1) = ($00:$01) to (D), putting result in D
DAA	Adjust sum in (A) to BCD, putting result in A
Subtract Instructions	
SBCA 5, +X	Subtract C and (M) from (A), put result in A
SUBA $12, Y	Subtract (M) = (Y+$12) from (A), put result in A
SUBD −3000, X	Subtract (M:M+1) = (X−3000) from (D), put result in (D)
Multiplication Instruction	
MUL (unsigned)	Multiply (unsigned) 8-bit in (A) by 8-bit in (B), put result in D
EMUL (unsignes)	Multiply (unsigned) 16-bit in (D) by 16-bit in (Y), put result in Y:D
EMULS (signed)	Multiply (signed) 16-bit in (D) by 16-bit in (Y), put result in Y:D
EMACS $3200	Multiply and accumulate (signed)
Division Instructions	
IDIV (unsigned)	Divide (D) by (X), put quotient in X and remainder in D
IDIVS (signed)	Divide (D) by (X), put quotient in X and remainder in D
EDIV (unsigned)	Divide 32-bit in (Y:D) by (X), put quotient in Y and remainder in D
EDIVS (signed)	Divide 32-bit in (Y:D) by (X), put quotient in Y and remainder in D
FDIV (unsigned)	Divide (D) by (X), put quotient in X and remainder in D

that uses register X ends with the letter X; replacing it with the letter Y defines the mnemonic for the equivalent Y-based instruction. The group supports addition and subtraction, multiplication, and division operations on 8-bit and 16-bit operands directly and can be extended to manipulate multiple-byte operands. Instructions in the table with no operand are INH mode instructions.

Addition and Subtraction

This subgroup performs direct addition and subtraction of unsigned binary numbers in registers and memory. It includes add and add-with-carry instructions and subtract and subtract-with-carry instructions. Add instructions using index registers X and Y are also provided. The DAA instruction handles addition of decimal numbers using binary-coded decimal (BCD) representation. Refer to the discussion on the H status bit in Section 6.2.7 for how this is handled. BCD arithmetic is useful when BCD displays are used, as discussed in Section 7.6.

Example 6.8: Addition of unsigned numbers–Two unsigned 16-bit numbers $101A and $4036 are stored in memory as four bytes at address $0100 as 10 1A 40 36. Give the instruction sequence that adds the two bytes using single-byte and double-byte additions.

Solution: The following code performs single-byte addition using the 8-bit B:

```
LDAB    $0100    ; Load $1A into B
ADDB    $0103    ; Add $1A to $36 (C = 1)
STAB    $0105    ; Store result at $0105
LDAB    $0101    ; Load $0101 into B
ADCB    $0102    ; Add to B $40 and C, put result in B
STAB    $0104    ; Store B at $0104
```

The same can be accomplished using double-byte addition and the 16-bit D:

```
LDD     $0100    ; Get first 2-byte number $101A
ADDD    $0102    ; Add it to second 2-byte number $4036
STD     $0104    ; Store the result at $0104 and $0105
```

The CPU uses B to add the least significant bytes first and then A to add-with-carry the most significant bytes.

Example 6.9: Subtraction of unsigned numbers–Subtraction is similar to addition, except the carry bit is used for borrow operations. Write the code to subtract $30 from $1A.

Solution

```
LDAA    #$1A   ; Load $1A into A
SUBA    #$30   ; Subtract $30 from $1A and put the result in A
               ; Condition codes: C = 1, N = 0, H = 0, Z = 0,
               V = 0.
```

The result is a signed 2s-complement $0A, which represents negative $F6.

Example 6.10: BCD addition–A temperature sensor reads a value of 42°C. The calibration curve of the sensor indicates a bias of −5°C. Write a sequence of instructions to determine the correct temperature.

Solution: The following code adds 5 to 42 to yield the correct temperature:

```
LDAA    #$05
ADDA    #$2E   ; A = $05 + $2E = $33
DAA            ; Adjust A = $33 + $06 = $39
```

Multiply Instructions

The CPU12 supports multiplication using four instructions. The MUL instruction multiplies the two unsigned bytes in A and B and puts the 16-bit unsigned result in D in one clock cycle. The EMUL instruction multiplies two unsigned 16-bit operands, one in D and the other in Y, and stores the 32-bit result in Y and D. The EMULS instruction performs the same on signed numbers. The MUL instruction affects the C bit only, whereas EMUL and EMULS affect N, Z, and C status bits. The arithmetic shift left instruction ASL multiplies an operand by powers of 2 but requires longer execution times. The MUL instruction sets the C bit if bit 7 of the result in B is 1. This enables the use of the ADCA or ADCB instruction to round the 16-bit result in D to an 8-bit result in A. This is possible because a binary system can be scaled to represent decimal fractions between 0 and 1. The MSB represents 0.5, or 2^{-1}, the next significant bit represents 0.25, or 2^{-2}, and so on. For example, the binary number 10110111 is converted to a decimal fraction as $1 \times 2^{-1} + 0 \times 2^{-2} + 1 \times 2^{-3} + 1 \times 2^{-4} + 0 \times 2^{-5} + 1 \times 2^{-6} + 1 \times 2^{-7} + 1 \times 2^{-8} = 0.7148$.

The extend multiply and accumulate instruction EMACS multiplies two signed 16-bit numbers to produce a 32-bit result. This instruction is useful in implementing simple digital filters and in performing defuzzification operations in fuzzy logic applications involving 16-bit operands.

Example 6.11: Multiplication—A temperature sensor outputs 36 μV/°C. What would the output voltage be when the temperature reaches 24°C?

Solution: This example requires the multiplication of the two hex numbers $24 \times \$F0$:

```
LDD    #$24F0        ; Load the two bytes $24F0 into D
MUL                  ; Multiply them
ADCA  #$00           ; 8-bit result in A = $03, equivalent
                     to a 0.011719 decimal fraction.
```

The MUL operation results in $0360 and stores it in D and C = 0. This result is equivalent to a decimal value of 864 μV if the bytes were integers. If the bytes being multiplied represent decimal fractions, the result is interpreted as 0.013184 decimal. After ADCA is executed, the 8-bit result in A remains unaltered at $03, with the equivalent decimal fraction of 0.011719. A simple way to convert an 8-bit hex number is to convert it to decimal and divide the result by 256. For a 16-bit hex number, divide by 65536.

Division Instructions

The five division instructions facilitate division operations on a variety of operands. The integer divide instruction IDIV divides a 16-bit unsigned binary integer in D (dividend) by another 16-bit binary unsigned integer in X (divisor). The 16-bit result (quotient) is placed in X and the remainder in D, as depicted in Fig. 6.25a. For a division by powers of 2, the arithmetic shift right instruction ASR may be used instead of IDIV, because it executes faster. The ASR instruction, however, assumes signed integers.

The fractional division instruction FDIV divides the 16-bit fraction in D by a larger 16-bit fraction in X and places the quotient in X and the remainder in D (see Fig. 6.25b). A decimal

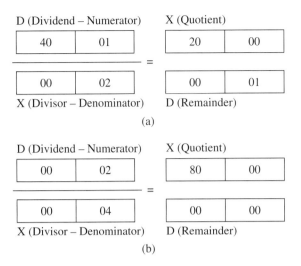

Figure 6.25 Division operations.

"fraction" in this context means that a 16-bit number is a binary-weighted fraction between 0 and 0.99998. The radix point, assumed to be in the same place for the numerator and the denominator, is to the left of the MSB of the result (bit 15 of the quotient). The fractional equivalent of the result is calculated by dividing the result by $10000. Thus, a result of $0001 implies $0001/$10000 = 0.00001, $80000 implies $8000/$10000 = 0.5, $C000 implies $C000/$10000 = 0.75, and $FFFF implies $FFFF/$10000 = 0.99998. If the denominator is less than or equal to the numerator, the V status bit will set. If a divide by zero is attempted, both IDIV and FDIV sets the C flag and the quotient in X to $FFFF. FDIV can be used to further resolve the remainder from an IDIV or FDIV operation. The Z and C condition codes are affected in the same way by IDIV and FDIV instructions. The Z flag is set only if all bits in the quotient are zero. Both IDIV and FDIV instructions execute in 12 clock cycles.

The E before IDIV or FDIV implies extended division of a 32-bit number by a 16-bit. The S at the end of a divide instruction implies signed numbers division. The V status flag will set if the result is greater than $FFFF for unsigned division or if greater than $7FFF for signed division. In all division operations the C flag will set if the divisor is $0000.

Example 6.12: Division operations—

a. Converting temperature from °F to °C requires division of 9 ($09) by 5 ($05). Assume that the two numbers are stored as 16-bit numbers at address $0000. Perform the required division using IDIV, and then use FDIV to further resolve the remainder from the IDIV operation.

b. Use the ASR Instruction to divide $12 by $8.

Solution

a. The following code accomplishes the required task:

```
LDD    $00    ; Load D with the dividend $0009
LDX    $02    ; Load X with the divisor $0005
IDIV          ; Quotient is 1 put in X as $0001 and
              remainder is 4 put in D as $0004
```

```
          STX   $04    ; Store integer result at $0004
          LDX   $02    ; Reload divisor
          FDIV         ; Divide remainder $0004 by 5 to get
                         fraction, X = $0DCC and D = $0000
          STX   $06    ; Store fraction
   here   BRA   here   ; Stop program
```

The memory starting at address $0000 will hold the following data after code execution is complete:

00   09   00   05   00   01   0D   CC

This represents the answer of $0001.0DCC. Other instructions could be added to round the result to a 16-bit integer or to an 8-bit integer without a fraction.

b. Since division by $8 is equivalent to three successive divisions by $02, using the AS R instruction three consecutive times accomplishes the required division:

```
          LDAA  #$12   ; Load the dividend $12 into A
          ASRA         ; Divide A by 2 three consecutive times
          ASRA
          ASRA
```

6.4.3 Special Math Instructions

Table 6.11 lists the instructions that perform special math operations. The TBL instructions perform the table lookup and interpolate operation on 8-bit data. Assume that the function that relates the data points x_i to y_i is approximated by successive linear segments and that the two endpoints of a segment are x_0, y_0 and x_1, y_1. If the data points are equidistant, then only the y_i values of the endpoints of the linear segments need to be stored in the lookup table. To

TABLE 6.11 Special Math Instructions

Mnemonic	Operand	Operation Performed
Table Interpolation Instruction		
TBL	A, X	Interpolate between two points in a lookup table of 8-bit entries; B and X are initialized before the instruction is used
ETBL	2, -Y	Interpolate between two points in a lookup table of 16-bit entries
Fuzzy logic Instructions		
MEM		Membership function evaluation - Fuzzification
REV		Evaluation of IF-THEN rules
REW		Weighted rule evaluation
WAV		Weighted average - Defuzzification

demonstrate the operation of the TBL instruction, it is desired to find the y_L corresponding to the lookup value x_L. The value of x_L is loaded into B, and a pointer then identifies the closest points x_0 and x_1 in the table that encompasses the exact lookup value. The value of x_L relative to the endpoint values of the segment that encompass x_L is determined as one of 256 possible result values by B = $(x_L - x_0)/(x_1 - x_0)$. The instruction then linearly interpolates between the two endpoint values y_0 and y_1 of the corresponding segment stored in a lookup table, using the value in B to compute the corresponding result value y_L, and stores it in A. The operation results in A = $y_L = y_1 + (B) \times (y_1 - y_0)$. The instruction executes in about 1 microsecond.

```
LDX    #table   ; Point to the data in table
LDD    data     ; Load 8-bit data in B
TBL    A, X     ; Interpolate and store result in A
```

If the x values of the line segments are not equidistant, then both x and y values need to be stored in two separate lookup tables.

The instruction set of the 9S12C includes four instructions to help implement fuzzy logic control using minimal MCU resources and with high speed. Section 13.6 covers fuzzy logic control in detail, where these instructions becoming very useful in implementation.

Example 6.13: Lookup table interpolation—The resistance (R, Ω) and temperature $(T, °C)$ data for a thermistor is given in the following table. Write a simple assembly code using the TBL instruction to determine the resistance corresponding to $T = 102°C$.

T (°C)	0	5	101	15	20	25	30
R (Ω)	27,280	2,205	17,960	14,680	12,090	10,000	8,313
T (°C)	40	45	50	55	60	65	70
R (Ω)	6,941	5,826	4,912	4,161	3,537	3,021	2,589
T (°C)	75	80	85	90	95	100	105
R (Ω)	2,229	1,669	1,451	1366	1,108	938	858

Solution: The following code determines the resistance corresponding to the desired temperature value at "temp":

```
;Thermistor lookup table
ORG     $C100
value   DS.B  1
table   DC.W  $6A90, $......   ; Determine the rest similarly

ORG     $C000
LDX     #table
LDD     value
TBL     A, X
```

TABLE 6.12 Logic Instructions

Mnemonic	Operands	Operation Performed
ANDA	#$7A	AND A with (M) = #$7A, putting result in A
ANDB	$05A1	AND B with (M) = $05A1, putting result in B
EORA	A, X	Exclusive-OR A with (M) = A + X, putting result in A
EORB	-120, Y	Exclusive-OR B with (M) = Y − 120, putting result in B
ORAA	[D, X]	Inclusive-OR A with (M) = D + X, putting result in A
ORAB	[$C400, SP]	Inclusive-OR A with M = SP + $C400, putting result in B

6.4.4 Logic Instructions

Table 6.12 gives the instructions in this group that are used directly to perform the Boolean logical operations AND, inclusive-OR, and exclusive-OR. Each accumulator has an instruction for each of these logical operations. The contents of A (or B) are ANDed, ORed, or exclusively-ORed bit by bit with another byte, and the result is placed back into A (or B).

Example 6.14: Use of logic instructions–Indicate the result of executing the given logical instructions. The instructions operate on three data bytes, $A1, $8B, and $D2, stored at address $C100.

Solution: The results are indicated in the corresponding comment fields.

```
LDD    $00    ; Load $A18B into D; A = $A1 and B = $8B
ANDA   #$C6   ; AND $A1 with $C6 and put the result $80 into A
ORAB   #$C6   ; OR $8B with $C6 and put the result $CF into B
EORB   $02    ; Exclusive-OR $CF with $D2 and put the result
              ; $11 into B
BITA   #$01   ; AND $80 with $01 result is $00 and Z = 1
```

6.4.5 Data-Compare and -Testing Instructions

The group of instructions listed in Table 6.13 can operate on single-bit operands or on any combination of bits of a data byte in memory. With compare instructions, the CPU subtracts two operands internally and updates the condition code bits without altering either operand. The bit test instruction BITA (B) logically ANDs the contents of A (or B) with a byte in memory to update the N, Z, and V condition code bits. Compare instructions involving D, X, Y, or SP as a base register compare the 16-bit number in the register with two consecutive bytes in memory without altering the base register or memory. Test instructions are available but hardly used, because almost all other operations lead to automatic updating of the condition code bits.

TABLE 6.13 Data-Testing and Bit-Manipulation Instructions

Mnemonic	Operands	Operation Performed
BITA	[D,X]	AND A with (M) = D + X, put result in A
BITB	[$1200,Y]	AND B with (M) = Y + $1200, put result in B
CBA		Compare A with B; (A) − (B)
CMPA	1200, X	Compare A with (M) = X + 1200; (A) − (M)
CMPB	-370, Y	Compare B with (M) = Y - 370; (B) − (M)
CPD	B, X	Compare D with (M) = B + X; (D) − (M)
CPX	1, +SP	Compare X with (M) = SP + 1
CPY	3, SP-	Compare Y with (M) = SP, postdecrement SP by 3
CPS	$7F01	Compare SP with (M) = $7F01; (SP) − (M)
TST	[$4000, X]	Test (M) = X + $4000 for 0 or minus
TSTA		Test A for zero or minus
TSTB		Test B for zero or minus

6.4.6 Condition Code Register Instructions

This group of instructions are listed in Table 6.14. They are used to manipulate bits in the CCR. Set and clear instructions are available for the C, I, and V bits only. ORCC and ANDCC instructions can be used to set or clear any desired bit or bits in the CCR, respectively. The instruction

$$\text{ORCC} \qquad \#\$01$$

inclusively-ORs the contents of the CCR with the immediate byte $01 and modifies the CCR accordingly. Similarly, the instruction

$$\text{ANDCC} \qquad \#\$B5$$

ANDs the contents of the CCR with the immediate byte $B5 and changes the CCR bits accordingly.

The transfer TAP instruction transfers the contents of A to the CCR, and the TPA instruction does the reverse.

TABLE 6.14 Condition Code Register Instructions

Mnemonic	Operand	Operation Performed
CLC		Clear carry bit C
CLI		Clear interrupt mask bit I
CLV		Clear overflow bit V
SEC		Set carry bit C
SEI		Set interrupt mask bit I
SEV		Set overflow bit V
TAP		Transfer A to CCR
TPA		Transfer CCR to A
ORCC	#$01	Inclusive-OR CCR with #$01, put result in CCR
ANDCC	#$B5	AND CCR with #$B5, putting result in CCR

6.4.7 Program-Control Instructions

This group includes instructions used to control program flow. It is divided into three categories: branch instructions, subroutine-handling instructions, and interrupt-handling instructions.

Branch Instructions

Branch instructions are divided into three categories, as indicated in Table 6.15. Short and long branch instructions, branch as a result of logical operation, and loop primitives. When executing a branch instruction, the CPU polls the associated condition code bits, and, if the branch condition is true, it redirects program execution to the destination address accordingly.

Short/Long Branch Instructions Short and long branch instructions perform similar operations. A long branch is implied if the instruction begins with the letter L. An asterisk indicates that the corresponding branch instruction works with signed arithmetic. In this case a CMPA instruction is inherently executed prior to the branch. Refer to the discussion on relative addressing mode in Section 6.3.5 for more details on short and long branch instructions. Note that all operands (relative offset) associated with the mnemonics are labels, because the assembler will assign the relative offset based on the destination address. Similar labels may be used with the long branch instructions, but the offset may be a 16-bit relative address.

TABLE 6.15 Program-Control Instructions

Short Branch		Long Branch	Operation Performed
BCC	repeat	LBCC	Branch if Carry Clear
BCS	halt	LBCS	Branch if Carry Set
BEQ	done	LBEQ	Branch if equal zero
BGE	finish	LBGE	Branch if A >= M
BGT	stop	LBGT	Branch if A > M
BHI	down	LBHI	Branch if A > M
BHS	full	LBHS	Branch if A >= M
BLE*	count	LBLE	Branch if A <= M
BLO	up	LBLO	Branch if A < M
BLS	increment	LBLS	Branch if A <= M
BLT*	pumpon	LBLT	Branch if A < M
BMI	nochange	LBMI	Branch if r = negative
BNE	continue	LBNE	Branch if A <> M
BPL	change	LBPL	Branch if r = positive
BVC*	false	LBVC	Branch if r = sign OK
BVS*	loop	LBVS	Branch if r = sign ERROR

Conditional Branch Instructions

BRCLR	$320A, #2B, near	Branch if bit(s) clear
BRSET	A, #40, done	Branch if bit(s) set

Loop Primitive Instructions

DBEQX	loop	Decrement counter X and branch to loop IF X = 0
DBNE	Y, fill	Decrement counter Y and branch to fill IF Y $\neq$ 0
IBEQ	A, paint	Increment counter A and branch to paint IF A = 0
IBNE	B, step	Increment counter B and branch to dry IF B $\neq$ 0
TBEQ	D, $-$60	Test counter D and branch to (PC + $3 + $40) IF D = 0
TBNE	S, $40	Test counter SP and branch to (PC + $3 - $60) IF SP $\neq$ 0

BRCLR and BRSET Instructions These instructions are useful for polling interrupt status flags and for making program decisions based on bit(s) values. BRCLR instruction logically ANDs an operand with a mask byte, and, if the specific bits in the result are clear, a branch to a destination address occurs. The instruction's argument contains the operand mask and the relative address offset. As an example of its use, consider two switches connected to pins PM[1] and PM[3] of port M at address $0250. If a switch is closed, the corresponding port B bits will set. When this happens, a subroutine at address "pumpstop" is executed. Assuming X is pointing to $0000 and port M bits are initially clear, the instructions

```
noinput   BRCLR   portm,#$0A, noinput
          JSR     pumpstop
```

ANDs port M with the mask byte $0A. If neither switch is closed, the result will be $00, and the branch condition is satisfied. The program then executes the instruction again. If the branch condition is not true, that is, one or both switches were closed, then the CPU executes the JSR instruction, directing program execution to the "pumpstop" subroutine.

The BRSET instruction functions in a similar way, with one exception: It logically ANDs the complement of the operand with the mask byte and makes decision to branch if the selected bits are set. The following instruction demonstrates its operation:

```
BRSET   $025F, #$01, htron
```

This ANDs the value at $025F with $01, and, if bit 0 is set, execution then resumes at "htron".

Loop Primitives Loop primitive instructions use A, B, D, X, Y, or SP as a counter to decide to continue or abort execution of a program segment at the end of a program loop if either the loop counter has counted to zero or it has not. The first letter in the mnemonic, I, D, or T, causes the counter to increment, decrement, or be tested, respectively, before a BEQ or BNE instruction is executed. The branch offset is nine bits, allowing for a branch range between -512 and $+511$. If the counter value does not satisfy the branch condition, execution resumes at the following instruction. Test and branch instructions are used when the counter adjustment is other than 1. The adjustment is defined earlier using autoincrement/-decrement instructions.

Example 6.15: Move block of data–Write a code that moves a block of ASCII characters "Think Community. Recycle." stored in "table" at $C100 to RAM at $3800.

Solution: The following code accomplishes the required task:

```
        ORG   $C100            ; Originate table at $C100
table   DC.B  "Think Community."
        DC.B  "Recycle."
        DC.B  $00              ; Last character to be sent

ram     EQU   $3800            ;

        LDX   #table           ; Initialize source pointer to
                               table
```

```
        LDY    ram            ; Initialize desitination pointer
                              to ram
loop    LDAA   1, X+          ; Read table entry pointed to by
                              X, then increment X
        CMPA   #$00           ; Check if last byte is moved
        BEQ    done           ; End program if all data in block
                              is moved
        STAA   1, Y+          ; Store data Y, then postincrement
                              Y
        BRA    loop           ; Execute loop again
done    END                   ; End program execution
```

The same task can be performed using MOVW and DBNE instructions, as follows:

```
        LDAB   #20            ; Use B as a counter, load it with
                              the number of bytes to be moved
loop    MOVW   1, X+, 1, Y+   ; Move data
        DBNE   B, loop        ; Loop again if counter is not
                              zero
        END
```

Forming Loop Constructs

The two basic forms of program flow structures are the IF-ELSE action and the LOOP action, such as the DO-WHILE loop. Both types can be implemented with conditional branch instructions, as Fig. 6.26 indicates. Consider controlling the temperature of an aluminum plate to 43°C, 20°C above ambient, using a resistive heater and a temperature sensor. This may be implemented as an IF-THEN-ELSE construct as follows: IF the temperature T from the sensor is below 43°F, THEN the MCU turns power to the heater ON, ELSE it keeps the heater power OFF. Another example that demonstrates the LOOP action is controlling the filling station in a bottling operation. The loop may be organized either to fill a predetermined number of bottles or to continue filling until no more bottles are delivered to the filling station.

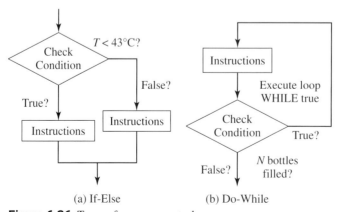

(a) If-Else   (b) Do-While

Figure 6.26 Types of program controls.

Example 6.16: IF-THEN-ELSE operation–A level sensor monitors the level of fluid in a tank. A pump is controlled to run as long as the fluid level is below the threshold or desired level. Develop a simple code to manage the operation of the pump. Assume the sensor output "level" is stored at $0090 and that the desired fluid level is $C1 cm. The tank is initially empty. Figure 6.27 shows a flowchart for the operation.

Solution: This is an example of an IF-THEN-ELSE operation. It is interpreted as follows: IF the "level" is LESS THAN $C1, THEN keep the pump on, ELSE stop the pump.

```
          ORG     $C100       ; Start address of the following
                                code
          LDAB    #$C1        ; Load B with the desired level
          JSR     pump_on     ; Turn pump on and start filling the
                                tank
not_full  LDAA    $0091       ; Read "level" from address $0090
          CBA                 ; Compare "level" with the desired
                                value
          BGE     full        ; If A ≥ B, tank is full
          BRA     not_full    ; If A ≠ B, tank is not full
full      JSR     pump_off    ; Turn pump off
here      BRA     here
```

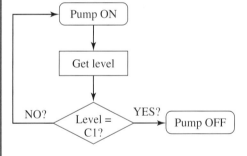

Figure 6.27 Flowchart for the operation of Example 6.16.

Example 6.17: Do-While operation–A bottling station is to be controlled to fill *N* bottles. The bottles are delivered to the bottling station by an appropriate mechanism. If a bottle is properly delivered to the filling station, it closes a switch hooked up to PB[4] of port B at address $0001. When this happens, the CPU initiates the filling operation. The process continues until all *N* bottles are filled. Write a code for the MCU to control the bottling operation. Figure 6.28 shows a flowchart for the operation.

Solution: This is an example of a DO-WHILE loop, interpreted as follows: WHILE not all *N* bottles are filled, execute (the DO part) the filling operation LOOP.

```
N         EQU     40          ; Number of bottles to fill
          ORG     $C000       ; Start address of the
                                following code
```

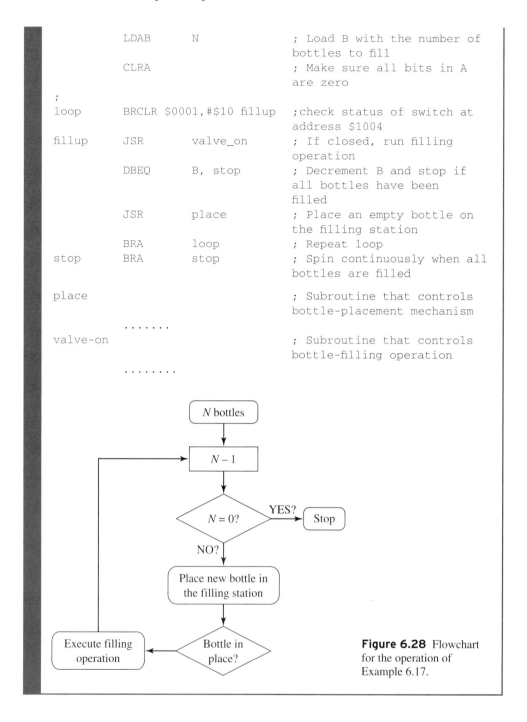

```
            LDAB      N              ; Load B with the number of
                                     bottles to fill
            CLRA                     ; Make sure all bits in A
                                     are zero
;
loop        BRCLR $0001,#$10 fillup  ;check status of switch at
                                     address $1004
fillup      JSR       valve_on       ; If closed, run filling
                                     operation
            DBEQ      B, stop        ; Decrement B and stop if
                                     all bottles have been
                                     filled
            JSR       place          ; Place an empty bottle on
                                     the filling station
            BRA       loop           ; Repeat loop
stop        BRA       stop           ; Spin continuously when all
                                     bottles are filled

place                                ; Subroutine that controls
                                     bottle-placement mechanism
        . . . . . . .
valve-on                             ; Subroutine that controls
                                     bottle-filling operation
        . . . . . . . .
```

Figure 6.28 Flowchart for the operation of Example 6.17.

Jump and Subroutine Call/Return Instructions

This group of instructions is listed in Table 6.16. The jump instruction JMP is used to redirect program execution to any absolute address within the standard memory space without any set condition. The following instructions demonstrate its use:

TABLE 6.16 Jump and Subroutine Call and Return Instructions

Instruction		Operation Performed
JMP	$200, X	Jump to address (X + $200)
BSR	fire	Branch to subroutine fire
JSR	lighton	Jump to subroutine lighton
RTS		Return from subroutine
CALL	[D, SP]	Call Subroutine in Extended Memory at (D + SP)
RTC		Return from Call

```
JMP    $C015    ; Jump to address $C015
JMP    there    ; Jump to address label "there"
JMP    [D, X]   ; Jump to address D + X
```

The use of subroutines facilitates structural programming. A large program is divided into smaller modules that are easier to debug and read. Program size may also be reduced if frequently used segments are coded in subroutines. The main program can call a subroutine more than once, and a subroutine can call other subroutines, including itself. The CPU uses the stack automatically for temporary storage of return addresses during subroutine calls. Figure 6.29 shows program

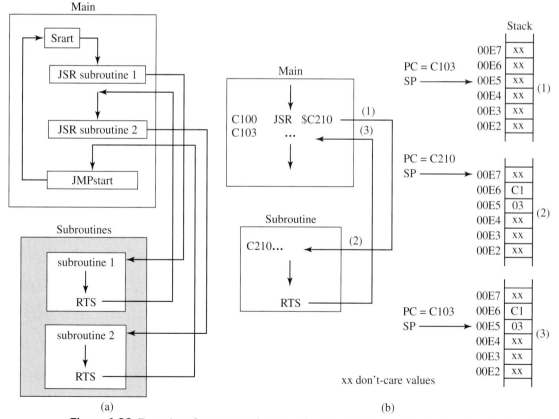

Figure 6.29 Execution of a program with subroutine (a) and stack operation during subroutine jump (b).

flow using subroutines and the stack operation during a subroutine call. Example 6.18 demonstrates the use of subroutines.

The short branch instruction BSR may be used if the subroutine address is at 8-bit relative offset from the current position. The JSR instruction, on the other hand, can cause execution of a subroutine anywhere in the standard 64K address space. The CALL instruction can call subroutines located in an expansion memory page. A subroutine terminates with an RTS or RTC instruction to return program execution to the instruction following the one that called the subroutine. The following traffic light control example includes a subroutine to generate time delays.

Example 6.18: Traffic light control–Consider the control of traffic lights at an intersection of roads bound east–west (EW) and north–south (NS) (see Fig. 6.30). A set of three LEDs is used to indicate the traffic signals for each direction: one green, one orange, and one red. The EW lights are connected to pins PM[0] (green), PM[1] (orange), and PM[3] (red) of port M at address $0250, whereas the NS lights are interfaced to PM[3] (green), PM[4] (orange), and PM[5] (red) pins. Develop a simple assembly code for this application (C-code implementation is given in Example 6.20).

Solution: The following code allows for a long time delay for EW (NS) traffic to pass as EW green (NS red) is on, and vice versa. It also provides for a short time delay for an orange light. The values for the "longt" and "shrtt" variables are set to provide the required time delay. Since the code runs continuously, the segment of code responsible for generating the short time delays and the long time delays are organized in subroutines.

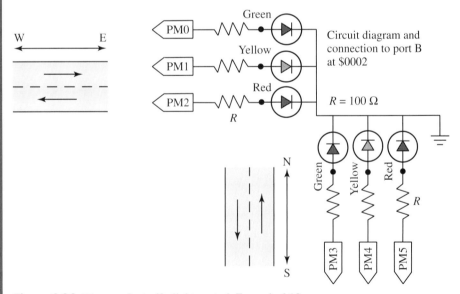

Figure 6.30 Diagram for traffic light control, Example 6.18.

```
; Traffic light control. The Assembler adds a few lines of com-
ments that are removed.
   INCLUDE 'mc9s12c32.inc'    ; include 9S12C specific macros
```

```
            ORG     $C100
table       DC.B    $21, $22, $0C, $41 ;Traffic Light Control
longt       DC.W    $FFFF
shrtt       DC.W    $4000
            ORG     $C000
main        ; Main code begins here
            MOVB    #$FF, DDRM  ; Configure port M for output
            LDX     #table
loop        LDAA    1, X+       ; Load A with $21
            STAA    PTM         ; Turn on EW green and NS red
            JSR     longdly     ; Long delay
            LDX     #table
            LDAA    2, X+       ; Load A with $22
            STAA    PTM         ; Turn on EW yellow and NS red
            JSR     shrtdly     ; Short delay
            LDX     #table
            LDAA    3, X+       ; Load A with $0C
            STAA    PTM         ; Turn on EW red and NS green
            JSR     longdly     ; Long delay
            LDX     #table
            LDAA    4, X+       ; Load A with $41
            STAA    PTM         ; Turn on EW red and NS yellow
            JSR     shrtdly     ; Short delay
            BRA     loop        ; Repeat loop

longdly:
            LDX     longt       ; Long-delay subroutine
loop1       LDY     #$0800
wait1       DBNE    Y, wait1
            DBEQ    X, done1
            BRA     loop1
done1       RTS

shrtdly:
            LDX     shrtt       ; Short-delay subroutine
loop2       LDY     #$0080
wait2       DBNE    Y, wait2
            DBEQ    X, done2
            BRA     loop2
done2       RTS
```

Interrupt-Handling Instructions

When the CPU receives an *interrupt request* from a source, a signal that warrants immediate attention, it completes the execution of the current instruction and then vectors off (jumps) to execute a special *interrupt* or *reset service routine* (ISR). This action is called an *interrupt operation*. The interrupt mechanism provides the MCU with the capability for *real-time control* of external events and hardware connected to I/O lines. This section focuses on the software interrupt-handling instructions. Chapter 7 covers hardware interrupts.

TABLE 6.17 Miscellaneous Instructions

Instruction		Operation Performed
BRA	poll	Branch Always to *poll*
BRN	fall	Branch Never to *fall*
NOP		No Operation (1-cycle delay)
STOP		Stop clocks

The software interrupt instruction SWI behaves like an interrupt, in that it causes the CPU to save the contents of all working registers on the stack, set the I bit in the CCR, and fetch the contents of its vector address. SWI is useful in simulating hardware interrupts during system development. A SWI service routine is used to implement *break point* functions in debugging software, such as simulators. A break point marks the location of an instruction where it is desired to halt the MCU execution to allow the developer to assess the performance of the program up to that point.

Some applications do not require continuous program execution; rather, pausing for a while and then resuming the execution at a later time may be desirable. The wait-for-interrupt instruction WAI puts the MCU in a wait mode until some interrupt occurs. This *standby state* reduces power consumption by the MCU while the clock continues to tick. When a WAI instruction is executed, the CPU stacks the registers, suspends execution, and waits for an unmasked interrupt to occur. WAI can also be used to reduce some latency time to some important interrupt.

6.4.8 Miscellaneous Instructions

This group of instructions is listed in Table 6.17. The no-operation instruction NOP serves several useful purposes. It can be used to introduce a small time delay into the flow of a program to meet time requirements of slow peripherals. NOP instructions may be executed in nested loops to generate larger time delays into loops. The branch-never instruction BRN is equivalent to a three-cycle NOP. BRA instructions will always branch to a specified address. Another way to modify program execution time is to use instructions in sequence with alternate addressing modes without changing the program's function. NOP may also be used to replace unwanted instructions during program development to effectively remove them without having to rearrange the rest of the program.

The effect of executing the STOP instruction depends on the S bit in the CCR. If the S bit is set, the CPU treats the STOP instruction as a NOP instruction. If S is clear, the STOP instruction freezes the oscillator and all MCU clocks and puts the MCU in the stop mode, halting program execution and reducing power consumption. RESET, XIRQ, or IRQ signals can be used to wake up the MCU. Sections 7.8–7.10 discuss the operation of these functions.

6.5 ASSEMBLER DIRECTIVES

The assembler program that will convert the source assembly code to the machine code executable by the 9S12C MCU is the Macro Assembler, which is part of the CodeWarrior IDE Studio from Metrowerks. The Macro Assembler supports a specific set of assembler directives the user may include in the source code to tell the assembler how to convert the program instructions into machine code during the assembly process. The assembler directives occupy the instruction field on the assembly lines as statements, but the assembler does not convert them into machine code during assembly. The Macro Assembler supports integers and string constants. The integer constants can be decimal, hexadecimal, octal, or binary. A string constant consists of printable characters enclosed by single quote marks or double quotes.

The most commonly used assembler directives supported by the Macro Assembler are briefly introduced in this section; examples of their use are available in the source code of previous examples. The HC12 assembler document available at www.freescale.com describes all Macro Assembler directives in detail and provides the equivalent directives used by other assemblers. The directives are categorized according to their use, as presented next.

6.5.1 Section Definition Directives

The ORG directive sets a location counter that defines the start address of the code or the absolute segment that follows it. For example:

```
        ORG     ram     ;Start the absolute segment at address
                        $0380
timer   DS.B    2       ; The "timer" occupies $0380 and $0381
                        locations
queue   DS.B    10      ;"queue" occupies $0382 - $038C locations
```

The ORG directives tells the assembler to assign two space bytes for the "timer," followed by 10 spaces for the "queue" at the beginning of ram section.

The SECTION directive is used to specify a relocatable section of code of at least one instruction, or a section of constants with only DC or DCB directives, or a section of data with at least a DS directive. It facilitates modular programming by specifying sections of code. The directive may be assigned a name defined in the label field. The first time the SECTION directive is used, the location counter for the section is set to zero. If SECTION directives are subsequently invoked for the same section, the location counter will begin with the value that follows the address of the last code in the section.

```
position   SECTION         ; Define a data SECTION called
                           position
angle      DS.B 1          ; Allocates one space for angle
motion     SECTION         ; Defines a code SECTION called
                           motion
right      LDAA    portb   ; Read limit switch connected to
                           Port B
           STAA    angle   ; Store A in angle
           STAA    porta   ; Display the switch status on a
                           LED connected to port A
```

6.5.2 Constant Definition Directives

The EQU, or *equate*, directive defines a name in the label field only once and assigns a value to it. For example,

```
tmax    EQU     $40     ;assign the value $40 to the name
                        tmax
```

6.5.3 Data Allocation Directives

The DC, or *define constant*, directive assigns constants in memory for each operand in the directive. Postletter B is used to specify 1 byte to be allocated for each constant, letter W to specify two bytes, and L to specify four bytes. In Example 6.7, the assembler assigns $20 to the label "count" at address $3801 and the 20 bytes to the "table" starting at address $3802.

```
first       DC.W   $3A2C,$0110,$42A2,$7D2F,$6E81
pmax        DC.B   20
display     DC.W   "High Pressure"
```

The DCB, or *define constant block*, directive tells the assembler to reserve a memory block of specific length, each of a specific size, starting form the specified initial value.

```
tower   DCB.B   3, $FF     ; allocates a 3*1 bytes memory block
data    DCB.W   3, $FFFF   ; allocates a 3*2 bytes memory block
area    DCB.L   3, $FFFF   ; allocates a 3*4 bytes memory block
                             starting at $FFFF
```

The DS, or *define space*, directive tells the assembler to allocate memory space for variables. It does not assign data to the addresses. In Example 6.7, the assembler allocates one byte for the label "minbyte" at address $3800. Other examples:

```
maxt    DS.B  2
period  DS.W  6
```

6.5.4 Assembly Control Directives

The INCLUDE directive instructs the assembler temporarily to insert into the current source code the contents of another specified file before assembly begins. These files are often known as *header files*. They are generally long files that contain a section that would appear in many source codes, for example, the directive statement

```
INCLUDE  "equates.asm"  ; Include register equates code
                          "equates.asm" in the current source
                          file during assembly.
```

The file must be enclosed in quotations.

The END directive marks the end of the source code processed by the assembler. All subsequent statements will be ignored.

The XREF directive tells the assembler to refer to an external source file for information.

The XDEF directive directs the assembler to make the code public for use by other source codes.

6.5.5 Listing File Control

The LIST directive tells the assembler to list the source text that follows the LIST directive until a NOLIST directive or an END directive is reached.

```
NOLIST                  ;Do not list "equates.asm" file
  INCLUDE               "Equates.asm"
  LIST                  ;list the rest of source text
```

The LLEN <n> directive sets the number of characters from the source file to be included on the listing line to <n>.

The TITLE directive tells the assembler to print the title of the program on the top of each listed page.

The PAGE directive inserts a page break in the assembly listing. The NOPAGE directive does the opposite.

6.5.6 Conditional Assembly

The IF . . . ELSE . . . ENDIF construct is a conditional assembly directive that instructs the assembler to conditionally assemble portions of the source code. It provides flexibility by allowing the reuse of source code sections. The statement between IF and the corresponding ELSE is assembled if the condition is true; otherwise the statement between ELSE and the corresponding ENDIF will be assembled. For example, if a large box moving on a conveyor belt is detected, it should be routed to the bin on the right; but if a small size box is detected, it should be routed to the left bin. The following example demonstrates the use of this directive.

```
IF      large = 1
        STAA  rightbin  ;Assemble if a large box is detected
ELSE
        STAA  leftbin   ;Assemble if small box is detected
ENDIF
```

6.5.7 Macro Control

The MACRO directive allows the user to define a macro and create new instructions from a sequence of MCU instructions. A MACRO instructs the assembler to perform some task before it starts the assembly process. The MACRO directive is a label that replaces a block of instructions. Consider how the macro named "pump_on" is created:

```
pump_on  MACRO
         LDAA    #$FF
         STAA    portb
         ENDM
```

If the macro is invoked in the source code, the assembler substitutes the macro reference "pump_on" with the sequence of instructions it contains. In this case the assembler loads $FF into A and then stores A into portb. ENDM is used to indicate the end-of-macro definition. Though macros are faster than subroutines, they do tie up more memory.

6.6 DEVELOPMENT OF AN ASSEMBLY LANGUAGE PROGRAM

An assembly language program must follow the rules of the associated assembler to avoid possible error messages from being generated during assembly. An assembled assembly language program will eventually contain three main sections: A code section contains the instructions dedicated to the operation of the target system; a data section contains the data processed by the instructions; and a stack section stores information temporarily during execution.

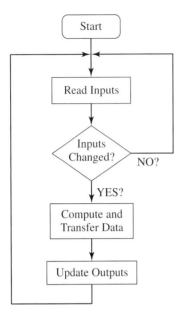

Figure 6.31 Polling programming strategy.

6.6.1 Program Strategies

One of several programming strategies can be used to organize the application code and construct algorithms for control applications. Commonly used methods include simple loop polled I/O, interrupt driven, and state machines. Depending on the application at hand, a practical program may combine all programming structures to provide acceptable performance; each has its advantages and disadvantages.

Polled I/O

Figure 6.31 shows the structure of the *polling* scheme in which the CPU runs in a tight loop, repeatedly executing a sequence of instructions and continuously checking inputs and modifying outputs accordingly. Since all I/O transfers are controlled by the program, the CPU is blind to changes in the input data that might occur. Additionally, CPU is tied at all times, even when no external events are occurring. This approach may work well for simple I/O operations; but as the complexity of the I/O requirements increase, it becomes inefficient and may render some events undetected when they occur. An improved version of the polling scheme is to allow the loop to respond only to changes in the inputs, as indicated in Fig. 6.31.

Interrupt Driven

Significant performance improvements over the polling scheme can be realized if program execution is based on interrupts, especially in real-time control applications. In this type of program, various events called *interrupts* are allowed to interrupt the CPU operation as it executes a foreground task and to redirect program execution to any desired destination. This allows real-time events to leverage the flow of program execution and provides the I/O devices a means to notify the CPU when its attention is needed.

The 9S12C MCU handles interrupts by means of software that recognizes input changes when they occur and responds to them accordingly. It provides a wide range of CPU hardware/software interrupts. Detailed treatment of interrupts is presented in Chapter 7.

State Machines

A *finite-state machine* (FSM), or simply *state machine*, is a simple, abstract way of representing a sequential process that evolves in time. A state machine has a set of valid states and a set of transition rules to move the machine between those states. The transition from a state to the next depends on the current state and present inputs. The action taken during a transition may be simply to change the state number or may lead to executing a task. Vending machines, combination locks, and elevator control are examples of sequential machines. The user must deposit the right amount of money in the vending machine before the selected item is dispensed. A security combination lock will open only if the user enters the right sequence of numbers. The elevator system must remember requests made by passengers on different floors while in motion to make smooth transition between floors (states). A state machine can be implemented with hardware, or it can be realized using software. In a software state machine, the past history of the process defines the current state. The program decides to move the process from one state to another based on both the current inputs and the present machine state. Before the code is developed, state machines are first expressed as either state transition diagrams or state tables.

Figure 6.32 shows a flowchart of a simple state machine dedicated to the operation of an elevator (refer to Example 7.9). At any given time the program is in one of a finite number of valid states (floor position). The gate keeper in the main program receives a signal, possibly from an interrupt source, and executes the code dedicated to updating the outputs necessary to make the transition to the destination state (floor). Once a transition is complete, control is returned to the main program to make updates at regular times and to limit the rate and circumstances under which transition to another state (floor) is made.

6.6.2 Source Code Structure

A typical source code is organized in several sections according to the following hierarchy.

The programs begin with the *header comments* section, which describes the purpose of the program and its organization. This is followed by comments on the *program hierarchy*, which

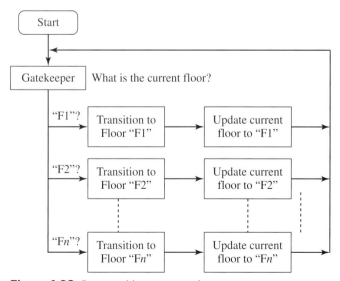

Figure 6.32 State machine programming strategy.

is like the table of contents of the program. It gives, in order indicated by indentation, the name of the mainline program, the name of each mainline subroutine it calls, and the list of subroutines called by each mainline subroutine.

The *equates* section defines constants used in the program as symbolic labels using the EQU directive. Symbolic labels make code easier to understand and debug.

The *allocation of variables* section allocates a specific memory location for each variable name used in the program. This section begins with an ORG directive to point to the addresses to be used by the assembler.

The *mainline program* section contains the application code responsible for the operation of the target device for which it was developed.

The *subroutines* section includes the codes of all subroutines. A subroutine begins with header comments that describe its purpose and define parameters transferred to and from it. Before a subroutine terminates with an RTS or RTC instruction, it should restore the registers to their original values, except those that are used to pass parameters back to the calling routine.

The *interrupt service routines* section includes a listing of all service routines developed to service interrupt requests from corresponding sources.

The *tables* section contains tables of data essential to the operation of the application code. For example, for a temperature measurement using a thermistor, a lookup table might include temperature values corresponding to thermistor's output voltage. The CORDIC algorithm for computing trigonometric functions is organized as a table.

Strings of ASCII characters is a list of the character strings that form messages to be presented to a user on a display or sent to a computer via universal asynchronous receiver-transmitter (UART) output. The ASCII code is given in Appendix F.

Interrupt and reset vectors are specified that require exact positioning in the memory space, as indicated in Table 7.5.

6.6.3 Conversion from Assembly Code to Machine Code

The prototype hardware, or the *target system*, is the focus of both hardware and software development work. The application code represents the value added to the MCU, and it should be carefully developed to support all pertinent hardware functions. Eventually the code will reside in the MCU memory for execution.

The development chain of the machine code that controls the operation of the MCU begins with the source code. The source code is a text file created by a text editor or a word processor that contains all instructions required to carry out target system tasks. The source code is converted into the required machine code format for the target CPU by the *assembler*. The assembler generates the *object code*. The object is a *relocatable code*, in the sense that it can start anywhere in memory without affecting its operation. The object code contains the machine code instructions, plus other information, but does not necessarily contain the final address assignment of all instructions. A large application code may consist of several separately generated object code sections. A program called the *linker* combines two or more object code sections to produce the hex code that contains the final machine code, including final address assignments, ready to load into the MCU memory. The assembler and linker are usually run on a host computer and not on the MCU itself. Finally, a *loader* program is used to convert the hex code into the executable binary machine code.

The Integrated Develeopment Environment provided by Metrowerks CodeWarrior Development Studio integrates all the steps involved in the chain of machine code generation for Freescale 68HC(S)12 MCUs.

6.6.4 Debugging Tools

After assembly, the machine code is debugged to verify its performance. Debugging tools include simulators and emulators. A *simulator* is a software program that resides on a PC in the form of simulated MCU. A simulator allows the user to step through the code and monitor the variables, registers, ports, and memory to see if they are changing as they should. The simulator can run in the single-step mode, or it can run from reset until it reaches a *break point* inserted by the user at desired locations to check the contents of the MCU's simulated memory and registers. If execution errors are detected, the user corrects them and continues execution, either in single steps or until another break point is detected. The simulator is limited to checking the application code only; it provides no means for simulating the target system in which the MCU will be embedded. This latter function is provided by a device called the *emulator.*

The emulator is a piece of hardware that connects to a PC on one end and has a cable that connects to a socket where the MCU will ultimately reside on the target board. The target system thinks of the emulator as if it were the real MCU sitting in its socket. The emulator allows the user to develop complex codes on a PC and to monitor and control the internal state of the MCU, look at registers, stop execution, and modify memory locations.

Once the code is bug-free, it is downloaded from the PC into the EEPROM of the MCU in a process known as *burning the code* into the chip. It is accomplished by means of a *device programmer*. One end of this device hooks to a PC, while the other is equipped with a socket into which the MCU is inserted.

6.7 HIGH-LEVEL LANGUAGE

The source code of an application program may ultimately include a mix of assembly language and high-level language code segments. High-level language provides a more user-friendly platform for program development, allowing the programmer to express programs in a way that makes no reference to registers or instruction sets of the target microprocessor. However, knowledge of bits and registers, as well as available memory resources, is required. High-level languages are therefore intended to be machine independent, thus portable. The programmer is only required to obey the programming rules of the language. For large and complex applications, such as database applications, operating systems, graphics applications, and artificial intelligence, bulk programming is usually done in high-level language, with time-critical functions handled in assembly.

Programming in C is a common practice in developing MCU-based systems because it blends the power and portability of the high level with the efficiency and flexibility of assembly. The source codes of a C-language program conforms to ANSI (and other) standards. A C-based source code is translated into equivalent assembly code for the 9S12C by a dedicated compiler. The assembler is then used to convert the assembly code to the hex machine object code that will eventually be linked and download to the MCU as a machine code for execution. Although the C-compiler usually optimizes the equivalent object code, the machine code generated from a well-written assembly language program that performs the same function is generally shorter and executes faster. However, programming in C is relatively easier than assembly, and the machine code derived from a well-written C code is usually compact in size and runs faster than assembly while allowing direct control over I/O devices for real-time control applications. Example 6.19 highlights some essential features of a C-language program.

Example 6.19: A C-code example—Write a C source code to add two arrays, array1 and array2, used in Example 6.5.

Solution

```
/* Refer to Example 6.5, where an assembly code version for
the same example was developed.
   main ( )
{
   int count;
   static int array1[16] = {4, 7, 5, 23, 65, 9, 7, 74, 52, 3,
   12, 82, 71, 27, 14, 8 };
   static int array2[16] = {7, 12, 9, 30, 44, 2, 1, 8, 34, 7,
   19, 61, 15, 37, 49, 73};
   count = 16;
   while (count--)
   {array1[count] = array1[count] + array2[count];}
}
```

Every C program, which may include many files, begins with the label "main". The variable "count" has been declared as an integer variable. The arrays are declared as having 16 integer elements, and each element may be referenced by an index so that the first value of array1 is "array[0]" and the last is "array[15]". The statement "while (count--)" evaluates the variable count within its parentheses as a Boolean variable, the result of which will be false if the value of "count" is zero, or it will be true if "count" has a value other than zero. The "--" causes "count" to be decremented immediately after the evaluation so that the index value is correct for the arrays. The term "static" in the declarations of the two arrays is used to indicate to the compiler the type of integer to be used. The term "static" is referred to as a "storage class identifier" and determines how functions, or subroutines, access those variables. Every statement terminates with a semicolon ";". A group of single statements enclosed in braces { } represents a combined statement.

6.7.1 C-Programming for the 9S12C MCUs

The CodeWarrior Development Studio from Metrowerks includes an ANSI-compliant C compiler for the MC9S12C32 MCU. When a new 9S12C project is started, the C compiler provides the user with a sample C code to use as a template, with the following lines at the top:

```
#include  "hidef.h"          /* common defines and macros */
#include  "mc9s12c32.h"      */ derivative information */
#pragma   LINK_INFO DERIVATIVE  "Mc9s12c32"
```

The first two lines make reference to the header files that allows the application program to access the register bits of the 9S12C MCU and other related information using their standard mnemonics. "include" is a preprocessor directive that tells the compiler to insert the header files in quotes, "hidef.h" and "mc9s12c32.h", into the current source that will be executed. The header files and the source code must reside in the same directory at the time of execution. The third line, beginning with `#pragma`, defines how information is passed from the front end (C-language dependent) to the back end (target-processor dependent) . The particular pragma LINK_INFO instructs the compiler to pass information to the linker from the contents of the "mc. . ." The comipler will include the DERIVATIVE "mc. . ." pair into the ELF file.

Before variables are used, the program must first *declare* them using reserved words that specify their types. The following statements demonstrate how various types are declared:

```
char      title[ ] = "System Response:";
unsigned  count;           /* Declare count as an unsigned
                              number */
short     pmax;            /* Declare variable "pmax" as
                              unsigned 8-bit integer
long      hperiod 100;  /* Declare "hperiod" as a long data
*/
Int       temp [6] = {10, 20, 25, 18, 54, 29}
                           /* Define integer array with 6 entries
                              */
```

The C program must begin with a "`main`" statement such as

```
void main (void)
{
.... }               */ End of main */
```

`main` is a primary C function. All statements within a main must be included within braces. The main code defines the ports and initializes them, call functions, and so on. To initialize the ports, the following statements are included in main:

```
DDRT = 0b11111111  /*  set up Port T pins as outputs */
PTT = 0xFF         /*  initialize Port T bits */
DDRM = 255         /*  Set up Port M as outputs (255 = 0xFF
                       = 0b11111111) */
PTM = 0x5A         /*  Send 0x5A byte to Port M*/
```

The "0b" prefix indicates a binary number, the "0x" prefix indicates a hex number, and no prefix means a decimal number.

The C code may read any port and store its contents in a RAM variable. The statement

```
temp = PTM  /* Store port M at temp */
```

reads port M and stores the result in a RAM location defined by the label "`temp`".

Specific bits of a port may also be assigned the name of the device connected to it. For example, if the code includes the statements

```
#define switch PTT_PTT0  /* Assigns the name "switch" to bit 0
                            of port T
#define motor PTM_PTM1   /* Assigns the name "motor" to bit 1
                            of port M
```

then anytime the names *switch* and *motor* appear in the code, they will refer to pin PT[0] of port T and pin PM[1] of port M, respectively.

The actual program begins after initialization and bit definitions are complete. Pins are referenced by their names to affect their states. To turn the motor connected to pin PM[1] of port M on or off, the source code includes the statement

```
motor = 1  /* Turn motor on   */
motor = 0  /* Turn motor off  */
```

Conditional constructs are used to perform operations that depend on a certain condition's being satisfied. For example, to turn the motor on if the switch is closed, the code will include

```
If (switch = = 1)
{
    motor = 1;
}
```

These statements drives pin PM[1] of port M high and turn the motor on if Pin PT[0] of port T becomes high when the switch connected to it is closed. The condition for execution may also be constructed using any of the recognizable logical bit operators, namely "|" for OR, "&" for AND, and "~" for NOT. Furthermore, two logical bit reassignment operations are possible: "|=" and "&=". The following two statements demonstrate the use of these two operators and their meaning:

```
tflg1 | = $03   /* tflg1 = tflg1 OR $03 */
tflg1 & = $08   /* tflg1 = tflg1 AND $08 */
```

The C code could be organized using functions that are similar to subroutines. The following is an example of a function that would start the motor when a switch is closed.

```
void motoron (void)
short start (void)

main
{
    while (true)
    {
        If(start( )) motoron ( );
    }
}                          /* end main   */

short start ( )            /* function checks switch status,
                              return true if switch is closed
{
    if switch == 1
        return (true)
    else
        return (false)
}
motoron ( )                /* function to turn motor on
{
    PTM_PTM1 = 1;
}
```

Function "`motoron (void)`" is defined as a void, meaning that it performs an action only and does not return a value to the calling function. Function "`start`" is defined as short because it returns a signed 8-bit integer value to the calling function.

Example 6.20: A C code for traffic light control–Write a C source code to control the traffic light for Example 6.18.

Solution

```
/* Refer to Example 6.18 for assembly code version for this
example

#include <hidef.h>      /* common defines and macros */
#include <mc9s12c32.h>  /* derivative information */

#pragma LINK_INFO DERIVATIVE "mc9s12c32"

void longdly (void)
    {   int i;
        for(i=0;i<100;i++) {
            for(i=0;i<1000;i++) {
            }
        }
    }
void shrtdly(void)
    {   int i;
        for(i=0;i<10;i++)  {
            for(i=0;i<1000;i++)  {
            }
        }
    }

void main(void) {
    int n[4] = {0x21, 0x22, 0x0c, 0x41};  /* Traffic light
                                          sequence */
    DDRM = 255;
    while (1)             /* Keep traffic light running */
        {
        PTM = n[0];    /* Send first traffic code to port M */
        longdly();     /* Execute long time delay */
        PTM = n[1];    /* Send second traffic code to port M */
        shrtdly();     /* Execute short time delay */
        PTM = n[2];    /* Send third traffic code to port M */
        longdly();     /* Execute long time delay */
        PTM = n[3];    /* Send fourth traffic code to port M */
        shrtdly();     /* Execute short time delay */
        }
}
```

6.8 DEVELOPMENT TOOLS FOR THE MC9S12C

Various platforms are available from many sources to develop 9S12C-based embedded systems. The MCU Student Learning Kit (MCUSLK) from Freescale Semiconductors is a simple stand-alone, low-cost suite that includes everything the user needs to develop skills in interfacing concepts and writing application codes for the MC9S12C MCU and verifying their performance. Figure 6.33 shows a picture of the MCUSLK kit. It includes a CSM12C32 application module with a 48-pin QFP MC9S12C32 MCU core, a prototyping platform (Project Board 2) with interface connection to the CSM12C32 module, CodeWarrior Integrated Development Environment (IDE) Studio, power adaptors and connecting cables, and supporting documents.

The CSM12C32 module includes an MC9S12C32 MCU, a 40-pin connector to allow access to almost all I/O signals and to connect the module to the Project Board 2, a 16-MHz ceramic resonator, a background DEBUG port compatible with the 9S12C BDM interface cables and software, a monitor preloaded into the MCU flash and accessible through the COM connector for user code development. The MC9S12C32 User Manual and Reference Manual documents describe those features in detail.

The Project Board 2 provides bidirectional communication between the CSM12C32 module and the debugger. It contains a USB BDM POD to connect the target board with a host PC and to aid in program development and debugging via the BDM. The board has a PCI–edge connector for use with the NI-ELVIS, National Instrument Educational Laboratory Virtual Instruments workstation. The Project Board 2 may be powered from one of three available sources: a wall plug transformer, the integrated USB-BDM, or the NI ELVIS workstation. The onboard voltage regulator provides four different voltage levels, with an LED indicator for

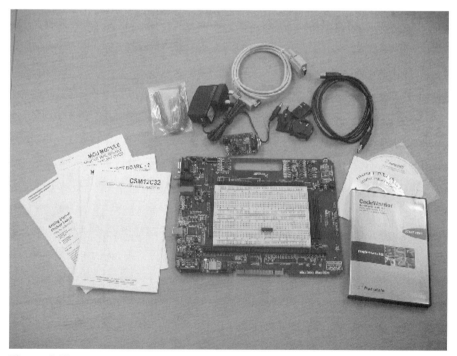

Figure 6.33 A photo of the components of the MCUSLK kit.

each level: 5 V_{DC} @ 500 mA, -3.3 CDC @ 500 mA, and ± 15 V_{DC} @ 50 mA when powered from the USB-BDM. The board provides socket header's breakout connectors to allow for user-generated I/O signals to access the following auxiliary components available onboard: an 8-pin connector for a 12-key or 16-key keypad; one single 5-kΩ trim pot for use in prototyping; an LCD module interface with a serial-to-parallel shift register connector configured to add an-character $\times$ 2-line LCD (liquid crystal) display; a COM port with a 9-pin DSUB connector; an RS-232 interface configured as a DCE (Data Communication Device) with an option to isolate the transceiver; a socket to add either a 14-pin or a canned 8-pin optional crystal oscillator; eight common-cathode green LED (light-emitting diode) displays; eight active low push-button switch inputs and eight active-high DIP switches; a large solderless breadboard area for adding components and prototyping. Refer to the MC DOC-0368-050 REV A document for details (available from Axiom manufacturing at www.axman.com).

The CodeWarrior Integrated Developoment Environment (IDE) Studio includes a project manager, a search engine, a source browser, a build system, and a debugger to facilitate user code development. The user code may be assembly, C, or combination of the two languages.

6.8.1 Background Debug Mode (BDM)

The role of the BDM module of the 9S12C is to synchronize serial communication between a host system and the 9S12C target processor for application development, debugging, and testing of target systems.

The BDM is accomplished via a single wire that connects the host to the BKGD pin on the MCU. The BKGD pin becomes the serial interface dedicated to the BDM operation after the MCU comes out of reset. The BDM is available in all operating modes of the MCU if it is enabled by setting the ENBDM bit in the BDM status register using a hardware command such as WRITE_BD_BYTE. Once enabled, the BDM is activated by a hardware BACKGROUND command received on the serial BKGD pin, executing a CPU BGND instruction, BDM instruction tagging mechanism. When active, the BDM control registers and a small ROM are mapped to addresses FF00–FFFF. Three registers are dedicated to the BDM function. The *BDM status register* (BDMSTS) occupies address $FF01, the *BDM CCR holding register* (BDMCCR) occupies address $FF06, and the *BDM internal register position register* (BDMINR) occupies address $FF07. The BDM registers are used by the BDM logic and not by the CPU. The ROM holds a lookup table of the firmware that manages the interaction between the host and the MCU and provides a full set of debug options.

The BDM control logic does not reside in the CPU; it communicates serially (via the USB on Project Board 2) with an external host via the single BKGD pin. There are nine BDM hardware commands used to read and write all target system memory locations that are accessible by the CPU and to enter active BDM by executing a hardware command such as WRITE_BD_BYTE. Hardware commands can execute while the CPU is operating normally, taking advantage of free cycles, when available, if the operation can be completed in a single cycle; otherwise the CPU clocks are frozen until operation is completed. The read commands are READ_BD_BYTE, READ_BD_WORD, READ_BYTE, and READ_WORD. Replacing the READ with a WRITE forms the four hardware write commands. The other hardware commands are BACKGROUND, ACK_ENABLE, and ACK_DISABLE.

Background firmware commands require the CPU to be in the active BDM to execute. Those commands are used to read and write to the CPU registers D, X, Y, PC, and SP and to exit from the BDM. Program execution can be resumed using the command GO. These instructions include the TRACE1 instruction and return BDM, read or write next byte/word, and the CALL instruction. The firmware consists of six read commands: READ_NEXT, READ_x (x = PC, D, X, Y, SP). Six write commands are formed by replacing READ with WRITE; GO and GO_UNTIL; and TRACE1 and TAGGO. The opcodes for all BDM commands are eight bits long.

The TAGHI and TAGLO pins are used to tag the high byte and the low byte of an instruction word being read in the instruction queue. An active BDM does not interfere with the program execution taking place in the CPU.

Some on-chip modules, such as the timer, include a BDM control bit that allows the user to suspend module operation during BDM and reenable it when program execution resumes.

BKGD is a pseudo–open-drain pin that can be driven by an external controller or by the MCU. It can receive a high or low level or transmit a high/low level. A falling edge received by the serial interface on the BKGD pin indicates the start of each bit time. An external controller generates this falling edge, whether data is transmitted or received. Data is transferred MSB first at 16 E-clock cycles/bit. The interface times out and hardware clears the command register if 256 E-clock cycles occurs between two successive falling edges from the host. Refer to Chapter 6 in the MC9S12C Reference Manual for more details.

6.9 16-KBYTE FLASH MODULE

The 9S12C MCU contains 16 Kbytes fixed and 16 Kbytes paged flash EEPROM modules. The 16-Kbyte flash array module is organized into 256 rows of 64 bytes each. A group of eight rows form a sector of 512 bytes for erasure. The flash array is used to hold the program code and frequently data such as lookup tables for fast execution. It allows for field programming and erasing without the need for an external voltage source, because the high voltage required for those operations is generated internally. The flash array occupies the $C000–$FFFF address space and is ready for use after reset in the single-chip mode. Meanwhile, the flash array is mapped into the $0000–$7FFF address space but is not readily available in the expanded mode. The flash array may later be mapped to an alternate address range.

The logic state of an erased or unprogrammed flash bit is 1 and that of a programmed bit is 0. Flash is arranged in units of 16 bits that may be accessed either in one bus cycle as a byte or as an aligned 16-bit word or in two bus cycles as a misaligned 16-bit word. The flash may be programmed one byte, or aligned word, at a time, whereas erasure is done either one byte or one sector (512 bytes) at a time. Flash EEPROM has hardware safeguards to protect the stored data from being accidently tampered with.

Flash operations are controlled by the bit settings in the six registers shown in Fig. 6.34. The FPROT and FSEC registers are loaded during a reset sequence from flash addresses FF0D and FF0F, respectively.

6.9.1 Security

Access to the flash array is allowed only if the MCU is unsecured by launching a specific sequence of commands to enable backdoor access. The sequence begins by setting the KEYEN[1:0], *backdoor key security enable* bits in the FSEC, *flash security register* at address $0101–1:0, followed by setting the KEYACC, *enable key security writing,* bit 5 in the FCNFG, *flash configuration register* at address $0100. The next step is to write four words of specific code to the backdoor key registers at addresses $FF00–FF07, the first word to

FCLKDIV: Flash Clock Divider Register ($0100)

FDIVLD	PRDIV8	FDIV5	FDIV4	FDIV3	FDIV2	FDIV1	FDIV0

FSEC: Flash Security Register ($0101)

KEYEN1	KEYEN0	NV5	NV4	NV3	NV2	SEC1	SEC0

FCNFG: Flash Configuration Register ($0103)

CBEIE	CCIE	KEYACC	0	0	0	0	0

FPROT: Flash Protection Register ($0104)

FPOPEN	NV6	FPHIDS	FPHS1	FPHS0	FPLDIS	FPLS1	FPLS0

FSTAT: Flash Status Register ($0105)

CBEIF	CCIF	PVIOL	ACCERR	0	BLANK	0	0

FCMD: Flash Command Register ($0106)

0	CMDB6	CMDB5	0	0	CMDB2	0	CMDB0

Figure 6.34 Registers associated with the flash array.

FF00–FF01 and the last word to FF06–FF07, in that order, followed by clearing the KEYACC bit. A match between the written code and the backdoor key code stored in the FF00–FF07 flash addresses prompts the MCU to become unsecured and forces the SEC[1:0], *flash security bits* in the FSEC register to assume the unsecured state of 1:0. The user then will have full control of the four backdoor key words in the $FF00–FF07 flash addresses that can configure.

The user code stored in the flash array needs to retrieve the backdoor key access code through any serial line on the chip. The NV[5:2] bits in the FSEC register are available to the user as nonvolatile flag bits.

6.9.2 Flash Protection

The flash array is protected out of reset to prevent accidental corruption of its contents. However, the out-of-reset default protection may be altered in normal operating modes, allowing for re-programming in single-chip mode. The FPOPEN, *flash protection function for program or erase* bit in the FPROT, *flash protection register at address* $0104 is set to allow the flash address range in the upper sector specified by the FPHS[1:0] bits to be protected when the FPHIDS bit is 0 or to allow no protection at all if the FPHIDS is set. For FPHS[1:0] = 00, 01, 10, and 11, the upper sector specified for protection is $F800–FFFF, $F000–FFFF, $E000–FFFF, and $C000–FFFF, respectively. When FPOPEN is clear, either the whole flash array is fully protected if FPHIDS = 1 or the higher address range is unprotected if FPHIDS = 0. In all, four protection options are possible, depending on the value settings of the FPOPEN and FPHIDS bits:

- *Option* 0 (FPOPEN = 0, FPHIDS = 0): The high range of flash array is unprotected. But to keep it that way the contents of the FSEC register must be changed directly by writing to the flash address $FF0F; otherwise the MCU returns to its secure operating mode after a reset.
- *Option* 1 (FPOPEN = 0, FPHIDS = 1): Full array protection.
- *Option* 2 (FPOPEN = 1, FPHIDS = 0): The higher range sector specified by the FPHS[1:0] is protected.

- *Option* 3 (FPOPEN = 1, FPHIDS = 1): No protection is provided, and all flash commands can be executed.

As a rule, protection can be added but cannot be removed. Therefore, protection mode can change from Option 0 to Option 1, from Option 1 to Option 2, and from Option 3 to Options 0,1, or 2 only. If an attempt is made to program an address or to erase in a protected area of the flash array or if a mass erase command is launched with protection enabled, the PVIOL, *protection violation*, flag in the FSTAT register will set, and the flash command controller will lock until the flag is cleared.

6.9.3 Flask Clock

Any write sequence to the flash array will be ignored unless the FDIVLD, *clock divide loaded* bit 7 in the FCLKDIV, *flask clock divider register*, at address $0100, is set, which will occur when the FCLKDIV register is written to after reset. The purpose of writing to the FCLDDIV is to set the values of the PRDIV8 and FDIV[5:0] bits to set the flash clock frequency (FCLK) to a value between 150 kHz and 200 kHz. This will be satisfied if (1/FCLK + 1/ECLK) > 5 μs. An ECLK frequency < 1 MHz renders the flash array inaccessible, and an FCLK < 150 kHz may destroy the flash array.

6.9.4 Flash Configuration (FCNFG)

The CBEIE, *command buffer empty interrupt enable* bit in the FCNFG, register enables an interrupt request from the CBEIF flag in the FSTAT register at address $0105 when it is set to indicate that the address, data, and command buffers are empty and a new write sequence may be initiated. The CCIE, *command complete interrupt enable* bit, is set to enable an interrupt request from the CCIF flag, which sets after all flash command sequences have been processed. The KEYACC, *enable security keywriting*, bit is set to cause any write attempt to the flash to be treated as a backdoor key and any read attempt to return invalid data. If the MCU's stop mode is activated, the CCIF and ACCERR flags will set and any active flash command will be aborted.

6.9.5 Flash Operations

The flash array may be programmed, erased, or erase verified by executing a strict three-step sequence of several instructions each. Bulk erase and sector erase are also supported. Flash registers and array reads are allowed during a command write sequence, but writes are not possible during a write command sequence.

No flash command can commence unless the ACCERR and PVIOL flags in the FSTAT register are clear. The CBEIF flag must also be set, indicating that the address, data, and command buffers are empty and ready for a command sequence to be issued. Here are the three steps for all possible command sequences.

Step 1. Write to a valid address in the flash array memory. The address will be stored in the FADDR(H and L) registers and the data will be stored in the FDATA(H and L) registers.

Step 2. This step is different for each possible flash command. Only one command type is issued in this step.

    I. Write an *erase verification command* ($05) to the FCMD register. Address and data written will be ignored.

    II. Write a *program command* ($20) to the FCMD register. The data written will be programmed to the specified flash array address.

    III. Write a *sector erase command* ($40) to the FCMD register. The specified flash address determines the sector to be erased, while ADDR[8:0] and the data written are ignored.

    IV. Write a *mass erase command* ($41) to the FCMD register. Address and data are ignored. This is the only command allowed in the special single-chip mode.

Step 3. Clear the CBEIF flag in the FSTAT register by writing a 1 to its bit location (bit 7) to activate the command issued in Step 2. When the CBIEF flag clears, the CCIF flag also clears automatically to indicate that the command has been successfully activated.

The flash module implements the concept of pipelining that queues any command and its associated address and data until any command in progress is executed. Pipelining helps increase the flash speeds.

After all active and pending write command sequences of the same type have been processed, the CCIF flag will set.

Example 6.21: Programming the flash array sector–Write a code to manage the erasure and then programming of a flash sector.

Solution

```
                        XDEF Entry,      ; export symbols
                        main
                        INCLUDE '        ; include derivative specific
                        mc9s12c32.inc'   macros
MY_Variables:           SECTION          ; variable/data section
MyCode:                 SECTION          ; code section begins here
main:
Entry:
data        DC.B        ......           ; insert the data to be
                                         written
            DC.B        $00              ; End of data to be loaded
            JSR         init
            JSR         erase
            JSR         program
done        BRA         done

init        ; Start of the subroutine to initialize flash erasing
            and programming
            MOVB        #$FF, FCLKDIV    ; Write to the FCLKDIV
                                         register to establish FCLK
here        BRSET       FSTAT, #$80, here ; ensure CBEIF flag is set
            LDAA        FSTAT
            ORAA        #$30
            STAA        FSTAT            ; Clear PVIOL and ACCERR flags
            RTS
```

```
erase                                   ; Start of subroutine to
                                          erase the flash array
            LDY     #$C000              ; Flash address sector to
                                          erase
            STAA    1, Y+
            MOVB    #$CC,0,Y            ; Write any data
            MOVB    #$40, FCMD         ; Must immediately follow
                                          the write of data instruc-
                                          tion
            LDAA    FSTAT
            ORAA    #$08
            STAA    FSTAT              ; Clear CBEIF flag
here2       BRCLR   FSTAT, #$40, here2 ; Poll CCIF flag if set to
                                          exit or wait until clear
            RTS
program     ; Start of subroutine to program the flash array
            MOVB    #$FF, FCLKDIV      ; Write to the FCLKDIV reg-
                                          ister to establish FCLK
here3       BRSET   FSTAT, #$80, here3 ; ensure CBEIF flag is set
            LDAA    FSTAT
            ORAA    #$30
            STAA    FSTAT              ; Clear PVIOL and ACCERR
                                          flags
            LDX     #data
byte        LDAA    1, X+
            BEQ     done1
            LDY     #$C000             ; Flash address byte to be
                                          programmed
            STAA    1, Y+
            MOVB    #$80,FCMD           ; Write program command
            LDAA    FSTAT
            ORAA    #$08
            STAA    FSTAT              ; Clear CBEIF flag
here4       BRCLR   FSTAT, #$80,       ; Poll CBEIF flag if set to
                    here4                exit or wait until clear
            BRA     byte
done1       RTS
```

6.10 MICROCHIP PIC MICROCONTROLLERS

The purpose of this section is to highlight the main features of the 8-bit MCUs from Microchip that are referred to as peripheral interface controllers (PICs). A complete list of all MCU versions in all PIC families with available options can be found on the Microchip website at www.microchip.com. The website also includes other, related resources to develop real-world MCU-based designs.

All PIC MCUs are based on the RISC processor with a simple instruction set. They are based on the Harvard architecture, which separates program memory from data memory to allow simultaneous access to both program instructions and data in one instruction cycle. The PIC18F452 MCU is used as a basis for the following discussion.

6.10.1 Architectural Overview of the Microchip PIC 18F452 MCU

The 16F452 MCU has many features that suit a wide range of applications. It is available in three packages: the 40-pin DIP, the 44-pin PLLL, and the 44-pin QFP. Figure 6.35 shows the pinout of the 40-pin DIP version. The PIC18F452 is built around a hybrid RISC/accumulator-based architecture.

The PIC18F452 contains an 8-bit ALU and accumulator that perform arithmetic and logic operations and operate on the working register W and data memory. The CPU supports up to 32 level-deep stacks.

6.10.2 Instruction Set

The RISC processor has a 75-word instruction set. Three instructions are 32-bit (double-word) instructions, and the rest are all 16-bit single-word instructions. The instructions are grouped as data access and movement, arithmetic, logic, compare, shift and rotate, bit manipulation, program flow, control flow, and others. Some instructions involve two operands (one operand is the W register and the other operand is a file register or a constant) and others involve one-operand (the operand is either the W register or a file register). The W register is implicitly specified and cannot be accessed directly.

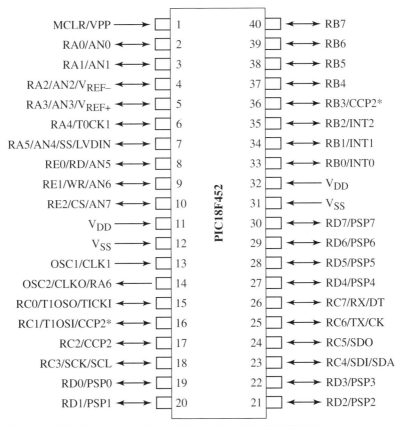

Figure 6.35 Pinout of the 40-pin DIP version of the PIC18F452.

6.10.3 Pipelining

The PIC18F452 is based on the Harvard architecture, which separates instruction and data busses, allowing for a 16-bit-wide instruction word with a separate 8-bit-wide data. The PIC core implements a two-stage pipeline that overlaps the fetch and execute cycles, allowing an instruction to be executed during the cycles following its fetch, for faster execution throughput.

6.10.4 Clocking Scheme

The PIC18F452 MCU is compatible with clock speeds of 4–40 MHz. The external oscillator clock is internally divided by 4 to produce four quadrature clocks—Q1, Q2, Q3, and Q4—which constitute one machine cycle. An instruction cycle consists of two machine cycles: a fetch cycle (four clock cycles) and an execute cycle (four clock cycles). During the fetch cycle, instructions are fetched from instruction memory; during the execute cycle, instructions are latched into the IR, decoded, and executed. The overlapping of the fetch/execute cycles allows for a new instruction to be executed in one machine cycle encompassing four clock cycles (1 ms for a 4-MHz oscillator), with the exception of branch instructions, which execute in two machine cycles (eight clock cycles).

6.10.5 Memory Organization

The 18F452 has a 21-bit-long PC allowing access to 2 MB of program memory. The PIC18F452 deploys separate program and data memory spaces with separate program and data memory buses. Program memory consists of 32 Kbyte flash occupying addresses $0000–7FFF. Data memory includes 4096 bytes of RAM and 256 bytes of EEPROM. Addresses $0000 and $0004 are reserved for reset and interrupt vectors, respectively.

Access to data memory is made by a 12-bit-long address encompassing a space of 4096 bytes that can be accessed directly, using absolute addresses, divided into 16 banks of 256 spaces each. Banks 0–14 are used as general-purpose registers (GPRs) used by the application program for general data storage. The upper half of bank 15 (F80h–FFFh; h means hex) is reserved for the special-function registers (SFRs). A complete list of all SFRs are available in the PIC18452 User's Manual. The SFRs are used to configure and control most MCU features and status functions. SFRs are divided into two groups: The core group is associated with the CPU operation (bank 0), and the peripheral group is dedicated to I/O functions (bank 1). All special-function registers, including the PC, are also mapped to data memory and treated as memory addresses. The status register includes a carry (C) flag, a digit carry (DC) flag, and a zero (Z) flag.

6.10.6 Addressing Modes

The instructions can access data in one of two addressing modes: direct addressing mode and indirect addressing mode. The data memory address is not included in the instruction in the indirect addressing mode but pointed to by the contents of the FSR. In the direct addressing modes, the register file (or data memory) can be accessed directly or indirectly.

6.10.7 I/O Ports

The 18F452 has five ports, called ports A, B, C, D, and E. Their associated pins are designated, respectively, RA[7:0], RB[3:0], RC[7:0], RD[7:0], and RE[7:0]. The operation of each port is controlled by three registers. The TRI*x*, *data direction register*, is used to set a pin to act as an input if the TRI*x*[*i*] bit is clear or as an output if the TRI*x*[*i*] bit is set. Port *x*, *data register*, is used to read input data and to write output data. The LAT*x*, *output latch register*, is used to manage read-modify-write operations of I/O data.

6.10.8 Timers

The MCU supports four timer modules, called TIMER 0–3. Timer 0 can be selected by software to be either an 8-bit or a 16-bit timer. Timer 1 is a 16-bit timer. Timer 2 can be used as either an 8-bit timer or for 8-bit period measurement. TIMER 3 is a 16-bit timer/counter.

The timer operation is configured by the following registers: INTCON, interrupt configuration register; TxCON configures timer *x* (*x* = 0–3; TRISSA . . .; TMRxL and TMRxH are two 8-bit register pairs that make up the 16-bit timer/counter register.

6.10.9 Compare/Capture/PWM (CCP) Module

The MCU also has two CCP, *capture/compare/ PWM*, modules, CCP1 and CCP2, brought out on pins RC[2] and RC[1], respectively, which perform the identical operation. Each may be configured to perform one of three functions—output compare, input capture, or pulse-width modulation—as determined by bit settings of the 8-bit CCP1CON and CCP2CON configuration registers: 16-bit data registers CCPR1 and CCPR2. Timer 2 or timer 3 may be used as timer sources for capture or compare events, and timer 2 serves as the timer source for the PWM. CCP1 and CCP2 provide the 10-bit PWM output, and register PR2 holds the PWM in microseconds. The T2CON register is used to select the prescaler value of timer 2 for the PWM function.

6.10.10 Analog-to-Digital Conversion (ADC) Module

The PIC18F452 MCU has a 10-bit ATD that caters to eight multiplexed channels. ADCCON0 and ADCCON1 registers configure the ADC and control its operation. ADRESH and ADRESL are two 8-bit registers that hold the ADC result in a 10-bit format. When a conversion completes, the conversion-complete flag is indicated by clearing bit 2 of the ADCCON0 register, and a conversion-complete interrupt is generated if it had been enabled.

6.10.11 Interrupt Structure

The PIC18F452 MCU supports 18 internal and external sources that could trigger an interrupt on either a rising or a falling edge: three external interrupt pins INT[2:0] shared with RB[2:0]; four interrupts associated with a change of state on port B pins RB[7:4]; ADC conversion complete; four timer overflow interrupts associated with TIMER0–TIMER3; two interrupts associated with input capture and output compare events on CCP1 and CCP2; and one interrupt associated with the operation of the USART, one interrupt for the parallel slave port, and one EEPROM write complete interrupt. Each interrupt source could be assigned one of two priorities: a high-priority vector at $000008 or a low-priority vector at $000018. Only one high-priority interrupt could be active at one time. Ten registers are dedicated to controlling the interrupts. Associated with each interrupt source are three bits, for enable/disable, priority setting, and status flag.

INTCON is a special-purpose register used to enable interrupts. Global interrupt enable (GIE)—bit 7—enables interrupts when set and disables interrupts when clear (the default at power-up). When GIE is set and an interrupt occurs, it is automatically cleared to mask further interrupts. Executing the RETFIE (return from interrupt) instruction sets GIE again. When an interrupt occurs in the PIC16F84, the PC is saved on the stack and then loaded with address 0x0004. This is called the *interrupt vector*. Typically, a CALL instruction that branches to the interrupt service routine is stored at that address. Alternatively, the interrupt-service routine (ISR) itself can begin at that address. The interrupt-service routine is used to save the processor state (e.g., STATUS and W registers), identify the interrupt source if several interrupt types are enabled, handle the interrupt, restore the processor state, and return from the interrupt.

6.10.12 PIC Development Suite

The following is a list of low-cost simple tools to help the user develop experience with the PIC18F452 MCU.

The PICDEM 2 plus demo board has a PIC 18F452 MCU onboard, a 9-V AC/DC adapter, 4-MHz clock, an RS-232 interface, four LEDs, and a prototyping area.

The MPLAB ICD2 in-circuit debugger has a common cable to connect a PC with the PICDEM 2 or a target system to allow debugging of PIC 18F452 MCU code.

The MPLAB IDE includes an editor, an assembler MPASM, a linker MPLINK, a debugger, and a simulator for PIC MCUs.

The MPLAB C18 ANSI-compliant C compiler operates under MPLAB IDE. It allows the same source code to contain assembly language and C-language instructions.

Additionally, the user can use the microchip's free simulator to check the program as to whether variables, ports, and registers change as intended.

6.11 SUMMARY

This chapter presented the brain of mechatronics products. The microcontroller inner workings and programming were covered in detail. In order to stress the details involved in deploying a microprocessor-based IC in a system, a specific MCU (9S12C) was adopted and discussed from various aspects: CPU and internal architecture, registers, instruction set, addressing modes, memory, programming, and other, associated development tools. This complete coverage of a specific MCU was necessary in order to overview the various functions and utilization of the facilities and tools that come with an IC. Another family of MCUs available in the markets was presented, specifically the PIC Microchip microcontrollers. Programming of these ICs was covered using assembly and C codes.

RELATED READING

S. F. Barrett and D. J. Pack, *Embedded Systems: Design and Applications with the 68HC12 and HCS12.* New York: Prentice Hall, 2004.

H.-W. Huang, *HC12/9S12: An Introduction to Hardware and Software Interfacing.* Clifton Park: Thomson Delmar Learning, 2005.

J. B. Peatman, *Design with Microcontrollers.* New York: McGraw-Hill, 1988.

E. Ramsden, "Embedded Microcontrollers, or Making Your Sensors Smart," *Sensors,* June 1999.

P. Spasov, *Microcontroller Technology: The HC12,* 2nd ed. New York: Prentice Hall, 1996.

A. K. Stiffler, *Design with Microprocessors for Mechanical Engineers.* New York: McGraw-Hill, 1992.

The following Freescale Semiconductor (www.freescale.com) documents related to the 9S12C MCU were used as the main references for the contents of this chapter.

• *Code Warrior Development Tools: C Compiler Reference 3.2* (CCOMPILERM.pdf)

- *HCS12 Microcontrollers: MC9S12C128 Data Sheet*, 2005 (MC9S12C128_V1.pdf)
- *MC68HC12 & HCS12 Microcontrollers: CPU12-Rev 3.0 Reference Manual*, 2002 (INSTRUCTION_ CPU12RM.pdf)
- *Freescale HC12 Assembler* (HC12ASMRM.pdf)

QUESTIONS

6.1 Name the basic characteristics of a computer.

6.2 Name the basic parts of a microcontroller.

6.3 What is the difference between a byte and a word?

6.4 Why does a CPU require a clock?

6.5 Why does a CPU require a reset?

6.6 What is held in the program counter?

6.7 What part of an instruction tells the CPU what to do?

6.8 Describe the role of the accumulators.

6.9 Describe the function of the stack pointer

6.10 What is the function of the ALU?

6.11 What is the meaning of each flag bit in the CCR, and how each is used?

6.12 How does the carry flag differ from the overflow flag?

6.13 What does *stack overflow* mean, and what are the complications it causes?

6.14 Explain the operating modes of the 9S12C MCU.

6.15 Describe the memory map of the 9S12C MCU.

6.16 Explain the difference between polled I/O, interrupt, and state machine programming techniques.

6.17 Explain the difference between machine language, assembly language, and high-level language.

6.18 Describe the process of converting an assembly code into a machine code.

6.19 What function does a linker program provide?

6.20 What is a compiler?

6.21 What is the difference between using IX and IY instructions?

6.22 Why must the stack pointer be initialized as one of the first things done in a program?

6.23 Explain the operation of the memory stack.

6.24 Name possible modes of addressing an operand.

6.25 Discuss the differences and similarities between a subroutine and an interrupt-service routine.

6.26 What is the difference between the DB, DW, and DS directives?

6.27 What is a macro?

6.28 Describe the various sections of an assembly program.

6.29 Explain the role of tables.

6.30 Discuss the process of erasing and programming the EEPROM.

6.31 Compare and contrast the main features of the PIC18F452 with those of the 9S12C.

PROBLEMS

6.1 Determine the status of the CCR flags after the execution of the following program statements. Before each statement, accumulator A contains $5A, and the conditions flags are N = 1, C = 1, Z = 0, and V = 0.

```
1. ADDA    #$5A
2. ADDA    #$A5
3. LSLA
4. ASRA
5. SUBA    #$AE
6. TSTA    #$A6
```

6.2 Write down the object code for the following program. Give also the address and indicate the addressing mode of each instruction.

```
ORG     $E000
LDAB    #$40
SUBB    $40
STAB    $0040
```

6.3 Write the machine code for the following assembly language source code.

```
SEG     page0
```

```
        DS.B   10
        DC.W   %01110110
        DC.B   "Team Work \t\n"
        ORG    $C000
begin   LDS    #$FF
        LDX    $4B
        LDD    $3000
        STD    $C102
        MUL
        XGDX
        RORA
        LDX    #$30
        ABX
        ANDA   #$A1
        ORAB   #$6A
        NOP
here    BRA    here
```

6.4 Write the execution history of the code of Problem 6.3. The initial conditions in hex are:

```
0048  10  B4  32  66  A1  E2  90  B2
0050  89  72  30  01  03  2D  33  16
3000  33  5D  28  46  2D  00  01  02
PC    CCR A   B   IX IY SP
C000  B2  72  5D  0020  00D5  00FF
```

6.5 The following code may be used to create a time delay in an application code. Assuming a clock frequency of 2 MHz, determine the time delay in seconds.

```
        LDX    #25000
loop    NOP
        NOP
        NOP
        DEX
        BNE    loop
        END
```

6.6 Determine the number of times the following loop will be executed and the time in seconds it takes to execute it.

```
        LDAA   %10000000
loop    LSRA
        STAA   $00
        ADDA   $00
        BMI    loop
```

6.7 What are the memory assignments for the following assembler directives?

```
        ORG    $0F
        DC.W   $34A0, $0041
```

```
        DC.W   #65535
        DC.W   '0'
```

6.8 Write assembler directives to reserve 50 bytes starting at $00 for a lookup table to hold the cosine of an angle between 0 deg and 90 deg.

6.9 Write assembler directives to build a jumpt table to hold the ASCII codes of the lowercase letters a–z beginning at address $00.

6.10 Write assembler directives to store the message "THINK COMMUNITY" starting at $00 in memory.

6.11 Write a program to compute the sum $BC02 and $841A, and store the result at $C000–$C001.

6.12 Write a program to subtract the two-byte number stored at $00–$01 from the two-byte number stored at $02–$03, and save the difference at $03–$04.

6.13 Write a program to compute the mean of an array of 16 8-bit numbers stored at $C100–$C110, and save the result at $00.

6.14 An array contains 20 8-bit numbers stored at $00 in ascending order. Write a program to rearrange the numbers in descending order.

6.15 An array has 20 elements stored at $00–$13. Write a program to divide each element by 8.

6.16 Write a program to determine how many elements in the array of Problem 6.14 are divisible by 4, and store the result in $30.

6.17 Write a program to multiply the contents of two arrays, each contain 10 8-bit numbers, and store the result at start address $D0. The two arrays are stored at addresses $00–$0A and $0B–$14.

6.18 Write a program to find the largest unsigned byte in a block of data, and store the result at address $01. The address of the first block element is $10; the address that contains the number of bytes in data block is $00.

6.19 Using a look-up table, write a program to find the square root of a single hex digit. The result is an 8-bit representation of the square root as a 4-bit integer with 4-bit fraction. For example, the square root of 6 is 2.449. The program would return a value of $25, representing $2.5 = 2 + 8/16.

6.20 Write a program that replaces all lowercase letters with uppercase letters. Initially, IX points to the start address of a block of ASCII character bytes. Assume that all bytes have bit 7 reset (no parity). A byte of $04 marks the end of the block. Note that this type of block is often known as a *character string*. The byte $04 is the ASCII code for control D. It is often used to mark the end of a string.

6.21 Given the 8-bit, signed numbers A, B, C stored in memory locations A, B, C. For each of the logic statements involving A, B, and C, give the appropriate 9S12C code to set the CCR and to branch to the ELSE part of an IF-THEN-ELSE operation.

a. IF A $\leqslant$ C

b. IF A + B $\geqslant$ 1

c IF C $<$ A − B

d. IF (A $>$ C) OR (B $<$ C)

6.22 A microcontroller system controls a robot arm. Write a program to determine the arm projection on the horizontal plane. The system has sensors to measure the arm length (radial) and its angle with respect to the horizontal plane. The program has the following input information:

```
A   arm length   zero extension
                  = $00
                  Full extension
                  = $FF
B   angle   0°  = $00
           89.6°  = $FF
            90° = $100,  have to
            use $FF!
```

You may use a cosine lookup table that is scaled to contain only 64 entries for the cosine of 0–90°. The program has the following output after execution:

```
A horizontal projection  zero
                            = $00
                   full = $FF
B cosine of angle minimum = $00
                  maximum = $FF
```

6.23 Write a section of 9S12C code to implement the following pseudo-code given. Assume x1, x2, and x3 are signed, 8-bit integer numbers stored in memory locations x1, x2, and x3.

```
while x1 = x2
  loop
    If x3 > x2
      Then x2 = x1
    else
      x2 = x3
    endif x3 > x2
    x1 = x1 + 1
  endloop
endwhile x1 < x2
```

LABORATORY PROJECTS

6.1 Write a C code to maintain the water in a tank to within two threshold levels, LOW and HIGH. When the level sensor indicates a low threshold, a pump is turned on; when the level reaches the high threshold, the pump turns OFF. Prepare a simple setup, with a small tank and a pot connected to a mechanical device for sensing the water level, to demonstrate the level-control operation.

6.2 Implement the setup of Example 6.2.

6.3 Modify the traffic light control code of Example 6.18 to control traffic at a three-way intersection.

7

Parallel I/O and Interrupt Mechanism

OBJECTIVES

Thoughtful engagement with the material presented in this chapter will enable the student to:

1 Write assembly programs to use parallel I/O ports
2 Interface different types of switching devices and displays
3 Understand resets and interrupts and apply them accordingly
4 Describe interrupt priorities
5 Write interrupt service routines

7.1 INTRODUCTION

What seems to be a microcontroller's magic is actually derived from its ability to do something in the real world: sense the environment, listen to the user, and act out an appropriate response. The interplay between a computer program devised to implement an idea and the real world is called *interface*. Interface encompasses all software and hardware that makes the interplay between the MCU and the real world possible, including analog-to-digital (A/D) and digital-to-analog (D/A) converters, communication protocols such as the RS232, and so on.

What makes the interface possible is an array of devices, called *peripherals*, that are connected to the I/O ports. Commonly interfaced input devices include switches, keypads, keyboards, touch screens, and output devices such as visual displays, LEDS, seven-segment displays, and LCD panels.

In a real-time environment, the MCU interacts with its target system as events occur, independent of the operation of the CPU. Proper operation of a target system hinges on the timely coordination of all relevant tasks requiring the MCU to handle all I/O operations. If the application program is required to manage all I/O transfers, the CPU will always be tied, running a loop to continuously check on inputs and change outputs accordingly. In this polling I/O scheme, the CPU will not know of the changes in input data at the time they actually occur. This approach may be suitable for simple I/O operations. But as the I/O requirements increase or the loop becomes very long, data manipulations increase and occurring events may not be detected in a timely manner, and real-time control is compromised. The MCU provides a vectored interrupt mechanism to manage I/O transfers and maintain its real-time control ability.

This chapter describes in detail the parallel input and output ports and provides examples on how they are used to interface input and output devices. The chapter also provides an overview of the interrupt mechanism of the 9S12C. Many MCU modules operate using the same I/O port pins. Subsequent chapters deal with the management of port functions associated with peripheral modules, such as the timer, PWM, serial communication, and A/D, and the operation of associated interrupts.

7.2 PARALLEL INPUT/OUTPUT (I/O)

The I/O section is the means by which the MCU communicates with the outside world. Signals received by the MCU from the outside world are referred to as *input data*, and signals transmitted from the MCU to the outside world are called *output data*. The point at which the I/O device connects to the MCU is called an I/O *port*. An I/O port is a collection of I/O pins on the MCU chip that represents a unit of data. Input information may be analog or digital, and it may come from devices such as bar code readers, mechanical switches, keyboards, other computers, and sensors. Output information may be sent to devices such as displays, printers, speakers, and actuators, and it may be analog or digital. Many devices, such as the touch screen, are I/O devices because they can serve as both an input and an output.

The 9S12C MCU interfaces with the outside world through eight I/O ports named ports A, AD, B, E, J, M, P, S, and T, as shown in Fig. 6.3. A set of data and control registers are dedicated to the operation of all I/O functions. Each I/O register is treated as a memory location to which a specific address is assigned, allowing all I/O transfers to be treated as memory transfers. Refer to Section 6.2.5 for memory mapping. The interface of I/O ports with external devices follow a *standard*, a protocol agreed to by the communicating devices. The I/O interface circuit on the MCU typically consists of an address decoder, an output or input latch, and buffers or drivers. Both the data and address busses connect to the I/O interface. The *address decoder* monitors the address bus, and it includes the interface circuitry for a particular device only when its proper address is detected. It also prevents different I/O devices from interfacing with each other. The *output latch* is used to store the output data for an output operation. Because the data from the CPU will appear on the data lines for only an instant, holding the data in a latch allows the I/O devices more time to examine and respond to the data. *Buffers* or *drivers* are necessary when several circuits share the same bus. Buffers provide isolation between the I/O device and the CPU bus.

The number of pins on the 9S12C MCU dedicated for I/O functions is 31 on the 48-pin device, 35 on the 52-pin device, and 60 on the 80-pin device. While this seems large, the small number of I/O pins is the most limited resource for an MCU application. Accordingly, the I/O ports share their pins with peripheral modules available on the chip. The port integration module (PIM) synchronizes the interface between the I/O port pins and the associated modules.

TABLE 7.1 Addresses of Ports and Registers Associated with the General-Purpose I/O Operation

	A	B	E	AD	J	M	P	S	T
Port x	$0000	$0001	$0008	$0270	$0268	$0250	$0258	$0248	$0240
PTIx	—	—	—	$0271	$0269	$0251	$0259	$0249	$0241
DDRx	$0002	$0003	$0009	$0272	$026A	$0252	$025A	$024A	$0242
RDRx	$000D	$000D	$000D	$0273	$026B	$0253	$025B	$024B	$0243
PERx	—	—	—	$0274	$026C	$0254	$025C	$024C	$0244
PPSx	—	—	—	$0275	$026D	$0255	$025D	$024D	$0247
MODRRx	—	—	—	—	—	—	—	—	$0247
WOMx	—	—	—	—	—	$0256	—	$024E	—
PIEPx	—	—	—	—	$026E	—	$025E	—	—
PIFx	—	—	—	—	$026F	—	$025F	—	—

7.2.1 Common Port Features

All I/O ports involve many registers that perform similar functions; they will be discussed first by referring to a port as port x and to pin i on port x as Px[i]. Port-specific features will also be indicated. Table 7.1 gives the registers responsible for setting up the I/O ports configured for general-purpose I/O operations. Registers dedicated to setting up the operation of specific modules that share their pins with the I/O ports are covered in subsequent chapters. Hereafter, the addresses used for registers conform to the mapping of the register block into the $0000–03FF memory space.

Each pin Px[i] can be used as a general-purpose input or output or as an input to or output from a peripheral module. The specific use is selected by configuring a set of associated bits in I/O registers dedicated to the operation of the associated port x. The following list details these registers and their functions.

> The DDRx, *data direction register*, determines whether Port x pins acts as inputs or outputs. Setting bit DDRx[i] in this register makes the corresponding Px[i] pin an output. Clearing the DDRx[i] bit makes Px[i] pin an input. The DDRx[i] bits are ignored if the Px[i] is controlled by a peripheral module.

> The port x data register holds the I/O data. A read of this port provides the status of the input pins, and a write to it sends its contents to the corresponding output latch register. Meanwhile, the PTIx, *port x input register*, always reads back the status Px[7:0] pins, and a write has no effect on it.

> The RDRx, *reduced drive register*, controls the drive strength of an output signal when Px[i] acts as an output. The reduced-drive option reduces power consumption and helps reduce the effect of radio frequency interference (RFI). Setting the RDRx[i] bit causes the output drive to be one-third of its full strength and should be used when light loads are involved. The output drive capability of any port pin is 25 mA sinking or sourcing.

> The PERx, *pull device enable register*, is used to connect any Px[i] pin that is configured to act as an input pin to an internal *pull-up* resistor. This feature is enabled by setting the PPSx[i] bit in the PPSx, *polarity select register*. If PPSx[i] is clear, the Px[i] is a pull-down input pin. The pull-up feature is very useful for interfacing switches and keypads to the input ports. The internal pull-up holds the pin high until the connected switch is closed.

The following pins are not available on the 52-pin package: PA[7:3], PB[7:5, 3:0], PE[6,5,3,2], PP[7:6, 2:0], PJ[7:6], and PS[3:2]. These pins are not bonded out and should be initialized to be inputs with enabled pull-up resistance to avoid excess current consumption.

Example 7.1: Configure I/O functions—Write a program segment to configure pin PTT[0] of port T as an input pin with pull-up and PT[1] as an output pin with reduced drive.

Solution: The following code sets the desired configurations. The associated registers and their addresses are port T, DDRT, RDRT, PERT, and PSPT. Refer to Table 7.1 for register addresses.

```
LDAA   #$02   ; Load %00000010 into A
STAA   $0242  ; Configure PTT[0] as input and PTT[1] as output
STAA   $0243  ; Set PTT[1] for reduced drive
LDAA   #$01   ; Load %00000001 into A
STAA   $0244  ; Set PTT[0] device for pull-up or pull-down
CLRA          ;
STAA   $0247  ; Select polarity of PT[0] device as pull-up
```

7.2.2 Specific Port Features

Ports A and B

Ports A and B at addresses $0000 and $0001 perform similar functions. In the single-chip mode, pin PA[i] of port A (or pin PB[i] of port B) may act as an input pin if the DDRA[i] (or DDRB[i]) bit is clear or as an output pin if DDRA[i] (or DDRB[i]) bit is set. In the expanded mode, ports A and B and their associated registers are not in the address map because they are used for the multiplexed address and data bus. Port A will carry the high address and data bytes, ADDR[15:8] and DATA[15:8], and port B carries the lower address and data bytes, ADDR[7:0] and DATA[7:0].

Setting bit RDPA (or RDPB) in the RDRIV register forces reduced drive strength on all PA[7:0] (or PB[7:0]) pins that are configured as outputs; otherwise, full-drive option is assumed. Setting the PUPAE (or PUPBE) bit in the PUCR register connects the internal pull-up resistor to the port A (or port B) pins that are configured as inputs.

Port E

Any port E pin PE[7:2] may act as an input pin if DDRE[i] is clear or as an output pin if DDRE[i] is set. Pins PE[0] and PE[1] can be used as input pins only. It is recommended to initialize port E before its pins are enabled for outputs. The settings of port E and DDRE are not mapped into memory when the MCU operates in the peripheral mode or in the expanded mode unless the EME, *emulate port E*, bit in the MODE register at address $000B is set.

Port E pins may also be configured to serve alternate functions as assigned by the PEAR, *port E assignment register* at address $000A. The settings of the PEAR bits override the DDRE settings. The PEAR register contains five bits PEAR[7,5:2] that determine the alternate function of port E pins: NOACEE (bit 7), PIPOE (bit 5), NECLK (bit 4), LSTRE (bit 3), and RDWE (bit 2). After reset, the values of these bits in the normal single-chip mode are 0, 0, 1,0, 0, respectively, and all port E pins act as general-purpose I/O. In the normal-expanded mode, the corresponding bit values are 0, 0, 0, 0, 0, and 1, configuring all port E pins for input or output, except PE[4], which is used as an alternate ECLK pin.

The NOACEE, *CPU no access output enable*, bit defines the function of PE[7]. If NOACCE = 0, PE[7] is a general-purpose I/O pin; and if NOACCE = 1, PE[7] is configured as an output to indicate whether the CPU cycle is a free cycle or not. The NOACEE bit is irrelevant in the single-chip and special-peripheral modes.

The PIPEOE, *pipe status signal output enable*, bit defines the operating function of PE[6:5]. If PIPEOE = 0, PE[6:5] pins act as general-purpose I/O pins; and if PIPEOE = 1, PE[6:5] pins are configured as outputs to indicate the state of the instruction queue PIPE1 and PIPE0 (refer to Section 6.2.8). The PIPEOE bit is irrelevant in the single-chip and special-peripheral modes.

The NECLK, *no external E clock*, bit defines the operating function of PE[4]. If NECLK = 0, PE[4] pin acts as a general-purpose I/O pin; and if NECLK = 1, PE[4] is configured as the external ECLK pin. If the ESTR, *E clock stretches*, in the EBICTL, *external bus interface control register*, at address $000E is clear, the ECLK is a free-running clock; and if ESTR is set, the ECLK becomes a bus control signal that will be active only for external bus cycles. The stretched-bus-cycle concept provided by the ESTR bit allows slow external memory resources to be accessed by the fast internal bus. The ECLK is available as output in all modes.

The LSTRE, *low strobe* ($\overline{\text{LSTRB}}$) *function*, bit defines the operating function of PE[3]. If LSTRE = 0, PE[3] pin acts as a general-purpose I/O pin; and if LSTRE = 1, PE[3] functions as the $\overline{\text{LSTRB}}$ bus control output used during external writes. If the BDM tagging is enabled, $\overline{\text{TAGLO}}$ is multiplexed on this pin on the rising edge of ECLK, and $\overline{\text{LSTRB}}$ is driven out on the falling edge of the ECLK. The LSTRE bit is irrelevant in the single-chip, special-peripheral, and expanded-narrow modes.

The RDWE, *read/write enable*, bit defines the operating function of PE[2]. If RDWE = 0, PE[2] pin acts as a general-purpose I/O pin; and if RDWE = 1, PE[3] functions as the R/$\overline{\text{W}}$ pin used for external writes.

Port E pins PE[1] and PE[0] may be used for the external interrupt functions $\overline{\text{XIRQ}}$ and $\overline{\text{IRQ}}$, respectively, which are discussed in Sections 7.10 and 7.11, respectively.

Port T

Port T data register at address $0240 shares its pins with the timer module functions. The same pins may also be used to generate PWM, *pulse-width modulation*, waveforms. If the TEN bit in the TSCR1, *timer system control register* 1, is enabled then port T pins are dedicated to the timer functions, *input capture*, *output compare*, and *pulse accumulator*. The MODRR, *port module routing register*, at address $0247 selects the module to be connected to Port T. If MODRR*i* bit is set, then PT[*i*] is connected to PWM channel *i*. The operations of the timer module and PWM are discussed in detail in Chapter 9.

Port P

When port P is not used for general-purpose I/O, port P pins are dedicated to the PWM, *pulse-width modulation*, channel output module. Only three pins of port P are available on the 52-pin version of the 9S12C MCU, pins PP3, PP4, and PP5, which are shared with PWM3, PWM4, and PWM5 channels, respectively.

The PPSP register is also used to select the polarity of the active interrupt edge. Interrupts are discussed in Section 7.8. If PPSP[*i*] is clear, a falling edge on PP[*i*] causes the associated flag bit PIFP[*i*] in the PIFP, *port P interrupt flag register*, to set. If PPSP[*i*] is also set, a rising edge on PP[*i*] sets the PIFP[*i*] flag and sets the device connected to PP[*i*] pin as a pull-down.

If the PWM module is not enabled, the general-purpose I/O signals of port P can be set to operate under interrupt control. The interrupt mechanism associated with port P operation can be used to synchronize data transfer on inputs or outputs. To enable an interrupt associated with PP[*i*], the PIEP[*i*] bit in the PIEP, *port P interrupt enable register* at address $025E, must be set. An interrupt is generated when an active edge is detected on PP[*i*] pin. The active (asserted) edge can be set up to be a falling edge if bit PSPP[*i*] has been cleared or to be a rising edge if PPSP[*i*] has been set. Recall that the PPSP register is also used to set the polarity of the

PP[*i*] input pin as a pull-up or pull-down. Whenever an active edge is detected on PP[*i*] pin, the corresponding PIFP[*i*] flag in the PIFP, *port P interrupt flag register* at address $025F, will set, causing an interrupt if the PIEP[*i*] had been set earlier. A digital filter on each pin prevents interrupts from occurring unless the pulse continues for a specific time duration.

The 80-pin version of the 9S12C contains port, J which has similar registers and functional features as port P.

Example 7.2: Writing a data byte to port P—Give instructions to transfer $6A to port P when a rising edge is received on pin PT[3].

Solution: The following instructions may be used to handle the operation.

```
        INCLUDE    'mc9s12c32.inc'     ; Bit definitions and
                                       registers addresses
        MOVB       #$00, ddrp          ; Set port P for input
        MOVB       #$08, ppsp          ; Set PP[3] polarity
                                       for rising edge
        MOVB       #$FF, ddrt          ; Configure port T for
                                       output
wait  BRCLR        pifp, #$08, wait    ; Wait (polling) for
                                       PIFP[3] flag to set
        MOVB       #$6A, portt         ; Write the byte to
                                       port T at address $0258
```

The same code may be used under interrupt control.

```
        INCLUDE    'mc9s12c32.inc'     ; Bit definitions and
                                       registers addresses
        MOVB       #00,ddrp            ; Set port P for input
        MOVB       #$08, plep          ; Enable interrupt on
                                       pin 3 of port P
        MOVB       #$,08, ppsp         ; Set PTP[3] polarity
                                       for rising edge
        MOVB       #$FF, ddrt          ; Configure port T for
                                       output
here  BRA          here                ; Wait for Interrup to
                                       occur and execute
                                       pt3_isr subroutine
pt3_isr                                ; Interrupt service
                                       routine for PT[3]
        MOVB       #$6A, portt         ; Write the byte to
                                       port T at address $0240
        RTI
```

Port AD

Port AD at addresses $008F shares the pins with the analog-to-digital (ATD) converter module on the chip. Chapter 10 presents the ATD converter in detail. The ADPU, *ATD enable*, bit in the ATDCTL2, *ATD control register* 2 at address $0082, is set to enable the ATD function. While any number of lines may be used for analog inputs, the remaining lines are available as general-purpose digital inputs or outputs, as determined by the bits in the DDRE register. Any Port AD pin PAD[*i*] may be used as a digital input while the ATD converter is enabled. This is accomplished by setting the IEN[*i*], *ATD digital input enable on channel* AN*i* in the ATDDIEN, *ATD input enable register* at address $008D. However, power consumption may increase if the IEN[*i*] bit is enabled and the AN*i* channel is being used as an analog input channel.

Port S

The WOMS, *port S wired-OR mode register*, at address $024E, affects only the pins that are configured as outputs. Setting the WOMS[*i*] bit configures pin PS[*i*] to operate in the wired-OR mode, forcing the output on the buffers to change from push–pull outputs to open-drain output (active low only).

The 52-pin version of the 9S12C has two port S pins, PS[0] and PS[1]. These pins act as either input or output if they are not configured for used with the serial communication module. If the *serial communication interface* function (SCI), also called *universal asynchronous receiver-transmitter* (UART), is enabled, pin PP[0] becomes the receive pin RXD and PP[1] pin acts as the transmitter pin TXD. The operation of the SCI module is discussed in Chapter 8. The SCI function is disabled after reset and must be enabled thereafter.

Port M

The WOMM, *port M wired-OR mode*, register at address $0256 only affects pins configured as outputs. Setting the WOMM[*i*] bit configures pin PM[*i*] to operate in the wired-OR mode, forcing the output on the buffers to change from push–pull outputs to open-drain output (active low only). Connecting pull-up resistors to these output pins causes current to be drawn through them instead of the MCU outputs.

The port M data register shares its pins with the SPI, *serial peripheral interface*, and MSCAM, *controller area network*, modules on the chip. The pins act as outputs if the SPI or MSCAN transmit channels are enabled, and they are forced to be inputs if the SPI or the MSCAN receive channels are enabled. The SPI operation is discussed in detail in Chapter 8.

> **Example 7.3: Using port M to drive a solenoid**–Write a program segment to drive a solenoid connected to PM[0] of port M.
>
> **Solution:** The following code performs the desired operation.
>
> ```
> MOVB    #$01, ddrm    ; Configure PM[0] for output
> MOVB    #$01, ptp     ; Write a 1 to PM[0] to drive the solenoid
> ```

7.3 MECHANICAL SWITCHES

Data can be entered to the MCU through mechanical switches, manual or automatic. Manual switches may be the hand-tripped type, such as a toggle switch, or the object-tripped type, such as a limit switch. Automatic switches are usually the ON-OFF type commonly used in process control. An automatic switch is caused to trip when the specific variable reaches a threshold value. Examples are the flow switch, pressure switch, thermoswitch, light switch, and humidity-controlled switch (humidistats).

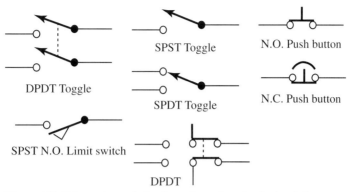

Figure 7.1 Schematic symbols of a variety of mechanical switches.

A mechanical switch consists of one or more pairs of contacts, each forming a separate circuit. A switch is characterized by its number of *poles* and *throws*. The pole is a movable piece that connects one contact of each circuit. Each pole can close a contact at a *single* or *double* position called the *throw*. Thus switches are designated as SPST (single-pole single-throw), SPDT (single-pole double-throw), DPDT (double-pole and double-throw), and so on. Limit and toggle switches can be used in a normally open (NO) or normally closed (NC) mode. In the former mode, if the switch is touched by an object, the contacts close and a signal is sent to the MCU indicating the presence of an object. Figure 7.1 shows several switch arrangements.

7.3.1 Interfacing Binary Switches

The signal from an ON-OFF switch can be easily interfaced to any pin of an input port of the MCU. A switch connected to a MCU forms what is called a *dry circuit*, because the voltage and current levels are very low (microvolts and microamps). This would extend the life of the switch to the order of 10^4–10^6 activations, depending on the mechanical quality. The electrical signal is made compatible with logic circuits by using a *pull-up resistor*. Setting the input pin for pull-up eliminates the need to add a pull-up resistor externally. Figure 7.2 shows the resistor for an SPST switch interfaced to pin PB4 of port B. When the switch is closed, the output voltage is zero, or LOW. When the switch is open, the pull-up resistor provides a voltage $V_0 = 5\ V - IR$, where I is the HIGH input current to the port; for CMOS this is very small.

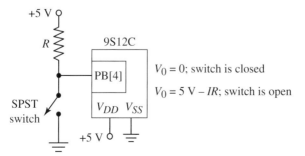

Figure 7.2 Interfacing an SPST switch to the 9S12C.

Figure 7.3 Switch bounce for a break-and-make of a mechanical switch.

The resistance should be large enough to limit power dissipation when the switch is closed while satisfying the HIGH voltage requirement. A typical pull-up resistor is 1 kΩ.

7.3.2 Switch Debounce

The inherent mass and flexibility of the moving part in a switch, exacerbated by the lack of sufficient damping, cause the switch to exhibit oscillatory motion when it is activated. These oscillations cause the contacts to open and close rapidly (2–10 times) over a period of 5–25 milliseconds before finally coming to rest and the "break" or "make" is complete. This behavior, depicted in Fig. 7.3, is called *switch bounce*.

To prevent an erroneous decision by the MCU, switch bounce must be avoided. A common practice is to provide for switch *debounce*, in hardware, using flip-flops and Schmitt triggers, or in software. Hardware debounce becomes impractical if the number of switches in a given application is large, as in keypads. Software debounce relies on a time-delay routine executed after initial contact is detected to kill time while waiting for the switch to stop bouncing. Figures 7.4 a–c show ways to debounce SPST switches using a Schmitt trigger and D and SR flip-flop circuits. The D and CLK inputs are ignored in Fig. 7.4b.

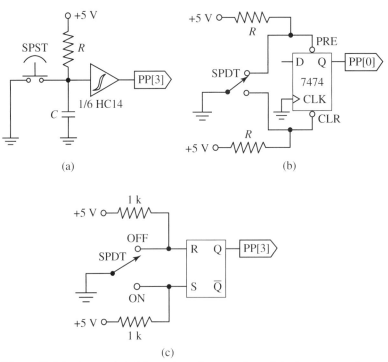

Figure 7.4 Various switch debouncing schemes using a Schmitt trigger and flip-flops.

Example 7.4: Interface binary switches to port B—Write a code fragment that can be used to detect the closing of a switch interfaced to pin PB[4].

Solution

```
        INCLUDE   "mc9s12c32.inc"    ; Include 9S12C32
                                     macros
switch  EQU       $10                ; Define mask byte
                                     ;
        BCLR      ddrb, $FF          ; Configure port B
                                     for input
        BSET      ptb, $01           ; Clear port B
        BRCLR     ptb, switch,       ; Take appropriate
                  closed             action if switch is
                                     closed
        BRA       open               ; Otherwise do some-
                                     thing else
closed
        ....                         ; Include code for
                                     the switch-closed
                                     condition here
open
        ....                         ; Include code for
                                     the switch-open con-
                                     dition here
```

7.4 INTERFACING KEYBOARDS

A keyboard with few keys may be arranged as a linear array of switches with an input pin for each switch. However, keyboards with large number of keys are generally arranged as a matrix of switches wired like the 3 × 4 telephone keypad (a keyboard with 16 or fewer keys) illustrated in Fig. 7.5.

Each key is associated with a row and a column connection which identifies the keycode or scan code and the character associated with it. When a key is pressed (*make*) the corresponding row and column connects and form a short circuit. An open circuit exists if the key is released (*break*). Figure 7.5 shows the keyboard connected to Port B pins. To ascertain the key code of a pressed key, the MCU drives one line connected to the corresponding row (or column). If the input line connected to the corresponding column (or row) is also low, the key is pressed, otherwise an open-circuit causes the pull-up resistor to drive the input line high. A look up table constructed from the table in Fig. 7.5 can be used to determine the keycode. The MCU compares the data on Port B with the expected data if a specific key is pressed. Keycode of 00 may indicate that no key is pressed.

Some keyboards have an extra common terminal, which causes another short circuit to occur between the column and row of a key and the common line whenever a key is pressed. This is called the 2-out-of-7 style of a 12-key matrix. The common line may be either an input or an output, and the rest are outputs or inputs, respectively. If the common line is configured as output and is driven low, one of the row inputs and one of the column inputs will also be low when a key is pressed. The application code would determine which key is pressed and get the code from a lookup table.

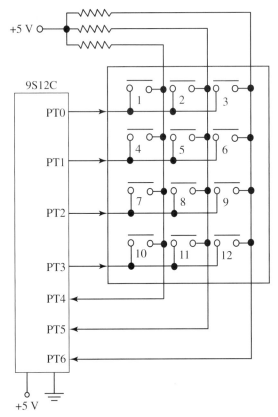

Key code	Row	Column	Port C data (binary/hex)
1	1	1	11101110/EE
2	1	2	11011110/DE
3	1	3	10111110/BE
4	2	1	11101101/ED
5	2	2	11011101/DD
6	2	3	10111101/BD
7	3	1	11101011/EB
8	3	2	11011011/DB
9	3	3	10111011/BB
10	4	1	11100111/E7
11	4	2	11010111/D7
12	4	3	10110111/B7
None	—	—	xxxx1111/0xF

Figure 7.5 A 3×4 matrix keyboard interfaced to port B (a), and port B inputs for corresponding key codes (b).

7.4.1 Hardware Decoding

Software decoding can be simplified by using hardware decoding and selecting devices. This is illustrated in the circuit of Fig. 7.6, in which an 8×8 keyboard is interfaced to the 9S12C. The 74HC151 IC is a CMOS analog data selector/multiplexer used to select the column. The three address selector inputs A, B, and C select which of the eight data inputs, D0–D7, is transferred to the chip's output Y whenever the strobe line is low. The 74HC138 is a 3-to-8 decoder used to select the row. Only one output can be low at one time, depending on the level of A, B, and C inputs. If the input is %011, Y3 goes low. PT[0:6] of port T are configured as outputs and used to specify the column and row numbers for each of the 64 keys. PT7 is configured as an input and is used to detect whether a key has been pressed. Both ICs are enabled by driving the PT6 line low.

The data inputs of the 74HC151 are pulled high. When a key is pressed, its associated row and column are shorted. With the decoder output low, the common output of the 74HC151 will also be driven low, which is detected by the MCU. For example, an output of %01100100 sent to port C will cause Y3 to be low. If key 29 is pressed, D4 line will also be low and a logic low is sensed at PC7.

A condition called *rollover* occurs when two adjacent keys are pressed simultaneously. This may damage the decoder because its output pins will be shorted together. Practical keyboard circuits include diodes to protect the interface circuitry from being damaged as well.

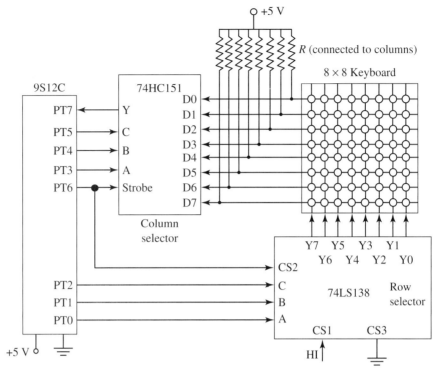

Figure 7.6 Hardware decoding of an 8 × 8 keyboard.

Dedicated ICs

The user can choose from among many ICs available to perform keypad scanning and debouncing. One example is the National Semiconductor (www.national.com) LM74C922 and LM74C923 ICs. The LM74C923 IC is a 20-key keyboard scanner. The AY-5-2376 keyboard encoder is a fully encoded keyboard manufactured by General Instrument (www.gi.com). The combination of logic and ROM it contains provides the means to generate a standard code to match each labeled key.

7.5 DISPLAYS

Light indicators are widely used in instruments to provide visual status. Common displays and comparative characteristics are given in Table 7.2.

TABLE 7.2 Specifications of Common Displays

Color	Voltage (V)	Current (mA)	Brightness (nf)	Color	Life (h)	Comments
Neon	$\approx$120 AC/DC	0.5–3	10^2	Red	10^3	Expensive
Incandescent	1–30 DC	>20	10^4	White	10^4	Bright, expensive
LED	1.5–3	5–25	10^2	R, Y, G	10^5	TTL compatible, least expensive
LCD	12–30 AC	None	Reflected	Reflected	10^5	CMOS compatible, expensive

TABLE 7.3 Specifications of LED Displays

| Color | Equivalent Circuit Parameters of LEDs | | | | Design Currents for Seven-Segment Displays (mA) | | |
	R_D (Min)	R_D (Typical)	R_D (Max)	V (Turn On)	Dim (Home)	Moderate (Office)	Bright (Outdoors)
Standard red	—	3	7	1.60	8	12	20
High-efficiency red	17	21	33	1.55	4	8	10
Yellow	15	25	37	1.80	6	10	16
Green	12	19	29	2.10	8	12	20

7.5.1 Light-Emitting Diodes (LEDs)

LEDs are the most commonly used indicators in microcontroller-based systems because of their power efficiency and long life. Since they can be turned on and off rapidly, high-power LEDs are used to drive digital data as pulses of light through fiber-optics. Specifications of LED displays are given in Table 7.3. A LED can be interfaced to any pin of an output port. Figure 7.7 shows how LEDs or lamps can be interfaced to the 9S12C. In a common-cathode connection, the cathode is tied to a common ground, each LED is turned ON by a logic HIGH on the driver bit, and the driver supplies a source current through a current-limiting resistor to the anode. In a common-anode connection, the anode is connected to a 5-V supply, each LED is activated by a logic LOW on the driver bit, and the LED is driven by current sinking through the current-limiting resistor. The current-limiting resistor R must be included when interfacing LEDs. If R is omitted, a forward-biased condition will short-circuit the LED and cause it to draw high current, damaging the LED and the output driving it. The resistor is selected to satisfy

$$V_{CC} - V_{OL(MAX)} = V_{Turn-ON} + I_D(R_D + R)$$

where V_{CC} is the supply voltage and $V_{OL(max)}$ is the maximum output voltage when the output is LOW. From Table 4.6, $V_{OL(max)} = 0.5$ volts for a CMOS output LOW. Also, $V_{Turn-ON}$ is the

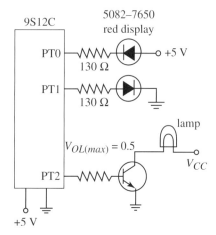

Figure 7.7 Interfacing common-cathode LEDs, common-anode LEDs, or lamps to port T.

turn-ON voltage, R_D is the resistance, and I_D is the design current. For a 5082-7650 standard-efficiency red display, $V_{Turn-ON} = 1.55$ volts, $R_D = 21\ \Omega$, and $I_D = 20$ mA. Thus,

$$R = \frac{(5 - 0.5 - 1.55)\,\text{volts}}{20\ \text{mA}} - 21\ \Omega = 126.5\ \Omega$$

Example 7.5: Interfacing eight LEDs to port M–Write an assembly code to control the status of eight common-anode LEDs interfaced to port M.

Solution: The following program fragment lights the LEDs from port M.

```
INCLUDE  "mc9S12c32.inc"
BSET  ddrm, #$FF   ; Configure port M for output
CLRA               ; Clear PM[7:0]
STAA  ptm          ; Write the content of A to port M to light
                     the LEDs
```

The CLRA and STAA instructions may be replaced with one single MOVB instruction:

```
MOVB    #$00, porta
```

If a common-cathode connection is used, a write of %11111111 to the port would light the LEDs.

The LEDs may be arranged in many ways to produce a desired lighting effect. They may also be controlled to generate rhythmic patterns that add a visually pleasing effect to any instrument.

Example 7.6: Coordinating switches and LEDs–An array of eight switches is interfaced to port M while a bank of eight LEDs is interfaced to port T as shown in Fig. 7.8. An LED should light when the corresponding switch is activated; that is, LED[i] lights up when S[i] is pressed. Write a code to handle this task.

Solution: The following code is one way to achieve the coordination between the LEDs and the switches.

```
        INCLUDE   "mc9S12c32.inc"
        MOVB      #$00,ptm    ; Configure port M as input
        MOVB      #$FF, ptt   ; Configure port T as output
loop    LDAA      ptm         : Load the value on port M
                                into A
        STAA      ptt         ; Transfer A to port T
        BRA       loop        ; Hold the CPU in a tight
                                loop
```

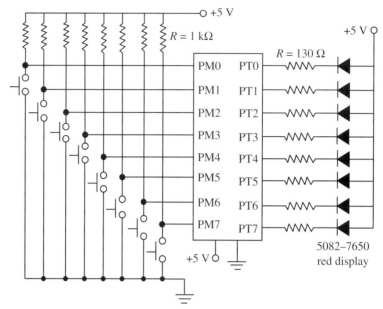

Figure 7.8 Coordinating switch states with LED displays.

Many possible I/O scenarios may be implemented without changing hardware. For example, inserting a COMA in the code would light the LEDs in a pattern that reflects the 1s-complement of the binary number set of the switches:

```
Loop    LDAA    ptad
        COMA                ; Take 1s-complement A
        STAA    ptt
        BRA     loop
```

Executing the ASLA instruction twice causes a display pattern that is four times the number set by the switches:

```
loop    LDAA    ptad
        ASLA                ; Multiply A by 2 twice
        ASLA
        STAA    ptt
        BRA     loop
```

The switches/LEDs combination may be used to provide a useful outcome in many applications. The switches may be an array of IR emitter-detector pairs that are placed on the periphery of a mobile robot to detect obstacles in eight different directions. If an IR switch detects an obstacle, the corresponding LED lights up. The switches may also be Hall switches arranged in a manner to indicate the direction in which wind blows in a weather station. Thus, this simple setup has a wide range of useful applications.

Example 7.7: C code to manage switches and LEDs–For the switch array of Example 7.6, write a C code to light the LEDs in a pattern that reflects the 1s-complement of the binary number set of the switches.

Solution: The following C code, developed using CodeWarrior C compiler, manages the required operation.

```
#include <hide.h>          /* Header file for common defines and
                              macros */
#include <msc9s12c32.h>    /* Header file for information specific
                              to the 9S12C32 */
#pragma LINK_INFO DERIVATIVE "mc9s12c32"

void main(void)
  { int i = 0;
    unsigned number;
    DDRM = 0xFF;            /* Configure port M for output */
    while (i == 0)
    { number = PTIT;        /* Get Port T value */
      PTM = ~number;        /* Write Port T value to port M */
    }
  }
```

7.6 INTERFACING LED DISPLAYS

Numeric and alphanumeric LED displays are available in one of two basic types: seven-segment and dot matrix. Both types use LEDs as the display elements. Depending on the physical arrangement of the display elements, several fonts are possible. The seven-segment display is most commonly used to display BCD digits. However, it is limited to numeric and a small number of alphabetic characters, including hexadecimal characters. Duplicating all alphabet letters requires a 16-segment display. The 5×7 dot matrix display provides full alphanumeric capability but requires a more complex circuitry and/or software.

Seven-segment displays are available in two versions: common anode and common cathode. Common anode is driven by current sinking and common cathode by current sourcing. Common anode is usually preferred because TTL is readily available with the appropriate range of current sinks. Using open-collector buffers, such as the 7404, or buffer drivers, such as the 74LS244, can source 15 mA at 5 V, sufficient to drive a common-cathode display.

LED displays are also available with a left-hand decimal, right-hand decimal, and $\pm$ overflow sign packages. Each segment is assigned a letter, as shown in Fig. 7.9. The top segment is assigned letter a. The other segments are assigned $b–g$, in clockwise order from a. To display a specific character, an appropriate 7-bit pattern is applied to the display pins. Table 7.4 gives the code for all possible characters.

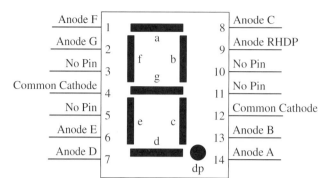

Figure 7.9 A common-cathode seven-segment (a-g) display with a decimal point (dp).

TABLE 7.4 Hexadecimal Display Representations

| Display | Hex Representation | | Display | Hex Representation | |
	Common Cathode	Common Anode		Common Cathode	Common Anode
0	3F	C0	L	38	C7
1	06	F9	O	3F	C0
2	5B	A4	P	73	8C
3	4F	B0	U	3E	C1
4	66	99	Y	66	99
5	6D	92	b	7C	83
6	7D	82	c	58	A7
7	07	F8	d	5E	A1
8	7F	80	h	74	8B
9	67	98	n	54	AB
A	77	88	o	5C	A3
C	39	C6	r	50	AF
E	79	86	u	1C	E3
F	71	8E	—	40	BF
H	76	89	?	53	AC
I	06	F9	.	80	7F
J	1E	E1			

7.6.1 Software Decoding

Driving a display by means of the MCU may be accomplished in many ways. The *software decoding* method uses a lookup table that contains data relevant to the application and type of display being used.

Example 7.8: Software decoding–Figure 7.10 shows a common-cathode seven-segment display driven by port B through the a 74244 buffer driver that can provide up to 3 V on output. The segment is used to display one BCD digit. Write a code to display the word "HELLO" with a 2-s delay between the letters.

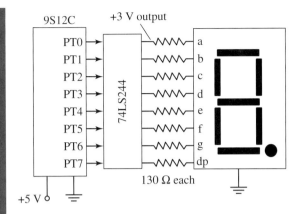

Figure 7.10 A single common-cathode display drive for software decoding.

Solution: The following code accomplishes the required task.

```
          INCLUDE  "mc9s12c32.inc"
tdly      EQU      200              ; Define time delay
          ORG      $00
table     DC.B     $76,$79,$38,$38,$3F,$00    ;bytes for the
                                              letters H, E, L,
                                              L, and O.
                                   ;
          ORG      $3800
          MOVB     #$FF, ptt       ; Configure port T as output
          LDY      #table          ; Point to data table
loop      LDAA     0,Y             ; Get the BCD number to be dis-
                                   played
          BEQ      end             ; Program ends if last byte is
                                   $00
          STAA     ptt             ; Display letters
          JSR      delay           ; Pause for a while
          LDAA     1,Y             ; Load next byte into A
          JSR      delay           ; Pause again
          IBNE     Y, loop         ; Increment Y and go to loop if
                                   Y not = 0
end       BRA      end
;Delay Subroutine
delay     LDAB     #tdly
Loop      LDX      #200000
tenth-s   NOP                      ;1/10 s delay loop
          NOP
          DBNE     X, tenth-s
          DBNE     B, loop         ;repeat the 1/10 s delay
          RTS
```

The following is a C code that implements the same task.

```
#include <hidef.h>         /* common defines and macros */
#include <mc9s12c32.h>     /* 9S12C32 specific information */
```

```
#pragma LINK_INFO DERIVATIVE "mc9s12c32"
void delay(void)     {
      int x;
      for(x=0 ; x ,< 1000 ; x++)         {
      }
}
void main(void)
{
      int i=0,j;
      int n[6]={118 , 121 , 56 , 56 , 63 , 0};
      DDRT=0xFF;
      while (i==0)          {
      for (j=0;j<6;j++)  {
          PTT=n[j];
          delay();
      }
    }
}
```

Example 7.9: A simple elevator control–Develop a 9S12C MCU-based elevator system that services a three-story building. The system should include all circuits and components to allow the user to request service from any floor and to command the elevator to travel to any floor from within the car. The current location of the elevator should also be displayed inside the car. Whenever the car reaches the desired destination, the elevator door should open automatically.

Solution: Figure 7.11 shows the main circuits for the required elevator operation. A push-button switch is placed beside the elevator door on each floor to allow the user to request service. A DC motor driven by the L298 H-bridge IC (refer to Section 12.4) controls the UP/DOWN motion of the passenger car. A DC motor driven by a BJT controls the opening/closing of each door. A seven-segment display indicates the current location of the elevator. Whenever the elevator reaches the destination floor, it causes a switch to close, signaling the MCU to stop the car motor and open the door. The following code manages the operation of the elevator.

```
        INCLUDE      'mc9s12c32.inc'
        ORG    $C100
        MOVB   #$00, ddrt    ;Configure port T as input
        MOVB   #$FF, ddrm    ;Configure port M as output
        MOVB   #$FF, ddrad   ;Configure port AD as output
        MOVB   #$FF, portad  ; Clear the seven-segment display
        MOVB   #$18, portm   ; Lock the motor of elevator car
start   LDAA   ptt
        ANDA   #$01          ; Check if first button is pressed
        BEQ    two
        JSR    first         ; If pressed go to first subroutine
two     LDAA   ppt
        ANDA   #$02          ; Check if second button is pressed
```

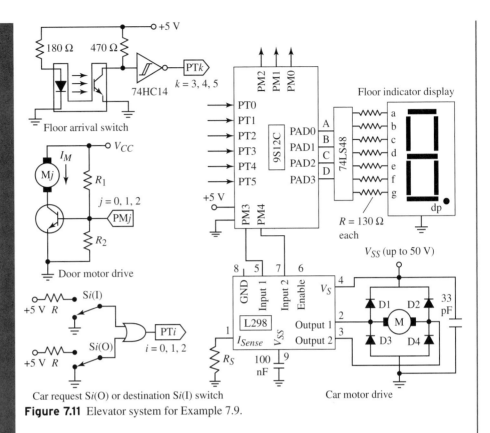

Figure 7.11 Elevator system for Example 7.9.

```
        BEQ     three
        JSR     second          ; If pressed go to second subroutine
;
three   LDAA    ppt
        ANDA    #$04            ; Check if third button is pressed
        BEQ     there
        JSR     third           ; If pressed go to third subroutine
there   BRA     start
; Subroutines Section
; Subroutine to manage a move to the first floor
first   MOVB    #$01, ptad      ; Display 1 on the 7-segment
        LDAA    #$10
        STAA    ptm             ; To go down
then    LDAA    ptt
        ANDA    #$08
        BEQ     then            ; Check if the elevator reached first
                                  floor
        JSR     stopr1
        RTS
; Subroutine to manage a move to the second floor
second  MOVB    #$02, portad    ; Display 2 on the 7-segment
```

```
            LDAA    portad
            ANDA    #$20            ; Check if elevator is on third floor
            BNE     down            ; If it is, go down
            LDAA    ptm
            ANDA    #$08            ; Check if elevator is on first floor
            BNE     up              ; If it is, go up
            LDAA    ptm
            ANDA    #$10            ; Check if elevator is on second
                                    floor
            BNE     stopr2          ; If it is there, open the door
up          MOVB    #$08, ptm
            BRA     when
down        MOVB    #$10, ptm
when        LDAA    ptad
            ANDA    #$10
            BEQ     when            ;check if the elevator reached
                                    second floor
            JSR     stopr2
            RTS
; Subroutine to manage a move to the third floor
third       MOVB    #$03, ptad      ;display 3 on the 7-segment
            MOVB    #$08, ptm       ;to go up
soon        LDAA    ptt
            ANDA    #$20
            BEQ     soon            ;check if the elevator reached third
                                    floor
            JSR     stopr3
            RTS
;Subroutine to mange a stop on the first floor
stopr1      MOVB    #$19, ptm       ;lock the elevator car, and open
                                    first door
            JSR     delay
            MOVB    #$18, ptm       ;lock the elevator car, and close
                                    first door
            RTS
; Subroutine to mange a stop on the second floor
stopr2      MOVB    #$1A, ptm       ;lock the elevator car, and open
                                    second door
            JSR     delay
            MOVB    #$18, ptm       ;lock the elevator car, and close
                                    second door
            RTS
; Subroutine to mange a stop on the third floor
stopr3      MOVB    #$1C, ptm       ;lock the elevator car, and open
                                    third door
            JSR     delay
            MOVB    #$18, ptm       ;lock the elevator car, and close
                                    third door
            RTS
; Subroutine to enforce a time delay
delay       LDX     #$014D ;typical delay routine
loop1       LDY     #$0F94
```

```
loop2    DBNE    Y, loop2
         DBNE    X, loop1
         RTS
```

7.6.2 Multiplexed Displays

To display multiple BCD digits simultaneously, the time *multiplexed display* technique is commonly used. In this technique the MCU drives, or *refreshes*, each display separately for a very short time while the other displays are turned off. As the MCU continually cycles the ON time from one display to the next, the eye perceives the character as though it were ON constantly. To produce a flicker-free display, the refresh frequency should be at least 1 kHz. For N displays refreshed at f Hz, the ON time for each character would be simply $t_0 = 1/Nf$. For $N = 3$ at $f = 1$ kHz, $t = 3.333$ ms, whereas the response time of an LED is $\tau \approx 0.1$ μs.

Figure 7.12 shows three-digit multiplexed common-cathode displays. In this circuit, port C pins are normally high. The common cathode of each display is connected to an *npn* 2N2222 transistor. To display a BCD digit, the program pulses the corresponding port B pin high. This will drive the connected transistor into saturation and pulls the common cathode low (0.2 V), which lights the display.

Each transistor must be selected to sink the peak current for a maximum of eight LEDs (seven segments plus a decimal point). The turn-on current for an LED, $I_{turn\text{-}on}$, for efficient multiplexed display is about 7 mA. Since each segment will be lit for $1/N$ of the time, the peak current through each LED reaches $I_{peak} = N \times 7$ mA $= 21$ mA for $N = 3$ displays. Therefore, each transistor must be able to supply 8 (7 segments + dp) $\times$ 21 = 168 mA. The *npn* 2N2222 can sink between 100 and 300 mA.

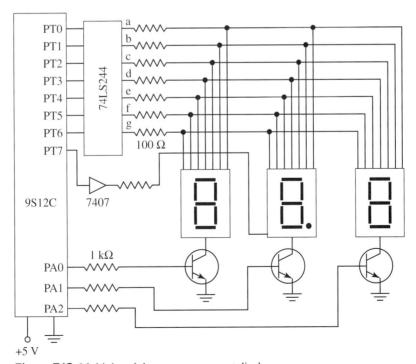

Figure 7.12 Multiplexed three seven-segment displays.

Example 7.10: Multiplexed displays—Write an assembly code to display three digits using the multiplexed connection shown in Fig. 7.12.

Solution: The following code provides for the required display.

```
            INCLUDE      'mc9s12c32.inc'        ; Include 9S12C32 spe-
                                                       cific macros
            ORG     $3800
; Dat definition
  display   DC.B    $04, $06
            DC.B    $02,$5B
            DC.B    $01,$6D            ; Digits to be displayed
; Code section
            ORG     $C000
            MOVB    #$FF, ddra
            MOVB    #$FF, ddrad
            LDY     #display          ; Point to digit
            LDAB    #3
  loop      LDAA    0, Y+             ; Turn on the driver pin for
                                        the right display
            STAA    porta
            LDAA    0,Y+             ; Output the digit code
            STAA    portad
            JSR     delay
            DBNE    B, loop
here        BRA     here
delay       LDX     #1600            ; Delay
            LDAA    #1B
count       DBNE    A, count
            RTS


The C-code version

#include <hidef.h>          /* common defines and macros */
#include <mc9s12c32.h>      /* 9s12c specific information */
#pragma LINK_INFO DERIVATIVE "mc9s12c32"
void delay(void)
   {   int x;
       for(x=0 ; x < 100 ; x++)
          {
          }
   }
void main(void)
   {   int i, j;
       int n[6] = {6 , 88 , 96}; /* Data to be displayed */
       int p[6] = {1 , 2 , 4};   /* Transistor drive data */
       DDRT = 0xFF;              /* Set port T for output*/
       DDRP = 0xFF;              /* Set port P for output */
            for (i = 0;i < 1; I = 0)
                { for (j = 0; j < 3; j++)
                     {
```

```
                                    PTT = n[j];
                                    PTP = p[j];
                                    delay();
                          }
                 }
        }
```

7.6.3 Hardware Decoding

Software decoding increases program length and execution time. Hardware decoding is an alternate option. Some ICs are specially designed to decode/drive seven-segment displays. The decoder receives a BCD logic number, decodes it for the display, and provides ample current drive for the segments. A commonly used decoder/driver for seven-segment displays is the 7447, manufactured by Motorola and Texas Instruments. This IC provides 40 mA of sink current for common-anode displays. Similar to the 7447, the 7448 IC sources current to a common-cathode display. The 8857 decoder/driver IC by National Semiconductor provides 60-mA source current to common-cathode displays. The MC14495 IC, from ON semiconductor, is a hex-to-seven-segment latch/decoder ROM/driver with built-in current-limiting resistors. The MC14499, also from ON semiconductor, provides a serial interface (see Chapter 8).

The Intersil/Harris ICL7107 CMOS IC combines a 31/2 digit A/D converter, reference voltage, timing clock, and decoder/driver circuit for common-anode, seven-segment displays. The 7106 is the liquid crystal display (LCD) version, which incorporates a backplane driver.

7.7 LCD DISPLAYS

Being simple, inexpensive, and low power, optical actuators are the main reason for the widespread use of liquid crystal displays (LCD) in instrument front panels and many battery-operated devices, such as cell phones and calculators. Liquid crystal is an anisotropic liquid material whose molecules can be aligned by surface interactions or electric fields. A typical LCD consists of a layer about 6-μm-thick nematic liquid crystal (LC) material sandwiched between two polarizers (analyzers) of buffed glass (with an alignment layer), with a reflective surface placed behind the assembly. The polarizers act to dictate the direction of the LC molecules in direct contact. The polarizers allow incident light to pass through if it is polarized in the same direction. A transparent electrode (usually indium tin oxide, ITO) is evaporated on the inner surface of each polarizer. When no electric field is applied, an incident light polarized in the same direction as the outer polarizer is allowed to pass. Guided by the twist of the LC molecules, the light is then aligned to the inner polarizer, passes through the inner polarizer, and is then reflected back out in reverse, keeping the associated pixel of the LCD bright and the display "off." To turn "on" a segment, an AC electric field with sufficient RMS value (3–12 V) is applied across the two ITO electrodes. This will align the molecules in a manner that will prohibit the LCs from being able to rotate the polarization of the incident light. Reflected light cannot go back out of the display, and the activated segment of the LCD appears "dark," i.e., "on."

The type of LCD just described is the reflective, front-lit version. Transmissive LCDs are back-lit; that is, then transmit light from the back. LCDs are available in many segment layouts, including the seven-segment format, which is similar to seven-segment LEDs, the dot matrix LCDs, and the flat-screen LCD panels. The MC14543B is designed to drive a seven-segment LCD. The 4543 IC driver takes in a BCD code and generates the code for a seven-segment LCD display. The IC driver MC145000 is designed to control a 5 × 8 dot matrix

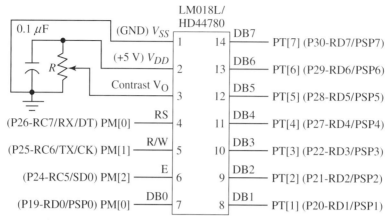

Figure 7.13 The LM018L/HD44780 LCD drive module with pin connections to the 9S12C and PIC16F87X (bracketed).

LCD display. Flat-screen LCD panels are available in various sizes and in many varieties: STN, thin-film transistors (TFTs), monochrome or color, or with 4-to 18-bit color depths.

LCDs have slow response time, on the order of 200 ms, and limited operating temperatures, and the drive voltage must not have a DC component. A constant DC excitation polarizes and destroys the crystals.

The Hitachi family of LCD modules can handle one to four alphanumeric rows with 16–40 characters in each row. All versions include a rear-mounted PCB that contains a drive circuit with a dedicated, embedded Hitachi MCU to implement the many control commands. For example, the Hitachi LM018L, 40-character $\times$ 2-line reflective-type LCD module includes the HD44780 controller. Figure 7.13 shows the interface of the LM018L to the 9S12C. The LM044L version is a 20 $\times$ 4 module. A detailed description of the module and associated commands for its operation can be found at www.doc.ic.ac.uk/~ih/doc/lcd/index.html.

The S1D13708 embedded-memory LCD controller by EPSON is designed for PDA and cell phone applications and can interface directly to numerous TFT panels. This controller can interface directly to various CPU busses and adds various display features, such as picture-in-picture and overlays. It has onboard 80-kB stack RAM for display purposes and data storage. For example, the Epson ND-TFD 160 $\times$ 160 panel in eight color modes uses 25 kB of RAM. It has seven I/O lines, eight output lines, and a PWM line. The interface bus relies on chip-select (CS) signal. The indirect-mode 68 interface (also 80, 6800 bus interface with only one address line (e.g., A0)) uses commands and data bus cycles to access the internal components of the S1D13708, with speed penalty for access. Interface requires four more lines to act as control signals, for a total of 12 lines needed on the MCU.

7.8 INTERRUPT MECHANISM

The 9S12C interrupt module uses vectored interrupt to manage I/O operations in real time. Real-time events can control the flow of program execution, and an I/O device has the means to notify the CPU when it needs attention. Physically, an *interrupt* takes the form of a single input from an external device signaling the CPU that it needs servicing. A special type of interrupt called *reset* could also trigger the MCU to restart if for some reason it stopped operating properly.

The 9S12C handles interrupts via software that can read input changes when they occur and change outputs as soon as possible thereafter.

In support of the extensive on-chip hardware I/O features (timer, SPI, SCI, etc.), the 9S12C provides one software interrupt using the SWI instruction and an extensive set of CPU hardware interrupt sources, listed in Table 7.5. The top five sources including SWI are reset interrupts, followed by two external interrupt sources, the $\overline{\text{XIRQ}}$ and $\overline{\text{IRQ}}$ inputs that share the PE[1] and PE[0] pins with port E (refer to Fig. 6.3). The following hardware interrupts are caused by internal interface hardware associated with the I/O operation generating the interrupt.

7.8.1 Maskable and Nonmaskable Interrupts

Distinction is made in Table 7.5 between maskable and nonmaskable interrupts. Some interrupts can be ignored for a few seconds while the CPU is performing a more important task. This type of interrupt is called a *maskable interrupt* because it can normally be enabled and disabled by the CPU. A *nonmaskable interrupt*, on the other hand, cannot be ignored and therefore cannot be masked. It is an interrupt to which the CPU will respond immediately. Its function is identical to the maskable interrupt, but it is normally reserved for emergency situations. Hardware reset interrupts and software interrupts using the SWI instruction are nonmaskable. The $\overline{\text{XIRQ}}$ pin interrupt is a nonmaskable interrupt used to detect power failure or for external emergency conditions that require immediate attention.

7.8.2 Interrupt Process

Each maskable interrupt has two associated bits that determine whether the corresponding interrupt source is enabled or disabled: the I bit in the condition code register and a specific *source enable bit*. As indicated in Table 7.5, the I bit has control over all maskable interrupts. When the I bit is set, all maskable interrupt sources are disabled. When the I bit is clear, a maskable interrupt is enabled (or disabled) if the corresponding source enable bit is set (or clear). The SEI and CLI instructions may be used anytime in the program to set or clear the I bit as desired. The X bit in the CCR is the mask bit for the external interrupt request input XIRQ (see Section 6.2.7).

The CPU knows an I/O device has requested an interrupt when the corresponding interrupt source flag is set. When an interrupt request is received, the I bit is automatically set to mask further interrupts. Meanwhile, the CPU finishes executing the present instruction in the mainline program, stacks the registers, and begins executing a special subroutine called the *interrupt-service routine* (ISR) from a predefined address. The ISR is written specifically to take any necessary actions to service the interrupt source. At the end of an ISR, the CPU executes a *return-from-interrupt* RTI instruction and transfers control back to the opcode address in the interrupted program held by the PC prior to servicing the ISR. The ISR is like a subroutine, but it is invoked by a hardware signal to the CPU instead of the JSR instruction. To enable future interrupts from the same source, the corresponding flag bit must be cleared within the ISR before the RTI instruction is executed. The process of clearing a flag differs from one I/O subsystem to the other and will be discussed with each system separately. In general, writing a 1 to the bit location associated with a flag clears the flag.

Nested interrupts are possible, provided the I bit is cleared accordingly. However, to avoid running an infinite loop or a locked-up program condition, the interrupting source should be disabled or the interrupting flag source cleared before the I bit is cleared, to prevent another interrupt from immediately being generated.

TABLE 7.5 9S12C Interrupts Listed in Order of Priority (Highest at Top, Lowest at Bottom)

Interrupt Source (Mask Bit)	Vector Location	Source Enable Bit	Enable Bit Address	Source Flag	Source Flag Address	HPRIO Priority Value
System reset pin (NM*)	FFFE-FF	None	—	None	—	—
Clock failure reset (NM)	FFFC-FD	CME	1039/3	None	—	—
Watchdog timer (NM)	FFFA-FB	NOCOP	100F/2	None	—	—
Illegal opcode trap (NM)	FFF8-F9	None	—	None	—	—
Software interrupt (SWI)	FFF6-F7	None	—	None	—	—
BDM vector request (NM)	FFF6-F7	None	—	None	—	—
XIRQ signal (X)	FFF4-F5	None	—	None	—	—
IRQ signal (I)	FFF2-F3	IRQEN	001E/7	—	—	00F2
Real-time clock (I)	FFF0-F1	RTIE	0038/7	RTIF	0037/7	00F0
Timer channel 0 (I)	FFEE-EF	C0I	004C/0	C0F	004E/0	00EE
Timer channel 1 (I)	FFEC-ED	C1I	004C/1	C1F	004E/1	00EC
Timer channel 2 (I)	FFEA-EB	C2I	004C/2	C2F	004E/2	00EA
Timer channel 3 (I)	FFE8-E9	C3I	004C/3	C3F	004E/3	00E8
Timer channel 4 (I)	FFE6-E7	C4I	004C/4	C4F	004E/4	00E6
Timer channel 5 (I)	FFE4-E5	C5I	004C/5	C5F	004E/5	00E4
Timer channel 6 (I)	FFE2-E3	C6I	004C/6	C6F	004E/6	00E2
Timer channel 7 (I)	FFE0-E1	C7I	004C/7	C7F	004E/7	00E0
Timer overflow (I)	FFDE-DF	TOI	004D/7	TOF	1023/1	00DE
Pulse accumulator overflow (I)	FFDC-DD	PAOVI	0060/1	PAVOF	0061/1	00DC
Pulse accumulator input edge (I)	FFDA-DB	PAI	0060/0	PAF	0061/0	00DA
SPI serial transfer complete (I)	FFD8-D9	SPIE	00D8/7	SPIF	00DB/7	00D8
		SPTIE	00D8/5	SPTF	00DB/5	00D8
SCI serial port—transmitter (I)	FFD6-D7	TIE	00CB/7	TDRE	00CC/7	00D6
		TCIE	00CB/6	TC	00CC/6	00D6
SCI serial port—receiver (I)	FFD6-D7	RIE	00CB/5	RDRF	00CC/5	00D6
		ILIE	00CB/4	IDLE	00CC/4	00D6
ATD (I)	FFD2-D3	ASCIE	0082/1	ASCIF	0082/0	00D2
Port J (I)	FFCE-CF	PIEP7-6	0250/7-0	PIFP/7-0	025F/7-0	00CE
CRG PLL (I)	FFC6-C7	LOCKIE	0038/4	LOCKIF	0037/4	00C6
CRG self-clock mode (I)	FFC4-C5	SCMIE	0038/1	SCMIF	0037/1	00C4
Flash (I)	FFB8-B9	CCIE	0103/6	CCIF	0105/7	00B8
		CBEIE	0103/7	CBEIF	0105/7	00B8
CAN wake-up (I)	FFB6-B7	WUPIE	0145/7	WUPIF	0144/7	00B6
CAN errors (I)	FFB4-B5	CSCIE	0145/6	CSCIF	0144/6	00B4
		OVRIE	0145/1	OVRIF	0144/1	00B4

*NM: nonmaskable.

A *reset* interrupt is a special type of interrupt. When a reset signal is received, the CPU suspends its current operation and immediately jumps off, before finishing the execution of the current instruction, to execute the reset-service routine. When executed, the reset routine does not return control to the interrupted program.

7.8.3 Vectored Priority Interrupt

The 9S12C employs the vectored-priority-interrupts concept, in which each interrupt and reset is assigned a vector and a priority level. Resets have the highest priority, followed by the nonmaskable $\overline{\text{XIRQ}}$ interrupt source. If two or more nonmaskable interrupt or reset sources request service at the same time, the CPU will respond first to the one with the highest priority.

While the priority of the 9S12C maskable interrupts is fixed, the programmer can elevate any one of the I-bit maskable interrupt sources to the highest priority at initialization time. This is done by programming the PSEL[7:1], *highest priority I interrupt select*, bits in the HPRIO, *highest priority I interrupt register* at address $001F. PSEL[7:0] forms the least significant byte of the associated interrupt vector. The HPRIO register can be written to only if the I mask is set, but it can be read anytime. Interrupts are listed in Table 7.5 in terms of their default priority, with the $\overline{\text{IRQ}}$ interrupt source having the highest priority. Writing a value higher than $00F2 indicates either an unimplemented vector or a non-I-bit masked vector address. In this case the $\overline{\text{IRQ}}$ vector $FFF2 will be the default high-priority interrupt. If while one interrupt source is getting service other interrupt sources request service, the interrupt priority decoder evaluates the validity and priority of pending interrupts. If a pending interrupt has a higher priority than the one currently being serviced, the CPU goes to service it; otherwise it completes the execution of the current service routine and then goes immediately to the service the highest-priority interrupt among pending sources. This is illustrated in Fig. 7.14, where a higher-priority source, routine 2, requests service while the lower-priority source, routine 1, is receiving service. Rather than making the higher-priority source wait, the CPU puts the lower-priority routine on hold, executes the higher-priority service routine, and, when finished, continues the execution of the lower-priority interrupt.

Three limiting factors must be examined when using interrupts for real-time control: interrupt latency, interrupt density, and waiting-time limit. *Interrupt latency* is the period from the time an interrupt source requests service to the time the CPU actually begins executing the corresponding interrupt-service routine. This time period includes the time it takes the CPU to finish executing the current instruction and the time the CPU spends executing a critical region of the program where the interrupts are disabled. *Interrupt density* is the percentage of time the CPU spends servicing interrupts. *Waiting-time limit* is the time an interrupt can wait before it is too late for its service to be meaningful. For any interrupt, if the latency time is greater than the maximum allowable waiting time, the system fails.

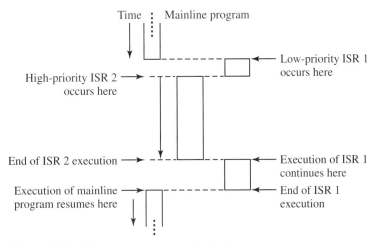

Figure 7.14 CPU response to vector priority interrupts.

7.8.4 Interrupt and Reset Vectors

Each interrupt source has a vector location or address, given in Table 7.5. A *vector* is a memory location that contains the start address for the ISR. Each vector has its own address in memory. The *vector address* is the two-byte memory location, where the vector is stored. Vector addresses are fixed, but the programmer can load the address vector with a vector value that addresses any location in memory. The *vector table* is the memory block where the vectors of interrupt sources are stored. The vector table occupies the highest memory addresses and usually ends up in ROM to become a permanent part of the chip hardware. The chip manufacturer puts in those addresses the vector values provided by the application developer. The 9S12C is designed with the top 256 addresses, $FF00–$FFFF, reserved for the vector table. If the BDM is active, the vector space will be used by the background module.

The sequence of operations performed by the CPU when a specific interrupt occurs is shown in Fig. 7.15. In this case the vector addresses for the real-time clock, FFF0–FFF1, contain $E190. When the real-time clock interrupt occurs, the CPU responds by loading the PC with $E190, the start address of the corresponding ISR.

7.8.5 Stacking the Registers

When an interrupt request is made from an enabled source, the CPU pushes the values of all registers onto the stack before the start address of the corresponding interrupt-service routine is fetched. The process is called *stacking the registers*, and the CPU pushes the registers in the following order: PC, Y, X, D, CCR. The lower byte of a 16-bit register is pushed first. The last instruction executed in an interrupt-service routine is the RTI instruction. When executed, the CPU pulls the data from the stack, restores them to the CPU registers in reverse order, and then transfers control back to the interrupted program. However, the CPU does not stack the registers if a reset interrupt occurs, because a reset ISR does not normally return control back to the interrupted program.

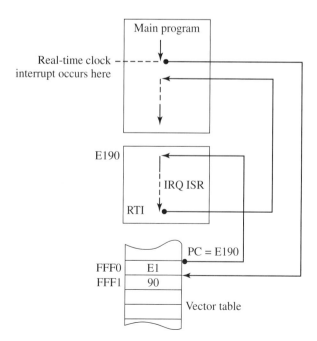

Figure 7.15 Vector interrupt processing.

Example 7.11: Effect of an interrupt request on the stack–After the execution of the current instruction, the CPU registers contain the following data:

```
ACCD    $5C76
IX      $1020
IY      $B67F
CCR     $81
PC      $E915
SP      $00D6
```

If an interrupt request is received while executing the current instruction, what would be the stack contents?

Solution: The stack will hold the following data after the given statements are executed:

```
00C4    XX   XX   XX   XX   XX   XX   XX   XX   XX   XX
00CE    81   5C   76   10   20   B6   7F   E9   15   XX
```

7.9 RESETS

In a typical 9S12C-based target system, the MCU operates under the control of the application program, also called the *monitor program*, in flash or ROM. The monitor program controls all CPU functions at all times. When the CPU is initially turned on, it starts by executing a short subroutine to initialize all vital facilities, such as the stack pointer and displays. Five possible sources can cause the CPU to execute the reset function at any later time: external pin RESET, power-on reset (POR), low-voltage reset (LVR), computer operating properly (COP), Watchdog timer (WDT) reset, and clock monitor reset (CMR). In many ways these resets are like interrupts, because, when a reset occurs, the CPU suspends its current operation and jumps off to the address of the reset ISR, where program execution resumes. The main goal of the reset function is to load into the PC the start address of the monitor program from addresses $FFFE–$FFFF (see Table 7.5).

7.9.1 External Pin RESET

Pin 18 on the 52-pin version of the 9S12C is an active low RESET control pin. When a reset event is initiated by an external circuit connected to this pin, an internal reset generator circuit drives the RESET pin LOW for 128 system clock cycles and then allows it to rise to HIGH. After waiting for 64 additional system clock cycles, the circuit samples the reset pin to determine the source that caused the reset and to fetch the corresponding reset vector for service. The CPU then executes a reset sequence to initialize the system to a start-up state, regardless of what the CPU is doing at the time. The RESET line is bidirectional and is pulled LOW internally as a result of a CMR, an LVR, or a COP failure.

In a given application, the RESET line may be tied to the V_{DD} line or generally connected to an external circuit, such as a switch, to provide the operator the means to pull the reset line LOW and to restart the system if it locks up or runs aimlessly for some reason. The external reset circuit is required to pull the reset line LOW at reset time and to provide a high-impedance pull-up state at all other times if any one of the five reset conditions manifest itself. The circuit connected to the reset pin should not include a large capacitance; otherwise the signal may not return to the valid HIGH state within the 64 cycles after the LOW drive is released, making the CPU misinterpret the type of reset that had occurred. Thus an external power-up delay circuit, such as the *RC* circuit shown in Fig. 7.4a, should not be used with the reset pin. A better approach is to use a *voltage-threshold circuit* designed to handle power glitches and power-on resets. Dallas Semiconductors (www.maxim-ic.com) manufactures a line of internally trimmed microprocessor supervisory ICs to provide necessary reset requirements.

7.9.2 Power-On Reset

At the onset of a power turn-on, the POR reset sequence provides the automatic means for the system to start operating in an orderly manner. When power is first turned on, the reset input to the MCU is held LOW until the voltage on the power supply pin V_{DD} reaches the level required for the MCU to operate properly before it asserts an POR, an LVR, or both. At this point, a positive transition is detected on the V_{DD} pin and the reset pin is pulled LOW, indicating that the CPU can operate properly and initiating a power-on reset. Both power-on and external pin resets have the highest priority over any other reset or interrupt, and both cause the CPU to execute the reset ISR from the address fetched from $FFFF–$FFFE in the vector table (see Table 7.5). POR only initializes the internal circuitry during cold starts and cannot be used to force a reset as system voltage drops. When a POR reset occurs, the PORF flag in the CRGFLG, *CRG flag register* at $00x3, is set. Similarly, the LVRF flag in the CRGFLG sets when an LVR occurs.

The power-on circuitry includes a built-in clock circuit that checks the quality of the incoming OSCCLK signal for a 4095 clock rising edge. This window provides ample time after the first oscillator operation for the clock oscillator to stabilize and internal device registers to be properly loaded. If a valid clock signal is detected, the reset sequence begins under the control of the OSCCLK; otherwise the reset sequence starts using the self-clock mode. Clock quality check is triggered anytime a POR, LVR, or CMR occurs.

The clock watchdog (COP) reset and the clock monitor reset pull the $\overline{\text{RESET}}$ pin LOW for four clock cycles when they occur, whereas an external reset pulls the line LOW for six clock-cycles. Hence the six-clock-cycle delay enables the CPU to distinguish an external reset from the clock-related reset.

7.9.3 COP Failure Reset

The computer-operating-properly (COP) reset is a safety mechanism against software failures. This reset occurs when the CPU gets "hung up" in a software *runaway* for a period of time. This means that the CPU is not executing certain sections of the application code within an allotted time period, which could have an adverse effect on the control system. For instance, this happens when the stack gets corrupted somehow. The 9S12C has a *watchdog timer* (WDT) that must be continually reset by the application program (to start timing all over again) before it times out; otherwise a COP reset occurs.

The time interval between two consecutive watchdog timer resets is called the *timeout period*. This period is chosen to be longer than the time required to execute the longest loop in the application program. The COP counter is enabled and starts counting when the CR[2:0], *COP timer rate select*, bits in the COPCTL, *COP control register* at address $003C, are written to. The timeout period may range from 2^{14} when CR[2:0] are all cleared, to 2^{24} clock cycles when CR[2:0] are all set. This range corresponds to a time from 1.04 s to 1.0486 s with a 16-MHz clock. If the timer is not reset, it will time out, resulting in a COP failure reset.

Three registers are associated with the operation of the watchdog timer. To enable COP failure reset, the RESET routine must clear the WCOP, *window COP mode bit*, bit 7 in the

COPCTL register. The watchdog timer may be reset by the program in a two-step process: a write of a $55 to the COP reset register COPRST at address $103A, followed by a write of $AA (complement of $55) to ARMCOP, *COP arm/reset register*, at address $003F. This reset sequence will not be executed if for some reason the CPU fails to properly execute the program, causing the watchdog timer to time out and to internally drive the reset pin LOW. The CPU will then load the PC with the reset vector address from $FFFA and $FFFB.

7.9.4 Clock Monitor Reset (CMR)

This reset occurs if the CPU clock fails to operate properly. The CMR is enabled if the CME, *clock monitor enable*, bit and the SCME, *self-clock enable mode*, bit in the PLLCTL, *PLL control register*, at address $003A, are clear. The CME bit is set after reset, but it may be cleared or set anytime thereafter.

7.9.5 Reset Sequence

When a reset is triggered, the CPU executes the following sequence of events (see Fig. 7.16): (1) Disables all interrupt sources by setting the interrupt mask bits I and X and the stop disable bit S in the CCR to prevent any I/O interrupt request source from interfering with reset

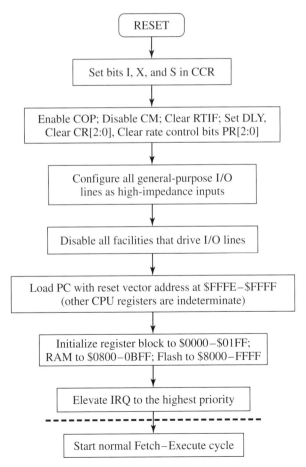

Figure 7.16 Reset interrupt sequence.

sequence; (2) changes all registers and control bits to their start-up states; (3) configures all I/O lines as input; (4) disables all interrupt sources (IRQE, CME) and all facilities that can drive the I/O pins; and (5) fetches the restart vector from the corresponding address in the vector table and loads it onto the PC. The reset sequence is then terminated and the monitor program takes control over subsequent CPU activities. The RAM is not automatically initialized out of reset.

7.10 NONMASKABLE INTERRUPT ($\overline{\text{XIRQ}}$)

The $\overline{\text{XIRQ}}$/PE0 interrupt line brought out to pin 26 of the 52-pin package provides a means to generate an external nonmaskable interrupt of highest priority after reset initialization. It is used to handle situations that cannot be ignored, such as a power-loss-detect interrupt. It may also be used with limit switches to prevent erroneous operation of a machine or to shut down the operation of a manufacturing cell if someone breaks through the safety screen. Because it is level sensitive, multiple sources may be connected to the $\overline{\text{XIRQ}}$ input as a wired-OR

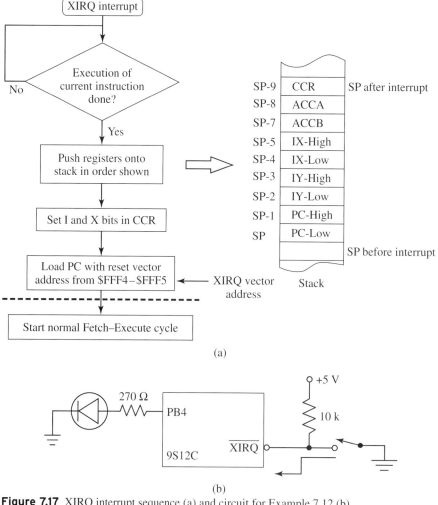

Figure 7.17 XIRQ interrupt sequence (a) and circuit for Example 7.12 (b).

network. The pin is always an input and can be read anytime. An active internal pull-up is forced on this pin during and immediately after reset, which can be turned off by clearing the PUPEE bit in the PUCR register.

This line operates in conjunction with the X bit in the CCR. By default, the X bit is set and the $\overline{\text{XIRQ}}$ disabled after a reset sequence. Many instructions, such as the TAP instruction, can be executed thereafter to clear the X bit and enable $\overline{\text{XIRQ}}$ interrupts. Once enabled, the $\overline{\text{XIRQ}}$ cannot be masked because the X bit cannot be set again by software. However, $\overline{\text{XIRQ}}$ will be masked during a reset operation and if a prior interrupt is being serviced.

The $\overline{\text{XIRQ}}$ input is level sensitive, and, if the internal pull-up resistor is active, it holds the pin HIGH until the connected switch is closed. If the internal pull up is not used, an external pull-up resistor (≈ 100 kΩ) tied to the supply may be used for proper operation. If the X bit is clear and a HIGH-to-LOW transition on the XIRQ line is detected, the CPU, after completing the current instruction, initiates the interrupt sequence shown in Fig. 7.17a.The CPU begins the sequence by pushing the contents of the PC IX, IY, accumulators, and CCR onto the stack, in that order. It then sets the I and X bits to ignore interrupts while it services the current one. Then it executes the corresponding ISR, starting at the address it loads to the PC from $FFF4–$FFF5 locations in the vector table. At the end of the ISR, the CPU executes the RTI instruction and restores all registers to their values prior to the interrupt.

Example 7.12: XIRQ interrupt–A machine includes a push-button switch as a safety feature. If an emergency condition occurs, the operator pushes the switch to halt machine operation and light up an LED. Write an assembly code to handle this situation by means of the 9S12C.

Solution: Figure 7.17b shows one possible solution. The switch is connected to the $\overline{\text{XIRQ}}$ pin through a pull-up resistor, and the LED is interfaced to pin PB[4] of port B. Closing the switch initiates the CPU to execute the "rxirq" ISR to light the LED.

```
        INCLUDE  'mc9s12c32.inc'  ; 9S12C specific macros
        ORG    $C000            ; Start address of the code
        TPA                     ; Transfer CCR to A
        ANDA   #$BF             ; Clear X bit (bit 6 in CCR)
        TAP                     ; Transfer A back to CCR
loop    BRA    loop             ; Spin around the loop while waiting
                                  for interrupt
;Service routine rxirq outputs 'H' when a high-to-low transi-
tion occurs at pin XIRQ
rxirq
        MOVB   #$10, ptb        ; Write #01 to port B
        RTI                     ; Return from interrupt
```

7.11 MASKABLE INTERRUPTS

The location of the *source-enable* or *mask* bits and the *source flag* for all maskable interrupts are given in Table 7.5. All I-bit maskable interrupt sources are masked if the I bit is set, and the CPU will ignore any associated interrupt request. When a maskable interrupt occurs, the corresponding source flag will set, informing the CPU that a pending interrupt needs service.

The CPU will respond to the interrupt request only if the I bit is clear and the corresponding source-enable bit is set; otherwise the interrupt request will go unnoticed. The interrupt-source flag must be cleared by the service routine before the RTI instruction is executed to enable future interrupts from the same source; otherwise the corresponding ISR will execute indefinitely.

The CPU automatically sets the I bit whenever an interrupt sequence is initiated. This prevents any I-bit maskable interrupt from interrupting the CPU while a current $\overline{\text{XIRQ}}$ or $\overline{\text{IRQ}}$ interrupt is being serviced.

Executing the CLI instruction clears the I bit, and writing a 1 into the source-enable bit location of the associated register sets the local enable bit. As an example, consider the interrupt source associated with PTP[0] pin of port P. If the active edge assigned by the PPSP[0] bit occurs on the PTP[0] pin, the corresponding PIFP[0] source flag sets, and an interrupt is generated if the source-enable bit PIEP[0] had been set. The following example demonstrates the interrupt operation.

Example 7.13: Drive a solenoid under interrupt control—Write a program segment to drive a solenoid connected to the PP[0] pin of port P when a push-button switch connected to the PP[1] pin is pressed. The circuit is shown in Fig. 7.18.

Solution: The following code performs the desired operation.

```
        INCLUDE  'mc9s12c32.inc'
        MOVB   #$01, ddrp    ; Configure PP[0] for output and
                               PP[1] for input
        MOVB   #$02, ppsp    ; Enable rising edge on PP[1]
        MOVB   #$02, piep    ; Enable interrupt on PP[1]
here    BRA    here          ; Wait for interrupt
ptp_isr
        MOVB   #$01, ptp     ; Write a 1 to PP[0] to drive the
                               solenoid
        RTI
```

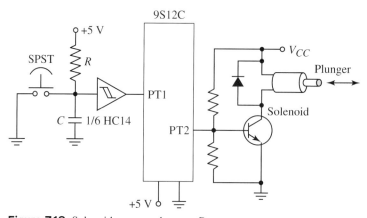

Figure 7.18 Solenoid connected to port P.

When an a maskable interrupt source requests service, the CPU finishes executing the current instruction and then checks the I bit in the CCR. If the I bit is set, the CPU ignores the interrupt request and executes the next instruction; but if the I bit is clear, the CPU pushes the registers onto the stack, PC first and CCR last. Next, it sets the I and X bits to ignore other interrupt requests until the present interrupt has been serviced. The CPU then executes the corresponding ISR. At this point, the CPU could clear the I bit as it executes the ISR to allow itself to be interrupted; if not only the nonmaskable interrupt $\overline{\text{XIRQ}}$ or resets could interrupt the execution of the ISR. Once the interrupt is serviced, the CPU executes an RTI instruction, restores the registers to their original state, and proceeds with the normal operation.

IRQ Interrupt Source

The IRQ line provides a means for an asynchronous interrupt request. The function of this line is very similar to the function of the $\overline{\text{XIRQ}}$, except the $\overline{\text{IRQ}}$ interrupt is maskable. The $\overline{\text{IRQ}}$ line is always an input and can be read anytime. The asserted state of the $\overline{\text{IRQ}}$ pin, that is, the trigger that prompts an interrupt, may be set up as either a low-level sensitive or as a falling-edge sensitive input, depending on the value written to the IRQE bit in the INTCR register at $001E. At reset, the $\overline{\text{IRQ}}$ is enabled and the IRQE bit is cleared for low-level sensitive triggering. Clearing the IRQE disables the $\overline{\text{IRQ}}$ interrupt source. To set up the triggering on a falling edge, the IRQE bit must be set. The CPU initiates the interrupt sequence depicted in Fig. 7.17a if the interrupt mask I bit in the CCR and the $\overline{\text{IRQ}}$ source-enable bit IRQEN in the INTCR register are both set. If more than one source forming a wired-or network is tied to the $\overline{\text{IRQ}}$ pin, the asserted state must be set to low-level sensitive operation. An active internal pull-up is connected to this pin during and immediately after reset, which can be turned OFF by clearing the PUPEE bit in the PUCR register. The internal pull-up resistor holds the pin HIGH until the connected switch is closed. The sequence of events that take place when an $\overline{\text{IRQ}}$ interrupt occurs is similar to that associated with the $\overline{\text{XIRQ}}$ interrupt, except that if the I bit is set, the $\overline{\text{IRQ}}$ interrupt request will be ignored.

7.12 SUMMARY

Parallel interfacing data to and from the MC9S12C was presented in this chapter. The usefulness of the MCU in mechatronics products is based on processing input data from the various events and broadcasting results to the outside world through different output devices (displays: LCD, LED, etc.). This chapter covered the parallel I/O facilities of the 9S12C microcontroller and the MCU various interrupts. It was clear that the adopted MCU has many interrupt levels; some are software selectable and others are based on hardware choices. The chapter showed the handling priorities, interfacing, and servicing of the interrupt requests. Practical examples were given, with the complete programming code (assembly and C) for realizing I/O operations.

RELATED READING

S. F. Barrett and D. J. Pack, *Embedded Systems: Design and Applications with the 68HC12 and HCS12*. New York: Prentice Hall, 2004.

H.-W. Huang, *HC12/9S12: An Introduction to Hardware and Software Interfacing*. Clifton Park: Thomson Delmar Learning, 2005.

G. Kovacs, *Micromachined Transducers Sourcebook.* New York: McGraw-Hill, 1998.

J. B. Peatman, *Design with Microcontrollers.* New York: McGraw-Hill, 1988.

P. Spasov, *Microcontroller Technology: The 68HC11*, 2nd ed. New York: Prentice Hall, 1996.

A. K. Stiffler, *Design with Microprocessors for Mechanical Engineers.* New York: McGraw-Hill, 1993.

The following Freescale Semiconductor (www.freescale.com) documents related to the 9S12C MCU were used as the main references for the contents of this chapter.

- *Code Warrior Development Tools: C Compiler Reference 3.2* (CCOMPILERM.pdf)
- *HCS12 Microcontrollers: MC9S12C128 Data Sheet*, 2005 (MC9S12C128_V1.pdf)
- *MC68HC12 & HCS12 Microcontrollers: CPU12-Rev 3.0 Reference Manual*, 2002 (INSTRUCTION _CPU12RM.pdf)
- *Freescale HC12 Assembler* (HC12ASMRM.pdf)

QUESTIONS

7.1 Describe the common features of the I/O ports of the 9S12C.

7.2 What are the output drive capability and the maximum input loading of each I/O port?

7.3 Explain the role of the data direction registers.

7.4 Explain the role of the pull-up feature of an input port pin and how to enable it.

7.5 Describe the role of each pin on port E.

7.6 Explain the various types of switching mechanisms.

7.7 Explain contact bounce.

7.8 How does a thermoswitch work?

7.9 Explain how a humidistat works.

7.10 What does *break-before-make* mean?

7.11 Explain how a switch matrix is scanned?

7.12 Explain the difference between software and hardware decoding of keyboards.

7.13 Explain the difference between software and hardware decoding of displays.

7.14 Explain how an LCD works.

7.15 List five possible applications of interrupts.

7.16 Explain the levels involved in enabling and disabling an interrupt.

7.17 What is the difference between maskable and nonmaskable interrupts?

7.18 What is the difference between an interrupt and a reset?

7.19 Explain the role of a vector and a vector address in the interrupt and reset operations.

7.20 What address does the 9S12C use to find the address of an interrupt-service routine for a timer overflow?

7.21 Describe the actions the CPU takes between the time it receives an interrupt request and when it begins executing the interrupt-service routine.

7.22 What would you do to raise an interrupt to the highest priority?

7.23 Define *interrupt latency.* On what does it depend?

7.24 Describe the process of enabling an interrupt.

7.25 Explain the process of clearing a flag.

7.26 How does the CPU react to an interrupt-service request while servicing another interrupt request?

7.27 What is meant by *interrupt latency* and *interrupt density*?

7.28 Give an example of a system failure in the case of latency time exceeding the maximum allowable interrupt waiting time.

7.29 Explain the five reset functions handled by the CPU of the 9S12C.

7.30 What is meant by the timeout period related to the watchdog timer?

7.31 Describe the various functions provided by the MAX817 IC.

7.32 Explain the difference between IRQ and XIRQ interrupts.

7.33 What is the role of the SWI instruction?

7.34 What instructions can be used to save power when waiting for an interrupt to occur?

PROBLEMS

7.1 Use port T pins PT[3:0] to drive blue, green, red, and yellow LEDs. Light the blue LED for 2 seconds, then the green LED for 4 seconds, then the red LED for 8 seconds, and finally the yellow LED for 16 seconds. Repeat the operation indefinitely.

7.2 A pulse generator and an oscilloscope can be used to observe the actions of an interrupt on any port P pins. Toward that end, the output of a pulse generator is connected to pin PP[0] and to one of the scope's input channels, whereas the PP[1] is connected to a second scope channel. The square wave of the pulse generator serves to drive PP[0] periodically through its active edge. Write the assembly code to mange the process, and observe the oscilloscope to see the reaction on the PP[1] line if the ISR drives it.

7.3 A keyboard is connected to port T lines. It sends a byte to port T and a KESTROBE to a port P pin whenever a key is depressed. Write a program segment to store $(80)_{10}$ bytes in the buffer from $C100 to $C14F.

7.4 Connect a photocell, an infrared emitter/detector pair, an LED, and a buzzer to the 9S12C. Write a program so that when it is dark, the LED turns ON and when the light path between the emitter and the detector is broken, the buzzer sounds OFF. Add as many features as you would like to explore various possibilities.

7.5 Write a program to control a traffic light at a three-way intersection. The lanes are bound N–S, SE–NW, and SW–NE. Refer to Example 6.18 for a start.

7.6 Write a program to read the time of day from an eight-DIP switch, and use ports M and T to drive six seven-segment displays.

7.7 Write a program to control a sump pump. A sump well collects water runoff. The sump pump pumps out the well whenever it fills up. When the level rises to a *high limit*, the pump turns on. It stays on until the level drops to a *low limit*. Then the pump shuts off. It does not turn on again until the well fills up to the *high limit*. Data is stored in the following addresses:

Address	Data
$3800	high limit
$3808	level reading
$380F	pump control
$3810	low limit

The level has a range of $00–$FF, representing empty to full. You can set the low and high limits to values within the range (e.g., $10 and $F0). Sending a value of $00 to address 380F turns on the pump. Sending a value of $FF shuts it off.

7.8 Obtain the specs for the LM74C923 keyboard encoder, and write a program to interface it with the 9S12C.

7.9 Obtain the specs for the AY-5-2376 keyboard encoder, and write a program to interface it with the 9S12C.

7.10 Obtain the specs for the Motorola MC14499 seven-segment decoder/driver, and write a program to interface it with the 9S12C. Using interrupts, make the program flash the message HELP 10 times per second.

7.11 Obtain the specs for the EPSON S1D13708 LCD controller, and write a program to interface it with the 9S12C.

7.12 Suppose that the 9S12C is executing the following program fragment when the IRQ interrupt occurs while the TSY instruction is being executed. What will be the contents of the top 10 bytes in the stack?

```
        ORG     $0100
        LDS     #$FF
        CLRA
        LDX     #$1000
        BSET    10, X $48
        LDAB    #$40
        INCA
        TAP
        PSHB
        TSY
        ADDA    #10
```

7.13 Write a code to control the number of times a push-button switch is triggered and to set off an alarm after 10 consecutive switch activations.

LABORATORY PROJECTS

7.1 Connect a debounced switch that can generate a negative-going pulse to the IRQ pin and an LED to pin PB0. Also connect two solenoids to two output port pins. Then write a program with an IRQ service routine to manage the interplay between the switch, the LED, and the operation of the solenoids. Specifically, the main program initializes the variable "icr-cnt" to N, stays in loop, and keeps checking the value of "irq-cnt". When "irq-cnt" decrements to 0, the main program jumps the IRQ service routine, turns on the LED, activates one solenoid for 3 seconds, and activates the other solenoid for 3 more seconds before returning. The system may be used to count N bottles moving on a conveyor belt to a storage box. When the box is full, one solenoid activates a mechanism to push it off the conveyor, and the other solenoid engages another mechanism to place an empty one. Select the components, build a model of the process, and control it from the MCU.

7.2 Develop a code to manage the operation of the 3×4 telephone keypad interface shown in Fig. 7.5 with an LCD display. Implement the code on actual hardware such that the number entered on the keypad is displayed on the LCD.

7.3 Design and build a system to retrieve boxes stored on shelves as shown in Fig. 7.19. The boxes contain chemicals and are arranged on shelves as shown. Each shelf contains a distinct product that could be used to retrieve. On command from the user, a small cart moves to the required shelf, retrieves a box, and moves back to the delivery station. The cart is guided by a reflective tape or other means to stay on course. When a box is retrieved and delivered to a bin accessible by the user, the bin door should open; once the user retrieves the box, the bin door should close again. The user should be alerted whenever the last box from a shelf is retrieved so that supplies could be added.

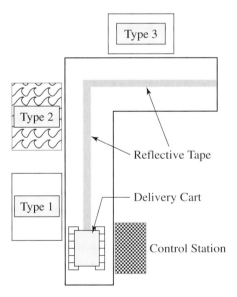

Figure 7.19

7.4 Build a small model of the elevator of Example 7.9 that can be operated by a remote control.

8

Serial Interface Facility

OBJECTIVES

Thoughtful engagement with the material presented in this chapter will enable students to:

1 Explain the basic concepts of serial communications
2 Describe the SCI and SPI facilities on the 9S12C MCU
3 Write assembly codes to manage SCI communications
4 Understand the interface between the 9S12C and the RS232 port
5 Write assembly codes to interface peripheral devices to the SPI facility

8.1 INTRODUCTION

Serial data transfer between two communicating devices is an alternative technique to a parallel interface. Parallel interface of equipment is simpler and requires less hardware, whereas serial interface is much more versatile. Transmission of data with a serial interface requires only one data wire, as opposed to the eight wires normally used to transmit a data in a parallel configuration. Therefore a serial interface is more economical, especially if the data is to be transferred to a distant peripheral. Additionally, serial transmission enables data equipment to use commercial communication facilities such as regular telephone or data lines. A serial link requires more time to transmit a character than a parallel link. However, this does not pose a problem in most sensor and control applications because achievable data exchange rates using either technique is far greater than necessary. The Controller Area Network (CAN) automotive standard by Bosch is an example of a serial standard accommodated by the 9S12C NCU.

The 9S12C provides for serial interface through three integrated subsystems: the *serial communications interface* (SCI), the *serial peripheral interface* (SPI), and the MSCAN I/O. The SCI subsystem is an on-chip equivalent to the *universal asynchronous receiver-transmitter* (UART) device, which converts data into a serial stream, and vice versa. A number of dedicated UART ICs are available, including the General Devices 1013 or 1015, the Maxim MAX3140, and the

Intel 8251. The SCI subsystem enables the 9S12C to exchange data with other UART-equipped devices. UART devices are identified by an ISO standard as *data terminal equipment* (DTE), such as a terminal or a computer, and *data communications equipment* (DCE), such as modems. The SCI can communicate with only one DTE or with a network of one or more DCEs or other MCUs in a multidrop network environment. The I/O serial port SPI is a high-speed software-driven synchronous protocol used for serial data transfer between master and slave units in a serial network. It also allows many SPI-equipped MCUs or devices to be interconnected, expanding the I/O capability of the 9S12C. The MSCAN scalable communication controller implements the CAN 1.0 A/B protocol. The CAN serial protocol was originally developed by Bosch automotive to reliably handle serial data communications in the EMI-rich automotive environment.

The balance of this chapter is dedicated to the operation of the serial facilities onboard the 9S12C MCU, namely, the SCI and SPI facilities. An overview of the CAN facility and example programs on its use are available on the text website.

8.2 SERIAL COMMUNICATION INTERFACE (SCI)

The SCI asynchronous I/O subsystem of the 9S12C plays the role of a UART device that converts bytes of data into a serial data stream, and vice versa. With this capability, the 9S12C can use industry-standard cables, telephone lines, or radio transmitters to communicate data with remote devices. The SCI subsystem consists of a transmitter and a receiver that function independent of each other, but both use the same data format and bit transfer rate. The SCI can transmit and receive data simultaneously without the need for any chips external to the 9S12C itself. This ability to transmit and receive data simultaneously is called full-duplex communications. Communication devices can use one of two other protocols for data transmission along a communication channel: simplex and half-duplex. Transmission using simplex protocol is possible in one direction only, analogous to traffic flow on a one-way street. In half-duplex, the communication channel can be used to transmit and receive data, but not simultaneously. This is like a one-lane street where two-way traffic is allowed in different directions at different times.

From only two pins, PS[0] and PS[1] of port S, any data communication device (DCE) or data terminal device (DTE), such as a terminal or a host computer, can easily be interfaced. The transmit line, TxD sends data to a remote receiver via pin PS[1] of port S. The receive line RxD receives data from a remote transmitter via pin PS[0] of port S. When the SCI is not in use (i.e., is disabled), port S lines can be used as general-purpose I/O lines as determined by the data direction register DDRS of port S. The SCI facility is configured by the bits in the SCICR1 and SCICR2, SCI control registers 1 and 2, at addresses $00CA–CB. When enabled, the SCI takes control of the PS[0] and PS[1] lines. For the transmit output to make it out of the chip, the DDRS register must be initialized with bit 1 set. The 9S12C does not have any of the modem control signals, which include RTS (request to send), CTS (clear to send), DTR (data terminal ready), DSR (data set ready), and TI (ring indicator). These signals are part of the RS232 standard discussed in Section 8.5.

8.2.1 Communications Protocol (Framing)

Serial communication requires a handshaking protocol, a set of standard procedures that coordinates information transfer between the transmitter and the receiver. Asynchronous communication allows the data bits for one character (ASCII uses seven bits per character) to be transferred at one time, as a stream one after the other, on a single line. However, a delay may occur between transmitted bytes. Therefore, the receiver needs to know when a new data byte

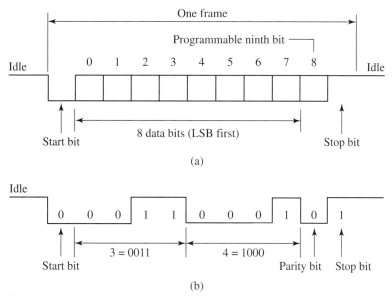

Figure 8.1 UART communication protocol: 11-bit frame (a) and asynchronous transmission of the letter C (b).

is about to be received. To achieve this, the transmitting and the receiving devices must agree on a communication protocol. One popular protocol used by asynchronous communications is a process called *framing*. The 9S12C supports the protocol shown in Fig. 8.1. Each data byte is framed by a start and a stop bit in order for the receiver to know when a data byte starts and stops. The signal line is considered to be *idle*, that is, no data is being transmitted, if the transmitter continuously transmits the voltage level corresponding to logic 1. A 1-to-0 transition indicates that a bit stream is about to be sent. A character ends with transmission of one logic 1 stop bit. Thus a start bit is always zero and a stop bit is always 1. The SCI hardware automatically creates the framing bits, and the receiver hardware automatically removes them and converts each framed serial character into a parallel byte. A parity is used as part of a frame to detect single bit errors by counting the number of the 1 bits in a binary character. Still others use two stop bits. Figure 8.1b shows the output produced when character C ($43) is transmitted in an 8-bit asynchronous format with odd parity. *Parity* refers to how many 1s the transmitted binary number contains: If the number of 1s is odd, the binary number has an odd parity; and if the number of 1s is even, the binary number has an even parity. A parity that is not what the receiver and transmitter agree on indicates that the received data is wrong.

8.2.2 Data Transfer (Baud) Rate

For a device to receive data reliably, it must understand not only the data frame configuration, but also the time duration of a bit. Thus, the receive and transmit devices should operate at the same clock frequency, called *baud rate*. The baud rate is the serial data speed, defined as the number of transmitted bits per second (BPS), including the start, parity, and stop bits. The SCI baud rate generator uses a 13-bit modulus counter to set the baud rate for both the transmitter and the receiver to an integral fraction of the bus clock frequency. The SCI baud rate is determined by

$$\text{Baud rate} = \frac{ECLK}{16 \times BR}$$

where *BR* is any value between 1 and 8191, depending on the code in the SBR[12:1], *SCI baud rate bits* in the SCIBDH and SCIBDL registers at addresses $00C8–$00C9, for a baud rate range between 122.08 kHz and 1 MHz. If the target baud rate is 9600, the *BR* value coded in SBR[12:1] should be 104, resulting in an error of 0.16%. The transmitter is driven at the baud rate frequency, whereas the receiver uses the internal RT clock to sample each bit at a frequency 16 times the baud rate; that is, it takes 16 RT clock ticks (RT1–RT16) to sample each bit. This will help verify the integrity of data transfer. The baud rate generator is enabled by setting the TE, *transmit enable*, or the RE, *receive enable*, bits 3 and 2 in the SCICR2, *SCI control register* 2. It is disabled if BR = 0, regardless of the RE or TE settings.

8.3 SCI REGISTERS

The SCI manages the transmit and receive operations by using the registers given in Fig. 8.2, (which also shows their corresponding bits). A brief description of the registers' bits follows.

8.3.1 Data Register

The SCIDR, *SCI data register*, at addresses $00CE–CF, handles data-transmit and data-receive operations using two internal shift registers, an 11-bit write-only *transmit shift register* (TSR) and an 11-bit read-only *receive shift register* (RSR). To transmit data, a write to the SCIDR register is executed, which automatically transfers the data to the TSR buffer. To receive a byte, a read of the SCIDR register is initiated, which prompts the transfer of the data received on the RSR. A write instruction prompts the CPU to write to the TSR buffer when the TDRE, *transmit data register empty flag*, bit 7 in the SCISR1 register, is set, and a read instruction causes

SCIBDH: SCI Baud Rate Registers ($00C8)

0	0	0	SBR12	SBR11	SBR10	SBR9	SBR8

SCIBDL: SCI Baud Rate Registers ($00C9)

SBR7	SBR6	SBR5	SBR4	SBR3	SBR2	SBR1	SBR0

SCICR1: SCI Control Register 1 ($00CA)

LOOPS	SCISWAI	RSRC	M	WAKE	ILT	PE	PT

SCICR2: SCI Control Register 2 ($00CB)

TIE	TCIE	RIE	ILIE	TE	RE	RWU	SBK

SCISR1: SCI Status Register 1($00CC)

TDRE	TC	RDRF	IDLE	OR	NF	FE	PF

SCISR2: SCI Status Register 2 ($00CD)

0	0	0	0	0	BK13	TXDIR	RAF

SCIDRH: SCI Data Register High ($00CE)

R8	T8	0	0	0	0	0	0

SCIDRL: SCI Data Register Low ($00CF)

R7/T7	R6/T6	R5/T5	R4/T4	R3/T3	R2/T2	R1/T1	R0/T0

Figure 8.2 Pins and data registers of the SCI facility.

the CPU to read the RSR register and automatically clears the RDRF, *receive data full flag*, bit 5 in the SCICR1 register. Having a separate receive-data register instead of reading the input shift register directly is called *double buffering*.

The SCIDR register occupies two bytes in the register block: the SCIDRH, data register high at $00CE, and SCIDRL, data register low at address $00CF. The SCIDRL is used to transmit T[7:0] data bits with a write instruction or to receive R[7:0] data bits with a read instruction. By convention, the LSB (start bit) is transmitted first.

The 9S12C can transmit or receive nine bit characters instead of the standard eight data bits. The SCIDRH holds the ninth data bit (MSB) **R8**, *receive data bit* 8, during a receive operation if the SCI had been configured to receive nine data bits. The extra, ninth bit can be used to "wake up" or enable the SCI for a receive operation. It also holds the ninth data bit for a transmit operation **T8**, *transmit data Bit* 8, when the SCI had been configured to transmit nine data bits.

8.3.2 SCI Control Registers

The SCI includes two control registers, SCICR1 at address $00CA and SCICR2 at address $00CB. These registers provide the means for the CPU to set up the SCI protocol. They also enable the SCI to generate interrupts to inform the CPU when it is ready to transmit or receive another character. The registers also enable the circuitry that connects the SCI to the physical pins on the chip. A reset operation clears all the bits in the control registers.

SCI Control Register 1 (SCICR1)

The SCICR1 register contains eight bits for control. A brief overview of their role follows.

> The **LOOPS**, *loop select*, bit 7 is set to enable a loop operation in which the RXD pin is disconnected from the SCI and connected internally to the transmitter output, provided that the RXD and TXD are both enabled. In an enabled loop operation, the **RSRC**, *receiver source*, bit 5 determines the source for the receiver shift register input.

> The **M**, *data format mode*, bit 4 is used to select whether the length of the word to be transmitted or received is eight or nine bits, M is set to 0 for eight bits and M is set to 1 for nine bits. This allows the framing format to use two stop bits or eight data bits plus a parity. To transmit nine bits (M = 1), the CPU writes the ninth bit, T8, to bit 6 in the SCIDRH register and the remaining bits to the SCIDRL. To receive nine bits (M = 1), the CPU reads R8, the MSB of the received character, from the SCIDRH and the remaining bits from the SCIDRL.

> The **WAKE**, *wake-up condition*, bit 3 is normally used when multiple 9S12Cs are connected to a common asynchronous bus. A receiver of an SCI can be put to sleep if the RWU, bit 1 in the SCICR2 register, had been set. The wake-up feature will only wake up the sleeping receiver that is destined to receive the message and respond to it. The WAKE bit sets one of two available wake-up protocols: idle-line wake-up and address mask wake-up. If the WAKE bit is 0, idle-line wake-up is selected. In this mode, the receipt of an IDLE character causes the RWU bit to reset and the receiver to wake-up. If WAKE is set to 1, address mask wake-up protocol is selected. This protocol generally uses 9-bit characters; a 1 in the MSB of the character indicates that the character is an address. When a receiver is awake and the receive buffer is full, the receive-buffer-full flag RDRF in the SCISR1 register will set, and the application program polls the RDRF or respond to an interrupt, if the RIE bit in SCICR2 is enabled.

> The **ILT**, *idle-line type*, bit 2 controls when the receiver starts counting idle characters (logical 1s). Counting starts after the start bit if the ILT bit had been cleared. In this mode,

an idle character may be falsely recognized if a stream of 1s precedes the stop bit. Meanwhile, the receiver starts counting after the stop bit if the ILT bit is set to 1. This mode avoids false recognition of idle characters but requires properly synchronized transmission.

The **PE**, *parity enable*, bit 1 is set to insert a parity bit in the MSB position of a transmitted word. The parity function is disabled if the PE bit is clear.

The **PT**, *parity type*, bit 0 forces the SCI to generate (or check for) an odd parity if it is set to 1 or to generate (and check for) an even parity if it is cleared to zero. If the parity is even, an even number of 1s clears the PE bit and an odd number of 1s sets the PE bit. The inverse is true if the parity is odd.

The **SCIWAI**, *SCI stop in wait mode*, bit 6 controls how the SCI functions in the wait mode. If the SCIWAI bit is clear, the SCI continues to operate normally; and if it has been set, the SCI clock generator stops. Ongoing SCI operation is halted and will resume if an interrupt source brings the MCU out of the wait mode.

SCI Control Register 2 (SCICR2)

The **TIE**, *transmit interrupt enable*, bit 7 is set to enable an interrupt when the TDRE flag is set. This will occur as soon as the contents of the SCIDR have been transferred to the TSR buffer. The CPU responds to the interrupt request by writing another byte to the SCIDR data register.

The **TCIE**, *transmit complete interrupt enable*, bit 7 is set to enable the SCI to generate an interrupt when the TC flag sets as soon as the contents of the TSR have been shifted out of the SCI, with no new data in the transmit buffer TSR waiting for transfer. This interrupt may be used for some protocols to confirm an end of message.

The **RIE**, *receive interrupt enable*, bit 6 is set to enable an interrupt when the RDRF flag sets, which will occur when a received byte has been transferred from the RSR buffer to the SCIDR register or when an overrun error has occurred. The CPU responds to the interrupt request by reading the SCIDR register before it receives another byte.

The **ILIE**, *idle line interrupt enable*, bit 4 is set to enable when the IDLE flag sets, which will occur when an idle condition is detected on the receive line RxD for a time corresponding to at least one character of 10 consecutive 1s (11 if M bit is 1). This feature may be used to end the communication for some protocols.

The **TE**, *transmit enable*, bit 3 is used to enable the transmit line of the SCI. When enabled, it connects the external transmit data line TxD/PS[1] to the TSR buffer. If TE is clear, pin PS[1] acts as a general-purpose I/O pin as configured by the DDRS register.

The **RE**, *receive enable*, bit 2 is used to enable the receive line of the SCI. When enabled, it connects the external receive data RxD pin PS[0] to the internal TST buffer. If the RE bit is clear, pin PS[0] acts as a general-purpose I/O pin as configured by the DDRS register.

The **RWU**, *receiver wake-up*, bit 1 determines the wake-up status of a receiver in a multiple receiver communication network. If the RWU bit is clear, the wake-up feature of the SCI is disabled and the associated receiver operates in its normal awake mode. If the RWU bit is set, the wake-up feature is enabled and the receiver is forced to sleep, ignoring the current message because the message is presumably intended for another device.

The **SBK**, *send break*, bit 0 is set to force the SCI to transmit continuous 10- or 11-bit blocks of logical 0s, signaling a break in the flow of a serial data. A break is the opposite of an idle. As soon as the SBK bit is cleared, the SCI will transmit at least one logical 1 bit after at least one break block so that a subsequent START bit can be recognized by any receiving device.

8.3.3 SCI Status Registers

Two status registers are associated with the operation of the SCI. The registers hold the bits that indicate the status of the SCI system at any time. Also, the bits provide inputs to the SCI interrupt logic circuits for generating previously enabled SCI interrupts. All the SCISR bits are cleared after reset.

SCI Status Register 1 (SCISR1)

The SCISR1 is an 8-bit register located at address $00CC. The role of each bit is described next.

The **TDRE**, *transmit data register empty flag*, bit 7 is set when a transmit operation shifts the contents of the SCIDR to the TSR. An interrupt is generated if the TIE bit in the SCICR2 register has been set. A read of the SCIDR followed by a write to the SCIDRL register clears the TDRE flag.

The **TC**, *transmit complete flag*, bit 6 sets when the entire 10- or 11-bit word has been shifted out of the SCI. An interrupt is generated if the TCIE bit in the SCICR2 register has been set. A read of the SCISR1 followed by a write to the SCIDRL clears the TC flag.

The **RDRF**, *receive data register full flag*, bit 5 sets when a serial word is transferred from the RSR buffer to the SCIDR register, signaling to the CPU that a word has been received and is ready to be read from the SCIDR. At the same time, an interrupt is generated if the RIE bit in the SCICR2 has been set. A read of the SCISR1 followed by another read of the SCIDR clears the RDRF flag.

The **IDLE**, *idle line detect flag*, bit 4 sets when the SCI receiver line remains idle for at least 10 consecutive 1s. An interrupt is generated if the ILIE bit in the SCICR2 register has been set. The IDLE bit is cleared by reading the SCISR register and then the SCIDRL register.

The **OR**, *overrun error flag*, bit 3 is set when a new serial word is about to be received before the CPU had read the current word in the SCIDR1. An interrupt is also generated if the RIE bit in the SCICR2 register is set. The OR bit is cleared by reading the SCISR1 register and then the SCIDRL register.

The **NF**, *noise flag*, bit 2 indicates if any noise has been received on any of the serial bits, including the START and STOP bits. This flag cannot initiate an interrupt, but it can be polled by the CPU to determine if any noise has been detected on the serial word. The NF flag is cleared by reading the SCISR1 register and then the SCIDRL register. The sampling clock used by the receiver is 16 times the baud frequency. Once the receiver has defined the bit boundaries, it samples the bits during the 8th, 9th, and 10th clock cycles; if these samples do not match, noise is suspected and the NF flag sets.

The **FE**, *framing error*, bit 1 is set when no stop bit is detected in the received data string. The FE bit cannot be used to generate an interrupt, but it can be polled by the CPU to determine if a framing error has occurred. The FE bit is cleared by reading the SCISR1 register and then the SCIDRL register.

The **PF**, *parity error flag*, bit 0 is set if the PE bit is set and the parity of the received data does not match the parity-type bit PT. This bit is set simultaneously with the RDRF but does not get set in the case of an overrun. The PE bit is cleared by reading the SCISR1 register and then the SCIDRL register.

SCI Status Register 2 (SCISR2)

This register is located at address $00CD and contains three nonzero bits. The role of each is described next.

The **BK13**, *break transmit character length*, bit 2 is set to make the SCI transmit 13 or 14 logical 0s to indicate a break in the flow of a serial data (refer to the SBK bit from earlier).

The **TXDIR**, *transmit pin data direction in single-wire mode*, bit 1 is set to force the use of the TXD pin as an output in single-wire operation; otherwise it will be used as an input pin.

The **RAF**, *receive active flag*, bit 0 is set to indicate that reception is in progress after the receiver has detected a logical 0 within the RT1 time period of the start-bit search. RAF will clear if an idle character is detected.

8.4 SCI OPERATION

This section includes assembly and C examples on setting up the SCI and managing the transmit and receive operations.

8.4.1 SCI Configuration

Before the SCI is ready to communicate with other devices, it has to be set up via the following procedure.

1. Program the SBR[12:0] bits in the SCIBDH and SCIBDL registers to select the required BR and to start the baud rate generator.
2. Configure the control bits in the SCICR1 register to enable/disable the loop operation (LOOPS), select SCI operation in wait mode (SCIWAI), define the receiver source bit when LOOP = 1 (RSRC), select data format (M), select address mark or idle-line wake-up (WAKE), select the receiver option to start counting for idle character bits of 1s (ILT), enable/disable parity (PE), and select the type of parity (PT).
3. Configure the control bits in the SCICR2 register to enable transmitter interrupt (TIE), enable/disable interrupt after transmission is complete (TCIE), enable/disable interrupt when RSR is full (RIE), enable/disable idle-line-flag interrupt (ILIE), enable transmit function (TE), enable receive function (RE), enable/disable the wake-up function (RWU), and allow/forbid sending break characters (SBK).
4. Ensure that flags are cleared.

The following example implements the required configurations.

Example 8.1: SCI setup—Write a code to set up the SCI operation.

Solution: The following code configures the SCI subsystem to operate at a baud rate of 9600 (SBR[12:00] = $30). The frame is set as one start bit, eight data bits, and one stop bit, idle-line wake up feature (SCICR1 = $00). The receiver and transmitter are enabled to operate with no interrupt and no send-break option (SCICR2 = $0C).

```
        INCLUDE    'mc9s12c32.inc'

sci_setup
            TPA                      ; Transfer CCR to A
```

```
ORAA    #10              ; ORAA with #$10 to clear
                         the I bit
TAP                      ; Transfer A back to CCR
MOVW    #$34, SCIBD      ; Set BR = 9600
MOVB    #$00, SCICR1     ; 8-bit data format, Loop
                         mode, parity disabled
MOVB    #$08, SCICR2     ; No interrupts TX is
                         enabled
LDAA    SCISR1           ; First step to cleat
                         TDRE Flag
STD     SCIDRL           ; Second step to clear
                         TDRE flag
RTI
```

8.4.2 Transmit Operation

Figure 8.3 depicts the transmit operation performed by the SCI. The CPU transmits a character to the SCIDR whenever the TDRE flag becomes set. The CPU knows the TDRE is set by either polling the SCISR1 register or by receiving a TDRE flag-interrupt request. In the polling mode, the CPU monitors the TDRE flag; when set, the CPU writes the 8- or 9-bit character to be transmitted to the SCIDR register and then goes back to monitoring the TDRE flag. This step is repeated until all characters have been transmitted. A write to the SCIDR register automatically clears the TDRE flag and prompts the SCI to transfer the contents of the SCIDR to the TSR, with the LSB first; it starts shifting it out immediately, and sets the TDRE flag to indicate that the SCIDR is empty and ready to accept new data from the CPU. The transmit-driver routine may respond by writing another character to the SCIDR register while the

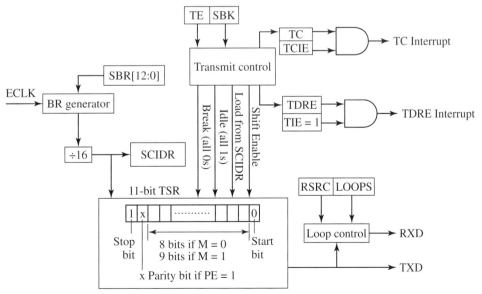

Figure 8.3 SCI transmit operations.

previous data byte is being shifted out. The TSR buffer automatically shifts out a start bit, followed by the data bits, and then automatically adds a stop bit at the end. The value settings of the baud rate register determines the shift rate.

Using the interrupt-driven mode, the SCI can be set up to interrupt the CPU each time the TDRE flag is set by setting the SCI interrupt enable bit TIE in SCICR2. This mode frees the CPU to do other things, knowing that it will be informed when it can send another byte to the SCI.

Example 8.2: Transmit a message via the SCI—Write a subroutine to handle the transmission of the indicated message through the SCI subsystem using polling.

Solution: The following code manages the required transfer. The output is first passed in accumulator B. Because each ASCII character consists of only seven bits, the code clears bit 7 of B first.

```
                INCLUDE 'mc9s12c32.inc'

                ORG     $3800
char_blk  DC.B    "Life is short!"
          DC.B    $0D, $0A          ; Codes for Carriage
                                      Return and Line Feed
          DC.B    "Live Love and Leave a Legacy."
          DC.B    $0D, $0A
          DC.B    "......... Steven Covey."
Lastchar  DC.B    $04
          JSR     sci_setup
          JSR     sci_txd
here      BRA     here
sci_txd
          LDY     #char_blk         ; Point to start of data
                                      to be transmitted
chech_tdr BRCLR   scisr1, #$80,     ; Is TSR empty?
                  check_tdr
          MOVB    1, Y+, scidrhl    ; Transmit a new
                                      character and
                                      increment Y pointer
          CPY     #lastchar         ; Check if the last
                                      character had been
                                      transmitted (TSR empty)
          BNE     check_tdr
          RTS
```

8.4.3 Receive Operation

Figure 8.4 shows a block diagram depicting the receive operation performed by the SCI. If the receiver is awake and the idle line is at logical 1, the receiver waits until a 1-to-0 transition in the signal occurs, indicating the arrival of the start bit. On the first low-level bit, the receiver shifts the next eight or nine bits into the RSR buffer. The rate at which data is sampled and shifted in is determined by the setting of the SCIBD register. When all bits have been received,

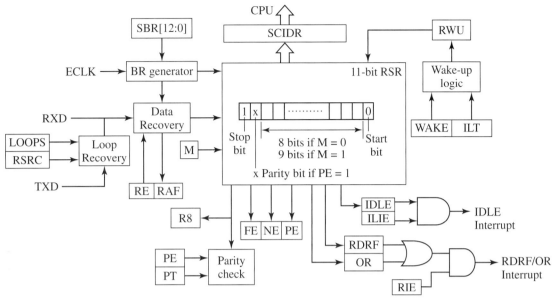

Figure 8.4 SCI receive operations.

the SCI automatically transfers a full byte from the RSR to the SCIDRL register and a ninth bit (if M = 1) to the R8 bit location in the SCIDRH register. If the SCI detects that a previously received byte of data has not yet been read by the CPU, it sets the OR bit in the SCISR1 register. It will also set the FE flag in the SCISR1 if it reads a 0 when a stop bit is expected. When the received data is transferred to the SCIDR, SCI sets the receive-data-ready flag RDRF to inform the CPU that the RSR is full and ready to be read. The SCI also generates an interrupt if the RIE bit is set. The receiver samples the stop bit and waits for another start bit before it shifts in the next group of data bits.

When the RDRF flag is set (by polling or by interrupt), the CPU first checks the framing error bit FE, the overrun error bit OR, and the noise bit NF in the SCISR1 register to make sure that reception has gone smoothly. If all of these flags are clear, the CPU then reads the byte from the receive data register. Reading the RSR register automatically clears the RDRF bit.

Example 8.3: Receive a character via the SCI—Write a subroutine to receive a character through the SCI subsystem using polling.

Solution: The following code manages the required transfer. The input is first passed in B before it is stored for later use.

```
            INCLUDE    'mc9s12c32.inc'
            ORG        $3800
char_blk    DS.B       100

            ORG        $C000
            JSR        sci_setup
```

```
                JSR      sci_rxd
      here      BRA      here

sci_rxd
                LDX      #char_blk
chech_rdr       BRCLR    scisr1, #$20, check_rdr  ; Is RDR full?
                LDAB     scidr    ; Read SCIDR
                STAB     1, X+
                RTS
```

8.5 INTERFACING THE 9S12C WITH THE RS232 PORT

Almost all computers use the RS232 (EIA 232) standard for serial communications. Most RS232 interfaces use the 25-pin mechanical connectors DB-25P (plugs) and DB-25S (socket), shown in Fig. 8.5a. Interfaces with fewer signals use the DB-9 connector, shown in Fig. 8.5b, instead. Although the RS232 standard defines 25 signal levels, only a few are ever used, and very often three signals are required for a connection: receive data RxD, transmit data TxD, and signal ground SG. Figure 8.5c gives the most commonly used signals. The EVBoard uses only three signals. The RS232 voltage levels are $+3$ to $+12$ V for logical 0, referred to as *space*, and -3 to -12 V for logical 1, or *mark*. Voltage levels between -3 and $+3$ V is for no logic signal. The most common RS232 voltages are ±9 and ±12 V. The high voltages used provide for better noise margins and allow logic signals to be transmitted over greater distances.

The RS232 standard uses two types of equipment for data communications: data terminal equipment (DTE), such as terminals and computers, and data communications equipment (DCE), such as line drivers and modems for telephone line communications. The DTE is the initial source or the final recipient of data. The DCE provides the function to establish,

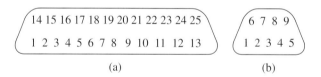

(a)   (b)

Pin number		EIA			
25-pin	9-pin	designation	Direction	Name	Function (as seen by DTE)
2	3	BA	OUT	TXD	Transmitted data
3	2	BB	IN	RXD	Received data
4	7	CA	OUT	RTS	Request to send (DTE ready)
5	8	CB	IN	CTS	Clear to send (DCE ready)
20	4	CD	OUT	DTR	Data terminal ready
6	6	CC	IN	DSR	Data send ready
8	1	CF	IN	DCD	Receive line signal detect
22	9	CE	IN	RI	Ring indicator
1	-	AA	-	FG	Frame ground (chassis)
7	5	AB	-	SG	Signal ground

(c)

Figure 8.5 RS232-D 25-pin connector (rear view of male or front view of female) (a), 9-pin connector (b), and circuit definition of mostly used pins (c).

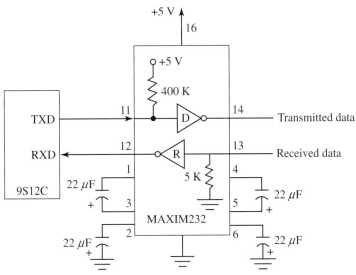

Figure 8.6 Interfacing the 9S12C to an RS232 port via a MAXIM232 IC.

maintain, and terminate connection, and it codes and decodes signals as required. By convention, DTE uses a plug and DCE uses a socket. Mechanical connectors are usually made for DTE-to-DCE connection. However, some applications require DTE-to-DTE or DCE-to-DCE connections, for example, connecting a computer (DTE) to a target system, a single-board computer (another DTE). In these situations, the *null modem* scheme is required for proper communication. The null modem scheme is implemented by crossing some connecting wires. For connections that require only three signals, wires 2 and 3 have to be crossed.

The RS232 electrical signal convention for logical 1 and 0 differ from the standard TTL logic levels of 0 and 5 V for the SCI waveforms generated within the 9S12C. Therefore, interfacing the SCI to a device that employs an RS232 interface requires that voltage levels be transformed. This is achieved by using ICs specially developed for this purpose. A popular IC is the MAX232, from MAXIM. This IC operates from a single +5-V supply to convert a 0- and 5-V to +10- and −10-V output using an internal charge pump. The ±10-V signals are sufficient according to the RS232 standard. The MAX232 is a 16-pin DIP that contains two DC-to-DC converters as well as two drivers and two receivers. Figure 8.6 illustrates how to interface the 9S12C to an RS232 port using the MAX232 IC.

8.6 SERIAL PERIPHERAL INTERFACE (SPI)

Synchronous communication between the MCU and external devices is carried out by the SPI facility onboard the 9S12C. The SPI offers full-duplex serial transfer of messages composed of many characters or bits in one continuous bit stream. The SPI subsystem is an I/O serial port through which data can be transferred at lower rates than through a parallel I/O. When one or more SPI devices communicate, one of the devices is configured as a master and the others are configured as slaves. The master initiates data transfer, whereas the slave can only react. The master and slaves must use the same clock signal, furnished by the master, to synchronize data transfer. Although the SPI is usually used as an I/O serial port, it can also be used to facilitate multiprocessor communications, I/O port expansion, and interface with peripheral devices. A list of over 190 SPI-compatible devices is available at www.mct.net/faq/spi.html. Examples of these devices are MAX5544 14-bit DAC from MAXIM; ADS7835, 12-bit ADC from Analog Devices; DS1267 digital potentiometer and DS1722 digital thermometer from Dallas Semiconductor; KP100 pressure sensor from Infineon; MAX3140 YART from MAXIM; and MAX7219 LED display drive from MAXIM.

The SPI subsystem communicates with external devices through four bidirectional pins (SS, SCK, MISO, and MOSI) that are shared with pins PM[5], PM[4], PM[3], and PM[2] of port M, respectively. The DDRM bits corresponding to the output lines of the SPI must be set, whereas the input lines of the SPI will act as input regardless of the DDRM bit settings. Master-in slave-out pin MISO/PM[2] serves as an input in a master device and as output in a slave device. The MISO line in a slave device is forced to a high-impedance state if the corresponding device is not selected. Master-out slave-in pin MOSI/PM[3] is configured as output in a master device and as an input in a slave device. Both MOSI and MISO lines are used to transfer data in one direction, MSB first. The serial clock SCK synchronizes data transfer along the MISO and MOSI lines. A byte of data is exchanged along those lines during eight SCK cycles. The SCK clock cycles are initiated by the master and received as input by the slave. The slave-select SS signal is used to select a slave device with which to transfer data when the SPI is configured as a master, and it is used as an input to receive the signal from another master when the SPI is configured as a slave. This line must be LOW prior to and during data transfer in a slave device. It must be tied HIGH in a master device; otherwise a *mode fault flag* (MODF) in the SPISR register is set. If bit 5 in the DDRM register is set to 1, the SS line is configured as an output line, thereby disabling the mode-fault-detection circuit.

8.6.1 Port M Data Direction Register (DDRM)

When the SPI is enabled, the SPI input lines act as inputs, regardless of the bit settings in the DDRM (Refer to Section 7.2). However, to configure the necessary lines as outputs, the corresponding data direction bits are set to 1. Table 8.1 summarizes the settings of the DDRM bits and their effect on port M pins for an SPI master and slave.

8.6.2 SPI Baud Rate Register (SPIBR)

The SPR[2:0], *baud rate preselection bits*, and the SPPR[2:0], *baud rate selection bits*, in the SPIBR, *SPI baud rate register* at address $00DA, are used to determine the *baud rate divisor* (*BRD*), which is the value by which the ECLK is divided to yield the SCK clock rate and thus the SPI bit-transfer time. The *BRD* is given by

$$BRD = (SPPR + 1) \times 2^{(SPR+1)}$$

TABLE 8.1 Configuration of DDRM Bits and Effect on the Port M Pins

	PM[3]/SS	PM[5]/SCK	PM[4]/MOSI	PM[2]/MISO
Master	1, output 0, input	1, output	1, output	x, input
Slave	x, input	x, input	x, input	1, output

x = does not matter

for a baud rate range of 7.8125 kHz $< BR <$ 8 MHz for a 16-MHz ECLK. The SPI clock frequency or baud rate is then determined as $SCK = ECLK/BRD$. The default transfer time of 0.125 μs is normally chosen unless data is being transferred to a slow CMOS chip.

8.7 SPI REGISTERS

An overview of the registers involved in the operation of the SPI is shown in Fig. 8.7. The corresponding bits and their role are briefly discussed next.

8.7.1 SPI Data Register (SPIDR)

The 8-bit SPIDR register at address $00DD serves as the register for the exchange of input and output data between a master and its slaves. When a master SPI initiates a write of the data to be transmitted to the SPIDR register, the data is queued and then transferred to the shift register immediately after it becomes empty, as indicated by the SPTEF flag, and begins shifting it out on the MOSI pin, under the control of the serial clock. A read of the SPISR with SPTEF = 1 followed by a write to the SPIDR puts data into the TDR.

When new data has been received and moved into the SPIDR register, the SPIF flag sets, and the data can be read anytime thereafter but before the end of the next transfer. If the SPIF flag is not serviced by the end of a data transfer, the data bytes received on subsequent transfers will be lost because the SPIDR register retains the first byte until the SPIF is serviced.

8.7.2 SPI Control Registers

The SPI operation is controlled by several bits contained in two control registers. The registers and their associated bits are introduced next.

SPI Control Register 1 (SPICR1)

The SPICR1 contains eight different bits.

> The **SPIE**, *SPI interrupt enable*, bit 7 enables SPI interrupt requests when it is set and disables them when it is clear.

SPIBR: SPI Baud Rate Register ($00DA)

0	SPPR2	SPPR1	SPPR0	0	SPR2	SPR1	SPR0

SPICR1: SPI Control Register 1 ($00D8)

SPIE	SPE	SPTIE	MSTR	CPOL	CPHA	SSOE	LSBFE

SPICR2: SPI Control Register 2 ($00D9)

0	0	0	MODFEN	BIDIROE	0	SPISWAI	SPC0

SPISR: SPI Status Register ($00DB)

SPIF	0	SPITEF	MODF	0	0	0	0

SPIDR: SPI Data Register ($00DD)

Bit 7							Bit 0

Figure 8.7 Registers and bits for the SPI operation.

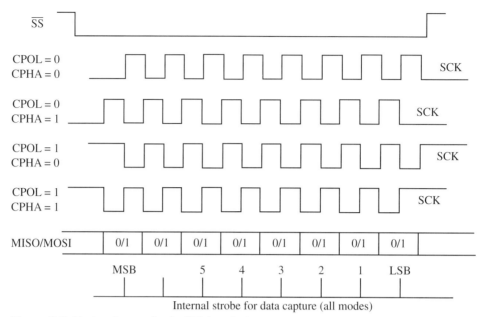

Figure 8.8 Timing diagram for the SPI data clock.

The **SPE**, *SPI enable*, bit 6 is set to enable the SPI facility to have control over port M pins. The lines intended for use as outputs should be configured by the corresponding DDRM bits.

The **SPTIE**, *SPI transmit interrupt enable*, bit 5 is set to enable the SPI to generate an interrupt if the SPITEF flag in the SPISR register becomes set. The SPITEF flag sets when the TDR becomes empty.

The **MSTR**, *master/slave select*, bit 4 is set to configure the SPI as a master device; it is cleared if the SPI is intended to be used as a slave.

The **CPHA**, *SPI clock phase*, and **CPOL**, *SPI clock polarity*, bits 2 and 3 specify the phase and polarity of the SPI clock (SCK), which controls data transfer between the master and its slaves. The choices available are shown in Fig. 8.8. When CPOL is 0 (1), the SCK pin of the master device will idle LOW (HIGH). When CPHA is 0, the shift clock output becomes the logic OR of the SCK with SS, and sampling of data occurs at odd edges of the SCK clock. As soon as the SS is pulled LOW, transfer of data begins with the first edge on the SCK signal. SS must go HIGH between successive characters in the message. When CPHA is 1, the SS pin may be left LOW for several SPI characters, and sampling of data occurs at even edges of the SCK clock. This setting requires fewer instructions during the data-transfer phase.

The **SSOE**, *slave select output enable*, bit 1 is set to enable the slave-select (SS) output feature, provided the MODFEN bit in the SPICR2 register is also set. If the MODFEN bit is clear and the SSOE bit is 1, the SS is an input with the MODF feature enabled.

The **LSBFE**, *LSB-first eanble*, bit 0 is set to allow data to be transferred with the LSB first. This will have no effect on the LSB and MSB positions in the SPIDR.

SPI Control Register 2 (SPICR2)

The SPICR2 contains four nonzero bits. The role of each bit is discussed next.

The **MODFEN**, *mode fault enable*, bit 4 is used to enable a mode fault detection, which occurs when more than one master is attempting system control. If the MODFEN bit is set, the SPI will then be able to detect a MODF failure if the SPI is configured as a master and the SS pin as an input line. The MODFEN has no effect on a slave SPI, since the SS line can only be used as an input.

The **SPC0**, *serial pin control*, bit 0 is set to enable the use of the MOSI line or the MISO line as bidirectional lines.

The **BIDIROE**, *output enable in the bidirectional mode of operation*, bit 3 is set to control the MOSI output buffer in the master mode or the MISO output buffer in the slave mode for the bidirectional operation of the MOSI line (master I/O) or for the bidirectional operation of the MISO line (slave I/O). If the state of this bit changes while the SPI is in the master mode while the SPC0 bit is set, transmission will be aborted and the SPI is forced into an idle state.

The **SPISWAI**, *SPI stop in wait mode*, bit 1 is set if it is desired to stop the SPI clock when the MCU is in the wait mode.

8.7.3 SPI Status Register (SPISR)

The SPISR register contains three flag bits used to synchronize the SPI operation. The other five bits are always 0.

The **SPIF**, *SPI transfer complete flag*, bit 7 is set whenever data transfer between the MCU and the external device is complete. The SPIF flag is cleared by executing a read of the SPISR register, followed by a read of the SPIDR register.

The **SPITEF**, *transmit empty interrupt flag*, bit 5 is set when the SPIDR register is empty. After it sets, the SPITEF flag is cleared by executing a read of the SPISR register and then a write to the SPIDR register

The **MODF**, *mode fault flag*, bit 4 is set when a master with an enabled MODF feature detects an error on the SS line. The MODF flag indicates that a conflict between more than one master attempting system control has occurred. Normally the MODF flag is clear and will set only if the SS pin on the master device is pulled LOW, signaling a change of status from master to a slave. The following occurs when the MODF flag sets: The SPI generates an interrupt if the SPIE bit has been set; the SPE bit becomes clear and the SPI is disabled; the MSTR bit becomes clear, forcing the device into a slave mode; and the DDRM bits associated with the SPI also become clear. After a MODF is set, the user must restore the affected bits to their original state by means of software.

8.8 SPI TOPOLOGIES

The SPI facility in the 9S12C can be configured to interface directly with other peripheral IC devices that support the SPI interface protocol, including other 9S12Cs. One of the interconnected devices is configured as a master (9S12C) and the other devices serve as slaves. There are two ways to connect multiple slaves to a master SPI. In the *bus connection* scheme, shown in Fig. 8.9a, the MISO lines on all devices are tied, as are the MOSI lines, and each can be selected as either an input line or an output line, depending on whether the associated chip is a master or a slave. The SCK line on a slave becomes input for receiving the clock signal from the master, which controls timing. The SS pin on the master is tied to a logic HIGH. The

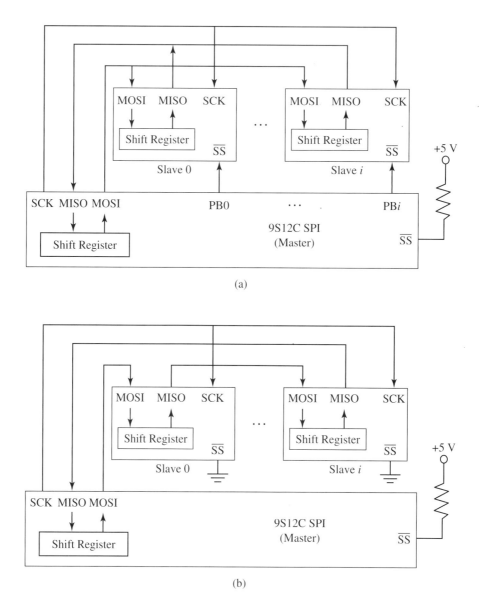

Figure 8.9 Single-master-to-multiple-slave bus connection (a) and cascade connection (b).

master uses output lines to drive the SS input on the slaves. When the master drives the SS line on a slave LOW, the slave knows that it is being selected for data transfer and treats the next eight bits transferred as one byte. This connection scheme is recommended even if only one slave is used to allow the master and slave to resynchronize without resetting the chips.

The other method of connecting multiple slaves is the *cascade connection*, shown in Fig. 8.9b. In the cascaded connection, the MOSI line of the master is wired to the MOSI line of slave 0. The MISO line of each slave is wired to the MOSI line of the subsequent slave. The MISO line of the last slave connects to the MISO of the master. Additionally, all slaves are enabled by tying their SS lines to ground. This connection enables the 9S12C to control a large number of devices with fewer pins.

8.9 SPI OPERATION

To operate the SPI facility, it must be first configured by the following procedure.

1. Write to the DDRM to program the bits corresponding to output pins to 1.
2. Write to the SPICR1 register to enable/disable SPI interrupts (SPIE); enable the SPI subsystem (SPE); enable/disable transmit SPTEF flag interrupt (SPTIE); set the SPI as master or slave (MSTR); set up the SCK clock phase and polarity (CPHA, CPOL); enable/disable SS output feature for an SPI master (SSOE); select first bit to be transmitted as the LSB or the MSB (LSBFE).
3. Write to the SPICR2 register to allow/inhibit MODF detection (MODFEN); control mode of MOSI/MISO in bidirectional mode (BIDIROE); select SPI operation in wait mode (SPISWAI); enable/disable bidirectional pin (SPC0).
4. Write to the SPR[2;0] and SPPR[6:4] bits in the SPIBR register to set the transfer baud rate.
5. For each byte to be transferred, the SPI does the following: Writes to the SPIDR, waits for the SPIF flag bit to set, and then reads a byte from the SPIDR.

The foregoing process applies to both a master and a slave SPI, but only the master controls the clock. The master can also assert the SS line LOW after it writes to the SPIDR register or deassert the SS HIGH after it reads a byte from the SPIDR register. The following examples demonstrate the SPI operation.

Example 8.4: SPI setup–Write a code to set up the SPI operation.

Solution: The following code configures the SPI subsystem to operate with the following configurations: SS line HIGH to prevent glitches; port M I/O levels: MOSI, SCK, SS* = output, MISO = input; SCK < 100 kHz. The SPI is configured as master, CPOL = 0, CPHA = 0, output drivers operate normally with active pull-up devices, no SPI interrupts.

```
            INCLUDE    'mc9s12c32.inc'

spi_setup
            BSET   PTM, #$80        ; SS High
            MOVB   #$E0, ddrm       ;
            MOVB   #$12, SPICR1     ; Enable the SPI
                                    subsystem
            MOVB   #$08, SPICR2     ; No interrupts TX
                                    is enabled
            MOVB   #$16, SPIBR      ; Set BR = 9600
            LDAA   SPISR            ; First step to cleat
                                    SPIF Flag
            STD    SPIDR            ; Second step to clear
                                    SPIF flag
            BSET   SPICR1, #$40     ; Enable the SPI
            RTI
```

Example 8.5: Master-to-slave transfer—Write a code to transfer one byte from a master SPI to a slave SPI. The slave is always enabled by pulling its SS line LOW.

Solution: The following code uses polling to manage the required transfer.

```
                INCLUDE   'mc9s12c32.inc'

                ORG       $00
data            DC.B      "American University of Beirut'
                DC.B      $0D, $0A      ; Carriage Return and
                                         Line Feed Codes
                DC.B      $00

                ORG       $C000
                JST       spi_setup
                JSR       transmit
here            BRA       here

transmit
                LDX       #data
                LDAA      1, X+         ; Get a byte to be sent
                BEQ       done
                BCLR      ptm, #$80     ; Assert SS line to start
                                         data transmission
                STAA      spidr         ; Start SPI transfer
chkflag         BRCLR     spisr, #$80,  ; Check bit 7 of SPISR
                          chkflag         to see if the SPI
                                          transfer is
                BSET      ptm, #$80     ;Deassert SS line
done            RTS
```

Example 8.6: Slave-to-master transfer—Write a code to for a master SPI to read a byte from a slave SPI. The slave is always enabled by pulling its SS line LOW.

Solution: The following segment of a larger code performs the required transfer.

```
;Define the offset addresses for the registers involved

        INCLUDE 'mc9s12c32.inc'

        ORG     $C000

        LDX     #data
        STAA    spdr, X    ; Start the SPI transfer
here    TST     spsr, X    ; Wait until a character has been
                             shifted in
        BPL     here
        LDAA    spdr, X    ; Place the character in acc A
loop    BRA     loop
```

Example 8.7: Seven-segment display—Write a code to emulate a BCD counter, display the count on a seven-segment display connected to port T, and transfer it to another 9S12C through its SPI slave.

Solution: The following code accomplishes the required task.

```
                INCLUDE 'mc9s12c32.inc'

                ORG   $C000
                JSR   delay       ; Execute a time-delay
                                  routine (Example 6.18)
                LDS   #4001       ; Initialize the stack pointer
                MOVB  #FF, ddrt   ; Configure port T for output
                JSR   spi_setup
continue        CLRA              ; Start BCD count at 00
loop            LDAB  spisr
                STAA  ptt         ; Output count to the
                                  seven-segment display
                STAA  spidr       ; Transfer counter value via
                                  the SPI
                JSR   delay       ; Pause the display
                INCA              ; Increment BCD count
                CMPA  #$09
                BGT   continue    ; Have not displayed 9 yet
around          BRA   around      ; Loop around always

delay
                .......           ; Include delay subroutine
spi_setup
                .......           ; Configure SPI registers
```

8.10 I/O EXPANSION OF THE 9S12C

The SPI subsystem provides a platform to expand the I/O capability of the 9S12C. The way to achieve this expansion using shift registers is discussed in the following sections.

8.10.1 Output Port Expansion

A simple way to expand the number of output ports from the 9S12C is to connect the SPI system to shift registers equipped with onboard SPI facility. The expansion requires three lines: a serial data output line, the serial clock line, and a line from an output port. Figure 8.10 shows the SPI connected to the 8-bit serial-in/parallel-out MC74HC595ADT shift register manufactured by ON Semiconductors (www.onsemi.com). As indicated, multiple 595 ICs can be connected to further increase the number of output pins. The 595 receives a character from the MOSI serial line of the 9S12C through pin A. The data received through pin A is shifted to the shift register whenever a LOW-to HIGH transition is detected at the shift clock (SC) pin, which is tied to the SCK line on the 9S12C. The 595 in turn provides two outputs: a serial data output through pin SQ_H and an 8-bit parallel output. The 595 also contains an 8-bit buffer

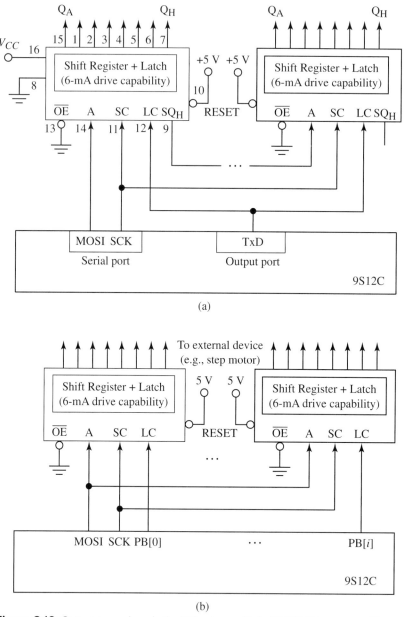

Figure 8.10 Output expansion via the SPI using multiple HC595 ICs connected in series (a) and in parallel (b).

between the parallel outputs of the shift register and the eight parallel output lines, labeled Q_A–Q_H. The transfer between the shift register and the buffer that drives the output lines occurs when the CPU writes a 0 and then a 1 to the output line labeled TxD, which is connected to the latch clock pin (LC) on the 595. The active low *output enable* (OE) pin is tied to ground (LOW) to allow data from the latches to be presented at the output. The active low reset pin resets the shift register only but has no effect on the buffer.

Example 8.8: Output port expansion using the MC74HC595 IC–Write a code to transfer four data bytes from the SPI to four 595 shift registers connected to the 9S12C as shown in Fig. 8.10.

Solution: The following code accomplishes the required task.

```
              INCLUDE  'mc9s12c32.inc'

              ORG      $3800
data          DC.B     $09,$18,$36,$72
count         EQU      $4
              ORG      $C000
              LDAA     #$3A           ;set DDRM bits
              STAA     ddrm
              JSR      spi_setup
              LDAB     #count         ; Initialize B with the number
                                      of bytes to shift out
              LDY      #data          ; Load the data buffer base
                                      address
continue      LDAA     0,Y            ; Get one character and
              STAA     spidr          ; Shift it out
chkflag       LDAA     spisr
              BPL      chkflag        ; Check flag if character has
                                      been transferred
              INY                     ; Point to the next character
                                      data
              DBNE     continue       ; Decrement B and continue until
                                      all bytes are shifted out
                                      ; Force a rising edge on TxD to
                                      shift data to the output latch
              BCLR     portm, #$02    ; Write a 0 to TxD pin
              BSET     portm, #$02    ; Write a 1 to TxD
              END

spi_setup                             ; Configure SPI
              ...........             ; Include spi_setup code here
```

8.10.2 Input Port Expansion

The number of input ports from the 9S12C may further be expanded by connecting the SPI system to SPI-based shift registers. The expansion requires three lines: a serial data output line, the serial clock line, and a line from an output port. Figure 8.11 shows the SPI connected to the 8-bit parallel-in/serial-out MC74HC589AN shift register available from ON Semiconductors. As indicated, multiple ICs can be connected to further increase the number of input lines. The 589 receives two inputs: a serial data input through pin S_A and 8-bit parallel inputs A–H. The serial inputs are shifted into the shift register on the rising edge of the shift clock (SC) input, which is tied to the SCK line on the 9S12C if the serial-shift/parallel-load SS/PL line is HIGH. Data on the S_A is ignored when the SS/PL is LOW. The parallel data inputs are stored in the data latch when the CPU writes a 0 and then a 1 to the

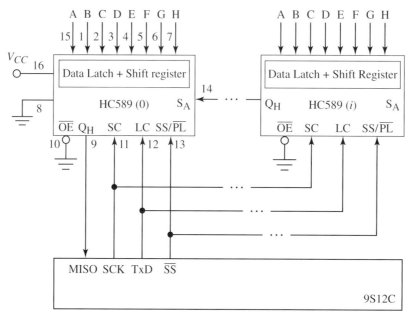

Figure 8.11 Input port expansion via SPI using multiple HC589 ICs connected in series.

output line labeled TxD, which is connected to the latch clock input (LC) on the 589. A character from the Q_H line is shifted in the 9S12C through the MISO serial line if the output-enable line on the 589 is LOW.

Example 8.9: Use of the HC589 IC for input port expansion—Write a code to retrieve four data bytes through the SPI from four HC589 shift registers connected to the 9S12C as shown in Fig. 8.11.

Solution: The following code accomplishes the required task.

```
            INCLUDE     'mc9s12c32.inc'

count       EQU         4
            ORG         $3800
data        DS.B        4

            ORG         $C000
            JSR         spi_setup
            BCLR        portm, #$02    ; Pull TxD (LC) line low
            BSET        portm, #$02    ; Pull TxD (LC) line high
            BCLR        portm, #$20    ; Force a low on SS for
                                       parallel load mode and trans-
                                       fer the; contents of the data
                                       latch into the shift register
            BSET        portm,#$20     ;pull SS line high to select
                                       serial shift mode
```

```
          LDY      #data        ; Load the base address of
                                the data buffer
          LDAB     #count       ; Load the number of bytes to
                                be retrieved
continue  STAA     spidr        ; Transfer a character and
holdon    LDAA     spisr        ; Wait until it has been
                                transferred
          BPL      holdon       ;
          LDAA     spidr        ; Retrieve one character and
          STAA     0,Y          ; Store it in the data buffer
          INY                   ; Point to the next location
                                in data buffer
          DBNE     continue     ; Decrement the loop count
                                until four characters are
                                received
          END
```

8.11 SUMMARY

Serial communication protocols are very useful in automation applications. This chapter presented two out of three available serial facilities in the MC9S12C, specifically the SCI and SPI. The CAN serial facility is left for the reader to obtain from the Freescale Semiconductor website. The covered SCI and SPI facilities included corresponding protocols, registers, and operations. Examples were given to highlight the various dedicated instructions needed when using the serial facilities of the adopted MCU. In addition, the input and output expansion capacity was covered as an important feature for integrating the 9S12C in various demanding I/O applications.

RELATED READING

S. F. Barrett and D. J. Pack, *Embedded Systems: Design and Applications with the 68HC12 and HCS12*. New York: Prentice Hall, 2004.

HCS12 Microcontrollers: MC9S12C128 Data Sheet, 2005 (MC9S12C128_V1.pdf)

H.-W. Huang, *HC12/9S12: An Introduction to Hardware and Software Interfacing*. Clifton Park: Thomson Delmar Learning, 2005.

M. B. Histand and D. G. Alciatore, *Introduction to mechatronics and Measurement Systems*, 2nd ed. New York: McGraw-Hill, 2002.

J. B. Peatman, *Design with Microcontrollers*. New York: McGraw-Hill, 1988.

P. Spasov, *Microcontroller Technology: The 9S12C*, 2nd ed. New York: Prentice Hall, 1996.

QUESTIONS

8.1 Explain the role of the SCI and SPI serial ports.

8.2 Explain the meaning of *simplex*, *half-duplex*, and *full-duplex transmission protocol*.

8.3 Describe the framing process.

8.4 Explain how to set up the SCI baud rate.

8.5 What is the role of parity?

8.6 Provide the necessary steps to configure the SCI system

8.7 Describe the SCI operation for transmitting data.

8.8 Describe the SCI operation for receiving data.

8.9 List and explain all SCI interrupt sources and the conditions for their occurrence.

8.10 List all SCI status bits, and explain their roles.

8.11 Describe the RS232C communication protocol.

8.12 Provide the necessary steps to configure the SPI subsystem.

8.13 Describe the SPI operation as a master.

8.14 Describe the SPI operation as a slave.

8.15 Describe SPI interrupt sources and the conditions for their occurrence.

8.16 List all SPI status bits, and explain their roles.

8.17 Explain the process by which a slave station SPI sends data to the master station.

8.18 Explain possible ways that the SPI facility can be used to expand the MCU ports.

PROBLEMS

8.1 Determine the code setting for the SBR[12:1] bits corresponding to a baud rate target of 4800 if the 9S12C MCU is clocked with a 16-MHz crystal.

8.2 Write an assembly code to send the message "HOLD ON" over the SCI port.

8.3 A device, while receiving data, cannot start transmission of its own data until 40 ms has elapsed after it has completely received its data. Write a program segment to manage the start of data transmission.

8.4 An MCU is set up to receive 8-bit characters with even parity and two STOP bits. Write program segments to detect:

    a. A parity error on reception.

    b. A framing error if the first STOP bit is 0.

    c. A framing error if the second STOP bit is 0.

8.5 Determine the code setting for the SPR[2:0] and SPPR[2:0] bits for a target SPI transfer time of 0.4 μs if the 9S12C MCU is clocked with a 16-MHz crystal.

8.6 The MAX7219 from MAXIM is an LED display driver with a serial interface port. Obtain the specs for this device, and write a code to manage its interface to the 9S12C.

8.7 The DS1267 is an SPI-compatible digital potentiometer. Obtain its specs, and write a code to interface it to the 9S12C.

8.8 Obtain the specs for the PCx3311 DTMF generator by Signetics, and write a code to interface it to the 9S12C.

8.9 The IEEE-4888, known as the general-purpose instrumentation bus, is a popular standard very widely used by instrument makers. Write a brief description of this standard.

8.10 The USB is becoming the standard communication of choice for computer peripherals. Write a brief description of this standard, and devise a method to interface the 9S12C with the outside world via the USB (e.g., USB9602 from National Semiconductor).

LABORATORY PROJECTS

8.1 Two 9S12Cs are needed to navigate and operate firefighting accessories of a firefighting robot. Devise a list of sensors and actuators needed to accomplish the task, and devise a plan to hook it all up to the two 9S12Cs. Write a C code to manage its operation. Implement your design.

8.2 Investigate interfacing the 9S12C to peripherals using wireless communication. Select a wireless transceiver, and provide a brief description of the associated communication protocol.

8.3 Obtain the specs for the MCP2510 CAN controller from Microchip. Investigate the interfacing of this IC to the 9S12C.

9

Programmable Timer Facility

OBJECTIVES

Upon completion of this chapter, the student familiar with the 9S12C MCU will be able to:

1 Describe the various capabilities of the programmable timer facility

2 Explain the role of timer interrupts in managing real-time control

3 Write assembly code to perform timing measurements using the input capture facility

4 Write assembly code to generate output signals using the output compare and PWM facilities

5 Use the pulse accumulator facility for counting operations

9.1 INTRODUCTION

The programmable timer subsystem provides the main interfacing environment for applications for which control decisions are required to be made in real time. Once set up by the CPU, the timer subsystem (and associated interrupts) can handle some timing tasks on its own, freeing the CPU to cater to other needs instead of executing time-delay loops to manage timing requirements. In a time-delay routine, the CPU generates specific time intervals by continuously executing a set of instructions around a loop for a specified number of times. If an interrupt occurs while a time-delay routine is executing, the interrupt duration will add to the intended time delay, upsetting any precise measurement. However, interrupts have no effect on the timer.

The programmable timer is the enabling source for the CPU to determine when it should interact with the I/O devices and to control events requiring precise timing. These attributes are valuable in machine tools, automobiles, robotics, and process-control applications, where various sensing and control tasks require profound coordination. Important timing tasks performed by the timer include:

- Generate specific time delays.
- Generate precisely timed control signals.
- Generate a single pulse of a specific duration or a stream of pulses at different frequencies and pulse widths to control actuators.
- Measure the time between external events.
- Measure the frequency or duration (pulse width) of an external pulse.
- Provide a real-time clock to generate periodic real-time interrupts.
- Provide the timing for the watchdog timer interrupts.
- Act as an internal timer without an output to start and stop a task at set times.

9.2 TIMER MODULE IN THE 9S12C MCU

The 9S12C MCU can accomplish the aforementioned timing tasks using three functions provided by its integrated *timer module* (TIM): the *input capture* (IC), the *output compare* (OC), and the *pulse accumulator* (PA) functions. These functions are brought out on pins PT[7:0] of port P. Any of these pins may act as a general-purpose input or output if it is not tied to a timer function. The OC function can generate outputs at precisely controlled times to drive any PT[i] pin LOW or HIGH. The IC function facility is used to determine the time of occurrence of an edge on any PT[i] channel and allows the MCU to measure the width of an input pulse or to determine the frequency of an input pulse train. The output compare and input capture functions use the same channels and share many registers. The application program can set each channel individually to act as an output compare channel or an input capture channel. The 16-bit PA can be used to accumulate gated time or to count events occurring on pin PT[7] of port T. The selected timer function depends on the value settings of corresponding bits in associated registers.

Figure 9.1 shows the bits in the TSCR1, *timer system control register* 1, at address $0046, which determines the operation settings of the timer module on the 9S12C. To use any timer function, the application program must first enable the timer module. This is accomplished by setting the TEN, *timer enable*, bit in TSCR1. The value settings of the TSWAI and TSFRZ bits in the TSCR1 determine the operating condition of the timer counter when the MCU operates in the wait and freeze modes, respectively. If either bit is set, the timer counter continues to run when the MCU operates in the associated mode. When the MCU operates in the STOP mode, the timer module is OFF because both PCLK and ECLK are stopped.

TSCR1: Timer System Control Register 1 ($0046)

TEN	TSWAI	TSFRZ	TFFCA	0	0	0	0

TEN = 1, Enable the main timer
TSWAI = 1, Timer stops when MCU in wait mode
TSFRZ = 1, Timer stops when MCU in freeze mode
TFFCA = 1, Timer fast flag clear all

Figure 9.1 Control bits of the TSCR1 register.

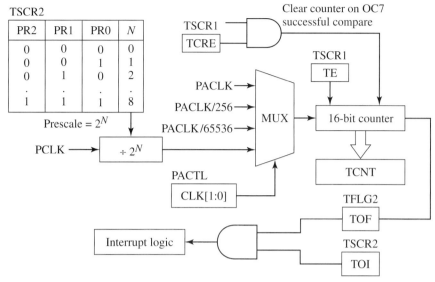

Figure 9.2 Block diagram depicting the essential features of the TCNT.

9.2.1 Free-Running Counter (TCNT)

When the TEN bit is set, the 16-bit TCNT, *free-running timer counter*, register at address $0044–0045 starts counting from $0000, and when it reaches $FFFF (full count) it automatically rolls over to $0000 and goes on counting. The TCNT is the main timer counter that provides timing reference for most of the timer module operations. Figure 9.2 shows a block diagram depicting the essential features of the TCNT. Figure 9.3 shows the associated registers.

The application program can read the TCNT at any time without affecting its value, but only the system bus (ECLK) can write to it. The ECLK increments the TCNT at a fixed rate unless a reset is enacted or power is removed from the MCU. Each increment is equivalent to a number of clock pulses, which is equal to the ECLK frequency divided by a prescale factor from 1 to 128, depending on the value settings of the PR[2:0], *timer prescaler select*, bits in the TSCR2, *timer control register* 2, at address $004D. The prescaler is simply 2^N, where N is the decimal equivalent of the binary code in PR[2:0]. For example, if PR[2:0] contains 110, then $N = 6$ and the prescaler is $2^6 = 64$; for a 16-MHz ECLK frequency, this corresponds to

TCNT: Timer Counter Register ($0044)

Bit 15			•••			Bit 0

TSCR2: Timer Control Register 2 ($004D)

TOI	0	0	0	TCRE	PR2	PR1	PR0

TOI = 1, Hardware interrupt generated when TOF is set
TCRE = 1, Enable timer counter reset on successful compare
PR[2:0], Timer prescale select bits

TFLG2: Timer Flag Register 2 ($004F)

TOF	0	0	0	0	0	0	0

TOF: Timer overflow flag sets when TCNT counts from $0000 to $FFFF

Figure 9.3 Registers associated with the operation of the TCNT.

a timer clock frequency of $16/2^N$. Because the function of PR[2:0] is critical in real-time control applications, they are time-protected bits, which means that they can only be modified once during the initialization phase after reset.

9.2.2 Timer Overflow

The TCNT can increment up to $FFFF (65,535) counts before it overflows. This is equivalent to an actual time that ranges from 4.1 to 524.3 ms with a 16-MHz ECLK and a prescale factor of 1–128, respectively. Each time the TCNT reaches full count, it rolls over and automatically sets the TOF, *timer overflow flag*, in the TFLG2, *timer flag register* 2, at address $004F. When the TOF is set, an interrupt will be generated; and the corresponding interrupt sequence is executed if the TOI, *timer overflow interrupt enable*, bit in the TSCR2 had been set. After reset, the TCNT is initialized to $0000, the TOI is disabled, the TOF is cleared, and the prescaler is set to 1.

9.2.3 Clearing the Timer Flag

The TOF and other flag bits can be cleared by writing a "1" to the bit location to be cleared. Many instruction sequences may be used to accomplish this. The flags would automatically clear if a read from an input capture channel or a write into an output compare channel is made while the TFFCA, *timer fast flag clear action*, bit in the TSCR1 register is set.

9.3 OUTPUT COMPARE

The output compare facility of the timer module enables the CPU to generate timed output waveforms to control external events that require precise timing, such as firing control of spark plugs. In combination with the TCNT, the OC subsystem can act as an internal timer to start and stop an event at set times without generating an output signal, independent of what the CPU is doing at the time. For instance, one output can be set to generate timed pulses to the engine ignition system while another can generate time delays without tying the CPU down with delay routine execution.

Figure 9.4 shows the type of pulses that can be generated by the timer output compare on any PT[i] channel. The output can be continuous pulse trains at different frequencies and pulse

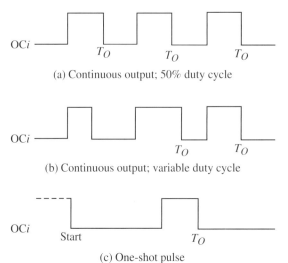

(a) Continuous output; 50% duty cycle

(b) Continuous output; variable duty cycle

(c) One-shot pulse

Figure 9.4 Output compare operation.

widths, as shown in Figs. 9.4a and 9.4b, or it may be a single pulse of a specific duration, as shown in Fig. 9.4c. Continuous pulse trains is useful in many applications, such as the speed control of a DC motor and causing a light to flash on an instrument display panel to indicate the occurrence of a pertinent condition.

9.3.1 Output Compare Registers

If the timer module is enabled, any PT[i] pin can be individually selected to act as either an output compare channel or an input capture channel, depending on the value setting of the IOSi bit in the TIOS, *timer input capture output compare select register*, at address $0040. If the IOS$i$ had been set, the PT[i] channel acts as an output compare; if the IOSi had been cleared, the PT[i] channel acts as an input capture. The input capture facility is discussed in Section 9.4.

Three other types of registers are involved in an output compare operation: control registers, data registers, and status registers. The mnemonic, the full name and addresses, and the associated bits in these registers are shown in Fig. 9.5. Two additional registers, OC7D and OC7M, are dedicated to the operation of the OC7 function and are introduced in Section 9.3.3.

Data Registers

In combination with the TCNT, each output compare channel has an associated 16-bit TCi, *timer counter register*. Each register occupies two memory locations, one for the high byte,

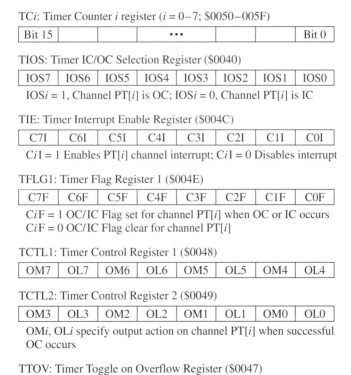

TCi: Timer Counter i register (i = 0–7; $0050–005F)

Bit 15			•••			Bit 0

TIOS: Timer IC/OC Selection Register ($0040)

IOS7	IOS6	IOS5	IOS4	IOS3	IOS2	IOS1	IOS0

IOSi = 1, Channel PT[i] is OC; IOSi = 0, Channel PT[i] is IC

TIE: Timer Interrupt Enable Register ($004C)

C7I	C6I	C5I	C4I	C3I	C2I	C1I	C0I

CiI = 1 Enables PT[i] channel interrupt; CiI = 0 Disables interrupt

TFLG1: Timer Flag Register 1 ($004E)

C7F	C6F	C5F	C4F	C3F	C2F	C1F	C0F

CiF = 1 OC/IC Flag set for channel PT[i] when OC or IC occurs
CiF = 0 OC/IC Flag clear for channel PT[i]

TCTL1: Timer Control Register 1 ($0048)

OM7	OL7	OM6	OL6	OM5	OL5	OM4	OL4

TCTL2: Timer Control Register 2 ($0049)

OM3	OL3	OM2	OL2	OM1	OL1	OM0	OL0

OMi, OLi specify output action on channel PT[i] when successful OC occurs

TTOV: Timer Toggle on Overflow Register ($0047)

TOV7	TOV6	TOV5	TOV4	TOV3	TOV2	TOV1	TOV0

TOVi = 1 causes OC pin PT[i] to toggle on overflow

Figure 9.5 Registers and bits involved in the output compare operation.

TCiH, and one for the low byte, TCiL. The TCi ($i = 0, 2, \ldots, 7$) register is used to control the length of a timed interval or the period of an output waveform on the associated PT[i] channel. TC0–TC7 also latch the current TCNT value during input capture events.

Control Registers

The TCTL1 and TCTL2, *timer action control registers* 1 and 2, at address $0048–$0049 contain a set of eight pairs of control bits (OMi, OLi; $i = 0, 1, \ldots, 7$) used to control the automatic actions that occur on the OCi pin when a match between the TCNT and TCi occurs. With this control register, the user selects the logic level generated on the timer output pins to be either low or high or to toggle when a successful compare occurs.

The TIE, *timer interrupt enable register*, at address $004C, is used to enable or disable the interrupt associated with an output compare or input capture event. It contains one bit associated with each of the eight output compare/input capture channels. If the CiI bit had been set, the timer interrupt feature is enabled for the TCi output compare or input capture operation.

Status Registers

The TFLG1, *timer interrupt flag register* 1, at address $004E, has eight flag bits C0F–C7F associated with the eight counter registers TC0–TC7. The OCiF flag bit will set on a successful compare between the TCNT and the TCi during an output compare operation. The flag bits are used in conjunction with input capture events.

The TTOV, *timer toggle on overflow register*, at address $0047, contains one bit associated with each output compare channel. If the TOVi bit had been set, the output on the OCi pin will toggle every time an overflow occurs. These bits are not relevant for input capture events.

9.3.2 General Setup for the Output Compare Operation

To carry out a successful output compare action on an OCi channel, the application program should perform the following sequence of events.

1. Use the SEI instruction to set the I bit in the CCR, and disable all interrupts while the CPU is setting up the timer, to ensure that no extra time is inadvertently inserted by any interrupting source while the CPU reads the TCNT, writes to the output compare register, and sets the output-level bit.

2. Configure the TSCR1 register to set the prescale select bits PR[2:0] in the TSCR2 to the desired prescale value and, if desired, to set the TOI bit to enable interrupt whenever an overflow occurs.

3. Configure port T pin PT[i] to act as an OCi channel by setting the IOSi bit in the TIOS register.

4. Configure the OMi and OLi bits in the TCTL1 or TCTL2 control registers to set the output logic level generated on the OCi line when a successful compare occurs, that is, when TCNT = TCi.

5. Enable the timer module by setting the TEN bit in the TSCR1 register.

6. If desired, set the CiI bit in the TIE register to enable an interrupt from the OCi channel when a successful compare occurs.

7. Acquire the current time T_0 from the TCNT, add to T_0 the appropriate time period to set the exact time when the desired output compare event is to occur, and store the result in TCi.

8. Use the CLI instruction to clear the I bit in the CCR and reenable interrupts.

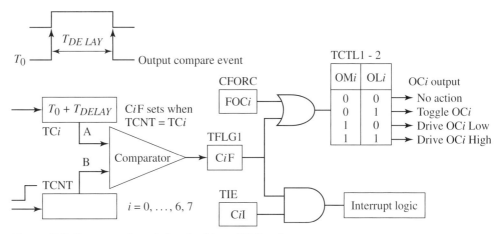

Figure 9.6 Sequence of events involved in an OC function.

Figure 9.6 shows the sequence of events involved in an output compare function on any OCi channel following the setup procedure. The value stored in the TCi register is continuously compared to the current TCNT value; when the two values match, successful compare occurs and the CiF flag sets and the selected output logic level bit is transferred to the OCi line, depending on the value settings of the OMi and OLi bits in TCTL1 or TCTL2. Furthermore, if the CiI bit in the TIE register had been set (step 6), an interrupt will be generated when the CiF flag sets, and appropriate action is enacted by executing the corresponding interrupt-service routine. Once the appropriate action is completed, the application program should clear the CiF flag to enable the next compare event. The sequence is repeated as needed.

Example 9.1: Generating a one-shot pulse—Write an assembly code to generate a pulse of 20-ms duration on the OC5 channel. This is equivalent to a pulse width of 20 ms/(0.5 µs/pulse) = 40,000 pulses if the TCNT counts with a prescale factor of 8 (PR[2:0] = 011).

Solution: The following code generates the required pulse.

```
            INCLUDE     'mc9s12c32.inc'   ;Include 9S12C macros

pls_width   EQU 40000                     ;equivalent to 20-ms time
                                          delay
;
            ORG $8000

pls_start   MOVB        #$20, ddrt        ;Configure PT[5] as
                                          output
            BSET        ptt, #$20         ;Drive PT[5]/OC5 high
                                          before enabling OC5
            JSR         tim_setup
            LDD         tcnt              ; Load TCNT into D
```

```
            ADDD        #pls_width-17       ; Add [(pls_width) -30]
                                              to D.
            STD         tc5                 ; Store D into TC5
waitflg                                     ; Wait for the C5F to
                                              set on successful
                                              compare
            BRCLR       tflg1,#$20,waitflg
clrflag
            LDAA        tflg1
            ORAA        #$20
            STAA        tflg1               ; Clear C5F flag
            END
; Timer setup subroutine

tim_setup
            SEI
            MOVB        #$08, tctl1         ;Set to drive OC5 low
                                            upon successful compare
            MOVB        #$20, tios          ;Set OC5 to act as an
                                            output compare channel
            MOVB        #$83, tscr1         ;Enable timer, select
                                            timer prescaler 8 for a
                                            frequency = 2 MHz
            MOVB        #$00, tie           ;No interrupt on
                                            successful compare
            CLI
            RTS
```

Note that 30 cycles have been executed from the time the OC5 is driven HIGH to the time the TCNT is read.

Generating short pulses: The minimum pulse width that could be generated must be longer than the time it takes the CPU to execute the instructions between the "pls_start" and "waitflg" labels; otherwise, preparing for an output compare operation would interfere with the action itself. The user must count the corresponding number of cycles to determine the minimum possible pulse width.

Example 9.2: Generating a square wave–Develop a code to generate a 2-kHz square waveform on the OC3 line. The period of the signal is 500 μs for a 0.5-μs clock period.

Solution: Since the high time and the low time are equal, the code is set up to configure an output compare function to toggle every half period, equivalent to 250 μs, encompassing 500 clock cycles ($0200) with 16-MHz E clock and a prescaler of 8. Subroutine *setup_oc3* initializes timer output OC3 to be interrupt driven. Once the main program sets up the variable *hitime* to $0200 and executes the *setup_oc3* subroutine, the CPU will be free to do other things. The interrupt-service routine `oc3_isr` schedules time delay for the next edge to be toggled.

```
; Program for generating a square wave at frequency =
1/(2*hitime); f = 1/(2×T_H)

                    INCLUDE    'mc9s12c32.inc'

        hitime      EQU     $0200          ;half-cycle delay (in
                                            0.5-μs increments)

        ; Mainline program
                    ORG     $C000

        stacktop    LDS     #$4000
                    JSR     setup_oc3      ;jump to subroutine
                                            setup_oc3
        clrflg      LDAA    tflg1
                    ORAA    #$08
                    STAA    tflg1          ; Clear C3F in TFLG1
        here        BRA     here           ;Interrupt driven from
                                            here
                    END                    ; End of assembly
```

;Subroutine *setup_oc3* initializes the timer for output compare on OC3

```
setup_oc3
        SEI
        MOVB    #$40, tctl2            ;Configure OM3 and OL3 to
                                        toggle output on OC3
        MOVB    #$08, tios            ;Set OC3 to act as an output
                                        compare channel
        MOVB    #$83, tscr1           ;Enable timer and select
                                        timer prescaler 8 for a
                                        freqency = 2 MHz
        MOVB    #$08, tie             ;Enable interrupt on
                                        successful compare
        MOVW    #hitime,tc3h          ;Load TC3 register with
                                        initial compare value
        CLI
        RTS                           ;Return
```

;Interrupt Service Routine *oc3_isr* schedules time delay for next edge to be toggled

```
oc3_isr
        LDD     tc3h                  ;Schedule next edge by
                                        loading current TC3 into D
        ADDD    #hitime               ;Add hitime to D
        STD     tc3h                  ;Update TC3 to schedule the
                                        next interrupt
```

```
clrflg  LDAA    tflg1
        ORAA    #$08
        STAA    tflg1                   ;Clear C3F
        RTI                             ;Return from oc3_isr
```

The program may be used to generate square waves at different frequencies by modifying the half-period time duration *hitime*. The half period cannot be less than the number of clock cycles required to execute the instructions in the `oc3_isr` routine and the instructions required to enter and exit it.

9.3.3 Operation of the OC7

The OC7 operation uses the *output compare mask register* OC7M at address $0042 and the *output compare data register* OC7D at address $0043 to control an output action on any combination of output pins when a match between the contents of the TC7 and TCNT registers occurs. The bits involved in the OC7M and OC7D registers are shown in Fig. 9.7. Bit OC7Mi in the OC7M register and bit OC7Di in the OC7D register ($i = 0, 2, \ldots, 7$) corresponds with OCi output compare channel. On a successful OC7 compare, the value stored in the OC7Di will appear on OCi if the corresponding OC7Mi bit is set. The values of the bits in the OC7D register are irrelevant if the corresponding bits in the OC7M register are zeros. Usually the user has to write to the OC7M register once to establish which pins will be controlled by the OC7 event. The instructions

```
MOVB    #$03,   oc7m
MOVB    #$03,   oc7d
```

sets the OC7M[2:0] bits so that the OC0, OC2, and OC3 will be driven to logic HIGH as indicated in the corresponding OC7D[2:0] bits when a successful compare on the OC7 line occurs.

OC7 is useful in situations when it is required to drive the same external device by several outputs. OC7 may also be used with another OCi line to drive the same pin and generate successive edges within very short durations, even for one E clock cycle.

The TCNT will reset whenever successful compare occurs on the OC7 channel if the TRCE bit in the TSCR1 register had been set. This feature facilitates the generation of PWM waveform on any OCi line. For example, TCi may be used to hold T_H (high time) of the PWM signal, while the TC7 holds the period T, with OC7M configured to drive the OCi HIGH on successful OC7 compare. Successful compare on TCi drives the OCi line LOW, whereas successful compare on TC7 drives the OCi HIGH again and reset TCNT. This idea is implemented in Example 9.3.

OC7M: Output Compare 7 Mask Register ($0042)

OC7M7	OC7M6	OC7M5	OC7M4	OC7M3	OC7M2	OC7M1	OC7M0

OC7D: Output Compare 7 Data Register ($0043)

OC7D7	OC7D6	OC7D5	OC7D4	OC7D3	OC7D2	OC7D1	OC7D0

Figure 9.7 Action mask and data registers for OC7 operation.

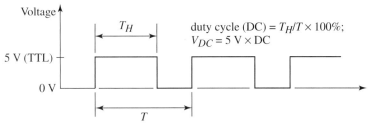

Figure 9.8 Pulse-width modulating signal (PWM).

Pulse-Width Modulation (PWM)

A PWM waveform is shown in Fig. 9.8. The *duty cycle* is the ratio of the time the signal is high (T_H) to the period for one cycle. PWM is a simple form of digital-to-analog conversion (DAC) used to obtain high-resolution control of a slowly changing variable. For example, a PWM output from the MCU can drive a heating coil for temperature control. The duty cycle correlates to the percentage of time the power is delivered to the heater. If the supply voltage to the coil is V_S, the average DC value of the output voltage is proportional to the duty cycle according to

$$V_{DC} = V_S \times (\text{duty cycle}) = V_S \times \frac{T_H}{T}$$

The period T of the output waveform for this application can be very long, on the order of 0.1 s, relative to the period of the MCU clock. However, PWM can also be used in applications that require fast-switching control, such as exciting speakers to generate a variety of sound effects, controlling the speed and position of stepper and DC motors, and controlling light intensity.

Example 9.3: Generating PWM output–Develop a code to generate a 2.5-kHz PWM waveform with a 75% duty cycle to regulate the speed of a DC motor. The motor control input is connected to IOC4 line on the 9S12C.

Solution: The following code provides the desired PWM signal. The main program determines the high time and period of the PWM signal from the "pwm1p" and "pwmdc" and stores them at addresses labeled "hitime" and "period". "pwm1p" is the number of cycles corresponding to 1% of the signal period, whereas "pwmdc" is the duty cycle of the PWM output. Then it calls routine "timsetup" to initialize OC4 and OC7 to act as output compare and schedule the PWM output. When a successful compare occurs on OC4, the output on OC4 goes high and when the OC7 compare occurs, the OC4 is driven high.

```
#include <hidef.h>        /* common defines and macros */
#include <mc9s12c32.h>     /* derivative information */

#pragma LINK_INFO DERIVATIVE "mc9s12c32"

void oc4setup(int hitime, int period)
          {  TCTL1 = 0x82;
             TIOS = 0x90;
```

```
            TSCR1 = 0x80;
            TSCR2 = 0x0B;
            OC7M = 0x10;
            OC7D = 0x10;
            TC4Hi = hitime;
            TC7Hi = period;
            TFLG1 = TFLG1 | 0x90;
        }

void main(void)
        {   int pwmdc = 75;
            int pwmlp = 5;
            int hitime, period;
            DDRT = 255;
            hitime = pwmdc*pwmlp;
            period = pwmlp*100;
            oc4setup(hitime, period);
            while(1)    {
            }
        }
```

This program may not properly generate duty cycles that are too close to 0% or 100%. Also, the motor speed may be increased by simply changing the duty cycle. If a 50% duty cycle is used, the output will be a square wave.

9.3.4 Forced Output Compare

The user can force a timer output action to occur immediately without waiting for a successful compare to happen. An example of this situation is spark-timing control of an automotive engine control. The forced output compare mechanism is designed to handle such situations. It is configured by simply writing 1s to the FOCi bits in the CFORC, *timer compare force register*, at address $0041, corresponding to the OCi channels to be forced. The contents of the CFORC register are shown in Fig. 9.9. FOCi bits are cleared out on reset. A sample instruction to set FOC3 bit to force a compare action on OC3 is

$$\text{BSET} \quad \text{CFORC, \#\$08}$$

After the write to the CFORC register, the compare action will be forced on the OC3 pin on the next TCNT count. The forced actions are synchronized to the TCNT clock rather than the E clock. The forced output compare does not affect the associated CiF flag, and it does not generate an interrupt if the CiI source had been enabled. The force mechanism should not be

CFORC: Timer Compare Force Register ($0041)

FOC7	FOC6	FOC5	FOC4	FOC3	FOC2	FOC1	FOC0

Figure 9.9 Timer compare force register (CFORC) at address $0041.

used on an OC*i* pin if it had been configured to toggle when an output compare occurs, that is, if the OM*i* and OL*i* bits had been set to 1.

9.4 INPUT CAPTURE FACILITY

The *input capture* function of the 9S12C MCU timer module performs tasks that involve precise timing measurements on input signals, such as those shown Fig. 9.10. The input capture action responds to the active edge of the input signal tied to the input capture line IOC*i* by reading the TCNT value and storing it in the TC*i*, *timer input capture register*. The active capture edge may be set to be a rising edge (low-to-high), a falling edge (high-to-low), or either edge. By measuring the time interval between signal transitions (two successive active edges) occurring on an input pin, the period and frequency of the input signal or any other related quantity, such as speed, can be accurately determined by the program.

9.4.1 Input Capture Pins and Registers

The pins, vectors, and controls associated with the output compare function are also used to manage the input capture function, provided that the associated ISO*i* bit in the TIOS register is cleared. The only exception is that the control registers associated with the input capture are the TCTL3 and TCTL4 registers, shown along with their associated bits in Fig. 9.11. These two registers contain eight control bit pairs (EDG*i*B, EDG*i*A; $i = 0, 1, \ldots, 7$) used to set the active transition on the corresponding IC*i* line to be a rising edge, a falling edge, or either edge (rising or falling) and with that the time of occurrence of the active edge.

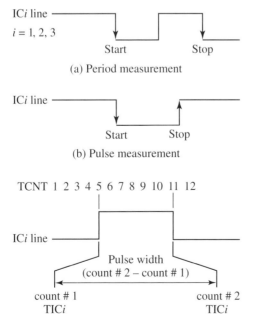

Figure 9.10 Typical timer input capture measurements.

TCTL3: Timer Control Register 3 ($004A)

EDG7B	EDG7A	EDG6B	EDG6A	EDG5B	EDG5A	EDG4B	EDG4A

TCTL4: Timer Control Register 4 ($004B)

EDG3B	EDG3A	EDG2B	EDG2A	EDG1B	EDG1A	EDG0B	EDG0A

EDGiB	EDGiA	Function of ICi input	
0	0	Capture disconnected from ICi pin	—
0	1	Capture TCNT$\rightarrow$ TICi on ICi rising edge	
1	0	Capture TCNT$\rightarrow$TICi on ICi falling edge	
1	1	Capture TCNT$\rightarrow$TICi on any ICi edge	

Figure 9.11 Registers involved with the input capture operation.

Figure 9.12 shows the sequence of events for an input capture action. When an active edge is detected on an ICi channel, the CiF flag in the TFLG1 register sets and the current value in the TCNT is copied into the corresponding timer input capture register TCi. Meanwhile, an interrupt will also be generated if the corresponding interrupt enable bit CiI in the TIE register had been set. The application program reads the TCi register and clears the CiF flag to schedule for capture on the next edge.

Pulse-Width Measurement

Figure 9.10c demonstrates how the input capture action on an ICi channel can be used to determine the time between two input edges of a pulse. The current TCNT value is captured and copied to the associated TCi register on the first active edge and again on the subsequent active edge. The pulse width is simply the difference between the two captured times.

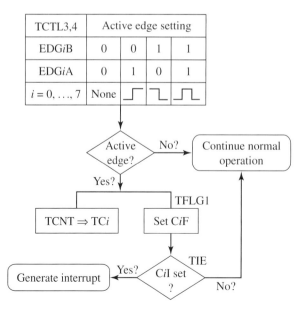

Figure 9.12 Operation of an input capture action.

Example 9.4: Measuring pulse width–A process application requires the measurement of the duration of a pressure spike inside a tank. The spike exceeds a certain threshold for a very short time. The output from the pressure sensor is connected to the input capture pin IC3 through a comparator. The output of the comparator will be HIGH if the pressure is above the set threshold; otherwise the output will be LOW. Develop a code to measure the width of the incoming pulse. Assume the spike occurs for a duration within one full TCNT count.

Solution: The following code polls IC3 instead of using interrupts.

```
                INCLUDE      'mc9s12c32.inc'

time1           EQU     $3800
pls_width       EQU     $3802                              ;

                ORG     $8000

                JSR     tim_setup
                LDAA    tflg1
                ORAA    #$08
                STAA    tflg1          ; Clear C3F in TFLG1
risedge                                ; Wait until rising edge is
                                         detected
                BRCLR   tflg1, #$08,risedge
                LDD     tc3            ; Capture the rise time
                STD     time1
                LDAA    tflg1
                ORAA    #$08
                STAA    tflg1          ; Clear C3F in TFLG1
faledge                                ; wait until falling edge is
                                         detected
                BRCLR   tflg1,#$08,faledge
                LDD     tc3            ;Capture falling edge time,
                                         time2
                SUBD    time1          ;Calculate pls_width = time2 -
                                         time1
                STD     pls_width      ;Store pls_width
done            BRA     spin           ;Done
tim_setup                              ;Timer setup subroutine
                SEI
                MOVB    #$08, tios
                MOVB    #$C0, tctl4    ;Set to capture on either
                                         edge
                BCLR    tios, #$08     ;Set IC3 to act as an input
                                         capture channel
                MOVB    #$80, tscr1    ;Enable timer
                MOVB    $#03,tscr2     ;Select timer prescaler 8 for
                                         a frequency = 2 MHz
                CLI
                RTS
```

Measuring Short Pulse Durations If the pulse width is extremely short, the trailing edge of the pulse may occur before the timer has the chance to set up for the next capture. One way to resolve this situation is to connect the input signal to two input capture lines, say IOC1 and IOC2; IOC1 can be set up to capture on a rising edge while IOC2 is configured to capture on the falling edge.

Measuring Long Pulse Duration If the pulse is longer in duration than the full count of the free-running counter, that is, more than 65,536 TCNT counts, counter overflow will occur and the result will be wrong. The program will have to increment a variable in RAM each time the timer-overflow flag sets, to count the number of timer cycles during the measurement of the pulse.

Period Measurement

If the signal is a periodic waveform, the period is simply the time between two identical transitions and its frequency is the inverse of the period. If the signal is that from a shaft encoder or a turbine flow meter, the frequency may easily be converted into speed of rotation and flow rate, respectively. The code in Example 9.5 is used to make these measurements by simply changing the active transition to be on falling or rising edge instead of both.

Example 9.5: Period measurement–Develop a C code to measure the period of a digital waveform brought into the MCU at the timer capture pin IC3.

Solution: The following code measures the period between two successive rising edges of a pulse train on the IC3 pin. The program ignores overflow; therefore, the results are valid for a period that falls between ~27 cycles and 65,535 TCNT counts.

```c
#include <hidef.h>              /* common defines and macros */
#include <mc9s12c32.h>          /* derivative information */

#pragma LINK_INFO DERIVATIVE "mc9s12c32"

void ic3setup(void)            /* Initialize input capture on
                                  IOC3 */
    {
        TCTL4 = 0x80;           /* Capture time on rising edge */
        TIOS = 0x08;
        TSCR1 = 0x80;
        TSCR2 = 0x03;
    }

void main(void)
    {
        int time1, time2, period;
        DDRT=0x00;
        ic3setup();
        TFLG1 = TFLG1 | 0x08;
        while((TFLG1 & 0x08) == 0) {
        }
        time1 = TC3;            /* Time on first rising edge */
```

```
        TFLG1 = TFLG1 | 0x08;
        while((TFLG1 & 0x08) == 0) {
        }
        time2 = TC3 - time1;      /* Time on second edge
        period = time2 - time1;   /* Calculate period */
    }
```

9.5 PULSE ACCUMULATOR

The pulse accumulator (PA) facility of the 9S12C timer is used in many applications that involve pulse counting, precise timing measurements of the input pulse train, discrimination between long and short pulses, and so on. The PA operates on input signals received on PT[7] pin of port T if it had been configured for use as the PA input line, hereafter referred to as PAI. An overview of the registers involved in the PA function are given in Fig. 9.13.

9.5.1 Pulse Accumulator Count Register (PACNT)

The 16-bit PACNT, at address $0062-63, is used by the PA as a counter. Unlike the TCNT, the PACNT can be read and written to, allowing the pulse accumulator to start counting from a preset count instead of zero. This provides for a specific number of events to be detected without having to compare two numbers.

When the PACNT counts from $FFFF to $0000, an overflow condition occurs that sets the PAOVF, *PA overflow flag*, bit 1 in the PAFLG, *PA flag register*, at address $0061. An interrupt will be generated if the PAOVI, *PA interrupt overflow interrupt enable*, bit 1 in the PACTL,

PACNT: Pulse Accumulator Counter Register ($0062)

Bit 15			•••		Bit 0

PACTL: Pulse Accumulator Control Register ($0060)

0	PAEN	PAMOD	PEDGE	CLK1	CLK0	PAOVI	PAI

PAEN = 1   Enables pulse accumulator system

PAMOD = 0 Selects event-counting mode

PEDGE = 1   Selects rising edge on PAI input to increment PACNT

PAOVI = 1   Enables interrupt on PACNT overflow

PAI = 1   Enables interrupt on active edge on PAI line

CLK1	CLK0	
0	0	Timer prescaler clock is used as timer counter clock
0	1	PACLK is input to the timer counter clock
1	0	Timer counter clock frequency = PACLK/256 as
1	1	Timer counter clock frequency = PACLK/65536

PAFLG: Pulse Accumulator Flag Register ($0061)

0	0	0	0	0	0	PAOVF	PAIF

Figure 9.13 Registers involved in the pulse accumulator operation.

PA control register, at address $0060, had been set. The PAOVF flag may be cleared by writing a 1 to it.

9.5.2 PA Enable and Active Edge Detection

The PAEN, *PA enable,* bit 6 in the PACTL must be set to 1 to enable the pulse accumulator function. When enabled, the PA increments the PACNT every time an active edge is detected on the PAI line. The active edge to trigger counting is selected to be the falling edge if the PEDGE, *PA edge control,* bit in the PACTL register is clear or to be the rising edge if the PEDGE bit is set. An active edge on the PAI line sets the PAIF, *PA input edge flag,* bit 0 in the PAFLG register. Furthermore, an interrupt will be generated if the PAI, *PA interrupt enable,* bit 0 in the PACTL register had been set.

9.5.3 PA Operating Modes

The PA may be configured to operate in one of two modes: event-counting mode and gated-time-accumulation mode, depending on the PAMOD, *PA mode select,* bit 5 in the PACTL register. If PAMOD = 0, the PA is set for event-counting mode. In this case the counter increments on a falling edge if PEDGE is zero or on a rising edge if PEDGE = 1. The setup for the PA operating in the event-counting mode is depicted in Fig. 9.14. This mode is used to count

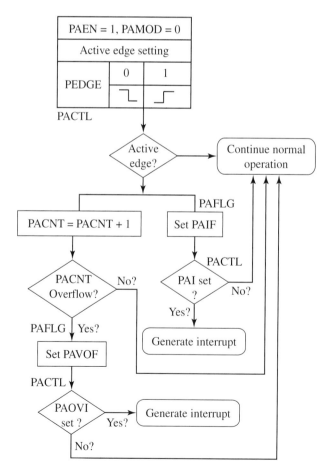

Figure 9.14 Pulse accumulator operation in event-counting mode.

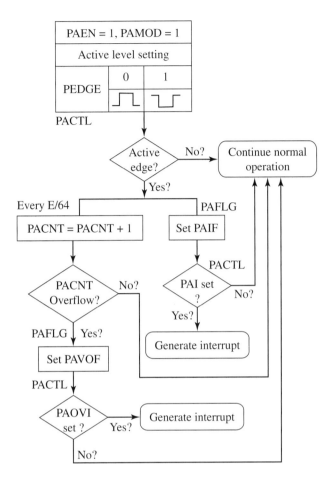

Figure 9.15 Operation of the PA in the gated-time-accumulation mode.

signal transitions of external events. It can be configured to generate an interrupt when a desired number of events has occurred. Typical pulse-counting applications include monitoring tape position in a VCR, monitoring an automobile odometer, monitoring total volume of fluid flowing through a pulse-generating flowmeter, such as a turbine meter, to name a few.

The PA is configured to operate in the gated-time-accumulation mode if the PAMOD bit is set to 1 (see Fig. 9.15). In this mode, the PEDGE bit is used to specify the active state of the input signal on the PAI input line; that is, the input that will inhibit time accumulation. If PEDGE is zero, the counter will not increment as long as the input on PAI is LOW. If PEDGE is set to 1, counting is inhibited if the input signal on the PAI is HIGH. The PACNT increments once every 64 E-clock cycles when an active edge occurs on the PAI line.

Example 9.6: Event counting—Consider the use of the 9S12C MCU to automate a box-labeling operation on an assembly line where a specified number of boxes are to be labeled using a labeling pad. A sensor such as an IR emitter/detector pair is placed at the end of the labeling line to provide an input signal to the IOC[7] pin whenever a box passes. Develop an assembly code to automate the labeling operation.

Solution: The following code manages the main labeling operation. The user needs to provide the missing code. The MCU sets up the labeling tool, loads the PACNT with the 2s-complement of the preset number of boxes to be labeled, and configures the pulse accumulator to produce a PACNT overflow interrupt. Every time a box passes the sensor, the IOC[7] pin goes HIGH and the PACNT increments. Once the preset number of boxes have been labeled, the PACNT overflows and an interrupt is generated.

The code uses the PA to label 60 boxes via subroutine *start_label*. This subroutine uses accumulator A to hold the label code. After returning from *start_label*, the labeling operation continues until subroutine stop_label is called by the overflow ISR, *paovi_isr*. The source code for subroutines *start_label* and *stop_label* are not provided.

```
        INCLUDE   'mc9s12c32.inc'


        JSR    pa_setup             ;initialize pulse accumulator
        JSR    start_label
wait    BRA    wait                 ;Wait until 60 labels done

;*Subroutine pa_setup initializes the pulse accumulator to
count the rising edges without interrupt, loads
;* the pacnt with the preset number of boxes held in
accumulator A, and enables interrupt when pacnt
;* overflows

pa_setup
        CLR                         ;Mask PAOVF interrupt
        BSET   pactl, #$5A          ;Set PAEN and PEDGE (bits 6
                                    and 4 of PACTL) to 1
        LDAB   #60                  ;Load A with the number of
                                    times the label is applied
        NEGB                        ;Take the 2s-complement of A
                                    and
        STAB   pacnt                ;Store to the pulse accumula-
                                    tor counter pacnt
        BCLR   paflg, #$02          ;Clear PAVOF flag
        RTS                         ;return
; Interrupt-service routine paov-isr handles pulse accumulator
overflow

paov_isr
        BCLR   paflg,#$02           ; Ensure PAVOF flag is clear
        JSR    stop_label           ; Stop labeling when all
                                    labels are executed
        RTI                         ; return

start_label
        .....                       ; Subroutine to manage the
                                    labeling tool
stop_label
        .....                       ; Subroutine to stop the
                                    labeling process
```

Example 9.7: Speed measurement–Develop a scheme to measure the rotational speed of a 120-tooth gear using the pulse accumulator.

Solution: Figure 9.16 shows a setup that uses the pulse accumulator to measure the speed of rotation of the 120-tooth gear by calculating the time the gear takes to make a single revolution. The IR detector generates a pulse each time a tooth of a gear interrupts the light path. The Schmitt trigger produces a square waveform that drives both IC2 and IC7/PAI. Both IOC2 and PAI can be set up to respond to rising edges of the input signal. The time of occurrence of the next rising edge is saved in RAM from TC2. Then the PACNT is preset to its maximum count (255) minus 120: $256 - 120 = 136$. An interrupt is enabled to occur when the PACNT overflows. Interrupts are not enabled from IC2. The TC2 register will record a new value on each rising edge of the input waveform, regardless of the fact that its C2F flag is not being cleared each time. After exactly $120 + 1$ rising edges, the PACNT will overflow and cause an interrupt. The interrupt-service routine reads the current TC2 value and subtracts the previous value (stored in RAM) from it to determine the time between 121 edges, or one revolution of the gear. The speed of the gear, in rad/sec is, then found by simply dividing 2π by this time.

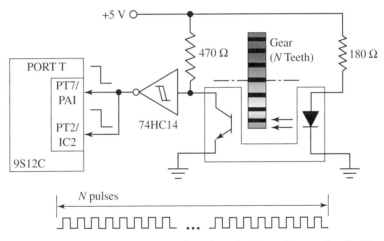

Figure 9.16 Measuring the speed of rotation of an N-tooth gear using the PA.

Although the pulse accumulator can be used for pulse-width measurement, it is not as accurate as the input capture function. However, it can identify a wide pulse from a narrow pulse much more easily. This makes it useful for decoding signals that use pulse width as part of their code. Thus a common use of this mode is to discriminate a wide pulse from a narrow pulse.

9.6 REAL-TIME CLOCK

The 9S12C MCU provides a means to implement a software-based real-time clock (RTC). The function of the RTC is to generate periodic interrupts when a specified real-time interrupt (RTI) period t_{RTI} has elapsed. The t_{RTI} is determined by dividing the frequency of the oscillator clock

(OSCCLK) by the product of the *modulus counter target value* (*MCV*) and a *prescaler factor* (*PF*). The *MCV* is the code value of the RTR[3:0], *RTI modulus counter select bits* in the RTICTL, *RTI control register* at address $003B. The *PF* is the code value of the RTR[6:4], *RTI prescale rate select bits* in the RTICTL. Thus, the RTI period is given by

$$ t_{RTI} = \frac{1}{OSCCLK/(MCV + 1)(2^{9 + PF})} $$

except when the RTR[6:4] code is 000, $t_{RTI} = 1/[OSCCLK/(MCV + 1)]$. For example, if $OSCCLK = 16$ MHz and the RTR[3:0] code is set to 0110 and the RTR[6:4] code is set to 110, then the corresponding RTI period is $t_{RTI} = 1/[(16 \text{ MHz})/(6 + 1)(2^{9+6})]$, or 14,336 ms. A 16-MHz OSCCLK yields $128 < t_{RTI} < 65,536$ ms. Peripheral devices such as the LM58274, available from National Semiconductor, gives hours, minutes, seconds, and tenth of seconds as well as month and day of the week.

The RTI mechanism is controlled by the RTIE, *RTI enable*, bit 7 in the CRGINT register at address $0038. As soon as the RTI time-out period ends, the RTIF, *RTI flag*, bit 6 in the CRGFLG register at address $0037 sets to 1, immediately triggering a new time-out period to start. The RTIF flag can be cleared only if a 1 is written to the associated bit location. A write to the RTICTL register restarts the RTI time-out period. The RTI mechanism is useful in many industrial control applications, such as food processing, where periodic monitoring of vital parameters such as temperature, pressure, and humidity is essential.

9.7 PULSE-WIDTH MODULATION (PWM)

The PWM module on the 9S12C MCU supports the generation of PWM output waveforms on port P pins on the chip. The number of PWM channels available on the 80-pin, 52-pin, and 48-pin versions of the 9S12C MCU are 6, 3, and 1 brought out on PP[5:0], PP[5:3], and PP[5] pins, respectively. The following discussion applies to any PWM channel if it is available on the chip.

Each PWM channel can be used individually to produce an 8-bit PWM output waveform. An existing pair of PWM channels may be concatenated to generate a 16-bit PWM output. If the CON*jk*, *concatenate channel jk*, bit in the PWMCTL, *PWM control register*, at address $00E5 is set, channels *j* and *k* are concatenated and channel *j* becomes the high byte while channel *k* provides the low byte for the 16-bit PWM output, where $j = 0, 2, 4$ and $k = 1, 2, 3$. The following instruction concatenates channels 4 and 5 to act as a double-byte PWM channel:

```
MOVB  #$40,pwmctl
```

The registers and associated bits dedicated to the operation of the PWM function are listed in Fig. 9.17. These registers are used to enable the PWM channels, select the PWM clock source and scaling, and define the start polarity of the output waveform. An overview of the bits' settings and the effect they have on the operation of the PWM is presented in the following sections.

PWM Enable

To enable a PWM channel for an 8-bit output, the *PWM enable* bit PWME*i* in the PWME, *PWM enable register*, at address $00E0, must first be set to 1. The PWM waveform becomes available and is automatically generated on the next source clock cycle without software

PWME: PWM Enable Register ($00E0)

0	0	PWME5	PWME4	PWME3	PWME2	PWME1	PWME0

PWMPOL: PWM Polarity Register ($00E1)

0	0	PPOL5	PPOL4	PPOL3	PPOL2	PPOL1	PPOL0

PWMCLK: PWM Clock Select Register ($00E2)

0	0	PCLK5	PCLK4	PCLK3	PCLK2	PCLK1	PCLK0

PWMPRCLK: PWM Prescale Clock Select Register ($00E3)

0	PCKB2	PCKB1	PCKB0	0	PCKA2	PCKA1	PCKA0

PWMCAE: PWM Center Align Enable Register ($00E4)

0	0	CAE5	CAE4	CAE3	CAE2	CAE1	CAE0

PWMCTL: PWM Control Register ($00E5)

0	CON45	CON23	CON01	PSWAI	PFRZ	0	0

PWMSDN: PWM Shut-Down Register ($00FE)

PWMIF	PWMIE	PWMRSTRT	PWMLVL	0	PWM5IN	PWM5INL	PWM5ENA

Figure 9.17 Registers associated with the operation of the PWM module.

intervention thereafter. PWME[4], PWME[2], and PWME[0] have no effect if CON45, CON23, or CON01 bits in the PWMCTL register have been set. The following instruction enables the PWM channel 4:

```
MOVB    #$40, PWME
```

PWM Polarity

The PWM channel may be configured to start the output waveform LOW or HIGH. If the PPOLi, *Pulse width channel i polarity* bit in the PWMPOL, *PWM Polarity Register*, at address $00E1, has been set to 1, the output pulse starts HIGH and then snaps to a LOW at the end of the duty cycle count. The opposite will occur if PPOLi has been cleared. The following instruction sets the starting polarity of the PWM waveform on Channel 4 to a high:

```
MOVB  $#10, PWMPOL   ; Set polarity on PWM Channel
                       4 to start high
```

PWM Period and Duty Cycle

Figure 9.18 shows the edges of all possible PWM waveforms that could be generated on a PWM channel and the timing of their occurrence. The heart of the operation of a PWM channel i is the PWMCNTi counter register. The PWMCNTi is an UP/DOWN counter that controls the occurrence of the edges of a PWM waveform. The PWMCNTi counts UP or DOWN every clock cycle, but it is forced to reset to $00 if it is written to. The PWMCNTi is configured as an UP counter after reset. The specific period and duty cycle on channel i are controlled by two registers: The PWMPERi register controls the desired period, and the

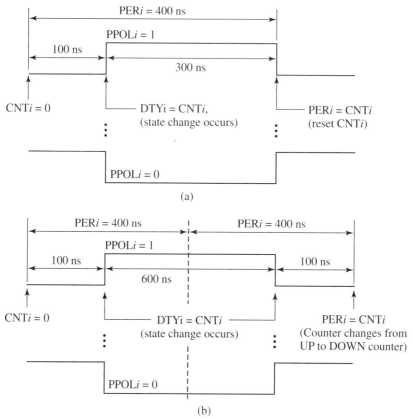

Figure 9.18 Formats of PWM waveforms.

PWMDTYi register controls the duty cycle resolution. The PWMPERi is loaded with PERi, an integer from 0 to 256 that corresponds to the desired period of the PWM waveform to be generated on channel i. The PWMDTYi register is loaded with the DTYi, an integer between 0 and 256 that corresponds to the desired duty cycle on channel i (DCi). The corresponding duty cycle is simply DCi = DTYi/PERi. If PERi = 256 and DTYi = 64, DCi = 25%. The following two instructions set the period and duty cycle on PWM channel 4 for a 25% duty cycle:

```
MOVB    $#64, PWMDTY4
MOVB    #$256, PWMPER4
```

The period and duty cycle registers are double buffered to ensure that any changes to their values can occur only when the PWMCNTi counter rolls over or when the associated PWM channel is disabled. Immediate change in the duty cycle or the period may be forced if a new value is written to the PWMPERi or to the PWMDTYi registers followed by a write to the PWMCNTi. This forces the counter to reset and prompts the duty cycle and/or the period to be latched.

When a match between the DTYi value and the value in the PWMCNTi counter occurs, the level of the PWM waveform changes during a period from LOW to HIGH (or HIGH to LOW, depending on the polarity setting). Moreover, when a match between the PER*i* value and the value in the PWMCNT*i* counter occurs, the level of the PWM waveform changes again at the end of the period, from LOW to HIGH or HIGH to LOW.

PWM Output Form Selection

A PWM output waveform may be configured as one of two forms: left-aligned (LA) or center-aligned (CA). The parameters associated with each form are indicated on Fig. 9.18 a and b, respectively. If the CAE*i*, *center align enable bit* in the PWMCAE, *PWM center align enable register*, at address $00E4, has been cleared, the corresponding PWM output will be LA. In this case, the PWMCNTi counter is configured as an UP counter only and will count from 0 to (PER*i* − 1). To select the CA form for the PWM output, CAE*i* must be set. In this case the PWMCNT*i* counter is configured as an UP-and-DOWN counter and will count from 0 to PERi and then back down to zero.

PWM Clock Setting

The operation of the PWM counters may use any one of four clock sources. PWM channels 0, 1, 4, 5 may use clock A or clock SA (scaled A), and channels 2 and 3 may use clock B or clock SB (scaled B). Figure 9.19 shows the possible settings for each clock source. The following discussion outlines the configurations of clock A and clock SA. The same applies to clocks B and SB as well, provided the associated bits are used.

The frequency of clock A is determined by dividing the bus (ECLK) frequency by a prescale *PR* determined by the settings of PCKA[2:0], *prescaler select for clock A* bits in the PWMPRCLK, *PWM prescale clock select register*, at address $00E3; $PR = 2^N$, where *N* is the value of the code bits PCKA[2:0]. Clock SA frequency is generated by dividing the frequency of clock A by twice the value written into the 8-bit PWMSCLA, *PWM scale A register* at address $00E8. If the value of the code bits in the PWMSCLA register is $00, the PWMSCLA

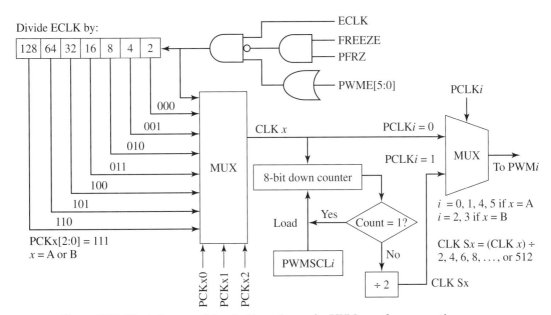

Figure 9.19 Block diagram of the clocking schemes for PWM waveform generation.

value is set to its full scale value of 256 and clock A is divided by 512. Succinctly, the frequency of the PWM clock A (f_{PWMCLK}) is determined by

$$f_{PWMCLK} = ECLK/PR$$

and for Clock SA it is

$$f_{PWMCLK} = ECLK/(2 \times PWMSCLA \times PR)$$

Reloading of the PWM counter register PWMCNTi with the prescale value in PWMSCLA occurs when the counter counts down to $0001 or when a write to the PWMSCLA is launched. If a write to PWMSCLA is attempted while the PWM channel is operating, the integrity of the generated waveform will be compromised. The PCLKi, *pulse width channel i clock select* bit in the PWMCLK, PWM *clock select register* at address $00E2, selects the desired clock. For Channel i, $i = 0, 1, 4,$ or 5, PCLK$i = 0$ selects clock A and PCLK$i = 1$ selects clock SA. The following instructions may be used to establish the desired clock settings:

```
MOVB  $#03, PWMPRCLK    ;Define PR = 8 for clock A
MOVB  #$00, PWMSCLA     ;Select full scale for PWMSCLA
MOVB  #$10, PWMCLK      ;Select clock SA for channel 4
```

The actual time corresponding to the PERi value written into the PWMPERi register is determined by

$$t_{PWMi} = PERi/f_{PWMCLK}$$

The frequency of the generated PWM waveform f_{PWMi} for the left-aligned option is determined by

$$f_{PWMi} = \frac{f_{PWMVCLK}}{PERi} \qquad \text{or} \qquad t_{PWMi} = PERi \times t_{PWMCLK}$$

For the center-aligned option, the frequency is determined by

$$f_{PWMi} = \frac{f_{PWMVCLK}}{2 \times PERi} \qquad \text{or} \qquad t_{PWMi} = t_{PWMCLK} \times (2 \times PERi)$$

The duty cycle for the LA or the CA output is determined by

$$DUTY\,Cycle = ((PERi - DTYi)/PERi) \times 100 \qquad \text{if} \quad PPOLi = 0$$
$$DUTY\,Cycle = (DTYi/PERi) \times 100 \qquad \text{if} \quad PPOLi = 1$$

Emergency Shutdown

The PWM module has an emergency shutdown mechanism that is enacted under the control of the PWMSDN register at address $01FE. The mechanism is enabled by setting the PWM5LNA bit. An emergency shutdown occurs if the active level defined by the PWM5INL bit is detected (1 for HIGH, 0 for LOW) on the PWM5 line. The output level forced on all active PWM channels in case of an emergency is the level (1 or 0) programmed into the PWMLVL bit. When an emergency shutdown condition occurs, the PWMIF flag sets and an interrupt is generated if the PWMIE, *PWM interrupt enable*, bit had been set.

PWM Operation Steps

Setting up and operating a PWM channel is achieved by means of the following steps.

1. Enable the PWM function of channel i by setting the PWM[i] bit in the PWME register.
2. Write to the PWMPOL register to configure the PPOL[i] bit, and select the polarity of the PWM output waveform.
3. Write to the PWMCLK register to configure the PCLK[5:0] code bits for the PWM output.
4. Select prescaler, PCKxi ($i = 0, \ldots, 5; x = A, B$) PWMPRCLK register.
5. Set the CAEi bit in the PWMCAE register if the CA option is desired.
6. Write to the PWMSCLA (B) register to set the prescale code bits.
7. Write PERi to the PWMPERi register to set the period.
8. Write DTYi to the PWMDTYi register to set the duty cycle.

Refer to the earlier discussion for instructions that may be used to execute these steps.

Example 9.8: PER and DTY settings–A heater rated at <20 V is driven by ON Semiconductor's MMSF5N02HD power MOSFET as shown in Fig. 9.20. The MOSFET has the following specifications: $R_{ON} < 25$ m Ω, VGS = 4.5 V, $I_D = 5$ A, and a drain-source resistance in the micro-ohm range. Thus the MOSFET may be directly controlled by the PWM waveform generated on the PWM0 channel, as shown. The duty cycle of the PWM waveform determines the percentage of time the I_D is on and thus the temperature of the heater. Determine the PER0 and DTY0 values to generate a PWM waveform of a periof of 1 ms and a duty cycle of 75%. Write a code to manage the operation. Make the PWM output waveform left-aligned and starting HIGH. The bus (ECLK) clock frequency is 16 MHz.

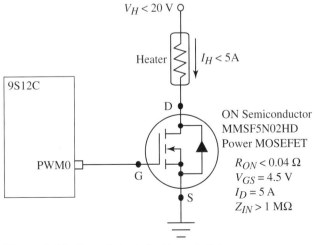

Figure 9.20 Control circuit for Example 9.8.

Solution: The PER0 and DTY0 values that should be programmed into the PWMPER0 and PWMDTY0 registers are determined as:

$$PER0 = t_{PWM0} \times \frac{ECLK}{2 \times PWMSCLA \times PR}$$

$$= (1 \times 10^{-3}) \times \frac{16 \times 10^6}{16 \times 2 \times 5}$$

$$= 100$$

$$DTY0 = PER0 \times Duty\ Cycle = 100 \times 0.75$$

$$= 75$$

The following code manages the desired heater temperature control.

```
        INCLUDE   'mc9s12c32.inc'

per0    EQU    100
dty0    EQU    75

        ORG    $C000
main    ; Start of the main code
        JSR    pwm_setup
here    BRA    here

pwm_setup     ; Start the subroutine to setup PWM output on
              channel 0

        MOVB   #$01, pwme        ;Enable PWM0
        MOVB   #$01, pwmpol       Select polarity at PWM0 at 1
        MOVB   #$00, pwmcae      ;Select left-aligned option
        MOVB   #$01, pwmclk      ;Select SA clock for PWM0
        MOVB   #$04, pwmprclk    ;Set PR = 16
        MOVB   #$05, pwmscla     ;Set clock SA prescaler to 5
        LDAA   per0
        STAA   pwmper0           ;Load PRE0 into PWMPER0
                                  register

        LDAA   dty0
        STAA   pwmdty0           ;Load PWMDTY0 into PWMDTY0
                                  register

        RTS
```

9.8 SUMMARY

This chapter presented the timer facilities of the MC9S12C. This programmable facility supports the mechatronics product's many useful functions, from real-time clock to PWM. This chapter covered the various registers associated with the adopted MCU timer facility. Programming examples were given for the input capture and output compare facilities needed in timing applications. Dedicated registers were presented, and related examples were explained. Pulse-width modulation facilities were covered, with attention given to their very useful applications in interfacing the digital data of the MCU to the driven analog circuits that exist in many applications.

RELATED READING

S. F. Barrett and D. J. Pack, *Embedded Systems: Design and Applications with the 68HC12 and HCS12.* New York: Prentice Hall, 2004.

HCS12 Microcontrollers: MC9S12C128 Data Sheet, 2005 (MC9S12C128_V1.pdf)

H.-W. Huang, *HC12/9S12: An Introduction to Hardware and Software Interfacing.* Clifton Park: Thomson Delmar Learning, 2005.

J. B. Peatman, *Design with Microcontrollers.* New York: McGraw-Hill, 1988.

P. Spasov, *Microcontroller Technology: The 68HC11*, 2nd ed. New York: Prentice Hall, 1996.

QUESTIONS

9.1 Explain how the 16-bit TCNT value is read.

9.2 Explain how to program the TCNT clock prescaler.

9.3 When is the timer overflow flag set?

9.4 Explain the interrupt mechanism associated with the output compare function.

9.5 Explain the interrupt process associated with the TCNT overflow.

9.6 How is the timer overflow flag reset?

9.7 What timing resolution can be achieved with the output compare?

9.8 Give the name, the register, the address, and the default state for the following bits:

   a. The bit indicating successful comparison on output compare 2.

   b. Timer output compare 2 interrupt enable bit.

   c. The bit to set the output compare 3 I/O HIGH on a successful comparison.

9.9 Give the two registers that control which data bits are output when the output compare flag is set.

9.10 What is the process of selecting the active edge for input capture 2?

9.11 What bits in what registers must be set to enable the pulse accumulator input edge interrupt?

9.12 List advantages of PWM-generated signals over other digital outputs.

PROBLEMS

9.1 Write a program to output a square wave at a frequency 10 kHz. Use the routine in Example 9.2.

9.2 Use the 9S12C MCU to generate a PWM signal with a duty cycle of 60% at a frequency of 5 kHz. Use the routines in Example 9.3.

9.3 Write a routine to measure the width of a HIGH pulse. Use the routines in Example 9.4.

9.4 Write a program to produce two PWM signals at pins PT[0] and PT[1]. Function OC7 controls the two pins in conjunction with OC0 and OC1. OC7 drives the period and the scheduling of OC0 and OC1.

9.5 The international tuning standard for musical instruments is "A above middle C" at a frequency of 440 Hz. Write a code to generate this tuning frequency, and sound a 440-Hz tone on a loudspeaker (such as the one found in PCs or children's toys) connected to OC2. Due to rounding of values placed in TC2, a slight error in the output frequency is expected. What is the exact output frequency, and what is the percentage error? What value of clock frequency would yield exactly 440 Hz with the code you have written?

9.6 Write an interrupt-driven program that continually plays a musical scale of your choice (e.g., A major).

9.7 Write a C code to implement a 4-bit counter at the counting rate of 50 kHz.

9.8 Write a code to implement a speed control of a DC motor by generating a PWM waveform using the available facilities in the 9S12C MCU without deploying an external DAC device.

LABORATORY PROJECTS

9.1 Design and build a vending machine to dispense 75-cent soda cans. The machine should be able to accept nickels, dimes, and quarters. When the right amount is supplied, the machine drops a can in the collection bin. It also counts the number of dispensed cans. When the can holder becomes empty, the machine sends a signal to the managing center requesting to be filled agin.

9.2 Design and implement the bottling operation described in Example 6.17.

9.3 Implement the measurement system to measure the speed of rotation of a gear as described in Example 9.7.

9.4 Develop a code using the RTI mechanism to announce the time by causing an alarm to sound off a number of times that is equal to the current hour.

10

Analog-to-Digital (A/D) and Digital-to-Analog (D/A) Conversion

OBJECTIVES

Thoughtful engagement with the material in this chapter will enable students to:

1 Understand fundamental concepts and operation in A/D and D/A conversion processes

2 Learn how various A/D and D/A converters interface in a mechatronic system

3 Describe and use the ATD facility of the 9S12C MCU

4 Write programs to manage sensor interface with the A/D converter

10.1 INTRODUCTION

Typical mechatronics applications involve sensors to measure relevant physical parameters, such as pressure, speed, and temperature. The measurement chain is shown in Fig. 10.1. The time-varying analog signal generated by the sensor, usually voltage or current, in response to a stimulus is presented to the system computer as a series of equivalent digital values sampled at discrete time intervals for further processing. The rate at which a signal is sampled, called the *sampling frequency* f_S (Hz), affects the accuracy of the discrete-time representation of the analog signal. The device that samples the analog signal and encodes it as a binary number for the processing computer is called an analog-to-digital (A/D) converter, or ADC. The necessity of converting analog signals to their binary equivalent makes the ADC an essential device in mechatronics applications, allowing a wide variety of sensors to be used directly with the MCU.

Mechatronics also involve controlling actuators to maintain the desired performance of a machine or a process according to the inputs received from sensors. Actuators such as DC motors, heaters, and pumps are analog devices. To control them from a computer requires a device to translate the digital control signal to its equivalent analog value. The device that accomplishes this task is called a digital-to-analog (D/A) converter, or DAC. Recall that the 9S12C can generate a PWM signal, which is a simple form of a DAC. This chapter deals with ADC and DAC systems. A major portion of the chapter focuses on the operation of the ADC facility onboard the 9S12C MCU.

10.2 FUNDAMENTALS OF A/D CONVERSION

A measurement signal evolves through various stages, as indicated in Fig. 10.1. The input to the ADC receives the analog signal v_i, compares it with its reference signal V_R, converts it into a fraction of all possible outputs, and presents it at its output as an unsigned binary value N.

The digital output of an ADC depends on two parameters that are fundamental to a converter's operation: the low and high reference voltages, V_{RL} and V_{RH}, and the number of bits k the input signal is coded into. A k-bit ADC generates 2^k output levels, also called *quanta*. Any input signal will be converted to one of these possible levels. The input signal v_i falls between the minimum and maximum values of the specified range. The minimum value is called the *offset*, and the difference between the maximum and minimum values is called the *range*, *span*, or *full scale*, *FS*. The minimum and maximum analog values, V_{MIN} and V_{MAX}, are translated to all zeros and all 1s in the binary code, respectively. For an 8-bit ADC, the number of levels is $2^8 = 256$, V_{MIN} maps to $N = 0$ (decimal), %00000000 (binary), and $00 (hex) values, and V_{MAX} maps to $N = 256$ (decimal), %11111111 (binary), and $FF (hex).

If V_{RL} and V_{RH} have the same polarity, the ADC is a *unipolar* device; otherwise it is a *bipolar* device. The reference voltages (V_{RL}, V_{RH}) for a unipolar ADC may be $(0, +5)$, $(0, -5)$, and so on, and for a bipolar device they may $(-2.5, +2.5)$, $(-5, +5)$, and so on. The full scale of the ADC is $FS = V_{RH} - V_{RL}$.

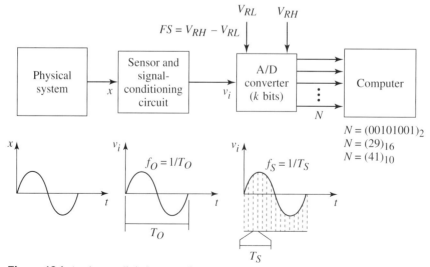

Figure 10.1 Analog-to-digital conversion process.

10.2.1 Resolution

Resolution is the smallest variation in the analog input signal that would cause the ADC output code to change by one level, or quantum. It is usually expressed in terms of the full scale and the number of output code bits k:

$$V_Q = \frac{FS}{2^k - 1} \qquad (10.1)$$

The resolution represents a quantum (or step size) and corresponds to the value of the LSB in the output code. Equation (10.1) also defines the resolution in terms of the input signal x, such as pressure. In this case V_Q is interpreted in the units of x (e.g., N/m^2) and $FS = x_{MAX} - x_{MIN}$.

10.2.2 I/O Mapping

Any voltage level v_i within the input signal range is translated to its decimal equivalent N as

$$N = INT\left(\frac{2^k - 1}{FS} \times [v_i - V_{RL}]\right) \qquad (10.2)$$

N is truncated to its closest integer value and then converted to its binary and hex values. Refer to Appendix G. To find the analog voltage corresponding to a specific output N, the following relation is used:

$$v_i = N \times \left(\frac{FS}{2^k - 1}\right) + V_{RL} \qquad (10.3)$$

Equation (10.3) indicates that A/D conversion is basically a ratiometric operation. If N is expressed in binary as $b_{k-1}b_{k-2}\,b_{k-3}\ldots b_1b_0$, the analog voltage is determined by

$$v_i = (b_{k-1} \times 2^{-1} + b_{k-2} \times 2^{-2} + \cdots + b_1 \times 2^{-k+1} + b_0 \times 2^{-k}) \times FS + V_{RL} \quad (10.4)$$

where $2^{-1}, 2^{-2}, \ldots, 2^{-k}$ are the *weights* of the corresponding bits. Note that the weight of the MSB is $\frac{1}{2}$ and the weight of the LSB is the resolution. If a unipolar 8-bit ADC with $FS = 5$ V generates a binary code of %01001110, the corresponding voltage value, by Eq. (10.4), is

$$v_i = (0 \times 2^{-1} + 1 \times 2^{-2} + 0 \times 2^{-3} + 0 \times 2^{-4} + 1 \times 2^{-5}$$
$$+ 1 \times 2^{-6} + 1 \times 2^{-7} + 0 \times 2^{-8}) \times 5 + 0 = 1.5234\,\text{V}$$

Figure 10.2 shows the attributes of an ideal 3-bit ADC. The LSB size is $\frac{1}{8}FS$, and the input range is quantized into eight distinct levels between 0 and $\frac{7}{8}FS$. It is obvious that the highest binary code, %111, represents not the FS but, rather, $\frac{7}{8}FS$, $FS - V_Q$. For example, with a 5-V reference, the resolution by Eq. (10.1) is 0.625-V. Any change in the input voltage less than this value will not cause a change in the output code of the converter.

Conversion Error

Analog values within a quantum level generate the same output code. If the output represents the midrange value or the *threshold* of a quantum, the output signal within a code involves a

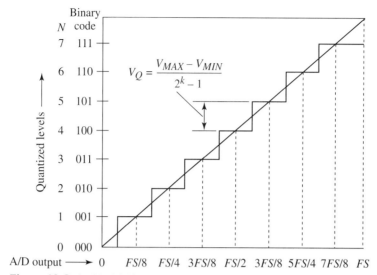

Figure 10.2 An ideal 3-bit conversion process.

conversion uncertainty equivalent to a $\pm\frac{1}{2}$ LSB. The error V_E between the digitized voltage v_i and the input voltage v_i is estimated by

$$V_E = v_i - NV_Q \tag{10.5}$$

This error is shown in Fig. 10.3 as a function of v_i. The maximum error in converting the input voltage is V_Q.

The error associated with the conversion process can only be minimized by increasing the number of bits in the converter's output code. The minimum number of bits required in the ADC for a specific allowable error and input voltage range can be found by

$$k = \frac{\log([V_{MAX} - V_{MIN}]/V_Q + 1)}{\log(2)} \tag{10.6}$$

The result must be rounded up to the next-highest integer. The theoretical root mean square (RMS) signal-to-noise ratio (SNR) for a k-bit ADC is estimated by

$$SNR = 6.02 \times k + 1.76 \text{ dB} \tag{10.7}$$

Table 10.1 provides the resolution, LSB for a 5-V reference voltage, and the dynamic range of 4- to 16-bit ADCs.

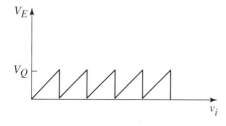

Figure 10.3 Conversion error.

TABLE 10.1 Number of Bits, Resolution, LSB Value, and Dynamic Range of ADCs

k-Bits	V_Q (% FS)	LSB Voltage for FS = 5 V	Dynamic Range (dB)
4	6.25	300 mV	24.08
8	0.3906	19.5 mV	48.16
10	0.0977	4.90 mV	60.12
12	0.0244	1.20 mV	72.25
14	0.00610	305 μV	84.29
16	0.00153	75 μV	96.33

Example 10.1: Offset, span, and resolution of ADC–A temperature sensor with a gain of 10 mV/°C is used to measure the temperature of a process within the range of -50 to $+200$°C. An 8-bit ADC with a range from -5 to $+5$ V is used. A signal-conditioning circuit is needed to match the limits of the sensor output v_S with the input voltage v_i of the ADC.

a. Determine a linear relationship for the signal-conditioning circuit between v_S and v_i.

b. Determine the offset, full scale, and resolution of the measurement in terms of voltage and temperature.

c. Determine the output of the A/D if the temperature is $+50$°C, and express it in decimal, binary, and hex equivalents.

Solution

a. The range of sensor outputs for the temperature range of interest is $-0.5\text{ V} \leq v_S \leq +2\text{ V}$. The sensor voltage V_S is related to the input voltage to the ADC by $v_i = mv_S + b$, where m is the slope and b is the intercept. Applying corresponding voltage limits gives the two equations $-0.5m + b = -5$ V and $2.0m + b = +5$ V. Solving these two equations simultaneously yields $m = 4$ and $b = -3$ V. The final relation is $v_i = 4v_S - 3$.

b. The voltage and temperature offset and full scale are, respectively, $(-5.0\text{ V}, -50°\text{C})$ and $(10\text{ V}, 250°\text{C})$. The resolution from Eq. (10.1) in terms of the ADC output is

$$V_Q = \frac{10}{2^8 - 1} = 0.0390625 \text{ V}$$

And the temperature resolution is

$$V_Q = \frac{250°\text{C}}{2^{k-1} - 1} = 0.98°\text{C}$$

c. The sensor output voltage when the temperature $T = 50$°C is

$$v_s = [(0.01 \text{ V/°C}) \times 50°\text{C}] \times 4 - 3\text{V} = -1.0 \text{ V}$$

The equivalent output code from Eq. (10.2) is

$$(-1.0 \text{ V} - (-5 \text{ V}))/(0.039 \text{ V}) = (102)_{10} \text{ decimal}$$

$$= (66)_{16} \text{ hexadecimal}$$

$$= (01100110)_2 \text{ binary}$$

Example 10.2: ADC resolution–An ADC is needed to sample the output voltage of a pressure sensor. The output of the sensor is 0 volts when the pressure is 0 kPa and 10 volts when the pressure reaches 10 kPa. If the sensor error is not to exceed 0.010 kPa, determine the number of ADC bits to resolve the sensor output.

Solution: The minimum number of bits is found from Eq. (10.6) to be

$$k = \frac{\log\left(\dfrac{10}{0.01} + 1\right)}{\log(2)} = 9.97$$

A 10-bit ADC can be used to sample the pressure output.

Example 10.3: Amplification and A/D conversion–An IC temperature sensor with a gain of 10 mV/°C is used to measure the temperature of an object up to 100°C. The sensor output is sampled by an 8-bit ADC with a 5-V reference. Design the signal-conditioning circuit to interface the sensor signal to the ADC, and determine the temperature resolution.

Solution: The output of the ADC at maximum temperature must be 11111111. This number corresponds to a maximum A/D input voltage of $v_i = 10 \times (2^{-1} + 2^{-2} + \cdots + 2^{-7} + 2^{-8}) = 9.9609375$ V. Meanwhile, the maximum output generated by the sensor is $(5 \text{ mV/°C}) \times 100°C = 0.5$ V. Bridging the difference requires amplification with a gain of $A = 9.9609375/0.5 = 19.92$. A noninverting op-amp with $R_f = 18.92$ kΩ and $R_i = 1$ kΩ (see Fig. 4.7). The temperature resolution that can be mitigated will be

$$\Delta T = \frac{5/(2^8 - 1)}{19.92} \times \frac{1}{5} = 0.19°C$$

10.2.3 Aliasing

The analog signal received at the input of the ADC is a combination of the usually low-frequency useful signal and the usually high-frequency noise either generated by the components in the measurement chain or picked up as EMI (see Section 2.15). If the sampling frequency of the ADC is not high enough, a phenomenon known as *aliasing* may occur. To demonstrate aliasing, assume that a sinusoidal signal of frequency f is being sampled by an ADC at a sampling frequency $f_S < 2f$, as shown in Fig. 10.4. The circles on the figure indicate the samples, and when connected they form a sinusoid with an apparent frequency f_a that is less than the frequency of the original signal, f. What happened here is that the digitized version of the original signal has been misrepresented by the ADC as a lower-frequency signal. This is known as *aliasing*. Succinctly, aliasing is the misrepresentation of a high-frequency signal as a lower-frequency one. With aliasing, sampling a signal of frequency $f, f + f_S, f - f_S, f + 2f_S, f - 2f_S$, or, in general, $f \mp nf_S$, where n is an integer at a sampling frequency f_S, yields the same result. The *apparent* low frequencies f_a due to aliasing are found by

$$f_a = |f_0 - nf_s| \tag{10.8}$$

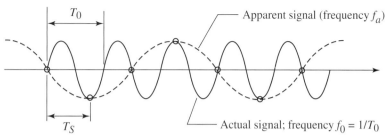

○ Sampling times

Figure 10.4 Apparent low-frequency signal as a result of aliasing due to $f_0 \geq f_S/2$.

for integer n and f_a. The lowest frequency can be found by choosing n such that

$$f_a \leq \frac{1}{2}f_S \tag{10.9}$$

Aliasing causes erroneous results and wreaks havoc on any control attempt, especially when high-frequency noise is involved. Digital systems are prone to aliasing if the sampling frequency f_S is less than twice the highest frequency of the signal.

Aliasing may be prevented by blocking signal frequencies that are higher than the *Nyquist frequency* $f_S/2$ from reaching the input of the ADC. Because it may not be convenient or practical to eliminate high-frequency noise, an analog *antialiasing prefilter* is often placed between the sensor and the ADC. In many cases a simple low-pass filter will suffice. The purpose of this filter is to attenuate the effect of higher-frequency noise components in the analog signal and prevent the noise from being aliased as a lower frequency by the sampling process. The filter should remove (ideally) all frequencies that are not less than half the sample frequency.

10.2.4 Amplitude Uncertainty

The ADC requires some time between the start and end of the conversion process, called *conversion time* T_C. If the value of the analog signal changes during this time, the output generated by the DAC will not correspond to the value of the input signal that existed at the start of the conversion process. If the change is faster than T_C, a possible conversion error might occur. This conversion error is referred to as *amplitude uncertainty*, or aperture error. If the input is a DC voltage or a slowly varying AC signal, the aperture error is less of a concern in comparison with the quantization error. However, if the input voltage changes rapidly, the aperture error can be significant, particularly if the period of the input signal is of the same order of magnitude as T_C. Figure 10.5 illustrates amplitude uncertainty and conversion errors. If the rate of change of the input voltage is dv_i/dt, the conversion error is

$$\Delta v_i = \frac{dv_i}{dt}T_C \tag{10.10}$$

Full conversion accuracy can be achieved only if the uncertainty is less than the ADC resolution; that is,

$$\left.\frac{dv_i}{dt}\right|_{MAX} = \frac{v_{MAX} - v_{MIN}}{(2^k - 1)T_C} \tag{10.11}$$

Conversion time has the same effect on closed-loop system stability as time delays.

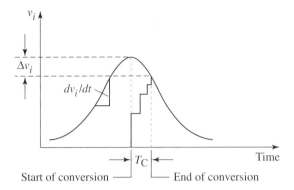

Figure 10.5 Conversion error due to a fast-changing input signal.

Example 10.4: Maximum frequency of input signal—Consider the sinusoidal input signal $v_i = v_0 \sin(2\pi ft)$ being converted by an 8-bit monolithic ADC that has a conversion time $T_C = 16$ μs. Determine the maximum frequency f of the signal for which an 8-bit representation can be found.

Solution: The rate of change of the input signal is

$$\frac{dv_i}{dt} = 2\pi f v_0 \cos(2\pi ft) \tag{10.12}$$

The maximum rate of change of the input voltage when $\cos(2\pi ft) = 1$ would be

$$\left(\frac{dv_i}{dt}\right)_{MAX} = 2\pi f v_0 \tag{10.13}$$

If we have a signal with full scale $FS = v_{max} - v_{min} = 2v_0$, then

$$2\pi f v_0 = \frac{2v_0}{(2^k - 1)T_C} \tag{10.14}$$

which gives the maximum frequency that could be processed as

$$f_{max} \leq \frac{1}{(2^k - 1)\pi T_C} \approx 78.02 \, \text{Hz}$$

This indicates that the 16-μs converter can find an 8-bit representation for signals with frequencies that are limited to less than 78 Hz. However, this restriction is eliminated if the signal is passed through a sample-and-hold circuit prior to its arrival at the input of the ADC.

10.2.5 Sample and Hold (S/H)

Decreasing conversion time is one way to reduce conversion error in the output code of a rapidly changing input signal. An alternative approach to this costly solution is to add the *sample-and-hold* (S/H) circuit shown in Fig. 10.6 in the signal path prior to the ADC input. The S/H circuit is an analog device controlled by a digital signal. It consists of an input op-amp follower, an electronic switch that can be a FET, BJT, or a diode bridge, a capacitor,

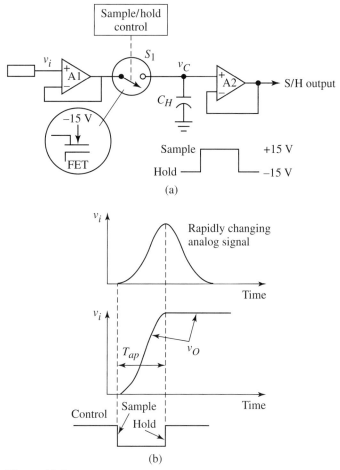

Figure 10.6 Sample-and-hold (S/H) circuit for an ADC.

and an output op-amp follower. The S/H circuit captures the analog input voltage at the instant when the conversion is required and holds it constant during the conversion process. This ensures that the ADC output value corresponds to the value of the input signal at the time the conversion began, regardless of the conversion time of the ADC. S/H circuits are also needed if the same ADC is used to convert multiple sensor signals. The S/H circuit holds a sample from one channel for conversion while a multiplexer proceeds to sample the next one.

When the electronic switch is closed (binary 0), the capacitor charges to the input voltage. The first op-amp follower A1 provides a high-impedance input path from the input signal to the capacitor and prevents loading the signal source. It also increases the corner frequency of the effective LPF formed from a series combination of the source resistance and the on-resistance of the FET switch, effectively increasing the signal frequencies that can be followed. The output impedance of A1 is extremely low and can drive a considerable load, including capacitive loads. Meanwhile, when the switch is open (binary 1), an open circuit exists between the A1 and the capacitor, and the voltage across the capacitor is held constant. Op-amp A2 provides a high-impedance output for the hold capacitance and has little effect on its charge. However, if the hold time is long, the capacitor will discharge gradually, or "droop," through the parallel combination of the off-resistance of the switch and the high input impedance of the follower.

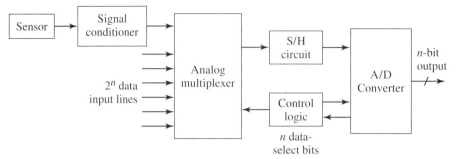

Figure 10.7 Multichannel data-acquisition system using an analog multiplexer and a single ADC.

This is not significant if the hold time is short, because the charge on the capacitor varies little. The output of the A2 will simply be identical to the original input.

With a S/H circuit used, the minimum sampling period becomes

$$T = T_C + T_{aq} + T_{ap} \tag{10.15}$$

T_{aq} is the acquisition time and T_{ap} is the aperture time. T_{aq} is the time needed for the S/H circuit to change from the hold to the sample mode to reacquire the signal. Aperture time refers to the elapsed time between a hold command and the actual holding of the signal.

A whole line of ADC ICs that can easily be interfaced with the MUC are available. Analog Devices' AD585 S/H IC has an internal capacitor C. It is characterized by an acquisition time of $t_{aq} = 3$ μs for a 0.01% accuracy in response to 10-V step. The LF398 IC contains all S/H components except the hold capacitor.

10.2.6 Multiple Sensor Inputs

A multiplexer circuit is employed when a single ADC is used to sample data from multiple sensors, as shown in Fig. 10.7. The multiplexer allows switching from one input signal to another. The n data-select lines of the multiplexer select the specific data output line to be routed to the input of the ADC from 2^n input lines. An input register or a binary counter usually handles channel selection. If n channels are sampled without an S/H circuit, the converted value of the last input may have varied considerably from its value when conversion from the first input started. This conversion error, also called *slewing error*, becomes n times that given by Eq. (10.10). Slewing error would render any control calculation inaccurate if all inputs are required to be sampled at precisely the same time. Slewing error may be eliminated either by using a separate ADC for each input or by adding an S/H circuit on each input.

10.3 A/D CONVERSION TECHNIQUES

10.3.1 Integrating ADCs

These converters work on the principle of averaging the input over fixed time intervals. The dual-slope and delta-sigma methods are two types of converters that are based on this technique. This technique provides accuracy, high resolution, and good noise rejection. However, it is relatively slow.

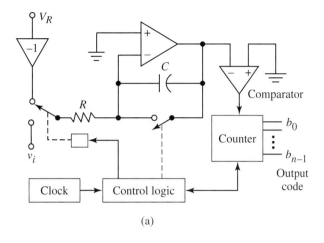

(a)

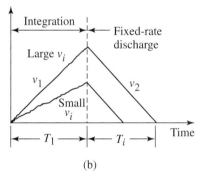

(b)

Figure 10.8 Block diagram of a dual-slope ADC (a) and circuit waveforms (b).

Dual-Slope Technique

The circuit in Fig. 10.8 shows the essential components of a dual-slope converter. In addition to control circuitry, the circuit includes a counter, an analog integrator, a voltage comparator, and a reference voltage. The conversion process begins by integrating the input voltage v_i for a fixed interval of time T_1, which is generally limited by the maximum count of the counter. Once the integration period completes, the output reaches $v_1 = (1/RC)T_1 v_i$ and the controller resets the counter and switches the input of the integrator to the negative reference voltage V_R.

The counter is reactivated to count again at the instant the output of the integrator begins to decrease linearly according to $v_2 = v_1 - (1/RC)t V_R$. The capacitor discharges completely to $v_2 = 0$, and the counter stops within a time interval T_i. Because v_i and V_R are constant,

$$T_1 v_i = T_i V_R \tag{10.16}$$

Therefore,

$$\frac{T_i}{T_1} = \frac{v_i}{V_R} \tag{10.17}$$

The time interval ratio represents the binary count relative to the counter's full count. Therefore the count at the end of T_i is the output word of the ADC.

Integrating over a fixed period T_1 eliminates noise at frequencies that are multiples of $1/T_1$ imposed on the input signal. Consequently, the dual-slope technique could be made to have excellent *normal mode rejection*, essential for measuring small DC voltages (e.g., thermocouple output) while ignoring a much larger 50- or 60-Hz noise signal impressed on the DC signal. However, achieving normal mode rejection comes at the expense of speed. For example, to suppress the 50- or 60-Hz line frequency noise, T_1 has to be at least 16.67 ms, too slow for typical fast data-acquisition applications. For this reason, dual-slope converters are generally used in applications requiring low sample rates, such as digital panel meters (DPMs), digital multimeters (DMMs), and temperature sensing.

Sigma-Delta Technique

Sigma-delta (Σ-Δ) converters are ideal for realizing high resolution. The main components of a first-order Σ-Δ ADC are shown in Fig. 10.9: They a modulator, a digital low-pass filter, and a decimation filter. The modulator consists of an integrator, a comparator, and a 1-bit D/A converter in a negative feedback loop. The difference between the analog input signal and the output of the 1-bit DAC is applied to the integrator. When the integrator's output voltage equals the comparator's reference voltage, the comparator output switches from high to low, or vice versa, depending on its original state. The comparator's output is clocked into both the 1-bit DAC and the digital filter stage. When the comparator changes its state, the 1-bit DAC changes its analog output voltage to the difference amplifier on the next clock pulse. This in turn changes the output voltage of the difference amplifier, causing the integrator's output to change in the opposite direction.

The modulator samples the analog input at a frequency many times the Nyquist rate (Eq. (10.9)) and converts the signal into a binary weighted digital output. The digital filter uses an oversampling and an averaging algorithm to achieve higher resolution. The combination of the digital and decimation filter stages also removes out-of-band quantization errors and reduces the sample rate and the amount of data for subsequent transmission, and the oversampling rates used reduce antialiasing requirements considerably.

State-of-the-art Σ-Δ designs contain a programmable gain amplifier, a multiorder Σ-Δ converter, a calibration microcontroller with on-chip static RAM, a clock oscillator, a programmable digital filter, and a bidirectional serial communications port. Examples of such devices are the Analog Devices AD7705 and AD7711 and Burr Brown ADS1210 and ADS1211 converters. They are used in process control primarily as direct transducer interface for pressure, temperature, flow, weigh scales, and force sensors. They are well suited for portable applications such as thermometers and gas and blood analyzers.

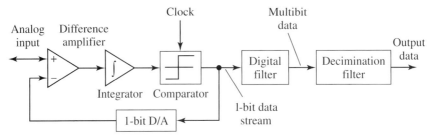

Figure 10.9 Components of a generic Σ-Δ ADC.

10.3.2 Successive-Approximation ADC

Figure 10.10 shows the basic components of a successive-approximation ADC and its operation for a 3-bit ADC. The circuit employs a D/A converter, a voltage comparator, reference voltage, and control circuitry to carry out fast conversion. An k-bit converter of this type sets all bits in the code successively, one bit at a time, each during one clock cycle, beginning with the MSB bit b_{k-1}. When the current bit b_k ($n = k - 1, \ldots, 0$) is set, the comparator compares the current estimated D/A output $V_F = (b_{k-1}2^{-k+1} + \cdots + b_n2^{-n}) \times V_R$ with the input signal voltage v_i. If $v_i > V_F$, the controller clears the current bit k in the code; otherwise it keeps it set. The next bit becomes current and the process continues until the LSB b_0 is tested. At that point, the best binary approximation of the input signal is achieved and constitutes the output digital word. Once a bit is tested, the appropriate level (0 or 1) is used in determining V_F when subsequent bits are tested. Thus the successive-approximation technique performs k comparisons before the best approximation is found. This process is demonstrated for a 3-bit converter in Fig. 10.10b. When the MSB is set, the initial estimate of the signal is ½FS of the input signal, corresponding to %100, as shown in Fig. 10.10c.

Converting a fast-changing input signal using the successive-approximation technique without an S/H circuit produces, at the time conversion is completed, an output signal that corresponds to the value of the signal at an earlier instant in time. Consequently, the samples appear to have been with different time intervals between them, as depicted in Fig. 10.11a. This phenomenon, called *jitter*, is more profound if the conversion time T_C is a significant portion of the sampling period T_S. An S/H circuit eliminates jitter, as illustrated in Fig. 10.11b.

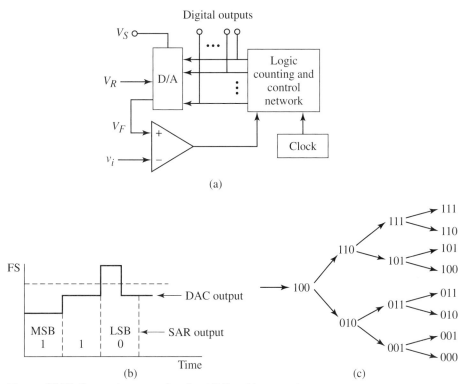

Figure 10.10 Successive-approximation ADC and its operation.

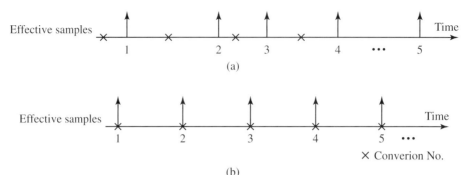

Figure 10.11 Effective sampling with S/H (a) and without S/H (b).

The successive-approximation ADC does not have high resolution (16 bit or less) but is much faster than the dual-slope technique. It can convert an analog signal to a 16-bit digital word in less than 10 μs.

Example 10.5: Successive A/D conversion—The output of a sensor is connected to a 4-bit successive-approximation ADC, with a reference of 10 V. What would be the ADC output if the sensor signal is 7.125 volts?

Solution: According to the procedure outlined, conversion is performed in four steps ($V_R =$ 10 V, $v_i =$ 7.125 V, V_F is from Eq. (10.4)):

Step 1: Set $b_3 = 1$, $V_F = 10 \times (2^{-1}) = 5$ V; since $v_i > 5$ V, keep $b_3 = 1$.
Step 2: Set $b_2 = 1$, $V_F = 10 \times (2^{-1} + 2^{-2}) = 7.5$ V; since $v_i < 7.5$ V, reset $b_2 = 0$.
Step 3: Set $b_1 = 1$, $V_F = 10 \times (2^{-1} + 2^{-3}) = 6.25$ V; since $v_v > 6.25$ V, keep $b_1 = 1$.
Step 4: Set $b_0 = 1$, $V_F = 10 \times (2^{-1} + 2^{-3} + 2^{-4}) = 6.875$ V; since $v_i > 6.875$ V, keep $b_0 = 1$.

The best approximation for $v_i = 7.125$ is the 4-bit digital word $(1011)_2$.

10.3.3 Flash ADC

The components of a typical flash ADC are shown in Fig. 10.12. An k-bit flash ADC employs 2^k comparators generating k outputs that are 1 LSB apart. All comparators are biased by equal voltages across a precision resistive network that divides the reference voltage V_R into 2^k equal parts. Conversion is completed in one step by comparing the input signal v_i with the threshold levels of all comparators simultaneously. The comparators that are biased above the input signal are turned off, whereas those biased below are left on. The output from the 2^k comparators are converted to an n-bit digital word by a $2k$-line-to-k-line priority encoder. The resolution of available flash ADCs is limited to six to eight bits because of the large number of required comparators, 255 for an 8-bit converter. However, they have the fastest speeds of all three ADC types, limited only by the propagation delays through the circuit. With 8-bit resolution, conversion rates up to 100 MHz are achievable, which makes flash ADCs suitable for applications requiring very high-speed conversions, such RF, video, radar, and digital oscilloscope applications. The relatively large amount of circuitry consumes large amounts of power.

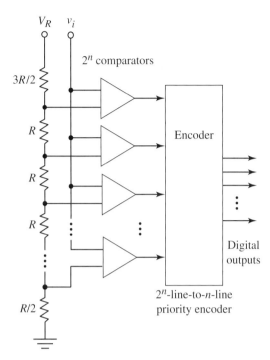

Figure 10.12 Flash ADC.

10.4 ADC FACILITY OF THE 9S12C MCU

The components of the ADC module onboard the 9S12C MCU are shown in Fig. 10.13. The ADC receives inputs through eight channels, labeled AN0–AN7, and shares pins 30–37 with bits PAD[7:0] of port AD at address $008F. The internal circuit that services the eight channels includes the following: a 10-bit successive-approximation ADC, an analog input multiplexer to route only one signal from the eight sensors at one time, and an S/H circuit to reduce the slewing error and to eliminate jitter in the sampling process, that is, to provide outputs that are equally spaced in time. The ATD may also be configured to operate at an 8-bit resolution to reduce power consumption. The 9S12C User's Manual uses ATD to refer to analog-to-digital, but ADC is a commonly used acronym to refer to analog-to-digital conversion. Both acronyms will be used hereafter.

10.4.1 Voltage References

The ADC requires a precise and stable reference voltage to accurately convert the input signal to a ratio. The high and low analog reference voltage levels are supplied through the V_{RH} (V_{ref}) and V_{RL} (V_{GND}) pins on the chip; V_{RL} is bonded to V_{SSA} on the 52-pin chip. An input voltage on any ANi channel that is equal to V_{RH} will be converted by the ADC configured for 8-bit conversion to $FF unsigned or to $7F signed output, and an input voltage level that is equal to V_{RL} will be converted to a $00 unsigned or to $80 signed data. The corresponding outputs of a 10-bit ADC are $0000–$FFC0 for unsigned data and $8000–$7FC0 for signed data.

The two reference voltages may fall within the range 0 to +5.12 volts, with $V_{RH} > V_{RL}$. The internal circuitry delivers the reference input voltages to the ADC, accounting for the voltage drop that may have occurred on the lines supplying power to the chip, for a more accurate

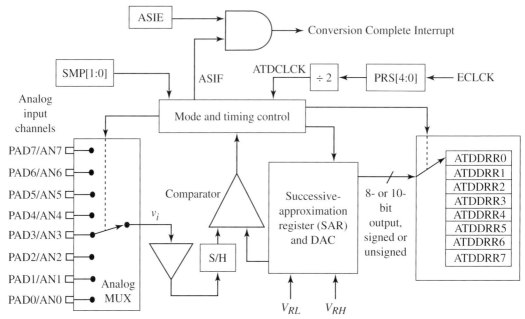

Figure 10.13 ADC facility onboard the 9S12C MCU.

conversion. The analog circuitry of the ADC is powered from the external power supply connected to the V_{DDA} and V_{SSA} pins on the chip.

10.4.2 ATD Registers

An overview of the registers involved in the ADC process is given in Fig. 10.14. Eight ATD result registers ATDDRR0–ATDDRR7 located at addresses \$0090–\$009F are used to store the conversion results from the eight channels; four ATD control registers ADCTL2–5 located at addresses \$0082–85 contain various bits that are used to configure different features of the conversion process: two ATD status registers, ATDSTAT0 and ATDSTAT1 at addresses \$0086 and \$008B, and one ATD interrupt enable register at \$008D.

10.4.3 ATD Setup

The four control registers ADCTL2–4 determine the operation of the ADC. The registers contain various bits that control associated features of the conversion process as outlined next.

The ADPU, *ATD power up*, bit 7 in the ADCTL2 register must be set once after reset to enable the ADC function. The MCU should then wait at least 100 μs for the ADC to stabilize before it is commissioned.

The SRES8, *ATD resolution select*, bit 7 in the ADCTL4 register determines the conversion resolution. A 10-bit resolution is intended if SRES8 is set to 1, and an 8-bit accuracy if SRES8 is set to 1.

Eight ATD *data result registers*, ATDDRR0–7, are available to store the ADC outputs. The way the result data is stored in a register depends on the value settings of the DJM and DSGN control bits in the ADCTL5 register. The DJM, *result register data justification*, bit 7 forces the data to be left justified if it is 0 or right justified if it is set. If the left-justified option is chosen, the 8-bit result, or bits 2–9 of a10-bit result will be stored in

ADCTL2: ATD Control Register 2 ($00xx)

ADPU	AFFC	AWAI	ETRIGLE	ETRIGP	ETRIGE	ASCIE	ACSIF

ADCTL3: ATD Control Register 3 ($00xx)

0	S8C	S4C	S2C	S1C	FIFO	FRZ1	FRZ0

ADCTL4: ATD Control Register 4 ($00xx)

SRES8	SMP1	SMP0	PRS4	PRS3	PRS2	PRS1	PRS0

ADCTL5: ATD Control Register 5 ($00xx)

DJM	DSGN	SCAN	MULT	0	CC	CB	CA

ATDSTA0: ATD Status Register 0 ($00xx)

SCF	0	ETORF	FIFOR	0	CC2	CC1	CC0

ATD STAT1: ATD Status Register 1 ($00xx)

CCF7	CCF6	CCF5	CCF4	CCF3	CCF2	CCF1	CCF0

ATDDIEN: ATD Data Enable Register ($00xx)

IEN7	IEN6	IEN5	IEN4	IEN3	IEN2	IEN1	IEN0

ATDDRR*x*H: ATD Result Register H (8-bit or Left Justified,10-bits) ($00xx)

BIT9	BIT8	BIT7	BIT6	BIT5	BIT4	BIT3	BIT2

ATDDRR*x*L: ATD Result Register L (Left Justified, 10-bits) ($00xx)

BIT1	BIT0	0	0	0	0	0	0

ATDDRR*x*H: ATD Result Register H (Right Justified, 10 bits) ($00xx)

0	0	0	0	0	0	BIT9	BIT8

ATDDRR*x*L: ATD Result Register L (8-bit or Right Justified, 10 bits) ($00xx)

BIT7	BIT6	BIT5	BIT4	BIT3	BIT2	BIT1	BIT0

Figure 10.14 Pins and registers dedicated to the operation of the ADC.

the ATDDRR*x*H, high byte of the ATD data register. Bits 0 and 1 of a 10-bit conversion will be stored in bit locations 6 and 7 of the ATDDRR*x*L, low byte of the data register. If the right-justified option is selected, the ATDDRR*x*L register will hold the 8-bit result of an 8-bit conversion or bits 0–7 of a 10-bit conversion. Bits 8 and 9 of a 10-bit conversion will be stored at bit locations 0 and 1 of the ATDDRR*x*H.

The DSGN, *result register data signed or unsigned representation*, bit 6 defines the polarity of the result data. If DSGN is set, the data result is signed; if clear, the data result is unsigned. Signed data is stored in 2s-complement format. This option is available only in the left-justified format.

The AFFC, *ATD fast flag clear*, bit 6 in the ADCTL2 register forces the CCF[7:0], *conversion complete flags* in the ATDSTAT1 register, and the SCF, *sequence complete flag*, bit 7 in the ATDSTAT0 register, to clear automatically after the result register is read. The AFFC bit is clear for normal flag clearing.

The AWAI bit is set to suspend ATD operation in the wait mode. If it is clear, the ATD continues to run in the wait mode.

The value settings of the FRIZ[1:0], *background freeze enable*, bits in the ATDCTL3 register determine the operating mode of the ATD system in the BDM; the value 00, 10, or 11 for FRIZ[1:0] causes the ATD either to continue to run, finish the current conversion sequence, and then freeze or to freeze immediately in the BDM.

The IENi, *input enable*, bit in the ATDIEN, *ATD input enable register*, at address \$008D is used to control the digital input buffer from the ANi channel to PTADx data register. Setting IENi while the ANi is used as an analog input causes potentially higher power consumption.

10.4.4 Conversion Time

The conversion time consists of two phases. In the first phase, the input sample is transferred to the ATD node via the buffer amplifier in two ATD clock cycles. In the second phase, the ATD places the external analog signal onto the storage node for final charging and increased accuracy. The second phase takes 2, 4, 8, or 16 clock cycles if the coded value in the SMP[1:0], *sample time select*, bits in the ADCTL4 register is, respectively, 00, 01, 10, or 11. The ATD clock may further be divided by a prescale (*PRS*) value between 2 and 128, depending on the value settings of the PRS[4:0], *prescale select,* bits in the ADCTL4 register. For a PRS[4:0] code between 00000 and 11111, the prescale setting is between 1 and 32, respectively. The frequency of the ATD clock is determined by dividing the bus (or ECLK) frequency by $[PRS + 1] \times 2$; that is,

$$ATDCLK = 0.5 \times \frac{ECLK}{PRS + 1}$$

The allowable ECLK frequency is $([PRS + 1] \times 2 \div 2) < ECLK < ([PRS + 1] \times 2 \times 2)$ MHz. The *PRS* value out of a reset is 5, dividing the *ECLK* by 12, for an allowable range of $6 < ECLK\ 24 <$ MHz. If the nominal bus frequency is 16 MHz, the default ATD clock frequency is 4/3 MHz for a period of 750 ns. However, $2 < ATDCLK < 5$ MHz. If the *ATDCLK* is slower than 5 kHz, charge leakage in the converter begins to affect conversion accuracy.

10.4.5 Channel Selection

The ATD module in the 9S12C MCU carries out a sequence of *NC* consecutive conversions, from 1 to 8, depending on the value settings of the S8C, S4C, S2C, and S1C, bits 6, 5, 4, and 3 in the ADCTL3 register. As indicated in Fig. 10.15, the *NC* samples can be

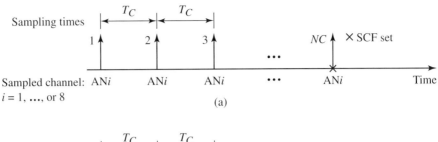

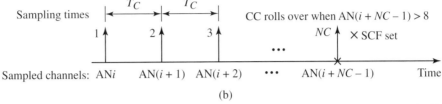

Figure 10.15 Conversion from a single-channel AN*i* (a) and from four different channels (b).

either from a single input channel or from multiple channels, depending on the value of the MULT, *multiple channel control*, bit 4 in the ADCTL5 register. If MULT = 0, the ADC operates in the single-channel mode, converting all *NC* samples from the channel selected by the value settings of the CC, CB, and CA, bits 2, 1, and 0 in the ADCTL5 register, as indicated in Fig. 10.16.

If the MULT bit is set to 1, the ADC operates in the multichannel mode, converting *NC* samples from *NC* channels. The first sampled analog channel is selected by the CC, CB, CA code bits. Subsequent channels are determined by incrementing the channel selection code bits. For example, if MULT = 1, CC = 0, CB = 0, CA = 0, and the S8C-S1C are set for *NC* = 4, the sampled channels will be AN0, AN1, AN2, and AN3, in that order.

10.4.6 Channel Sampling and Conversion Results

The ADC system may be configured to carry out the *NC* conversions only once and stop or to continuously scan the *NC* input channels as determined by the SCAN, *continuous scan control*, bit 5 in the ADCTL5 register. If the SCAN bit is clear, the ADC stops after *NC* conversions; otherwise the conversion process continues, as shown in Fig. 10.17. After each conversion from channel AN*i*, the associated CCF*i* flag in the ATDSTAT1 register is set and the result is placed in the ATDDRR*x* register. At this point the CPU may poll the CCF*i* flag; if the flag is set,

CC	CB	CA	Channel select
0	0	0	AN0
0	0	1	AN1
0	1	0	AN2
0	1	1	AN3
1	0	0	AN4
1	0	1	AN5
1	1	0	AN6
1	1	1	AN7

Figure 10.16 ADC channel select options.

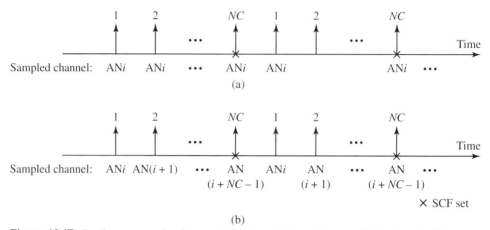

Figure 10.17 Continuous scanning from a single channel (a) and from multiple channels (b).

the program reads the result register. The CCFi flag clears under one of three conditions: a write of 1 to the CCFi location, a read of the ATDSTAT1 register followed by a read of the ATDDRRi register with the AFFC bit set to 0, or a write to the ADCTL5 register with AFFC = 1.

If the ATDDRRx register is overwritten before it is read, the FIFOR, *FIFO overrun flag*, bit 4 in the ATDSTAT0 register sets, indicating an overrun condition has occurred. The FIFOR flag may be cleared in one of two ways: by writing 1 to the FIFOR bit location or by starting a new conversion sequence.

At the end of a conversion sequence, the ADC sets the SCF, *sequence complete flag*, in the ADCTL2 register. The CPU may poll the SCF flag and, if set, directs the program to retrieve the result from the result registers. Alternatively, an interrupt may be enabled to occur when a conversion sequence completes by setting the ASCIE, *ATD sequence complete interrupt enable*, bit 1 in the ADCTL2 register. If ASCIE is set to 1, the ASCIF, *ATD sequence complete interrupt flag*, bit 0 in the ATDCTL2 register sets concurrently with the SCF flag. The CPU must process the results stored in the ATDDRx register before they are overwritten. Because the program has no control over the sampling rate if the SCAN bit is 1, the maximum sampling rate should be used or the program should retrieve the *NC* samples in a timely manner. The SCF flag clears under one of three conditions: a write of 1 to the SCF location, a read of the ATDDRRi register, or a write to the ADCTL5 register with AFFC = 1.

The conversion results are placed in the result registers in sequential order under the control of the conversion counter (CC). When the conversion sequence completes, the CC resets, and the same result registers are used all over again if the FIFO bit in the ATDCTL3 register is clear. Setting the FIFO bit to 1 configures the ATD to operate in the FIFO mode. In this mode the CC does not reset when a conversion sequence completes. Consequently, the result of the first conversion of a new sequence will be placed in the next available data result register pointed to by the contents of the CC[2:0] bits in the ATDSTAT0 register. For example, if CC[2:0] contains 110, the next conversion result will be placed in the ATDDRR6 register. The CC[2:0] rolls over from 111 to 000 when the *NC* result register block is used up. If a conversion sequence is aborted or a new conversion starts, the CC[2:0] bits will clear, even if FIFO = 1, and the result of the first conversion is placed in the ATDDRR0 register. FIFO mode is intended for use with continuous-scan or external-trigger conversion modes. The FIFOR flag will set if a result register is overwritten before the CCFx flag is cleared.

External ADC Trigger

The ATD operation may be synchronized to occur when an external trigger input is received on ATD channel 7. Channel 7 becomes unavailable in this mode. The external-trigger mode is enabled by setting the ETRIGE, *external trigger mode enable*, bit 2 in the ATDCTL2 register. The sensitivity of the external trigger is controlled by the ETRIGLE, *external trigger level/edge control*, bit 4 in the ATDCTL2 register. If ETRIGLE is set, the trigger is level sensitive; if it is clear, the trigger is edge sensitive. Furthermore, the polarity of the trigger is set by the ETRIGP, *external trigger polarity*, bit 2 in the ATDCTL2 register. If the bit codes of the ETRIGLE and ETRIGP bits are 00 or 01, trigger will occur on a falling edge or on a rising edge, respectively; whereas bit codes of 10 and 11 set the trigger to a low level or to a high level, respectively. If a new active trigger is detected while a conversion is in progress, the ETORF, *external trigger overrun flag*, bit 5 in the ATDSTAT0 will set.

Example 10.6: ATD setup–The following subroutine may be added to any code that requires the use of the ADC facility. The subroutine powers up the ADC facility and selects the clock source.

```
atd_setup
            LDAA    #$80
            STAA    adctl2      ; Normal ATD function, normal
                                flag clearing, run in wait
                                mode,   ; no external trigger-
                                ing, and no interrupts
            JSR     delay       ; Execute time delay for
                                100 µs for the ATD to become
                                stable
            LDAA    #$20
            STAA    adctl3      ; Conversion sequence NC = 4,
                                result placed in data result
                                registers ; sequentially,
                                continue to run in BDM
            LDAA    #$87
            STAA    adctl4      ; Set conversion accuracy to
                                8 bits, use 2 ATDCLCKs for
                                the second ; conversion
                                phase, set PRS to 7 to divide
                                ECLK by 16.
            RTS
                                ; Subroutine to generate the
                                100-µs delay
delay
            LDAA    #$C8
loop        DECA
            BNE     loop
            RTS
```

Example 10.7: Multiple-channel, continuous-scan operation—Write a code to manage four conversions of four unsigned analog signals, connected to the AN0, AN1, AN2, and AN3 channels, into digital form. The ATD operates in multiple-channel continuous-conversion mode.

Solution: The following program handles the required conversion.

```
              INCLUDE "mc9s12c32.inc'  ; include 9S12C macros
              ORG      $C000
nsets         DC.B     16
result        DS.B     64                 ; Reserve 64 bytes to store
                                          conversion result
;
              ORG      $8000
              JSR      atd_setup          ; Set up the ADC (see
                                          Example 10.6)
start         LDAA     #$30               ; SCAN = 1, MULT = 1, and
                                          select AN0-AN3 channels
              STAA     adctl5             ; Start conversion
              LDAB     nsets              ; Initialize B as a
                                          counter
              LDX      #result            ; Point to the start of
                                          results block
clrscf        LDAA     atdstath
              ORAA     #$10
              STAA     atdstath           ; Clear SCF flag
wait          BRCLR    atdstath, #$80, wait; Poll SCF flag to set
              LDAA     atddrr0h           ; Read conversion result
                                          from ATDDRR0H and
              STAA     1, X+              ; save it at address
                                          result
              LDAA     atddrr1h           ; Read conversion result
                                          from ATDDRR1H and
              STAA     1, X+              ; save it at address result
                                          + 1
              LDAA     atddrr2h           ; Read conversion result
                                          from ATDDRR2H and
              STAA     1, X+              ; save it at address result
                                          + 2
              LDAA     atddrr3h           ; Read conversion result
                                          from ATDDRR3H and
              STAA     1, X+              ; save it at address result
                                          + 3
              DECB
              BNE      clrscf
              END

atd_setup
              .........                   ; Include ATD setup code
                                          here (Example 10.6)
```

Example 10.8: C-code version–Write a code to store 20 samples from channels AN0–AN3 into RAM for later use.

Solution

```
#include <hidef.h>                /* common defines and macros */
#include <mc9s12c32.h>           /* derivative information */

#pragma LINK_INFO DERIVATIVE "mc9s12c32"
void delay (void)
   {
     int dlytime = 0x80;         /* Generate time delay */
     while ((dlytime = dlytime-1) !=0) {
           }
   }
void atdsetup (void){
          ATDCTL2 = 0x80;
          delay();               /* Call delay */
          ATDCTL3 = 0x20;
          ATDCTL4 = 0x87;
        }
void main(void)
        {
          int nsets = 16,result[64], i = 0, k;
          atdsetup();
          ATDCTL5 = 0x30;
          while ((nsets = nsets − 1) ! = 0)
          {
            k = i*4;
            ATDSTAT1 = ATDSTAT1 | 0x10; /* clear SCFFag */
            while ((ATDSTAT1 & 0x80) = = 0)   {
            }
            result[k]     = ATDDR0H;
            result[k + 1] = ATDDR1H;
            result[k + 2] = ATDDR2H;
            result[k + 3] = ATDDR3H;
          }
          i++;
        }
```

10.4.7 Input Signal Range

Obviously the ADC onboard the 9S12C converts unipolar analog levels between V_{RL} and V_{RH}. Usually, the input signal does not span the full ADC input range. To take full advantage of the converter's dynamic range and reduce the relative effect of the converter's error on the output, the input signal is brought to within the ADC full scale by inserting an appropriate interface circuit between the signal source and the ADC input. The analog input signal v_S and the ADC input are assumed to have a linear relationship, as outlined in Example 10.1. The relation is then implemented by an appropriate op-amp circuit similar to that in Example 4.2.

Applications that require polarity information in the digital output should use a bipolar converter, since the ADC in the 9S12C is unipolar.

Example 10.9: Utilizing the full range of an ADC—A sensor generates an output that varies within the range $-5 \leq v_S \leq +5$ V. The sensor signal is to be processed by the unipolar ADC of the 9S12C. Design a signal-conditioning circuit to be placed between the sensor output and the ADC input to utilize the full ADC range.

Solution: The linear relation between the sensor output and the ADC input in this case will be $v_i = 0.5(v_S + 5)$ V. This relation can be realized by a summing op-amp with two inputs and a gain of 0.5, as shown in Fig. 10.18; one input is the sensor signal and the other is a $+5$-V offset.

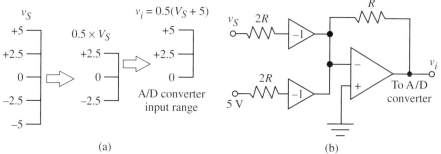

(a)   (b)

Figure 10.18 Op-amp circuit to convert a bipolar signal using a unipolar ADC.

Example 10.10: Light meter—It is desired to develop a 9S12C-based light meter to measure the intensity of ambient light using a photocell. The output from the photocell is interfaced to AN0 of port AD. Develop a source code that reads the light intensity from the ATDDRR0H register and displays it on two seven-segment displays connected to port T, as in Fig. 10.19.

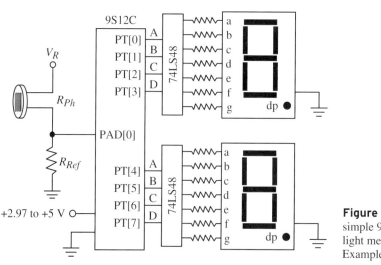

Figure 10.19 A simple 9S12C-based light meter for Example 10.10.

Solution: The following code uses a lookup table containing 128 bytes corresponding to the full range of the ADC. The 128 bytes corresponds to 64 possible displays on two seven-segment displays. The code continuously reads the output of the ADC stored at ATDDRR0H and updates the displays as hex digits accordingly. Note that 0 V corresponds to bright light and $FF represents complete darkness.

```
          INCLUDE    'mc9s12c32.inc'  ; Include register equates
          INCLUDE    'atd_setup.asm'  ; Include ATD setup routine

          ORG        $C000
          MOVB       #$FF, ddrt    ; Configure all port T pins as
                                   outputs

          JSR        set_up        ; Set up ADCTL2-4 (Refer to
                                   Example 10.1)
          LDAA       #$20          ; SCAN = 1, MULT = 0, conversion
                                   at AN0
          STAA       adctl5        ;
start     LDAA       atddrr0h      ; Load the contents of ATDDRR1H
                                   into A
          COMA                     ; Take its complement to increase
                                   display number with light
                                   intensity
          LSRA                     ; The two LSRA instructions
                                   convert the 8-bit number to a
                                   6-bit number,
          LSRA                     ; since 256 ADC outputs are repre-
                                   sented by 64 displays.
          LDAB       #$02          ; Since the 64 display codes are
                                   stored in 128 memory locations,
                                   two ; for each code, we need to
                                   define a LOW byte to indicate the
                                   location ; of the number on page
                                   $C1 to display.
          MUL                      ; B will hold the address on Page
                                   C1 to display
          LDAA       #$C1          ; Loads page address $C1 of the
                                   lookup table into A
          XGDX                     ; Exchange contents of X with D
          MOVW       X, ptt        ; Load two display code bytes and
                                   write them to port T
delay     LDY        #$FFFF        ; The next three lines provide
                                   time delay for displays to be read
loop      DBNE       Y, loop
          BRA        start

; Lookup table for the 65 display codes stored at address $C100
          ORG        $C100
table     DC.B       $ED,$B7,$ED,$D5,$ED,$76,$ED,$57,
                     $ED,$C5, $ED,$D3            ;00 - 05
```

```
        DC.B        $ED,$F3,$ED,$15,$ED,$F7,$ED,$D7,
                    $A0,$B7,$A0,$D5                          ;06 - 11
        DC.B        $A0,$76,$A0,$57,$A0,$C5, $A0,$D3,
                    $A0,$F3,$A0,$15                          ;12 - 17
        DC.B        $A0,$F7,$A0,$D7,$6E,$B7,$6E,$D5,
                    $6E,$76,$6E,$57                          ;18 - 23
        DC.B        $6E,$C5, $6E,$D3,$6E,$F3,$6E,$15,
                    $6E,$F7,$6E,$D7                          ;24 - 29
        DC.B        $EA,$B7,$EA,$D5,$EA,$76,$EA,$57,
                    $EA,$C5,$EA,$D3                          ;30 - 35
        DC.B        $EA,$F3,$EA,$15,$EA,$F7,$EA,$D7,
                    $A3,$B7,$A3,$D5                          ;36 - 41
        DC.B        $A3,$76,$A3,$57,$A3,$C5,$A3,$D3,
                    $A3,$F3,$A3,$15                          ;42 - 47
        DC.B        $A3,$F7,$A3,$D7,$CB,$B7,$CB,$D5,
                    $CB,$76,$CB,$57                          ;48 - 53
        DC.B        $CB,$C5,$CB,$D3,$CB,$F3,$CB,$15,
                    $CB,$F7,$CB,$D7                          ;54 - 59
        DC.B        $C7,$B7,$C7,$D5,$C7,$76,$C7,$57,
                    $C7,$C5                                  ;60 - 64
; End of look-up table
```

10.5 DIGITAL-TO-ANALOG CONVERSION (DAC)

Figure 10.20 shows a block diagram of an MCU-based control application. The ADC generates a digital word corresponding to a physical external condition perceived by the sensor. Based on the level of the physical quantity, the MCU makes a control decision and sends it in the form of a control word to the DAC. The DAC converts the digital word into an equivalent analog voltage proportional to that number, as illustrated in Fig. 10.21.

A DAC can be used to generate any kind of waveform. However, the DAC output is a staircase approximation of the desired continuous waveform and must be filtered to reconstruct the analog signal. A common application of DACs is to generate signals to drive various types of actuators, including DC motors, heating coils, valves, and any type of device requiring *proportional* output rather than on–off control.

The 9S12C MCU does not provide a DAC output. However, the PWM module can generate high-resolution waveforms (refer to Section 9.7) that represent a special form of the DAC.

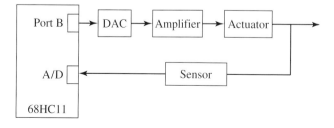

Figure 10.20 Block diagram of a simple MCU-based control application.

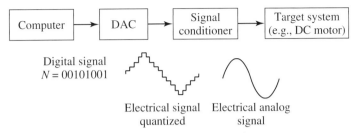

Figure 10.21 Digital-to-analog conversion (DAC) process.

10.5.1 Components of a D/A Converter (DAC)

Figure 10.22 shows the block diagram of a typical k-bit DAC. Its basic elements include a set of k latches, a precise voltage reference, a resistor network, switches, and an op-amp. The latches hold the binary number to be converted to the analog level. The binary output of a latch controls a corresponding single-pole, double-throw switch connected to a particular resistor in the resistor network. The switches are usually implemented using BJTs or FETs. Each input resistor is twice as big as the one preceding it. This construction is called *binary weighted ladder.* A switch is ON or OFF, depending on the corresponding bit in the input digital word. The reference voltage V_R connected to the resistor network (usually $+2.56$V, $+5$V, or $+10$ V) controls the range of the output voltage. The accuracy of a DAC output depends on maintaining a precise reference voltage. The output op-amp is a summing amplifier to add the signals through the activated switches. The op-amp also functions as a current-to-voltage converter because it sums (and converts to voltage) the binary weighted currents flowing through the resistors.

10.5.2 Output Voltage

Figure 10.23 shows the basic design of a k-bit unipolar binary weighted-resistor network DAC. The values of the op-amp input resistors are selected to ensure that the output voltage is

$$v_O = V_R\left(\frac{R_f}{R/2^{k-1}}b_{k-1} + \frac{R_f}{R/2^{k-2}}b_{k-2} + \cdots + \frac{R_f}{R/2}b_1 + \frac{R_f}{R}b_0\right) \qquad (10.18)$$

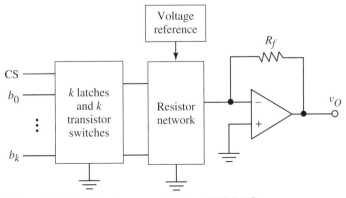

Figure 10.22 Block diagram of a typical k-bit DAC.

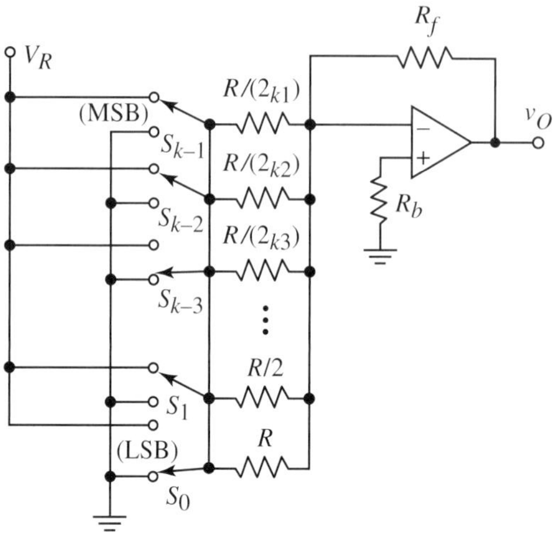

Figure 10.23 Basic weighted-resistor network DAC.

where V_R is the output voltage of the D flip-flops corresponding to logic-1 (nominally +5 volts) and $b_{k-1}, b_{k-2}, \ldots, b_1, b_0$ are the 0 or 1 binary outputs of the flip-flops. Equation (10.18) may also be written as

$$v_O = V_R \frac{R_f}{R} N \tag{10.19}$$

where

$$N = b_{k-1} 2^{k-1} + b_{k-2} 2^{k-2} + \cdots + b_1 2^1 + b_0 2^0 \tag{10.20}$$

N in Eq. (10.20) is the binary number provided by the CPU as input to the DAC. Therefore, the output of the DAC is proportional to N. The gain of the DAC can be adjusted as desired by changing R or R_f. The values of the smaller input resistors to the op-amp must have higher precision to produce the desired linear relation in the output voltage. In general a k-bit converter requires k resistors, and the LSB input resistor must be 2^{k-1} times greater than the MSB resistor. Switch resistance must also be selected to be smaller than $1/2^k$ times the smallest input resistor.

10.5.3 Range

The range of the DAC output is the difference between the maximum output voltage V_{MAX} and the minimum output voltage V_{MIN}. The output voltage of the DAC is maximum when all bits are 1. In this case,

$$N = 2^k - 1 \tag{10.21}$$

and

$$V_{MAX} = V_R \frac{R_f}{R} (2^k - 1) \tag{10.22}$$

The output voltage of the DAC is minimum when all bits are 0. In this case $N = 0$ and $V_{MIN} = 0$.

10.5.4 Resolution

The minimum nonzero voltage that can be generated by a DAC is obtained with $N = 1$; that is,

$$V_Q = V_R \frac{R_f}{R} \tag{10.23}$$

where V_Q represents the resolution of the DAC. The minimum number of bits required from the DAC for a specified resolution and range is

$$k = \frac{\log\left(\dfrac{V_{MAX} - V_{MIN}}{V_Q} + 1\right)}{\log(2)} \tag{10.24}$$

The result must be truncated to the next highest integer value because k must be an integer.

10.5.5 Accuracy

The accuracy and linearity of conversion by the circuit shown in Fig. 10.24 are highly dependent on the precision of the op-amp input resistors. A weighted-resistor DAC requires resistor values encompassing the range from R to $(2^{k-1})R$. This range may be too large (2048:1 for a 12-bit DAC) where high accuracy is difficult to achieve. The *R–2R ladder network* shown in Fig. 10.24 overcomes this disadvantage, at the expense of adding a resistor for each bit. The actual values of the resistors are not critical, but precise resistor matching is. It can be shown that Eq. (10.18) equally applies to this configuration. The *R–2R* ladder network results in improved accuracy and low cost and is commercially available in sizes from 8 to 16 bits.

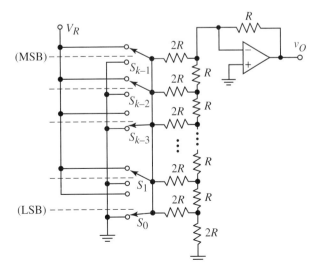

Figure 10.24 Basic $R-2R$ DAC converter.

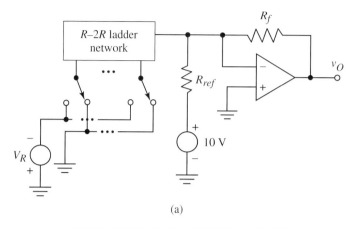

(a)

Unipolar		Bipolar	
V_O	Binary	2's compliment	V_O
0	000	100	$-V_{FS}$
	001	101	
	010	110	
	011	111	
	100	000	
	101	001	
	110	010	
$+V_{FS}$	111	011	$+V_{FS}$

(b)

Figure 10.25 Block diagram of a bipolar DAC (a) and corresponding codes (b).

10.5.6 Bipolar DACs

Many applications require the DAC to provide signed outputs, both positive and negative voltage values. This can be achieved by adding to the unipolar DAC an extra resistor, R_{ref}, and an additional reference power source, V_{ref}, as shown in Fig. 10.25. The 10-V reference voltage V_{ref} supplies a constant current through R_{ref} to the inverting input of the amplifier. This current is adjusted so that its value is equal to the current associated with the most significant bit (MSB). This in effect adds a bias voltage of $-V_R(R_f/R)(2^{k-1})$, which represents the MSB and indicates polarity to the output voltage v_O.

Negative numbers are represented by their 2s-complement counterparts (see Appendix G). In this case, the 2s-complement number zero (10000000 for 8-bit DAC) produces the expected $V_O = 0$. Also,

$$V_{MAX} = V_R \frac{R_f}{R} (2^{k-1} - 1) \tag{10.25}$$

and

$$V_{MIN} = -V_R \frac{R_f}{R} (2^{k-1}) \tag{10.26}$$

The 3-bit binary and the 2s-complement binary codes for a unipolar and a bipolar DAC are given in Fig. 10.25b.

Example 10.11: D/A conversion—The 10-bit converter constructed as shown in Fig. 10.23 outputs an unsigned decimal number 261. The reference voltage is 10 volts, and $R_f = R/2^k$, one-half the value of the smallest input resistor, $R/2^{k-1}$. Determine the output voltage, the maximum voltage, the resolution, and the number of bits required to enhance the resolution to 0.0025 volts.

Solution: The 10-bit binary equivalent of 261 decimal is $(0100000101)_2$. Hence from Eq. (10.18), the DAC output is

$$v_O = 10\left(\frac{1}{2} \times 0 + \frac{1}{4} \times 1 + \frac{1}{8} \times 0 + \cdots + \frac{1}{128} \times 0 + \frac{1}{256} \times 0\right.$$
$$\left. + \frac{1}{512} \times 0 + \frac{1}{1024} \times 1\right)$$
$$= 2.55 \text{ volts}$$

The maximum voltage, from Eq. (10.22), is

$$V_{MAX} = 1 \times \frac{1}{1024} \times (2^{10} - 1)$$
$$= 9.99 \text{ volts}$$

The range of the output voltages is therefore from 0 to 9.99 volts, and the resolution can be found, from Eq. (10.23), as

$$V_Q = 10 \times \frac{R}{2 \times (2^9 R)}$$
$$= 0.00977 \text{ volts}$$

To improve the resolution to 0.0025 volts, the number of bits required in the DAC can be found from Eq. (10.24):

$$k = \frac{\log\left(\dfrac{10}{0.0025} + 1\right)}{\log(2)} = 11.97 \text{ bits}$$

A 12-bit converter would therefore be needed to satisfy the required range and resolution.

10.5.7 DAC ICs

Analog Devices, Burr-Brown, and other manufacturers provide DAC ICs that contain all necessary circuitry to complete the conversion process for a wide range of applications. The user is required to provide a reference voltage for some DAC ICs and op-amps to shift voltage levels or to convert current to voltage.

The DAC-08 is an 8-bit $R–2R$ ladder DAC that generates two unipolar current outputs, I_O and its complement, with a full-scale output current of 2 mA. A precision reference voltage source and a precision resistance with a low temperature coefficient of resistance (TCR) may be used to generate the required reference current I_{ref}. For example, with $V_{ref} = 10$ V and $R_{ref} = 5000\ \Omega$, the DAC output is made TTL compatible. In this case, a 0- to 0.8-V output represents a LOW and a $+2.4$- to $+5$-V output represents a HIGH. Alternatively, a constant-current source may be used. Current-to-voltage ($I–V$) conversion may be achieved by connecting a resistor between the output and ground or by using an inverting follower op-amp with a feedback resistor. A shunt capacitor across the feedback resistor provides limited filtering (on the order of 6 dB/octave) and smooths out the output signal.

10.6 SUMMARY

Signal conversion is needed in mechatronics applications that typically control a continuous process via a digital controller. This chapter presented the concepts and operations of various analog-to-digital and digital-to-analog conversions. The material covered the necessary practical interfaces of ADC and DAC. The onboard MC9S12C A/D facilities were discussed in detail, with examples of programming the MCU for operating the ADC ports. The 9S12C does not have DAC facilities. However, the available PWM was discussed as a useful interface between the digital and analog parts of a system. In addition, this chapter presented many commercial DAC ICs, with their specs and part models. The chapter also included many examples of ADC and DAC applications integrated with the 9S12C programming.

RELATED READING

D. G. Alciatore and M. B. Histand, *Introduction to Mechatronics and Measurement Systems*, 2nd ed. New York: McGraw-Hill, 2002.

S. F. Barrett and D. J. Pack, *Embedded Systems: Design and Applications with the 68HC12 and HCS12*. New York: Prentice Hall, 2004.

J. Bollinger and N. A. Duffee, *Computer Control of Machines and Processes*. Reading, MA: Addison-Wesley, 1988.

HCS12 Microcontrollers: MC9S12C128 Data Sheet, 2005 (MC9S12C128_V1.pdf).

H.-W. Huang, *HC12/9S12: An Introduction to Hardware and Software Interfacing*. Clifton Park:. Thomson Delmar Learning, 2005.

C. D. Johnson, *Process Control Instrumentation Technology*, 7th ed. New York: Prentice Hall, 2002.

J. B. Peatman, *Design with Microcontrollers*. New York: McGraw-Hill, 1988.

P. Spasov, *Microcontroller Technology: The 68HC11*, 2nd ed. New York: Prentice Hall, 1996.

QUESTIONS

10.1 What is meant by quantization and quantization error?

10.2 Define *offset, span, step size, resolution*, and *full scale*.

10.3 Explain what aliasing is, what causes it, and how it can be avoided.

10.4 Explain amplitude uncertainty.

10.5 Describe the components of an ADC.

10.6 What is the role of the multiplexer in the ADC process?

10.7 What is the role of the sample-and-hold circuit in the ADC process?

10.8 Describe the operation of the successive-approximation ADC technique.

10.9 Describe the operation of the dual-slope ADC technique.

10.10 Explain the procedure to power-up the ADC facility on the 9S12C.

10.11 How long must the program delay before using the ADC after power-up?

10.12 The A/D converter is programmed to convert four channels in continuous-conversion mode. What is the maximum-frequency signal on PE0 that can be converted without aliasing? (Ignore the aperture-time effect).

10.13 Describe the operation of the weighted-resistor network DAC.

10.14 Describe the R–$2R$ ladder network DAC.

PROBLEMS

10.1 Give the digital outputs of an 8-bit ADC ($k = 8$) to the analog values $V_X = -5, -3.75, -2.5, -1.25, 0, 1.25, 2.5, 3.75$, and 4.961 for an analog range of -5 to $+5$ volts.

10.2 Give the bit weights for converting from -5 to $+5$ volts analog to 8-bit digital values.

10.3 A process is to be manipulated using a DAC with a range of ±10 volts and a resolution 0.05 volts. How many bits are needed in the DAC?

10.4 A DAC is to be used to deliver velocity commands to a motor. The maximum velocity is to be 3000 rpm, and the minimum nonzero velocity is 2 rpm. How many bits must be present in the DAC for bidirectional operation?

10.5 A pressure sensor generates 0.25–4.75 volts of output over a full-scale pressure of 15 psi. Determine the sensor resolution if the sensor is interfaced to the ADC of the 9S12C.

10.6 The output of a buffer op-amp is connected to a 12-bit ADC with a 3-V range. If the op-amp's source resistance is 1 MΩ and its bias current is 10 nA, determine the number of A/D conversion counts contributed by the input bias current.

10.7 A rotary potentiometer is to be used as a remote rotational displacement sensor. The maximum angular displacement to be measured is 180°, and the potentiometer is rated for 10 volts and 270° of rotation.

a. What voltage increment must be resolved by an ADC in order to resolve an angular displacement of 0.75°? How many bits would be required in the ADC for full-range detection?

b. The ADC requires a 10-volt input voltage for full-scale binary output. If an amplifier is placed between the potentiometer and the ADC, what amplifier gain should be used in order to take advantage of the full range of the ADC?

c. Design the circuitry needed to interface the potentiometer to the 9S12C.

d. What is the displacement if the corresponding digital output is $49?

10.8 The ADC of the 9S12C is monitoring a sinusoidal signal with a 3-V peak amplitude. Determine the maximum frequency that can be tracked.

10.9 The rotational velocity of a motor is controlled with a proportional digital controller using the output voltage of a tachometer sampled at a rate of 120 Hz with an ADC. Motor velocity control voltage is issued using a DAC. The motor operates over a speed range of 2000–3000 rpm in one direction only.

a. It is necessary to achieve a velocity resolution of 1 rpm. How many bits are required in the ADC?

b. It is observed that the motor velocity oscillates at a frequency of about 1 Hz during operation at 2220 rpm. The tachometer is known to have 13 commutator bars. What is likely to be the source of the oscillation? How can it be corrected?

10.10 A dual-slope ADC operates with a reference voltage of 5 V. The integrator employs a 50-kΩ resistor and a 0.01-μF capacitor. Determine the conversion time for a 3.5-V input signal if the fixed integration time is 10 ms.

10.11 The position of an object manipulated by a Cartesian robot is ascertained from the measurement of the x-, y-, and z-position sensors. The output from the sensors is to be interfaced to a computer and sampled at a rate of 100 Hz via an ADC. The sampled data will be used to find $x = f(z)$ and $y = g(z)$. The maximum x, y, and z velocities are 0.02 m/s, the ADC has one analog input, and its conversion time is 70 μs. Complete the A/D conversion system shown in Fig. 10.26, and explain the functional role of each component that you added.

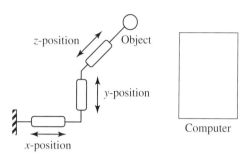

Figure 10.26

10.12 A process involves the measurement of temperature from 40 to 100°C scaled at 40 mV/°C, pressure from 1 to 100 psi scaled by 100 mV/psi, and flow from 3 to 90 gal/min scaled at 150 mV/(gal/min). Develop a complete signal-conditioning system so that these signals can be connected as *inputs* to the ADC of the 9S12C. Specify the resolution of each measurement in terms of the change of each variable that corresponds to an LSB change of the A/D output.

10.13 A 50-Hz signal is sampled at 1.5 Hz. What is the lowest apparent frequency in the sampled data that arises from aliasing? What range of sample frequencies ensures that aliasing does not occur when a 50-Hz signal is being sampled?

10.14 A 10-bit ladder resistor DAC is built, with $V_R = 5.12$ V and $R = 50$ kΩ. Determine:

a. The required value of R_f for a gain of 0.01 V/bit.

b. The maximum possible output voltage.

10.15 An 8-bit bipolar DAC has a 5.0-V reference.

a. Determine the output voltage generated for the input digital words $A2, $B6, and $9D.

b. What is the closest digital word if an output of 3.856 V is desired? What is the percentage error?

c. If a 50-mV RMS high-frequency noise corrupts the output, how many output bits are obscured by this noise?

10.16 Obtain the specs for the National Semiconductor DAC0808 or DAC0801 DAC IC, and write a program to interface it with the 9S12C.

LABORATORY PROJECTS

10.1 Interface a thermistor and two seven-segment displays to the 9S12C MCU, and write a program to measure and display the temperature.

10.2 Interface a photocell and two seven-segment displays to the 9S12C MCU, and write a program to measure and display the light intensity.

10.3 Interface a potentiometer that reads the angular position of a robotic joint between 0° and 90° and two seven-segment displays to the 9S12C. Write a program to measure and display the angular position of the robotic joint using the MCU.

10.4 Build a data-acquisition system using the 9S12C and the Burr Brown ADS1210 Σ-Δ ADC, and interface it with a thermistor and a display.

Sensors and Their Interface

11.1 INTRODUCTION

The measurement of physical quantities by appropriate sensors is an integral feature of any mechatronics system. The quantity to be measured is referred to as the *input signal*, *measurand*, or *stimulus*. Figure 11.1 shows the main components involved in a typical measurement, which includes the sensor, also called the transducer, the signal-conditioning (SC) circuit, and the processing unit, such as the computer or the MCU. The sensor/transducer converts the usually analog, nonelectrical, input signal, such as displacement, into a useful electrical signal in the form of voltage, current, frequency, train of pulses, or phase shift. The electrical sensor output is sent through the SC circuit to the processing unit for further use. A list of typical

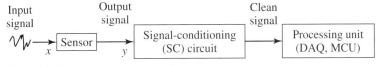

Figure 11.1 Typical components of a measurement system.

TABLE 11.1 Sensors for Measuring Various Physical Quantities

Measurand	Type of Sensor
Position and displacement	Potentiometer, LVDT, resolver, optical encoder, Hall effect, capacitive, inductive, eddy current, magnetoresistive
Level	Capacitive, inductive, Hall effect, ultrasonic
Proximity	Capacitive, eddy current, IR emitter/detector, ultrasonic, reed switches, Hall effect, magnetoresistive
Velocity	Tachometer, potentiometer, optical encoder, piezoelectric
Acceleration	Piezoresisitive accelerometers, capacitive accelerometers
Strain	Strain gauge, piezoelectric, moire interferometry
Pressure	Mercury, bellows, piezoresistive, capacitive, moire interferometry, ultrasonic
Force	Strain gauges, tactile
Temperature	Thermometer, RTDs, thermistor, thermocouple, thermodiode, thermotransistor, thermochromic paints, surface acoustic wave (SAW) devices
Light	Photoresistor, phototransistor, photodiode
Flow	Ultrasonic, thermal transport, capacitive flow, resonant bridge
Acoustic	Resistive, capacitive, fiber-optic, piezoelectric, electret, ultrasonic
Chemical	Ion-sensitive electrodes (ISEs), ISFET, resistive, capacitive, chemotransistors (CHEMFET), chemodiode, thermochemical, pellistor, organic gas

physical quantities encountered in machine and process control applications and the types of sensors used to measure them are given in Table 11.1.

To successfully implement a measurement of any kind, the user must choose the right sensor to measure the physical variable of interest, within pertinent environmental conditions and cost considerations. Keen understanding of sensor characteristics and interfacing issues are fundamental requirements. Abundant sensing schemes are available, and the designer always has to balance between often-conflicting requirements, such as accuracy and cost. For example, measuring the angular position of a rotating object may be accomplished using Allen Bradley's 2500-line incremental encoder (www.ab.com), which costs hundreds of dollars, or by a rotary Inductosyn transducer by Farrands Controls (www.ruhle.com), which may cost thousands of dollars. The resolution of the former is 0.144°, whereas the accuracy of the later is better than ±0.5 arc-seconds.

11.2 CLASSIFICATION OF SENSORS

Sensors are commonly classified according to one of the following attributes:

- The *physical quantity* the sensor can measure, such as, pressure, temperature, velocity, color.
- The *physical principle* on which the sensor is based, such as magnetoresistive, optoelectronic, Hall effect, shape memory alloys, piezoelectric.
- The *technology* used to fabricate the sensor, such as semiconductor, electromechanical, fiber-optic.
- The *spatial relationship* between the sensor and the object on which measurement is performed: contact, noncontact, or remote.
- The main form of *energy* involved in sensing, such as electromagnetic, thermal, solar, mechanical. This is the most commonly used classification.

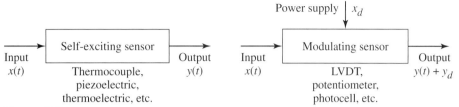

Figure 11.2 Self-exciting and modulating sensors.

Sensors are further classified in other ways.

> *Digital vs. analog sensors.* Sensors are overwhelmingly analog. An analog sensor produces an analog signal, typically voltage, proportional to the measured variable, which must be converted to a digital equivalent via an analog-to-digital (ATD) converter, or ADC, before it can be interfaced to a computer. Digital sensors, on the other hand, generate a digital signal, which can be interfaced directly to a computer, with some signal preprocessing to obtain a crisp digital output.

> *Simple vs. complex.* A simple sensor generates a single value proportional to the measurand. For example, a tachometer, outputs a voltage proportional to the speed of a rotating shaft. Whereas a complex sensor, such as a tactile array generates a matrix of digital values.

> *Self-generating vs. modulating.* These two types are shown in Fig. 11.2. A self-generating, or self-exciting, sensor is passive, because it does not need an external supply of power to operate; it draws signal energy from the signal source itself. A thermocouple draws energy from the thermodynamic system with which it is in contact to produce an output voltage signal proportional to the temperature difference across its junctions. The output signal of a self-generating sensor has low output power and needs considerable amplification to reach a useful level. A modulating sensor is an active device because it requires an external power source to operate. Photoconductive devices and thermistors are examples of modulating sensors. For example, the forward-bias current of a photodiode modulates the photoinduced electrons. Modulating sensors are more efficient and generate higher output energy as compared with self-exciting sensors. Further processing of the output signals from modulating sensors is usually required to obtain standard electrical signals, such as 0–100 mV of voltage or 4–20 mA of current (see Section 11.25).

11.3 SMART SENSORS

Although universal agreement as to what constitutes a smart sensor is lacking, the IEEE-P1451 family of standards provides protocols for interfacing sensors on wired networks. The IEEE-P1451.2 standard, *Transducer to Microprocessor Communication Protocol and Transducer Electronic Data Sheet* (TEDS) *Formats*, defines a smart sensor as a sensor that "provides functions beyond those necessary for generating a correct representation of a sensed or controlled quantity" at least. "This function typically simplifies the integration of the transducer into applications in a networked environment." Inherent in this definition are the necessary features that render a sensor smart. Succinctly, a smart sensor developed for a particular application will integrate many of the following hardware and software features.

- An onboard microcontroller to provide computing power, onboard memory for data storage, and circuits to enable digital and analog outputs. These features include signal conditioning, data processing, calibration algorithm, and conversion from the secondary variable (resistance of a thermistor) to the primary variable (temperature) or from volts to useful engineering variables, such as temperature.

- Circuitry to facilitate interactive serial communication between the sensor and the control computer, so the sensor can identify and describe itself in TEDS format, and to communicate that information to a larger, distributed sensing network. Moreover, the communication protocol provides the means to configure the sensor remotely when it is installed in a network and to adjust its parameters easily to accommodate changes in the process.
- Storage of initial calibration data in resident memory to correct for measurement anomalies, such as temperature compensation, drift, and bias for improved performance.
- Compatibility with other components on the network so as to react to instructions from application software resident on a network and to interact with the data being collected.
- Direct storage of process data to a hard disk over a period of time. This data can quickly be imported into a spreadsheet or database for further use.

Silicon-based microfabrication technologies are the enabling medium for smart sensors. Through this medium the transducer can be integrated on a single chip with microcontrollers, DSPs, application-specific ICs (ASICs), and other technologies. Using microelectromechanical system (MEMS) platforms has rendered many economically viable smart sensors, such as pressure sensors, accelerometers, and IR temperature sensors, among many others.

11.4 SENSOR MODELS AND RESPONSE CHARACTERISTICS

In general, the sensor consists of many integrated components. The signal generated by the sensor represents the physical quantity being measured, modulated by the behavior of sensor components. Knowing the effects of sensor dynamics is essential for extracting the useful signal. Also, to determine the performance of a control system accurately, the mathematical model must include the effects of sensor behavior.

The behavior of most sensors can be described mathematically by a zero-order, first-order, or second-order model. Zero-order sensors represent ideal or perfect dynamic performance. The response characteristics of a *zero-order* sensor, shown in Fig. 11.3, are governed by the linear algebraic relation

$$y(t) = Kx(t) \tag{11.1}$$

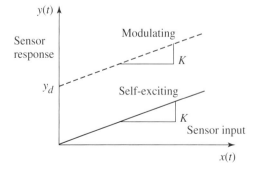

Figure 11.3 Response of a zero-order sensor.

where K is the sensor gain. The transfer function of a zero-order system is

$$G(s) = \frac{Y(s)}{X(s)} = K \tag{11.2}$$

where s is the Laplace operator. Sensors that exhibit zero-order behavior include potentiometers, tachometers, and LVDTs, to name a few.

First-order behavior is governed by the first-order differential equation

$$\tau \frac{dy(t)}{dt} + y(t) = u(t) \tag{11.3}$$

where $u(t)$ is the excitation signal and τ is the time constant. The response to a step input, $u(t) = K$, is given by

$$y(t) = K(1 - e^{(-t/\tau)}) \tag{11.4}$$

and is shown in Fig. 11.4. The transfer function of the system is given by

$$H(s) = \frac{Y(s)}{U(s)} = \frac{K}{\tau s + 1} \tag{11.5}$$

Systems that exhibit first-order behavior include a parallel spring damper in linear (rotary) displacement, a series mass damper subject to linear (rotary) velocity, $\tau = m/c$; electrical charge flow in a series RC circuit, $\tau = RC$; current flow in a series LR circuit, $\tau = L/R$; fluid flow through a restriction; and quenching of a rod in an infinite bath.

Second-order behavior is represented by the second-order differential equation

$$\frac{d^2y(t)}{dt^2} + 2\zeta\omega_n\frac{dy(t)}{dt} + \omega_n^2 = Kx(t) \tag{11.6}$$

ζ is the nondimensional damping ratio and ω_n is the natural frequency. The transfer function of the system is given by

$$H(s) = \frac{K}{s^2 + 2\zeta\omega_n s + \omega_n^2} \tag{11.7}$$

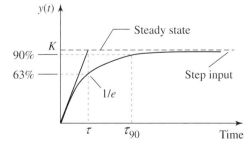

Figure 11.4 Response of a first-order system to a step input.

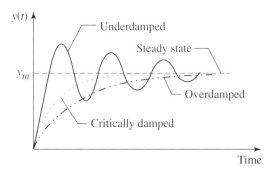

Figure 11.5 Response of a second-order sensor to a step input.

Typical response of a second-order system to a step input is shown in Fig. 11.5. Systems that exhibit this behavior include band-pass filters, accelerometers, and others.

11.5 SENSOR CHARACTERISTICS

Selecting the right sensor for a specific application requires properly matching the performance of the sensor to the characteristics of the measured signal. Sensor performance is characterized by the following criteria.

Sensitivity S is the smallest detectable change in output of the sensor y to a change in the input x, evaluated at a specific input x_0. It is expressed as

$$S_0 = \frac{dy}{dx}\bigg|_{x_0} \tag{11.8}$$

For a linear sensor, S is constant and represents the slope of the input–output curve. Figure 11.6 shows the relation between S and the input signal of ideal and real sensors. High and constant sensitivity over the entire working range is highly desirable. If the output signal varies with time, the sensitivity also varies with time, and the changing sensitivity is referred to as *sensitivity drift*.

Gain K is defined as the absolute value of the output signal relative to the absolute value of the input signal:

$$K = \frac{y}{x} \tag{11.9}$$

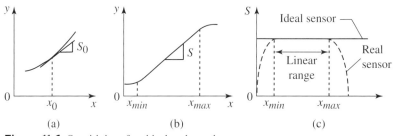

Figure 11.6 Sensitivity of an ideal and a real sensor.

The gain is equal to the sensitivity for a linear sensor.

Full-scale input (FS) specifies the working range of the measured variable that can be converted by a sensor; $x_{max} - x_{min}$. For instance, an angular-position sensor may be limited to one revolution, or a range of $0°$ to $360°$. An infinite working range is a desirable sensor feature.

Span or **full-scale output** (FSO) is the algebraic difference between the electrical sensor output values corresponding to maximum and minimum values of the input signal; $y_{max} - y_{min}$.

Threshold refers to the minimum value of the input signal needed to stimulate or activate sensor output.

Resolution refers to the incremental change in the input signal from a nonzero arbitrary value that would cause a corresponding change in the sensor output. For example, a thermometer with a display that reads to three decimal places would have a resolution of $0.001°C$.

Response Δy is the measured change in the output signal. It is expressed as

$$\Delta y = y(t) - y_0 \tag{11.10}$$

where y_0 is called the *bias* (*baseline*, or *zero offset* value). It represents the output of the unperturbed sensor under a specified condition, such as the emf generated by a thermocouple at room temperature. Usually the bias is a systematic offset voltage or current that may be removed by placing the sensor in a bridge circuit. A sensor with a zero bias is desirable. If the output varies with time, the zero drift also varies with time, and the changing zero offset is referred to as a *zero drift*.

Response time τ is a measure of the speed of response of a sensor to a changing input. It is the time duration it takes the sensor output to rise to a specified percentage of its steady-state value as a result of a step change of input signal. Response time is measured as time constant τ, rise time t_r, fall time t_f, settling time t_s, etc. The response time of an ideal sensor is $\tau = 0$.

Linearity is a measure of how close the sensor response curve is to being a straight line. A linear response means that the output is proportional to the input signal. Linear response is always a desirable sensor feature. However, all physical devices exhibit some form of nonlinear behavior due to phenomena such as saturation and hysteresis. Saturation may be due to magnetic saturation, plasticity in mechanical components, or nonlinear deformations in springs. Hysteresis is manifested by a change in the input/output curve if the direction of measurement is reversed. Examples of hysteresis are backlash in gears and nonlinear damping, such as coulombic friction.

Repeatability expresses the deviation in the output values for repeated measurements of the same input signal under identical conditions. For example, the width of the variation in the output of a position sensor that is observed when the input being measured is returned to exactly the same position several times is a measure of repeatability, w in Fig. 11.7a. Repeatability error δ_r may also be expressed as the maximum difference between output readings as determined between two calibrating cycles, one varying the input from zero to full scale and the other in reverse, as depicted in Fig. 11.7b, unless otherwise specified. It is usually represented as % of FS:

$$\delta_r = \frac{\Delta y}{FS} \times 100\% \tag{11.11}$$

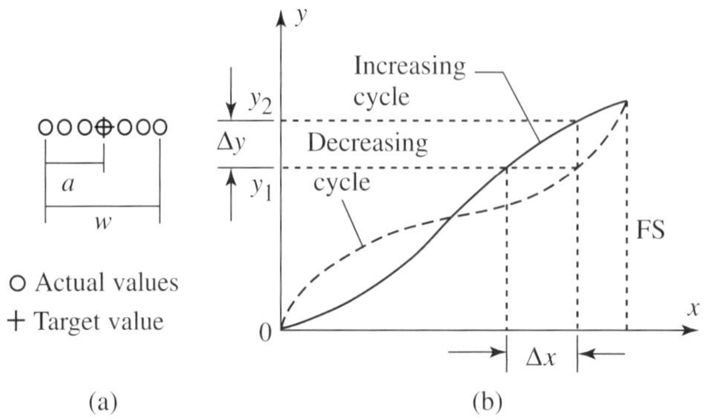

Figure 11.7 Repeatability error: two output signals corresponding to incrementing the same input signal in ascending and descending orders.

Accuracy is defined as the closeness of agreement between the true value and the sensor output of a measurand (Fig. 11.7a). It is sometimes expressed in terms of the error that can be expected between the measurement and the measurand. If the distribution of error has a known standard deviation σ, then the accuracy is frequently expressed as $\pm 2\sigma$ or $\pm 3\sigma$. For some sensors, accuracy is more commonly expressed as a percentage of the full scale. Various errors, such as hysteresis and friction, adversely affect sensor accuracy. Existing standards, such as those of the National Institute of Standards and Testing (NIST), classify accuracy as a qualitative concept not to be used quantitatively.

Precision is the closeness of agreement between independent measurements obtained under controlled conditions. Precision is used in the context of repeatability and not accuracy, as a qualitative rather than a quantitative term.

Error is the difference between the sensor output and the true value of the measurand.

Uncertainty is the estimated possible deviation of the result of measurement from its actual value. Measurement uncertainty is ascertained by combining the uncertainties of the individual components in the measurement using the root-sum-of-squares method.

Dynamic response refers to the maximum sinusoidal frequency of change in a measured variable for which agreement can be maintained between the measurement and the measured variable. It is usually limited by the electrical and mechanical characteristics of the sensor. It may be expressed in terms of the 3-dB attenuation bandwidth of the sensor's frequency response.

11.6 SIGNAL CONDITIONING

The raw electrical signal generated by a sensor might not be useful because it may be too small or too noisy, contain wrong information due to poor transducer design or installation, have a DC offset usually due to the transducer and instrumentation design, or be incompatible with the input requirements of the processing device, such as the MCU. Therefore, the raw sensor signal requires conditioning before it can serve a useful purpose. The circuit that is used to interface the signal properly to the computer is called a signal-conditioning (SC) circuit. Table 11.2 gives typical signal range and output impedance for common sensors. A signal-conditioning circuit performs one or more functions, depending on the characteristics of the signal being measured. The most common functions of the SC circuit are briefly discussed next.

TABLE 11.2 Typical Characteristics of Common Sensors

Sensor	Typical Signal Range	Output Impedance (Ω)
Thermocouple	0–50 mV	—
Thermistor	1–100 mV	10^3–10^7
Platinum RTD	—	10^2–10^3
Strain gauge	0.1–10 mV	10^4
Piezoelectric crystal	1–100 mV	10^4
photodiode	1 nA to 1 mA	10^2
photomultiplier	1–1000 nA	10^8
pH electrode	−1 to +1 V	10^9

11.6.1 Amplification

A major role for the SC circuit is to amplify a weak raw signal to a particular level. Amplification means that the signal is multiplied by a constant, called *gain*. The raw signal generated by common transducers may be on the order of millivolts or picoamps. This is true for most passive sensors. Meanwhile, standard data-processing electronics, such as ATD converters and frequency modulators, require sizable input signal values—on the order of volts (V) and milliamps (mA). Voltage gains up to 1000 and current gains up to 1 million are not uncommon. The output from a thermocouple made from one iron arm (high emf at +3.54 mV) and one nickel arm (low emf at −3.1 mV) would be 6.64 mV for a temperature gradient 200°C. This voltage is too small and serves no practical use if not adequately amplified. Some sensors are packaged on an IC chip with built-in amplifiers. Amplification is accomplished by using a combination of inverting and noninverting op-amps.

11.6.2 Conversion

The output of most sensors is an analog (continuous) signal in the form of voltage or current. The MCU is a digital device that can read only binary numbers (0s and 1s). Before the signal from an analog sensor reaches the MCU, it needs to be converted to a digital signal by an ATD converter circuit. As discussed in Chapter 10, the ADC onboard of the 9S12C MCU contains an 8-bit successive-approximation ADC, a multiplexer, and a sample-and-hold circuit. An analog signal applied to any ATD channel (pins PAD[7:0]) is converted to a digital word in the range of 0 to 255 and stored in the ATD result registers ATDDRR[7:0].

11.6.3 Filtering

A very important role of the SC circuit is filtering in order to reduce the effect of noise that corrupts the useful signal. Refer to Chapters 2 and 4 for more details. Noise filtering may require the addition of low-pass, high-pass, band-pass, or notch filters. Filtering limits the dynamic range of the sensor output. Various filter functions can be implemented via software. A simple moving-average filter may be implemented, using a number of samples that is an integer power of 2, allowing the division to be implemented using bit-shift operations. General-purpose finite impulse response filters, which allow for low-pass, high-pass, and band-pass filtering functions can also be implemented.

11.6.4 Impedance Buffering

In many applications, the output impedance of a signal source is of the same order as the input impedance of the transducer. However, the output impedance of a sensor can be very high (refer to Table 11.2); if the input impedance of the SC circuit is low, it could load the sensor circuit and reduce the level of the signal it is trying to measure. Loading errors occur because a device with low input impedance extracts a high level of power from the preceding output device. Therefore, the signal-conditioning circuit should have a considerably larger input impedance Z_{IN} in comparison with the output impedance of the sensor Z_{OUT} in order to reduce loading errors. A *unity-gain* op-amp used as a *buffer* amplifier is usually installed between the sensor and the input stage of the SC circuit to eliminate loading errors. Refer to Sections 2.10.2 and 4.6.3 for more details.

11.6.5 Modulation/Demodulation

Modulation is the process of altering the original signal baseband by means of another signal, called the *carrier*. This SC technique can be one of three forms: amplitude modulation, frequency modulation, or phase modulation. Amplitude modulation and demodulation, depicted in Fig. 11.8, is commonly used in bridge circuits, measurement with LVDTs, and other sensors. In a strain-gauge bridge excited by an AC voltage, the carrier frequency is the frequency of the AC excitation voltage and the modulation frequency is the frequency of the measured signal. The modulated signal is not suitable for processing until the transducer signal is separated from the carrier signal. The separation process is called *demodulation*. Performance of modulating sensors often hinges on the ability of the SC circuit to provide a stable power/or carrier signal. This concern is less important when a bridge circuit is used because first-order fluctuations of the power supply are irrelevant.

11.6.6 Linearization

All physical devices exhibit nonlinear input/output relations. A linear output from a sensor is highly desirable. Only a few sensors exhibit a linear relationship within a limited range of their input–output curve. A role of the SC circuit is to linearize the output of the sensor against the input, within the desired operating range.

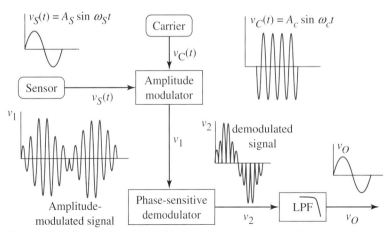

Figure 11.8 Components of amplitude modulation and demodulation.

In many applications it is necessary to pass the sensor output raw signal through a customized function to remove measurement offsets, gain errors, and nonlinearities. With low- to moderate-resolution ATD converters (8 to 12 bits), straight lookup tables in which "corrected" data points are stored as an array can be used to accomplish linearization. In this case, the result of the ATD conversion represents an index into the array. A k-bit input requires a table of 2^k entries. If memory is not available to store the lookup table data, linearization may be achieved via *linear interpolation*. In this scheme, the correction function is represented as a connected series of linear segments requiring storage of endpoints only.

11.6.7 Grounding and Isolation

If two or more ground points in a measurement system are at different voltage levels, *ground loops* are formed. Ground loops provide a path for currents to circulate and carry noise signals that may be many times larger than the actual signal. A measurement system involves two important ground references: signal ground and measurement system ground. The purpose of system ground is to ensure that the measurement is taken with respect to a zero-voltage reference and that the enclosure does not carry a voltage. System ground is achieved by connecting the casing of a device to a stable ground rod by one or more heavy copper conductors. On the other hand, signal ground provides a reliable reference for measuring all low-level signals. Proper grounding of measurement signal and measuring equipment is essential to ensure measurement accuracy, user safety, and equipment protection. Figure 11.9 shows an incorrect and a proper grounding of a signal circuit.

The harmful effects of ground loops may be eliminated by a signal-conditioning technique known as *isolation*. A device known as an isolator separates and isolates the signal ground from the system ground. Refer to Section 2.16 for more details on grounding and isolation techniques.

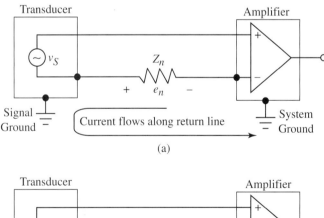

Figure 11.9 Grounding signal circuit: a wrong way (a) and a proper way (b).

11.7 POTENTIOMETER SENSORS (Pot)

A potentiometer (or simply pot) is a resistive transducer used to measure the linear and angular displacement of a moving object. Figure 11.10 shows two types of potentiometers. A pot consists of a single-wire resistive element and a sliding contact known as the wiper or brush. The *resistive element* is normally wire-wound or conductive plastic. Cermet and hybritron elements are also available. Conductive plastic and hybritron elements are applied in voltage divider circuits only. The resistance per unit length of the resistive element is usually constant along the element. The *wiper* connects mechanically to the object whose position is to be measured, while the rest of the pot is attached to a stationary object. The wiper is usually electrically isolated from the sensing shaft.

When an excitation voltage V_R is applied across the resistive element, an output voltage v_O proportional to the position of the wiper that slides along the resistive element is produced. The pot is essentially a voltage divider, and Eq. (2.18) governs the relation between V_R and v_O as long as the interface circuitry does not load the sensor. Assuming the output load resistance R_L to be infinite, a carefully fabricated pot exhibits a linear relationship between the output voltage and the displacement, as indicated in Fig. 11.10a. This relation is expressed as

$$\frac{v_O}{V_R} = \frac{x}{x_M} \quad \text{(linear)}$$

$$\frac{v_O}{V_R} = \frac{\theta}{\theta_M} \quad \text{(angular)}$$

(11.12)

from these relations we can write

$$x = Kv_O = \frac{x_M}{V_R} \times v_O \quad \text{(linear)}$$

$$\theta = Kv_O = \frac{\theta_M}{V_R} \times v_O \quad \text{(angular)}$$

(11.13)

where x (or θ) is the displacement corresponding to v_O, x_M (or θ_M) is the full-scale displacement when $v_O = V_R$, and K is the gain.

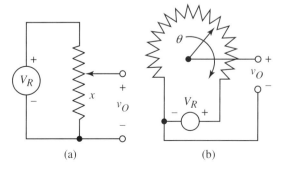

(a)   (b)

Figure 11.10 Single-turn potentiometer for measuring linear (a) and angular (b) displacements.

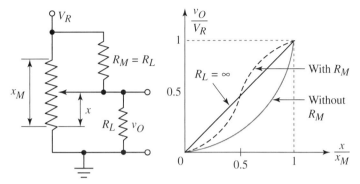

Figure 11.11 Circuit diagram and output of a potentiometer with and without a shunting resistor.

The nonlinearity of the pot increases as the output load resistance R_L decreases from the ideal infinite value. One way to reduce this effect is to shunt the upper arm of the pot with a resistance R_M, as shown in Fig. 11.11. This reduces the error to zero at $x = 0.5x_M$. Another way to achieve linearity is to use the pot as a feedback resistor in an inverting op-amp circuit. The output voltage of the op-amp is linearly dependent on the feedback resistor.

Potentiometers are simple and useful in many applications. However, they need physical coupling with the moving object, and their performance is hindered by many anomalies, which include noticeable mechanical load (friction), low operating speeds, heating caused by friction and excitation voltage, and low environmental stability.

Figure 11.12 shows one problem associated with a wire-wound pot. As the wiper moves across the winding, it will make contact with either one or two wires, producing an uneven output voltage steps, that is, variable resolution. Therefore, only the average resolution can be considered in a wire-wound pot. The resolution is dependent on the coil diameter and can be improved by using a very fine wire. A linear pot with N turns/mm (50 turns/mm is the practical limit) has a resolution n determined by

$$n = 1/N \tag{11.14}$$

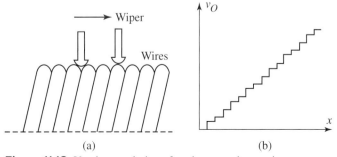

Figure 11.12 Varying resolution of a wire-wound potentiometer.

Example 11.1: Interfacing a potentiometer to the 9S12C–A pot is used to measure the angular position of a robot joint. The pot is rated for a 270° rotation and 5-V range. However, the range of motion of interest of the robotic joint is 90°. Provide a scheme to map the varying output voltage from the sensor into the range of digital values between 0 and 255 available to the MCU.

Solution: Figure 11.13 shows the circuit connection and required mapping for this example. The output signal from the pot is processed by an amplifier before it is received at pin PAD[0] of port AD on the 9S12C MCU configured to operate in the ADC mode. To take advantage of the full range of the ADC, the amplifier's gain should be

$$K_A = \frac{90 \times 5}{270} = \frac{5}{3} \tag{11.15}$$

The voltage that the ADC receives is $K_A v_O$. The result is a linear mapping where the joint angle of 0° is mapped to number 0 and an angle of 90° to 255.

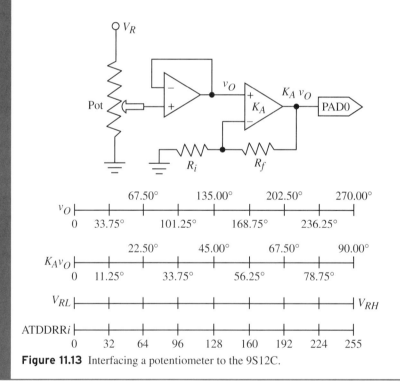

Figure 11.13 Interfacing a potentiometer to the 9S12C.

11.8 LIGHT DETECTORS

Electromagnetic radiation propagates through a region in free space in the form of a large amount of discrete units, or quanta, called *photons*. Photon propagation is characterized by the frequency of oscillation f (Hz) or wavelength λ (m), but all waves in the electromagnetic spectrum travel at the speed of light $c = 3 \times 10^8$ m/s. Regardless of the light source, the wavelength and frequency of a wave are related by

$$\lambda = c/f \qquad (11.16)$$

The energy of a photon is $E_p = hf$ (J), where $h = 6.63 \times 10^{-34}$ (J-s) is Planck's constant. The energy of a photon is usually expressed in electron-volts, where 1 eV $= 1.602 \times 10^{-19}$ J. The number of photons emitted by a light source at a given frequency is the ratio of the net energy of all emitted photons to the energy of one photon. Optical or radio frequency energy requires no medium to propagate. The photons transmit most effectively through vacuum, and earth's atmosphere impedes their propagation.

Light refers to the electromagnetic radiations that occupy the region between 0.02 μm and 100 μm of the electromagnetic spectrum. This range encompasses the following light components:

ultraviolet (UV)	$0.002 \leq \lambda \leq 0.38$ μm
visible	$0.38 \leq \lambda \leq 0.78$ μm,
near-infrared (NIR)	$0.78 \leq \lambda \leq 1.7$ μm,
middle-infrared (MIR)	$1.7 \leq \lambda \leq 6$ μm,
far-infrared (FIR)	$6 \leq \lambda \leq 1000$ μm.

The units used to measure optical output are either radiometric or photometric. *Radiometric* units measure, in watts, pure energy flow, independent of wavelength, whereas *photometric* units, called *lumens*, measure energy flow within the spectrum of human vision. A lumen (lm) is a measure of the brightness of light. The lumen is the luminous flux of light emitted within a solid angle of 1 steradian from a point source having a uniform intensity of 1 candela. A *candela* (cd) is the luminous intensity of light emitted from a monochromatic source in a given direction at a frequency of 540×10^{12} Hz (peak sensitivity of human visual system) such that the radiant intensity in that direction is $1/683$ watts/steradian. Thus, 1 watt at 555 nm is equivalent to 683 lumens. The illuminance produced by a luminous flux uniformly distributed over a surface of 1 m^2 is called the *lux* (lx), and 1 lx $= 1$ lm/m^2.

Detectors of the electromagnetic radiations between ultraviolet and infrared regions are referred to as light detectors. Several light detectors are discussed in the following sections. Table 11.3 gives the characteristics of a few of them.

TABLE 11.3 Characteristics of Various IR Sensors

Device (Manufacturer)	Package	Type	Range	Output	Response Time
P394A (Hamamatsu)	TO-5	PbS PC (cooled)	1–2.5 μm	5×10^4 V/W @λ_{peak}	100–400 μs
P791 (Hamamatsu)	TO-5	PbSe PC (uncooled)	2–5 μm	8×10^2 V/W @λ_{peak}	2–5 μs
B2538-01 (Hamamatsu)	TO-8	Ge PD (cooled)	1–1.8 μm	800 mA/W @λ_{peak} $I_{SC} = 1.4$ μA @ 100 lx	1 μs
G3476-01 (Hamamatsu)	TO-18	InGaAs PD (uncooled)	0.8–1.7 μm	900 mA/W @ 1.3 μm	0.3 ns
SD5491-3 (Honeywell)	TO-18	Silicon PT	0.5–1.1 μm	2.0–5.0 mA	2 μs
DP-2101-101 (Sentel)	TO-5	Pyro	7–14 μm	1800 V/W	100 ms

11.8.1 Materials for Light Detectors

Light detectors are usually made of semiconductor material. Under an incident beam of light, the semiconductor material converts the optical energy (photons) into an electrical signal (electrons). When a semiconductor crystal is exposed to light of the proper wavelength, the concentration of charge carriers (holes and electrons) in the crystal increases, causing the conductivity of the crystal to increase. Commonly used semiconductor materials include silicon, cadmium sulphide (CdS), and mercury-cadmium-telluride (MCT). Silicon is not suitable as an IR detector, whereas MCT material covers a wide range and is practically independent of wavelength.

11.8.2 Types and Modes of Operation of Light Detectors

Light detectors are classified into two major groups: *quantum* and *thermal* detectors. Quantum detectors are devices that convert photons of electromagnetic radiation directly into charge carriers. They are generally produced in the form of photodiodes, phototransistors, and photoresistors. Quantum detector operation is based on either the *photoconductive* effect or the *photovoltaic* effect. The photoconductive effect is the change of the resistance of a semiconductor material caused by the incidence of light. The photovoltaic effect is the generation of a voltage across a semiconductor *p-n* junction when the junction is exposed to electromagnetic radiation. Quantum detectors operate between the ultraviolet and mid-infrared regions and exhibit a strong correlation between wavelength and faster response.

Thermal detectors are most useful in the mid- and far-infrared regions, where their efficiency at room temperature exceeds that of the quantum detectors. They are mainly used for noncontact temperature measurement. The class of light detectors used to detect infrared radiations between 0.75 μm and about 1000 μm, which lie between the visible light and microwave regions, are referred to as *IR detectors*.

11.8.3 Applications of Light Detectors

Light detectors are generally used to perform one of two functions: to indicate light or dark conditions or to measure light intensity. Light detectors are typically used in the measurement of light intensity, temperature, position, velocity, level, and flow; in the detection of human or animal movement, color, proximity, and fire; in smoke-density monitoring, and in fiber-optic communication and imaging.

11.9 PHOTORESISTOR (PHOTOCELL)

The photocell (PC) is a *photoconductive* device whose internal resistance falls as the amount of incident light increases. It behaves as a variable resistor, in many ways similar to the potentiometer, except the change in the resistance is caused by a change in light level rather than by moving a wiper. Photocell packages, symbol, structure, and resistance vs. light of a plastic CdS PC are shown in Fig. 11.14.

Photocells are more sensitive to light than are any other type of light-sensitive devices. Their sensitivity depends on the wavelength of the incident light. The resistance of a typical cell might be as high as several hundreds of megaohms in complete darkness and as low as few hundred ohms in normal room light. The biggest disadvantage of photocells is their slow response to light change, on the order of several milliseconds. This response, however, is not influenced by temperatures between 0° and 50°C. The color response of a PC is roughly

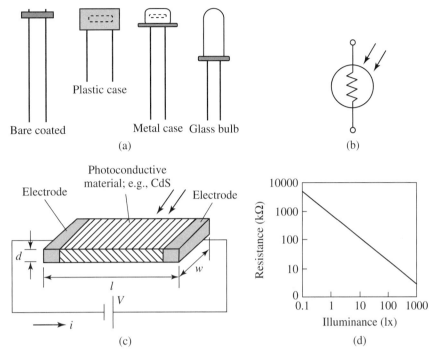

Figure 11.14 Photocell packages (a), symbol (b), basic structure (c), and resistance of a CdS cell with illuminance (d).

equivalent to that of the human eye. Exposure to high humidity or intense UV light for long periods shortens a device's life. Typical applications of common photoconductive sensors include automatic dimmers, light switches for outdoor day/night lighting, robotic light-level sensors, photoelectric servos, electronic camera shutters, melody greeting cards, and photoelectric relays.

11.9.1   Materials for Photocells

Depending on the material used in their manufacture, photocells can be made to detect light between the visible and infrared regions. The most commonly used materials for visible-light photocells are cadmium sulphide (CdS) and cadmium selenide (CdSe) semiconductors. The spectral range of CdS is essentially visible light, from 0.4 μm to 0.8 μm, with a peak response at 0.55 μm.

Photocells made from thin films of lead sulfide (PbS), lead selenide (PbSe), and mercury-cadmium-telluride (MCT) semiconductors can detect electromagnetic radiation in the infrared region. The PbS cells have a spectral range between 1 and 3 μm, with a peak response at 2.2 μm. PbSe cells have a spectral range between 1 and 4.5 μm, with a peak response at 3.8 μm (both at 25°C). Depending on the composition ratio of mercury-telluride (HgTe) to cadmium-telluride (CdTe), MCT cells can detect light in the spectral range between 2 and 12 μm.

11.9.2   Interfacing a Photocell to the 9S12C

A photocell is easily interfaced to the 9S12C MCU. Two approaches are shown in Fig. 11.15. The output of the voltage divider in Fig. 11.15a changes in proportion to light intensity. The

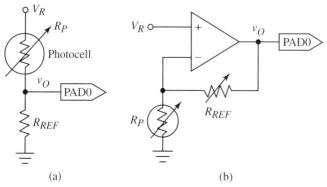

Figure 11.15 Interfacing a PC to the 68HC11 via a voltage divider (a) and a noninverting op-amp (b).

resistance R_{REF} is selected so that its value is the same as the resistance of the photocell when exposed to a light level in the middle of the range. This approach is a compromise between sensitivity and range. In Fig. 11.15b, the cell is used as part of the resistive-feedback network of a noninverting op-amp. The gain of the op-amp is controlled by the resistance of the cell. As a result, the op-amp output is a function of light intensity.

11.10 PHOTODIODE

A photodiode (PD) is a device that converts light energy (photons) into electrical energy (charge carriers), roughly one electron and one hole per photon. The output photocurrent I_L is proportional to the incident light. Photodiodes are fabricated from a variety of semiconductor materials, the most popular of which are silicon and germanium, whose spectral sensitivities range from 300 nm to 1800 nm. The wavelength of maximum sensitivity for Si is ~950, Ge ~1500, InGaAs ~1600, and GaAlAs ~880 nm. Indium-arsenic (InAs) and indium-antimony (InSb) are also used as sensors in the NIR region.

11.10.1 Photodiode Types

Photodiodes are available in one of four types: the *p-n* photodiode ($0.19 \le \lambda \le 1.1$ μm), the *p-i-n* photodiode ($0.32 \le \lambda \le 1.1$ μm), the *Schottky*-type photodiode ($0.19 \le \lambda \le 0.68$ μm), and the *Avalanche photodiode* ($0.4 \le \lambda \le 0.8$ μm). A standard photodiode is constructed from a *p-n* junction, with a transparent nonreflective silicon dioxide window placed on the *p*-layer to allow light to enter the device and a metal back-plate attached to the *n*-layer to allow electrical contact to this region. The *p-i-n* photodiode is constructed by placing a high-resistance intrinsic *i*-layer between the *p*- and *n*- regions of a standard *p-n* diode. *p-n* and *p-i-n* diodes are often mounted on an insulating substrate and sealed within a metal case. A glass window is provided to allow light to enter and strike the diode. Windows in IR sensors are typically made of Si ($\lambda = 5$ μm), irtran ($\lambda = 4.3$ μm), and SnSe ($\lambda = 10.6$ μm). Photodiodes are packaged in many styles, e.g., TO-46 flat-window and TO-46 dome-lensed metal packages, and side-looking plastic packages. Specifications for several photodiodes made by Hamamatsu are given in Table 11.3. Refer to Fig. 3.13 for the symbol and typical appearance of the photodiode.

11.10.2 Photodiode Characteristics

The overall characteristics of a photodiode are captured in the relation

$$I_L = I_S\left[\exp\left(\frac{q_e V_D}{k_b T}\right) - 1\right] - \frac{\eta q_e P}{hf} \qquad (11.17)$$

where,

I_L = electrical output photocurrent
I_S = reverse saturation current attributed to the generation of electron–hole pairs
V_D = voltage applied to the diode
k_b = Boltzman constant (1.3805×10^{-23} J/°K),
T = absolute temperature
q_e = charge of an electron (1.6021×10^{-19} C)
P = optical power of an incident beam
h = Plank's constant (6.6256×10^{-34} J-s),
f = propagation frequency
η = probability that a photon of energy hf will produce an electron in a detection

The second term on the right-hand side of Eq. (11.17) is the short-circuit current, I_{SC}, proportional to the optical power incident on the detector. Advantages of photodiodes over photocells include higher sensitivity, faster response time to changes in illumination, smaller size, better stability, excellent output photocurrent linearity over several magnitudes of light intensity, and a wider range of color response.

11.10.3 Operating Modes

The photodiode can operate in one of two modes: the photovoltaic mode and the photoconductive mode. Figure 11.16 shows the circuit for these two modes. In the *photovoltaic mode* (zero biased), the photodiode generates a small voltage across its semiconductor junction when exposed to light. This voltage increases with light intensity. Typical values of output diode current I_L are around 10 μA when illuminated and around 10 nA dark current. The photodiode is not an ideal current source, because its frequency response is seriously limited by an output impedance that consists of a parallel combination of a finite resistance R_O that varies over voltage and a parasitic capacitance C_O of about 100 pF. For example, 100 MΩ in combination with 100 pF results in a low-pass filter with a roll-off frequency of only 1600 Hz. Consequently, it is not adequate to measure the small output current by simply measuring the voltage drop in a high resistor (1 MΩ) placed across the photodiode. A far better solution when dealing with low-level signals is to employ the I–V converter shown in Fig. 11.16a. This circuit balances the incoming current with a separately generated current, detects the difference, and outputs a large voltage. Practically, the photodiode (and any transducers generating a low-level signal) is interfaced to a trans-impedance amplifier (e.g., OPA627 FET input), which provides a solution to the precision of I–V conversion problems. Photovoltaic devices are commonly known as *solar cells*. A typical 4-in. -diameter solar cell produces about 1 watt of power. Individual solar cells are usually grouped into panels and then into arrays to produce several kilowatts of power.

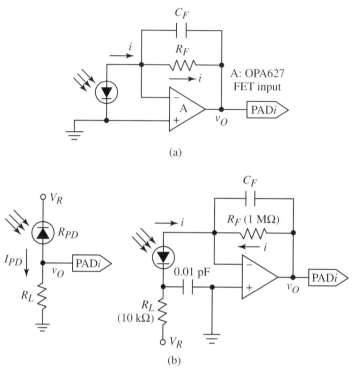

Figure 11.16 Interfacing a photodiode to the 68HC11: photovoltaic mode (a) and photoconductive mode (b).

The photodiode operates in the *photoconductive mode* when a reverse-biased voltage is applied to its junction. With a given reverse-biased potential and under dark conditions, the diode resistance is so high that little or no reverse current flows into the circuit. The reverse current of the photodiode varies with illumination. When light strikes the photodiode, the increase in the photocurrents widens the depletion region around the junction, causing the resistance across the junction to decrease and more reverse currents to flow through it. Thus the junction resistance is proportional to illumination. Furthermore, the capacitance across the junction decreases and the speed of response improves. The photoconductive mode offers a linear response over a wide range of illumination and bandwidth in the range of hundreds of megahertz. However, a design trade-off for the increase in speed of response is an increase in dark current (a DC error) and in signal-to-noise ratio as compared with the photovoltaic mode.

11.10.4 Applications

Photodiodes are widely used in photometers, spectrometers, videocassette recorders, IR pyrometry, chromatography, and medical instrumentation, such as CT scanners and blood analyzers. They are also used, in communication systems for detecting modulated light, in solar cells for satellites, and as sensors to detect the presence of an object. They can also be used to measure the presence, intensity, and wavelength of UV radiation to near-UV radiation. Two photodiode applications are shown in Fig. 11.17.

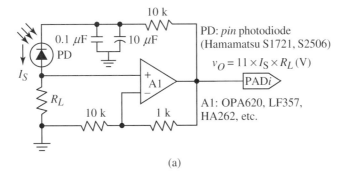

(a)

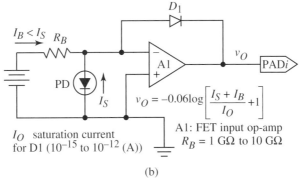

(b)

Figure 11.17 Examples of PD applications: a high-speed light sensor (a) and a light-to-log voltage converter (b).

11.11 PHOTOTRANSISTOR

A phototransistor is a photojunction device with similar construction and appearance to that of an ordinary BJT transistor, but it is often packaged in a metal case with a glass window, much like a photodiode. Although the phototransistor has a base lead as well as emitter and collector leads, the base lead is seldom used. The flow of current through a phototransistor is directly proportional to the amount of incident light. Any increase in light intensity results in an increase in the base current, which causes a large increase in the collector current due to the amplification function of the transistor.

11.11.1 Phototransistor Characteristics

Phototransistors are used in the same way as a photodiode operating in the photoconductive mode. They are more sensitive to light variations than a photodiode, making them more suitable for a wider variety of applications. However, phototransistors do not respond as quickly to changes in light intensity as a photodiode, and, therefore, they are not suitable for applications where extremely fast response is required. Phototransistors are characterized by an impedance that varies from 1000 Ω to 1 MΩ in most low-level DC circuits and a response time that is inversely proportional to incident light—moderately high intensity levels yield response well under 1 millisecond. Phototransistors have poor response to green and blue but acceptable response to red and near-infrared light. Specifications of a typical phototransistor are given in Table 11.3.

11.11.2 Applications

Figure 11.18 shows the electronic symbol for a phototransistor and three phototransistor applications. In Figure 11.18b, a phototransistor (Panasonic MT2 7935) circuit is used as a light meter to control the amount of light incident on a light-sensitive area, e.g., light-sensitive plants. It is powered by a 5-V power supply to keep the output of the op-amp within the operating range of the ADC on the 9S12C. The circuit is biased at the appropriate operating point by a 10-kΩ resistor that was suggested empirically. A unity-gain op-amp acts as a buffer between the emitter voltage and the ADC.

Figure 11.18c shows a phototransistor used in a fire detector circuit. Being sensitive to infrared or near-infrared light, a phototransistor can detect flames with little visible light, such as those produced when many gases burn, including hydrogen and propane. In this application, an opaque infrared filter should be used to block any light except infrared light from reaching the phototransistor. The 10 kΩ variable resistor is used to adjust the sensitivity of the phototransistor. Low sensitivity is desirable in this application so that ambient infrared light does not trigger the comparator. This circuit does not work when the background light has excessive infrared content (e.g., sunlight, incandescent lamp).

Figure 11.18d shows a phototransistor being used with a light source to measure the rotational speed of an object (gear, fan, propeller, etc.). The phototransistor is connected to one input of an op-amp comparator, whose output is connected to the PAI pin on the 9S12C. Each time the rotating object breaks the light beam, the comparator circuit generates an output pulse. These pulses can be counted by the pulse accumulator to determine the rpm of the object.

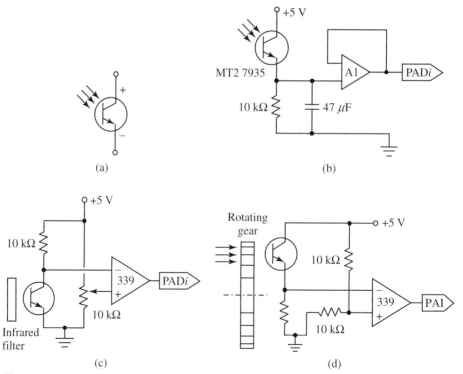

(a)   (b)   (c)   (d)

Figure 11.18 Phototransistor symbol (a), and applications as a light meter (b), a fire detector (c), and a speed meter (d).

11.12 IR EMITTER/DETECTOR PACKAGES

Phototransistors and photodiodes are well matched within the light spectrum to infrared-light-emitting diodes, or IR LEDs. The TIL 901-6 IR LED has a high relative output power, which is very important for long-range requirements, such as in remote control applications. An optical sensor circuit that is very useful in a variety of applications is formed by combining an IR LED emitter and a phototransistor or a photodiode detector. The emitter/detector pair can be arranged in one of two modules: the *optical interrupter* (also known as *optical switch*) and the *optical reflector*. In both modules, the IR emitter is constantly illuminated and the detector operates as an optical switch that turns on and off with the presence or absence of impinging light. Optical interrupters and reflectors generally operate on voltage levels up to 20 V, which makes them compatible with voltage levels of digital control circuits.

Combining the light source and the light sensor in one package offers the following advantages: (1) The light source and light sensor are matched within the light spectrum for maximum efficiency; (2) the physical parameters, such as alignment and source-to-sensor distance, are optimized; (3) the sensor can be properly shielded from ambient light; and (4) the package is standardized and mass-produced for cost effectiveness.

11.12.1 Optical Interrupter

The circuit diagram of an optical interrupter is shown in Fig. 11.19a. The device consists of an IR LED (e.g., SFH484 by Siemens) and a phototransistor or a photodiode (e.g., MTH320 by GI) mounted on opposite sides of a physical slot. The constantly energized IR LED shines a focused beam of light on a phototransistor through a carefully shaped aperture (the opening of the phototransistor). The phototransistor is ON and its output is LOW as long as the light beam is not interrupted. When an object is placed within the slot, it breaks the light beam, causing the phototransistor to turn OFF because it stops receiving light. Figure 11.19c shows a typical connection of an optical interrupter circuit. The pull-up resistor raises the output voltage to +5 V.

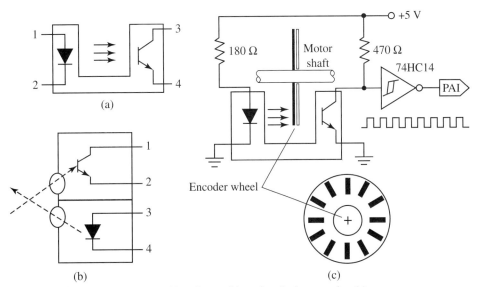

Figure 11.19 Optical interrupter (a), reflector (b), and typical connection (c).

The high-impedance output from the photodiode drives the high-impedance input circuit, such as the CMOS Schmitt trigger inverter shown. The Schmitt trigger sharpens up the output; even if the output should change slowly, its output will snap from high to low or from low to high without dithering between 1 and 0. The resulting phototransistor output can then be used to stop or start machinery, trigger an alarm, supply count pulses, or perform other functions dictated by the device's application. The width of the slot is usually less than 0.375 in., with 0.1 and 0.125 in. being very common. This limits the size of the object that can be detected by the device. Usually an encoder wheel is connected to the moving object.

Slotted optical interrupters are made for a variety of operating conditions. The user can choose from such available options as aperture size, mounting configuration, and housing material. The unit in Fig. 11.19c, made by Honeywell, has a slot width of 0.125 in. and a maximum trigger current of 20 mA. The device contains an IR LED, a photodiode, an amplifier, a voltage regulator, a Schmitt trigger, and an *npn* output transistor with 10-kΩ pull-up resistor. The output rise and fall times of this device are independent of the rate of change of incident light. Optical interrupters are used in many applications, such as to measure the angular (linear) position and the velocity of a rotating shaft (translating object), count objects, and obtain a zero-reference position for a motor, among others.

Example 11.2: Object detection using the IR interrupter–Explain how the IR interrupter can be used to detect objects within the gripper of a robot.

Solution: An IR emitter/detector pair can be used effectively to detect the presence of an object within the grasping area of a robot gripper. In this application, the IR LED is mounted in one finger and the phototransistor is mounted in the other finger of the gripper. When an object breaks the light path, the transistor switches OFF. The control computer detects the change and sends a command to an actuator to close the gripper.

11.12.2   Optical Coupler (Optical Isolator)

The optical interrupter circuit forms the basis of a solid-state optical coupler, which is equivalent to a solid-state relay contact (SPST normally open) except that normally open currents of a few milliamps can flow through the phototransistor. Optical coupling provides complete electronic isolation between two circuits. Since the output and input communicate by light, the main use is to interface two systems without need for interconnecting lines. The result is exceptional noise immunity. Optical couplers are of great benefit when interfacing systems that include PLCs, computers, or instrumentation equipment. Refer to Section 3.14 for more details.

11.12.3   Optical Reflectors

The circuit diagram of an optical reflector module is shown in Fig. 11.19b. The emitter and the detector are mounted side by side, and their optical axes are aimed to converge at a focal point just beyond the surface of the module. When the light from the constantly illuminated IR LED strikes a target surface present within the field of view near the focal point, a burst of light is reflected back to the detector and causes its output to change. A number of control circuits, such as a comparator, feed the ON/OFF output from the phototransistor to the computer. The sensing distance of a reflector module is usually limited to less than 0.5 in.

Example 11.3: Obstacle avoidance using the IR reflector–Explain how an IR reflector can be used in robots to avoid obstacles.

Solution: The basis for using the optical reflector as a proximity sensor for obstacle detection is that the light emitted by the IR LED bounces off the obstacle and the reflected light causes the detector output to change. The computer receives the ON/OFF output from the detector and commands the robot to steer away from the obstacle. For successful obstacle detection, the emitter/detector pair must be properly aligned and the detector must be blocked from both ambient room light as well as direct light from the IR LED.

Example 11.4: Application of an optical reflector in a line-tracing robot–Explain how the IR reflector can be used to guide a robot along a path.

Solution: Guiding a robot to move along a prescribed path can be accomplished by placing a strip of white or reflective tape on a dark and hard floor, such as wood or concrete. One or more IR pairs are placed on the robot. Based on the ON/OFF states of the detectors caused by the light reflected off the tape, the computer commands the drive motors to turn in the proper directions and to steer the robot along the desired path. Performance integrity of this application hinges on properly shielding the detector from ambient light.

11.12.4 NIR Receiver/Demodulator Sensors

Many vendors integrate IR detectors with the necessary signal-conditioning circuit in a compact package, for reliability and cost-effectiveness. The GP2D02 detector from Sharp Electronics (www.sharpelectronics.com) is such a device; it can be used to measure the distance to an obstacle. The onboard circuitry returns an 8-bit signal that is proportional to the distance from an obstacle located within a range of 10 to 80 cm. The G1U52X is another device widely used as a detector in a variety of applications, such as televisions, VCRs, line-of-sight wireless communications, and audio components, and for obstacle avoidance in robots. It operates from a 5-V source, which simplifies the connection of the sensor to TTL or CMOS ICs. The detector module consists of (1) a built-in filter, a lens with visible-light cutoff resin, designed to block visible light and reduce or eliminate false operation caused by ambient light sources; (2) a *pin* photodiode that has a peak sensitivity in the NIR range; (3) a preamplifier/limiter circuit to sharpen the output signal from photodiode; (4) a band-pass filter to reject all signals outside the passband range of 40 kHz $\pm$ 4 kHz; (5) a demodulator, an integrator, and a schmitt trigger stage to produce a clean digital waveform without the carrier.

The detector responds to an emitter signal that is modulated by a 40-kHz carrier frequency generated by an oscillator circuit. The emitter is programmed to blink at 40 kHZ for 600 μs and then to go off for 600 μs. The oscillator circuit to blink the emitter can be implemented in a variety of ways, using a 555 timer circuit, a crystal circuit, or a PWM signal generated by the MCU. Figure 11.20 shows a simple *RC*-based oscillator circuit using a CMOS 74HC04 (and no other TTL version) inverter powered from a +5-V source. Also shown is the detection protocol. Optimum emitter sensitivity occurs when the signal's peak wavelength is 0.88 μm and if the oscillator's frequency is very near the center frequency of the detector's band-pass filter.

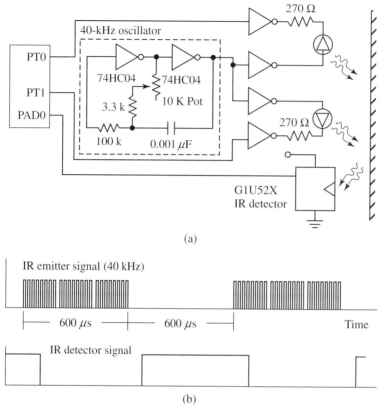

(a)

(b)

Figure 11.20 Obstacle-detection circuit using the GP1U52X NIR detector: circuit diagram (a) and timing protocol (b).

Figure 11.20 shows the interface of the G1U52X to the 9S12C as a robotic obstacle detection sensor. Each emitter receives a signal from a port T pin (PT[1:0]). A 74HC04 invertor and a current-limiting resistor are placed between the port T pin and the emitter. The 40-kHz oscillator runs constantly, but the emitter blinks only when the output at a pin is a logic HIGH. The application program is responsible for driving PT[0:1] pins HIGH and then LOW for 600 μs. As depicted in Fig. 11.20b, the digital output will be LOW (0 V) when the G1U52X detects reflected light and HIGH (5 V) when it does not, and thus pin PAD[0] of the 9S12C is used as normal digital input channel. The operating distance of the detector is limited to a few meters.

11.13 OPTICAL ENCODER

Optical encoders are noncontact, digital sensors used to sense the linear or angular position and velocity of mechanical objects. Rotary encoders are used to measure the angular position and velocity of a rotating shaft, whereas linear encoders are used to measure the linear position and velocity of a translating object. Linear encoders are more expensive than their rotary counterparts. Rotary encoders are the most widely used position sensors in robotic applications because they provide acceptable resolution with good noise immunity at low cost. The following discussion focuses on rotary encoders, but it applies equally to linear encoders.

Optical encoders are available in two distinct types: *absolute* and *incremental*. Both types use similar optical scanning techniques, which are implemented with five basic components

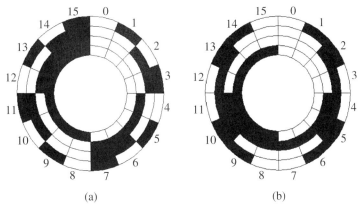

(a)   (b)

Figure 11.21 Absolute encoder wheel: binary code (a) and gray code (b).

(refer to Fig. 11.19c): a light source, a light sensor, an encoder disk, a mask, and interface electronics. The light source is usually an infrared-emitting diode (IR LED) and the light sensor is an IR phototransistor or photodiode. The emitter and detector are spectrally matched to operate at wavelengths in the region from 820 nm to 940 nm. The rotating encoder disk is attached to the moving body and placed between the light source and the light sensor. The disk is made from a thin laminate of chrome on glass, etched metal, or Mylar. The code disk is formed of opaque and transparent segments. The opaque lines are produced by a photographic process. The stationary mask is employed in high-resolution encoders to block extraneous light. The light emitted by the IR LED passes through the encoder disk and the stationary mask to produce an electronic output from the detector array according to the pattern on the disk. As the shaft rotates, the IR detector would be either on or off. When the opaque section of the disk breaks the light beam, the sensor is turned off, indicating digital zero; when the light passes through a transparent section, the detector is on, indicating digital 1. The output from the light sensor is usually a weak sinusoidal or triangular signal. The role of the interface circuitry is to amplify the signal and pass it through a Schmitt trigger to obtain a TTL signal compatible with digital code.

The laminated disk of an absolute encoder takes the form of N separate, equally spaced, concentric *tracks* with 2^N precisely placed opaque and transparent radial *sectors*. Figure 11.21 shows a binary-coded and a gray-coded wheel consisting of 4 tracks and 16 sectors. In the gray code, only one bit in the output changes as a result of wheel rotation. The basic resolution unit for a rotary absolute encoder is 2^N. The corresponding angular resolution r_θ for a single complete rotation is $r_\theta = 2\pi/2^N$.

The incremental encoder is simpler and has a less complex code disk and fewer components than the absolute encoder. The disk, shown in Fig. 11.22, is formed from a single track, with N

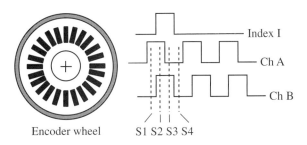

Encoder wheel   S1 S2 S3 S4

Figure 11.22 Encoder wheel, and index and quadrature counting.

opaque radial lines equally spaced around the rim of the disk. The resolution of an incremental encoder, which is the angular position it can resolve, is $360°/N$. The number of lines N is typically 100, 128, 200, 256, 500, 512, 1000, 1024, 2048, or 2540. The incremental encoder employs the principle of *phase quadrature sensing* to acquire information simultaneously on incremental steps and on direction of motion. This is accomplished by using two IR sensor pairs forming two output channels, A and B. The two output stages are separated by a quarter of a cycle to generate two identical pulse train signals (N pulses per revolution) that are approximately 90° out of phase. The direction of rotation (CW or CCW) depends on whether the output signal from channel A (channel B) leads or lags the signal from channel B (channel A). The 90° shift of the two signals divides each encoder cycle into four quarters, each called a quadrature count, as shown in Fig. 11.22. This increases the sensor's resolution in such a way that an encoder with N cycles/rev produces $4N$ quadrature counts per revolution. The resulting encoder gain would be $K_F = 4N/2\pi$. To obtain information on the number of complete rotations, the encoder provides a signal through a third channel, called the index I channel.

Rotary and linear shaft encoders are available from various vendors, such as the PED differential linear probe encoder series and the S5D rotary encoder series from US Digital (www.-usdigital.com). The PED linear plunger-style encoder is available with 1-in. and 2-in. travel versions and a range of cycles per inch (CPI) for a wide range of resolutions between 0.00210 and 0.00069 in. The PED tracks from 0 to 400 in./s and generates TTL-compatible quadrature digital outputs. The encoder can sustain a plunger force of 3–6 oz (1-in. travel) and 3–9 oz (2-in travel), a slide load of 1 lb, and vibrations up to 20 g. It has an internal spring to return the plunger to its fully extended position. The PED may be mounted in one of three different configurations. The PED has an internal line driver (26C31) that can source and sink 20 mA at TTL levels. Three twisted-pair cables plus power and ground should connect to the PED.

The S5D rotary series may be used for position feedback or for manual interface. The S5D operates from a single $+5$-V_{DC} supply, generates TTL-compatible quadrature outputs and an index output (optional), and tracks from 0 to 100,0000 cycles/s. Internal driving and connecting cable arrangements are similar to those of the PED series.

The HEDS series of optical encoders can detect linear position (when used with a linear encoder strip) or rotary position (when used with an encoder wheel). It provides two TTL-compatible quadrature outputs and an index pulse (optional). It operates from a single $+5$-V_{DC} supply and offers resolutions up to 2048 cycles per revolution.

Quadrature decoding is a time-intensive task if accomplished via software. Shifting this burden to hardware decoding relieves the microcontroller for other functions and improves overall performance of digital closed-loop motion-control systems. The HCTL 20xx family by Agilent Technologies (www.agilent.com) is a popular series of decoder/counter CMOS ICs designed to interface quadrature incremental encoders or digital potentiometers to the microcontroller. Each member of the family consists of a 4X quadrature decoder, a binary UP/DOWN state counter (12-bit for the HCTL 2000 and 16-bit for the HCTL 2016 and 2020), and an 8-bit bus interface. The HCTL 20xx units operate at frequencies up to 14 MHz.

11.14 PYROELECTRIC SENSOR

A pyroelectric sensor is a heat-flow detector that generates electric charge in response to a change in temperature. The active element of the sensor is essentially a capacitor formed from a thin slice or film of a pyroelectric crystal with electrodes deposited on opposite sides of the crystal to collect the thermally induced charge. When exposed to a heat source, the temperature of a pyroelectric material rises by ΔT. The sensor outputs a corresponding change in voltage Δv or charge Δq according to the relation

$$\Delta v = \frac{\Delta q}{C} = p_q \frac{A}{C} \Delta T = p_q \frac{\varepsilon_r \varepsilon_0}{h} \Delta T \qquad (11.18)$$

where p_q is the pyroelectric charge coefficient, ε_r is relative permitivity, ε_0 is the dielectric constant, A is the sensor's area, and h is its thickness. Lithium tantalate ($LiTaO_3$), TGS, lead-zirconate-titanate (PZT) ceramics, or plastic materials such as PVDF are commonly used pyroelectrics.

In a typical construction of a pyroelectric sensor, the sensor elements are placed inside a metal TO-8 or TO-39 can, to shield them. The can has a silicon window to protect the detector from outside moisture and pollutants. The inner space of the can is filled with dry air or nitrogen, and the sensor requires no cooling to produce a useful signal. Applications of pyroelectric sensors include radiation thermometry, gas analysis, human body detection (e.g., burglar alarms), and fire detectors. Pyroelectric sensors used in burglar alarm systems are optimized to detect radiation in the 8-to10-μm range, which is the range of infrared energy emitted by humans.

A pyroelectric sensor typically includes a focusing lens, usually a Fresnel plastic lens, to focus a thermal image of a body onto the surface of the sensor element. The charge generated on the electrodes is the sum of the charge induced by heating the crystal and the charge induced by the thermal expansion of the heated side. To capture the thermally induced charge and to compensate for rapid thermal changes, two symmetrically opposite identical pyroelectric elements, arranged in parallel or in series, are used. Alternatively, the dual element is often made from a single crystal, with two pairs of electrodes deposited on both sides, as shown in Fig. 11.23a. The upper electrodes are connected to form one electrode, while the two bottom electrodes are separated to form two series-connected capacitors C_1 and C_2.

The thermal image of an object is focused on the pyroelectric side covered by the continuous electrode. The charge is generated across the electrode pair subjected to the heat flux. As the thermal image moves from one electrode to the other, the pyroelectric current i_P through the sensing element changes from zero, to positive, back to zero, to negative, and again to zero, as Fig. 11.23c shows.

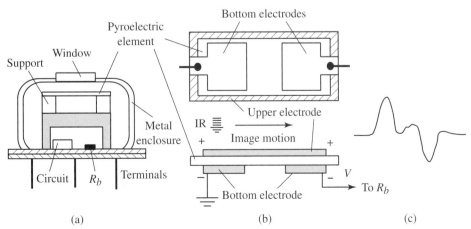

(a) (b) (c)

Figure 11.23 Pyroelectric sensor package (a), dual sensing element and electrodes (b), and voltage across R_b as the thermal image moves from left to right (c).

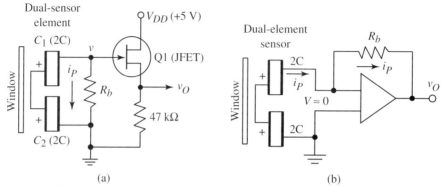

Figure 11.24 Impedance converters for pyroelectric sensor: voltage follower with JFET (a) and current-to-voltage converter with op-amp (b).

11.14.1 Signal Conditioning

The pyroelectric element behaves as a capacitance C in parallel with a leakage resistance of a value on the order of 10^{12}–10^{14} Ω. The value of the AC current generated by the element in response to a moving person is on the order of 1 pA. Practically, the output current flows through a very high-bias resistance R_b. The output impedance of the sensor, which is equivalent to the parallel combination of the capacitance C and R_b ($C\|R_b$), is very high. This high impedance is buffered by an impedance converter to prevent loading the sensor. The impedance converter is usually integrated with the sensor in the same package. Figure 11.24 shows two possible converter solutions; one is a voltage follower formed of a JFET transistor and a resistor, and the other is a current-to-voltage (I–V) converter. The voltage follower in Fig. 11.24a converts the high output impedance of the sensor into the output resistance of the follower, which is a parallel combination of the transistor's transconductance and the 47 kΩ. This low-cost circuit is simple and has low noise. But it has many disadvantages, which include (1) a response time that depends on the electrical time constant of the sensor $\tau_e = CR_b$, and (2) a large offset voltage across the output resistor that could reach several volts. The I–V converter circuit, on the other hand, has a faster time response, is insensitive to the capacitance of the sensor element, and has a low offset output voltage. However, because it has a broad bandwidth, it may suffer from high noise. The op-amp used for the I–V converter should have a very low input bias current, on the order of 1 pA. CMOS op-amps, such as the OP-41, the AD546, and the TLC271BCP, are preferable. At very low frequencies, the JFET circuit or the I–V converter transforms the pyroelectric current i_P into an output voltage according to

$$v_O = i_P R_b \tag{11.19}$$

11.15 THERMAL DETECTORS

Thermal sensors are used to measure heat-related variables, most fundamental of which is temperature. Knowing the temperature of an object, other important quantities, such as heat flux, heat capacity, and thermal energy, can be ascertained. Thermal sensors are contacting sensors in which the sensing element physically touches the heat source, such as thermocouples,

TABLE 11.4 Characteristics of Various Temperature Sensors

Device	Range (°C)	Resolution (°C)	Response Time	Sensitivity
Thermocouple (thin foil, type T)	−160 to +370	±3	5 ms	40 μV/°C
PRT (thin foil)	−200 to +1000	±0.1	1 ms	0.4Ω/°C @ 100 Ω
Thin-film platinum RTD (Honeywell HEL-700)	−200 to +540	±0.3	$\tau < 0.15$ s (in water) $\tau < 1$ s (on metal surfaces) $\tau < 4$ in air at 10 ft/s	
Thermistor (mini bead, oxide)	−270 to +450	±3	5 ms	Variable as NTC, PTC
Si thermodiode	−50 to 200	±3	10 ms	−2 mV/K
Si thermotransistor	−50 to 200	±1	10 ms	1 μA/K @ 300 μA

thermistors, RTDs, thermodiodes, and thermotransistors. Noncontacting thermal sensors using semiconductor elements, thermopiles, and pyroelectric sensors are considered radiation sensors used to detect electromagnetic waves emitted by a body. Infrared temperature measurement uses the 0.7- to 14-μm band of the electromagnetic spectrum. Table 11.4 gives the characteristics of various temperature sensors, many of which are discussed in the following sections.

11.15.1 Thermocouple

Seebeck Effect

The Seebeck effect is the generation of a measurable electric field due to temperature gradients across an element. The generated field tends to oppose the flow of charges attributed to the temperature imbalance. The measured voltage, or thermoelectric emf, across the element is determined by

$$\Delta v = \alpha(T_X - T_R) + \gamma(T_X^2 - T_R^2) \qquad (11.20)$$

where α is the Seebeck coefficient (or thermopower), γ is a constant, and T_R and T_X are the temperatures at the ends of the element. Table 11.5 gives the thermoelectric emf and Seebeck coefficient for various metals and thermocouple alloys. The Seebeck effect forms the basis for the design of a thermocouple to measure temperature and other, related quantities. It can also be exploited to generate thermoelectric power if a temperature difference across a junction is maintained.

A thermocouple consists of two dissimilar materials joined together, as shown in Fig. 11.25a, to form what is called the *sensing junction*. Neglecting the second term on the right-hand side of Eq. (11.20), the voltage captured between the leads is the sum of the voltages across the two thermocouple wires:

$$\Delta v = \alpha_1 \times (T_R - T_X) + \alpha_2 \times (T_X - T_R) = (\alpha_2 - \alpha_1) \times (T_X - T_R) \qquad (11.21)$$

The emf for any thermocouple may be obtained by subtracting the pertinent values listed in Table 11.5. A thermocouple with one arm made of iron and the other made of nickel has, for a 200°C temperature gradient, an emf of 6.64 mV, or a thermopower of 33.2 V/K.

TABLE 11.5 Thermoelectric emf's ΔV_R and Seebeck Coefficient (Thermopower) α of Various Metals and Alloys at 200°C Relative to Platinum at 0°C

Element	ΔV_R (mV)	α (μV/K)	Element	ΔV_R (mV)	α(μV/K)
Antimony	+10.14	+50.7	Pt—10% Rh	+1.44	+7.20
Chromel	+5.96	+29.8	Aluminum	+1.06	+5.3
Iron	+3.54	+17.7	Tantalum	+0.93	+4,65
Molybdenum	+3.19	+16.0	Platinum	0.00	0.00
Tungsten	+2.62	+13.1	Calcium	−0.51	−2.55
Cadmium	+2.35	+11.8	Palladium	−1.23	−6.15
Gold	+1.84	+9.20	Alumel	−2.17	−10.85
Copper	+1.83	+9.15	Cobalt	−3.08	−15.40
Silver	+1.77	+8.85	Nickel	−3.10	−15.50
Rhodium	+1.61	+8.05	Constantan	−7.45	−37.25
Pt—13% Rh	+1.47	+7.35	Bismuth	−13.57	−67.85

Chromel: 90% Ni, 10% Cr; constantan: 55% Cu, 45% Ni; nisil: 95% Ni, 4.5% Si; alumel: 95% Ni, 2% Al, 2% Mn, 1% Si; nicrosil: 71–86% Ni, 14% Cr, 0–15% Fe.

Thermocouples Types

Thermocouples are fabricated as thick wires, thin wires, or thin foils. A thick-wire thermocouple is very robust, usually consisting of a single wire of 200-μm diameter. A thin-wire thermocouple has lower thermal mass and responds rapidly ($\tau < 0.1$ s) to temperature change. A thin-foil thermocouple consists of a typically 50-μm-thick foil supported by a polymide film. It occupies a small surface space and provides extremely fast response ($\tau \sim 5$ ms). Semiconductor thermocouples (Si and Ge) are used to make thermoelectric microsensors. The Seebeck coefficients of a 0.38-μm-thick polysilicon films relative to Pt are–100 for n-polysilicon (30 Ω/x), −450 for n-polysilicon (2600 Ω/x), and 270 for p-polysilicon (400 Ω/x).

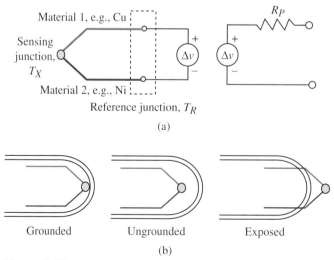

Figure 11.25 Thermocouple junction (a) and junction types (b).

TABLE 11.6 American Standard ANSI MC 96 (1975) Thermocouples

Junction Material	Designation	Sensitivity (μV/°C @ 25°C)	Temperature Range (°C)	Accuracy
Copper/constantan	T	40.69	−270 to +600	±1.5°C; ±2.5%
Iron/constantan	J	51.71	−270 to +1000	±2.2°C; ±0.75%
Chromel/alumel	K	40.46	−270 to +1300	±2.2°C; ±0.75%
Chromel/constantan	E	60.93	−200 to +1000	±1.7°C; ±0.5%
Pt /Rh—10% Pt	S	6.02	0–1550	±1.5°C; ±0.75%
Pt/Rh—13% Pt	R	5.93	0–1600	±1.5°C; ±1.5%

Commonly used thermocouples have been standardized according to color and type. Table 11.6 gives the type designation, material used, sensitivity, temperature range, and accuracy of the American Standard ANSI MC 96 (1975) thermocouples that offer reasonably linear output with temperature.

Figure 11.25b shows the possible thermocouple junction types. The sensing junction of a grounded type is attached directly to the probe wall, which provides good heat transfer between the outside and the junction through the probe wall. The response of the ungrounded type is slower because the sensing junction is detached from the probe wall. The sensing junction in the exposed thermocouple protrudes outside the sheath and is exposed to the environment. This construction provides for the fastest response time, but it is not suitable for measurements in a corrosive or pressurized environment.

Cold-Junction Compensation

To capture the output from the thermocouple, its wires are connected to the measuring device using lead wires of dissimilar materials, usually copper. This will inevitably introduce two other undesirable junctions, known as parasitic thermocouple junctions. According to the law of intermediate metals, these junctions will have no effect if they are isothermal, that is, held at the same temperature. Meanwhile, standard thermocouple tables provide the thermoelectric voltage corresponding to a reference junction temperature of 0°C. Thermocouples tables are available from many sources, such as www.iseinc.cm. If the absolute temperature of the process is needed, *cold-junction compensation* (CJC), to account for a nonzero junction temperature, becomes necessary. Many methods could be used to accomplish CJC. One approach is to measure the reference junction temperature T_R with another sensor, convert it into a thermoelectric voltage V_R, and add it to the thermoelectric voltage of the thermocouple Δv. V_R is equivalent to a voltage that would be generated if the sensing junction were heated to the reference temperature while the reference junction was held at 0°C. The resultant thermoelectric voltage V_X (T_R = 0°C) = $\Delta v + V_R$ corresponds to a 0°C reference junction. Another way to accomplish CJC is to use an IC with built-in CJC, such as the AD594/AD595 ICs from Analog Devices.

Thermocouple Interface

Figure 11.26 shows the essential components of a thermocouple-based temperature measurement. The thermocouple leads are connected to a reference junction via an isothermal terminal block. The terminal block maintains the leads at the same temperature and provides electrical isolation. It also couples the reference junction to a thermistor or IC temperature sensor, such as the LM335 from National Semiconductor (www.national.com) or the AD590 from Analog Devices (www.analog.com).

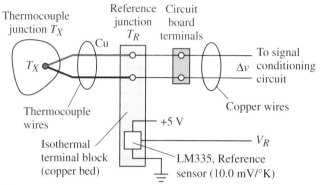

Figure 11.26 Temperature measurement using a thermocouple with a semiconductor reference sensor.

Because the emf output from a thermocouple is very small, it is susceptible to electrical noise generated by surrounding electrical machinery. Additionally, the thermocouple wires act as an antenna for picking up electromagnetic noise from various sources. Measures that could be taken to reduce the effect of noise on thermocouple measurements include the use of active filters, the use of a twisted-lead-wire shield in a grounded sheath, and differential measurements using an instrumentation amplifier with high CMRR. For long connecting wires, 4-to 20-mA current loop transmitters (refer to Section 11.25) provide the most reliable way to transmit thermocouple data.

The thermocouple output voltage is converted into an equivalent temperature value using a linear relation or a power series polynomial with type-dependent coefficients. The linear relation has a slope equal to the Seebeck coefficient and a zero offset. The power series polynomial offers higher accuracy. Polynomial coefficients for the thermocouple type used are obtained from thermocouple standards.

11.15.2 Thermopiles

Invented by James Joule, a thermopile is essentially a group of N thermocouples wired to be electrically in series and thermally in parallel, to produce a voltage N times that of a single thermocouple. The arrangement of the five identical thermocouples shown in Fig. 11.27 produces five times the output of a single thermocouple. Contemporary thermopiles belong to a class of PIR (passive infrared) detectors used to detect thermal radiation in the far-infrared region. The heated junctions are coated with an absorber (i.e., gold or bismuth black) to convert incident radiation into heat and to generate a voltage signal. Thin-film thermopiles (CU/constantan and Ni/Cr) have been realized using microfabrication technology. Figure 11.27b shows a typical layout of a radiation thermopile.

11.15.3 Theremoresisitive Devices

These devices operate on the principle that the resistance of a material changes with temperature. The relation between the resistance and the temperature of a material is called the *temperature coefficient of resistance* (TCR) α. Metals have positive temperature coefficients (PTC), because their resistance changes directly with temperature. Many semiconductors and oxides have negative temperature coefficient of resistance (NTC), because their resistance

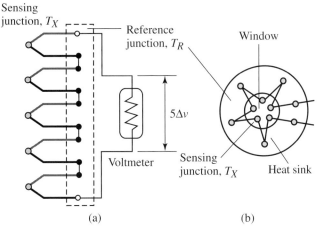

Figure 11.27 A basic thermopile circuit (a) and a typical layout of a thermopile radiation sensor (b).

varies inversely with temperature. Figure 11.28 shows the variation of resistance with temperature of PTC and NTC semiconductors and that of platinum and nickel. $\alpha(/°C)$ is 0.0067 for Ni, 0.0039 for Pt, 0.0038 for Cu, 0.0045 for W, and -0.0005 for C.

There are two types of thermoresisitve sensors: metal-based *resistance-temperature detectors* (or RTDs) and semiconductor based *thermistors*.

Resistance-Temperature Detectors

Although virtually all metals can be used to fabricate RTDs, platinum, copper, nickel, and nickel/iron (known by the trademarks Balco and Hytempco) are most common, with platinum by far the most common. Table 11.7 gives temperature range, sensitivity, and base resistance values for these metals. Platinum exhibits extremely linear resistance vs. temperature variations,

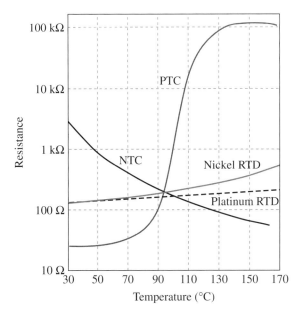

Figure 11.28 Variation of resistance with temperature for typical PTC and NTC thermistors and platinum and nickel RTDs.

TABLE 11.7 Properties of Various Metal RTDs

	Platinum	Nickel	Copper	Nickel/Iron (70%/30%)
Temperature range °C (°F)	−269 to 593 (−452 to 1100)	−80 to 320 (−112 to 608)	−73 to 149 (−100 to 300)	−46 to 343 (−50 to 650)
Temperature coefficient (Ω/Ω/°C)	0.00385–0.003923	0.0067	0.0043	0.0051
Base resistance value (Ω) @ 0°C (32°F)	100, 200, 500, 1000	120	10	—
Fabrication	Wire wound, thin foil	Wire wound	Wire wound, etched foil	Wire wound, etched foil

as depicted in Fig. 11.28. Platinum RTDs are also used for their accuracy, repeatability, predictable response, long-term stability, and durability. Reference-grade platinum is used for most industrial, commercial, laboratory, and critical RTD temperature measurements. Nickel RTDs are used for some older navy shipboard measurements and with older controllers. Copper RTDs are used for temperature measurement in electric motors and generators and in a few other industrial applications. Nickel/iron RTDs are used for windshield temperature measurements and with older controllers. Tungsten RTDs are usually applicable for measuring temperatures over 600°C.

RTDs are fabricated in two forms: wire-wound RTDs and thin-film RTDs. A typical *wire-wound* probe consists of a precision platinum element wound as a coil around high-temperature ceramic mica or glass core, sealed within a ceramic or glass capsule, and then inserted into a hollow metal (usually stainless steel) sheath of diameter of 1/8 to 1/4 in. to protect it from moisture and hostile environment. RTDs are made with two, three, or four lead wires protruding from the sensor casing. Wire-wound RTDs are widely used in industrial and scientific applications due to their high stability.

Thin-film RTDs are fabricated from a thin film of platinum or its alloys deposited on a flat ceramic substrate and encapsulated in silicone rubber to provide a moisture-proof seal. A thin-film RTD is as small as a small signal transistor. Thin-film RTDs provide excellent linearity, fast response (typically less than 0.5 s), high accuracy (tolerance less than 0.1%), and excellent stability.

The resistance versus temperature function for platinum is accurately modeled by the Callendar–Van Dusen equation:

$$R_T = R_0(1 + AT + BT^2 - 100CT^3 + CT^4) \tag{11.22}$$

where R_T (Ω) is the resistance at temperature T (°C) and R_0 is the resistance at 0°C. A, B, and C are the Callendar–Van Dusen constants, which are determined from the properties of the platinum used in the construction of the sensor. They are basically derived from α and other constants, δ and β, which are obtained from actual resistance measurements. These constants are related by:

$$A = \alpha(1 + \delta \times 10^{-2})$$
$$B = -\alpha \times \delta \times 10^{-4} \tag{11.23}$$
$$C_{T<0} = -\alpha \times \beta \times 10^{-8}$$
$$\alpha = (R_{100} - R_0)/100R_0$$
$$\delta = [R_0(1 + 260\alpha) - R_{260}]/(4.16R_0 \times \alpha)$$

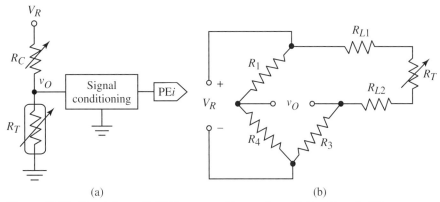

Figure 11.29 Signal from a Pt RTD processed by a voltage divider (a) and a Wheatstone bridge (b).

where R_{100} and R_{260} are resistance at 100°C and 260°C, respectively. Note that $\beta = 0.16$ for $T < 0°C$ and $\beta = 0$ for $T > 0°C$.

Figure 11.29 illustrates the interface between the 9S12C and a two-wire RTD using voltage divider and bridge circuits. Because RTDs have low absolute value and temperature coefficients, they are prone to measurement errors, which result in an increase in the output voltage v_O. One source for the error is the voltage drop across the resistance of the lead wires, R_{Li}. This error is significant when the sensor is remotely located. For example, if a 100-Ω RTD is located 100 ft away and the signal is transmitted by a conventional twisted-pair shielded cable, than a 5-Ω lead resistance would induce a 15°C error in a 0–100°C temperature measurement. The second source of error is due to increase in the resistance of the lead wires and the RTD caused by Joule's heating.

Two-wire RTDs are typically used with very short lead wires—when the sensor is connected directly to the receiver—or with a 1000-Ω element. A better measurement arrangement is shown in Fig. 11.30. In the three-wire RTD shown in Fig. 11.30, R_{L1} and R_{L3} carry the measuring current while R_{L2} acts as a potential lead. Ideally R_{L1} and R_{L3} are perfectly matched and their effect cancelled out. R_3 is chosen to be equal to R_T at a given T (usually the midpoint of the temperature range) where no current passes through the center lead. As the temperature increases, R_T also increases, causing the resistance to be out of balance. Current will flow in the center lead, indicating an offset temperature.

Figure 11.30b shows a four-wire RTD. This is the optimum arrangement because it removes the error of mismatched resistance of the lead wires. A constant current is passed through R_{L1} and R_{L4} while R_{L2} and R_{L3} measure the voltage drop across the RTD. With a constant current, the voltage is strictly a function of the resistance and a true measurement is achieved. This arrangement is the most expensive, but it provides the highest measurement accuracy.

Thermistors

A thermistor is a semiconductor device that exhibits a change in resistance with temperature. Thermistors are similar to resistance thermometers, but with much improved sensitivity because they exhibit a higher rate of change in resistance with temperature. As Fig. 11.31a demonstrates, the thermistor has nonlinear temperature-resistance characteristics, which are generally approximated by one of several different relations. In a negative temperature

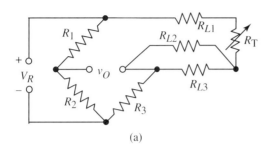

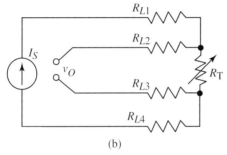

Figure 11.30 RTD measurement with three wires (a) and four wires (b).

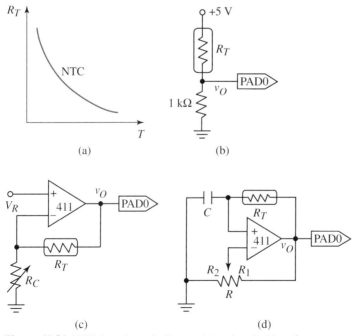

Figure 11.31 NTC thermistor: R–T curve (a) and typical interface methods (b–d).

coefficient (NTC) thermistor, the resistance decreases exponentially as temperature increases. The relation can be approximated as

$$R_T = R_{T_0}\exp[\beta(1/T - 1/T_0)]$$ (11.24)

where R_T is the resistance at temperature T measured in degrees Kelvin (K), R_{T_0} is the resistance at a known reference temperature T_0, and β (K) is a measure of the thermistor's sensitivity, referred to as a characteristic temperature of the thermistor. β can be determined by making two resistance measurements, R_1 and R_2, at two different temperatures, T_1 and T_2. Figures 11.31b and c illustrate the interface between the thermistor and the 9S12C. The output voltage V_{OUT} of the op-amp circuit in Fig. 11.31c is

$$v_O = V_R(1 + R_T/R_C)$$ (11.25)

R_T is found from Eq. (11.25) and then used in Eq. (11.24) to determine T. In addition to measuring temperature, thermistors are used in a wide variety of other applications, such as measuring the velocity of a moving fluid, measuring the level of fluid in a tank, and measuring altitude, to name a few. The op-amp circuit in Fig. 11.31d uses a thermistor for temperature-to-frequency conversion.

11.15.4 Thermodiode

When a voltage source is connected to a *p-n* junction diode, current flows continuously and the junction is said to be *forward-biased*, provided that the applied voltage V_D is greater than the forward-biased voltage V_b of the diode, which is 0.7 V for silicon and 0.25 V for germanium at 25°C. The forward current I that flows through a forward-biased *p-n* junction diode for a given applied voltage is given by the diode (Shokley) equation:

$$I = I_S\left[\exp\left(\frac{q_e V_D}{\eta k_b T}\right) - 1\right]$$ (11.26)

I_S is the constant reverse saturation current, which is related to the junction area A, the doping profile, and other junction parameters. For a Si diode, the current is about 25 nA at 25°C, but it could reach a level of 7 mA at 150°C. η is a nonideality multiplicative factor. $\eta = 1$ for the ideal diode and 1.5 typically for nonideal behavior. q_e is the electron charge (1.602×10^{-19}C), k_b is Boltzmann's constant (1.3085×10^{-23} J/K), and T is the absolute temperature (K). Rearranging Eq. (11.26) gives the voltage dependence on temperature as

$$v_O = (\eta k_b T/q_e)\ln(I/I_S + 1)$$ (11.27)

This equation indicates that the voltage is directly proportional to absolute temperature (PTAT) and provides the basis for using the diode as a temperature sensor. The forward junction voltage is approximately linear with temperature, with a slope of -2 mV/°C within a region between 50 and 300 K. Since diode voltage is a function of current, a stable constant-current source is essential. A diode operating under a constant-current source gives a temperature coefficient (dr_O/dT) that is theoretically independent of temperature and dependent only on current terms.

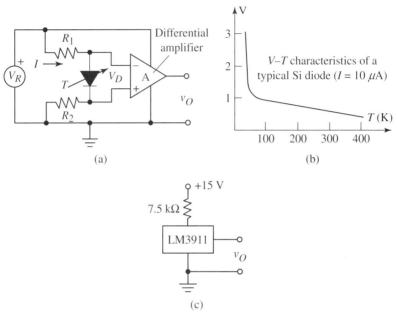

Figure 11.32 Basic circuit (a) and V–T characteristics (b) of a typical Si diode, and typical connection of the LM3911 Si thermodiode (c).

Diodes are more sensitive and exhibit higher linearity in most of their usable range but have less repeatability than a thermocouple or a resistance thermometer. Measurement accuracy within $\pm 1°C$ is possible. A high-precision GaAS diode thermometer may have an accuracy of ± 0.002 K within the range from 14 to 300 K. Diode thermometers should not be used in the presence of a magnetic field greater than 1 tesla. Advantages of using thermodiodes as temperature sensors include compatibility with IC technology and low manufacturing cost.

Figure 11.32 shows an amplifier circuit to measure the forward-biased voltage drop across a thermodiode. Resistors R_1 and R_2 limit the current I flowing through the junction. The output from the differential op-amp is linearly dependent on temperature. Integrated temperature sensors, such as the LM3911, manufactured by National semiconductors (www.national. com), include on a single device a thermodiode and an amplifier operating from a single voltage supply.

11.15.5 Thermotransistor

The voltage between the base and the emitter of a transistor depends on temperature. The governing relation is similar to Eq. (11.26), except I_S and other constants depend on the device geometry. The voltage–temperature relation is exploited in the manufacture of IC thermotransistor microsensors that are easily integrated into microelectronic circuits. The ability to include laser-trimmed resistors onboard the IC further improves the accuracy of the thermotransistor.

If two different collector currents are applied to a single transistor, a high current I_{C1} and a low current I_{C2}, the difference in base–emitter voltage V_{BE} would depend only on the collector currents rather on the geometrical or material factors, according to the relation (from Eq. (11.26))

$$\Delta V_{BE} = (V_{BE1} - V_{BE2}) = (kT/q_e)\ln(I_{C1}/I_{C2}) \qquad (11.28)$$

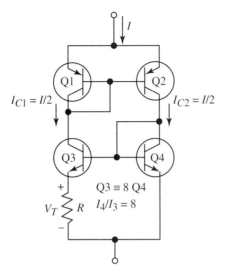

Figure 11.33 Integrated Si thermotransistor.

This equation indicates that the transistor is a proportional-to-absolute-temperature (PTAT) device.

Several thermotransitors are usually integrated in a single device to provide a more robust measurement. Figure 11.33 shows a circuit in which several thermotransitors are connected to form a temperature sensor. This is the basic circuit on which the Analog Devices AD590 sensor is based. Transistors Q1 and Q2 have the same base-to-emitter voltage V_{BE1}. They form a current mirror that splits the total current into two equal currents I_{C1} and I_{C2}. Q3 consists of eight transistors identical to Q4 connected in parallel. The collector current passing through Q4 is thus eight times that passing through each Q3. The voltage drop across the resistor R is the difference between the base–emitter voltages of Q3 and Q4; that is,

$$V_T = (V_{BE4} - V_{BE3}) = (KT/q_e)\ln(I_4/I_3) = \frac{KT}{q_e R}\ln 8 \qquad (11.29)$$

Since the current through R is $I/2$, the current is directly proportional to the absolute temperature and is given by

$$I = \frac{2KT}{q_e R}\ln 8 \qquad (11.30)$$

11.16 HEAT FLUX SENSOR

A heat flux sensor is used to measure the heat flux through a surface. A typical heat flux sensor employs a pair of thermocouples or a thermopile 5 sensing layers and a layer of thermal resistance material separating the sensing elements. When the junctions of the two thermopile layers are at different temperatures, a differential voltage proportional to the heat flux is generated. To ensure the existence of a proper thermal gradient, the thermal conductivity of the sensor has to be the same as or larger than that of the material to which the sensor is mounted. The response time of the sensor is improved by reducing the thickness of the thermal resistance layer.

Heat transfer through a medium occurs in three modes: radiation, convection, and conduction. Radiation heat sources are used to calibrate most flux sensors because they are repeatable sources. The relation of incident and absorbed heat flux for a radiation source is

$$q''_{absorbed} = \varepsilon q''_{incident} \tag{11.31}$$

where ε is the emissivity of the body, which is equal to the absorbtivity since the body is assumed gray. Sensor emissivity and the amount of radiation from the source that actually impinges on the sensor considerably affect the integrity of the measurement.

The heat flux equation for convection is

$$q''_{absorbed} = h \times \Delta T \tag{11.32}$$

where h is the convective heat transfer coefficient of the sensor, which is usually determined experimentally. It is a function of the fluid's thermal conductivity and the characteristics of fluid flow. ΔT is the temperature difference between the sensor and the fluid.

The heat flux equation for conduction is

$$q''_{absorbed} = q''_{incident} = -k\left(\frac{\partial T}{\partial \mathbf{n}}\right) \tag{11.33}$$

where k is the thermal conductivity of the sensor, $\partial T/\partial \mathbf{n}$ is the thermal gradient, with $\mathbf{n}$ the unit vector normal to the surface through which the heat flux is being measured. Flat layered sensors are usually mounted on a target surface with a thermally conductive adhesive to minimize thermal contact resistance.

11.17 MAGNETIC SENSORS

Magnetic sensors are used to detect the presence of a magnetic field. If a moving object to which a magnet is attached passes by a magnetic sensor, the sensor generates a logic signal that indicates the presence of the object. The magnetic reed switch and the Hall-effect device operate on this principle.

11.17.1 Magnetic Reed Switch

Figure 11.34 shows the essential components of a magnetic reed switch. The switch is made of a pair of slightly overlapped ferromagnetic reed contacts enclosed in a sealed glass tube approximately 1 in. long and with a $\frac{1}{8}$-in. diameter. The contact tips touch when exposed to an external magnetic field and close the circuit. Like a mechanical switch, the operation of a magnetic switch requires a pull-up resistor. Reed switches are employed as proximity sensors and as door- or window-closure sensors in security applications. They require no power to operate and have very high impedance in the open state and negligible impedance (150 mΩ) when closed. A reed switch can operate over 100 million or more cycles at levels of 50 V/100 mA or higher before it fails. The typical sensing distance of a reed switch is $\frac{1}{2}$ in.

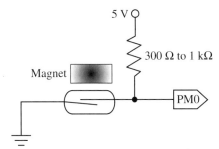

Figure 11.34 A Magnetic reed switch interfaced to the 9S12C.

11.17.2 Hall-Effect Device

The operation of this device is based on the Hall-effect principle, discovered by E. H. Hall in 1879. It is based on the Lorentz force that deflects flowing electrons to one edge of the conductor. The uneven lateral charge distribution gives rise to electric field **E**, which, in turn, exerts a force $\mathbf{F} = q\mathbf{E}$ in opposition to the Lorentz force. At equilibrium, the two forces balance each other.

If a constant-current I flows through a plate that is under the influence of an external magnetic field B perpendicular to the current, a very small voltage V_H, called the Hall voltage, is produced across the transverse direction, as shown in Fig. 11.35a. For a plate of length l, width w, and thickness t (cm), the voltage is linearly related to the magnetic flux according to

$$V_H = \frac{R_H \times I \times B}{t} \tag{11.34}$$

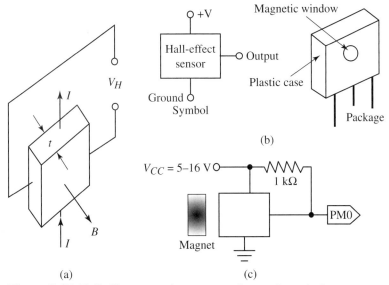

Figure 11.35 Hall-effect sensor. A current carrying conductor in the presence of a static magnetic field generates transverse voltage (a); a Hall-effect device (b) and its interface to the 9S12C (c).

where R_H is the Hall coefficient, a measure of the open-circuit product sensitivity (mV/mA-kG). The polarity of the Hall-effect voltage is a function of the polarity of the magnetic field. R_H values for bulk InAs, thin-film InAs, GaAs, and InSb are, respectively, 0.1, 1.0, 2.0, and 160.0.

As shown in Fig. 11.35a, a Hall-effect device consists of a semiconductor Hall plate with four leads attached to it. The current flows through two leads and the differential Hall voltage is generated across the other two leads. A Hall-effect device and its symbol are shown in Fig. 11.35b.

Figure 11.35c shows a Hall-effect sensor interfaced to the 9S12C. A pull-up resistor is inserted between the V_{CC} supply lead and the output lead. A third lead connects to the ground. The supply voltage varies between 5 and 16 V. The Hall-effect device can be used to sense proximity just as a reed switch does. However, it requires a stronger magnetic field than a reed switch. Typical sensing distance of a Hall-effect device is 1/16 to 1/4 in. Figure 11.36 shows a Hall-effect sensor driving various loads.

Linear Hall-effect sensors are generally packaged with a voltage regulator and an output amplifier, as shown in Fig. 11.37. The resulting package improves stability and increases the bandwidth of the sensor to over 100 kHz. Furthermore, adding a Schmitt-trigger threshold detector and a transistor output driver transforms the linear Hall-effect sensor into a digital Hall-effect switch that is insensitive to reverse magnetic polarity and provides for very rapid switching response times, typically in the 400-ns range.

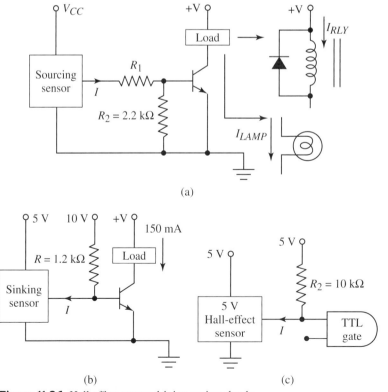

Figure 11.36 Hall-effect sensor driving various loads.

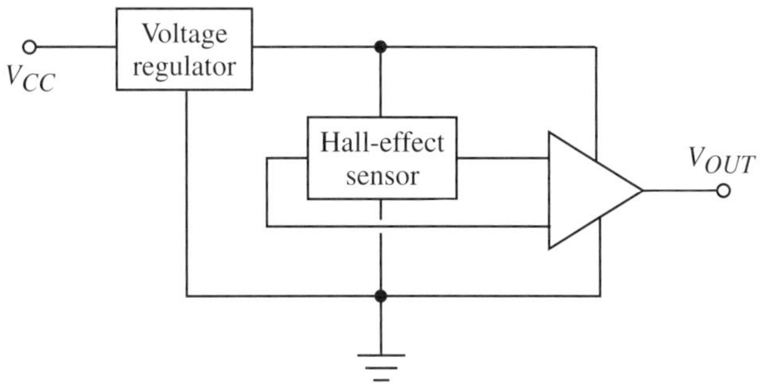

Figure 11.37 A Hall-effect sensor incorporating a voltage regulator and a stable DC output amplifier.

Position Sensing

Hall-effect sensors can also be used for position and speed sensing, motor commutations, guide-path following, and magnetic compasses. An array of reed switches or Hall-effect devices can be used to measure the linear or angular position of moving objects, as shown in Fig. 11.38. A magnet is embedded into the bottom of a moving object (rotating shaft). As the object moves laterally (or the shaft rotates), the magnet passes over several separate reed switches or Hall-effect sensors. Each sensor is located at a precise position so that the position of the table (shaft) can be determined from the sequential activation of sensors. The linear (angular) position resolution depends on the number and spacing of the sensors. Also, sensors must be located close enough to the table (shaft) to be within their given sensing range.

Current Sensing

Two effects may be utilized for current sensing: voltage drop and magnetic field generation. Voltage-drop measurement requires the placement of a highly linear and stable shunt resistor in the current path. The voltage drop across the sense resistor is measured, and Ohm's law is then used to determine current flow. Because the output of a voltage-drop measurement is typically in the millivolt AC range, further conditioning is required to affect contact closure or to yield a

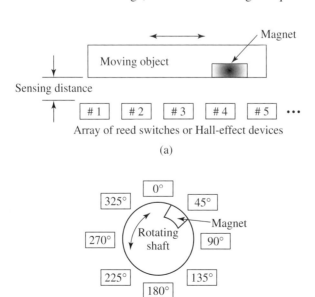

Figure 11.38 Sensing linear (a) and angular (b) positions using magnetic sensors.

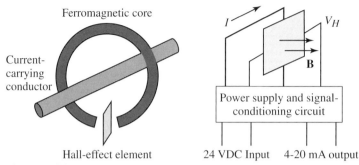

Figure 11.39 Current-sensing scheme using a Hall-effect device.

process level signal such as 4–20 mA. This technique ignores application and system requirements and is prone to errors emanating from nonideal component behavior. It is suitable only for DC current measurements and low-frequency AC measurements (<100 Hz).

Sensing current on the basis of magnetic field generation offers the advantage that no direct contact with the current being monitored is required. Possible principles that can be utilized include the transformer, the Hall effect, and the magnetoresistive effect. Figure 11.39 shows a Hall-effect sensor used to measure current through a conductor. Depending on the design, frequencies from DC to about several kilohertz can be measured. The current-carrying conductor is surrounded by a circular, magnetically permeable core to concentrate the magnetic field produced by current flow. The Hall-effect element is carefully placed in a small air gap in the core, at a right angle to the concentrated magnetic field. When excited by a constant current in another plane, the Hall-effect element generates a very small voltage, proportional to the measured current. The output voltage is then amplified into a process-level signal, such as 4–20 mA, or to close a contact in the case of a current switch. The sensitivity of the measurement is temperature dependent and requires adequate compensation. There is an inevitable offset, a small DC voltage at zero current.

The Hall-effect current sensor can be employed either open loop or closed loop. In open loop, the amplified output signal of the Hall-effect element is used directly as the measurement value. In the closed-loop arrangement, the output current from the amplifier flows through a secondary coil on the magnetic core. The amplitude of the magnetic field generated in the secondary coil has the same value but is opposite in direction to that of the primary current conductor. The result is that the magnetic flux in the core is compensated to zero. Closed-loop measurement offers better sensitivity, linearity, and less temperature dependence than open-loop measurement, but at a higher cost. However, the offset remains. Closed-loop current sensors can measure frequencies up to approximately 150 kHz.

11.18 STRAIN GAUGES

Strain gauges provide a simple and cost-effective way to measure strain and strain-derived quantities, such as stress, force, torque, pressure, displacement, and acceleration. The principle of operation of a strain gauge is based on the change in electrical resistance of a material when it is mechanically deformed. The fractional change in resistance of a strained wire can be expressed as a linear function of the applied strain, ε, by

$$\frac{\Delta R}{R} = (1 + 2\nu)\varepsilon + \frac{\Delta \rho}{\rho} = G\varepsilon \qquad (11.35)$$

TABLE 11.8 Composition, Gauge Factor, and TCR of Common Alloys and Semiconductors

Material	Composition %	Gauge Factor, G	TCR (C^{-1}10^{-5})
Advance or constantan	45 Ni, 55 Cu	2.1	±2
Karma	74 Ni, 20 Cr, 3 Al, 3 Fe	2.0	+2
Isoelastic	36 Ni, 8 Cr, 4 Mo, 4 Mn, 4 Si, 52 Fe	3.6	+17
Nichrome	80 Ni, 20 Cr	2.1	10
Platinum-tungsten	92 Pt, 8 W	4.0	+24
Armour D	70 Fe, 20 Cr, 10 Al	2.0	
Silicon	p-type	100 to 170	70–700
Silicon	n-type	−100 to −140	70–700
Germanium	p-type	102	
Germanium	n-type	−150	

where $\Delta R/R$ is the fractional change in resistance, $\Delta\rho/\rho$ is the fractional change in resistivity, v is the Poison's ratio, and ε is the strain, is equal to $\Delta L/L$. G is the *gauge factor*, or *strain sensitivity*, expressed as

$$G = \frac{\Delta R/R}{\varepsilon} = \frac{\Delta\rho}{\rho\varepsilon} + (1 + 2v) \qquad (11.36)$$

Strain gauges are fabricated as metal filaments, metal foils, or single-crystal semiconductor material. The first term in Eq. (11.35) corresponds to the dimensional change in the wire, which is dominant in metals. For most metals, $v \approx 0.3$ and G is at least 1.6. The second term results from the change in resistivity due to the change in the crystal lattice of the material under strain. This term is called the *piezoresistive effect*, which dominates in semiconductor materials. Semiconductor materials possess a higher gauge factor than metals. Single-crystal semiconductor strain gages are most suitable for dynamic measurements. However, they are sensitive to temperature variations because they have high temperature coefficient of resistance (TCR). Consequently, temperature compensation is required for measurements over a broad temperature range. Temperature compensation is best achieved using a four-gauge balanced Wheatstone bridge circuit. Table 11.8 gives gauge the composition, gauge factor, and TCR of commonly used strain gauge materials.

The variation in resistance in a strain gauge usually does not exceed 2%, and the resistance of the metallic wire may be expressed as

$$R = R_0(1 + \Delta) \qquad (11.37)$$

where R_0 is the stress-free resistance and $\Delta = G\varepsilon = \Delta R/R$. For semiconductors, the relation depends on the doping concentration.

11.18.1 Bridge Circuit

Bridge circuits are one form of signal conditioning that implements the ratiometric technique for the purpose of improving sensor sensitivity and accuracy. Bridge circuits are commonly used to process signals from resistive, capacitive, and inductive sensors. The four-arm resistive bridge shown in Fig. 2.13 is widely used in the measurement of temperature, pressure, strain, force, and magnetic fields. The excitation voltage V_R may be AC or DC. One or more arm of the bridge is replaced with a resistive sensor, such as a strain gauge, a thermistor, a resistance thermometer, or a piezoresistive pressure sensor. Many arrangements for locating active and passive sensors in the bridge circuit are possible. Refer to Section 2.6.2 for the null condition relation and equivalent circuit.

The Wheatstone bridge operates in one of two modes: the deflection mode and the null mode. In the *deflection (unbalanced) mode*, the bridge starts from a balanced state. If the sensor resistance R_S changes by Δ, an offset voltage across the bridge diagonal is generated, rendering $\Delta v \neq 0$. The output voltage Δv is a nonlinear function of disbalance and can be used to compute the cause of change. This method requires a calibrated meter and is suitable for small changes in resistors ($\Delta < 0.005$). In the *null (balanced) mode*, the bridge is always maintained at the balanced state by varying a calibrated variable resistor, say R_2, using a control circuit. Any change in the sensor resistance R_S causes the meter to deflect and the control circuit to generate an error signal that is used to modify R_2 until the effect of changing R_2 cancels that of R_S and the bridge is balanced again. This approach takes more time if R_2 is adjusted manually. The choice between the null and deflection methods in a given case depends on the speed of response, the drift, and other factors dictated by the application.

Bridge response is very nearly linear as long as the resistance variations Δ_i ($i = 1, 2, 3, 4$) are a small percentage of R_i. In strain gauges, Δ_i rarely exceed 1% R, or $\Delta << 1$, and, for this practical interest, the sensitivity of the bridge is given by

$$\frac{\Delta v}{v_R} = \frac{R_4 \Delta_1 - R_1 \Delta_4}{(R_1 + R_4)^2} - \frac{R_3 \Delta_2 - R_2 \Delta_3}{(R_2 + R_3)^2} \tag{11.38}$$

If the bridge has one active transducer with resistance $R_S = R_0(1 + \Delta)$ in arm 3 and the other three arms have a fixed baseline of resistance values R_0, then for $\Delta << 1$ the sensitivity reduces to

$$\frac{\Delta v}{v_R} = \frac{\Delta}{2(2 + \Delta)} \approx \frac{\Delta}{4} \tag{11.39}$$

Δ must be less than 0.02 if the error owing to nonlinear effects is to be kept to less than 1%.

Equation (2.24) may be used to assess loading errors associated with the bridge circuit. If $R_L = \infty$, then $\Delta v_L = \Delta v$; however, if $R_L \neq \infty$, then the bridge signal $\Delta v_L < \Delta v$. For example, if $R_L = 10R_B$, then $\Delta v_L / \Delta v = 0.9$, which means that the signal will lose about 9% of its ideal value, which reduces bridge sensitivity.

Temperature Compensation

Resistance-based sensors are prone to changes in resistivity with temperature. Unless properly compensated, the dependence of resistivity on temperature results in a highly undesirable effect. One way to compensate for temperature effects is to incorporate with the bridge circuit a temperature-compensation network, as shown in Figure 11.40. If the sensitivity α of all arms in the bridge to the stimulus s is

$$\alpha = \frac{1}{R} \frac{dR}{ds} \tag{11.40}$$

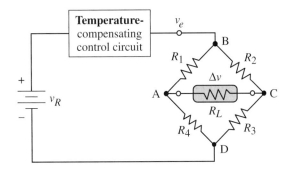

Figure 11.40 Four-arm bridge circuit with a temperature-compensation network and load resistance.

then the network generates a temperature-related signal, which is incorporated in a feedback control loop, to vary the excitation voltage of the bridge v_e with temperature at the same rate, and with opposite sign, at which the temperature coefficient of sensitivity α of bridge arms changes with temperature. Among many options for temperature compensation, a popular technique, which provides satisfactory compensation for a resistive Wheatstone bridge, is to use a fixed resistor R_C with low temperature sensitivity as the compensating network. It can be shown that for a bridge with four resistive arms R, the compensating resistor is selected according to

$$R_C = -R\frac{\beta}{\gamma + \beta}, \qquad |\beta| < \gamma \tag{11.41}$$

where β is the temperature coefficient of sensitivity (TCS) of the bridge arm. For a resistive bridge, it is given by

$$\beta = \frac{1}{\alpha}\frac{\partial \alpha}{\partial T} \tag{11.42}$$

γ in Eq. (11.41) is the temperature coefficient of bridge arm resistance (TCR), defined as

$$\gamma = \frac{1}{R}\frac{\partial R}{\partial T} \tag{11.43}$$

If β, γ, and R are known, then Eq. (11.41) is used to select R_C.

11.18.2 Strain-Gauge Measurement

Figure 11.41 shows two identical two-element strain gauges attached to opposite faces of a thin beam to measure its bending stresses, deflections, or vibrations under load P. A simple system to accomplish the measurement task is also shown. A similar arrangement is employed for strain-gauge measurement of pressure, acceleration, etc. The measurement system consists of a stable DC excitation, a bridge circuit with four active arms, and an instrumentation amplifier. The amplifier output signal is fed to the ADC of the 9S12C. If $v_R = 5$ V, $\Delta = 0.1\%$, and the gain of the instrumentation amplifier is 800, then the signal fed into the ADC is $v_i = (0.001)(5)(800) = 4$ V.

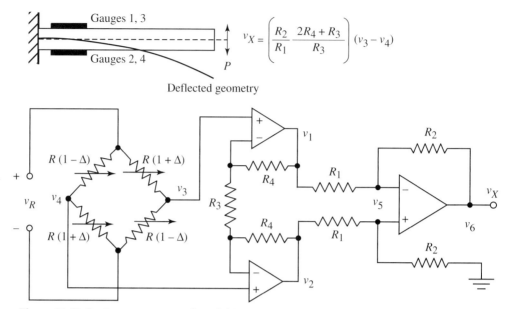

Figure 11.41 Strain measurement using a bridge circuit interfaced to an instrumentation amplifier.

11.19 ACOUSTIC MEASUREMENT

Mechanical vibrations in an elastic medium produce waves characterized as rapid changes in the air pressure and perceived by the brain as sound. *Sound* refers to the hearing (audible) range of the human ear, encompassing the frequency range from 20 Hz to 20 kHz. Sound waves with frequencies less than 20 Hz are referred to as *infrasound* and above 20 kHz are called *ultrasound*. The range of the audio spectrum is from 0 Hz to 30 kHz.

11.19.1 Properties of Wave Propagation

Sound waves cannot propagate through a vacuum, but can through an elastic medium, such as a solid, a liquid, or a gas. The wavelength λ (m) of an acoustical wave relates to the frequency f (Hz) through the longitudinal speed of propagation s (m/s):

$$\lambda = s/f \tag{11.44}$$

Sound waves behave very much like light waves at very short wavelengths.

The propagation of sound waves in a medium is longitudinal, whereas each particle in the medium oscillates near a stationary position in the direction of wave propagation. Sound waves cause compression in the medium in the form of a pressure change Δp, which results in a change in volume ΔV. The ratio of pressure change Δp to the corresponding relative change in volume $(\Delta V/V)$ defines the *bulk modulus of elasticity* K_M of the medium. The speed of wave propagation in a medium is given by

$$s = \sqrt{K_M/\rho} \tag{11.45}$$

where ρ is the density of the medium outside the compression zone. The speed of sound also depends on the temperature of the medium, since both K_M and ρ are functions of temperature.

TABLE 11.9 Acoustic Data for Several Common Materials

Material	Sound Velocity s (m/s)	Mass Density ρ (kg/m³)	Acoustic Impedance Z_a (kg/m²-s)
Air	332	1.281	425.3
Hydrogen	1269	0.090	114.2
Water	1461	998	1.46×10^6
Clay, ceramic	3000–5000	1500–2500	4.5×10^6 to 12.5×10^6
Wood	3048–4572	480–800	1.5×10^6 to 3.7×10^6
Aluminum	5102	2643	13.48×10^6
Brass	3499	8553	29.93×10^6
Copper	3557	8906	31.68×10^6
Iron	5000	7100	35.5×10^6

This dependence is the basis for the operation of the *acoustic thermometer*. Sound travels in 25°C air at approximately 1138 ft/sec at sea level. The actual speed is influenced by altitude, humidity, and barometric pressure. The speed of sound in seawater at 25°C at sea level is 5034 ft/sec. In solids, the speed of sound may be expressed as

$$s = \sqrt{\frac{E(1 - v)}{\rho(1 - v)(1 - 2v)}} \tag{11.46}$$

where E is the modulus of elasticity and v is the Poisson's ratio of the material. Acoustic data for several common materials is given in Table 11.9.

Acoustic Impedance

The acoustic impedance in an idealized (no loss) sound-propagating media is

$$Z_a = \rho s \quad \text{or} \quad Z_a = \sqrt{E\rho} \tag{11.47}$$

where Z_a is the acoustic impedance (kg/m²-s), ρ is the mass density of the medium (kg/m³), s is the velocity of sound in the medium (m/s), and E is the modulus of elasticity of the medium. For a freshwater medium with a density of 1000 kg/m³, the speed of sound through it is $s = 1490$ m/s and the acoustic impedance is $Z_a = 1.49 \times 10^6$ (kg/m²-s). For aluminum, $E = 7.3 \times 10^{10}$ N/m², $\rho = 2643$ kg/m³, and so $Z_a = 1.39 \times 10^7$ (kg/m²-s). In an acoustic measurement, if the transducer's impedance Z_T does not match the impedance of the medium Z_M through which sound is traveling, the use of an *impedance-matching transformer* becomes necessary. The transformer's material is selected to provide the coupling impedance Z_C according to

$$Z_C = \sqrt{Z_T Z_M} \tag{11.48}$$

Sound Intensity

The intensity of a sound wave refers to the acoustical power (W/m²) concentrated or transferred per unit area. It is expressed as the ratio

$$I = P^2/Z = Pv \tag{11.49}$$

TABLE 11.10 Sound Levels β Referenced to I_0 at 1 kHz

Sound Source (Distance from Observer)	Level (dB)
Rocket engine at 50 m	200
Threshold of pain	120+
Jet engine (3 m)	140+
Rock concert	120
Piezoelectric buzzer (12 in)	108
Subway train	102
Heavy-machine shop	100
Air Force T-38 (2500 ft overhead)	90
CO_2 pellet gun (12 in.)	90
Digital alarm clock (12 in.)	85
Vacuum cleaner (18 in.)	80
Air Force T-38 (1 mile)	70
Typical conversation	65
Paper clip dropped on desk (12 in.)	62
Telephone dial tone (1 in.)	56
Pencil eraser tapped on desk (12 in.)	54
Average residence	45
Soft background music	30
Quiet whisper	20
Threshold of sound	0

p is the *acoustic pressure*, defined as the difference between the instantaneous pressure at a given point in the media and the average pressure, and v is the instantaneous velocity of a vibrating particle. Sound pressure is usually quantified by *sound pressure level*, which is defined as

$$\beta = 10\log_{10}[I/I_0]$$ (11.50)

where $I_0 = 10^{-12}$ W/m^2 is the reference sound intensity that is just audible to the human ear. It corresponds to a reference pressure $p_0 = 2.9 \times 10^{-9}$ psi caused by sound propagating in air. I_0 is usually referenced to a pure tone at 1 kHz to which the human ear is most sensitive. The units of β is the decibel, or dB. An increase of 10 dB corresponds to an increase of 10 times the initial intensity. Table 11.10 gives β values for several sound sources.

11.19.2 Acoustic Sensors

Acoustic sensors are used to measure sound waves propagating at frequencies within a wide range, from low megahertz to very high gigahertz. An acoustic sensor can achieve high resolution when operating at shorter wavelengths. The design of a specific sensor depends on the frequency range of the signal to be measured and on the media through which sensed sound waves propagate. For example, ultrasonic diagnostic instruments used in medical applications operate below 1 MHz, much higher than the relatively low frequencies of sound waves traveling in air.

The two essential elements of an acoustic sensor are a diaphragm that moves under impinging sound wave and a displacement sensor to capture the movement. The diaphragm motion creates compression and expansion of the medium. The sensor converts the diaphragm deflection into an electrical signal. In addition to these two elements, an acoustic sensor may include an acoustic damping chamber (muffler) to prevent ringing from occurring following a transmitted pulse, a focusing lens to concentrate acoustic power at a specific location, an impedance-matching circuit, housing, and connecting wires.

Acoustic Lens

The acoustic lens helps concentrate the energy radiating from the vibrating source, such as a piezoelectric element, to the target area by guiding the sound waves to travel along a predictable path. The concave surface of a spherical lens produces a *spot focus*, whereas a conical lens produces a *line focus*. For a spherical lens, the exact location of the focal point relative to the source is found by

$$F_L = \frac{r}{1 - s_1/s_2} \tag{11.51}$$

where F_L is the focal length from the center of the radiating lens to the focal point (m), r is the radius of the concave lens surface (m), s_1 is the velocity of sound inside the lens (m/s), and s_2 is the velocity of sound in the target medium (m/s); s_2 must be larger than s_1 for Eq. (11.51) to yield a relevant location of the focal point.

11.19.3 Types of Transducer Element

Acoustic transducers are of two types: electrostatic and piezoelectric. The electrostatic transducer is similar to a parallel-plate electrical capacitor with one fixed plate. The electrostatic transducer generates small forces but yields relatively large displacement amplitudes. It also couples more effectively to a compressible medium such as air. Using a foil membrane allows for faster turn-on and turn-off due to the membrane's very low inertia compared to that of the plate. Membrane transducers are much more broadband because they are not limited to a specific resonant frequency; however, they are limited to an upper frequency in the kilohertz range.

Condenser Microphone

This capacitive-based sensor converts the distance d between two parallel plates into an electrical voltage according to the following relation:

$$V = \frac{d}{q\varepsilon_0 A} \tag{11.52}$$

where, $\varepsilon_0 = 8.8542 \times 10^{-12}$ C^2/Nm^2 is the permitivity constant, A is the plate area, and q is the charge. The displacement of the moving plate under the impinging acoustic pressure changes the distance d between the two plates. Thus, the electrical output of the microphone is proportional to the sound-pressure level. The sensitivity of the microphone depends on the magnitude of the charge, supplied either by a (20- to 200-V) external power supply or by an internally built-in electret layer. Silicon diaphragms are used instead of a capacitor plate in some designs. Larger static diaphragm deflections result in higher transducer sensitivities but

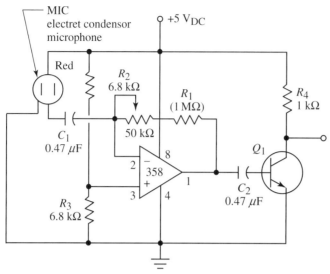

Figure 11.42 An electret microphone connected to a sound detector amplifier.

require higher bias voltages, which may result in reduced shock resistance and lower dynamic range. A smaller capacitor's air gap reduces the mechanical sensitivity of the microphone at higher frequencies.

Figure 11.42 shows a simple interface circuit for an electret condenser microphone. An LM567 tone decoder IC, shown in Fig 11.43, may be hooked to the output of the 358 op-amp to decode tones at any frequency of interest. To complete an acoustic system, a tone generator built from the 555 timer shown in Fig. 11.44 may be added to generate tones at any desired frequency.

Figure 11.45 shows the components of a piezoelectric microphone. An externally applied AC voltage excites the piezoelectric element to oscillate near its resonant frequency, where it is most efficient and sensitive. The resonant frequency depends on the geometry of the piezoelectric element, which is usually a disk with electrodes deposited on both sides. Signal wires are either

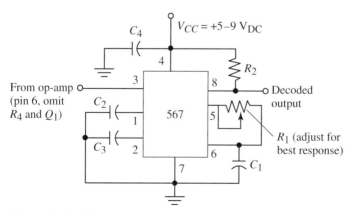

Figure 11.43 Schematic and connection of the LM567 tone detector.

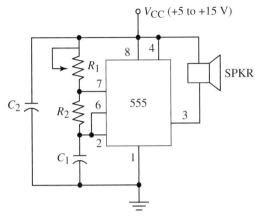

Figure 11.44 A tone generator using the LM5555 timer.

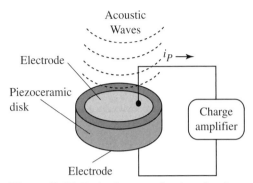

Figure 11.45 Basic elements of a piezoelectric microphone.

soldered to the electrodes or connected to them by electrically conductive epoxy. Piezoelectric transducers exhibit some latency due to mechanical inertia of the vibrating piezoelectric crystal. The force generated by the piezoelectric can be significant, but the amplitude of vibration is typically small. The output impedance of a piezoelectric sensor is very high, and processing the output signal requires a high-input-impedance amplifier. The charge amplifier shown in upcoming Fig. 11.50 is commonly used for this purpose. Piezoelectric transducers couple well to solids and liquids but rather poorly to low-density compressible media such as air.

Acoustic transducers used in certain measurements, such as level and time of flight, require a transmitter and a receiver. A single piezoelectric element can be used as a transmitter and a receiver if it is excited to emit a single pulse. If it is excited to transmit continuous waves, separate elements are employed for the transmitter and the receiver.

Electret

An electret is a crystalline material permanently polarized by electrical means to create a constant charge density σ on its surface. Figure 11.46 shows an electret acoustic sensor. It is an electrostatic transducer consisting of a metallized electret diaphragm and a back plate with an air-gap separation. The constant charge density on the surface of the electret sets an electric field E in the air gap. The resistor R connects the metallic layer bonded to the electret and the back plate. The deflection of the diaphragm caused by a sound wave reduces the air gap x_1 by

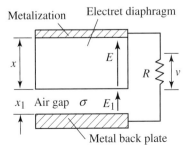

Figure 11.46 Components of an electret microphone.

a distance Δx and generates a voltage v across the electrodes in phase with the diaphragm deflection. The open-circuit amplitude of the variable portion of the output voltage is given by

$$V = \frac{x\Delta x}{\varepsilon_0(x + \varepsilon x_1)} \frac{2\pi fRC}{\sqrt{1 + (2\pi fRC)^2}} \tag{11.53}$$

where f is the frequency of the acoustic wave, ε_0 is the permitivity of air, x is the thickness of the electret, C is the capacitance, and ε is the permittivity of the electret. This voltage is amplified and used as the output signal.

The electret condenser microphone is the most sensitive acoustic sensor. Although it is as small as a watch battery, it can pick up sound several meters away. It is, however, susceptible to ambient noise, and, unlike piezoelectric elements, the electret microphone requires electricity to operate. Most electret microphones come with a built-in FET amplifier stage.

11.19.4 Types of Measurements

Acoustic sensors are used in the measurement of proximity, level, time of flight, velocity of moving objects, temperature, manufacturing defects, object detection, fluid flow, material properties, and many more. In medical applications, ultrasonic imaging is safer to use than X-ray machines, and, by varying the intensity of the transmitted waves, image penetration at different depths becomes possible. As a result, images of organs that would normally be transparent to X-rays may be obtained. Other applications of acoustic sensors are briefly introduced in the following sections.

Object (Proximity) Sensing

A typical ultrasonic proximity sensor consists of a transmitter and a receiver, although a single transducer could be used for both functions. The transmitter emits a longitudinal wave at a typical frequency between 20 and 200 kHz. Higher-frequency (short-wavelength) acoustic waves bounce off smaller surface areas far more efficiently than lower-frequency (longer-wavelength) waves. As a rule of thumb, to effectively detect an object using high-frequency waves, the object must be at least one wavelength in length or width.

Part of the emitted acoustic energy reflects back to the receiver if a target is present in the acoustic field. When the amplitude of the returned energy reaches a preset threshold, the state of sensor output changes, indicating detection. Ultrasonic proximity sensors are capable of detecting liquid and solid objects over distances of several feet. The maximum detection range depends on the power level and the direction of the emitted wave and on the cross-sectional area and reflectivity of the target.

Time of Flight (TOF)

The ultrasonic TOF ranging system commonly known as *sonar* (sound navigation and ranging) measures the round-trip time t required for a *single* pulse of emitted ultrasonic wave (initiation of a ping) to travel to a reflecting object and then to echo back to a receiver. The round-trip distance is given by

$$d = v \times t \tag{11.54}$$

where t is the elapsed time and v is the speed of propagation. TOF is a straight-line active sensor in which the returned signal follows essentially the same path back to the receiver, located coaxially with or in close proximity to the transmitter. The sensor does not depend on the plane property or orientation of the target surface, provided that reliable echo detection is sustained. As a consequence of Huygen's principle, spherical wavefronts will develop around objects lying in the path of the target. This will scatter the acoustic energy and reduce the energy received at the target. As a result, a weaker signal will be reflected back to the receiver. Thus, the detected echo represents only a small portion of the original signal.

The wavelength of sonar operating at 200 kHz would be approximately 0.3 m under water and 0.07 m in air. In theory, better resolution is expected with sonar in air than in water. However, the opposite is true, for the following reasons: Water is a better conductive medium; the mismatch of acoustical impedance between the transducer and air is much larger than that for water ($Z_{Water} > Z_{Air}$); and underwater operation involves locating larger discrete objects.

Ultrasonic sensors are used, rather than sonar, to detect small objects because a small surface area reflects shorter wavelengths (higher frequencies) far more efficiently than wave sounds with longer wavelength. To be detected effectively by an ultrasonic signal, the object must be one wavelength in length or width.

Transit-Time Flow Measurement

A transit-time ultrasonic flow meter is used to measure the flow of blood in vessels and of liquids (water, milk, pharmaceuticals, paper pulp slurry) in pipes. The sensor uses a transmitter and a receiver placed at a distance L either externally on opposite sides of a pipe or blood vessel or within the pipe along the flow direction. The former arrangement is preferred and is shown in Fig. 11.47. The sound wave path makes an angle θ with the direction of flow. The transit time t is determined by

$$t = \frac{L}{s \pm \hat{u}\cos\theta} \tag{11.55}$$

where L is the distance between the transducers, s is the speed of sound in the fluid, and $\hat{u}$ is the velocity of the fluid averaged along the ultrasonic path. $\hat{u} = 1.33\bar{u}$ for laminar flow and $\hat{u} = 1.07\bar{u}$ for turbulent flow, $\bar{u}$ being the velocity averaged over the cross-sectional area. The $(+)$ in the denominator is used when the sound waves travel along the flow (receiver is downstream) and the $(-)$ if the receiver is upstream.

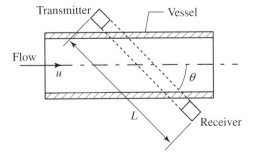

Figure 11.47 Ultrasonic transit-time flow meter.

Doppler Frequency

The radial direction and velocity of a moving object may be determined by measuring the frequency of the returned energy of a continuous-wave signal. If a source emits an acoustic wave at frequency f_e while moving at a speed v_s, the Doppler frequency f_r arriving at a fixed receiver is determined by

$$f_r = f_e(s/s \pm v_s) \tag{11.56}$$

where s is the speed of sound in air. The sign of v_s is negative for a source approaching the observer and positive as the source is moving away from the observer. Meanwhile, if the source is fixed but the observer is moving at a velocity v_0, the observed Doppler frequency is

$$f_r = f_e(s \pm v_0/s) \tag{11.57}$$

In the case of a reflected wave, it is more convenient to use the Doppler shift instead of the Doppler frequency f_r. The doppler shift is the change in frequency Δf, given by

$$\Delta f = f_e - f_r = \frac{2f_e v \cos\theta}{s} \tag{11.58}$$

where v is the velocity of the target object and θ is the relative angle between the direction of motion and the acoustic beam axis. The factor 2 in Eq. (11.58) indicates a round-trip path.

Doppler radar is also used to measure the speed of or distance to a moving target. It works on a similar principle, but it is based on radar signal transmission and receiving rather than on ultrasonics. Equation (11.58) applies in this case, except the speed of sound s is replaced with the speed of light c.

Example 11.5: Fluid velocity−A Doppler ultrasonic flow meter operating with a carrier frequency of 5 MHz is used to measure the velocity of fluid flow in a pipe. Given that the speed of sound in the fluid is 1500 m/s and the angle between the transducer and the direction of the flow is 60°, determine the velocity of the fluid if a Doppler frequency shift of 6 kHz is obtained.

Solution: Rearrange Eq. (11.58) as

$$v = \frac{\Delta f \times s}{2f_e \cos\theta}$$

Substituting the pertinent data yields

$$v = \frac{(6 \times 10^3 \, \text{Hz})(1500 \, \text{m/s})}{(2)(5 \times 10^6 \, \text{Hz})(\cos 60°)} = 1.8 \, (\text{m/sec})$$

Acoustic Pyrometry

The speed of sound is a strong function of the temperature of the medium through which sound waves travel. The speed of sound in a gas is related to gas temperature by

$$s = (\gamma RT/M)^{1/2} \tag{11.59}$$

where γ is the ratio of the specific heat of the gas at a constant pressure to that at a constant volume, R is the universal gas constant (8.314 J/K-mol), T is absolute temperature (K), and M is the molecular weight of the gas (kg/mol). This phenomena is exploited in acoustic pyrometry to measure the average temperature of the path traversed by the acoustic wave. If d (m) is the distance traveled by the sound wave in τ (s), then substituting $s = d/\tau$ into Eq. (11.59) yields the relation between gas average temperature T (°C), distance d, and flight time τ:

$$T(°C) = \left(\frac{d}{\tau B}\right)^2 - 273.16 \tag{11.60}$$

where $B = (\gamma R/M)^{1/2}$ is the acoustic constant. Experience has indicated that the practical frequency range of the acoustic pyrometer in measuring temperatures in large power plants is between 500 and 2000 Hz. With temperatures reaching as high as 3000°F, practical temperature resolution and accuracy are possible, provided the flight time is resolved to a fraction of one wavelength.

Detecting Manufacturing Defects

High-frequency ultrasound is highly suitable for detecting internal manufacturing defects such as cracks, delaminations, and disbonds in relatively small parts and assemblies such as microelectronic devices. Depending on the application, the transducer emits ultrasonic pulses of frequencies between 10 and 100 MHz thousands of times per second as the object of interest is being scanned. Lower frequencies can penetrate thick parts, but the image echoed back to the receiver has lower resolution. On the other hand, higher frequencies are more useful for thinner materials or more acoustically transmissive materials, but they achieve higher resolution. The characteristics of the return echoes provide clues to whether internal defects are present. Depending on the condition of the target, three possible manifestations occur. If the target material is homogeneous and defect free, the transducer receives no echoes. If the target consists of bonded layers of material, the acoustic impedance value at the layer interface changes, causing part of the signal to be reflected back, and the nonreflected part continues to travel deeper into the target to the next interface. If the target contains a defect such as a crack, disbond, delamination, or void, the entrapped air forces all the ultrasound to be reflected back to the transducer.

11.20 PIEZOELECTRICITY

11.20.1 Piezoelectric Effect

Piezoelectric materials are a special class of anisotropic crystals that exhibit coupling between mechanical and electrical behavior in a manner known as the *piezoelectric effect*. When subjected to a mechanical stress, a piezoelectric element generates electric charge and voltage. This is known as the *direct piezoelectric effect*, which is exploited for many sensors. If, on the other hand, an electric field is applied to the crystal, a mechanical stress is produced. This is known as the *converse piezoelectric effect*, which gives the crystal its usefulness as an actuator.

The piezoelectric effect occurs naturally in quartz (SiO_2), tourmaline, Rochelle salt, and ammonium dihydrogen phosphate. These crystalline materials are nonconducting and exhibit varying degrees of piezoelectricity, and their properties are easily controlled. Piezoceramics such as lead-zirconate-titanate, or PZT ($Pb(ZrTi)O_3$), are artificially synthesized piezoelectrics. They are manufactured by compacting a mixture of powder under high pressure and temperature, followed by a process referred to as *poling*. Curie temperatures of piezoceramics can be as high as 575°C. A piezoceramic material is dense, brittle, and hard, resists wear, and is chemically inert. It exhibits excellent electromechanical coupling and has good longitudinal piezoelectric coefficients, a high dielectric constant, a low voltage coefficient, and high acoustical impedance. However, it is susceptible to failure under impact or shock loads.

Piezopolymers such as polyvinylidene fluoride, or PVDF, are synthetic noncrystalline materials. Piezoelectricity is induced into them by a special stretching-and-poling treatment. They are light and flexible, have low acoustic impedance, a low dielectric permittivity, and a high voltage coefficient, require a large electric field to accomplish polarization, and can sustain much larger strains. Piezopolymers can be produced in large sheets and complex geometries at low cost. They can operate at temperatures up to 100°C, have a dynamic sensitivity in the range 10^{-8} to 10^6 N/m (286-dB range), and have a maximum response frequency in the gigahertz range, and their dielectric constant is about one one-hundreth that of ceramics.

11.20.2 Piezoelectric Use in MEMS

Piezoelectric effects are widely utilized in fabricating semiconductor microsensors and actuators. Since silicon does not posses a piezoelectric effect, a thin layer of crystalline material is deposited on a silicon substrate to provide for the piezoelectric effects. Zinc oxide (ZnO), aluminum nitride (AlN), and PZT are the most popular materials used in semiconductor microsensors. ZnO provides for ease of chemical etching. A thin layer of ZnO and a thin layer of PZT are usually deposited by sputter deposition. AlN has high acoustic velocity and can withstand humid and high-temperature conditions. Thin films of AlN are usually deposited by chemical vapor deposition (CVD) or reactive beam epitaxy (MBE) methods. Both techniques require the substrate to be heated to high temperatures (up to 1300°C). Thick films of PZT have also been deposited on silicon substrates using the sol-gel deposition method.

11.20.3 Constitutive Relations in One Dimension

Figure 11.48 shows the convention used to indicate direction in a polarized crystal. Axis 3 is the polar axis, (indicated by the arrow and P) because it is taken parallel to the direction of ceramic polarization. The directions of the applied stress and the poling axis identify the

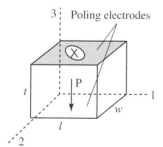

Figure 11.48 Axes of a polarized crystal.

mode of the ceramic material, which is described by the numbered axes 1, 2, and 3. If, for example, axis 3 is the poling direction and the stress axis is along axis 1, the crystal is said to operate in 31 mode; the first digit refers to the poling axis and the second refers to the stress direction. When the mechanical stress or strain is shear, number 5 is used as the second subscript. For shear operations, the poling electrodes are removed and replaced by electrodes deposited on a second pair of faces. Graphite and copper are two materials commonly used as electrodes for ceramic crystals. The electrodes are either plated or sprayed onto the crystal element's body. l, t, and w are, respectively, the length, thickness, and width of the piezoelectric element.

The constitutive relations in one dimension include one strain component S_1 (m/m), one stress component T_1 (N/m^2), one electric field component E_3 (V/m), and one electric displacement component D_3 (coul/m^2). They are expressed as

$$D_3 = d_{31}T_1 + \varepsilon^T_{33}E_3$$

$$S_1 = \frac{1}{Y_P}T_1 + d_{31}E_3$$

(11.61)

where d_{31} is a piezoelectric constant (Coul/N or m/V), ε_{33} is the dielectric permittivity constant (farads/m) and Y_p is the Young's modulus (N/m^2) of the piezoelectric material. The first subscript in d_{31} indicates the *electrical axis*, which is the direction of the electric field, and the second subscript indicates the *mechanical axis*, which is the direction of the piezoelectrically induced strain or applied stress. The electrodes are perpendicular to the direction indicated by the second subscript.

11.20.4 Piezoelectric Sensor

When used as a sensor, a piezoelectric element generates a charge proportional to the applied stress. The level and polarity of the induced charge depend on the placement of the electrodes in relation to the polling axis and on the direction of the applied force. For example, the polarity of the output voltage is opposite to the direction of the polarizing voltage if a tension force is applied along the polar axis. The piezoelectric sensor behaves as a capacitor; the crystal is the dielectric and the electrodes form the parallel plates on which charges are deposited. Figure 11.49 shows possible sensor configurations and the corresponding relations between the applied force and the voltage and charge they produce.

Example 11.6: Piezoelectric sensor–A crystal having a length of 5 mm, a width of 20 mm, and a thickness of 3 mm is sandwiched between compression plates. A 5-N force is applied on this crystal along the polar axis. Determine the amount of voltage generated if the voltage coefficient is $g_{33} = 26.1 \times 10^{-3}$ V-m/N.

Solution: Figure 11.49a represents the given crystal arrangement. Thus,

$$v = \frac{Fg_{33}t}{lw} = \frac{(5\,N) \times (26.1 \times 10^{-3}\,\text{V}.\text{m/N}) \times (0.003\,\text{m})}{(0.005\,\text{m})(0.020\,\text{m})}$$

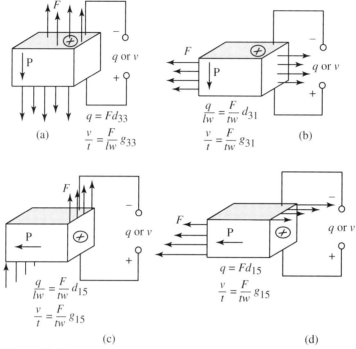

Figure 11.49 Possible piezoelectric sensor arrangements.

Amplification

The piezoelectric transducer is a dielectric with a high leakage resistance; i.e., it has a very high output impedance Z_{OUT}. The conditioning circuit must have a very high input impedance to prevent loading the sensor.

The AC *charge amplifier* shown in Fig. 11.50 was developed specifically for the high-impedance, piezoelectric crystal-type transducers used in the dynamic measurement of shock,

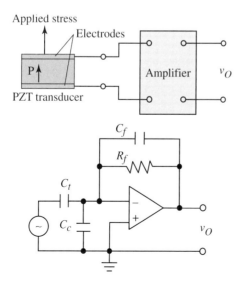

Figure 11.50 AC charge amplifier.

vibration, pressure, force, and sound. The amplifier converts to voltage the very low charges or currents signals generated by devices such as capacitive sensors, quantum detectors, pyroelectric sensors, and accelerometers. The gain of the AC amplifier is relatively independent of the shunt capacitances normally associated with the use of piezoelectric transducers, an advantage over voltage amplifiers. C_t in Fig. 11.50 represents the capacitance of the PZT element and C_c is the shunt capacitance of the cable. The effective input capacitance of the amplifier circuit is $C_i = (C_t + C_c - C_f)(A-1)$, where A is the open-loop gain of the voltage amplifier. The transfer function of the converter is

$$v_O = \frac{\Delta q \times A}{(C_t + C_c - C_f)(A - 1)} \qquad (11.62)$$

Assuming A to be very large, the effects of C_t and C_c are negligible and Eq. (11.62) reduces to

$$v_O = -\frac{\Delta q}{C_f} \qquad (11.63)$$

with an overall charge conversion gain of $-1/C_f$. The negative sign indicates that the output is 180° out of phase with the input. If C_f is measured in nanofarads, the charge conversion gain is in units of mV/pC. A large feedback resistance R_f must be used to prevent the nonzero op-amp bias current from developing a significant steady charge on C_f. However, it causes the op-amp to behave as a high-pass filter with a time constant $R_f C_f$. A high amplifier input resistor R_i is also needed, which is usually obtained by using an FET in the input stage. Consequently, the low corner frequency of the amplifier $f = 1/2\pi R_f C_f$ precludes DC operation. However, using a T-resistor network provides for a much higher effective feedback resistor than a single resistor and allows for very low operating frequencies, less than 2 Hz. Since the high-input-impedance FET amplifier is a virtual ground, the voltage across the equivalent impedances (resistances and capacitances) of the cable and transducer is essentially zero. Thus, the length of the cables has no effect on the sensitivity or frequency response of the system. The leakage resistance of the transducer must be substantially larger than the impedance of the capacitor at the lowest operating frequency to provide for good low-frequency response. When used with quartz-crystal transducers, the value of C_f is from 10 to 100,000 pF and R_f is from 10^{10} to 10^{14} Ω.

11.20.5 Piezoelectric Mass-Sensitive Chemical Sensor

A piezoelectric quartz crystal coated with a thin film of a chemically sensitive soft layer, such as spin-coated polymer, may be used as a chemical sensor. When a chemical species reacts with the coating, it results in a small change in the total mass Δm of the crystal. The change in crystal mass Δm is converted to a change in frequency Δf according to the Sauerberry equation,

$$\Delta f = -\frac{1}{\rho_m k_f} f_0^2 \frac{\Delta m}{A} \qquad (11.64)$$

where A is the crystal surface area, ρ_m is the density of the thin active coating (electrode), and $1/\rho_m k_f$ is a physical prefactor, approximately equal to 2.3×10^{-7} m²/Hz-kg for a quartz plate. The active coating ultimately determines the performance and chemical selectivity of the quartz mass. This device is called a *quartz crystal microbalance*, or QCM. It has high sensitivity but can only operate at relatively low temperatures ($<50°C$). The sensor is not suitable for use with high-temperature oxide materials, such as SnO_2.

11.21 RESOLVER

A resolver is a mutual-induction device used to measure the absolute angular position of a shaft. The construction of a resolver is similar to that of a small AC motor (refer to Section 12.4). It consists of a rotor and a stator, each having two coils that are mechanically displaced 90° relative to one another. The AC excitation may be applied to either the rotor or stator coils, and the signal generated at the other coils represents the output. Figure 11.51 is a commonly used configuration, in which one rotor winding is excited with the AC carrier signal

$$V_r = V\sin\omega_{ac}t \tag{11.65}$$

where V_r is the rotor voltage and ω_{ac} is the frequency of the excitation signal. The AC excitation is coupled to the rotor via slip rings and brushes. In brushless resolvers, a special cylindrical transformer is employed to couple the excitation signal to the rotor. The voltage output from the stator windings is proportional to the sine and cosine of the applied rotor excitation and the angular position of the rotor shaft θ. The amplitude-modulated voltage signals induced in the two stator windings are given by

$$V_a = V\sin\theta\sin\omega_{ac}t$$
$$V_b = V\cos\theta\sin\omega_{ac}t \tag{11.66}$$

where V is a constant that depends on the inherent characteristics of the resolver. Both signals in Fig. 11.51 are needed to capture the position and direction of rotation. The modulated stator signals are demodulated to filter out the carrier signal. The output is presented in the form of square voltages, which are converted to analog signals proportional to the rotor angle. The accuracy of commercially available resolvers range from 2 to 20 min of arc.

Resolvers offer a rugged and reliable alternative to measuring the absolute angular positioning of a shaft within a single turn, and with resolution limited only by the electronics used

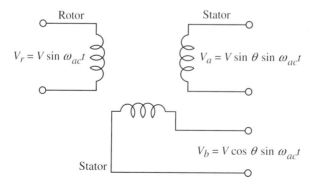

Figure 11.51 Electrical circuit of a simple resolver with input excitation through the rotor windings and outputs through the stator windings.

to decode their signals. Interface devices are currently available to provide excitation signals and decode the output to a digital signal similar to that of an incremental encoder. The EMR57 resolver manufactured by Empire Magnetics (www.empiremagnetics.com) is accurate to within $\pm 0.0463\%$ of full scale.

11.22 TACHOMETER

The tachometer is the most commonly used velocity sensor with DC motor control. A tachometer is a DC generator that converts the rotational motion of a motor into electrical energy. Although a DC motor can be used in this manner, DC tachometers have smaller volume and weight and are specially designed to produce a relatively ripple-free output voltage that exhibits linear variation with the speed of rotation of the armature over the entire operating range. The magnitude of the DC voltage V_{TACH} output from a permanent magnet (PM) tach is given by

$$V_{TACH} = K_G \omega \qquad (11.67)$$

where K_G is the tach gain, which is mainly dependent on the magnetic strength of the PM, and ω is the angular velocity of the shaft. The polarity of the output voltage depends on the direction of rotation.

The armature of a tachometer is formed from a set of copper or aluminum coils, which are wound longitudinally on a cylindrical piece of iron. The ends of the coils are connected to a commutator—a segmented ring with twice as many segments as the coils.

The output DC voltage has two signals superimposed on it: a modulation at shaft rotation frequency and a high-frequency ripple. The modulation is due to a slight eccentricity of the armature and is usually small. The ripple frequency is caused by the effects of commutation in switching from one coil in the armature to the next. It is given by

$$f_r = (100/6)ukn \qquad (11.68)$$

where u is the number of commutation segments, $k = 1$ if u is even and 2 if u is odd, and n is the shaft speed (rpm). The high-frequency ripple can be filtered out. A typical tachometer produces 5 V/1000 rpm, with a linearity of 1% and a root mean square ripple voltage of 5% of the mean DC level. Like potentiometers, tachometers usually have an accuracy of $\pm 0.5\%$.

Tachometers are also available with a moving coil used for an armature. This design allows for more coils, such as 19–23, to be used and results in a significantly reduced weight. In addition, the ripple voltage is on the order of 1% of the DC output, resulting in improved low-speed performance.

11.23 CAPACITIVE SENSORS

Capacitance formulas (Eqs. (2.26) and (2.28)) are very important for capacitive-based sensor design. Figure 11.52 shows possible capacitive sensor arrangements that are exploited to measure myriad physical quantities, including position, area, volume, pressure, force, liquid level, humidity, and chemical species concentration, among many more. They are used in the construction of microphones, proximity sensors, and intrusion sensors. The capacitance varies by changing d, A, or κ. The change in capacitance, ΔC, is measured and related to the value of the

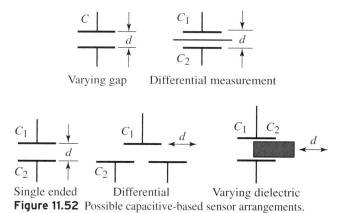

Varying gap   Differential measurement

Single ended   Differential   Varying dielectric

Figure 11.52 Possible capacitive-based sensor arrangements.

quantity being measured. Although the configurations shown in Fig. 11.52 indicate linear motion between the capacitor's plates, similar configurations can be constructed with angular relative motion between the plates.

Figure 11.53a shows a coaxial capacitor used as a displacement sensor. The outer cylinder plate is stationary, whereas the inner cylinder plate is attached to a moving object, such as a machine tool table. As the inner cylinder plate moves within the outer cylinder plate, the capacitance changes as indicated. A coaxial capacitive-level sensor immersed in a water tank is shown in Fig. 11.53b. Water has a high dielectric constant but significant leakage resistance. Consequently, the surface of each conductor is coated with a thin isolating layer to prevent an electric short-circuit through water. As the water level increases, the sensor's capacitance

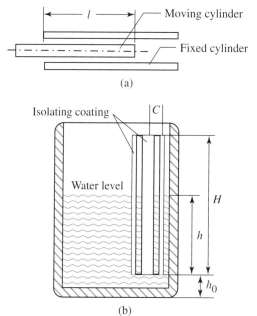

Figure 11.53 Capacitive displacement sensor (a) and a liquid-level sensor (b) with capacitance vs. liquid level (c).

changes as water fills more space between the conductors. Using Eq. (2.26), the total capacitance of the sensor can be found by adding the capacitances of the water-free and water-filled portions:

$$C_h = C_1 + C_2$$
$$= \frac{w\varepsilon_0}{d}[H - h(1 - \kappa)] \qquad (11.69)$$

where h is the height of the water-filled portion of the sensor and w is the width of the plates.

11.24 INDUCTIVE SENSORS

11.24.1 Motion-Detection Sensor

Equation (2.52) is the basic principle of the magnetic motion detector sensor shown in Fig. 11.54. The sensor works as follows. If a coil of N turns moves into a gap of a permanent magnet, the flux enclosed by the loop is $\phi = Blx$, where lx is the area of that part of the loop that entered the gap of the magnet. The voltage induced in the loop is

$$V = N\frac{d}{dt}(Blx) = NBl\frac{dx}{dt} = NBlv \qquad (11.70)$$

where v is the velocity of the coil. The output voltage depends on the coil cross section. The output voltages for rectangular and circular coils are shown in Fig. 11.54.

11.24.2 Linear Variable Differential Transformer (LVDT)

The LVDT is an electromechanical device that generates a signal proportional to the displacement of a movable core. The basic construction, corresponding circuit diagram, and interface implementation of an LVDT are shown in Figure 11.55. The LVDT consists of a primary coil energized by an external AC source and two series opposing secondary coils that generate voltages of opposite polarity. The difference between the output voltages of the secondary

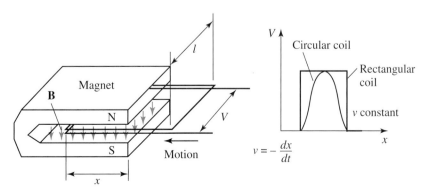

Figure 11.54 Magnetic motion detector and its transfer function.

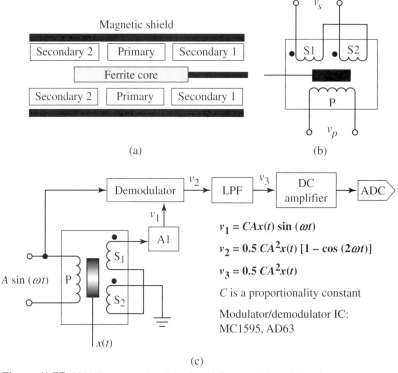

Figure 11.55 LVDT construction (a), circuit diagram (b), and interface implementation (c).

coils is proportional to the displacement of a separable noncontacting movable core. The excitation voltage of the primary coil is a precise sinusoidal input of the form

$$v_p = A \sin(\omega t) \tag{11.71}$$

The excitation signal has a frequency ω ranging from 60 to 20 kHz and a stable RMS amplitude of 3–15 V. The difference between the voltages of the secondary coils is a sine wave representing the net output of the transducer and varies linearly with the change in the core position $x(t)$:

$$v_1 = CAx(t)\sin(\omega t) \tag{11.72}$$

where C is a proportionality constant, partially provided by op-amp A1. The frequency of the output signal is the same as the excitation frequency, but its amplitude is modulated by the low-frequency movement of the core. For a given LVDT the output is generally not in phase with the excitation, except at a particular frequency; if the LVDT operates at this frequency, the phase shift between the primary and secondary coil signals is zero.

The excitation signal and secondary coil signal are multiplied (modulated) to yield

$$\begin{aligned} v_2 &= CA^2x(t)\sin^2(\omega t) \\ &= 0.5CA^2x(t)[1 - \cos(2\omega t)] \end{aligned} \tag{11.73}$$

To recover the core movement, the output signal has to be demodulated and then filtered. The output voltage experiences an abrupt $180°$ phase shift as the core moves from one side of null to the other. Thus, displacements of the core to the left or to the right of null would produce the same magnitude of output voltage as when no directional detection is performed. Therefore, the demodulator must be able to discern the direction of core movement. The Motorola MC1595 four-quadrant multiplier IC performs phase-sensitive demodulation. The demodulator produces a full-wave rectified output that must be fed to a low-pass filter. In practice, if the ratio of the excitation frequency to the core displacement frequency is 10:1 or more; a simple RC low-pass filter may be adequate. Setting the corner frequency of the LPF at 0.1 times the excitation frequency, the output from the filter will be

$$v_3 = 0.5CA^2x(t) \tag{11.74}$$

This signal is properly amplified before it is sent to the ADC. Figure 11.55(c) shows a complete implementation of LVDT measurement.

The noncontacting nature of the movable core provides for frictionless measurement, resulting in infinite resolution and infinite mechanical life. The symmetric construction of the LVDT and the lack of deformation eliminate mechanical hysteresis and help achieve a repeatable null position with no drift. The LVDT also provides for complete isolation between the excitation and output signals.

11.25 FOUR- TO 20-mA TRANSMITTERS

A commonly used standard in industrial processing to transmit analog signals over long distances is the 4- to 20-mA transmitter. The main component of the transmitter is a V–I converter that converts the sensor voltage corresponding to a measured signal such as pressure into a current signal ranging from 4 to 20 mA. The current signal can be sent without signal modification to a controller or a data-acquisition system for further processing. Transmitting signals of current instead of voltage eliminates the effects of supply voltage variations, voltage drop from line resistance, and random noise induced in the loop. The 4-mA signal indicates a "live" zero corresponding to the lowest measurand value. A "real" dead zero indicates a fault condition in the instrument or line.

There are three basic transmitter types, indicating the number of wires required to supply transmitter power: type 2, type 3, and type 4. Figure 11.56 shows the components and

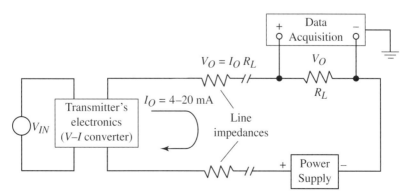

Figure 11.56 Two-wire 4- to 20-mA transmitter.

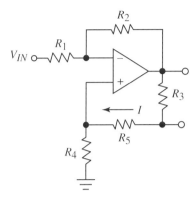

Figure 11.57 V-to-I converter.

connection of a typical 2-wire transmitter, in which the transmitter floats and the signal ground is in the data-acquisition system or receiver. The power and the signal are transmitted along the same wire. The loop is powered by an external supply included at the receiver. The standing loop current (4 mA) is used to power the remote instrumentation and the transmitter. The voltage available at the transmitter needs to overcome voltage drop along the line and across the receiver. A data acquisition (DAQ) device reads the voltage drop across a resistor, which converts the 4- to 20-mA range on the DAQ input channel into a voltage signal. The conversion resistor can be placed either on the supply ($+$) or return ($-$) lead of the circuit. Because current transmission requires a DC transmission path, it cannot be used over the public telephone system, which involves AC amplifiers in the repeater stations.

Many signal-conditioning ASICs are available to convert a voltage signal from a sensor to a 4- to 20-mA analog output. An example is the AM 422 integrated V–I interface from Analog Microelectronics (www.analog.com). The device is available in 2- and 3-wire versions. With a few external discrete components, the device can easily be hooked to a variety of sensors.

11.25.1 Voltage-to-Current Converter

Figure 11.57 shows an op-amp circuit that can sink a current into various loads while maintaining the voltage-to-current transfer characteristics. Additionally the current can be delivered in both directions, as the application requires. Using op-amp rules, the relation between the input voltage and output current is simply

$$I = -\frac{R_2}{R_1 R_3} V_{IN} \tag{11.75}$$

provided the resistances are selected to satisfy

$$R_1 R_5 = R_2 R_4 \tag{11.76}$$

11.26 SUMMARY

This chapter connected physical real-life phenomena and parameters to the scientific representation and manipulation of data. The chapter presented measurement and sensing concepts and characteristics. Various sensors were covered in detail, including commercial packaging and specifications that can measure light, temperature, magnetic, sound, strain, and piezoelectric.

The chapter covered the various conditioning circuits needed in interfacing the sensors to the 9S12C MCU. Examples were given that described the interfacing of the 9S12C with the conditioned sensors.

RELATED READING

C. Bergquist, *IC Projects*. Indianapolis: Prompt Publications, 1997.

W. deSilva, *Control Sensors and Actuators*. New York: Prentice Hall, 1989.

E. O. Doebelin, *Measurement Systems Application and Design*, 3rd ed. New York: McGraw-Hill, 1983.

H. R. Everett, *Sensors for Mobile Robots: Theory and Applications*. Wellesley: A. K. Peters, 1995.

J. Fraden, *AIP Handbook of Modern Sensors: Physics Design and Application*. New York: AIP, 1993.

R. Frank, *Understanding Smart Sensors*, 2nd ed. Norwood: Artech House, 2000.

J. W. Gardner, *Microsensors: Principles and Application*. New York: Wiley, 1994.

P. Horowitz and W. Hill, *The Art of Electronics*. Cambridge, UK: Cambridge University Press, 1989.

C. D. Johnson, *Process Control Instrumentation Technology*, 7th ed. New York: Prentice Hall, 2002.

J. L. Jones, B. A. Seiger, and A. M. Flynn, *Mobile robots, Inspiration to Implementation*. Natick: A. K. Peters, 1999.

S. Kamichik, *IC Design Projects*. Indianapolis: Prompt Publications,1998.

J. A. Kleppe, "High-Temperature Gas Measurement Using Acoustic Pyrometry," *Sensors*, vol. 13, no. 1, 1996.

G. Kovacs, *Micrmachined Transducers Sourcebook*. New York: McGraw-Hill, 1998.

D. Shetty and R. A. Kolk, *Mechatronics Systems Design*. Boston: PWS, 1997.

W. Tompkins and J. Websters, *Interfacing Sensors to the IBM PC*. New York: Prentice Hall, 1988.

J. G. Webster (ed.), *The Measurement, Instrumentation, and Sensors Handbook*. Boca Raton, FL: CRC Press, 1998.

J. Wheeler and A. R. Ganji, *Introduction to Engineering Experimentation*. New York: Prentice Hall, 1996.

QUESTIONS

11.1 List the main components of a measurement system, and describe each.

11.2 Explain the possible classifications of sensors.

11.3 What is the difference between self-generating and modulating sensors?

11.4 What are the essential elements to render a sensor smart?

11.5 List desirable sensor characteristics, and explain each.

11.6 What is loading error? How can its effect be reduced?

11.7 Explain the roles of interface electronics in measurement.

11.8 How can the nonlinearity error of a potentiometer be reduced?

11.9 What constitutes a light detector?

11.10 What is a lumen, and how is it quantified?

11.11 Describe the types and operating modes of light detectors.

11.12 Compare and contrast quantum detectors and thermal detectors.

11.13 Compare and contrast photoconductive and photovoltaic devices.

11.14 Prepare a table with various semiconductor materials used as light detectors, and give the wavelength of maximum sensitivity.

11.15 Write down the expression describing photodiode behavior, and explain each term.

11.16 How is a solar panel constructed?

11.17 Describe the operation of a phototransistor.

11.18 Compare and contrast the operation of an optical interrupter and an optical reflector.

11.19 Describe the essential components of an optical encoder.

11.20 What is the pyroelectric effect?

11.21 Describe the temperature-resistance characteristics of thermistors.

11.22 What is the Seebeck effect?

11.23 Why is a reference temperature needed when measuring temperature with a thermocouple?

11.24 How is a thermopile constructed?

11.25 Compare and contrast wire-wound and thin-film RTDs.

11.26 What is meant by the base ohm of an RTD?

11.27 Why is self-heating a major concern with RTDs, and why is it of lesser importance with thermistors?

11.28 Describe the disadvantages of using an RTD and ways to circumvent them.

11.29 Compare and contrast the temperature-measuring devices discussed in the chapter.

11.30 Describe the construction of a heat-flux sensor.

11.31 How can the thermal conductivity of a metal be measured?

11.32 How is the convective heat transfer coefficient measured?

11.33 Explain the operation of a magnetic reed switch.

11.34 Explain the Hall effect.

11.35 Compare and contrast current-sensing methods.

11.36 Explain how a Hall-effect device can be used to measure the current, position, and angular speed of a shaft.

11.37 How are very high-frequency currents measured?

11.38 Define the gauge factor for a strain gauge.

11.39 Explain the piezoresistive effect.

11.40 What are the main advantages of a semiconductor strain gauge?

11.41 Why is temperature compensation important for strain gauges, and how is it performed?

11.42 Describe the operating modes of a bridge circuit.

11.43 Compare and contrast acoustic and optical measurements.

11.44 Write the equation for sound pressure level, and explain the terms involved.

11.45 Describe the main components of an acoustic sensor.

11.46 What is meant by *acoustical impedance*, and how is it determined?

11.47 Why are high-frequency acoustic waves used to detect small objects?

11.48 List two advantages ultrasonic imaging has over X-rays.

11.49 What is an impedance-matching transformer?

11.50 Explain piezoelectric effects, and list the type of devices that exploit them.

11.51 Explain the role of the charge amplifier and its advantages over the current-to-voltage amplifier.

11.52 Explain how the crystal microbalance works.

11.53 List the components of an incremental shaft encoder, and describe the role of each.

11.54 What are the differences between an incremental and an absolute shaft encoder?

11.55 What is the difference between a binary-coded and a gray-coded absolute encoder wheel?

11.56 Explain the principle of quadrature count in an incremental encoder.

11.57 Describe the construction and the principle of operation of a resolver.

11.58 Describe the construction and the principle of operation of a tachometer.

11.59 Describe the construction and the principle of operation of the LVDT.

11.60 Describe the 4- to 20-mA transmitter.

PROBLEMS

11.1 A strain gauge is used to measure stresses in a cantilever beam. The strain gauge has a resistance of 120 Ω and a gauge factor of 2. It can sustain a maximum allowable current of 30 mA. This gauge is placed in a bridge circuit, with three other, equivalent dummy gauges forming the other arms of the bridge.

a. Determine the maximum allowable bridge voltage.

b. Consider the measurement of a stress level of 1000 lb/in.^{2} in a steel beam. Determine the equivalent change in strain. The relation between the stress s and strain ε is given by $\sigma = E\varepsilon$, where E is the modulus of elasticity (for steel $E = 30 \times 10^6$ lb/in.^{2}).

c. Determine the output voltage for the given bridge arrangement corresponding to the stress in (b).

d. The Johnson noise generated in a resistance (e.g., strain gage) is given by the relation

$$v_{noise,rms} = \sqrt{4kTR \, \Delta f}$$

where k is Boltzmann's constant (1.38×10^{-23} J/K), T is the absolute temperature of the resistor (K), R is the resistance (Ω), and Δf is the bandwidth (Hz). If the given strain gauge operates at 300 K over a bandwidth of 100,000 Hz, determine the corresponding noise and the *signal-to-noise ratio* corresponding to the signal due to the 1000 lb/in.^{2}.

e. Suppose a stress of 1 lb/in.^{2} is to be measured instead of 1000 lb/in.^{2}. Determine the corresponding signal-to-noise ratio.

f. What would you conclude based on the *signal-to-noise ratio* in the two stress levels?

11.2 The resistance of a CdS photocell varies according to $R(t) = R_i + (R_f - R_i)[1 - e^{-t/\tau}]$, where R_D is the dark resistance, R_i is the resistance under a light beam, and τ is the time constant of the cell. If $R_D = 65$ kΩ, $R_i = 25$ kΩ, and $\tau = 60$ ms, design a circuit that will trigger a 5-V comparator within 6 ms of interruption of the light beam.

11.3 A photovoltaic cell delivers 0.5–1.7 mA into a 100-Ω load when it is exposed to radiation of intensity from 5 to 12 mW/m^2. The output voltage across the cell ranges from 0.22 to 0.41 V.

a. Find the range of the cell's short-circuit current.

b. Develop an SC scheme to provide a linear voltage from 1.0 to 2.4 volts within the specified light intensity range.

11.4 A linearized RTD having a TCR $\alpha = 0.0075$/K and resistance of 2000 Ω at 25°C is connected in the psuedo-bridge circuit of Fig. 11.58. The maximum current the sensor can tolerate is 1 mA. The desirable temperature range to be measured is from 0°C to 40°C, with a corresponding output voltage from 0 V to 10 V. Assuming ideal op-amp behavior, select appropriate circuit components for the desired measurement range, and determine the temperature where the nonlinearity error is maximum $V_{CC} = 10$V.

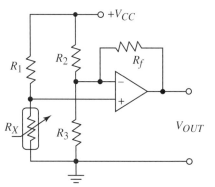

Figure 11.58

11.5 Figure 11.59 shows a voltage divider circuit in which a standard reference resistor is placed in series with a sensor whose unknown resistance R_x varies with a stimulus. For a linear resistance sensor, R_x is expressed as $R = R_0(1 + x)$, where R_0 is the baseline resistance and x is the fractional change in sensor resistance, which can be positive or negative. The range of x depends on the type of sensor used and largely influences the design of the interface circuit. It could take values between 0 and -1, and in many cases it may be much lower. Determine the gain and maximum sensitivity of the circuit.

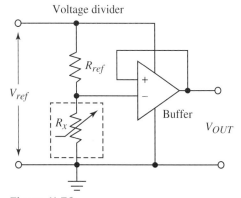

Figure 11.59

11.6 The signal from a differential capacitive sensor with grounded movable electrode is conditioned. Obtain a relation between the passive components in the circuit of Fig. 11.60 so that the output voltage is independent of the excitation frequency. Determine the maximum excitation frequency allowed by the finite slew rate if the maximum value of the output voltage is 10 V. Determine the value of the resistors and capacitors if the

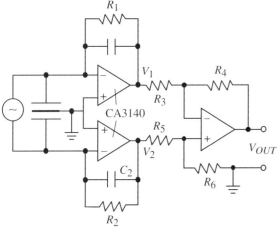

Figure 11.60

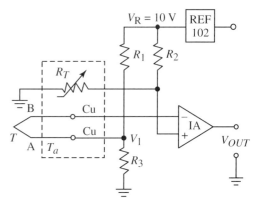

Figure 11.61

a. Derive a relation for the output voltage in terms of the thermocouple's sensitivity S_K and the value of R_T.

b. Determine the condition the resistors are required to satisfy to have a null output when $T_a = 0°C$.

c. Assume that R_2 and R_3 are chosen so that the output of the bridge circuit they form with R_1 and R_T is linear. Determine their values to compensate the cold junction in the entire range of T_a.

d. Determine the gain of the IA if the output voltage ranges between -1 V and $+1$ V.

e. Determine the maximum offset voltage for the IA to limit the maximum error to 0.1°C.

gap between the sensor's plates area is 1 cm and their area is 100 cm². Additionally, the input displacement varies between -1 and $+1$ mm and the peak excitation voltage is 10 V at 100 kHz.

11.7 Determine the acoustical impedance for air and the matching impedance for a sonic ranging device whose sonic emitter is made of iron.

11.8 Determine the Doppler shift at a receiver site of a sonar signal in water whose transmitted frequency is 250 Hz and whose transmitter is traveling toward the receiver at a velocity of 6.7 m/s.

11.9 A 350-Ω bridge is connected to an IA (e.g., AD524). One arm of the bridge is connected to a resistor sensor R_S (thermistor, photocell, etc.). The bridge is excited with a 10-V supply. The IA is jumped for a gain of 100. At this gain the minimum CMRR is specified as 100 dB up to 60 Hz. The maximum input offset for the IA is 100 μV. Determine:

a. The output signal from the bridge per 1-Ω change in R_S.

b. The total system response.

c. The maximum output error to common-mode input.

d. The error due to input offset voltage.

e. The total error due to common mode and input offset in terms of change in R_S.

11.10 A type K thermocouple is to be used to measure temperatures in the range $-100°C$ to $+100°C$(Fig. 11.61). Cold-junction compensation is achieved by using a Pt-100 RTD having a TCR of 0.385%/K at 0°C.

11.11 Search the web for an IR temperature smart sensor, and write a two-page summary of its characteristics. Show how it is interfaced to the 9S12C, and write a code to manage that interface.

11.12 Explain how you would set up a device to measure the density changes in liquids and solids in a testing facility by using a sounding device and precise timing equipment.

11.13 Design a sonic ranging system, using block diagrams only, that can measure distances of several hundred feet and is temperature compensated.

11.14 An acoustic pyrometer is used to measure average temperature across a 18.29-m-wide coal-fired utility boiler. The flue gas composition (%) dry volume is 82.0% N, 6.0% O, 12.0% CO_2, and moisture content per weight is 5% (kg moisture/kg dry gas). Calculate the path average temperature of the flue gas when the flight time of the acoustic wave is measured to be 25.94×10^{-3} s.

11.15 Search the web for a pressure smart sensor, and write a two-page summary of its characteristics. Show how it is interfaced to the 9S12C, and write a code to manage that interface.

11.16 Design a mechanism (see Appendix H) that connects a float for liquid-level measurement to an LVDT such that as the float moves from 0 to 1.5 m, the LVDT core moves over its linear range of 3 cm. Determine the resolution in the level measurement if the LVDT output is interfaced to a 10-bit ADC.

11.17 Search the web for a tilt sensor, and provide a summary of its characteristics. Show how it is interfaced

to the 9S12C, and write a code to manage that interface.

11.18 Search the web for a gyroscopic sensor, and write a two-page summary of its characteristics. Show how it is interfaced to the 9S12C, and write a code to manage that interface.

11.19 Polaroid manufactures a range of active ultrasonic ranging modules, series 6500, that is well cited in the literature. Prepare a two-page summary of its capability, and provide code for its interface with the 9S12C.

LABORATORY PROJECTS

11.1 Design and build a 9S12C-based climate control for an automobile using a thermistor and a photocell for temperature and light sensing. The device should have a display for temperature and light intensity. The device should also adjust the shades available for the back windshield.

11.2 Design and build a 9S12C-based weather vane to indicate the direction in which the wind is blowing. Use a set of eight reed switches or Hall-effect sensors to indicate the N, NE, E, SE, S, SW, S, and NW directions. Provide a display for the direction.

11.3 Design and build a 9S12C-based anemometer using an encoder consisting of an IR LED emitter, an IR detector, and an interrupt wheel. Provide a velocity display on a panel.

11.4 Design and build a 9S12C-based humidity-sensing device that generates a humidity-dependent frequency signal. Use a humidity sensor (e.g., Humirel HS11, Phillips 2322 691 90001) and a 555 timer or other oscillator circuit.

11.5 Design and build a humidity-sensing device that generates an output proportional to humidity. Use the Honeywell HIH 3605-A or other humidity sensor.

11.6 Obtain the data sheet for the Motorola MPX5100A pressure sensor, and design and build a device to measure pressure in a pipe. Display pressure on a panel.

11.7 Design and build a level-control device. Level detection is via an *RC*-based circuit. The device should open a gate for water to escape when the level is high.

11.8 Build a security acoustic sensor that would detect the breaking of window glass and engage an alarm when that happens.

11.9 Build an IR-based fire detector, and hook it up to the 9S12C to activate a sprinkler system when fire occurs.

Electric Actuators

Thoughtful engagement with the material presented in this chapter will enable the student to:

1 Understand the different types of electric actuators

2 Develop understanding of stepper motors and their types, construction, and interface to the MCU

3 Develop understanding of DC motors and their control and interface to the MCU

4 Select discrete components to control electric actuators

5 Develop knowledge on the operation of servos

6 Develop knowledge on the operation and control of AC induction motors

7 Select the motor type suitable for mechatronics driving applications

12.1 ACTUATORS

In a typical *closed-loop*, or *feedback*, control of a machine or a process, the *control computer* compares the actual sensor measurement with the desired value and adjusts the signal to the actuator accordingly. The *actuator*, or prime mover, converts signals into a physical quantity to initiate a motion, cause temperature change, regulate fluid flow in a pipe, regulate light intensity, and so on. Proper selection of the actuator hinges on many factors, such as power requirement, motion resolution, and operating bandwidth. These factors ultimately depend on the specific role of the actuator within the system.

Motion-generating actuators in particular are classified into three categories: electric, pneumatic, and hydraulic. Table 12.1 provides a qualitative comparison among various types. While many schemes can be used to control each family of motors, all control schemes may be implemented directly or indirectly using pulse signals generated by the MCU.

Electric actuators convert electrical power into mechanical power. Electric actuators are available in one of two types, direct current (DC) or alternating current (AC). AC induction

TABLE 12.1 Comparison of Pneumatic, Hydraulic, and Electric Actuators

Actuator Type	Advantages	Disadvantages
Pneumatic	High operating speed; availability of air; requires no return lines; clean; can stall without damage; relatively inexpensive	Air compressibility hampers speed control; less accurate; exhaust causes noise pollution; easy to leak; requires additional drying and filtering
Hydraulic	Moderate operating speeds; smooth running at low speeds; carry heavy loads; high power-to-weight ratio; incompressibility of fluid provides for accurate control; self-lubricated and self-cooled; can stall without damage; fast response; inflammable and safe in explosives areas	Leaks through seals; difficult high-speed cycling; requires return line and remote power source; cannot back drive links against valves; difficult to miniaturize due to high pressure and flow; expensive
Electric	Fast and accurate response; simple control; easy to implement in modular and new designs; relatively inexpensive.	Transmission drives may be needed to reduce speed and increase torque—introduces backlash and friction; electrical arcing may be a problem in flammable areas; overheats when stalled; requires brakes to lock in position

and synchronous motors are ideal for constant speed applications with little load variations. AC motors use line current to directly provide more power compared to DC motors of similar size. For position and speed control applications involving variable loads, DC motors are favoured. DC motors fall in one of three categories: conventional or brushed DC motors, brushless DC motors, and stepper motors. Servos are basically DC motors fitted with sensing and control components. In this chapter stepper motors and DC motors are discussed in detail with practical MCU-based control. A brief overview of servos and AC induction motors is also provided.

12.2 DC MOTORS

In conventional DC motors, large currents are delivered to the rotor through a pair of carbon brushes that maintain contact with a split ring commutator. The rotor current maintains a stationary position aligned with the stator magnetic field provided either by permanent magnets or by energized field coils. In a brushless DC motor, the phase windings are held in the stator and the permanent magnets are in the rotor. Stator windings are alternately energized using electronic switching circuits instead of mechanical commutation through brushes. Hall-effect sensors, encoders, or resolvers are used to sense the rotor position to achieve proper commutation. DC motors usually operate in a closed loop to achieve accurate response.

12.2.1 Principles of Operation of a DC Motor

A DC motor is an electromechanical device that converts direct current (DC) electrical energy into mechanical energy. The principle of operation of any electric motor is based on Ampere's

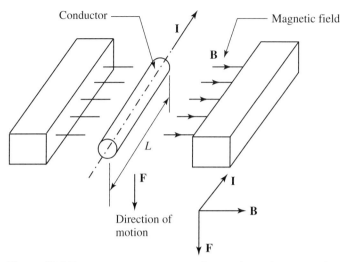

Figure 12.1 Force acts on a current-carrying conductor in a magnetic field.

law, which states that a conductor of length L will experience a force $\mathbf{F}$ if an electric current $\mathbf{I}$ flows through that conductor at right angles to a magnetic field having a flux density $\mathbf{B}$. Referring to Fig. 12.1, the force is determined by the cross product

$$\mathbf{F} = (\mathbf{B} \times \mathbf{I})L = BIL\sin\theta \qquad (12.1)$$

where θ is the angle between the current flow and the magnetic flux density and a boldface letter indicates a vector quantity. If $\theta = 90°$, vectors $\mathbf{I}$, $\mathbf{B}$, and $\mathbf{F}$ form a vector triad in which the force is normal to the plane formed by the $\mathbf{I}$ and $\mathbf{B}$ vectors. In this case Eq. (12.1) becomes

$$F = BIL \qquad (12.2)$$

If the conductor is free to move, the force F will cause it to move at some velocity v in the direction of the force. In turn, the motion of the conductor in the magnetic field B induces a *back electromotive force*, or *bemf*, E_b in the conductor. This voltage is obtained from Eq. (2.50) as

$$E_b = BLv \qquad (12.3)$$

According to Lenz's law, the flux due to E_b opposes the flux caused by the original current through the conductor and attempts to slow down the motor. This is the source of electrical damping in motors, where the back emf tends to oppose the voltage that produced the original current.

Based on the foregoing, a motor can be constructed from two basic components: one to produce the magnetic field, usually termed the *stator*, and one to act as the conductor, usually termed the *armature* or *rotor*. The stator magnetic field may be created either by *field coils* wound on the stator poles or by *permanent magnets* (PMs).

In field coil motors, stator windings are powered by a supply voltage V_f (subscript f indicates field) to provide the coils with an electric current I_f required to create magnetic poles on the stator. This is depicted in Fig. 12.2, where only one coil is shown. In a full motor, the stator

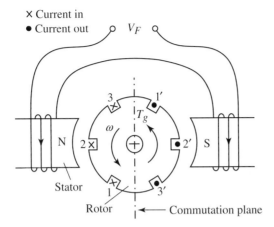

Figure 12.2 Stator windings in a field coil motor.

is made up of an even number of magnetic poles p, which are constructed from ferromagnetic sheets (laminated construction).

As shown in Fig. 12.3, the armature is composed of a laminated cylindrical core made from a ferromagnetic material, such as iron, which provides a low reluctance path and helps concentrate the magnetic flux toward the rotor. The rotor has many conductors (100 or less) embedded in closely spaced slots on its periphery and terminated by several of *commutator* segments. The rotor windings are powered by a supply voltage V_a (subscript a indicates armature) that supplies armature current I_a to the rotor *windings* through a pair of carbon "*brushes*" that maintain contact with the commutator as the rotor turns. Brushes are susceptible to wear and are a source of sparks and electrical transients.

A PM motor has no field coils. The source of the magnetic field is produced by permanent magnets held in the stator. Some PM motors include coils wound on the magnets simply to recharge the magnets if their strength fails. The material of the PMs are either *alnico* or *rare earth* (samarium cobalt) magnets. Alnico is used in relatively low-torque applications. Rare earth magnets are more expensive than alnico magnets, but, being three times stronger than alnico and almost impossible to demagnetize, they are used in most high-performance PM motors. Since PM motors do not use any field coils, the need for field power supplies is eliminated, which results in less heat dissipation and lower temperatures within the motor. This leads to lowered cooling requirements and higher current ratings and torque capacity. Thus, PM motors are more efficient and more reliable than field coil motors.

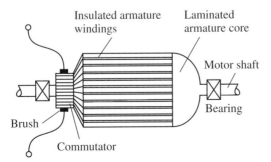

Figure 12.3 Rotor armature of a DC motor.

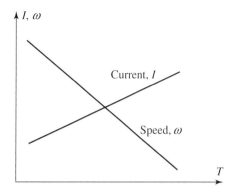

Figure 12.4 Current and speed vs. torque variations of a PM DC motor.

The stator magnetic flux of a PM motor is constant at all levels of armature current, and, as Fig. 12.4 shows, the speed–torque curve is linear over a wide operating range, which makes PM motors ideal for control applications.

Depending on the rotor construction, PM motors are classified into three types: *iron core*, *disk*, and *cup*. Although brushless DC motors employ PMs, they are not called PM motors because of their distinct characteristics. Torque motors are also PM motors designed for optimum torque. These are usually short and have larger diameters than other PM motors.

The armature of an iron-core PM motor is similar to the conventional armature structure described earlier. Iron-core motors deliver the highest power and have high starting torque. To overcome their typically high rotor inertia, the rotors are made long and with small diameters to decrease their inherent mechanical time constant.

As shown in Fig. 12.5, the armature wires of a disk motor are stamped from flat sheets of copper and laminated together with insulation. The laminations are connected to form a continuous wire. Current flows radially to the circumference and around radially placed field magnets.

The armature of a cup motor has a hollow-like structure, as shown in Fig. 12.6. The cup shell is formed of wires held together by epoxy or fiberglass. An extremely lightweight, small-diameter rotor gives the smallest inertia among the three types and thus the fastest acceleration. These motors are generally used in low-speed applications.

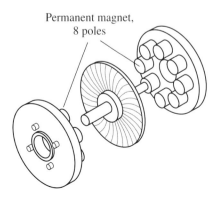

Permanent magnet, 8 poles

Figure 12.5 Disk motor construction.

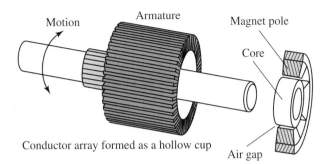

Figure 12.6 Cup motor construction.

In order to hold the force vector constant in magnitude and direction, it is necessary to have the **B** and **I** vectors fixed in space with respect to each other. This condition will not be met if the direction of current through the inductors is maintained as the rotor rotates about its axis. Therefore the direction of current in individual coils must be switched as the coils pass a fixed point in space. The switching is accomplished by means of a *commutator*.

The most commonly used commutator is a ring of many segments of conductive material, usually copper, that are electrically insulated from each other by insulating material, as shown in Fig. 12.3. The adjacent segments, however, are connected through the conductors in sequence, as shown in Fig. 12.7. The number of commutator segments is equal to the number of conductor slots in the rotor.

For the rotor position shown in Fig. 12.7, if the commutator rotates CCW 1/12 of a revolution, the current paths in conductors 1-1′ reverse while the remaining current paths remain unchanged. The successive passage of commutator segments under the brushes would also reverse the direction of current through them. Consequently, as the armature rotates, the current through its windings occupies a fixed position in space, which results in a fixed unidirectional torque. The commutator segments are arranged so that the current through the brushes is switched between the conductors to keep the torque steady. However, torque ripple and arcing associated with commutation influence the performance of the motor.

12.2.2 Modeling of DC Motor Behavior

Voltage and Torque Equations

Voltage and torque characteristics of a DC motor depend on design parameters, such as the number of conductors, magnet size, and the number of stator poles. Figure 12.8 shows the

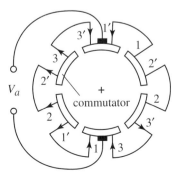

Figure 12.7 Sequential connection of conductors.

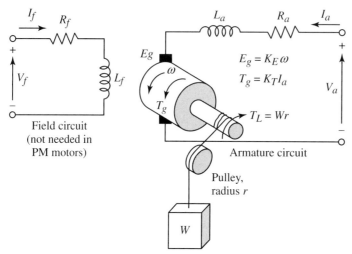

Figure 12.8 Equivalent electric circuit of a DC motor.

equivalent circuit diagram of a conventional field coil DC motor. An expression for the induced back electromotive force E_g due to armature rotation is derived from Eq. (12.3) to be

$$E_g = K_E\omega \tag{12.4}$$

where K_E is the *voltage constant* of the motor and ω is the angular velocity of the rotor. K_E is given by

$$K_E = K_2\phi = \frac{Zp}{60n'}\phi \tag{12.5}$$

where n' is the number of parallel conductor paths in the armature, p is the number of poles, Z is the number of conductors moving through the magnetic field, and ϕ is the radial magnetic flux. The 60 in the denominator of Eq. (12.5) is a consequence of converting ω (rad/s) into n (rpm), where $\omega = n\pi/30$.

The motor-generated torque T_g is linearly related to the armature current I_a by

$$T_g = K_T I_a \tag{12.6}$$

where K_T is the *torque constant* of the motor and is given by

$$K_T = \frac{Zp}{2\pi n'}\phi \tag{12.7}$$

Both K_E and K_T depend on the magnetic field, the geometry of the air gap, and armature construction. Depending on the system of units used, K_T and K_E are related as follows:

$$K_T(\text{oz-in/A}) = 1.3524 K_E(\text{V/krpm})$$
$$K_T(\text{N-m/A}) = 9.5493 \times 10^{-3} K_E(\text{V/krpm}) \tag{12.8}$$
$$K_T(\text{N-m/A}) = K_E(\text{V/rad-s}^{-1})$$

The last of Eqs. (12.8) will be assumed hereafter unless stated otherwise. For an ideal torque motor, K_T is constant. In reality, though, K_T varies with rotor position. An important performance index of a DC motor is the torque ripple, defined as

$$\text{torque ripple} = \frac{\text{torque deviation}}{\text{average output torque}} \tag{12.9}$$

If the input voltage frequency is much greater than 1 kHz, torque ripple and speed ripple are very small. The generated torque can also be related to the power P supplied to the motor. Using Eq. (12.6), power is evaluated as

$$P = V_a I_a = \frac{R_a}{K_T} T_g \times \frac{T_g}{K_T} = \frac{R_a}{K_T^2} T_g^2 \tag{12.10}$$

where V_a is the voltage applied to the armature and the armature resistance R_a includes the resistance of the brushes. From Eq. (12.10),

$$T_g = K_M \sqrt{P} \tag{12.11}$$

where the constant K_M is given by

$$K_M = \frac{K_T}{\sqrt{R_a}} (\text{Nm}/\sqrt{\text{W}}) \tag{12.12}$$

As Eq. (12.11) indicates, the output torque is proportional to the square root of the supplied power. The proportionality constant, K_M, is called the *motor constant*. It describes motor effectiveness in converting electric power to mechanical torque. Higher values of K_M mean the motor can generate higher torques with less power. For the same amount of power, a motor with a higher torque constant requires less current but a higher voltage.

Electrical Model of a DC Motor

Analysis of the circuit in Fig. 12.8 leads to the electrical equation of the DC motor:

$$V_a = L_a \frac{dI_a}{dt} + R_a I_a + E_g \tag{12.13}$$

where L_a is the armature inductance. Note that the resistance due to magnetic circuit losses is usually 5–10 times greater than R_a and its effect on motor operation insignificant, so it is neglected. Additionally, neglecting the effect of L_a (because its value is usually very small), Eq. (12.13) reduces to

$$V_a = R_a I_a + K_E \omega \tag{12.14}$$

This equation is also valid at constant load, that is, when dI_a/dt is constant.

Steady-State Performance

Performance of a DC motor under steady-state conditions can be characterized by three measures: stall torque, no-load speed, and constant-speed behavior. These measures can be studied by considering Eq. (12.14). Substituting Eq. (12.6) for I_a and Eq. (12.4) for E_g into Eq. (12.14) and making use of Eq. (12.12) yield

$$T_g = \frac{K_T}{R_a}V_a - K_M^2\omega \tag{12.15}$$

This linear relation is the basis for generating speed–torque curves of a DC motor. By setting $\omega = 0$ in Eq. (12.15), the blocked-rotor torque at the rated voltage V_a, called the *stall torque* T_s, is determined by

$$T_s = \frac{K_T}{R_a}V_a \tag{12.16}$$

The theoretical no-load speed is obtained by setting $T_g = 0$ in Eq. (12.15):

$$\omega_0 = \frac{K_T V_a}{R_a K_M^2} = \frac{V_a}{K_E} \tag{12.17}$$

If the speed of the motor is controlled by a PWM signal with a duty cycle dc, the applied voltage to the motor will be $V_a \times dc$ and Eq. (12.17) is simply multiplied by the dc to yield the corresponding speed. Equations (12.16) and (12.17) give

$$T_s = K_M^2\omega_0 \tag{12.18}$$

Since V_a (and V_f in case of field coil motors) is constant at steady state, Eq. (12.18) can be expressed as

$$\frac{\omega}{\omega_0} + \frac{T_g}{T_s} = 1 \tag{12.19}$$

A plot of Eq. (12.19) is shown in Fig. 12.9. Note that if V_a is increased, the curve translates to the right. If a field motor is used and V_f is increased, the curve tilts to the left.

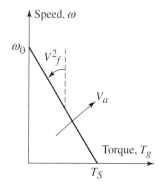

Figure 12.9 Variation of generated torque with angular velocity.

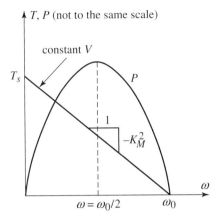

Figure 12.10 Power variation in a DC motor.

If the motor rotates at a constant speed, it produces a parabolic mechanical power curve, shown in Fig. 12.10 and given by

$$P = T_g \omega = \frac{K_T}{R_a} V_a \omega - K_M^2 \omega^2 \qquad (12.20)$$

Equation (12.20) reveals that the maximum power P_M is delivered at $\omega = {}^1/_2 \omega_0$ and its value is

$$P_M = \frac{K_M^2 \omega^2}{4} \qquad (12.21)$$

Dynamic Model

Figure 12.11 shows an equivalent dynamic model of a DC motor driving a load through a one-stage gear reduction. The equation of motion that governs the dynamic behavior of the system is easily established to be

$$T_g = T_f + T_L + T_D + J_T \frac{d\omega}{dt} + T_{gr} \qquad (12.22)$$

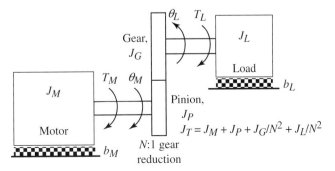

Figure 12.11 A motor driving a load via gear reduction.

T_f is a constant coulomb friction torque, T_L is the load torque, T_D is the viscous and friction torques, which are proportional to ω, and T_{gr} is a gravity torque. J_T is the sum of mass moments of inertia of the rotor J_M, the inertia of the pinion J_P, the equivalent inertia of the gear J_G, and the load J_L at the motor shaft (see Fig. 12.11). The viscous torque is determined by

$$T_D = B\omega \qquad (12.23)$$

where B is the equivalent viscous damping coefficient, which includes the damping in the motor b_M and the equivalent damping in the load b_L at the motor shaft.

Motor torque capability is expressed by two parameters: the continuous value and the peak value. The continuous torque is the torque that the motor can sustain, often at any speed, without overheating. The peak torque is the maximum torque that can be generated for short periods without causing mechanical damage or demagnetization. The peak torque is several times the continuous torque. The motor can generate any level of torque below the peak torque as long as the root mean square (RMS) value of the torque is within the continuous-torque level.

At this point, it is appropriate to provide parameters' values for a typical DC motor. The following are such values for the JDH-2250 Clifton Precision motor: $R_a = 2.7\ \Omega$, $L_a = 0.004$ H, $b_M = 0.0000093$ N-m-s/rad, $K_E = 0.105$ V-s/rad, $K_T = 0.105$ N.m/A, $J_M = 0.0001$ kg-m^2.

Electromechanical Model

Equations (12.13) and (12.22) represent the behavior of the DC motor and are expressed in state-space form as

$$\begin{bmatrix} dI_a/dt \\ d\omega/dt \end{bmatrix} = \begin{bmatrix} -R_a/L_a & -K_T/L_a \\ K_E/J_T & -B/L_a \end{bmatrix} + \begin{bmatrix} 1/L_a \\ 0 \end{bmatrix} V_a - \begin{bmatrix} 0 \\ 1/J_T \end{bmatrix}(T_f + T_L + T_{gr}) \quad (12.24)$$

Taking the Laplace transform of both sides of Eq. (12.13) while assuming zero initial conditions and solving for I_a yields

$$I_a(s) = \frac{V_a(s) - K_E\omega(s)}{L_a s + R_a} \qquad (12.25)$$

Similarly, taking the Laplace transform of Eq. (12.22) gives

$$T_g(s) = J_T s\omega(s) + B\omega(s) + T_L(s) + T_f(s) + T_{gr}(s) \qquad (12.26)$$

Figures 12.12a and b show block diagram representations of Eqs. (12.24) and (12.25), respectively. Solving Eq. (12.25) for $\omega(s)$ gives

$$\omega(s) = \frac{T_g(s) - (T_L(s) + T_f(s) + T_{gr}(s))}{sJ_T + B} \qquad (12.27)$$

Combining the block diagrams in Figs. 12.12a and b gives the block diagram for the velocity loop model of the DC motor shown in Fig. 12.12 c. Note that all previous relations can be expressed in terms of the angular position of the rotor θ, instead of its angular velocity, by substituting $s\theta(s) = \omega(s)$.

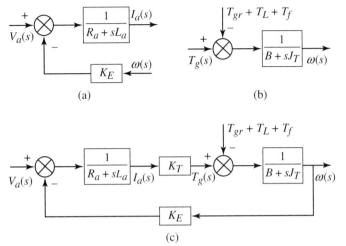

Figure 12.12 Block diagram of a DC motor velocity control.

Transfer Function of a DC Motor

The transfer function of a DC servo motor velocity loop is obtained from Fig. 12.12 as

$$G_m(s) = \frac{\omega(s)}{V_a(s)} = \frac{K_T/L_a J_T}{s^2 + ((R_a J_T + BL_a)/L_a J_T)s + ((R_a B + K_E K_T)/L_a J_T)} \qquad (12.28)$$

Note that the expression $T_l(s) + T_{gr}(s) + T_f(s)$ does not appear in $G_m(s)$ because these terms are treated as additional inputs. Also, $G_m(s)$ is an inherent characteristic of the system, and external disturbances are not part of it. The poles of the transfer function are the roots of the characteristic equation

$$s^2 L_a J_T + s(R_a J_T + BL_a) + (R_a B + K_E K_T) = 0 \qquad (12.29)$$

If damping is assumed negligible, i.e., if $B \approx 0$, the poles become

$$s_{1,2} = \frac{R_a J_T(-1 \pm \sqrt{1 - (4K_E K_T L_a/R_a^2 J_T)})}{2L_a J_T} \qquad (12.30)$$

In all practical motors, L_a is small, so

$$(R_a J_T)^2 - 4K_E K_T L_a J_T > 0 \qquad (12.31)$$

and the two poles are negative and real, given by

$$s_{1,2} = \frac{-R_a J_T \mp \sqrt{(R_a J_T)^2 - 4K_E K_T L_a J_T}}{2L_a J_T} \qquad (12.32)$$

When L_a is much smaller than $R_a^2 J_T/K_E K_T$, one can use the approximation (x is very small)

$$\sqrt{1 - x} = 1 - (x/2) \qquad (12.33)$$

which gives

$$R_a J_T \sqrt{1 - (4K_E K_T L_a / R_a^2 J_T)} = R_a J_T [1 - (2K_E K_T L_a / R_a^2 J_T)] \qquad (12.34)$$

and the two poles become

$$p_1 = \frac{-K_E K_T}{R_a J_T} \qquad \text{and} \qquad p_2 = \frac{-R_a}{L_a} \qquad (12.35)$$

With these two poles, the motor transfer function reduces to

$$G_m(s) = \frac{1/K_E}{(s\tau_m + 1)(s\tau_e + 1)} \qquad (12.36)$$

in which τ_m and τ_e are the mechanical and electrical time constants of the motor, respectively. They are given by

$$\tau_m = \frac{R_a J_T}{K_E K_T} \qquad \text{and} \qquad \tau_e = \frac{L_a}{R_a} \qquad (12.37)$$

Typical values for τ_e are around 1 ms, and τ_m varies within the range $1/10 < \tau_m < 1/20$ (sec/rad).

Note that the poles of $G_m(s)$ will be the negative reciprocal of τ_m and τ_e only if the two time constants are sufficiently different, or if $\tau_m > 10\tau_e$. Furthermore, if the motor inductance is large enough, yielding

$$4K_E K_T L_a J_T > R_a^2 J_T^2 \qquad (12.38)$$

the two poles of the $G_m(s)$ become complex. However, this is not the case for the majority of DC motors. The inductance of a DC motor is usually small and $L_a \approx 0$ may be assumed. In this case, the transfer function in Eq. (12.36) reduces to

$$G_m(s) = \frac{\omega(s)}{V_a(s)} = \frac{1/K_E}{1 + s\tau_m} \qquad (12.39)$$

which is the model that will be considered hereafter.

12.2.3 Heat Dissipation in DC Motors

The main factor that limits the performance of an incremental motion-control system is the heat dissipation in the motor armature. The energy dissipated in a motor W_C as it moves a load during a time of t_C seconds is given by

$$W_C = \int_0^{t_C} P(t)\, dt = R_a \int_0^{t_C} I_a^2(t)\, dt \qquad (12.40)$$

If the load torque T_L is constant, the energy dissipated in the motor during t_C seconds is

$$W_{C(T_L)} = I_a^2 R t_C = \frac{R}{K_T^2} T_L^2 t_C \qquad (12.41)$$

Because the energy dissipation is repeated f times per second, the average power is

$$P = fW_C \qquad (12.42)$$

The temperature in the motor rises proportional to the average power

$$\Delta T = R_{TH}P \qquad (12.43)$$

where R_{TH} is the motor thermal resistance. The temperature rise is also proportional to the RMS torque, T_{RMS}, given by

$$T_{RMS} = \left[\frac{1}{t_c} \int_0^{t_c} T_g^2 dt \right]^{1/2} \qquad (12.44)$$

Two important factors influence heat dissipation in a motor: the velocity profile and the coupling ratio. The *velocity profile* refers to the way the angular velocity of the motor is varied with time as the load moves, and the *coupling ratio* refers to the ratio of the angular velocity of the motor inertia J_M to the velocity of the load inertia J_L. The optimum velocity profile $\omega(t)$ and coupling ratio G_O may be determined to minimize W_C while satisfying Eq. (12.40) for a required torque and a rotor position given by

$$\theta = \int_0^{t_c} \omega(t)dt \qquad (12.45)$$

12.2.4 Velocity Profile Optimization

Three possible velocity profiles are investigated: parabolic, triangular, and trapezoidal. The effect of friction may be ignored in searching for an optimum velocity profile assuming friction is independent of velocity. The parabolic velocity profile shown in Fig. 12.13a is given by

$$\omega(t) = 6\theta \frac{t_C - t}{t_C^3} t \qquad (12.46)$$

In the triangular velocity profile, the load is accelerated at a fixed rate and then decelerated at the same rate, as shown in Fig. 12.13b. The trapezoidal profile is divided into three equal time zones: acceleration time, run time, and deceleration time, as shown in Fig. 12.13c.

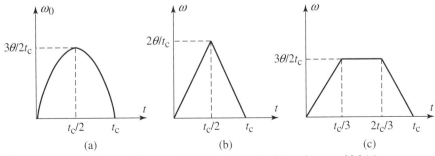

Figure 12.13 Velocity profiles: parabolic (a), triangular (b), and trapezoidal (c).

Substituting the three velocity profiles into Eq. (12.40), the energy dissipation during t_C can be expressed as

$$W_C = \frac{R}{K_T^2}\left[\lambda\frac{J_T^2\theta^2}{t_C^3} + T_L^2 t_C\right] \tag{12.47}$$

where $\lambda = 12/\eta$ and η is the velocity profile efficiency, expressed as

$$\eta = \frac{W_{C0}}{W_C} \tag{12.48}$$

and evaluated at $T_L = 0$. W_{C0} is the energy dissipation for a parabolic velocity profile, used as a reference for comparison because it is the optimal profile. $\lambda = 12$ for the parabolic profile, $\lambda = 16$ for the triangular profile, and $\lambda = 13.5$ for the trapezoidal profile with equal parts. Although the optimum velocity profile is the parabolic one, the trapezoidal profile is favored in practice for ease in control implementation.

12.2.5 Inertia Matching

Three possible mechanisms to couple a motor to a load are considered: gear transmission, belt-and-pulley drive, and the lead screw drive.

Gear Transmission

Figure 12.14 shows a DC motor driving a load via a gear reducer unit. The load parameters—angular rotation θ_L, load inertia J_L, and load torque T_L—are reflected to the motor shaft by the following relations:

$$\theta'_L = N\theta_L, \qquad J'_L = J_L/N^2, \qquad \text{and} \qquad T'_L = T_L/N \tag{12.49}$$

$N = N_G/N_P$, the ratio of the number of teeth of the gear to the number of teeth of the pinion. The equation for energy dissipation, derived from Eqs. (12.47) and (12.49), becomes

$$W_C = \frac{R}{K_T^2}\frac{12}{\eta}\frac{J_L^2\theta_L^2}{t_C^3}\left[N^2\left(\frac{J_M}{J_L} + \frac{1}{N^2}\right)^2 + \frac{\gamma}{N^2}\right] \tag{12.50}$$

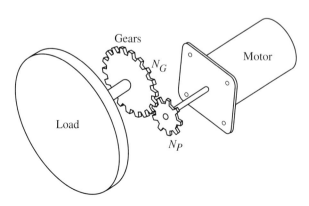

Load
Gears
N_G
N_P
Motor

Figure 12.14 A motor driving a load via a pinion–gear transmission.

where

$$\gamma = \frac{\eta}{12}\left[\frac{T_L t_C^2}{\theta_L J_L}\right]^2 \qquad (12.51)$$

In order to minimize the power dissipation in the motor for a given load power requirement, the bracketed term in Eq. (12.50) is differentiated with respect to N^2 and the result is equated to zero. This process gives the optimum gear ratio N_O that minimizes W_C:

$$N_0^2 = \frac{J_L}{J_M}\sqrt{1 + \gamma} \qquad (12.52)$$

In the absence of load torque, $T_L = 0$, so $\gamma = 0$, and Eq. (12.52) becomes

$$N_0 = \sqrt{J_L/J_M} \qquad (12.53)$$

which is known as *inertia matching*, since $J_M = J_L/N^2$.

Belt–Pulley Drive

The belt–pulley drive shown in Fig. 12.15 consists of a motor turning a pulley that pulls a belt to which the load is attached, thus converting rotary motion to translation. The coupling ratio G is the reciprocal of the radius r of the pulley connected to the motor. The load mass m is to be moved a distance x in time t_c, and the constant opposing force is F. Thus the energy dissipated in the motor armature during t_c

$$W_C = \frac{R}{K_T^2}\frac{12}{\eta}\frac{m^2 x^2}{t_C^3}\left[G^2\left(\frac{J_M}{m} + \frac{1}{G^2}\right)^2 + \frac{\beta}{G^2}\right] \qquad (12.54)$$

where

$$\beta = \frac{\eta}{12}\left[\frac{F t_C^2}{mx}\right]^2 \qquad (12.55)$$

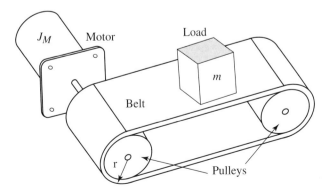

Figure 12.15 A DC motor driving a load via a belt–pulley mechanism.

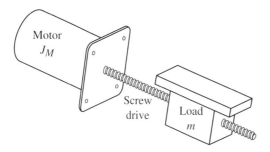

Figure 12.16 A DC motor moving a load via a lead screw drive.

The optimum coupling is easily determined to be

$$G_O^2 = \frac{m}{J_M}\sqrt{1 + \beta} \tag{12.56}$$

from which the optimum pulley radius is

$$r_O = \sqrt{J_M/m\sqrt{1 + \beta}} \tag{12.57}$$

If the load force is $F = 0$, then $\beta = 0$, and an exact inertia match is found to be

$$r_O = \sqrt{J_M/m} \tag{12.58}$$

Lead Screw Drive

A typical lead screw drive is shown in Fig. 12.16. Parameters x, θ, T_L, F, and m are the same as in a belt–pulley system. The problem in this case is to determine the optimum pitch, P_o, of the lead screw. The pitch is taken as the number of revolutions per unit length

$$P = G/2\pi \tag{12.59}$$

Since G is the motor rotation in radians per unit length, the optimum pitch is

$$P_{\text{opt}} = (1/2\pi)\sqrt{m/J_M} \tag{12.60}$$

12.2.6 Motor Selection

There is no step-by-step guide that would provide the ultimate answer to selecting a motor for a specific application. However, from among the many possible workable solutions, the engineer should be able to select the one that satisfies the desired motion requirements. A motor is selected according to the following criteria: (1) The motor must generate a torque T_g to accommodate peak torque T_{PEAK} and continuous torque T_C to drive the load along the desired trajectory, and (2) the motor should run at the maximum desired velocity without overheating. Peak torque is the maximum torque required, often needed during the acceleration phase of the motion, and it is computed from Eq. (12.22) using the maximum acceleration during start-up. Continuous torque is the torque needed to overcome friction and to drive the load. It is computed as the RMS torque from Eq. (12.44). Peak torque and continuous torque may also be measured directly. Overheating is not a problem if the rated continuous torque is greater than T_{RMS}.

Example 12.1: Inertia and torque calculations for a conveyor system—An application calls for selecting a DC motor to drive a conveyor system. The drive train and motion profile are shown in Fig. 12.17. The conveyer and load are to be accelerated from rest to 1.016 m/s in 0.5 s, driven at constant speed for 14 s, and decelerated to rest in 0.5 s. For worst-case analysis, the cycle is repeated continuously. The friction torque at the drive shaft is measured to be $T_f' = 54.23$ N-m. The system operates in ambient conditions at 25°C. The specifications of the drive train components are given in Table 12.2. Determine the RMS torque that will be used to select the motor.

Solution: The equivalent inertia of all components reflected at the motor shaft are summarized in Table 12.2.

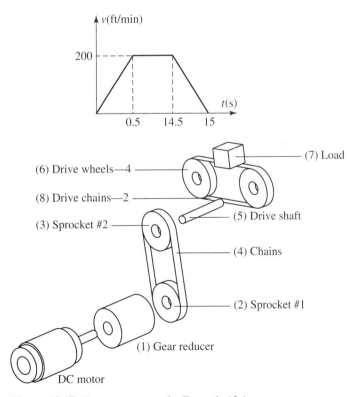

Figure 12.17 Conveyor system for Example 12.1.

TABLE 12.2 Specifications of the Conveyor Belt Components

Item	Part	Specifications	Inertia Reflected at Motor Shaft
1	Gear reducer	$N = 10$, $J_1 = 0.0002824$ (N-m-s^2)	$J_1 = 0.0002824$ (N-m-s^2)
2	Sprocket #1	$r_2 = 2$ in., $w_2 = 8.896$ N	$J_2 = 0.00001172$ (N-m-s^2)
3	Sprocket #2	$r_3 = 6$ in., $w_3 = 88.960$ N	$J_3 = 0.00011720$ (N-m-s^2)
4	Chain	$w_4 = 106.240$ N	$J_4 = 0.0002810$ (N-m-s^2)
5	Drive shaft	$r_5 = 0.0381$ m, $w_5 = 22.240$ N	$J_5 = 0.0000018$ (N-m-s^2)
6	Drive wheels (4)	$r_6 = 0.2032$ m, $w_6 = 88.960$ N	$J_6 = 0.0008332$ (N-m-s^2)
7	Load	$w_7 = 444.800$ N	$J_7 = 0.0020830$ (N-m-s^2)
8	Drive chains (2)	$w_8 = 667.200$ N	$J_8 = 0.0062420$ (N-m-s^2)

The total inertia reflected at the motor shaft is computed as follows:

$$J = \sum_{i=1}^{8} J_i = J_1 + \frac{m_2 r_2^2}{2N^2} + \frac{m_3 r_3^2}{2(N \times r_3/r_2)^2} + \frac{m_4 r_2^2}{N^2} + \frac{m_5 r_5^2}{2(N \times r_3/r_2)^2}$$

$$+ 4 \times \frac{m_6 r_6^2}{2(N \times r_3/r_2)^2} + \frac{m_7 r_6^2}{(N \times r_3/r_2)^2} + 2 \times \frac{m_8 r_6^2}{(N \times r_3/r_2)^2}$$

$$= 0.0002824 + 0.00001172 + 0.0001172 + 0.0002810$$
$$+ 0.0000018 + 0.0008332 + 0.0020830 + 0.0062420$$
$$= 0.00985(\text{N-m-s}^2)$$

The load torque during the acceleration phase (t_1: $t \leq 0.5$ s) is

$$T_1 = T_a + T_{Lf}$$

$$= J_T \frac{\omega_0}{t_1} + \frac{T'_{LF}}{N \times r_3/r_2}$$

$$= J_T \frac{1}{t_1} \frac{v}{r_6} N \frac{r_3}{r_2} + \frac{T'_{LF}}{N \times r_3/r_2}$$

$$= 0.00985(\text{N-m-s}^2) \times \frac{10}{0.5(\text{s})} \frac{1.016(\text{m/s})}{0.2032} \times \frac{0.1524(\text{m})}{0.0508(\text{m})}$$

$$+ \frac{54.23(\text{N-m})}{10 \times 0.1524(\text{m})/0.0508(\text{m})} = 2.955 + 1.8077$$

$$= 4.7627(\text{N-m})$$

Load torque during the constant-velocity phase (t_2: $0.5 \leq t \leq 14.5$ s) is simply the friction torque,

$$T_2 = \frac{T'_{LF}}{N \times r_3/r_2} = 1.8077(\text{N-m})$$

Load torque during the deceleration phase (t_3: $t \geq 14.5$ s) is simply

$$T_3 = T_a - T_{Lf}$$

$$= 2.955 - 1.8077 = 1.1473(\text{N-m})$$

The RMS load torque is then

$$T_{RMS} = \sqrt{\dfrac{\sum_{i=1}^{3} T_i^2 t_i}{\sum_{i=1}^{3} t_i}} = 1.963\,(\text{N-m})$$

Any motor capable of sustaining this torque while maintaining speed requirements can be used.

12.2.7 Servo Amplifiers

A servo amplifier is a power op-amp capable of supplying the required current and voltage to the motor. The amplifier typically receives an analog command signal within ± 10 V and amplifies it to the required level of current or voltage.

Power amplifiers are available in one of two types: linear amplifiers and PWM switching amplifiers. The principal output stage of either amplifier can be one of two basic designs: the "T"-type configuration, shown in Fig. 12.18a, and the "H"-type configuration, shown in Fig. 12.18b. The "H," or bridge output stage, consists of four transistors and a single unipolar power supply. It is wired in such a way that only two transistors are on at a time. The motor must be floated with respect to ground, and current and/or voltage feedback is not easily implemented. The "T" stage requires a *bipolar* power supply and two complementary transistors. The transistors cannot both be on at the same time, otherwise they would probably fail because the transistors would form a short circuit and all current will flow through them. Because the motor does not float with respect to ground, voltage and/or current feedback signals are easily implemented.

In linear amplifiers, the power transistors are operated in their active (or linear) range. Linear amplifiers supply constant output voltage, which is often a fraction of the total supply

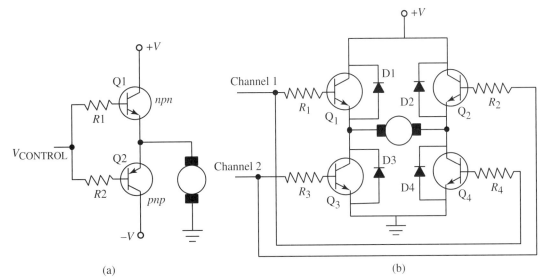

(a)   (b)

Figure 12.18 Servo amplifiers: T-type, requiring double-ended power supply with a complementary pair of transistors (a), and H-type, requiring unipolar power supply (b).

voltage. Consequently, the collector-to-emitter voltage drop of the conducting transistors is high and the power dissipated in the collector is significant, requiring large transistors and heat sinks. Because of power loss, the efficiency of linear amplifiers is low, and their use is restricted to low-power applications. A linear amplifier is easy to drive. Linear amplifiers use the "T" configuration, most often due to the ease of implementing voltage and/or current feedback.

A switching amplifier uses power transistors that can be switched at megahertz rates. The amplifier generates a voltage that switches between the high and low levels of the DC supply voltage. Switching amplifiers are more suitable for digital control. The most common method of amplifier switching is PWM. Here the transistors are switched at a constant frequency, and the resulting output voltage varies between the two extreme values. By varying the pulse width or the duty cycle, the average value of the output voltage can be changed in accordance with the desired drive. The "H" type is the more common form, especially when high voltage is required. Because power transistors are either off or in saturation, the power dissipated in the collectors is considerably less than in an equivalent linear amplifier. PWM is used almost exclusively when power requirements exceed 100 watts. The practical range of switching frequency is between 1 and 15 kHz. Frequencies higher than 15 kHz may cause the transistors to operate in the active region for a significant portion of the overall switching period, resulting in high power loss.

To compare the performance of linear and switching amplifiers, consider an application in which a 40-V supply delivers power to a 10-V, 5-A motor through an amplifier. A linear amplifier would output 10 V at all times, which means that the 30-V difference between the supply and output voltages falls across the transistors of the output stage. Drawing 5 A at 10 V, the motor absorbs 50 watts of power, while the 5 A of current flows through the 30-V voltage drop in the output transistors, resulting in 150 watts of power dissipation. The net efficiency of the amplifier will be only 25%. In comparison, a PWM amplifier delivers the required voltage at close to 98% efficiency. It switches the output voltage between zero and 40 V, so the voltage equals 40 V only 25% of the time and zero 75% of the time, with an average of 10 V. The driver is turned either fully on or fully off, which results in a significant reduction in power dissipation.

Power Amplifier Configurations

Power amplifiers can be configured as either current or voltage amplifiers. The type of amplifier used depends on the strategy employed to control the motor, which is dependent on the application and the environment in which the motor operates. Two alternative approaches can be used to control DC motors: torque control and velocity control. Torque control utilizes a current amplifier in the motor's drive unit to manipulate the motor current, since the motor's armature current is proportional to the generated torque. The torque generated overcomes gravity, friction, and torques due to load inertia. Speed control, on the other hand, utilizes a voltage amplifier in the motor's drive unit to manipulate the DC motor voltage.

The current-amplification strategy is commonly used when the system does not use a tachometer. A current amplifier receives an input voltage signal from the controller, commonly in the range between -10 and $+10$ V, and supplies a current to the motor that is proportional to the input voltage. This mode is also referred to as the torque mode, because it controls the motor output torque. The bandwidth of the power amplifier is finite. In most applications, the frequency response of the amplifier is adequately represented by a single-pole model. The transfer function of a current feedback amplifier is given by

$$T_A(s) = \frac{I_{OUT}(s)}{V_{IN}(s)} = \frac{A_I}{1 + s\tau_A} \tag{12.61}$$

where $1/\tau_A$ is the frequency bandwidth and A_I is the amplifier gain. In general, τ_A is very small and the amplifier is modeled as a gain A_I only, which indicates the output current in amperes for 1 V of command signal. The output current of a current amplifier is simply the product of the amplifier gain and the input voltage. A current amplifier should have a low output impedance.

A current amplifier may be configured in what is called the *velocity mode*. It includes a voltage-amplification stage, which compares the applied voltage with the motor velocity and amplifies the difference before the signal is converted to a proportional current in the current loop.

The voltage amplifier provides an output voltage proportional to the voltage signal it receives from the controller. It is capable of supplying the current required by the motor. The transfer function for a voltage amplifier is given by

$$T_A(s) = \frac{V_{OUT}(s)}{V_{IN}(s)} = \frac{A_V}{1 + s\tau_A} \tag{12.62}$$

where $1/\tau_A$ is the frequency bandwidth and A_V is the amplifier gain. In general, τ_A is very small and the amplifier is modeled as a gain A_V only. In the voltage mode, the output voltage is the product of the amplifier gain and the input voltage. This mode is used with motor velocity control.

Example 12.2: Amplifier selection–A DC motor drive system consists of a motor that has a moment of inertia $J_M = 10^{-4}$ kg-m^2, a resistance $R_a = 2\ \Omega$, and a torque constant $K_T = 0.2$ Nm/A. The moment of inertia of the load is $J_L = 2 \times 10^{-4}$ kg-m^2. The load is to be accelerated at $\alpha = 10{,}000$ rad/sec^2 to reach a slew velocity of $\omega_{MAX} = 100$ rad/sec against a friction load $T_f = 0.4$ Nm. Select an amplifier for the drive.

Solution: Selecting amplifier size for a given application requires knowledge of the peak values for the motor's current and voltage. The peak current is determined by

$$I_{MAX} = \frac{T_{PEAK}}{K_T} \tag{12.63}$$

where the peak torque T_{PEAK} is determined by Eq. (12.22) for the acceleration interval. Neglecting damping, the required peak torque is

$$T_{PEAK} = 10{,}000 \times (1 \times 10^{-4} + 2 \times 10^{-4}) + 0.4 = 3.4 \text{ Nm}$$

Meanwhile, the peak voltage is obtained by substituting $\omega = \omega_{MAX}$, and $I_a = I_{MAX}$ into Eq. (12.14) to yield

$$V_{PEAK} = K_T\omega_{MAX} + R_a I_{MAX} \tag{12.64}$$

For the current system,

$$I_{MAX} = 3.4/0.2 = 17 \text{ A}$$

and the required peak voltage from Eq. (12.64) is

$$V_{PEAK} = 0.2 \times 100 + 2 \times 17 = 54 \text{ V}$$

The results indicate that the amplifier must be able to deliver 17 A at 54 V. Since the system parameters may vary, a 25% margin on the amplifier ratings is recommended.

12.2.8 DC Motor Servo Drive

Load inertia and torques in robotic applications and the like can change by a factor of 10 over the required motion range. Furthermore, heat dissipation in the motor affects its constants. In those cases, closed-loop control is not a luxury but a necessity.

A typical closed-loop control system, also called a servo system, is shown in Figure 12.19. Its components include the controller, the power amplifier, the motor, and sensors. The control loop receives its command from a programmable controller, a terminal, and switches, but most commonly from a computer. The controller generates the proper signal and initiates a motion command to the driver. The power amplifier delivers the necessary current to drive the motor. For example, the driver takes a low-current electrical signal and amplifies it to a higher level of 10 A. The motor generates a torque proportional to the applied current to move the load. The sensors determine the velocity and position of the motor and report the results back and thus close the control loop. A tachometer is typically used to measure angular velocity, and either a shaft encoder or a resolver is used to measure angular position.

Position and Velocity Feedback

Position feedback is made available by using one of the commonly used position sensors, which include potentiometers, LVDT, resolver, and the encoder. An important factor to consider in choosing a position sensor is sensor resolution, which impacts the overall

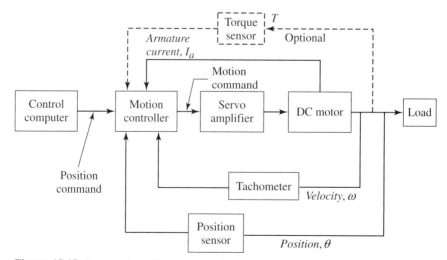

Figure 12.19 Typical closed-loop control of a servo system.

performance of the system. The increased resolution reduces the effect of *position quantization*. For example, the position quantization of a shaft encoder that generates N counts/rev is $360°/N$, which means that accuracy and repeatability below this level are not possible. Therefore, higher resolution means better potential accuracy and repeatability. If the same encoder is used to measure the speed of a motor running at ω (rev/s), the feedback sensor frequency f_S (Hz) would be

$$f_S(\text{Hz}) = \omega(\text{rev/s}) \times N(\text{counts/rev}) \tag{12.65}$$

This equation indicates clearly that the allowed sensor frequency based on the control loop bandwidth is a function of both maximum speed and the encoder's resolution.

Using velocity feedback affects system performance in many ways. It induces damping into the system and helps improve its stability, especially if the system structure is flexible enough to become susceptible to resonances and vibrations. However, a velocity sensor increases system cost and requires additional space. Additionally, velocity feedback produces position errors, which reduce positioning accuracy during a move. For these reasons, modern motion controllers achieve stability through the use of digital filters applied to the encoder counts instead of using separate velocity feedback.

Closed-Loop Control with DC Motors

The block diagram of a DC motor with velocity and position feedback is shown in Fig. 12.20. Neglecting disturbance torque, that is, assuming $T_d(s) = 0$, the transfer function of the system is easily determined to be

$$T(s) = \frac{\theta(s)}{\theta_d(s)} = \frac{A}{A(K_P + sK_G) + K_E(s)(1 + s\tau_A)(1 + s\tau_m)} \tag{12.66}$$

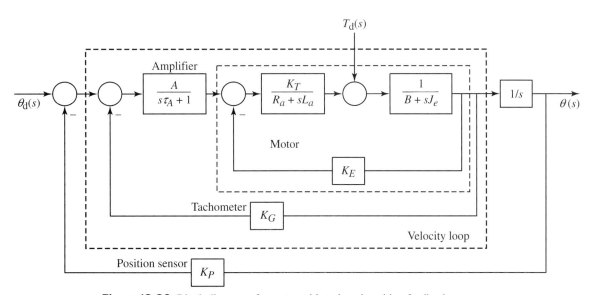

Figure 12.20 Block diagram of a motor with tach and position feedback.

To investigate the steady-state performance of the system, a step input of amplitude θ_d is applied. According to the *Laplace final-value theorem*, the steady-state value of a function $f(t)$ as $t \Rightarrow \infty$ is the same as the value of $sF(s)$ as $s \Rightarrow 0$; that is,

$$f(\infty) = \lim_{t \to \infty} f(t) = \lim_{s \to 0} sF(s) \tag{12.67}$$

In this case $F(s) = \theta(s) = \theta_d(s)T(s)$ and the final position is determined to be

$$\theta_{final} = \lim_{s \to 0}\left[s\theta_d(s)T(s)\right] = \frac{\theta_d}{K_P} \tag{12.68}$$

This result indicates that the actual position is the desired position divided by K_P. Therefore, multiplying the command signal by a position gain K_P forces the motor to reach the desired position θ_d with no error.

Effect of Tach Feedback Because the time constant of the amplifier τ_A is very small, the corresponding pole at $s = -1/\tau_A$ has little effect on the closed-loop response of the system. Under this condition, the joint position servo can be approximated by a second-order system and the closed-loop transfer function in Eq. (12.28) reduces to

$$T(s) = \frac{\theta(s)}{\theta_d(s)} = \frac{A/K_E\tau_m}{s^2 + s[(1 + AK_G/K_E)/\tau_m] + [(AK_P/K_E)/\tau_m]} \tag{12.69}$$

Comparing Eq. (12.70) to the transfer function of a standard second-order system gives the damping coefficient for the DC motor with position and tach feedback:

$$\zeta = \frac{0.5(1 + AK_G/K_E)}{\sqrt{AK_P\tau_m/K_E}} \tag{12.70}$$

Two conclusions can be inferred from Eq. (12.69). Increasing K_G increases the damping in the system and decreases settling times, and increasing K_P decreases system damping. This suggests that tach feedback enhances the stability of the servo system and reduces response oscillations. Though a small value of K_P seems desirable, other factors, such as the speed of response, influence the limit imposed on the minimum value of K_P.

Effect of Disturbance $T_d(s)$. To ascertain the effect of disturbance torque on the response of the servo system, with $T_d(s)$ as the input signal, we have

$$T'(s) = \frac{\Delta\theta(s)}{T_d(s)} = \frac{R_a(1 + s\tau_A)}{s(1 + s\tau_A)(K_EK_T + sR_aJ_e) + AK_T(K_P + sK_G)} \tag{12.71}$$

where $\Delta\theta$ is the disturbance-related error. The actual position of the motor becomes

$$\theta_a(s) = \theta(s) - \Delta\theta(s) \tag{12.72}$$

Substituting Eqs. (12.69) and (12.71) into Eq. (12.72) gives

$$\theta_a(s) = T(s)[\theta_d(s) - (T_d(s)R_a(1 + s\tau_a))/AK_T] \tag{12.73}$$

For simplicity, assume that the position of the motor is to be held constant at $\theta_d = 0$. If the disturbance torque is a step of amplitude $T_d(t)$, then $T_d(s) = T_d/s$, and applying the final-value theorem gives

$$\Delta\theta_{Final} = -\frac{T_d}{AK_TK_P} \tag{12.74}$$

This equation implies that the disturbance torque results in a steady-state error. If the power amplifier gain A is fixed, the right-hand side of Eq. (12.74) can be reduced by increasing K_P. If $\Delta\theta_{MAX}$ is specified, the minimum gain $K_{P(MIN)}$ to achieve this will be

$$K_{P(MIN)} = \frac{R_aT_L}{AK_T \times \Delta\theta_{MAX}} \tag{12.75}$$

For small $\Delta\theta_{MAX}$, K_P will be very large, which causes highly underdamped or unstable responses. Using tach feedback may improve the response, but such a solution is not always feasible. A practical solution adds derivative and integral control actions to eliminate the steady-state errors and achieve desirable damping levels. The overall control scheme is referred to as a PID controller. Chapter 13 presents classical, modern, and intelligent control schemes in detail.

12.2.9 Interfacing DC Motors to the 9S12C

Even small DC motors require relatively high current levels for operation. Since a microprocessor cannot supply the current required to drive the motor, a power stage is introduced between the motor and the microprocessor to deliver necessary power (V and I) to the motor.

A DC motor is easily controlled from the MCU through a DAC. The MCU generates the necessary control signals from an output port, and the analog output from the DAC controls the direction and speed of the DC motor. Recall that PWM is a special form of DAC, and the 9S12C can generate five PWM signals from its output compare line (see Section 9.3). In this case the PWM signal can directly control the speed of the motor without the need for a DAC by changing the duty cycle of the PWM signal. If the signal is applied to a DC motor circuit, the motor sees the average DC voltage level.

The control signal generated by the MCU is delivered to the interface circuitry of the power stage responsible for supplying necessary current and voltage levels required by the motor. The power stage is simply a power amplifier between the MCU and the motor.

The H-Bridge Circuit

The H-bridge is the most commonly used design to interface the MCU with the power stage of a motor drive. It merely consists of four transistor switches arranged in an H-pattern, with the motor floating as shown in Fig. 12.21. Relays could have been used as switches instead, but relays have slow switching speeds and are susceptible to wear.

The H-bridge works as follows. If transistors Q1 and Q4 conduct while Q2 and Q3 are turned off, the current flows through the motor from A to B and the motor will rotate in one direction. Meanwhile, if transistors Q2 and Q3 conduct while Q1 and Q4 are turned off, the voltage polarity is reversed and the current flows through the motor from B to A, causing it to spin in reverse. If the terminals are floating, the motor will freewheel; if the terminals are shorted, the motor will brake if it is a PM motor. If the transistors were ideal, there will be no voltage drop, and the voltage across the motor is always equal to the full magnitude of the supply voltage when opposite sets of transistors are on. However, transistors incur voltage drop.

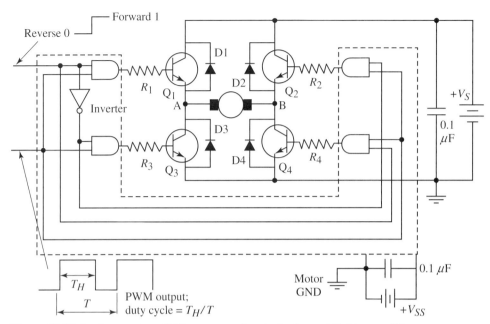

Figure 12.21 Implementation of DC motor speed control using an H-bridge amplifier.

Selecting Switching Transistors

The main criteria in selecting the switching transistors is that they be able to handle the current demands of the motor. Signal-level transistors are rated for no more than 500–600 mA. For most drive motors, however, continuous current demand is higher than this limit, on the order of 1–5 A, or more. High current demand requires the use of power transistors (refer to Section 3.12). In general, the p-type devices (pnp bipolar transistors and p-channel MOSFETs) have poorer performance than n-type devices. A more efficient H-bridge design is realized if all switches, including the high-side ones, are n-type devices. Power transistors must be used with suitable heat sinks. With most power transistors, the case is the collector (or drain) terminal.

The bipolar transistor is a current-controlled device, which requires significant amounts of base current to drive it into conduction. Because these currents cannot be delivered directly from the MCU or logic gates, another level of interface circuitry is often needed to excite the H-bridge into driving the motor. A biasing resistor is always needed for each BJT to prevent the transistor from pulling excessive current from the control port; otherwise the port could be destroyed from overheat. The resistor value depends on the voltage and current draw of the motor and the particular transistor used. This resistor is usually in the range from 1 to 3 KΩ. In addition to the added complexity involved in the bias network, the base current through the base resistor results in power losses.

In contrast with a bipolar transistor, a MOSFET is a voltage-controlled device, which requires very little current to operate and can be driven directly from the MCU or CMOS logic gates. The MOSFET acts as a switch when it operates in the linear region, where its on-resistance R_{ON} causes a finite voltage drop; a p-type MOSFET incurs a higher voltage drop because it has a higher R_{ON}. With an n-channel MOSFET driving a load that is connected between the supply and ground, the source voltage V_{SS} should be very close to that of the positive supply rail (V_{DD}) to fully open the MOSFET. Thus, to turn on an *n*-type MOSFET switch placed on the high side of the bridge, its gate voltage must be higher than that of the positive rail. Since the gate turn-on voltage $V_{GS\,(TH)}$ must be approximately 10 V higher than the source, the required gate voltage must be $V_{GS} = V_{SS} + V_{GS(TH)}$. Before this level is reached, the large stray capacitance between the gate and other channels of a power MOSFET must first be charged, which requires gate currents and charge-up time that may not be acceptable. A solution to these problems is to add a charge pump to the gate-drive network to create voltages higher than the supply voltage. The IR2110/IR2113(S) ICs from International Rectifier help overcome these problems. Each IC can provide a drive capability to high-voltage, high-speed power MOSFETs or IGBTs placed at the high and low sides of a one-half H-bridge; thus a full H-bridge requires two ICs.

The choice of MOSFETs or power BJTs depends largely on the power requirements of the motors and the choice of devices available. In general, power MOSFETs are preferred in designing the H-bridge, if they are available with low enough on-resistance and if, for the required current, the voltage drop incurred between the drain and the source is less than the saturation voltages of comparable bipolar devices. Recommended transistors to use for medium-duty applications are TIP31 *npn* or TIP32 *pnp* (TO-220 case) bipolar transistors, and for high-power applications the 2N3055 *npn* or 2N2955 *pnp* (TO-3 case) bipolar transistors and the IRF-511(TO-220 case) power MOSFET. A wide range of other choices is also available.

Open-Loop Speed Control

In applications, such as peripheral drive spindles, in which load inertia and steady-state load torque are relatively constant, open-loop control is adequate. Open-loop speed control of a DC motor can easily be achieved. Figure 12.21 shows this control scheme using the 9S12C MCU and an H-bridge amplifier circuit. The signal from port T pin PT[0] controls the direction of rotation of the motor. The AND gate is added to provide positive action control, with which only one pin of the MCU is used to control direction. A high control signal on pin PT[0] turns transistors Q1 and Q4 on, whereas a LOW signal on PT[0] turns transistors Q2 and Q3 on. The logic prohibits invalid operation. The PWM signal generated on the OC2 line controls the motor speed. Motor power and ground are separate. Using a separate power for the motor drive protects the controller from any possible drop in the supply. Bypass capacitors are placed close to the supplies to shunt high-frequency currents and prevent AC glitches from propagating to the motor commutation circuit, motor control and digital switches, and other circuits.

Commercially Available Full-Bridge ICs

Instead of using discrete components, one can employ H-bridge IC drivers that contain matched transistors and onboard inductive kick protection. Examples include the L293D from Unitrode Integrated Circuits (www.unitrode.com), the L298 IC from ST Microelectronics (www.st.com), the LMD18200(3 A, 55 V) from National Semiconductors (www.national. com), and many more.

Figure 12.22 shows the Multiwatt version of the L298 monolithic IC connected to the 9S12C. The L298 has dual full-bridge drivers, though only one is needed to drive one DC motor. The IC is designed to drive inductive loads such as relays, solenoids, DC motors, and stepper motors. The emitters of the lower transistors in each bridge are connected together, and the corresponding external terminal may be connected to an external sensing resistor R_S to control load current. Each bridge can handle up to 2 A maximum from a supply voltage up to 46 V. Logic inputs Input1–Input4 are TTL compatible, with an input voltage up to 1.5 V for a logic "0." The two inputs enable A and enable B are provided to independently disable bridge A and bridge B with a signal line separate from the logic control inputs. The L298 provides for an over-temperature protection and has an additional logic supply input to operate the logic at a lower voltage.

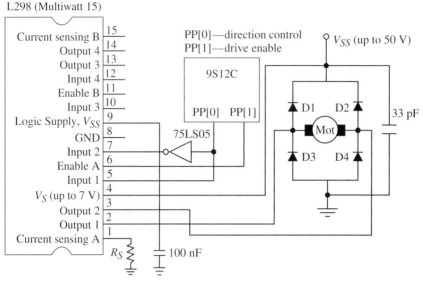

Figure 12.22 A DC motor driven by the 9S12C via an L298 driver IC.

Closed-Loop Control

Figure 12.23 shows the block diagram of a DC motor system with velocity feedback made available by a tachometer and position feedback by an optical encoder. The tachometer signal is fed directly back to an input channel of the ADC. The connection shows that a digital signal from a port T pin is transmitted through a DAC IC to deliver the proper supply through the amplifier. The MCU compares the tach signal with the desired velocity and adjusts the duty cycle in software accordingly. Additionally the signal from both the tachometer and the encoder may be used to provide for position and velocity control as needed. Other possible arrangements can be used.

PC-based closed-loop control of DC motor simulation and experimentation are covered extensively in practical examples, and using various control strategies, in Chapter 13.

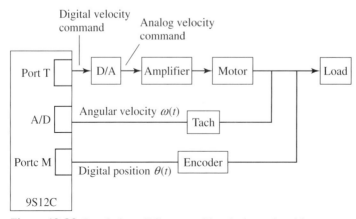

Figure 12.23 Interfacing a DC motor with velocity and position feedback to the 9S12C.

Controller ICs

High-performance digital motion control requires time-intensive, real-time computational tasks. To free the microcontroller from this computational burden and allow it to service other tasks of a target device, dedicated motion control processors are available for use with DC, brushless DC servo, and stepper motors. Two such driver ICs are the LM628 from National Semiconductor (www.national.com) and the HCTL-1100 from Agilent Technologies (www.agilent.com).

The LM628 provides an 8-bit output that can drive either an 8-bit or a 12-bit DAC. The LM629 provides an 8-bit PWM output for directly driving H-bridge circuits. Both ICs feature 32-bit position, velocity, and acceleration registers, programmable digital PID filter with 16-bit coefficients, a programmable derivative-sampling interval, and an internal trapezoidal velocity profile generator. The device can be configured to operate in either position or velocity mode. Velocity, target position, and filter parameters may be changed during motion. Additionally, the LM628 and LM629 have real-time programmable host interrupts, 8-bit parallel asynchronous host interface, and quadrature incremental encoder interface with index pulse input.

Figure 12.24 shows a complete 9S12C-based servo system built using the LM628. In addition to the LM628, the system employs a DC motor, an incremental encoder, a DAC, and a power amplifier. The 9S12C communicates with the LM628 through port T to facilitate programming the trapezoidal velocity profile and the digital PID filter. The profile generator calculates the required trajectory for either the position mode or the velocity mode of operation. The LM628 compares the actual (feedback) position to the desired (profile generator) position. The digital filter processes the error signal and computes compensated motor commands. Motor commands are made available as DAC output data, which is transformed by the DAC to a signal that is amplified and applied to the motor to drive it to the desired position.

The HCTL 1100 is a general-purpose controller IC from Agilent for velocity and position control of DC, DC brushless, and stepper motors. Adding a host processor such as the 9S12C, an amplifier, a motor, and an encoder to the HCTL-1100 provides for a complete servo system.

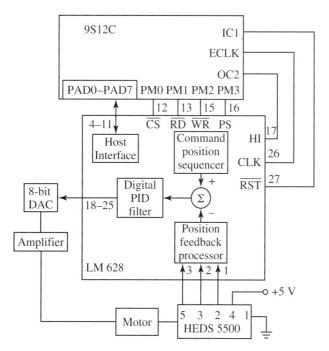

Figure 12.24 A complete servo system featuring the LM628 IC controller.

The user can select to operate the HCTL-1100 in one of four control modes: position control, proportional velocity control, trapezoidal profile control for point-to-point (PTP) moves, and integral velocity control with continuous velocity profiling using linear acceleration. The HCTL receives input signals from the host processor asynchronously along an 8-bit bidirectional multiplexed address/data (AD0/DB0–AD7/DB5 and DB6, DB7) bus. It also receives quadrature outputs (CHA, CHB, and INDEX pins) from a shaft encoder, decodes it, and makes it available as position information in a 24-bit counter. The profile generator calculates the required trajectory for the user-defined mode of operation. The digital filter processes the error between the actual position and the desired position and issues a control byte at the motor command port (MC0–MC7 pins) and a PWM signal at the PWM port (PULSE and SIGN pins). The HCTL-1100 also provides programmable electronic commutation (PHA–PHD pins) for DC brushless and stepper motors. Using the 9S12C in single-chip mode as a host processor requires the following connections to the HCTL-1100: PT[5:0] are connected to AD[5:0]; PT[6:7] to D[6:7]; PM[0] to OE; PM[2] to CS; PM[4] to ALE; PM[6] to R/W; and E clock to EXTCLK. A detailed description of the HCTL can be found at www.agilent.com.

12.2.10 DC Servos

A servo is a three-wire DC servomotor used in applications that require limited (noncontinuous) rotations, such as to adjust the position of the flaps and control surfaces on wings of model airplanes and to steer radio-controlled cars. Packaged into the housing of a servo are a DC motor, a gear train, limit stops beyond which the shaft cannot turn, a single-turn pot for position feedback, and an IC for closed-loop position control. A new generation of servos incorporates a microprocessor into the servo instead of an IC for control purposes (see Multiplex Servo Family mc/V2; www.multiplexrc.com.)

One of the three wires protruding from the case of a servo is connected to ground and another to power. The third wire is connected to the control input, which receives a train of *pulse-code modulated* (PCM) *signals*, as shown in Fig. 12.25. The pulse is repeated at a given period, which is typically set at 20 ms. The pulse width represents the code that causes the shaft to rotate to a specific position. For the neutral (center) position, the code width is usually

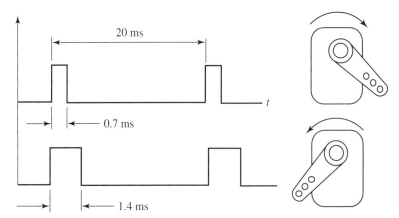

Figure 12.25 Depiction of a servo's pulse-code modulated signal.

1.3 ms. Pulse widths of 0.7 ms and 1.7 ms commands positions all the way to the left and all the way to the right, respectively.

Futaba (www.futaba-rc.com) also produces a line of analog and digital servos. The RS404PR robot servo has the following specifications: $40 \times 20 \times 37.6$ mm, 63 g, 0.13 s/60° at 6 V. It provides a torque of 13.8 kg-cm at 6 V, 180° rotational range, and PWM control.

12.3 STEPPER MOTORS

Stepper motors are basically brushless DC motors. However, they are treated as a separate category because they have a unique construction and functional features that distinguish them from conventional and brushless DC motors. Stepper actuators are available as stepper motors, which cause a shaft to rotate, and linear steppers, which cause a shaft to extend. Although the discussion to follow focuses on rotary stepper motors, it is equally applicable to linear steppers.

12.3.1 Characteristics of a Stepper Motor

A stepper motor is a digitally controlled device that moves incrementally, in response to a digital pulse, a fixed angular step, which is a fraction of the full 360° of shaft rotation. A stepper motor is easily interfaced and controlled by a microprocessor to rotate a specific number of steps at a desired speed and in a desired direction by using an appropriate pulse train. Under certain conditions, the motor may be used in an open-loop operation to eliminate the need for a position or velocity sensor and reduce cost. If the motor angular velocity is low enough so that steps are not missed, the motor responds to the pulse signal by moving the load to the specified position with an accuracy less than ± 1 step. More importantly, errors are not cumulative. However, stepper motors can slip if overloaded, and the error can go undetected. For this reason stepper motors use feedback control in high-precision applications. A stepper motor generates a relatively high torque at low velocities, which is used to accelerate the load to a desired speed. When energized by a DC signal, the stepper motor exhibits a relatively large holding torque that holds the rotor in a "self-locking" position. The stepper motor can stall during operation without causing damage (due to overheating). The rotor is set in motion when the voltage at the input terminals changes with time. Stepper motor construction is simple and rugged and on average has a long maintenance-free life. For this reason, and with improved permanent magnets and sophisticated drive circuits, stepper motors offer a cost-effective alternative in high-speed control applications involving light-to-medium loads. Stepper motors are used in numerous applications, including computer peripherals, machine tools, medical equipment, automotive devices, small business machines, and robots.

12.3.2 Classification of Stepper Motors

The stepper motor consists of two main components: the rotor, or the moving part, and the stator, or the stationary part. The stator holds several pairs of field windings called *phases* that can be switched on to create N and S electromagnetic poles. The rotor and stator are constructed in one of two schemes, shown in Fig. 12.26. In the first scheme (Fig. 12.26a), both the rotor and the stator have regular projections of teeth, with the stator holding the windings. In the second scheme (Fig. 12.26b), the rotor and stator have numerous teethlike projections to provide smaller step angles for higher accuracy. Accuracy for most stepper motors is on the order of 3% of the step angle, regardless of the number of steps.

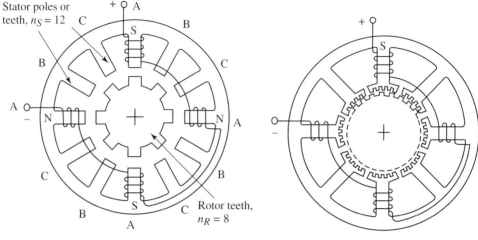

(a) 3-phase windings, single-stack VR motor   (b) 2-phase windings

Figure 12.26 Construction methods of stepper motors: single-tooth stator pole (a) and multiple-tooth stator poles (b).

According to Rotor Design

Depending on the construction of the rotor structure, stepper motors are classified into four types:

- variable reluctance (VR)
- permanent magnet (PM)
- hybrid
- linear

Variable Reluctance The structure of a typical VR motor is shown in Fig. 12.27. VR stepper motors have a nonmagnetized soft iron multiple-pole rotor and a stack of laminated wound stators. When the stator coils are energized, the rotor teeth are attracted to the stator teeth, causing the rotor to rotate. VR motors come in the form of three-or four-phase single-stator-stack constructions. Fundamental to the operation of this type of motor is that the rotor and stator do not have the same number of teeth. For the motor shown in Fig. 12.27, the number of teeth on

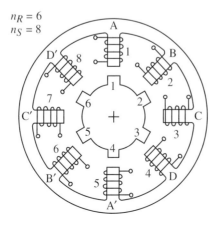

Figure 12.27 Basic construction of a single-stator-stack VR motor.

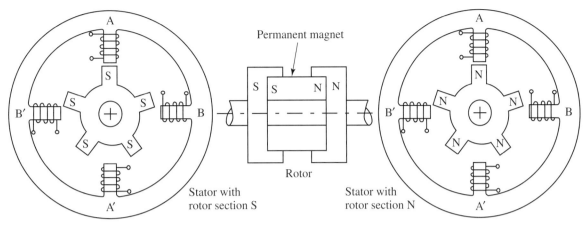

Figure 12.28 Schematic diagram of a PM stepper motor.

the rotor is $n_R = 6$ (located every 60°), and the number of teeth on the stator is $n_S = 8$ (located every 45°). In addition, each stator tooth has a coil wound on it, with oppositely placed coils (e.g., A and A′) grouped together and referred to as a *phase*. The number of phases in the motor of Fig. 12.27 is $p = 4$ phases, labeled A, B, C, and D. Step angles of VR motors are generally 7.5° or 15°. It has low rotor inertia, enabling faster response. A VR motor has no holding torque when it is not energized because the rotor has no permanent magnet.

Permanent Magnet This type of stepper motors is the most commonly used. A typical PM motor is shown in Fig. 12.28. It consists of a multiphase stator and a two-part PM rotor. The rotor and stator are also toothed. The opposite ends of the PM rotor form north and south poles, and their teeth are offset by half of a tooth pitch. The PM is made from alnico or a rare earth material (e.g., samarium cobalt). PM motors require less power to operate and provide a small holding torque when not energized. When magnets are also inserted into the stator, higher torques are achieved. The rotor inertia is higher than that of a VR motor, and thus the associated mechanical time response is slower. The rotor offset provides for a low inductance, leading to a faster current rise than that of a VR motor. Typical step angles of PM motors are 0.72°, 1.8°, 3.6°, 7.5°, 15°, 30°, 45°, and 90°.

Hybrid This type of stepper motor combines rotor features of the VR and PM types. It uses the construction scheme shown in Fig. 12.26b. A typical hybrid motor is shown in Fig. 12.29. The shaft is surrounded by a smaller PM. The rotor has two ends; one is a north pole and the other a south pole. Rotor teeth are cut into two iron-core cups that fit over each end. The hybrid de-sign can accommodate more teeth and has higher torque. Typical step angles are 7.5° and 1.8°.

Linear Steppers These are made flat instead of circular to generate linear instead of rotary motion. They are used in a variety of applications, including coil winding systems and trans-porting semiconductor wafer through a laser inspection station, among others. Linear steppers use the same switching and drive circuits as their rotary counterparts.

According to Phase Windings

Stepper motors are also classified according to the number of phases, or *independent* windings on the stator. Two-, three-, four-, and five-phase stepper motors are available. Figure 12.26a is a three-phase motor (only one drawn), and Fig. 12.26b is a two-phase motor. A *winding* refers to a coil section between the DC supply voltage and ground. If each pole has a single winding,

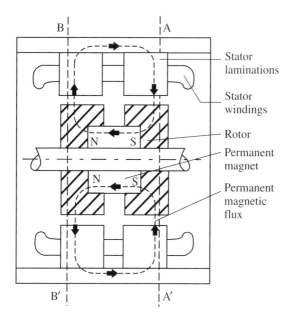

Figure 12.29 A hybrid stepper motor.

the arrangement is called *unifilar winding*; if each pole has two winding coils of opposite polarities, it is called *bifilar winding*. To advance the rotor, the polarity of each phase of winding must be reversed by switching supply voltage and ground ON and OFF in a specific sequence. The switching sequence varies with the motor and manufacturer. Bipolar or MOSFET switching transistors can be used to reverse the phase current. The number of connector wires protruding from the motor housing is not related to the number of phases. Various wiring schemes for two-, three-, and four-phase motors are shown in Fig. 12.30.

Two-phase motors have four-, six-, or eight-wire arrangements and five in some cases. Two-phase motors with four wires are called *bipolar* motors because generating a four-step sequence requires reversing the polarity of the windings. Four transistor switches and a bipolar ($\pm$) power source or eight switches and a unipolar ($+$) source provide the necessary switching pattern. If the power is applied to the center tap of a winding, the current can then be controlled to flow in either half of the winding, and the need for polarity switching is eliminated. This arrangement is called the *unipolar* drive, resulting in less torque at low step rates but providing for a better performance at high step rates as compared with the bipolar drive, as indicated in Fig. 12.31. The eight-wire unipolar motor uses half the number of switches and is not prone to short-circuit problems. Two-phase motors are normally used in light-duty applications using less expensive switching elements. A three-phase stepper motor has one four-wire arrangement, which requires only three switches and is used exclusively in VR motors.

Four-phase motors have two wires for each coil that may be wired externally in parallel or in series, as shown in Fig. 12.30. Parallel wiring provides for a better performance at high step rates, whereas the series arrangement has higher torque at low step rates. Four-phase motors are typically used in high-power applications.

12.3.3 Principle of Operation

A stepper motor can operate in one of three modes: *full-step*, *half-step*, or *micro-step*. In a full-step mode, all motor windings are energized simultaneously. An input pulse switches current

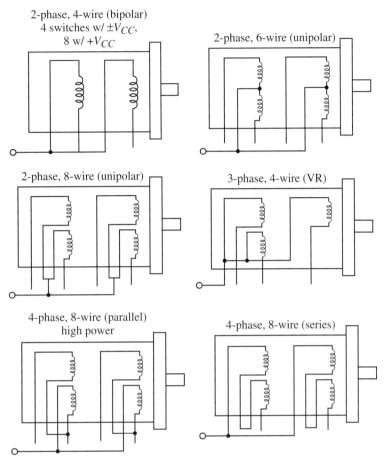

Figure 12.30 Motor phases and wire connection procedures.

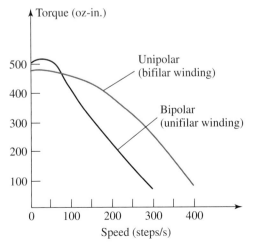

Figure 12.31 Torque characteristics of unifilar and bifilar steppers.

polarity in the windings and advances the rotor by its inherent step angle. In the half-step mode, the windings are energized so that within a three-step sequence only one winding is energized in the second step. A strong step followed by a weak step results in a lower torque at low step rates but in a slightly higher torque at higher step rates.

Operation of the VR Motor

The operation of the VR stepper is described for the motor shown in Fig. 12.27. It is based on the principle of "minimum reluctance," whereby the rotor always seeks to reorient itself at a stable position by finding the shortest magnetic path with the smallest air gap. Thus, the rotor of a VR stepper seeks a stable position. When phase A is energized, rotor teeth 1 and 4 align with stator teeth 1 and 5 and remain in this position as long as the coils in the same phase are energized. This is a stable equilibrium point, known as *detent position*, representing one motor step. With phase A-A' energized, the rotor is kept in position as long as the external torque is less than the holding torque. Increasing the current to the phase increases the holding torque.

If the excitation is removed from phase A-A' and placed on phase B, rotor teeth 6 and 3 should align with stator teeth 8 and 4, and the rotor moves CW 15°. This process can be repeated for phases C and D and then back to A. A complete sequence of phase excitation rotates the rotor 60°. Applying the same sequence in reverse forces CCW rotation.

If two adjacent phases (e.g., A and B) are energized simultaneously, an equilibrium point is created halfway between the two full step points acquired by exciting phases A and B separately, and the phase coils are identical, with the same excitation amplitude applied to both sets of coils. This process can be repeated for phases BC, CD, and DA. Consequently, if the phase excitation sequence A, AB, B, BC, C, CD, D, DA, A is executed, the rotor makes twice the number of CW moves as before. This operation is referred to as the *half-step mode*. Operating a stepper motor in the half-step mode reduces the overshoot as the rotor moves from one point to the next. However, the switching circuitry is more complicated than for the relatively simple full-step mode.

Operation of the PM Motor

Figure 12.32 shows a hypothetical stepper motor consisting of a PM rotor with a single N-S pole and a four-tooth stator that holds two windings, *A-A'* and *B-B'*. When *A, B* and *A', B'*

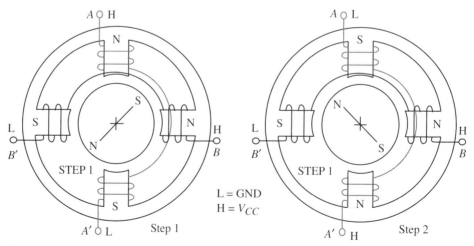

Figure 12.32 A hypothetical two-phase PM motor.

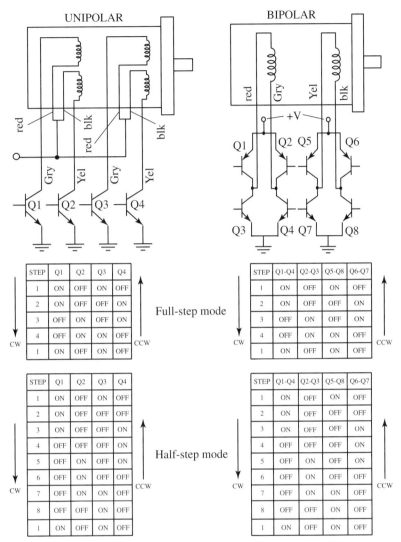

Figure 12.33 Switching transistors and sequence for unipolar and bipolar steppers.

terminals are connected to the DC supply and ground, respectively, the teeth on the supply side (top and right-side teeth) become N poles and the opposite teeth become S poles. This moves the rotor into a stable position at $+45°$. Reversing terminal connections to supply and ground also reverses polarity, and the rotor steps $90°$ to a new stable position at $-45°$. Executing the four-step sequence in Fig. 12.33 operates the hypothetical motor in the full-step mode (one phase on) and rotates it a full revolution. If the eight-step sequence in Fig. 12.33 is executed, the motor runs in the half-step mode and the motor rotates half a revolution. Reversing the step sequence causes the rotor to rotate in the opposite direction. Figure 12.33 shows typical switch connections using BJTs and the switching sequence for unipolar and bipolar PM motors for full-step and half-step modes.

12.3.4 Step Angle

The number of rotor and stator teeth (or poles), n_R and n_S, of a VR stepper motor with p phases in a single-stator stack and for $n_S > n_R$ are related by

$$n_S = n_R \pm n_{S/p} \tag{12.76}$$

where $n_{S/p}$ is the number of stator teeth per phase $n_{S/p} = n_S/p$. The rotor and stator pitch angles, θ_R and θ_S, are defined as

$$\theta_R = 360°/n_R \quad \text{and} \quad \theta_S = 360°/n_S \tag{12.77}$$

The number of steps s per motor revolution is related to n_R and n_S by

$$s = n_R p = |n_R n_S/(n_S - n_R)| \tag{12.78}$$

The step angle in degrees is then given by

$$\Delta\theta = 360°/s \tag{12.79}$$

For a p-phase stepper motor with $n_S \neq n_R$ (see Fig. 12.26a), the following relations hold:

$$
\begin{aligned}
\Delta\theta &= \theta_R - \theta_S \\
\theta_R &= p\Delta\theta = \theta_S + \theta_R/p \\
n_S &= pn_{s/p} = n_R + n_S/p \\
n_{s/p} &= \frac{s}{p(p-1)} \qquad n_s > n_R \\
n_{s/p} &= \frac{s}{p(p+1)} \qquad n_s < n_R
\end{aligned}
\tag{12.80}
$$

For the motor shown in Fig. 12.26a, $n_R = 8$, $n_S = 12$, $p = 3$, $n_{S/p} = 4$, $s = 24$, and $\Delta\theta = 15°$.

If the motor is constructed with many teeth on a stator pole, as in Fig. 12.26b, the step angle, the rotor rotation θ_R in p steps, and the number of stator teeth become

$$
\begin{aligned}
\Delta\theta &= \frac{n_S}{2p}\theta_R - \frac{180}{p} = (n_S/2p)(\theta_R - \theta_S) \\
\theta_R &= p\Delta\theta = \theta_S + 2\theta_R/n_S \\
n_S &= n_R + 2
\end{aligned}
\tag{12.81}
$$

In general, for a stepper motor with a PM rotor, the step angle is determined by

$$\Delta\theta = \frac{360}{n_S} = \frac{360}{2PQ} \tag{12.82}$$

where P is the number of pole pairs (e.g., 24) on the rotor and Q is the number of stator sections (e.g., 2).

$$R_{S(MAX)} = R(V_{CE(MAX)}/V) - 1$$

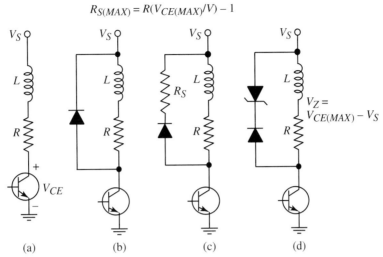

Figure 12.34 Equivalent circuit of a winding (a) and back-emf suppression schemes: diode (b), diode + resistance (c), and diode + Zener (d).

12.3.5 Electrical Model of an Energized Coil

The winding of a phase is modeled as a series combination of an inductance L and a resistance R, as shown in Fig. 12.34a. For a two-phase PM motor, the voltage in each phase is expressed by the first-order relations

$$L\frac{dI_a}{dt} + RI_a + E_a = V_a$$
$$L\frac{dI_b}{dt} + RI_b + E_b = V_b \tag{12.83}$$

where V_a and V_b are the voltages applied to the windings, I_a and I_b are the windings' currents, and E_a and E_b are the back emf induced by the rotation of the rotor. E_a and E_b are given by

$$E_a = K_e\omega\sin\theta$$
$$E_b = -K_e\omega\cos\theta \tag{12.84}$$

where K_e is the peak value of the flux leakage, or voltage constant of the motor, and ω is the angular velocity of the rotor. Assuming the current $I(0^+) = 0$, the solution to Eq. (12.83) is simply

$$I(t) = \frac{V}{R}(1 - e^{-t/\tau}) \tag{12.85}$$

where $\tau = L/R$ is the electrical time constant. Two important conclusions can be derived from Eq. (12.85). The first is that a large electrical time constant slows current buildup in a phase due to the inductance of the coil. This will result in a lower driving torque at the start of each step because the motor torque is proportional to the current flowing through its windings. For

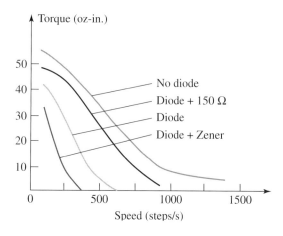

Figure 12.35 Effect of suppression methods on the torque–speed curve.

example, driving a motor at 2000 pulses per second would energize the windings for 0.0005 seconds. If windings need 0.01 seconds after power is switched ON to reach the rated current, only one-twentieth of the maximum torque can be realized.

The second conclusion is that it is not possible suddenly to turn off the current from an inductor ($V = L(dI/dt)$) because that would imply an infinite voltage generated across the inductor's terminals. When the phase is switched off, the sudden rise in the voltage across the inductor forces current to flow for some short time to dissipate the stored energy in the inductor. The instantaneous voltage can exceed the maximum breakdown voltage $V_{CE(MAX)}$ of the switching transistor, damaging it and other circuit elements. Therefore the harmful effect of self-induction voltage needs to be suppressed in order to protect the driver circuitry.

Back-emf Suppression

The back emf reduces the operating torque of the motor. Several methods may be used to suppress an inductor's back emf. Figure 12.35 shows the torque–speed curves associated with four suppression methods. The simplest method is to put "*flyback*" or a "*free-wheeling*" diode across the winding, as shown in Fig. 12.34b. When the winding is energized, the diode is reverse-biased and current cannot flow through it.

A better method is to connect a resistor and a diode in parallel with the phase winding, as shown in Fig. 12.34c. In this case the current will flow through R and R_S during the switching periods, which decreases the electrical time constant to $\tau = L/(R + R_S)$ and helps dissipate energy faster. Another effect is that the back emf will be higher and must not exceed the maximum breakdown voltage $V_{CE(MAX)}$ of the BJT. Therefore, the maximum resistance R_S is selected according to

$$R_{S(MAX)} = R(V_{CE(MAX)}/V_S) - 1 \tag{12.86}$$

During steady conditions, no current flows through R_S and the winding draws a larger current, which results in a higher torque.

Using a Zener diode and a diode as in Fig. 12.34d will damp the short-circuit current faster than the two previous approaches. If the Zener is selected with a breakdown voltage of $V_{CE(MAX)} - V_S$, it will conduct all current necessary to maintain its breakdown voltage, and most of the energy will be dissipated in it. Only one Zener may be used with all motor windings.

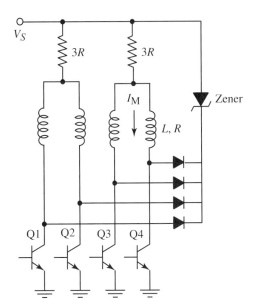

Figure 12.36 External resistance for an $L/4R$ drive.

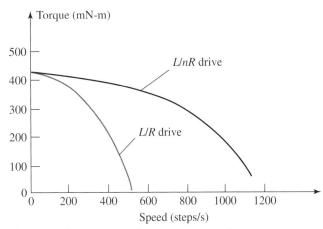

Figure 12.37 Effect of L/nR on torque–speed response.

12.3.6 Drive Methods

Generating high torques by a stepper requires fast current build-up in its windings. This can be accomplished in one of two ways. The *series resistance*, or L/nR *drive*, method requires adding an external resistance that is n times R in series with winding resistance R; hence the name L/nR drive. The larger effective resistance decreases the electrical time constant τ_e of the motor. Adding a $3R$ resistance to each winding of a motor results in the $L/4R$ drive circuit shown in Fig. 12.36; its effect on the torque is demonstrated in Fig. 12.37. The L/nR drive is simple and cheap but leads to a higher power dissipation, and the extra voltage drop across the series resistance requires a supply voltage that is n times the operating voltage of the motor.

The second approach is called the *chopper drive* method. A supply voltage much higher than the motor rated voltage is applied to the windings to speed up the current build-up in the windings and increase motor torque, as Eq. (12.85) indicates. The high voltage is applied until the current exceeds the rated current (V_S/R). A current-sensing network then switches the voltage OFF. When the winding current decays below the rated current, the same network switches the voltage back ON. The average of the resulting chopped winding current is equal to the rated current. Figure 12.38 illustrates the voltage–current relationship in a chopper drive. The

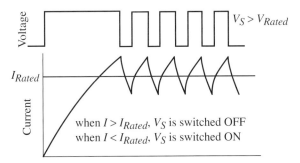

Figure 12.38 Typical voltage and current waveforms of a chopper drive.

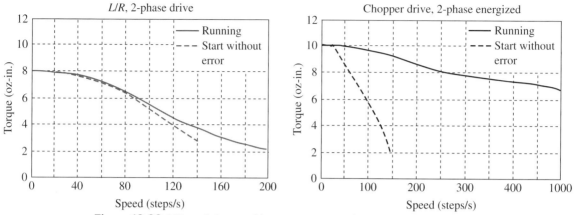

Figure 12.39 Effect of chopper drive on torque–speed response.

flywheel effect of the rotor inertia smooths out the chopped torque and provides for a smooth motion.

As illustrated in Fig. 12.39, the chopper drive results in a significant improvement in torque–speed characteristics over the L/nR drive. Although the chopper drive is more complex and more expensive than the L/nR method, more efficient and dedicated chopper drive circuits are available as IC packages.

Example 12.3: Selecting drive components—Determine the series resistance, select an appropriate BJT transistor, and determine the voltage, current, and power specifications for a Zener diode to drive a unipolar stepper motor with a step angle $\Delta\theta = 1.8° \pm 5\%$ and rated at $V_S = 12$ V. Also, each winding has a resistance $R = 104$ Ω and an inductance $L = 89$ mH and weighs 16 oz.

Solution: Use a series resistance that is three times the winding resistance, as in Fig. 12.36. Thus,

$$3R = 3 \times 104 \ \Omega = 312 \ \Omega$$

Since the transistor operates as a switch, thus in the saturated state, the steady-state current in each winding is then (Eq. 3.9)

$$I_{SS} = \frac{V_S - V_{CE(SAT)}}{R} = \frac{12 - 0.2}{104} = 0.113 \, \text{A}$$

The transistor is selected to handle a current higher than 0.12 A. Assuming $\beta = 150$, the required bias current is less than 1 mA. The drive capability of an output port on the 9S12C is

1.6 mA, which is sufficient to provide the transistor's bias. A 2N3904 transistor (see Table 3.4) may be used for this application. This transistor's breakdown voltage is $V_{CE(MAX)} = 40$ V.

The Zener diode is selected with a breakdown voltage that must be less than the transistor breakdown voltage minus the power supply voltage, or $V_Z = V_{CE(MAX)} - V_S = 40 - 12$ V $= 28$ V. Since only one transistor is switched from ON to OFF during any two-phase switching sequence, at the moment of switching, the current through the winding I_{SS} is diverted through the Zener diode circuit, where it will decay exponentially according to

$$I_Z = I_{SS}e^{-t/\tau} \cong \frac{V_S}{R}e^{-t/\tau} \qquad (12.87)$$

Since the energy produced in the inductor must all be dissipated by the Zener, we have

$$\frac{1}{2}LI_{SS}^2 = V_Z \int_0^\infty I_Z dt = V_Z I_{SS} \int_0^\infty e^{-t/\tau} dt = V_Z I_{SS}\tau \qquad (12.88)$$

and the time constant for current decay from this equation is

$$\tau = \frac{1}{2}\frac{LV_S}{RV_Z} \qquad (12.89)$$

where V_Z will remain constant as long as the Zener is conducting. Assuming that the current decays in 4τ s, the average dissipated power in the Zener diode will be

$$P_{Z(AVG)} = \frac{1}{2}\frac{LI_{SS}^2}{4\tau} = \frac{V_Z V_S}{4R} \qquad (12.90)$$

The Zener specifications will then be $V_Z = 28$ V and $P_{Z(avg)} = 1.6$ watts (using a safety factor of 2)

12.3.7 Stepper Motor Performance

Single-Step Operation

The static torque for one phase of the three-phase VR motor depicted schematically in Fig. 12.26a may be assumed to be a sinusoidal waveform, as shown in Fig. 12.40a. The motor has $p = 3$, $n_S = 12$, $n_R = 8$, and a step angle of 15°. Point D is the detent position when phase A is being energized. If the rotor is rotated 15° CCW by external means to

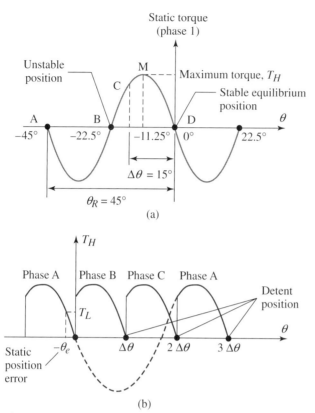

Figure 12.40 For the three-phase VR step motor of Fig. 12.26a: Static torque curve of phase 1 of Fig. 12.26a (a), and periodic torque distribution (b).

occupy position C, the previous detent position of the rotor before phase A was energized, a positive torque would act on the rotor, trying to rotate it back to position D. Point A, corresponding to a position at 45°, which is a rotor tooth pitch θ_R from position D, is also a detent point, but point B is unstable and a small torque would move the rotor in either direction. The torque at point C is positive but not maximum. The maximum torque, also called the *holding torque* T_H, occurs at point P corresponding to a rotor position 11.25° from point D. The static torque with maximum current applied to only one phase is expressed as

$$T = -T_H \sin n_R \theta = -T_H \sin(2\pi\theta/\theta_R) \tag{12.91}$$

For a motor with p phases, the static torque for each phase is periodic, with a period $p\,\Delta\theta = \theta_R$. This relation is shown in Fig. 12.40b for a three-phase VR motor. The static torque (at zero speed) in this case is expressed as

$$T = -T_H \sin(2\pi\theta/p\Delta\theta) \tag{12.92}$$

which is valid in the range $-\Delta\theta \le \theta \le 0$. If the motor windings are not energized, the torque required to rotate the stepper motor is called *detent torque*.

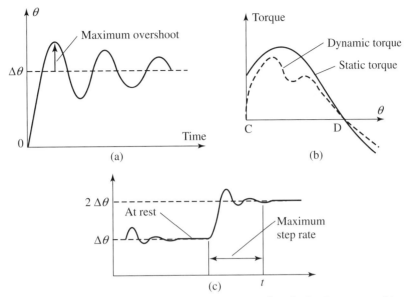

Figure 12.41 Single-step response (a) and corresponding single-phase torque (b), and typical response in the stepping mode (c).

The detent position is attained if no static load is applied to the shaft. In case a static load torque T_L is applied, the rotor will deviate from the detent position by an angle θ_e, termed the *static-position error*. This error can be found by substituting T_L for T and θ_e for θ in Eq. (12.91) to yield

$$\theta_e = \frac{p\,\Delta\theta}{2\pi}\sin^{-1}(T_L/T_H) = \frac{p}{s}\sin^{-1}(T_L/T_H) \tag{12.93}$$

where s is the number of steps per revolution, or the step rate. It is clear that as s increases, θ_e decreases.

When a single step is applied to an energized phase, the rotor turns through the step angle θ_0. Before the rotor comes to rest at the end of the step, it oscillates about the new stable equilibrium position, as shown in Fig. 12.41. These oscillations are caused primarily by the load inertia, which can be expressed, when damping and friction effects are neglected, by the second-order relation

$$J\ddot{\theta} = T \tag{12.94}$$

The time it takes for the oscillations to subside is called the *settling time t_s*, which is approximately equal to 4τ (τ is a time constant).

Slewing Operation

As the step rate increases, switching would be required before the rotor fully comes to rest in each step and the motion changes from discrete steps to a continuous motion, termed *slewing*

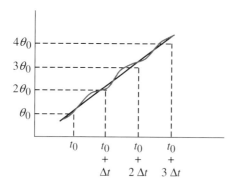

Figure 12.42 Typical slewing response.

motion. The time-displacement curve under steady-state slewing is shown in Fig. 12.42. The constant pulse rate s_R, or the slew rate (pulses/s), is expressed as

$$s_R = \frac{1}{\Delta t} \text{ (steps/s)} \tag{12.95}$$

where Δt is the time between successive pulses. If Δt is smaller than the settling time t_S, unavoidable periodic oscillations (or hunting) will result. The amplitude of these oscillations can be reduced by increasing damping. In addition, the upper limit of s_R depends on the inertia of the rotor and the load, the damping, and the load rating.

To drive the motor at a constant slew rate, the pulse rate is increased through accelerating the rotor from a lower speed by means of *ramping*. Assuming linear variation, the pulse rate increases according to

$$s_R(t) = s_0 + \frac{(s_R - s_0)t}{t_0} \tag{12.96}$$

where s_0 is the starting pulse rate (typically zero), s_R is the final pulse rate, and t_0 is the ramp time ($t_0 = n\,\Delta t$; n is the total number of pulses applied). The angular velocity of the motor, ω, is expressed in terms of s_R, as $\omega = k s_R$, where

$$k = \frac{2\pi \text{ rad/rev}}{(360°/\Delta\theta)\,\text{pulses/rev}} \tag{12.97}$$

The operating angular velocity ω_{OP} should not fall within the band of resonant frequencies given by

$$\omega_{N_1} = \sqrt{\frac{360 T_H}{J\Delta\theta}} \tag{12.98}$$

$$\omega_{N_2} = \sqrt{\frac{720 T_H}{\pi J\Delta\theta}} \tag{12.99}$$

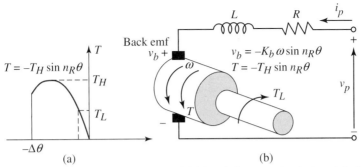

Figure 12.43 Generated torque model (a) and electromechanical model of a step motor (b).

Thus ω_{OP} should be $\omega_{OP} < \omega_{N1}$ or $\omega_{OP} > \omega_{N2}$ to avoid resonance and missed steps from occurring. Equations (12.98) and (12.99) are obtained from Eq. (12.94) using elementary kinematics.

Microstepping

A stepper motor may be driven such that each fundamental step $\Delta\theta$ is divided into a number of mini- or microsteps. Microstepping is accomplished by changing the phase currents incrementally in steps so that the currents in two adjacent motor phases are balanced, to force the rotor to assume a desired angular position between two adjacent stator poles. The vector sum of the magnetic fields generated in the adjacent phase windings defines the angular position along which the rotor will align. This driving approach improves the positional resolution of the motor, eliminates ripple in the output torque, and provides for an operating frequency that is higher than the resonant frequencies of the motor. However, the equilibrium points are not as well defined.

Dynamic Behavior

Figure 12.43 shows a model of a stepper motor connected to a load. The dynamic motion of the rotor is governed by

$$(J_M + J_L)\frac{d\omega}{dt} = T_a + T_b - B\omega - T_L \tag{12.100}$$

where J_M is the inertia of the rotor, J_L is the load inertia, B is the coefficient of viscous damping, T_L is the load torque, which may include coulomb friction torque and gravitational torque, and ω is the angular velocity of the rotor. The electromagnetic torques T_a and T_b generated by phases P-A and P-B are given by:

$$T_a = (-K_e \sin\theta)I_a(t)$$
$$T_b = (K_e \cos\theta)I_a(t) \tag{12.101}$$

A plot of steady torque output vs. speed is shown in Fig. 12.44. The plot outlines the regions in which the motor operates correctly. The holding torque is generated at standstill. As the step rate increases, phase switching takes place before the current in the inductor reaches its steady-state value and the torque decreases.

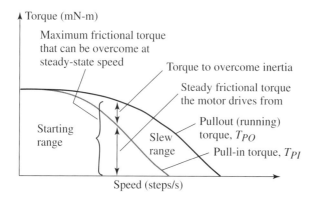

Figure 12.44 Relation between the starting T_{PI} and running T_{PO} torque ranges.

Figure 12.44 features two curves. The *pull-in torque* T_{PI}, or start-without-error torque, and the *pull out torque* T_{PO}, or running torque. The pull-in torque is the maximum torque the motor can produce starting from rest (or stop without loss of a step) and accelerating to operate at a given step rate. Since it includes the torque required to overcome rotor inertia, the pull-in torque represents the torque required to overcome constant friction and gravity torques.

Once the rotor reaches the steady-state speed, the acceleration becomes zero and no torque is needed to overcome inertia. The inertia torque (that determined from Eq. (12.94) will then be utilized to overcome higher friction and gravity torque. Pullout torque represents the maximum friction and gravity torque the motor can overcome at a steady rate. At a fixed step rate, the torque that overcomes inertia is the difference between the two curves. The region between the two curves is called the *slew range*.

If the friction and gravity torque is known and has a fixed value, its intersection with the pull-in torque curve gives the maximum step rate at which the motor can run while moving the load from rest, and its intersection with the pullout torque curve gives the maximum step rate possible after the motor reaches the pull-in step rate.

Another point of concern is that while the pullout torque curve is the same for any load inertia, the pull-in torque curve does not account for load inertia. For a stepper driving a load inertia, the pull-in torque at a given speed is

$$T_{PI(J_L \neq 0)} = \frac{J_L}{J_M} T_{PI(J_L = 0)} \tag{12.102}$$

Figure 12.45 shows the pull-in torque with $J_L = J_M$ compared to that with $J_M = 0$. Note that the distance between the curves at a given speed is the same because $J_L = J_M$.

Selection of a Stepper Motor

Proper selection of a stepper motor requires that the designers have complete and accurate knowledge of the motion and load characteristics of the application at hand. This includes speed and acceleration of motion, load torque, and natural frequencies. Armed with this information, the engineer should select the motor that best accommodates the operating conditions. The most important task in the selection process is the rating torque of the motor. The rating torque of the motor and the torque–speed characteristics must accommodate static torque and dynamic torque (slewing) requirements. This depends on many factors, such as the drive scheme of the motor and the back-emf suppression method used. The step angle must be of high enough resolution to provide the desired output motion increments. For example, the

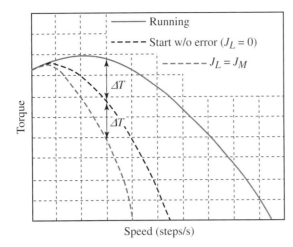

Torque (y-axis)

Speed (steps/s) (x-axis)

- Running
- Start w/o error ($J_L = 0$)
- $J_L = J_M$

ΔT

ΔT

Figure 12.45 Starting torque curves with and without equivalent load inertia.

step angle of a stepper motor needed to drive a lead screw with a lead $l = 0.5$ in. at increments of $\Delta x = 0.02083$ in. would be

$$\Delta\theta = (360 \times \Delta x)/l = (360 \times 0.02083)/(0.5) = 15°$$

Other important factors include the motor time constant, damping properties, size, mounting provision, thermal properties, and cost. Table 12.3 provides the specifications of the MS23C manufactured by US Digital. These specifications are typical for any stepper motor. The MS23C motor is compatible with the encoder series E5D, E5S, E6D, E6S, also by US Digital.

12.3.8 Interfacing Stepper Motors to the 9S12C MCU

Driving Stepper Motor Using Discrete Components

Interfacing a stepper motor to the 9S12C MCU is relatively simple. Figure 12.46 shows a unipolar stepper motor interfaced to port T of the 9S12C. The motor is rated at 5 V, and its step angle is 7.5°. Each winding is characterized by a resistance $R = 15.5\ \Omega$ and inductance $L = 13.5$ mH. The complete circuit is an $L/4R$ drive that includes an external $3R$ resistor connected in series with the winding resistance R to decrease the time constant and speed up the current buildup in the windings for higher torque throughput. Flyback diodes and a Zener are used for emf suppression. Each winding is connected to a DC supply voltage and ground. The current through a winding is reversed by switching between supply voltage and ground using a BJT

TABLE 12.3 Specifications of the MS23C Stepper Motor

Step Angle	Full-Step Current	Microstep Current	Resistance, R	Inductance, L	Power/Phase (max)
$1.8° \pm 5\%$	4.20 A	5.90 A	0.43 Ω	1.73 mH	7.65 W

Running torque, T_C	Detent torque, T_D	Holding torque T_H	Rotor inertia, J	Number of wires	Rated voltage
110 oz-in.	4.9 oz-in.	140 oz-in.	3.96×10^{-3} oz-in.-s^2	8 wires or 4-pin connector	24 V

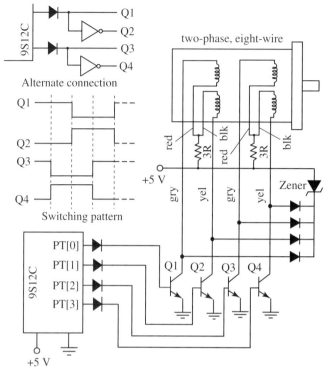

Figure 12.46 Interfacing of a unipolar step motor to the 9S12C.

switch, although a FET switch could also be used. The switching transistors are biased from port T, assuming that the bias current does not exceed the 1.6-mA drive throughput of the port. With transistors of high current gains (200 or more) there is an ability to drive motors with current ratings up to 0.4 A. Power MOSFETS or Darlingtons would be required for stepper motors with high current ratings. The circuit also includes diodes D1–D4 to protect the MCU in case a control transistor should short out. Resistors or optoisolators may also be used to achieve similar protection.

Example 12.4: Assembly code for a stepper motor drive–Write an assembly code to control the motion of the stepper motor drive shown in Fig. 12.46. The motor should rotate at 100 rpm CW in full-step mode.

Solution: The step rate corresponding to running the motor in full-step mode at a constant speed of 100 rpm will be: 100 (rev/min) $\times$ 360 (deg/rev) $\times$ 1 (min)/60 (s) $\times$ 1 (step)/7.5 (deg) = 80 (steps/s). This requires a $1/80$-s delay between steps. Using the switching pattern shown in Fig. 12.33, the following program segment would accomplish the required task.

```
INCLUDE    'mc9s12c32.inc'
tdly    EQU    2
;
```

```
;* Stepping Sequence: The program advances the motor shaft one
step at a time
;* A time delay is used between steps
;
        ORG     $C000
        MOVB    #11, ddrt

start   ; Main code begins here
        LDAA    #$0A
        STAA    ptt         ; Advance first step
        JSR     delay       ; Execute time delay
        LDAA    #$09
        STAA    ptt         ; Advance second step
        JSR     delay       ; Execute time delay
        LDAA    #$05
        STAA    ptt         ; Advance third step
        JSR     delay       ; Execute time delay
        LDAA    #$06
        STAA    ptt         ; Advance fourth step
        JSR     delay       ; Execute time delay
        BRA     start

delay   ; Time      delay subroutine
        LDAB    #tdly
again   LDX     #2
loop    NOP
        NOP
        DBNE    X, loop     ; Loop until 20000 done
        DBNE    B, again    ; Repeat as needed
        RTS
```

Example 12.5: Stepping sequence organized in a table–Write an assembly code to handle a stepper motor drive for any number of steps.

Solution: The following code will cause a stepper motor to move a specific number of steps.

```
;initializing registers should be done before the following
code
            INCLUDE   'mc9S12C32.inc'
            ORG       $C100
steptbl     DC.B      $C0,$80,$90,$10,$30,$20,$60,$40
```

```
nsteps          DS.B 2    ; Signed number to keep track of how
                            many steps remain to be taken
stepptr         DS.B 2    ; Pointer to point to the current posi-
                            tion in the switching sequence table
output          DS.B 1
;
                ORG       $C000
                MOVB      #3A, ddrm
                LDAA      #$0A           ; Define the number of
                                           steps
                STAA      nsteps         ; If positive, motor
                                           rotates CW, and if
                                           negative, it rotates CCW
                LDY       #steptbl
                STY       stepptr
                JSR       step
here            BRA       here

step            LDAA      #$00
                CMPA      nsteps         ; If steps taken, is
                                           complete
                BEQ       step_2         ; Return otherwise
                BMI       step_bkwd      ; If result is negative,
                                           rotate in CCW direction

step_frwd       DEC       nsteps         ; Rotate in CW direction
                INC       stepptr

                LDY       stepptr
                CPY       #steptbl+8     ; If pointer is > end of
                                           table, Y − M : M + 1
                BLO       step_1         ; Move a step
                LDY       #steptbl       ; Otherwise reset stepptr
                                           to top of table
                STY       strepptr
                JMP       step_1         ; And move a step

step_bkwd       INC       nsteps         ; Rotate in CCW
                                           direction
                DEC       stepptr
                LDY       stepptr
                STY       stepptr
                CPY       #steptbl_1     ; If pointer is < top of
                                           table, Y − M : M + 1
```

```
                    BHI       step_1          ; If no, move a step
                    LDY       #steptable+7    ; If yes, set stepptr to
                                                bottom of table

                    STY       stepptr

step_1              LDAA      stepptr
                    STAA      output
                    BCLR      output, #%00001111   ; clears upper
                                                     4 bits of output
                    BSET      output, stepptr      ; OR output with
                                                     data stored at
                                                     stepptr

                    LDAA      output
                    STAA      ptm, X                       ; move the step
step_2              RTS
```

Stepper Motor Driver ICs

Numerous ICs dedicated to stepper motor control are available from many manufacturers to suit a wide range of application requirements. Examples include the TPIC2406 driver IC from Texas Instruments, the MC1413 (up to 500 mA per winding), MC1416, and MC3479 driver ICs from ON Semiconductor, and the ULN200x (up to 500 mA per winding) series from STMicroelectronics. The SAA1027 by Signetics is suitable for small stepper motors with a winding current up to 350 mA.

Unipolar Motor Drive

Figure 12.47 shows the 16-pin DIP, 2559 protected Quad power IC driver from Allegro Microsystems driving a four-phase unipolar stepper motor. The inputs of the 2559 IC are compatible with TTL and 5-V CMOS. Each channel provides 700 mA of output current. Each output has its own integral output flyback diode and low output saturation voltage. Each driver is provided with an independent over-current protection, and there is integral thermal protection for the device and each driver. Full-step control is easily accomplished with signals from two pins of the 9S12C, PA[1] and PA[0], providing the logic to two sets of complementary control

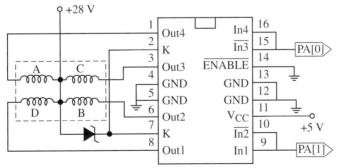

Figure 12.47 Driving a unipolar stepper motor via the 2559 driver IC.

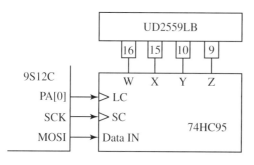

Figure 12.48 Driving a bipolar stepper motor using Allegro's A3966 IC driver.

inputs, (In1, In2) and (In3, In4), respectively. Active low ENABLE input can be used for chopper (PWM) applications. It is tied low if it is not used for this purpose. The external Zener is used to increase the flyback diode voltage, providing a much faster inductive load turnoff current decay.

Figure 12.48 shows a stepper motor driven from the serial port. A 74HC595 shift register receives the pulse train generated by the SPI output and transfers it to the motor through the 2559 driver IC. Four signals from the shift register provide the inputs to the 2559 IC.

Bipolar Motor Drive

Figure 12.49 shows the 16-pin DIP A3966SA Dual Full-Bridge PWM driver from Allegro Microsystems. The A3966SA includes two H-bridges to drive both windings of a two-phase bipolar stepper motor or two DC motors. The device can deliver continuous output currents of ± 650 mA while operating at 30 V. The IC has internal current-control circuitry that generates

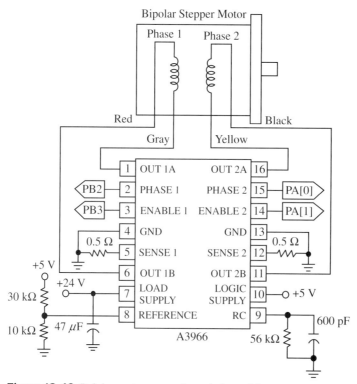

Figure 12.49 Driving a step motor through the serial port.

pulse-width modulated (PWM) signals at fixed frequency to control the current in the motor winding. The fixed-frequency pulse duration is set by a user-selected external RC timing network. The capacitor in the RC timing network also determines a user-selectable blanking window that prevents false triggering of the PWM current-control circuitry during switching transitions. The user can set the peak-load-current limit by selecting a reference voltage and current-sensing resistors. Two inputs are associated with each bridge, a PHASE input and an ENABLE input. The PHASE input controls load-current polarity by selecting the appropriate source-and-sink driver pair. Holding the ENABLE input high disables the output drivers. The IC features internal circuit protection, which includes thermal shutdown with hysteresis, ground-clamp and flyback diodes, and crossover-current protection.

Stepper motor translator/driver ICs are also available from various sources for applications where microprocessor control is not available or is overburdened. In addition to providing output drive capability, translator/driver ICs integrate the necessary logic to translate input signals to the proper sequence of output signals. The A3967SLB by Allegro Microsystems is an example of a motor driver with a built-in translator designed to operate bipolar stepper motors with no special power-up sequence required. Applying an input pulse to the STEP input pin, the motor will take a full-, half-, quarter-, or eighth-step, depending on two logic inputs. The A3967SLB has an output drive capability of 30 V and ±750 mA. It includes a fixed off-time current regulator that can operate in slow, fast, or mixed current-decay modes. This current-decay control scheme results in reduced audible motor noise, increased step accuracy, and reduced power dissipation. Internal circuit protection includes thermal shutdown with hysteresis, undervoltage lockout (UVLO), and crossover-current protection.

12.4 AC INDUCTION MOTORS

The simplest and most rugged electric motor, the AC induction motor consists of a wound stator and a rotor assembly. The AC induction motor is named because the electric current flowing in its secondary member (the rotor) is induced by the alternating current flowing in its primary member (stator). The power supply is connected only to the stator. The combined electromagnetic effects of the two currents produce the force that creates rotation (see Fig. 12.50). Induction machines are considered the most widespread machines produced and used in industry. They are simple in construction, robust, and cheap.

Alternating current (AC) motors are divided into two electrical categories, based on their power source: single-phase and polyphase motors. The three-phase type is presented in detail in the next section. The stator field in the single-phase motor does not rotate; it simply alternates polarity between poles as the AC voltage changes polarity. This type of motor does not differ in construction from the three-phase type except for the way windings are accommodated in stator slots. The single-phase induction motor with only one stator winding does not

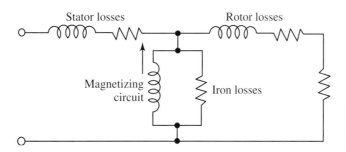

Figure 12.50 Typical torque/ speed characteristics for an induction motor.

develop any starting torque and therefore will not start to rotate if the stator winding is connected to an AC supply. However, if the rotor is given a spin or started by auxiliary means, it will continue to run. There are two theories as to how this happens: the "cross-field theory" and the "double-revolving-field theory." The methods for starting a single-phase induction motor are: the split-phase motor, capacitor-start induction-run, permanent-split-capacitor motor, capacitor-start capacitor-run (two-value capacitor) motor, or shaded-pole motor.

12.4.1 Three-Phase Motors

Construction

The induction machines main parts are the stator and the rotor. The stator comprises laminated cores, made of insulated sheet steel, pressed together. It is slotted on its inner cylindrical surface. The windings are made from embedded coils connected according to the number of poles and phases. On the other hand, the rotor carries the winding where current is induced. It is made also of laminated sheet steel; on its outer surface slots are milled. Slots in the rotor have different shapes and are in general of the following two types.

Squirrel-Cage Type The windings in this case are made of copper bars occupying rotor slots. These bars are brazed to end rings.

Slip-Ring Type This type of winding is used whenever a high starting torque with considerably less starting current is required. The leads of the rotor windings are brought out through slip rings, which can be connected to external resistances or short-circuited.

Theory of Operation

Motor torque is developed from the interaction of currents flowing in the rotor bars and the stator rotating magnetic field. In actual operation, rotor speed always lags the magnetic field speed, allowing the rotor bars to cut magnetic lines of force and produce useful torque. This speed difference is called *slip speed*. Slip also increases with load, and it is necessary for producing torque. Slip is the difference between the RPM of the rotating magnetic field and the RPM of the rotor in an induction motor. Slip is expressed in percentages and may be calculated by the following formula:

$$\text{Slip} = (\text{Synchronous Speed} - \text{Running Speed} \times 100)/\text{Synchronous Speed}$$

Equivalent Circuit

The induction motor can be treated essentially as a transformer, for analysis. As shown in Fig. 12.51, the induction motor has stator leakage reactance, stator copper loss elements as series components, and iron loss and magnetizing inductance as shunt elements. Similarly, the rotor circuit has rotor leakage reactance, rotor copper (aluminum) loss, and shaft power as series elements. The transformer in the center of the equivalent circuit can be eliminated by adjusting the values of the rotor components in accordance with the effective turn's ratio of the transformer.

From the equivalent circuit and basic knowledge of the operation of the induction motor, it can be seen that the magnetizing current component and the iron loss of the motor are voltage dependent and not load dependent. Additionally, the full-voltage starting current of a particular motor is voltage and speed dependent but not load dependent.

The magnetizing current varies with the design of the motor. For small motors, the magnetizing current may be as high as 60%; but for large two pole motors, the magnetizing current is more typically 20–25%.

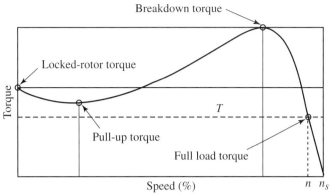

Figure 12.51 Equivalent circuit of an induction motor.

Dynamical Model

The following notations for dynamical parameters are used in the induction motor model.

A, B, C are the suffixes indicating stator variables.

a, b, c are the suffixes indicating rotor variables.

S is the suffix indicating stator quantity.

R is the suffix indicating rotor quantity.

q, d are the suffixes indicating the q- and d-axes, respectively.

i_{qs}, i_{ds} are the two-phase q, d axes stator currents.

i_{qr}, i_{dr} are the two-phase q, d axes rotor currents.

v_{qs}, v_{ds} are the two-phase q, d axes stator voltages.

v_{qr}, v_{dr} are the two-phase q, d axes rotor voltages.

R_s is the resistance of one stator phase.

R_r is the resistance of one rotor phase.

L_s is the apparent three-phase stator inductance per phase.

L_r is the apparent three-phase rotor inductance per phase.

M is the apparent three-phase mutual inductance per phase.

ω_r is the rotational frequency of the machine rotor.

θ is the angle between stator phase A and rotor phase a.

J is the rotor moment of inertia.

T_E is the electromagnetic torque developed.

T_L is the load torque.

T_D is the damping torque.

P is the operator d/dt.

(α, β) are the orthogonal rotating two-phase axes.

The study of the transient performance of three-phase induction motors and their behavior on nonsinusoidal supplies is much easier with transformation of the machine variables to a stationary reference frame. They are called the direct and the quadrature axes. Since the choice of time zero is arbitrary, it is most convenient to choose time zero at the instant when q, A, and a axes are all coincident, as shown in Fig. 12.52. Choosing an instant when the α-axis of the rotating two-phase axes (α, β) makes an angle θ with the stationary q-axis, and

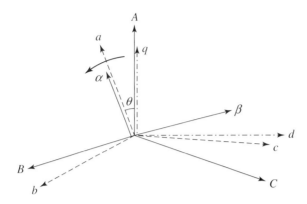

Figure 12.52 Different machine phase axes.

assuming no zero-sequence components, one form of transformation of rotor quantities between rotating three-phase (a, b, c) and stationary two-phase (d, q) is:

$$\begin{bmatrix} v_a \\ v_b \\ v_c \end{bmatrix} = \begin{bmatrix} \cos\theta & -\sin\theta \\ \cos(\theta+120°) & -\sin(\theta+120°) \\ \cos(\theta-120°) & -\sin(\theta-120°) \end{bmatrix} \begin{bmatrix} v_{qr} \\ v_{dr} \end{bmatrix} \qquad (12.103)$$

and

$$\begin{bmatrix} i_a \\ i_b \\ i_c \end{bmatrix} = \begin{bmatrix} \cos\theta & -\sin\theta \\ \cos(\theta+120°) & -\sin(\theta+120°) \\ \cos(\theta-120°) & -\sin(\theta-120°) \end{bmatrix} \begin{bmatrix} i_{qr} \\ i_{dr} \end{bmatrix} \qquad (12.104)$$

Also, we can define the (d, q) stator voltages as:

$$\begin{bmatrix} v_{qs} \\ v_{ds} \end{bmatrix} = \begin{bmatrix} 1 & 0 & 0 \\ 0 & -1/\sqrt{3} & 1/\sqrt{3} \end{bmatrix} \begin{bmatrix} v_A \\ v_B \\ v_C \end{bmatrix} \qquad (12.105)$$

These transformations do not give invariance of power, but this is not important in this case since variables are ultimately expressed in the three-phase reference frame.

Using per-unit quantities, the differential equations describing the behavior of the machine when referred to the stationary frame can be written in the form

$$\begin{bmatrix} v_{qs} \\ v_{ds} \\ v_{qr} \\ v_{dr} \end{bmatrix} = \begin{bmatrix} R_S & 0 & 0 & 0 \\ 0 & R_S & 0 & 0 \\ 0 & -\omega_r M & R_r & -\omega_r L_r \\ \omega_r M & 0 & \omega_r L_r & R_r \end{bmatrix} \begin{bmatrix} i_{qs} \\ i_{ds} \\ i_{qr} \\ i_{dr} \end{bmatrix} + \begin{bmatrix} L_S & 0 & M & 0 \\ 0 & L_S & 0 & M \\ M & 0 & L_r & 0 \\ 0 & M & 0 & L_r \end{bmatrix} p \begin{bmatrix} i_{qs} \\ i_{ds} \\ i_{qr} \\ i_{dr} \end{bmatrix} \qquad (12.106)$$

This is in the form $V = [R]i + [L]\, pi$, where pi is di/dt, and can be rewritten as $pi = [L]^{-1}\{V - [R]\{I\}$ from which we can get the following derivatives:

$$pi_{qs} = \frac{[L_r v_{qs} - M v_{qr} - \omega_r M^2 i_{ds} - \omega_r M L_r i_{dr} - L_r R_s i_{qs} + M R_r i_{qr}]}{L_r L_s - M^2}$$

$$pi_{ds} = \frac{\left[L_r v_{ds} - M v_{dr} + \omega_r M^2 i_{qs} + \omega_r M L_r i_{qr} - L_r R_s i_{ds} + M R_r i_{dr}\right]}{L_r L_s - M^2}$$

$$pi_{qr} = \frac{\left[-M v_{qs} + L_s v_{qr} + \omega_r L_s M i_{ds} + \omega_r L_s L_r i_{dr} + M R_s i_{qs} - L_s R_r i_{qr}\right]}{L_r L_s - M^2}$$ (12.107)

$$pi_{dr} = \frac{\left[-M v_{ds} + L_s v_{dr} - \omega_r L_s M i_{qs} - \omega_r L_s L_r i_{qr} + M R_s i_{ds} - L_s R_r i_{dr}\right]}{L_r L_s - M^2}$$

In per-unit quantities, the electromagnetic torque *TE* becomes

$$T_E = M[i_{dr} i_{qs} - i_{qr} i_{ds}]$$ (12.108)

Also, from the balance between input and output,

$$T_E = J p \omega_r + T_D + T_L$$ (12.109)

Equation (12.109) can be rearranged as

$$p\omega_r = [T_E - T_D - T_L]/J$$ (12.110)

$$p\theta = \omega_r$$ (12.111)

Substituting equations (12.110) and (12.111) into equation (12.107) and then substituting for i_{qr} and i_{dr} using Eq. (12.108) we have:

$$V_a = [R_r + pL_{eq}]i_a + [\omega_r(L_{eq} - L_r)i_{dr} - \omega_r M i_{ds} - (MR_s/L_s)i_{qs} + (M/L_s)v_{qs}]\cos\theta$$
$$- [\omega_r(L_r - L_{eq})i_{qr} + \omega_r M i_{qs} - (MR_s/L_s)i_{ds} + (M/L_s)V_{ds}]\sin\theta$$ (12.112)

$$V_b = [R_r + pL_{eq}]i_b + [\omega_r(L_{eq} - L_r)i_{dr} - \omega_r M i_{ds} - (MR_s/L_s)i_{qs} + (M/L_s)V_{qs}]\cos(\theta + 120°)$$
$$- [\omega_r(L_r - L_{eq})i_{qr} + \omega_r M i_{qs} - (MR_s/L_s)i_{ds} + (M/L_s)V_{ds}]\sin(\theta + 120°)$$ (12.113)

$$V_c = [R_r + pL_{eq}]i_c + [\omega_r(L_{eq} - L_r)i_{dr} - \omega_r M i_{ds} - (MR_s/L_s)i_{qs} + (M/L_s)V_{qs}]\cos(\theta + 120°)$$
$$- [\omega_r(L_r - L_{eq})i_{qr} + \omega_r M i_{qs} - (MR_s/L_s)i_{ds} + (M/L_s)V_{ds}]\sin(\theta - 120°)$$ (12.114)

where

$$L_{eq} = (L_r L_s - M^2)/L_s$$ (12.115)

12.4.2 Speed Control of the Induction Motor

An induction motor is considered a constant-speed motor whenever it is connected to a constant-voltage and constant-frequency power supply. Typical industrial applications require an adjustable range of speeds. There are many schemes to achieve variable speeds from the same supply and machine.

Changing the Number of Poles

The speed of an induction motor can be varied by changing the number of poles, because its operating speed is close to synchronous speed. Changing the coil connections of the stator winding can achieve the task. Two synchronous speeds can be provided by this method. The squirrel-cage motor is most commonly used for this method:

$$N_s = \frac{60f}{(p/2)} \text{ (rpm)} \qquad \text{or} \qquad \omega_s = \frac{2\pi f}{(p/2)} \text{ (rad/sec)} \qquad (12.116)$$

As we can see from Eq. (12.116), changing the number of poles will lead to a change in synchronous speed, where the synchronous speed of the machine is N_s [rpm], the number of poles is p, and the frequency of the supply is f [Hz].

Line Voltage Control (Stator Voltage Control)

The torque developed by the motor is proportional to the square of the applied voltage $T \propto V2$. Thus, the torque under running conditions changes with the change in supply voltage. The reduction in the operating speed of an induction motor can be achieved by reducing the applied voltage. This method is of limited use, since in order to achieve an appreciable change in speed, a relatively larger change in the applied voltage is required.

Power electronic components, such as thyristors, play an important role in speed control for line voltage control. This method is used in fans and other low-power applications.

Changing the Frequency of the Supply

The operating speed of an induction motor can be increased or decreased by increasing the frequency of the applied voltage. This method enables us to obtain a wide variation in the operating speed of the induction motor. This technique requires a frequency modifier or a variable frequency supply. The induced emf in the stator windings is directly proportional to the frequency. For this reason, the applied voltage should be varied in direct proportion to the frequency in order to maintain constant flux density.

When the frequency varies, the synchronous speed changes and so will the operational speed of the induction motor. Electronic switching devices are used to provide a variable frequency supply for the machine. Reducing the frequency compresses the horizontal speed scale, and the opposite is true as shown, in Figure 12.51.

Rotor-Resistance Control

This method in speed control is suitable only for wound-rotor induction motors. By modifying the speed–torque curve through altering the rotor resistors, the speed at which the motor drives a particular load can be altered. This has been used in winching-type applications. In this scheme, excessive heat is generated in the rotor resistors, and there is a consequential drop in overall efficiency. Because of these drawbacks, this method of speed control can be used only for short periods.

Scalar Controllers

The "voltage-frequency" (V/f) controller is the most widespread control technique among industrial applications. It is known as a scalar control, and it functions through imposing a constant relation between voltage and frequency. The structure is very simple, and it is normally used without speed feedback. However, this controller does not achieve acceptable accuracy in either speed or torque tracking.

Vector Controllers

These controllers deploy control loops for controlling both the torque and the flux. The most widespread controllers of this type are the ones that use vector transforms. The accuracy can reach values in the 0.5% range of speed and 2% range in torque. The main disadvantages are the huge computational complexity required and the required critical identification of the motor parameters.

Direct Torque Control (DTC)

Direct torque control (DTC) for AC drives allows accurate control of both motor speed and torque without pulse encoder feedback from the motor shaft. It delivers performance comparable to that of classical vector control, but with a simpler structure and control diagram. In DTC it is possible to control directly the stator flux and the torque by selecting the appropriate inverter state. The main advantage of DTC is the absence of coordinate transforms and of a voltage modulator block. Furthermore, the DTC yields a minimal torque response time that is faster in general than its counterpart among vector controllers. However, DTC requires torque and flux estimators, hence the need for the consequent parameter identification.

To demonstrate the DTC, an AC motor with the following dynamic parameters is considered: $L_r = 0.07131$; $L_s = 0.07131$; $M = 0.06931$; $R_s = 0.435$; $R_r = 0.816$; $J = 0.089$; $B = 0.006$. The following intermediate parameters were defined: $d = (L_r*L_s) - (M*M)$; $k_2 = (M*M)/d$; $k_3 = (M*L_r)/d$; $k_4 = (L_r*Rs)/d$; $k_5 = (M*R_r)/d$, $k_6 = M*L_s$; $k_7 = L_s*L_r$; $k_8 = M*R_s$; $k_9 = L_s*R_r$.

Figures 12.53–12.58 provide a SIMULINK model of the subject drive through critical subsystems and the corresponding response.

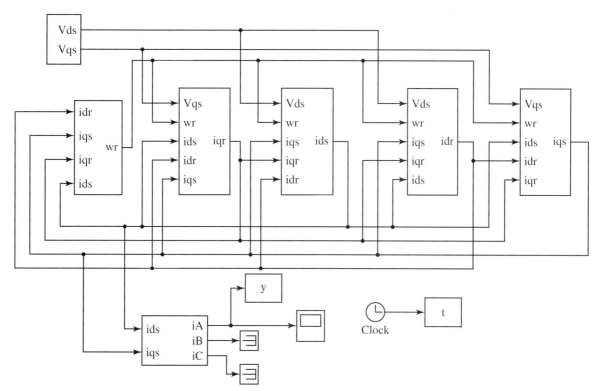

Figure 12.53 Simulink Diagram of DTC for 3 phase AC motor.

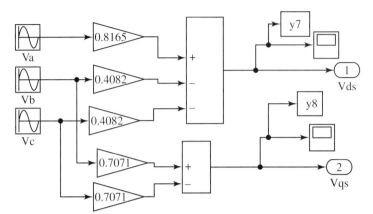

Figure 12.54 Simulink sub-block of DTC for stator voltage.

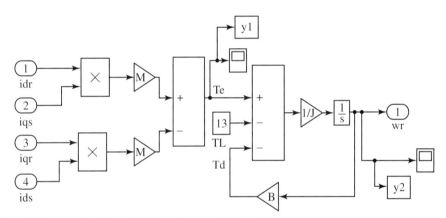

Figure 12.55 Simulink sub-block of DTC for AC motor output.

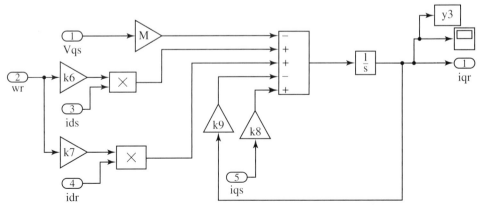

Figure 12.56 Simulink sub-block of DTC for controller loop.

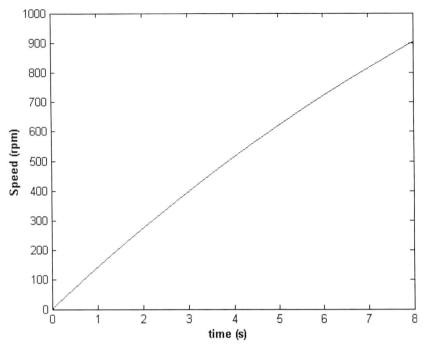

Figure 12.57 DTC response of AC motor speed.

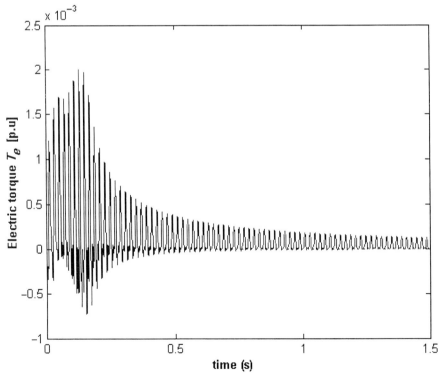

Figure 12.58 DTC response of AC motor torque.

12.5 SUMMARY

The useful motion-based mechatronics functions integrate one form of motors in order to actuate the mechanism. This chapter covered the basic operations, characteristics, modeling, construction, control, and driving circuitry of many types of electric motors. In addition, many commercial motors were listed and a comparative analysis among various types of motors was presented and used in selecting a type and size of a motor for a specific application.

RELATED READING

C. W. deSilva, *Control Sensors and Actuators.* New York: Prentice Hall, 1989.

Electrocraft Corporation, *DC Motors, Speed Controls, Servo Systems*, 1980.

G. Franklin, D. Powell, and A. Emami-Naeini, *Feedback Control of Dynamic Systems.* Reading, MA: Addison-Wesley, 1994.

M. Hidstand and D. Alciatore, *Introduction to Mechatronics and Measurement systems*, 2nd. ed. New York: McGraw-Hill, 2002.

J. Jones, B. Seiger, and A. Flynn, *Mobile Robots: Inspiration to Implementation.* Natick: A. K. Peters, 1999.

R. D. Klafter, T. A. Chmielewski, and M. Negin, *Robotic Engineering: An Integrated Approach.* New York: Prentice Hall, 1989.

B. C. Kuo, *Step Motors and Control System.* Champaign: SRL, 1979.

M. Mokhtari and M. Marie, *Engineering Applications of MATLAB 5.3 and SIMULINK 3.* New York: Springer, 2000.

J. Peatman, *Design with Microcontrollers.* New York: McGraw-Hill, 1988.

A. K. Stiffler, *Design with Microprocessors for Mechanical Engineers.* New York: McGraw-Hill, 1992.

J. Tal, *Step-by-Step Design of Motion Control Systems.* Galil Motion Control, 1994.

http://www.electricmotors.machinedesign.com/

http://www.iprocessmart.com/

http://www.itee.uq.edu.au/

http://www.lmphotonics.com/m_control.htm

http://www.abb-drives.com/

http://www.electricmotors.machinedesign.com/

QUESTIONS

12.1   Describe the various types of electric motors.

12.2   State the differences between a stepping motor and a DC motor.

12.3   What are the main advantages and disadvantages of brushless over brush-type DC motors?

12.4   Describe the various types of stepper motors.

12.5   Why is a stepper motor considered a digital actuator?

12.6   Explain the terms *winding* and *phase*.

12.7   Explain the terms *unifilar* and *bifilar windings*.

12.8   What are the main differences between unipolar and bipolar stepper motors?

12.9   State the major differences between a PM and an VR stepper motor.

12.10   State the various techniques used to protect control circuits for inductive loads, and explain the effect of each on motor performance.

12.11   How does adding an external resistance to a motor winding enhance the performance of a stepper motor?

12.12 What is meant by chopper drive?

12.13 What is meant by slewing?

12.14 How does motor torque change with speed?

12.15 What torque does a stepper motor generate when not energized?

12.16 Explain pull-in torque and pullout torque.

12.17 Describe the principle of operation of a DC motor.

12.18 What is the difference between a field coil motor and a PM motor?

12.19 Describe methods of PM motor construction.

12.20 List the various constants used to characterize DC motor behavior, and explain the meaning of each.

12.21 Define electrical and mechanical time constants, and explain the effect of each on motor performance.

12.22 How does the velocity motion profile affect motor performance?

12.23 What is the most efficient velocity profile?

12.24 What is meant by coupling ratio optimization?

12.25 What is the optimum coupling ratio for a gear drive at no load?

12.26 What are the most important considerations in selecting a DC motor for a given application?

12.27 Explain the difference between linear and switching amplifiers.

12.28 Explain the difference between a voltage amplifier and a current amplifier.

12.29 What should be determined before an appropriate amplifier is selected?

12.30 Describe the construction and principle of operation of a tachometer.

12.31 How does tachometer feedback affect controller performance?

12.32 Explain the operation of the H-bridge.

12.33 Discuss the important features of the LM628 motor controller IC.

12.34 How does the HCTL-1100 controller IC differ from the LM628?

12.35 List the major components of a complete servo drive, and explain the role of each.

12.36 Describe the most important feature in an induction motor torque curve.

12.37 List possible control strategies for AC motors.

PROBLEMS

12.1 Estimate the torque required on a 15-deg VR stepper motor used to accelerate a load with inertia 0.0002 oz-in.-s^2 from standstill to 10,000 steps/s in 10 msec.

12.2 A two-phase PM stepper motor has the following specifications: $R = 6\ \Omega$, $L = 0.02$ H, $k_e = 0.8$ Nm/A, $J = 0.00043$ Nm-s^2, $B = 0.35$ Nm-s, and $V_{DC} = 6$ V. The motor drives a load exerting a torque of $T_L = 0.001$ N-m. Determine v_a, v_b, T_a, and T_b. Also, determine the motor response for four steps over 1.2 s/pulse using a time increment of $dt = 0.1$ ms.

12.3 A 3-mm-pitch lead screw is to be driven by a stepper motor through a gear drive with a 2:1 ratio. The required axial motion increment is 0.0075 mm/step. Determine the step angle requirement of the motor.

12.4 Select the external winding resistance, Zener diode, and transistors for a stepper motor having a resistance per winding of 20.5 Ω, an inductance per winding of 45 mH, and an operating voltage of 12 V.

12.5 The pull-in torque curve of a 1.8°/step stepping motor can be approximated by

$$T = -0.1\omega^2 + 810$$

where T is the torque (gm-cm) and ω is the angular velocity (steps/sec). The motor is coupled to a lead screw of 5-mm pitch. The screw is connected to an NC work table that exerts the torque described by

$$T = 200 + 25v$$

where v is the load velocity (mm/s).

a. Draw a torque—velocity curve of the motor and load.

b. Determine the maximum starting frequency of the motor.

c. Determine the motor speed to drive the load at a linear velocity of 250 mm/s.

d. Is the motor capable of driving the NC work table?

e. What is the maximum allowable acceleration of the motor if the moment of inertia is 4.5 oz-in.-s^2.

12.6 Explain how to use flip-flops and logic gates to provide the proper sequence logic to sequentially energize the phase windings of a four-phase stepper motor. Draw a block diagram of such a circuit.

12.7 An open-loop control is needed for a unipolar stepper motor driving a single-axis NC milling machine. The control consists of an oscillator generating 1000 pulses/s (pps), a decelerator circuit, and a position down-counter. The counter initially contains the required motion stroke in basic linear increments. The counter decrements each time the position NC table increments. For a smoother motion, it is required that the stepper motor be decelerated from a velocity of 1000 steps/s to 125 steps/s about 20 ms before stopping. The motor stops when the counter reaches the zero position. Draw a block diagram of the control system, select all pertinent components, and implement the design.

12.8 Consider a stepper motor driven via the A2559 IC. Write the code to advance the motor in steps of 18° with a pause of 1 s after each step. At the 90° point, the motor reverses to the start point.

12.9 Obtain the specs for the A3967SLB driver/translator IC from www.allegromicro.com. Draw a block diagram showing its interface to the 9S12C, and write a code to manage this interface.

12.10 A PM servo motor has the following characteristics: voltage constant $K_E = 0.825$ V-s/rad, torque constant $K_T = 7.29$ lb-in./A, armature resistance $R_{a} = 0.41\ \Omega$, and armature inertia $J_M = 0.19$ lb-in.-s^2.

    a. Show that the voltage and toque constants are equal in SI units.

b. Calculate the mechanical time constant of the motor.

c. If the input voltage is 85 V, determine the steady-state speed at no load and when a 120 lb-in full load is being driven.

12.11 Obtain the specs for the A3967SLB driver/translator IC from www.allegromicro.com. Draw a block diagram showing its interface to the 9S12C, and write a code to manage this interface.

12.12 A company needs to design a DC motor servo to drive the table of an NC machine. The maximum axial velocity of the table is to be 2500 mm/min, and motion resolution is expected to be 0.02 mm. The table is mounted on a 10-mm-pitch lead screw, with a shaft encoder connected directly to it for position feedback. The desired motor speed is 1000 rpm. The maximum axial force accounting for friction and the cutting force generated during milling at a velocity of 600 mm/min is 400 N.

a. Determine the encoder gain and its maximum frequency.

b. Calculate the required gear ratio between the motor and the lead screw.

c. What is the maximum static load torque on the motor?

d. Derive a relation between the axial force F and motor torque T, gear ratio N and screw pitch P.

LABORATORY PROJECTS

    12.1 Design and build a stepper motor–based garden clock.

    12.2 Design and build an XY plotter to move a pen along a desired path.

13

Control Schemes

OBJECTIVES

Thoughtful engagement with the material presented in this chapter will enable the student to:

1 Become familiar with the history and background of control theory

2 Learn how to analyze and design popular classical control schemes needed in mechatronics

3 Develop skills in modern control design based on the state-space approach

4 Design, choose, and justify adaptive, fuzzy logic, and adaptive fuzzy logic controllers

5 Digitally implement control schemes as typically required in mechatronics projects

6 Use LabVIEW, MATLAB and Simulink to simulate various controllers

7 Develop a keen sense of the practicality of the developed control schemes by comparing experimental and simulation results

13.1 INTRODUCTION

13.1.1 History of Control

A control system is "an interconnection of components forming a system configuration, which will provide a desired system response as time progresses." Control theory was inspired by human behavior before a clear definition of control was devised. For example, a person in the shower senses the temperature of the water and adjusts the mixer knob to achieve the desired temperature level. The sense-adjust operation performed until a desired goal is reached is an example of a closed-loop control strategy. Man adopted the idea of control from nature, where control is inherent to most natural processes. The human body, for instance, maintains a 1% level of glucose in the blood. When the glucose level increases, the body transforms the extra quantity into glycogen for storage. In the event of a reduced glucose level, the body transforms glycogen back into glucose. Many other examples of feedback control are exhibited by the nervous system and other body functions.

Automatic control has played a key role in the advancement of engineering and science, such as space vehicle systems, missile guidance systems, autopiloting systems, robotics, automobiles, consumer electronics, and industrial processes in which regulating temperature, pressure, humidity, position, or flow is essential. Automatic control provides the means to attain consistent performance specifications of dynamic systems to improve productivity, regulate outputs to track desired reference values, and relieve humans from repetitive and tedious manual operations.

The initial work on feedback control took place in Greece in the period 300 to 1 BC. Float-regulator mechanisms were developed using feedback control. The float regulator was employed in the Ktesibiois water clock and in the regulatation of the fuel oil level in Philon's oil lamp. Heron of Alexandria authored his book *Pneumatic*, which outlined the use of float regulators in water level control. In 1572–1633, Cornelis Drebbel of Holland created a temperature regulator; and in 1647–1712, Dennis Papin created the first temperature regulator for steam boilers. The first historical feedback mechanism was a water-level float regulator invented by Poluznov of Russia in 1765. Significant work in the area of automatic feedback control in industrial processes was achieved in 1769, when James Watt invented a centrifugal governor for speed control of a steam engine. Moreover, important contributions to earlier development in control theory were made by Minorsky, Hazen, and Nyquist, among others. In 1922, Minorsky worked on automatic controllers for steering ships and showed how stability could be determined from the differential equations describing the system. In 1934, Hazen discussed the design of a relay servomechanism capable of closely following a changing input.

Prior to World War II, two main streams in control theory and practice had developed. The first stream was in the United States and western Europe, where the development of the telephone system and electronic feedback amplifiers by Bode, Nyquist, and Black at Bell Telephone Laboratory were the driving force in using feedback control. The frequency domain was used to describe the operations of control systems. The second stream originated in Russia, where time domain formulation using differential equations was utilized. In World War II, the need to design and construct autopiloting systems, radar antenna control systems, and other advanced weapon systems became urgent. As a result of the growing interest in control systems, control engineering became an engineering discipline in its own right. After World War II, frequency domain analysis, the Laplace transform, and the complex frequency plane were still in use. As modern plants with many inputs and outputs became more complex, the description of a modern control system required a large number of equations, thus making single-input, single-output (SISO) systems powerless. Since 1960, modern control theory based on the time domain using state variables was developed to cope with the increased complexity of modern plants and the stringent requirements on accuracy, weight, and cost in military, space, and industrial applications.

In the early 1950s, there was extensive research on adaptive control in connection with the design of autopilots for high-performance aircraft. This research was motivated by the problems created by ordinary constant-gain, linear feedback control, which could work well in one operating condition but not over a wide range of flight regimes. In the 1960s, research efforts culminated in the introduction of state-space and stability theories. In the 1970s and early 1980s, proofs for the stability of adaptive systems appeared, but under restrictive conditions. Research in the late 1980s and early 1990s gave new insight into the robustness of adaptive controllers.

Because modeling and analysis of nonlinear plants are very difficult, the need for control strategies not based on the plant transfer function or mathematical model emerged. Intelligent control aims to solve this problem. It was inspired by the fact that humans can control complex systems without knowing any mathematics about those systems. Three main projects provided the basis for today's intelligent control systems. In the seventeenth century, Gottfried Leibniz suggested developing a machine able to carry out whatever requested reasoning. In the nineteenth century, Charles Babbage invented an analytical engine capable of performing algebra. In 1936, Alan Turing suggested developing a machine that would think. In the first decades of the twentieth century, the nervous system and the brain as control devices were thoroughly studied. In the 1930s, *cybernetics* emerged from the studies performed on the brain. In the 1950s, cybernetics was integrated with computers by a group of scientists (McCarthy, Minsky, Newell, and Simon). The computer was able to solve geometric problems, to play chess, to recognize patterns, and to understand English. This introduced *artificial intelligence*. In the same year, the term *intelligent control* was proposed by K. S. Fu. The first neurocontroller was developed in 1963 by Widrow and Smith using a simple neural network trained to stabilize and control an inverted pendulum. Fuzzy sets were introduced by L. A. Zadeh in 1969.

The introduction of the digital computer facilitated rapid growth in the use of digital logic for the control of physical systems. Microcontrollers and other microprocessors have been the brains in large and small systems. Aircraft autopilots, oil refineries, paper-processing machines, hard disk drives, CD players, ABS braking in cars, and autofocusing cameras are examples. Digital control provides for flexibility, ease of change of control decision making or reprogramming, and more robustness in the face of the drift of electronic components and environmental changes.

13.1.2 Open-Loop Control

Determining adequate input for a process to achieve a desired response or output is the simple purpose of controllers. Hence, control engineering is concerned with understanding, modeling, analysis, and design of systems. Early attempts at control aimed at using *open-loop control* strategies, where an actuating device is stimulated to obtain the desired response using a well-known model of the process, as shown in Fig. 13.1.

In an open-loop system, the input is changed until a desired output is obtained. For example, in the case of a DC motor, the input voltage is changed manually until a desired speed of the shaft is attained. After that, this input voltage is supposed to keep that motor running at the desired speed. Open-loop control has many drawbacks: (1) If the motor comes in contact with a workpiece or any other load, its speed will be affected and the set control signal will no longer achieve the desired speed; (2) even if the motor runs at constant load, environmental effects such as temperature, drift in electronic components, and friction will change the operating parameters of the process, and again the designed input will no longer achieve the desired output.

13.1.3 Closed-Loop Control

Closed-loop control continuously and automatically monitors the output of the process being controlled; by comparing that output to a desired reference, an error signal is generated and used by the controller to adjust the input of the process to reestablish the desired output (Fig. 13.2).

Figure 13.1 Schematic diagram of an open-loop control system.

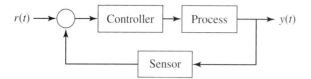

Figure 13.2 Schematic diagram of a closed-loop feedback control system.

Open-loop control is used for systems in which inputs and processes are known ahead of time and for certain environments. Closed-loop control has advantages when unpredictable disturbances and/or uncertain variations in system components are present. Closed-loop control systems are generally higher in cost and power consumption than their open-loop counterparts.

The following sections discuss in details the aforementioned types of closed-loop control.

13.2 CLASSICAL CONTROL

The term *classical control* refers to the techniques developed from the early days of control theory to the early 1960s, when the use of computers was very limited. The design was performed using transfer function–based techniques. Moreover, it was characterized mainly by the use of algebraic and graphical techniques applied to the single-input, single-output (SISO) systems.

The objective of designing a classical control system is to deliver desired closed-loop transient and steady-state specifications. This is usually accomplished by altering the system dynamics through compensation. Performance specifications are usually given either in terms of the time domain, such as rise time, maximum overshoot, and settling time, or in terms of the frequency domain, such as stability phase and gain margins, resonant peak value, and bandwidth. The two most popular strategies used to perform such tasks are the root-locus and the frequency response approaches.

13.2.1 Mathematical Modeling

Usually the mechatronics process to be controlled involves electromechanical and electronic components. Proper implementation of control strategies requires knowledge of the mathematical model of the involved hardware devices. The first step in the analysis of a dynamic system is to derive its mathematical model. It is important because all the calculations done afterwards are based on that model. One way to derive the mathematical model is to use the physical laws governing the systems, such as Newton's laws for mechanical parts and Kirchhoff's laws for electrical components, to represent the system as a set of differential equations. If time domain representation is difficult to analyze, taking Laplace transforms of signals and deriving a transfer function of the process is popular. Once a transfer function is derived, analysis and design take place in the s-domain before taking the inverse transforms back to the time domain for implementation.

Alternatively, if the transfer function of a system is unknown, it may be established experimentally by introducing known inputs and studying the corresponding outputs of the system. A mathematical model can be a set of differential equations that represent the dynamics of the system. In modeling, a more accurate representation is achieved at the disadvantage of more complexity. Thus, a compromise is most often the best choice.

Example 13.1: Experimental model determination of an unknown plant—A DC motor closes the gripper to grasp an unknown load of a pick-and-place robot. The model of the loaded end effector is not known. The Bode plot shown in Fig. 13.3 was experimentally obtained. Derive an estimate of the unknown transfer function.

Solution: The importance of frequency-response methods is that the transfer function of the plant, or other components of a system, may be determined by a simple frequency-response measurement. In this method, if the amplitude ratio and phase shift have been measured at a sufficient number of frequencies within the frequency range of interest, they may be plotted on the Bode diagram. Then the transfer function can be determined by asymptotic approximations. We build up asymptotic log magnitude curves consisting of several segments. With some trial-and-error juggling of the corner frequencies, it is usually possible to find a very close fit to the curve.

A sinusoidal signal generator is used to perform a frequency-response test. The required frequency range is from DC to 1000 Hz. The sinusoidal signal must be reasonably free of harmonics or distortion. To determine the transfer function, we first draw asymptotes to the experimentally obtained log-magnitude curve. The slopes of the asymptotes must be multiples of ± 20 dB/decade. If the slope of the experimentally obtained log-magnitude curve changes from -20 to -40 dB/decade at $\omega = \omega_1$, then the factor $1/[1 + j(\omega/\omega_1)]$ exists in the transfer function. If the slope changes by -40 dB/decade at $\omega = \omega_2$, then the transfer function must include a quadratic factor of the form

$$\frac{1}{1 + 2\xi\left(j\dfrac{\omega}{\omega_2}\right) + \left(j\dfrac{\omega}{\omega_2}\right)^2} \tag{13.1}$$

The undamped natural frequency of this quadratic factor is equal to the corner frequency ω_2. The damping ratio ξ can be determined from the experimentally obtained log-magnitude curve by measuring the amount of resonant peak near the corner frequency ω_2. Once the factors of the transfer function have been determined, the gain can be determined from the low-frequency portion of the log-magnitude curve. Since these factors become unity as ω approaches zero, at very low frequencies, the sinusoidal transfer function $G(j\omega)$ can be written as

$$\lim G(j\omega) = \frac{K}{(j\omega)^\lambda} \qquad \text{as } \omega \to 0 \tag{13.2}$$

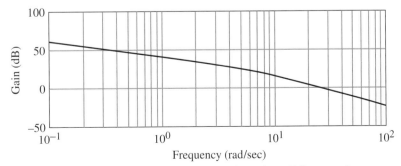

Figure 13.3 Experimental Bode plot for the robot gripper DC motor of Example 13.1.

In many practical systems, λ equals 0, 1, or 2.

- For $\lambda = 0$, or type 0 systems, $G(j\omega) = K$ for very small ω, and thus the low-frequency asymptote is the horizontal line at 20 log K dB. The value of K can thus be found from this horizontal asymptote.
- For $\lambda = 1$, or type 1 systems, $G(j\omega) = K/j\omega$ for very small ω, and thus the low-frequency asymptote has a slope of -20 dB/decade. The frequency at which the low-frequency asymptote (or its extension) intersects the 0-dB line is numerically equal to K.
- For $\lambda = 2$, or type 2 systems, $G(j\omega) = K/(j\omega)^2$ for very small ω, and thus the low-frequency asymptote has a slope of -40 dB/decade. The frequency at which this asymptote (or its extension) intersects the 0-dB line is numerically equal to $\sqrt{K}$.

Here are the steps followed to determine the transfer function of the system in this example.

1. Apply a sine wave signal to the input.
2. Vary the frequency of the applied signal over the range of interest.
3. At each frequency sample, record the output voltage.
4. Form the decibel gain by dividing the output voltage by the input voltage.
5. Plot the decibel gain values versus the frequency logarithm.
6. Approximate the -20-dB slope and get the poles of the system.

From the experimental Bode plot shown in Fig. 13.3, the gain and poles of the system are derived. Since the Bode plot shows a -20-dB/decade slope at low frequencies, the system type is 1 with a pole at zero. At $\omega = 6$ rad/sec, the slope becomes -40 dB/decade, and the system has a pole at $\omega = 6$ rad/sec.

Now, to get the gain, it is noted that this is a type 1 system and that the low-frequency asymptote extension will cross the 0-dB line at a frequency $\omega \approx 100$ rad/sec; thus gain = 100. After some manipulation in the fraction, the transfer function is estimated as

$$G(s) \cong \frac{670}{s(s + 6)} \tag{13.3}$$

13.2.2 Transfer Function

From a theoretical viewpoint, *system stability* means, "well-behaved inputs produce well-behaved outputs." By contrast, if any well-behaved input produces an undesirable, unbounded response, then the system is unstable. The most important characteristic of the dynamic behavior of a control system is its *absolute stability*; the system can be stable or unstable. A control system is in equilibrium if in the absence of any input or disturbances, the output stays in the same state. A *linear time-invariant control system* is stable if the output eventually comes back to its equilibrium state when the system is subjected to a disturbance. A linear time-invariant control system is unstable if oscillation of the output continues or the output diverges without bound from the equilibrium state when the system is subjected to a disturbance. Asymptotic stability depends on the system modes, i.e., poles $P_i (i = 1, \ldots, n)$, of the transfer function that were not cancelled with any common terms in the numerator. Three possibilities exist:

1. Real $(P_i) < 0$ for all i. In this case the output $y(t) \to 0$ as $t \to \infty$, and the system is asymptotically stable.
2. Real $(P_i) > 0$ for at least one value of i. Here, $y(t) \to \infty$ as $t \to \infty$, and the system is unstable.

3. Real $(P_i) = 0$ for more than one value of i. This means that the roots are pure imaginary and the system tends to oscillate. This is known as *marginal stability*.

To understand the concept of bounded-input, bounded-output (BIBO) stability, consider the output of a linear time-invariant system. This output is given in the time domain by the convolution of the system impulse response $g(t)$ with the input to the system $u(t)$:

$$y(t) = \int_{-\infty}^{\infty} g(\tau)u(t - \tau)\,d\tau \tag{13.4}$$

If the input is bounded, i.e., $|u(t)| \leq K < \infty$, then

$$|y(t)| = \left| \int_{-\infty}^{\infty} g(\tau)u(t - \tau)\,d\tau \right| \leq K \left| \int_{-\infty}^{\infty} g(\tau)\,d\tau \right| \leq K \int_{-\infty}^{\infty} |g(\tau)|\,d\tau \tag{13.5}$$

This implies that a system is BIBO stable if and only if

$$\int_{-\infty}^{\infty} |g(\tau)|\,d\tau < \infty \tag{13.6}$$

Consequently, it can be proven that a system is BIBO stable if and only if all poles of the system transfer function lie in the left half of the complex plane (LHP). If any pole lies in the right half-plane or on the $j\omega$ axis, the system is unstable. A popular analytical technique to ascertain the number of unstable poles without actually solving for them is *Routh's stability criteria*.

Finally, if a system is asymptotically stable, then it is BIBO stable, although the reverse is not true.

13.2.3 Transient and Steady-State Analyses

Once a mathematical model of the system is obtained, various methods are available for the analysis of its performance. In practice, the input signal to a control system is not fixed or limited to a certain class, so it cannot be expressed analytically. But in designing and analyzing controllers, there must be a basis of comparison of performance for those systems. This can be achieved by selecting a set of test input signals. Commonly used test input signals are the step, ramp, impulse, and sinusoidal functions. The test input used for a specific system will be chosen based on the inputs that the system most frequently experiences under normal operation. For example, if the system experiences sudden disturbances, step input may be a good choice. If it is subjected to shock inputs, an impulse function will be better. And if the inputs to the control system change frequently, then a sinusoidal function will be a suitable signal.

The *time response* of a control system consists of two parts: the transient and the steady-state responses. Transient response is the part where the output goes from the initial state to the final state and remains there. Steady-state response is the part of the output that proceeds the transient phase. Before reaching steady state, the *transient response* of a practical control system often exhibits damped oscillations. Here are the specifications for the step response of a second-order system, shown in Fig. 13.4.

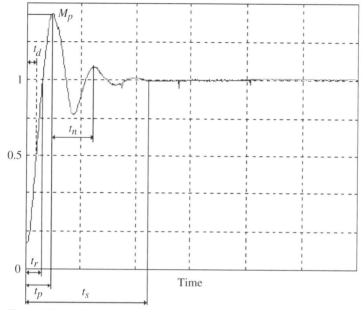

Figure 13.4 Unit step response of a second-order system.

1. *Delay time*, t_d, is the time required for the response to reach half the final value the very first time.

2. *Rise time*, t_r, is the time required for the response to rise from 10% to 90%, 5% to 95%, or 0% to 100% of its final value.

3. *Peak time*, t_p, is the time required for the response to reach the first peak of the overshoot.

4. *Maximum percent overshoot*, M_p, is the maximum peak value of the response curve, measured from the final value. It indicates the relative stability of the system and can be calculated by

$$M_p = \frac{y(t_p) - y(\infty)}{y(\infty)} \times 100\%$$ (13.7)

5. *Settling time*, t_s, is the time required for the response curve to reach and stay within a range of about 2% or 5% of its final value.

Most system specifications are often given assuming that the system is second order. In addition, for the analysis of higher-order systems, *dominant pole* techniques may be used to approximate the system with a second-order transfer function. The generalized form of the closed-loop second-order system is shown in Fig. 13.5; the transfer function is the following equation:

$$G(s) = \frac{y(s)}{u(s)} = \frac{\omega_n^2}{s^2 + 2\zeta\omega_n s + \omega_n^2}$$ (13.8)

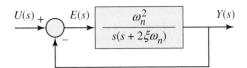

Figure 13.5 Unity-feedback closed-loop second-order system.

The dynamic behavior of the second-order system is described in terms of two parameters: the *damping ratio* ζ and the *natural frequency* ω_n. The closed-loop poles are given by

$$s_{1,2} = -\zeta\omega_n \pm j\omega_n\sqrt{1 - \zeta^2} \qquad (13.9)$$

The two poles can be real ($\zeta > 1$, overdamped), real and identical ($\zeta = 1$, critically damped), or complex conjugates ($0 < \zeta < 1$, underdamped). The transient response of critically damped and overdamped systems does not oscillate. If $\zeta = 0$ (undamped), the transient response does not die out. However, for the underdamped case, the step response $y(t)$ is given by

$$y(t) = 1 - \frac{e^{-\zeta\omega_n t}}{\sqrt{1 - \zeta^2}} \sin\left(\omega_d t + \tan^{-1}\frac{\sqrt{1 - \zeta^2}}{\zeta}\right) \qquad (t \geq 0)$$

$$\omega_d = \omega_n\sqrt{1 - \zeta^2}$$

$$(13.10)$$

where ω_d is called the *damped natural frequency*.

The rise time, peak time, maximum overshoot, and settling time can be derived for the underdamped case. The rise time t_r can be obtained by letting $y(t_r) = 1$:

$$t_r = \frac{1}{\omega_d} \tan^{-1}\left(\frac{\omega_d}{-\sigma}\right) \qquad (13.11)$$

where $\sigma = \zeta\omega_n$. The peak time can be calculated by differentiating $y(t)$ with respect to time and setting this derivative equal to zero. The maximum overshoot occurs at the peak time, or at $t = t_p$:

$$M_P = e^{-(\sigma/\omega_d)\pi} \qquad \text{at} \qquad t_p = \frac{\pi}{\omega_d} \qquad (13.12)$$

The settling time is obtained graphically. It corresponds to a $\pm 2\%$ or $\pm 5\%$ tolerance band. It may be measured in terms of the time constant $1/\omega_n\zeta$. The settling time is commonly defined by

$$t_s = \frac{4}{\sigma} = \frac{4}{\zeta\omega_n} \qquad (2\% \text{ criterion})$$

$$t_s = \frac{3}{\sigma} = \frac{3}{\zeta\omega_n} \qquad (5\% \text{ criterion})$$

$$(13.13)$$

In most control system designs, the interest is specifically in the final, or steady-state, output value. This is known as *steady-state accuracy*. Ideally, in the steady-state phase, the

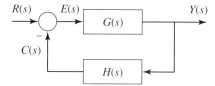

Figure 13.6 Closed-loop control system.

system output tracks the reference signal and the steady-state error is zero. However, this ideal situation is rarely met, and so the steady-state error for any system is determined. The steady-state error for a unity-feedback system is defined as

$$e_{ss} = \lim_{t \to \infty} e(t) = \lim_{t \to \infty} \left[r(t) - c(t) \right] \tag{13.14}$$

Consider the system shown in Fig. 13.6. The closed-loop transfer function is given by

$$\frac{Y(s)}{R(s)} = \frac{G(s)}{1 + G(s)H(s)} \tag{13.15}$$

$G(s)$ describes the model transfer function of the plant, and $H(s)$ represents the transfer function of a needed sensor that makes $y(t)$ compatible with $r(t)$ prior to generating the error signal $e(t)$. Meanwhile, the transfer function between the reference signal and the error signal is given by

$$\frac{E(s)}{R(s)} = \frac{1}{1 + G(s)H(s)} \tag{13.16}$$

Assuming that the closed-loop transfer function is stable, then it has all poles in the complex left half-plane, with a possible single pure imaginary pole at zero. A convenient and adequate *steady-state error* is obtained by application of the *final-value theorem* of the Laplace transform:

$$e_{ss} = \lim_{s \to 0} sE(s) = \lim_{s \to 0} \frac{1}{1 + G(s)H(s)} R(s) \tag{13.17}$$

For step, ramp, parabolic, and higher-order polynomial inputs of the form

$$r(t) = \frac{t^n}{n!} \to R(s) = \frac{1}{s^{n+1}}, \, n = 0, 1, 2, \ldots \tag{13.18}$$

the steady-state error is given by

$$e_{ss} = \lim_{s \to 0} \frac{1}{s^n + s^n G(s)H(s)} \tag{13.19}$$

For a unit-step input ($n = 0$),

$$e_{ss} = \lim_{s \to 0} \frac{1}{1 + G(0)H(0)} \tag{13.20}$$

TABLE 13.1 Steady-State Errors for Polynomial Inputs

	Step Input	Ramp Input	Acceleration Input
Type 0 system	$1/(1 + K_p)$	∞	∞
Type 1 system	0	$1/K_v$	∞
Type 2 system	0	0	$1/K_a$

It is obvious that the steady-state error depends on the structure of the *open-loop transfer function* $G(s)H(s)$. For example, if $G(s)H(s)$ has no poles at the origin, then $G(0)H(0)$ is finite, giving a finite step-response steady-state error. A *system type* is defined as the order of the input polynomial that the closed-loop system can track within a finite error. If $G(s)H(s)$ has no poles at the origin, the closed-loop system is type 0 and tracks a constant. If $G(s)H(s)$ has one pole at the origin, the closed-loop system is a type 1 system that can track a ramp. If $G(s)H(s)$ has two poles at the origin, the closed-loop system is type 2 and can track a parabola. Table 13.1 summarizes the steady-state errors for types 0, 1, and 2 systems when they are subjected to various inputs (for constants K_p, K_v, K_a).

13.2.4 Root Locus

The characteristics of the transient response of a closed-loop system are closely related to the location of the closed-loop poles in the complex plane. So when designing a controller for a system, one should know how the closed-loop poles move in the s-plane as the gain is varied. Sometimes, desired response can be achieved by only adjusting the gain of the controller. Since determining the controller's gain involves trial and error, the calculation should be repeated many times. If the system is of high order, solving the characteristic equation for the roots will be at best tedious. The root locus (RL) method, developed by W. R. Evans, provides a simple way to find the trajectories of closed-loop transfer function poles graphically in the s-plane as the gain is varying, by knowing only the open-loop transfer function poles and zeros. With a gain K added in series with the plant of Fig. 13.6, the roots of the characteristic equation of the closed-loop system are the poles:

$$KG(s)H(s) = -1 \qquad (13.21)$$

This implies that

$$|KGH| = 1 \quad \text{and} \quad \angle GH = \pm (2r + 1)\pi \quad \text{for } r = 0, 1, 2, \ldots \quad (13.22)$$

The basis of most stability studies may be deduced from the properties in this equation. For a point, s_0, in the s-plane to be part of the RL (i.e., a valid closed-loop pole location), the sum of angles from the poles and zeros of $G(s)H(s)$ to s_0 must be $\pm(2r + 1)\pi$. The gain K that corresponds to this point is found from $K = 1/|G(s_0)H(s_0)|$. A summary of the general rules and procedures for constructing the root loci of the system shown in Fig. 13.6 are as follows.

1. Loci parts start ($K = 0$) at poles and end ($K \to \infty$) at zeros of the open-loop transfer function $G(s)H(s)$, where poles and zeros at infinity are also included.

2. Loci segments exist on the real axis only to the left of an odd number of real poles and zeros of the open-loop transfer function.

3. Loci approach asymptotes with angles of $(2r + 1)\pi/(n - m)$, where n is the number of finite poles and m is the number of finite zeros of $G(s)H(s)$ (open-loop transfer function).

4. The asymptotes originate from the centroid on the real axis σ, where p_i are the poles and z_i are the zeros of the open-loop transfer function:

$$\sigma = \frac{\sum_{i=1}^{n} p_i - \sum_{j=1}^{m} z_j}{n - m} \tag{13.23}$$

Example 13.2: Generating the root locus using MATLAB–The electronic pacemaker regulates the speed of the human heart pump. The following second-order transfer function models the behavior of the pacemaker and the heart pump. Generate the root locus of the supplied system, where K is the design amplifier gain of the pacemaker:

$$KG(s) = \frac{K}{s(5s + 1)} \tag{13.24}$$

Solution: Using the RL function in MATLAB gives the root locus plot shown in Fig. 13.7.

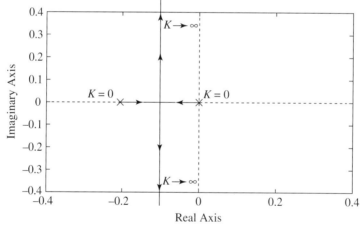

Figure 13.7 Root locus using MATLAB for the heart–pacemaker system of Example 13.2.

Example 13.3: Generating the root locus using LabVIEW–A hypersonic plane model is approximated by a fourth-order transfer function. Plot the root loci for the unity-feedback system with plane speed as output shown in Fig. 13.8.

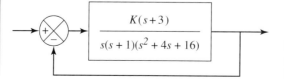

Figure 13.8 Hypersonic plane model with unity feedback for Example 13.3.

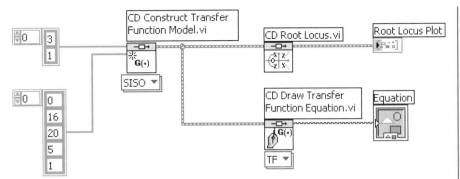

Figure 13.9 LabVIEW VI diagram for root locus, Example 13.3.

Solution: After installing the Control Design (CD) Toolkit, the LabVIEW block diagram shown in Fig. 13.9 is created.

To plot the root locus of the current system, two VIs are employed. The first is the CD Construct Transfer Function Model VI, which creates the following transfer function model of a SISO system from the sampling time (s), numerator, denominator, and delay:

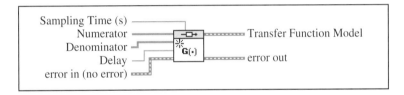

For the current example, only the numerator and the denominator are used to give the transfer function of the open-loop system:

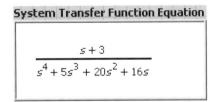

The second is the CD Root Locus VI, which plots the trajectory of the closed-loop poles of a SISO system as the feedback gain varies from zero to infinity:

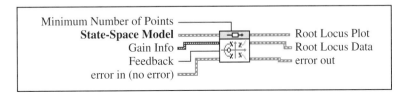

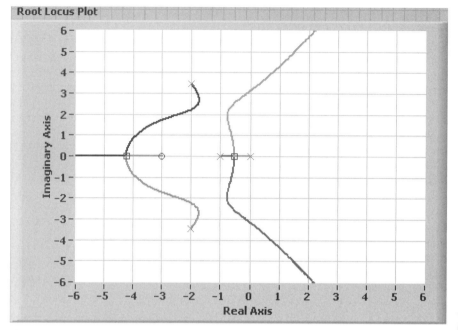

Figure 13.10 Root locus using LabVIEW in Example 13.3.

The root locus plot of the system is shown in Fig. 13.10.

Example 13.4: Control design using the root locus and LabVIEW–An automated camera-positioning servo system is modeled by the following third-order transfer function, as shown in Fig. 13.11. Using LabVIEW, plot the root loci, determine the amplifier gain K such that the damping ratio ζ of the dominant second-order closed-loop system is 0.5, and plot the unit-step response.

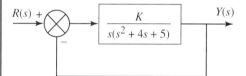

Figure 13.11 Camera-positioning system with unity feedback for Example 13.4.

Solution: The corresponding LabVIEW program and the generated root locus plot for the system are shown in Figs. 13.12 and 13.13, respectively. The dominant closed-loop desired poles have a damping ratio ζ of 0.5 and can be written as

$$s = x \pm j\sqrt{3}x \tag{13.25}$$

The characteristic equation for the closed-loop transfer function is

$$s^3 + 4s^2 + 5s + K = 0 \tag{13.26}$$

By substituting values of s (Eq. (13.25)) into Eq. (13.26), we get

$$(x + j\sqrt{3}x)^3 + 4(x + j\sqrt{3}x)^2 + 5(x + j\sqrt{3}x) + K = 0 \tag{13.27}$$

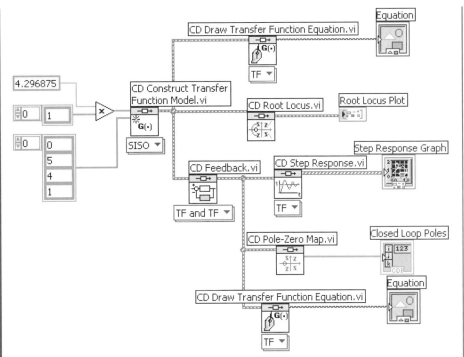

Figure 13.12 LabVIEW diagram for root locus, Example 13.4.

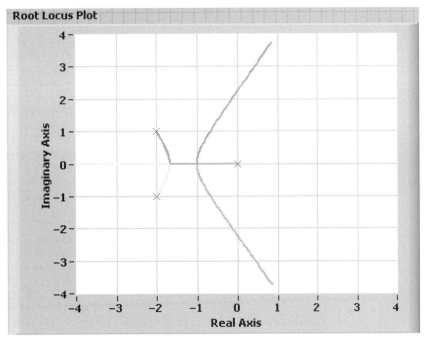

Figure 13.13 Root locus plot using LabVIEW for Example 13.4.

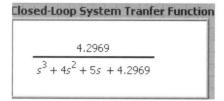

Figure 13.14 Closed-loop transfer function and poles for Example 13.4.

or

$$-8x^3 - 8x^2 + 5x + K + 2\sqrt{3}j(4x^2 + 2.5x) = 0 \tag{13.28}$$

By equating the real and imaginary parts to zero, respectively, we get

$$-8x^3 - 8x^2 + 5x + K = 0 \quad \text{and} \quad 4x^2 + 2.5x = 0 \tag{13.29}$$

Noting that $x \neq 0$, then $4x + 2.5 = 0$, or $x = -0.625$. Substituting $x = -0.625$ gives

$$K = 8x^3 + 8x^2 - 5x = 4.296875 \tag{13.30}$$

Thus, the closed-loop poles are located at $s = -0.625 \pm j\,1.0825$ and $s = -2.75$.

The closed-loop transfer function and the corresponding closed-loop poles of the system shown in Fig. 13.11 are shown in Fig. 13.14. The unit-step response curve of the system is shown in Fig. 13.15.

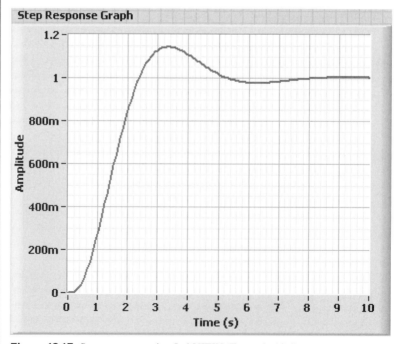

Figure 13.15 Step response using LabVIEW (Example 13.4).

$A \sin(\omega t)$ → $G(s)$ → $A|G(j\omega)|\sin(\omega t + \angle G(j\omega))$

Figure 13.16 Sinusoidal response of a linear system.

13.2.5 Frequency Response

Frequency response stands for the steady-state part of the response of a stable system subjected to a sinusoidal input. Consider the stable plant transfer function shown in Fig. 13.16. If the input is a sinusoidal signal, the steady-state part of the output will also be a sinusoidal signal of the same frequency, but with different magnitude and phase.

It is clear from the system contributions to the signal that the magnitude and timing of the output are functions of the transfer function and the frequency of the input (w). Bode plots, developed by H. W. Bode in the 1930s, consist of two plots, the $\log_{10}$ of the transfer function gain (multiplied by 20) versus the $\log_{10}$ of frequency, and phase (in deg) versus the $\log_{10}$ of frequency. These plots have two main advantages over linear plots: (1) The $\log_{10}$ of the frequency compresses the scale so that greater details are available over a very wide range of frequencies, and (2) the product of the transfer functions often analyzed can be converted to a sum by taking logarithms.

Example 13.5: Generating Bode plots using MATLAB—A laser eye surgery positioning controller with its camera is modeled by the following third-order system. Use MATLAB to generate the Bode plots for the supplied transfer function:

$$G(s) = \frac{1}{s(s + 1)(s + 2)} \tag{13.31}$$

Solution: Figure 13.17 shows MATLAB-generated Bode plots for the system in Eq. (13.31).

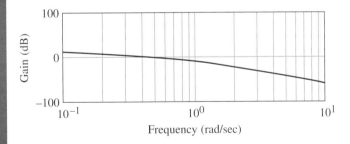

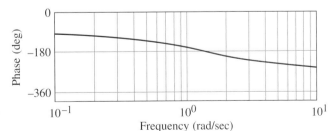

Figure 13.17 Bode plots using MATLAB for Example 13.5.

Example 13.6: Generating Bode plots using LabVIEW–In the pharmaceutical industry, an automatic system dispenses fluid into capsules that are precisely positioned horizontally below the nozzle using a linear motor whose transfer function is supplied in Fig. 13.18. Using LabVIEW, plot Bode diagrams for the closed-loop system for $K = 1$, 10, and 20.

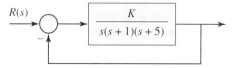

Figure 13.18 Fluid-dispensing positioning system with unity feedback (Example 13.6).

Solution: The closed-loop transfer function of the system is given by

$$\frac{C(s)}{R(s)} = \frac{K}{s(s + 1)(s + 5) + K} \tag{13.32}$$

The CD Bode polymorphic VI produces the magnitude and phase plots of a system on an XY graph. The VI accepts a state-space, transfer function, or zero-pole-gain model as input and converts it into a corresponding transfer function model before calculating the frequency response.

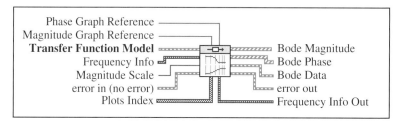

The CD Bode VI with transfer function model as input.

Figures 13.19 and 13.20 show the LabVIEW program and the resulting Bode diagrams, respectively.

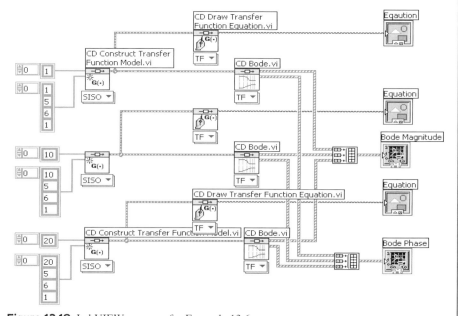

Figure 13.19 LabVIEW program for Example 13.6.

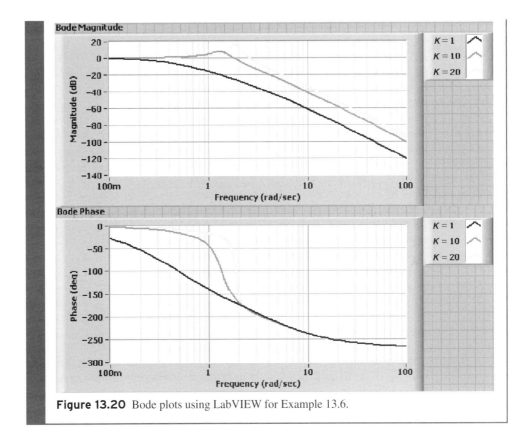

Figure 13.20 Bode plots using LabVIEW for Example 13.6.

Bode plots have several advantages. They are used to determine the stability and relative degree of stability, and hence they are useful in design schemes that aim to improve these specifications. If frequency is eliminated as the independent variable, then gain and phase can be shown on the same plot. There are many ways to present such data. Plots can be rectilinear or polar; gains can be linear or logarithmic.

H. Nyquist used linear magnitude polar plots in systems analysis. The Nyquist diagram plots the variations of magnitude and phase of a transfer function with frequency in the complex plane. At a given frequency, the gain determines the distance from the origin of the complex plane, and the phase determines the angle from the positive real axis.

Typically, systems are destabilized when their gain exceeds certain limits or if there is too much phase lag. These tolerances of gain or phase uncertainty are called *gain margin* and *phase margin*. The *phase margin* is defined as the additional phase lag of the $G(j\omega)H(j\omega)$ at unity magnitude that brings the system to the marginal stability with intersection of the $-1 + j0$ on the Nyquist plot. Let Ψ be the phase angle of the open-loop transfer function at the unity-magnitude frequency. Then the phase margin, *PM*, is

$$PM = \Psi - 180°$$ (13.33)

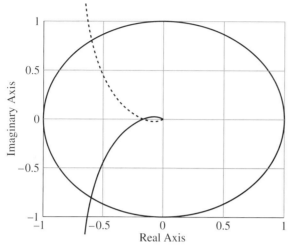

Figure 13.21 Using MATLAB, the Nyquist plot of the robot gripper DC motor (Example 13.1).

The *gain margin* is the increase in the system gain at the phase of $\pm 180°$ that results in a marginally stable system with the intersection of the $-1 + j0$ point on the Nyquist plot. If we let A_m denote the magnitude at this intersection, then the gain margin, *GM*, is defined as

$$GM = 1/A_m \tag{13.34}$$

The Nyquist plot of Example 13.1 estimated transfer function (Eq. (13.3)) is shown in Fig. 13.21.

Example 13.7: Generating a Nyquist plot using LabVIEW–An autonomous vehicle traveling in a linear direction is modeled as a second-order state model, with the position being the output. Generate the Nyquist plot for the system given by Eq. (13.35) using LabVIEW.

$$\begin{bmatrix} \dot{x}_1 \\ \dot{x}_2 \end{bmatrix} = \begin{bmatrix} 0 & 1 \\ -25 & -4 \end{bmatrix} \begin{bmatrix} x_1 \\ x_2 \end{bmatrix} + \begin{bmatrix} 0 \\ 25 \end{bmatrix} u$$

$$y = \begin{bmatrix} 1 & 0 \end{bmatrix} \begin{bmatrix} x_1 \\ x_2 \end{bmatrix} + [0]u \tag{13.35}$$

Solution: The Nyquist VI in LabVIEW produces the Nyquist plot of the system, in which the imaginary part of the frequency response is plotted against its real part. This data can be displayed in the CD Nyquist Plot indicator. This polymorphic VI accepts state-space, transfer function, and zero-pole-gain models as inputs. It converts state-space and zero-pole-gain models into transfer function models before determining the frequency response.

Figures 13.22 and 13.23 show the LabVIEW program and the generated Nyquist plot, respectively.

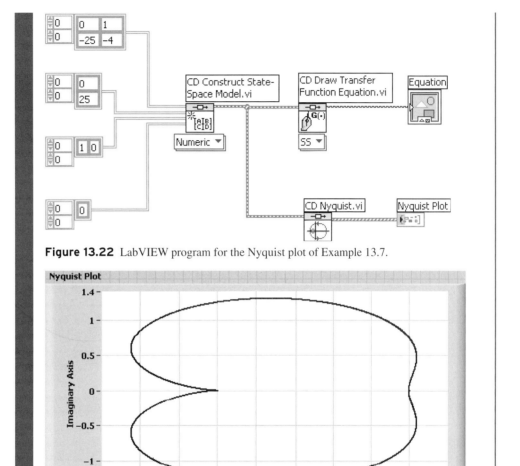

Figure 13.22 LabVIEW program for the Nyquist plot of Example 13.7.

Figure 13.23 Nyquist plot using LabVIEW (Example 13.7).

Example 13.8: Computing phase and gain margins using LabVIEW–A satellite-positioning system is modeled by the following fourth order transfer function, with the angular position being the output. Using LabVIEW, draw the Bode plots for the system shown in Fig. 13.24, with open-loop transfer function $G(s)$ given by

$$G(s) = \frac{20(s + 1)}{s(s + 5)(s^2 + 2s + 10)} \tag{13.36}$$

$$\frac{20(s + 1)}{s(s + 5)(s^2 + 2s + 10)}$$

$$G(s)$$

Figure 13.24 Satellite-positioning system with unity feedback (Example 13.8).

Determine also the system's phase margin and gain margin.

Solution: The LabVIEW VI program and the corresponding Bode diagrams are shown in Figs. 13.25 and 13.26, respectively.

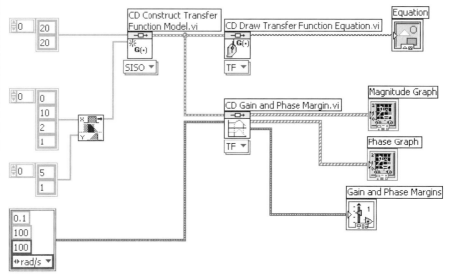

Figure 13.25 LabVIEW program for stability margin computation for Example 13.8.

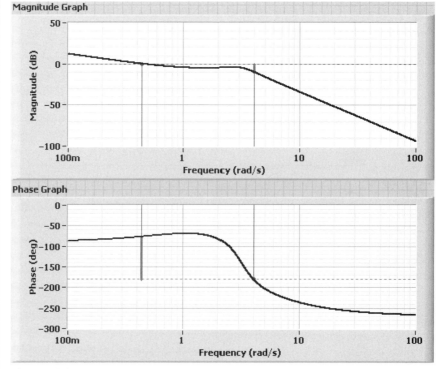

Figure 13.26 Bode plots with stability margins using LabVIEW (Example 13.8).

The corresponding phase margin, gain margin, phase crossover frequency, and gain crossover frequency are:

G.M. Frequency	Gain Margin
4.0148	9.9404

P.M. Frequency	Phase Margin
0.4426	103.6573

13.2.6 Lag-Lead Compensator

Control systems are designed to perform specific desired tasks. The requirements imposed are usually spelled out as performance specifications of the controlled process. The desired specifications may be given in terms of time domain performance measures, such as rise time, maximum overshoot, and settling time, or frequency domain performance measures, such as phase margin, gain margin, resonant peak value, and bandwidth. The design is usually based on root locus and/or frequency-response methods, as discussed earlier.

In designing a control system to alter the system response, an additional component is inserted within the structure of the feedback system. This additional block is the unit that compensates for the system performance deficiency. Various configurations exist for compensation. Among these are series, feedback, and a combination of both. In series compensation, the compensator is placed ahead of the plant. In feedback configuration, the controller is placed within the feedback path. Because feedback reduces the effect of parameter variations with respect to the elements in the forward path, the series configuration has better sensitivity properties. The series configuration, shown in Fig. 13.27, has traditionally been more popular.

The simplest and most common form of compensation is a transfer function with one zero and one pole. The transfer function of this compensator takes the general form

$$G_c(s) = K \frac{s + a}{s + b} \tag{13.37}$$

When $a > b > 0$, the corner frequency associated with the pole is smaller than that of the zero (hence an overall negative phase contribution), yielding a lag in the phase, and $G_c(s)$ is a lag compensator. When the order of corner frequencies of the zero occurs before that of the pole $(0 < a < b)$, $G_c(s)$ is known as a lead compensator, contributing a positive phase to the controlled system Bode plot. The maximum compensator phase contribution, lead or lag, occurs at a frequency in between the pole and the zero, specifically at $\omega = \sqrt{ab}$.

The choice of a lead or lag compensator depends on the desired system specifications. As an example, if the stabilization of the closed-loop system is the only requirement, any of the foregoing configurations can be used. Additional steady-state specifications may lead to the selection of one type of compensation over the other. For example, specifying desired phase and gain margins or a desired gain crossover frequency (where gain is 0 dB) ω_c will immediately determine

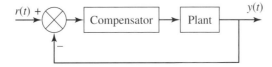

Figure 13.27 Block diagram of a series compensator.

the compensation type. If the plant ω_c is smaller then the desired ω_c, then lead compensation is needed. These outcomes can be deduced by examining the plant open-loop Bode plot.

Bode plot–based design techniques are used basically to restructure the open-loop transfer function of the plant. This change is supposed to deliver a desirable gain crossover frequency (for speed of response), a desirable low-frequency gain (for steady-state error or disturbance rejection properties), and adequate stability margins. On the other hand, the root locus design is based on re-shaping the root locus of the system through adding poles and zeros to the plant. This should force the loci to pass through a desired point (desired closed-loop pole location) in the complex plane.

The stability, steady-state error, and bandwidth of the closed-loop systems are affected by compensation. In general, lead compensation tends to increase the phase margin, which in turn increases the relative stability of the system and improves (makes faster) the transient response. Usually the steady-state error is decreased when using a lag compensator. Lag compensators also decrease the gain crossover frequency, ω_c, resulting in a slower step re-sponse settling time (i.e., increasing system damping). However, this results in a decrease in the closed-loop bandwidth, and a system with high disturbance rejection.

Example 13.9: Lead compensator design using MATLAB–The DC motor controlling the gripper of a manipulator was modeled in Example 13.1. Design a lead compensator for the system to achieve improved transient recovery from a disturbance.

Solution: The root locus method described earlier will be used for determining the locations of all closed-loop poles from a knowledge of the locations of the open-loop poles and zeros as the multiplying factor gain is varied. The root locus plot of the system indicates that a simple change in the gain of a controller will not be sufficient to get the desired performance. In this case, it is necessary to compensate the plant with a series transfer function or a compensator. Adding a pole to the system will lower the settling time. The addition of a zero to the open-loop transfer function has the effect of making the system more stable and speeding up the settling of the re-sponse. This is equivalent to a derivative control action. Lead compensation will yield an appre-ciable improvement in transient response and a small change in the steady-state accuracy.

The transfer function of the motor being controlled as obtained experimentally by frequency domain techniques in Example 13.1 is

$$G(s) = \frac{672}{s(s + 6.07)} \tag{13.38}$$

The unity-feedback closed-loop uncontrolled transfer function is

$$G(s) = \frac{672}{s^2 + 6.07s + 672} \tag{13.39}$$

The natural frequency, damping factors, and poles of the closed-loop transfer function are $\omega_n = 25.92$, $\xi = 0.1177$, and $p_i = -3.05 \pm j25.74$, respectively. The corresponding step response transient specifications are: final value = 1, time to peak = 0.125 s, overshoot = 68.7%, rise time = 0.0417 s, and settling time = 1.236 s.

If the desired overshoot of the controlled system is below 20% and the required settling time is less than 1 second, then the corresponding required parameters will be

$$\text{Overshoot} = (1 - \xi/0.6) \times 100 = 20 \Rightarrow \xi_{desired} = 0.48 \tag{13.40}$$

$$\text{SettlingTime} = (4.6/(\xi\omega_n)) \Rightarrow \omega_n|_{desired} = 9.58 \text{rad/s} \tag{13.41}$$

The design of the gain, the pole, and the zero locations of the lead compensator is discussed. The general form of the lead compensator is given in this equation:

$$G_c(s) = \frac{K_c(s + 1/T)}{s + 1/\alpha T}$$

(13.42)

1. The first step of the design is to determine the desired locations of the dominant second-order closed-loop poles:

$$\text{Desired characteristic polynomial} = s^2 + 9.2s + 92$$
$$\Rightarrow \text{Desired Roots are } -4.6 \pm 8.4j$$

(13.43)

2. In the complex plane, the damping ratio ξ of a pair of complex poles can be expressed in terms of the angle θ measured from the $j\omega$ axis to the segment joining the pole to the origin $\xi = \sin\theta$. Thus, lines of constant damping ratio are radial lines passing through the origin. The desired damping ratio is equal to 0.48, which implies that θ is equal to 29°.

3. Then find the sum of the angles at the desired location of one of the dominant closed-loop poles and zeros of the original system, and determine the necessary angle ϕ to be added so that the sum of all angles is $180°(2k + 1)$. The lead compensator must contribute this angle ϕ. In this system, the angle of $G(s)$ (open-loop transfer function) at the desired closed-loop pole is:

$$G(s) = \frac{672}{s(s + 6.07)}\bigg|_{-4.6+8.4j} = -7.785 + 2.646j$$

$$\angle -7.785 + 2.646j = -200°$$

(13.44)

4. Then a horizontal line passing through point P is drawn, the desired location for one of the dominant closed-loop poles. This is shown as PA in Fig. 13.28. Next, the bisect is drawn of the angle between PA and the segment joining P to the origin. Finally, two lines are drawn: PC and PD, which make angles $\phi/2$ with the bisector from each side. The intersection of PC

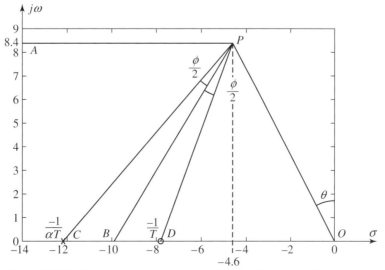

Figure 13.28 Lead pole and zero in the s-plane for Example 13.9.

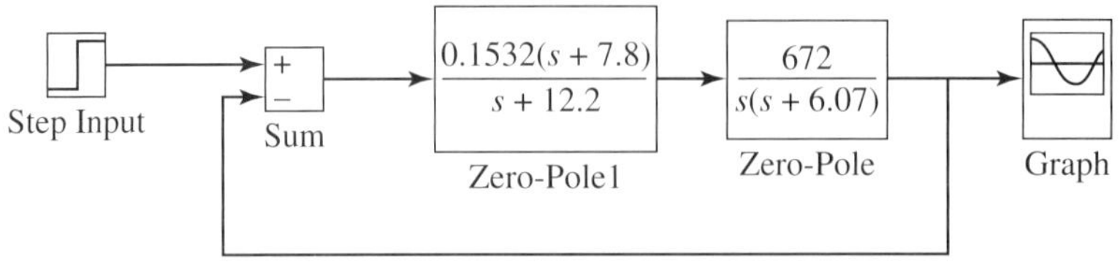

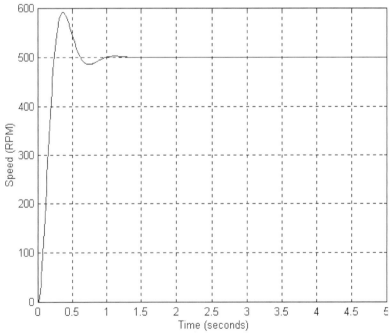

Step Input Sum $\dfrac{0.1532(s + 7.8)}{s + 12.2}$ $\dfrac{672}{s(s + 6.07)}$ Graph

Zero-Pole1 Zero-Pole

Figure 13.29 Block diagram of the lead compensator for Example 13.9.

and PD with the negative real axis gives the necessary location for the pole and zero of the lead compensator as

$$1/T = 7.8 \Rightarrow T = 0.128 \qquad (\text{zero}) \tag{13.45}$$

$$1/\alpha T = 12.2 \Rightarrow \alpha = 0.082 \qquad (\text{pole}) \tag{13.46}$$

The designed compensator will locate point P on the root locus of the compensated system.

5. Finally, the open-loop gain of the compensated system is determined from the magnitude condition. The magnitude of $G(s)G_c(s)$ evaluated at the desired pole location P will be equal to 1, and thus the gain of the lead compensator is calculated to be 0.1532. The transfer function of the lead compensator is

$$G_c(s) = \frac{0.1532(s + 7.8)}{s + 12.2} \tag{13.47}$$

Figures 13.29 and 13.30 show the SIMULINK block diagram and simulation results, respectively.

Figure 13.30 Simulation result of the robot gripper DC motor with a lead controller for Example 13.9.

Example 13.10: Lead controller design using LabVIEW–A machine tool is modeled by the following second-order system; it is typically controlled to track a predetermined desired path. Using LabVIEW, design a lead compensator for the system shown in Fig. 13.31 to achieve a static velocity error constant $K_v = 20\ \mathrm{s}^{-1}$, a phase margin at least $50°$, and a gain margin of at least 10 dB.

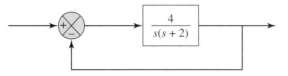

Figure 13.31 Machine tool model with unity feedback (Example 13.10).

Solution: The lead compensator has the form

$$G_c(s) = K_c\alpha\,\frac{Ts + 1}{\alpha Ts + 1} = K_c\,\frac{s + 1/T}{s + 1/\alpha T} \tag{13.48}$$

The compensated system will have the open-loop transfer function $G_c(s)G(s)$. Define

$$G_1(s) = KG(s) = \frac{4K}{s(s + 2)} \tag{13.49}$$

where $K = K_c\alpha$. The first step in the design is to adjust the gain K to meet the required static velocity error constant. Since the constant is given as $20\ \mathrm{s}^{-1}$,

$$K_v = \lim_{s\to0} sG_c(s)G(s) = \lim_{s\to0} s\,\frac{Ts + 1}{\alpha Ts + 1}G_1(s) = \lim_{s\to0}\frac{s4K}{s(s + 2)} \tag{13.50}$$

$$= 2K = 20 \Rightarrow K = 10$$

A value of $K = 10$ satisfies the steady-state error requirement. The Bode diagram of

$$G_1(j\omega) = \frac{40}{j\omega(j\omega + 2)} = \frac{20}{j\omega(0.5j\omega + 1)} \tag{13.51}$$

is shown in Fig. 13.32, using the LabVIEW program.

The phase and gain margins of the system with design constant gain $K = 10$ are $17°$ and $+\infty$ dB, respectively. The specification calls for a phase margin of at least $50°$. Thus an additional phase lead of at least $33°$ is necessary to satisfy the relative stability requirement.

The side effect of adding a lead compensator is the modification in the magnitude curve in the Bode diagram; in other words, the gain crossover frequency will be shifted to higher frequency values corresponding to an even smaller phase margin computation. To offset the additional phase lag of $G_1(j\omega)$ due to this increase in the gain crossover frequency, the maximum phase lead required from the controller, φ_m, is increased to $38°$ instead of $33°$.

Since

$$\sin\varphi_m = \frac{1 - \alpha}{1 + \alpha} \tag{13.52}$$

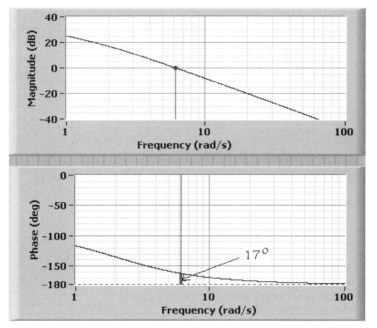

Figure 13.32 Bode diagram for $G_1(j\omega)$ for Example 13.10.

$\phi_m = 38°$ corresponds to $\alpha = 0.24$. The maximum phase-lead angle φ_m occurs at the geometric mean of the two corner frequencies of the controller, or $\omega = 1/(\sqrt{\alpha}T)$. The amount of the modification in the magnitude curve at $\omega = 1/(\sqrt{\alpha}T)$ due to the inclusion of the term $(Ts + 1)/(\alpha Ts + 1)$ is

$$\left| \frac{1 + j\omega T}{1 + j\omega\alpha T} \right|_{\omega = 1/(\sqrt{\alpha}T)} = \left| \frac{1 + j(1/\sqrt{\alpha})}{1 + j\alpha(1/\sqrt{\alpha})} \right| = \frac{1}{\sqrt{\alpha}} \tag{13.53}$$

Note that

$$\frac{1}{\sqrt{\alpha}} = \frac{1}{\sqrt{0.24}} = \frac{1}{0.49} = 6.2\,\text{dB} \tag{13.54}$$

And $|G_1(j\omega)| = -6.2\,\text{dB}$ corresponds to $\omega = 9$ rad/sec. At this frequency, it is most beneficial to measure the controlled system new phase margin; hence at this frequency the gain is forced to become zero, and this frequency shall become the new gain crossover frequency ω_c:

$$\frac{1}{T} = \sqrt{\alpha}\omega_c = 4.41 \qquad \text{and} \qquad \frac{1}{\alpha T} = \frac{\omega_c}{\sqrt{\alpha}} = 18.4 \tag{13.55}$$

Hence T is determined, and the lead compensator is

$$G_c(s) = K_c \frac{s + 4.41}{s + 18.4} = K_c\alpha \frac{0.227s + 1}{0.054s + 1} \tag{13.56}$$

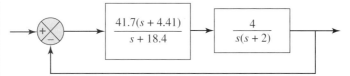

Figure 13.33 Machine tool system with a lead compensator for Example 13.10.

where the value of K_c is

$$K_c = \frac{K}{\alpha} = \frac{10}{0.24} = 41.7 \tag{13.57}$$

Hence the transfer function of the compensator is

$$G_c(s) = 41.7 \frac{s + 4.41}{s + 18.4} = 10\frac{0.227s + 1}{0.054s + 1} \tag{13.58}$$

Note that

$$\frac{G_c(s)}{K} G_1(s) = \frac{G_c(s)}{10} 10\, G(s) = G_c(s)\, G(s) \tag{13.59}$$

The compensated system has the following open-loop transfer function:

$$G_c(s)\, G(s) = 41.7 \frac{s + 4.41}{s + 18.4} \frac{4}{s(s + 2)} \tag{13.60}$$

The overall system shown in Fig. 13.33 meets the steady-state and relative-stability requirements.

The unit-step response and the unit-ramp response curves of the compensated and uncompensated systems will be determined to assess the response characteristics of the designed system. The LabVIEW programs shown in Fig. 13.34 generate the unit-step response

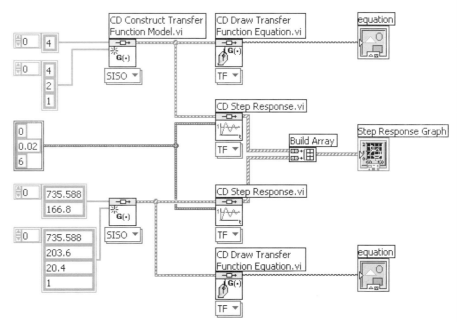

Figure 13.34 LabVIEW block diagram for Example 13.10.

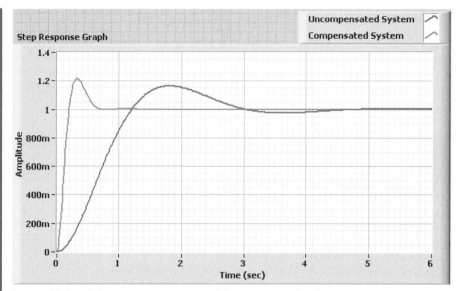

Figure 13.35 Unit-step response curves of the compensated and uncompensated systems using LabVIEW for Example 13.10.

and the unit-ramp response curves before and after compensation, as given in Figs. 13.35 and 13.36. The response curves indicate that the designed system is satisfactory.

Note that the achieved closed-loop poles for the compensated system are located at:

$$s = -6.9541 \pm j8.0592$$
$$s = -6.4918$$

(13.61)

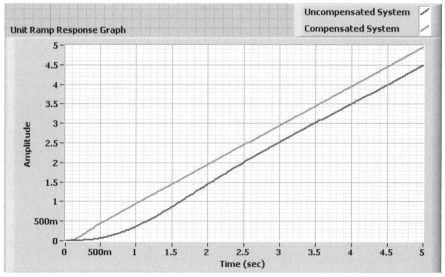

Figure 13.36 Unit-ramp response curves of the compensated and uncompensated systems using LabVIEW for Example 13.10.

13.2.7 Proportional-Integral-Derivative (PID) Controller Design

PID controllers are by far the most widely used controllers in mechatronics projects. The general transfer function of this compensator that uses the error signal between the actual response and the desired response is of the form

$$G_c(s) = K_P + \frac{K_I}{s} + K_D s \tag{13.62}$$

where K_P, K_D, and K_I are, respectively, the proportional, derivative, and integral gains.

The PID controller can take several forms: proportional only P, $K_D = K_I = 0$; proportional plus integral (PI), $K_D = 0$; proportional plus derivative (PD), $K_I = 0$; or full PID. The simplest control scheme is the proportional-only control. However, this one-degree-of-freedom structure allows designers to satisfy directly only one closed-loop specification, such as GM, PM, or steady-state error, among others. The addition of derivative control increases the response speed in the closed-loop system, while integral control increases the system type and, hence, decreases the steady-state error.

In the proportional control of a plant whose transfer function does not possess an integrator $1/s$, there is steady-state error in the response to a step input. It can be eliminated if the integral control action is included in the controller. Note that even though integral control action removes steady-state error, it may lead to an oscillatory response of slowly increasing amplitudes. Combining derivative and proportional control actions provides a controller with high sensitivity. An advantage of using derivative control action is that it responds to the rate of change of the actuating error and can produce a significant correction before the magnitude of the actuating error becomes too large. Thus, derivative control anticipates the actuating error, initiates an early corrective action, and tends to increase the stability of the system. Because derivative control operates on the rate of change of the actuating error and not on the actuating error itself, it cannot be used alone. Thus, it is always used in combination with proportional or proportional and integral action. In all cases, it is possible to force an actual characteristic polynomial to a desired polynomial reflecting desired poles or specs.

A PID controller can be designed to meet the transient and steady-state specifications via root locus– and Bode plot–based design techniques only if a mathematical model of the plant is available. However, if the plant is complicated, a mathematical model cannot easily be derived, and an analytical approach to the design of a PID controller is not possible. Experimental methods can be used. Ziegler and Nichols developed a method for tuning the three gain parameters of the PID controller. This method was based on a simple stability analysis. The procedure is summarized as follows. First, set $K_D = K_I = 0$, and increase the proportional gain until the system oscillates (i.e., the closed-loop poles on the $j\omega$ axis). The gains are then calculated by

$$K_P = 0.6 K_o \qquad K_D = \frac{\pi K_P}{4\omega_o} \qquad K_I = \frac{K_P \omega_o}{\pi} \tag{13.63}$$

where K_o is the proportional gain at which the system oscillates and ω_o is the oscillation frequency.

As a conclusion, it is advisable to consider classical control before addressing more advanced control methods, which usually result in more complex controllers. Time-varying, nonlinear, or unknown systems that cannot be adequately stabilized and desirably controlled using classical methods discussed in this section will be controlled by state-space, adaptive, or intelligent control methods.

Example 13.11: PID controller design using MATLAB—Design a PID controller for the robot loaded gripper DC motor described earlier by Eq. (13.3) from Example 13.1 in order to achieve in a step response a maximum overshoot of 20% and a settling time of less than 1 second.

Solution: Since the mathematical model $G(s)$ of the plant was experimentally approximated in Example 13.1, an analytical solution can be derived. For the required performance,

$$\text{Overshoot} = \left(1 - \frac{\xi}{0.6}\right) \times 100 = 20 \Rightarrow \xi_{desired} = 0.48 \tag{13.64}$$

$$\text{Settling time} = \left(\frac{4.6}{\xi\omega_n}\right) \Rightarrow \omega_n = 9.58 \text{ rd/s} \tag{13.65}$$

Hence,

$$\text{Desired characteristic polynomial} = s^2 + 9.2s + 92 \Rightarrow \text{Roots} = -4.6 \pm 8.4j$$

For a proportional controller with gain K_P, the closed-loop transfer function is given by

$$\frac{K_P G(s)}{1 + K_P G(s)} = \frac{672 K_P}{s^2 + 6.07s + 672 K_P} \tag{13.66}$$

Equating the desired-characteristic polynomial to the designed counterpart yields a proportional gain equal to 0.13. Figure 13.37 shows the block diagram of the closed loop P controller with the DC motor. Figure 13.38 shows the corresponding simulation results. Figure 13.39 shows the output speed for certain values of K_P, K_I and K_D.

It is clear that the addition of the derivative action increased the overshoot but improved the settling time and that the addition of the integral action increased the damping of the output.

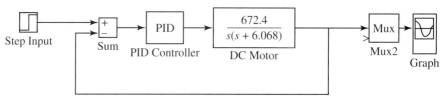

Figure 13.37 Block Diagram of PID-controlled DC motor for Example 13.11.

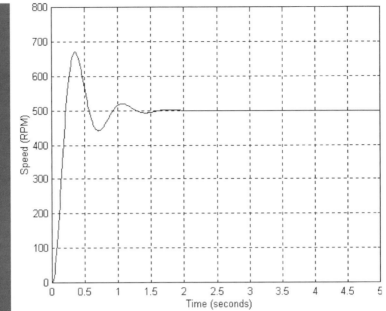

Figure 13.38 Simulation result of the speed response of the DC motor with a P controller.

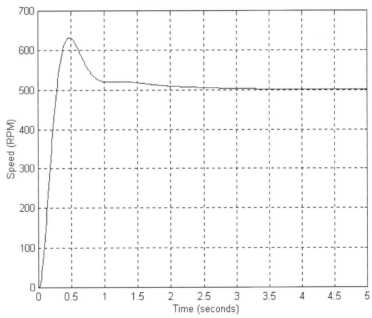

Figure 13.39 Speed response of the DC motor with a PID controller.
$K_P = 0.1; K_I = 0.1; K_D = 0.005$

Example 13.12: PID controller design using LabVIEW–A simplified linear model that describes the dynamics of a plane used in autopilot design consists of a third-order transfer function that relates the aileron deflection to the bank angle (Fig. 13.40).

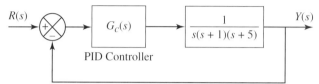

Figure 13.40 Autopilot plane model with PID controller, for Example 13.12.

The proposed PID controller has the general-purpose transfer function

$$G_c(s) = K_p\left(1 + \frac{1}{T_i s} + T_d s\right) \tag{13.67}$$

Apply the experimental technique of the Ziegler–Nichols tuning rule for the determination of the values of parameters K_p, T_i, and T_d. Once the PID design is selected, use LabVIEW to supply a step response of the controlled system

Solution: It is noted that the plant has a built-in integrator, so the experimental second method of Ziegler–Nichols tuning rule is recommended. Specifically, set $T_I = \infty$ and $T_D = 0$, and obtain the closed-loop transfer function (with P controller):

$$\frac{Y(s)}{R(s)} = \frac{K_p}{s(s+1)(s+5) + K_p} \tag{13.68}$$

Sustained oscillation occurs when the characteristic polynomial has roots on the jw axis. The corresponding roots can be obtained using Routh's stability criterion. Since the characteristic equation for the closed-loop system is

$$s^3 + 6s^2 + 5s + K_p = 0 \tag{13.69}$$

the Routh table becomes:

$$
\begin{array}{ccc}
s^3 & 1 & 5 \\
s^2 & 6 & K_p \\
s^1 & \dfrac{30 - K_p}{6} & \\
s^0 & K_p &
\end{array}
\tag{13.70}
$$

From the coefficients of the first column of the table, we can conclude that sustained oscillation will occur if $K_p = 30$. Thus, the critical gain K_{cr} is 30, and the characteristic equation becomes

$$s^3 + 6s^2 + 5s + 30 = 0 \tag{13.71}$$

The sustained oscillation frequency w is obtained by substituting $s = j\omega$ into the characteristic equation

$$(j\omega)^3 + 6(j\omega)^2 + 5(j\omega) + 30 = 0$$

or

$$6(5 - \omega^2) + j\omega(5 - \omega^2) = 0 \tag{13.72}$$

The corresponding period of the sustained oscillation is

$$P_{cr} = \frac{2\pi}{\omega} = \frac{2\pi}{\sqrt{5}} = 2.8099 \tag{13.73}$$

Based on the Ziegler–Nichols method:

$$K_p = 0.6K_{cr} = 18$$
$$T_i = 0.5P_{cr} = 1.405 \tag{13.74}$$
$$T_d = 0.125P_{cr} = 0.35124$$

The obtained transfer function of the PID controller is

$$G_c(s) = K_p(1 + 1/T_i s + T_d s) = 18(1 + 1/1.405s + 0.34124s)$$
$$= \frac{6.3223(s + 1.4235)^2}{s} \tag{13.75}$$

The controlled system closed-loop transfer function $Y(s)/R(s)$ shown in Fig. 13.41 is given by

$$\frac{Y(s)}{R(s)} = \frac{6.3223s^2 + 18s + 12.811}{s^4 + 6s^3 + 11.3223s^2 + 18s + 12.811} \tag{13.76}$$

The LabVIEW block diagram and generated unit-step response are shown in Figs. 13.42 and 13.43, respectively.

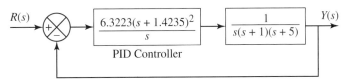

Figure 13.41 Block diagram of the plane model with PID controller (Example 13.12).

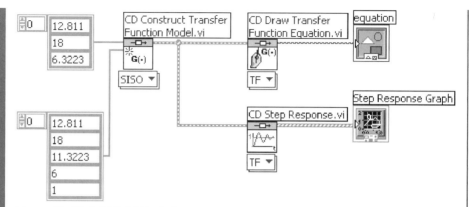

Figure 13.42 LabVIEW block diagram for Example 13.12.

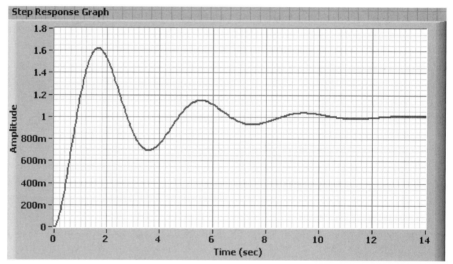

Figure 13.43 Unit-step response curve of a PID-controlled system design based on the Ziegler–Nichols tuning rule without iterations using LabVIEW for Example 13.12.

The achieved maximum overshoot in the unit-step response is large, about 62%. Fine-tuning the controller gains might improve the situation. After one iteration, keeping $K_p = 18$ and by moving the double zero of the PID controller to $s = -0.65$, we arrive at the new PID controller:

$$G_c(s) = 18\left(1 + \frac{1}{3.077s} + 0.7692s\right) = 13.846\frac{(s + 0.65)^2}{s} \tag{13.77}$$

The maximum overshoot in the unit-step response shown in Fig. 13.44 is reduced to about 18%.

In a second iteration, the proportional gain K_p is increased to 39.42, and the obtained PID controller is

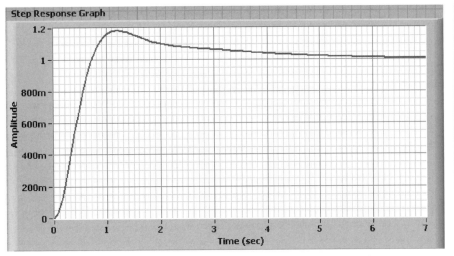

Figure 13.44 Unit-step response using LabVIEW with a PID controller having parameters $K_p = 18$, $T_i = 3.077$, and $T_d = 0.7692$ after the first iteration for Example 13.12.

$$G_c(s) = 39.42\left(1 + \frac{1}{3.077s} + 0.7692s\right) = 30.322\frac{(s + 0.65)^2}{s} \qquad (13.78)$$

This second iteration of PID gain settings speeds the response but increases the maximum over-shoot to about 28% (Fig. 13.45).

Hence, the tuned values of K_p, T_i, and T_d become, after two iterations,

$$K_p = 39.42, \qquad T_i = 3.077, \qquad T_d = 0.7692 \qquad (13.79)$$

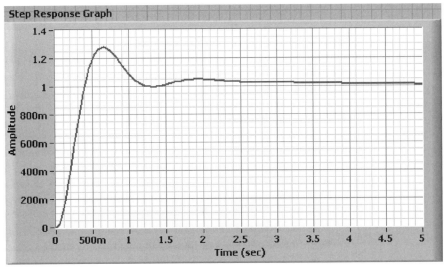

Figure 13.45 Unit-step response using LabVIEW with a PID controller having parameter. $K_p = 39.4$; $T_i = 3$; $T_d = 0.77$

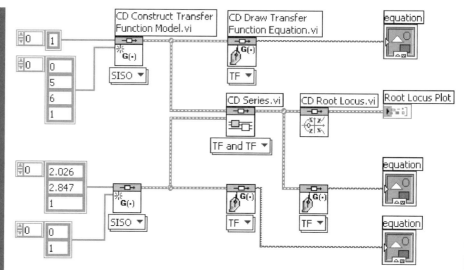

Figure 13.46 LabVIEW block diagram for the root locus for Example 13.12.

The second method of the Ziegler–Nichols tuning rule gives only suggested starting values for the gains (Fig. 13.46).

The root locus analysis in Fig. 13.47 corresponds to the system designed by the use of the second method of the Ziegler–Nichols tuning rule. Another root locus analysis is shown in Fig. 13.48.

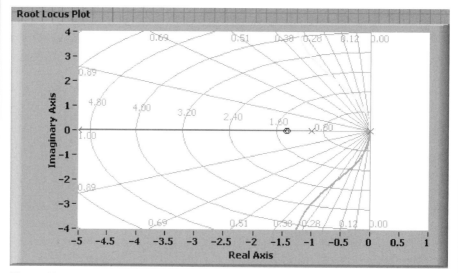

Figure 13.47 Root locus diagram with a double-zero PID controller (initial design) at $s = -1.42$.

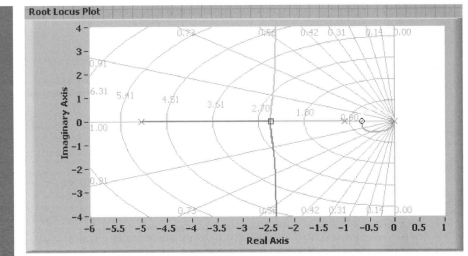

Figure 13.48 Root locus diagram of the system when the PID controller has double zero at $s = -0.65$, $K = 13.846$ corresponds to $G_c(s)$ using LabVIEW for Example 13.12.

13.3 STATE-SPACE-BASED CONTROL STRATEGIES

In classical control theory, the input–output relationship is described using a transfer function. Together with a variety of graphical techniques, such as root locus plots and Bode diagrams, the transfer function is used to analyze and design control schemes. Classical control techniques are simple and require a reasonable amount of computation. However, they are only applicable to linear time-invariant systems and mostly suitable for single-input, single-output plants. Therefore, they can not be used in the design of optimal and adaptive control systems, which are mostly time varying and/or nonlinear.

Modern control theory (state-space) utilizes the vector matrix analysis based in the time domain. Modern control theory founded on state-space representation of the control system compensates for the shortcomings of classical control. Using modern control techniques, it is possible to:

- Design control systems guaranteeing desired closed-loop poles.
- Design optimal control systems with respect to given performance indexes.
- Include nonzero initial conditions, if necessary.
- Design controllers for multiple-input, multiple-output (MIMO) systems
- Handle time-varing systems
- Design nonlinear controllers

Another advantage of state-space representation is the freedom of state variable choice. Variables do not need to represent physical parameters. Usually the large amount of mathematical manipulation in classical techniques tends to hide the physical interactions of the system elements. In state space, the states clearly reflect the status of the system parameters. The state-space model is a superior analytical tool. State space uses matrices in analysis and design, which allows the handling of large numbers of equations in an organized and efficient manner. All of the computations in a state model can be incorporated into a digital computer, which will make lengthy computations possible.

The state of a system at a time t_0 is the minimum set of internal variables which is sufficient to uniquely specify the system outputs given the input signal $[t_0, \infty)$. The state represents the effect of all past excitations and is fundamental in determining the future evolution of the system. The state model for a lumped linear time-invariant system is a set of four matrices A, B, C, and D, which define these first-order, degree-n vector differential equations:

$$\dot{x}(t) = Ax(t) + Bu(t)$$
$$y(t) = Cx(t) + Du(t) \tag{13.80}$$
$$x(t_0) = x_0$$

where $x(t)$ is the state vector $(n \times 1)$, $u(t)$ is the system input vector $(m \times 1)$, $y(t)$ is the system output vector $(r \times 1)$, A $(n \times n)$, B $(n \times m)$, C $(r \times n)$, and D $(r \times m)$ are real matrices, and $x(t_0)$ is the initial state vector or initial condition for Eq. (13.80).

Within the realm of all possible choices of state variables for a state model, a special set of choices leads to the so-called "canonical" state models. The most common of these are the controllable and the observable forms. In a controllable system, the poles of the system can be placed anywhere in the s-plane. The control signal is usually a function of the states. Sometimes, direct measurement of the states is not possible or, if measurable, will be noisy, and thus, an estimation of the states is needed. If a system is observable, then all the states can be estimated from knowledge of the inputs, outputs, and the A, B, C, and D matrices.

A system represented by its state model is said to be controllable, or, more commonly, reachable, if and only if there exists an unconstrained control $u(t)$ that can transfer any initial state $x(0)$ to any desired location $x(t)$. The test for *controllability* is: If rank $[B \; AB \; A^2B \; \ldots \; A^{n-1}B] = n$, then the system is said to be controllable.

A system is observable if and only if there exists a finite time t such that the initial state $x(0)$ can be determined from the observation history $y(t)$ given the control $u(t)$ as well as the pair (A, C). The system is observable if the following matrix has a rank of n:

$$\begin{bmatrix} C \\ CA \\ CA^2 \\ \vdots \\ CA^{n-1} \end{bmatrix} \tag{13.81}$$

Suppose a single-input, single-output process can be described by the state model of Eq. (13.80). In a control problem, the main objective is to determine a control function $u(t)$ such that the output $y(t)$ behaves in a predetermined desired manner.

A powerful control strategy is called the *state variable-feedback* strategy, where $u(t)$ can be calculated from a knowledge of the state vector $x(t)$. If the state vector $x(t)$ is not accessible or is corrupted by noise but is required for feedback, a state observer can be used to estimate it. When both the state vector and the model parameters are to be estimated, a combined state-and-parameter estimator can be used. By applying control techniques, any process can be given any desired closed-loop dynamic performance, within the limit of permissible input.

Assume a continuous process described by Eq. (13.80) is fully controllable. One can use state feedback to place the process poles anywhere in the complex plane. Called pole placement, this consists of replacing the input vector $u(t)$ by a linear combination of states, $-Kx(t)$, where K is a gain matrix to be designed. The closed-loop model becomes

$$\dot{x}(t) = Ax(t) - BKx(t) \tag{13.82}$$

This closed-loop model is a homogeneous vector differential equation yielding a regulator behavior or zero steady value of the solution $x(t)$. The equation can be Laplace transformed to yield

$$(sI - A + BK)X(s) = 0 \tag{13.83}$$

Suppose now that the closed-loop poles of the system are required to be at given desired locations $\alpha_1, \alpha_2, \ldots, \alpha_n$ in the s-plane. Control design then consists of choosing K so that

$$\det|sI - A + BK| = (s - \alpha_1)(s - \alpha_2)\cdots(s - \alpha_n)$$
$$= s^n + \beta_1 s^{n-1} + \beta_2 s^{n-2} + \cdots + \beta_n \tag{13.84}$$

We note that, for an nth-order process, a suitable matrix K always exists to allocate the poles to n arbitrary locations in the s-plane, provided the process is controllable. The algebra of finding the specific value of K is especially simple if the system matrices happen to be in the "controllable canonical form."

Most of the sophisticated and ambitious feedback controllers that can be designed using state-space techniques require knowledge of the state vector of the system that is to be controlled. Often the whole state vector is not available (inaccessible states), in which case an estimator is used to reconstruct the nonmeasurable states. Even when the state vector is available for measurement, it is more or less corrupted by noise and there is still a need for estimation of internal states of the system.

State-space methods are very powerful techniques, since they have expanded the range of problems that can be solved. Arbitrary pole placement is the main advantage of this technique, since it gives us a great deal of control over the time response of the system. Because the models of process are never perfect, state-space techniques may lack the robustness properties. Finally, if the state-space methods fail to give an acceptable response, other techniques, such as adaptive or optimal control, can be used.

Example 13.13: State-space control design using MATLAB—The DC motor of a robot gripper was experimentally determined in Example 13.1. Design a pole placement with a state feedback controller for a plant described by the transfer function given by

$$G(s) = \frac{Y(s)}{U(s)} = \frac{672}{s(s + 6.07)}$$

$$\Rightarrow s^2 Y(s) + 6.07sY(s) = 672U(s) \tag{13.85}$$

$$\Rightarrow \ddot{y}(t) + 6.07\dot{y}(t) + 0y(t) = 672u(t)$$

Solution: To transform this transfer function into a state-space model, let the first state be the speed of the motor $x_1 = y(t)$. Let the second state be the acceleration $x_2 = \dot{x}_1$,

$$\Rightarrow \dot{x}_2 = \ddot{x}_1 = \ddot{y} = 672u - 6.07x_2$$

Then the state-space model equivalent to $G(s)$ is given by this equation:

$$
\begin{bmatrix} \dot{x}_1 \\ \dot{x}_2 \end{bmatrix} = \begin{bmatrix} 0 & 1 \\ 0 & -6.07 \end{bmatrix} \begin{bmatrix} x_1 \\ x_2 \end{bmatrix} + \begin{bmatrix} 0 \\ 672 \end{bmatrix} u
$$
$$
y = \begin{bmatrix} 1 & 0 \end{bmatrix} \begin{bmatrix} x_1 \\ x_2 \end{bmatrix} + 0u
$$
(13.86)

This has the general form of Eq. (13.80). Now let the control signal u be a combination of the nonzero reference input u_c and a feedback of the states multiplied by a certain gain to give a desired performance according to

$$u \leftarrow Fx + u_c, \qquad F = \begin{bmatrix} f_1 & f_2 \end{bmatrix}$$
(13.87)

The closed-loop matrix becomes

$$\dot{x} = (A + BF)x + Bu_c$$

$$\dot{x} = \begin{bmatrix} 0 & 1 \\ 672f_1 & 672f_2 - 1 \end{bmatrix} *x + \begin{bmatrix} 0 \\ 672 \end{bmatrix} *u_c$$
(13.88)

$$\det(\lambda I - (A + BF)) = \det\left(\begin{bmatrix} \lambda & -1 \\ 672f_1 & \lambda + 6.07 - 672f_2 \end{bmatrix} \right)$$
$$= \lambda^2 + (6.07 - 672f_2)\lambda - 672f_1$$
(13.89)

In order to calculate matrix F, equate the preceding polynomial to the desired one. Assuming the desired $\xi = 0.5$ and the desired $\omega_n = 26$, the desired characteristic polynomial is:

$$\lambda^2 + 25.92\lambda + 672$$
$$f_2 = -0.03$$
$$f_1 = -1$$
$$u = u_c + f_1x_1 + f_2x_2$$
(13.90)

The MATLAB block diagram of the system is shown in Fig. 13.49 and the result of the simulation is shown in Fig. 13.50.

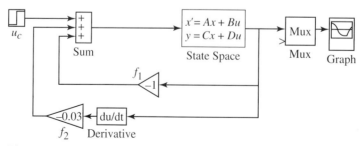

Figure 13.49 MATLAB block diagram of the state-space controller for Example 13.13.

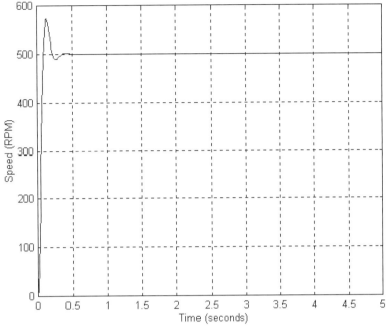

Figure 13.50 Speed response of the motor using MATLAB with a pole placement.

Example 13.14: Pole placement controller design using LabVIEW–The dynamics of a submarine are approximated by a third-order state model for controlling the depth:

$$\dot{x} = Ax + Bu$$

where

$$A = \begin{bmatrix} 0 & 1 & 0 \\ 0 & 0 & 0 \\ -1 & -5 & -6 \end{bmatrix}, \quad B = \begin{bmatrix} 0 \\ 0 \\ 1 \end{bmatrix} \quad (13.91)$$

Design using LabVIEW the gain vector **K** of the controller by using state feedback control $u = -Kx$ to deliver closed-loop desired poles at

$$\mu_1 = -2 + j4, \qquad \mu_2 = -2 - j4, \qquad \mu_3 = -10 \quad (13.92)$$

Solution: Determine the state feedback-gain matrix **K** with LabVIEW. Note here the use of the Tab control to switch between the cases where we used CD Ackermann VI and CD Pole Placement VI (Fig. 13.51).

The output of LabVIEW Program is a set of three gain values for the **K** vector that multiplies the state feedback vector **x**(*t*) (Figs. 13.52 and 13.53).

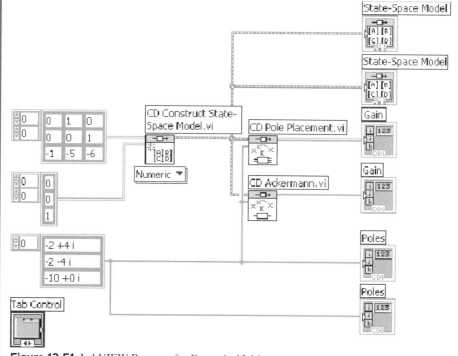

Figure 13.51 LabVIEW Program for Example 13.14.

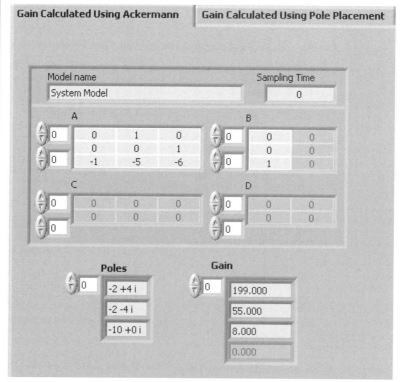

Figure 13.52 Output using CD Ackermann VI of LabVIEW pole-placement.

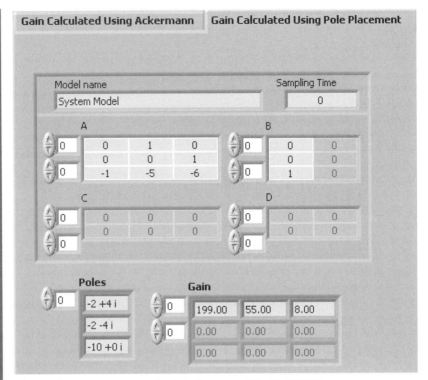

Figure 13.53 Output using CD Pole Placement VI of LabVIEW.

Note that the outputs of the two VI's are the same. But the CD Ackermann VI is used for SISO systems only, while CD Pole Placement can be used for MIMO systems.

13.4 ADAPTIVE CONTROL

In classical control systems, small deviations in system parameters from their nominal values will not alter the normal operations of the system. However, if plant parameters vary widely, the control may exhibit unsatisfactory response and may even cause instability. If we can anticipate the changing dynamic parameters and account for them, then we can compensate for the model variations by adjusting the controller gain parameters and thereby obtain satisfactory system performance under various environmental conditions. This strategy is "adaptive control."

An adaptive controller is thus a controller that can modify its behavior in response to changes in the dynamics of the process and the character of the disturbances. The system will have two loops; one is the regular feedback control loop and the other is the parameter adjustment loop. Several methods in adaptive control exist and are discussed in this section.

13.4.1 Gain Scheduling

It is well known that process dynamics may change according to the operating conditions of the process. Usually the nonlinear factors are the major contributors to the changes in the dynamics.

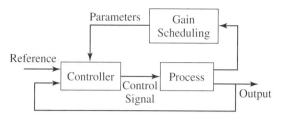

Figure 13.54 Block diagram of a system with gain scheduling.

By monitoring the process operating conditions, it is possible to change the parameters of the controller. In gain scheduling, the gain parameters of a linear controller are changed in a pre-programmed fashion as a function of the operating conditions or as a lookup table. Gain scheduling can thus be viewed as a feedback control system in which the feedback gains are adjusted. A block diagram of a system with gain scheduling is given in Fig. 13.54, which clearly shows two loops: the control feedback loop and the adaptive gain schedule loop.

Finding the suitable scheduling variables is a major problem in the design of systems with gain scheduling. When the scheduling variables are determined, the controller parameters can be calculated at a number of operating conditions by using any design scheme. The controller is thus tuned for each operating condition. The stability and performance of the system are typically evaluated by simulation.

One drawback of gain scheduling is that it is open-loop compensation. There is no feedback to compensate for an incorrect schedule. Another drawback of this technique is that the off-line design may be time consuming. The controller parameters must be determined for many operating conditions, and extensive simulations must check the performance. The advantage of gain scheduling is in the controller's fast response. This follows from not requiring any parameter estimation in the process.

13.4.2 Model-Reference Adaptive Control (MRAC)

Model-reference adaptive control (MRAC) is a popular adaptive controller in which the desired performance is expressed in terms of a reference model. The reference model produces the desired output for a given reference input. A block diagram of the system is shown in Fig. 13.55. The system has an inner loop, composed of the process and the ordinary fixed

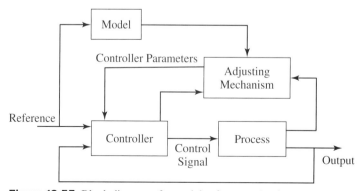

Figure 13.55 Block diagram of a model-reference adaptive system.

controller, and an outer loop, which changes the controller parameters. The adjusting mechanism is based on the difference between the output of the process and that of the reference model. The mechanism for adjusting the parameters in a model-reference adaptive system can be obtained in two ways: by using a gradient method, such as the MIT rule, or by applying stability theory, such as Lyapunov.

Consider a closed-loop system in which the controller has one adjustable gain parameter θ. The desired closed-loop response is specified by a model whose output is y_m. Let e be the error between the process output y of the closed-loop system and the output y_m of the model. One possibility is to adjust controller parameters in such a way that the loss function

$$J(\theta) = \frac{1}{2}e^2 \tag{13.93}$$

is minimized. To make J small, it is reasonable to change the parameters in the direction of the negative gradient of J:

$$\frac{d\theta}{dt} = -\gamma\frac{\partial J}{\partial \theta} = -\gamma e\frac{\partial e}{\partial \theta} \tag{13.94}$$

which is the MIT rule. The partial derivative $\partial e/\partial \theta$, which is called the *sensitivity derivative* of the system, tells how the error is influenced by the adjustable gain parameter.

13.4.3 Self-Tuning Regulators

In the adaptive schemes discussed earlier, the adjustment rules updated the controller parameters directly. These are called *direct* adaptive methods. In *indirect* adaptive methods, the estimation of the process parameters is done first. Then using the estimated dynamic parameters, the controller parameters are calculated. A block diagram of such a system is shown in Fig. 13.56. This control strategy is called a self-tuning regulator (STR). The STR can be thought of as being composed of two loops: the inner loop, which consists of the process and a feedback controller, and the outer loop, which estimates process parameters and adjusts the controller gains. The outer loop is composed of a recursive system-parameter estimator and a design calculation mechanism.

In the STR, the process parameters are estimated in real time and used in controller design. This is called the *certainty equivalence principle*. In brief, the STR attempts to automate the process of controller design.

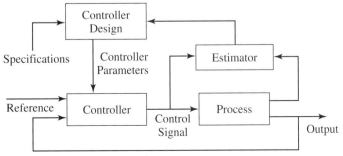

Figure 13.56 Block diagram of a self-tuning regulator.

13.5 DIGITAL CONTROL

Throughout this chapter, the design of the controller resulted in a compensation transfer function or a gain vector multiplying feedback states. The implementation or realization of such schemes can be analog through electrical, electronic, pneumatic, or mechanical components and circuits. However, in mechatronics and other applications, the microprocessor algorithm or subroutine with the adequate interfacing components composes the controller of the systems. This has become more common due to the affordability, packaging, and reliability of ICs. Robots, aircraft autopilots, oil refineries, countless electromechanical servomechanisms, various aspects of automobiles, and household appliances are among the many existing examples. A block diagram of a single-input, single-output digital control system is shown in Fig. 13.57. The digital computer receives the error in digital form. It performs the calculations and provides a digital output. The sensed data is converted from analog form to digital representation by means of the A/D converter. The control signal is in digital form and is converted to an analog equivalent signal using a D/A converter.

One advantage of a digital controller is that it can handle multiple-input, multiple-output systems easily, with flexibility of the control programs, whose complex formulas and control strategies are easy to implement. Furthermore, in digital control, drifts in components do not affect performance. Finally, it is much easier to implement necessary changes and to upgrade digital controllers than with the hardwired analog counterparts.

13.5.1 Discretization Techniques

A system is modeled using either transfer functions or state models. To analyze, design, and implement digital controllers, discretization is used, in one of two ways. The process is discretized from the point of view of the microprocessor, which issues samples out and receives samples in. Hence the design is done in the discrete domain. Or if the approximation is taken from the point of view of the continuous process, which is stimulated by analog inputs and generates corresponding continuous response outputs, then the design is done in the continuous time domain and, prior to implementation, the analog controller is discretized as an algorithm. Many techniques exist for discretization, some of which are presented here: the zero-order hold, the bilinear transform, and the zero-pole mapping.

Zero-order hold (ZOH) is the simplest and most commonly used D/A technique. In ZOH circuitry a D/A device holds the input sample constant for the period T duration, called the *discretization period*. The output will be the analog equivalent of the input.

Bilinear transform is one of the various numerical integration–based methods. In these techniques, s in $G(s)$ is replaced by a function of z. The *bilinear transformation*, also known

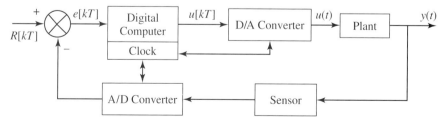

Figure 13.57 Block diagram of a basic digital control system.

as the trapezoidal or Tustin's transformation, will map the left half s-plane into the unit circle of the z-plane.

The *zero pole mapping* technique is based on mapping the poles and zeros of the transfer function $G(s)$ in the s-domain to the z-domain via the transformation $z = e^{sT}$. Note that zeros of $G(s)$ at ∞ should be represented as a zero at $z = -1$ in the equivalent $G(z)$.

All discretization techniques exhibit approximation errors. As in the continuous control systems, compensators should be designed for digital control systems to achieve a desired response. The compensator is then realized as a set of difference equations programmed into the computer. Design techniques for digital controllers parallel those for continuous ones.

13.5.2 Emulation

The first step in this method is to design a compensator using the classical or modern control design techniques in the continuous domain that were discussed in previous sections. Then using one of the various discretization methods, the designed controller is converted to its digital equivalent. At very high sampling rates, these techniques have proven very effective. However, at low sampling rates, this conversion usually creates problems, since the approximation errors introduced by the discretization process grow large with increased sampling periods. The advantage of this technique is that it can use all the knowledge acquired in analog control design.

13.5.3 Direct Digital Control

Direct digital control analysis and design are based in the discrete domain. The first step is to transform the plant model into its z-domain equivalent transfer function, and then to design a digital controller using the root locus, frequency, or state-space techniques.

The proportional-integral-derivative (PID) controller is as effective in digital systems as it is in continuous systems. One discretized equivalent PID controller in the z-domain is given by

$$G(z) = K_p + K_D \frac{(z - 1)}{zT} + K_I \frac{zT}{z - 1} \tag{13.95}$$

where $K_{(s)}$ are design gains to be determined and T is the constant sampling period, in seconds. Discrete lead or lag compensators can be designed. A discrete root locus technique follows the same mathematics as the continuous root locus. One difference is that the stability focus is now on the inside of the unit circle in the z-plane rather than on the left half of the s-plane.

The nth-order state-space model for a discrete plant is given by

$$\begin{aligned} x(k + 1) &= A_d x(k) + B_d u(k) \\ y(k) &= C_d x(k) + D_d u(k) \end{aligned} \tag{13.96}$$

where $k = kT$ is the kth sample, with a sampling period of T s, and the plant poles are the eigenvalues of A_d. Pole-placement techniques can be applied to discrete systems the same way they were applied to continuous ones. The control input is given by

$$u(k) = -F_d x(k) \tag{13.97}$$

The plant is discretized, and the desired poles are chosen inside the unit circle.

Various controllers will be implemented digitally in later sections containing simulation and experimental examples.

13.6 INTELLIGENT CONTROL

In the last few decades, several modern control techniques, such as state space, optimal, and adaptive controllers, have been developed that provide very good performance as compared to classical PID regulators. The design of effective adaptive control is based on mathematical modeling and is usually complex due to the computationally intensive algorithms.

Intelligent control is a classification given to control strategies that were invented, in particular, for solving control problems that could not be approached previously in any practical way. Most often, these control strategies try to model human behavior and characteristics. At present, control engineers are using intelligent instead of conventional control to get a more desirable performance from their systems. What is meant by conventional control is the type of control that once was called "modern control" and, upon the introduction of intelligent algorithms, has now been labeled "conventional."

The goal of intelligent process control is to emulate the behavior and structure of human biological systems. One of the most powerful tools, which can convert linguistic control rules based on expert knowledge, is fuzzy logic control. The knowledge is simply paraphrased in a set of IF-THEN rules together with an input-and-output interface to the real world. Fuzzy logic, based on the fuzzy set theory first formulated by Lotfi Zadeh in 1965, is a mathematical theory combining multivalued logic, probability theory, and artificial intelligence. Its design philosophy deviates from all previous methods by accommodating expert knowledge in controller design. Recently it has found wide popularity in various applications. Fuzzy logic has been successfully applied in the identification, modeling, and control of dynamic systems in different domains. In control systems, it can be considered an alternative to the conventional approach in the control of complex nonlinear plants, where process mathematical modeling is difficult or impossible. Fuzzy logic control is used in two application fields: in the design of controllers for ill-known systems, and in the design of nonlinear controllers for modeled systems.

13.6.1 Fuzzy Logic Control Design

When designing a fuzzy logic controller (FLC), the linguistic variables of the inputs and outputs of the dynamical system are created. With these variables, the membership function for each input/output is constructed. The membership function is the interface between the fuzzy logic scheme and the control process. The heart of the fuzzy controller is called the *fuzzy inference engine* (IF-THEN rules). Finally, a defuzzification technique will work on the output membership function to generate a crisp output value. A block diagram of the complete system is given in Fig. 13.58, which clearly shows three principal components:

1. A fuzzification interface that quantizes the inputs into a corresponding universe of discourse. The quantized input data are then converted into suitable linguistic variables, which may be viewed as labels of fuzzy sets. A membership function can take a bell, triangular, or trapezoidal shape. The choice is dependent on user preference. The same set of linguistic terms is assigned to the variables: positive small (PS), positive medium (PM), positive large (PL), zero (ZE), negative small (NS), negative medium (NM), and negative large (NL).

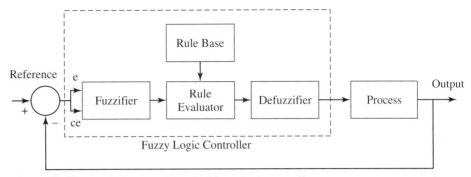

Figure 13.58 Block diagram of a typical fuzzy logic controller.

2. Knowledge base fuzzy control rules and an inference engine are developed based on expert experience and control engineering knowledge (Table 13.2)
3. A defuzzification interface is the mapping from a space of fuzzy control actions defined over an output universe of discourse into a space of nonfuzzy crisp control actions. The defuzzification strategy is aimed at producing a nonfuzzy control action that best represents the possibility distribution of an inferred fuzzy control action.

The control surface, a 3D plot showing the output corresponding to all possible combination of values of the inputs, can be used to facilitate the controller tuning. Control surfaces can be plotted for different combinations of membership functions and control rules.

FLCs are divided into three main categories:

1. The FLC where fuzzy inference and if–then rules are used without a precise math model.
2. The FLC with a structure based on a precise math model.
3. The FLC that assists in tuning a conventional controllers' parameters.

The proposed fuzzy controllers preserve the basic properties and merits of the general PID controller, but they have simple configurations similar to the PI and PD schemes. The resulting controller is a discrete-time fuzzy version of the conventional PD and PI counterparts. It has the same linear structure in the proportional, derivative, and integral parts but has varying gains that are nonlinear functions of the input signals that will enhance the self-tuning capabilities of the controller. The controller is designed based on a precise mathematical

TABLE 13.2 Sample Rules Table

ce	NL	NM	NS	ZE	PS	PM	PL
NL	NL	NL	NL	NL	NM	NS	ZE
NM	NL	NL	NL	NM	NS	ZE	PS
NS	NL	NL	NM	NS	ZE	PS	PM
ZE	NL	NM	NS	ZE	PS	PM	PL
PS	NM	NS	ZE	PS	PM	PL	PL
PM	NS	ZE	PS	PM	PL	PL	PL
PL	ZE	PS	PM	PL	PL	PL	PL

control model. Membership functions are triangular, with only four fuzzy logic if–then rules. Fuzzy PI and PD have the capabilities of the conventional PI and PD for linear systems and can handle nonlinear systems. On the other hand, the fuzzy PI and PD like controllers, are driven by a set of control rules rather than by two constant gains. Hence, the controller becomes self-tuning and capable of handling nonlinear systems.

PD-Like Fuzzy Logic Controller

The controller described in this section is a discrete-time fuzzy version of the conventional PD controller. The conventional continuous-time PD control law is given by

$$u(t) = K_p e(t) + K_d e(t) \tag{13.98}$$

where $e(t)$ is the error defined to be the difference between the reference and the actual sensed output. The corresponding digital PD obtained using the bilinear transformation is

$$U(z) = \left(K_p + K_d \frac{1 - z^{-1}}{1 + z^{+1}} \right) E(z) \tag{13.99}$$

The inverse z-transform of Eq. (13.99) is

$$\Delta u(nT) = K_p d(nT) + K_d r(nT) \tag{13.100}$$

In Eq. (13.100), n is the number of sample, $\Delta u(nT)$ is the incremental control, $r(nT)$ is the rate of change, and $d(nT)$ is the average of change of the error signal, given by

$$\Delta u(nT) = \frac{u(nT) + u(nT - T)}{T} \tag{13.101}$$

$$r(nT) = \frac{e(nT) - e(nT - T)}{T} \tag{13.102}$$

$$d(nT) = \frac{e(nT) + e(nT - T)}{T} \tag{13.103}$$

Thus the control signal fed to the plant is given by

$$u(nT) = -u(nT - T) + T \Delta u(nT) \tag{13.104}$$

$$u(nT) = -u(nT - T) + TK_p d(nT) + TK_D r(nT) \tag{13.105}$$

The block diagram of this controller is shown in Fig. 13.59(a).

In the fuzzy PD-like controller, the term $T \Delta u(nT)$ is replaced by a fuzzy control action $K_u \Delta u(nT)$, where K_u is a fuzzy control gain so that the control law is given by

$$u(nT) = -u(nT - T) + K_u \Delta u(nT) \tag{13.106}$$

The block diagram of the fuzzy PD is shown in Fig. 13.59(b), where it is clear that the controller has the same simple linear structure, except two gains, K_P and K_D, are not constant.

In the fuzzification step there are two inputs: the error signal $e(nT)$ and the rate of change of the error signal $r(nT)$. There is only one control output, $u(nT)$. The adopted memberships

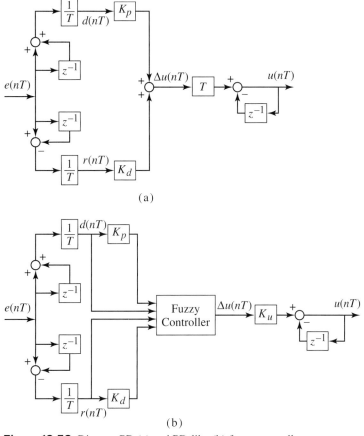

(a)

(b)

Figure 13.59 Discrete PD (a) and PD-like (b) fuzzy controllers.

are shown in Fig. 13.60, where, for simplicity, two memberships are used for the inputs and three memberships for the output. Also for simplicity, there are four rules only in the rule base:

R1: If error = ep and rate = rp, then output = oz.
R2: If error = ep and rate = rn, then output = op.
R3: If error = en and rate = rp, then output = on.
R4: If error = en and rate = rn, then output = oz.

Note that ep is error positive, rp is rate positive, en is error negative, rn is rate negative, op is output positive, on is output negative, and oz is output zero.

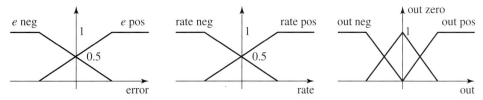

Figure 13.60 Sample memberships of the inputs and output.

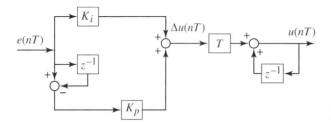

Figure 13.61 Discrete PI controller.

PI-Like Fuzzy Controller

The controller described in this part is a discrete-time fuzzy version of the conventional PI controller. The conventional continuous-time PI control law is given by

$$u(s) = K_p E(s) + \left(\frac{K_I}{s}\right)E(s) \tag{13.107}$$

where $e(t)$ is the error defined to be the difference between the reference and the sensed output. Using the bilinear transformation, the corresponding digital PI is obtained by

$$U(z) = \left[K_P + \frac{K_I}{1 - z^{-1}}\right]E(z) \tag{13.108}$$

Thus the following equations will be obtained:

$$u(nT) = u(nT - T) + T\Delta u(nT) \tag{13.109}$$

$$u(nT) = u(nT - T) + T\left[K_p\left[e(nT) - e(nT - T)\right] + K_I e(nT)\right] \tag{13.110}$$

The block diagram of this controller is shown in Fig. 13.61.

In the fuzzy PI-like controller, the term $T\,\Delta u(nT)$ will be replaced by a fuzzy control action $K_u\,\Delta u(nT)$, where K_u is a fuzzy control gain so that the control law will be given by Eq. (13.111). The block diagram of the fuzzy PI is shown in Fig. 13.62, where it is clear that the controller has the same simple linear structure, except that two gains, K_P and K_I, are not constant:

$$u(nT) = u(nT - T) + K_u\,\Delta u(nT) \tag{13.111}$$

Similar fuzzification and defuzzification membership functions (Fig. 13.60) were used in the PD-like FLC with the modified rules:

R1: If error = ep and rate = rp, then output = op.
R2: If error = ep and rate = rn, then output = oz.

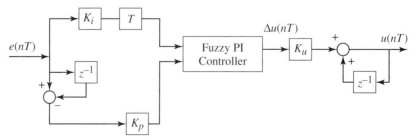

Figure 13.62 Discrete fuzzy PI-like controller.

R3: If error = en and rate = rp, then output = oz.

R4: If error = en and rate = rn, then output = on.

Example 13.15: Fuzzy logic controller design–MATLAB simulation–The loaded gripper of a robot driven by a DC motor does not have a known model as the load carried by the robot changes with various tasks. Develop fuzzy PI- and PD-like controllers to control this unknown system.

Solution: The controllers shown in Figures 13.59 and 13.62 are simulated using the DC motor transfer function (Eq. (13.3)). MATLAB block diagrams and simulation results are given in Figs. 13.63–13.68.

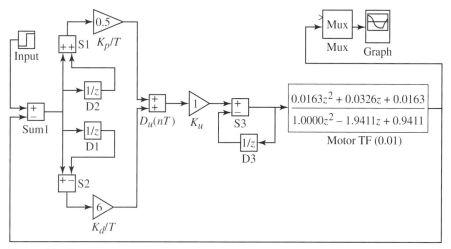

Figure 13.63 Simulink block diagram of the digital PD controller applied to the discretized motor transfer function for Example 13.15.

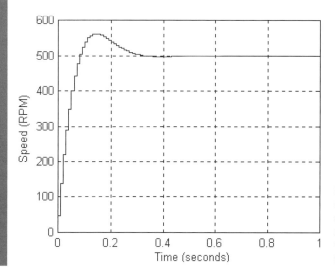

Figure 13.64 Simulation result of the digital PD controller applied to the discretized motor transfer function with 500-RPM reference for Example 13.15.

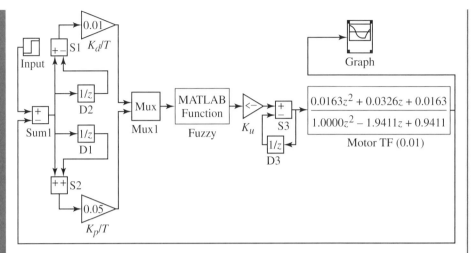

Figure 13.65 Simulink block diagram of the digital fuzzy PD-like controller applied on the discretized motor transfer function for Example 13.15.

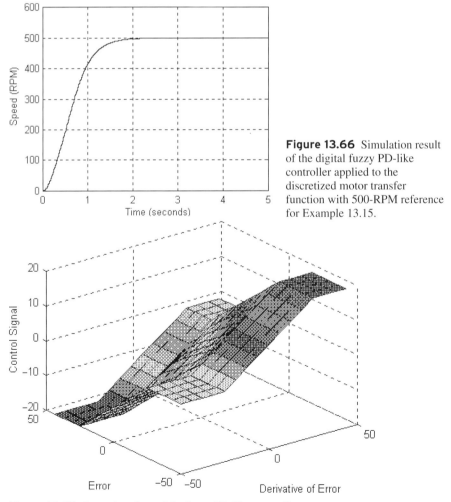

Figure 13.66 Simulation result of the digital fuzzy PD-like controller applied to the discretized motor transfer function with 500-RPM reference for Example 13.15.

Figure 13.67 Control surface of the fuzzy PD-like controller.

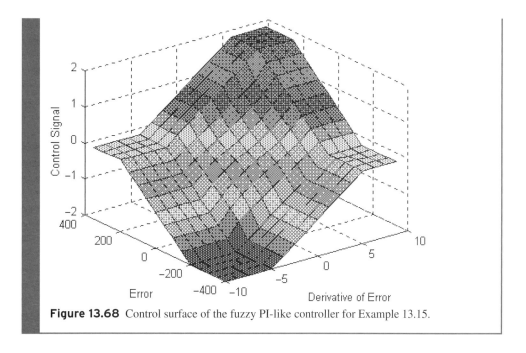

Figure 13.68 Control surface of the fuzzy PI-like controller for Example 13.15.

13.7 ADAPTIVE FUZZY LOGIC CONTROLLERS

13.7.1 Introduction

Compared to the conventional PID control schemes, the performance of fuzzy control of a poorly modeled system is substantially improved in terms of overshoot, steady-state error, load disturbance rejection, and variable speed. Fuzzy logic controllers (FLCs) have many advantages. However, FLCs with fixed parameters are inadequate in applications where large changes in the operating conditions are likely. The behavior of fuzzy controllers depends on the membership functions, their distribution, and the rules that influence the different fuzzy variables in the system. There is no formal method to accurately determine the parameters of the controller. It is, rather, a trial-and-error approach, unless reliable expert operator knowledge is available.

To preserve the controller effectiveness despite unreliable expert knowledge, the system response is observed and the controller parameters are accordingly adjusted or adapted to enhance the output signal. Observing the response of the controller and modifying the fuzzy sets in the universe of discourse of the input variables and output variable can improve response. This section discusses two adaptive fuzzy techniques, and their advantages over regular fuzzy techniques are highlighted. Simulation and experimental comparison of various controllers' implementation on the same system is included.

13.7.2 Fuzzy Model-Reference Adaptive Controller

The purpose of model-reference adaptive control (MRAC) is to tune control gains or parameters to account for the dynamically varying system parameters. The major components of a MRAC are an adjustable parameterized controller, a reference model, and an adaptation mechanism. The goal of the adjustable controller is to force the plant to respond similarly to the reference model. This is accomplished by having the adaptation mechanism change the adjustable controller gains or parameters based on the error between the output of the plant and that of the reference model. There are two feedback loops wrapped around the plant

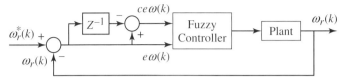

Figure 13.69 Fuzzy logic controller for a plant regulator.

(Fig. 13.55). The first, or inner, loop is the controller loop, and the second, or outer, loop contains the adaptation mechanism that updates the controller's parameters.

The reference model is used to specify the desired performance that satisfies the design criteria, such as transient response settling time and overshoot. In most cases, a first- or second-order system can be used as a reference model. The same reference input is applied to both the reference model and the fuzzy controller. A fuzzy logic adaptation loop is added in parallel to the fuzzy control feedback loop. In the nominal case, the model following is perfect and the fuzzy adaptation is idle. When significant parameter changes occur and/or expert knowledge is not reliable, an adaptation signal will be added to the output signal of the main fuzzy controller to preserve the desired model following performance.

The adaptation can be one of two types: *parameter adaptation*, where controller parameters are modified, or *signal adaptation*, where the controller output signal is corrected. In contrast to conventional adaptive systems, adaptive fuzzy controllers require no complex mathematical operations because most operations are simple and repetitive.

In the proposed system, the structures of the fuzzy controller and the adaptation mechanism are identical, so they can be implemented as two parallel fuzzy systems. A representation of the fuzzy controller is given in Fig. 13.69, where the two input variables $e\omega(k)$ and $ce\omega(k)$ are calculated at every sampling instant as

$$e\omega(k) = \omega_r^*(k) - \omega_r(k)$$
$$ce\omega(k) = e\omega(k) - e\omega(k - 1)$$

(13.112)

where $\omega_r^*(k)$ is the reference speed and $\omega_r(k)$ is the actual motor speed.

Figure 13.70 shows a fuzzy model-reference adaptive controller where, in addition to the main fuzzy controller, another error is generated from the difference between the actual output and the required output from an ideal reference model. This error is then fed to another fuzzy controller, which will generate a corrective signal to be added to the original control signal. This will ensure that the output signal approaches the desired values.

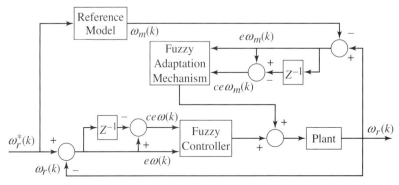

Figure 13.70 Fuzzy model-reference adaptive controller for a plant regulation.

Example 13.16: Adaptive FLC MRAC using MATLAB simulation—This example uses the previously designed PI-like fuzzy logic controller (Example 13.15) to control the speed of the loaded DC motor of a robot gripper. An MRAC is then designed such that the adaptation mechanism is executed using fuzzy logic based on the error, and its derivative, measured between the motor speed and the output of a reference model.

Figures 13.71 and 13.72 show, respectively, the Simulink block diagram and the control surface of the adaptation mechanism. Figure 13.73 shows the results of the simulation after the MRAC is applied.

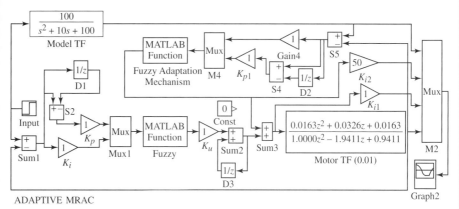

Figure 13.71 Simulink diagram of the digital fuzzy PI-like controller with the MRAC adaptive technique (Example 13.16).

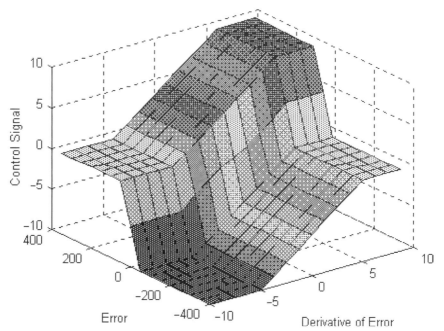

Figure 13.72 Control surface of the fuzzy adaptation mechanism (Example 13.16).

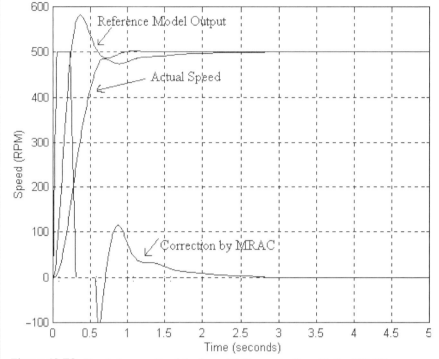

Figure 13.73 Simulation results of the fuzzy PI-like controller with the MRAC adaptation technique (Example 13.16).

Even though the computation time is long, the design and the operations performed are simple compared to the conventional adaptive controllers.

13.7.3 Membership-Tuning Adaptive Controller

This adaptive fuzzy scheme consists of three main components. The first component is a reference model describing the desired behavior of the control system. The second is a fuzzy controller that determines the adjustment to the control loop. The third is a fuzzy learning-and-adaptation mechanism that modifies the fuzzy controller according to the difference between the actual and the desired speed response. The modification is performed by shifting the membership functions of the fuzzy output sets associated with the consequent rules of the fuzzy controller. The proposed fuzzy controller is similar to the MRAC system described earlier, but the adaptation corrects the controller, as depicted in the block diagram of Fig. 13.74, instead of the control signal.

In the case of a fixed fuzzy controller, the membership functions of the consequent fuzzy output that sets $U(k)$ are fixed and can be called NL, NM, NS, ZE, PS, PM, and PL. A rule matrix can be devised to relate these control fuzzy sets to the input fuzzy sets. However, in the case of an adaptive fuzzy controller, the fuzzy sets $U(k)$ and the rule matrix are modified online by the adaptation-and-learning mechanism.

The purpose of the fuzzy adaptation block is to learn the environmental parameters and to modify the fuzzy controller accordingly so that the response of the overall system is close to

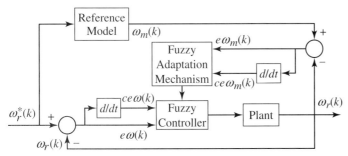

Figure 13.74 Membership-tuning adaptive fuzzy controller block diagram.

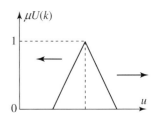

Figure 13.75 Fuzzification membership shifting.

the response of the reference model. The modification to the fuzzy controller is achieved by changing the membership functions of the control fuzzy set $\mu U(k)$. This change can take a number of forms, e.g., shifting, contracting, or expanding the membership functions. In the discussion to follow, shifting of the membership functions will be adopted.

Let the first input to the fuzzy adaptation mechanism be the error between the actual output and the output of the reference model, and let the second input be the derivative of that error. The output represents the shift to the center point of the membership function $\mu U(k)$, as shown in Fig. 13.75.

The error and the derivative of the error are fuzzified to obtain linguistic variables and the IF-THEN rules are used. The rules are:

1. **IF** the deviation and the derivative of the deviation are both close to zero, **THEN** do not shift the center point of the corresponding membership function.
2. **IF** deviation is positive but the derivative of the deviation is negative, or vice versa, **THEN** do not shift the membership function, allowing the opposing conditions to cancel out over time.
3. **IF** the deviation and the derivative of the deviation are both positive, **THEN** make a large shift in the positive direction. In this case the actual is below the desired response, and the system is responding too slowly; therefore, a large positive shift is needed to correct the situation.
4. **IF** the deviation and the derivative of the deviation are both negative, **THEN** make a large shift in the negative direction.
5. **IF** the deviation is positive but the derivative is zero, **THEN** make a small shift in the positive direction.
6. **IF** the deviation is negative but the derivative is zero, **THEN** make a small shift in the negative direction.

At each control time, the fuzzy values of the shift found by the fuzzy adaptation block are defuzzified to obtain the crisp values used to determine the center points of the fuzzy controller's membership function $\mu U(k)$. However, since not all fuzzy sets $U(k)$ contributed to the current response, the membership function of only those fuzzy sets that had activation level at the previous sampling time are shifted.

Example 13.17: Adaptive FLC with membership tuning using MATLAB simulation–
The same system used in Example 13.16 is the subject of this example. Figure 13.76 shows the Simulink block diagram of the fuzzy PI-like controller with the MRAC membership-tuning adaptive technique. Note that the output of the adaptation mechanism is not acting directly on the output signal, but it is fed to the regular fuzzy controller, where it adjusts the output memberships.

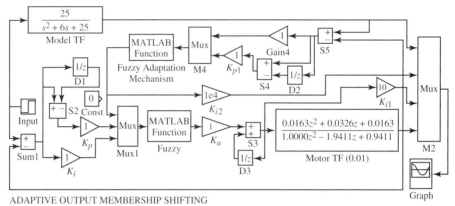

ADAPTIVE OUTPUT MEMBERSHIP SHIFTING

Figure 13.76 Simulink diagram of the fuzzy PI-like controller with the MRAC membership-tuning adaptive technique (Example 13.17).

Figure 13.77 shows the control surface of the adaptation mechanism. Figure 13.78 shows the simulation result of the fuzzy PI-like controller without adaptation and the output of the model. In Figure 13.79, after the adaptation to the output membership is made, it is obvious how the motor speed approached the model output before settling on the reference speed. This method is tested also for bad expert knowledge.

Figure 13.80 shows the output of a fuzzy controller with slow response due to wrong expert knowledge. Figure 13.81 shows the output that followed the model output.

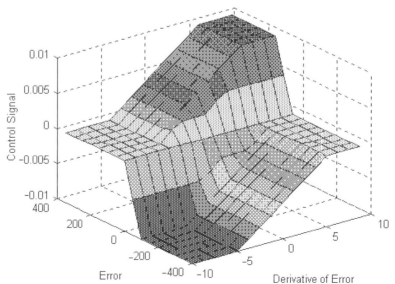

Figure 13.77 Control surface of the membership-tuning adaptation mechanism (Example 13.17).

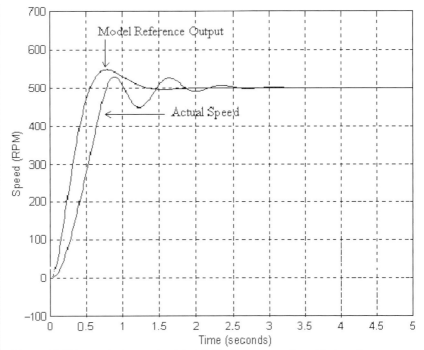

Figure 13.78 Simulation result of the PI-like fuzzy controller with the reference model output (Example 13.17).

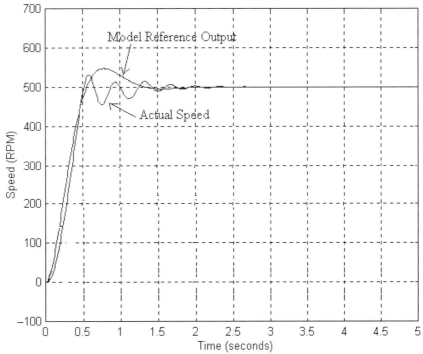

Figure 13.79 Simulation result of the PI-like fuzzy controller with the MRAC membership-tuning adaptation technique (Example 13.17).

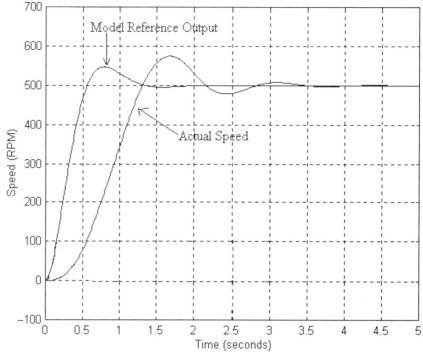

Figure 13.80 Simulation result of the PI-like fuzzy controller applied to the motor, with bad expert knowledge in the output membership (Example 13.17).

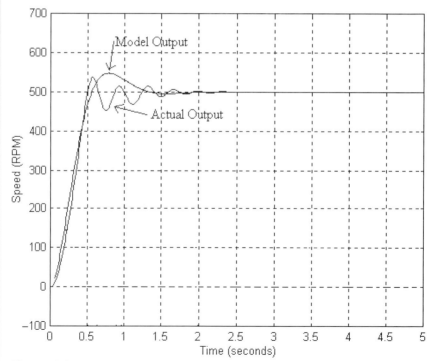

Figure 13.81 Simulation result of the PI-like fuzzy controller with bad expert knowledge and membership-tuning adaptation (Example 13.17).

13.8 EXPERIMENTAL COMPARATIVE ANALYSIS

13.8.1 Hardware Platform

Many advanced controllers were not widely applied until recently, due to the great advancement in digital realization techniques, which made the implementation of complex algorithms possible. For example, fuzzy logic controllers can be implemented using conventional processors, microcontrollers, reduced instruction set computing (RISC) processors, or digital signal processors (DSPs). It can also be implemented using fuzzy logic ASICs (application-specific integrated circuits). The processor selection depends on the complexity of the fuzzy controller and the desired sampling frequency. The major advantage of ASIC fuzzy logic chips is the high processing speed they can achieve. The American Neuralogix NLX230 fuzzy microcontroller is an example of such a device. It is a fully configurable fuzzy logic engine containing one of eight output selectors, 16 fuzzifiers, a minimum comparator, a maximum comparator, and a rule memory. Up to 64 rules can be stored in the on-chip 24-bit-wide rule memory. The NLX230 can execute 30 million rules per second.

On the other hand, conventional processors can provide more flexibility in both hardware and software implementation. Commercial processor boards based on microprocessors, microcontrollers, RISC processors, or DSPs are widely available with different standard bus formats. By using C language, the fuzzy control code can be developed and tested off-line on workstations or PCs. Once debugged, the same C code can be recompiled and downloaded to a specific processor board for real-time execution. The development process can be then accelerated.

13.8.2 Digital Control Workstation

In order to implement the different control schemes, the experimental station of Figure 13.82 is proposed. Many mechatronics case studies will be presented in Chapter 14. In this section, an experimental setup is presented for experimental evaluation of control strategies rather than the mechatronics technology integration functions required in useful products.

This station can supply a valid platform for most of the control schemes discussed in earlier sections. Specifically, the following components form the blocks of this station:

- A DC motor with an optical encoder generates a pulse train of frequency proportional to the speed of the shaft.
- A frequency-to-voltage converter accepts the speed as a pulse train and converts it to a voltage.

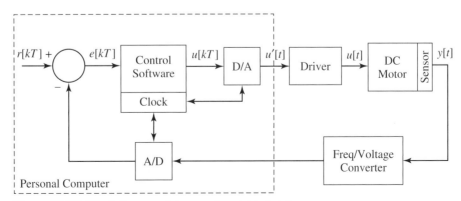

Figure 13.82 Block diagram of the experimental test station.

- A personal computer runs the control software algorithms and supplies the user interface.
- A data-acquisition card interfaces between the analog systems and the digital computer.
- A driver is supplied for the motor.

DC Motor and Sensor

The motor used in this station is a 48-W DC series motor with an armature resistance of 1.4 Ω and a rated voltage of 12 V. The field winding is connected in series with the armature winding. The series motor is a variable-speed motor, and its speed varies with the applied load. The speed torque characteristic is shown in Fig. 13.83. Closed-loop control of the DC motor uses an optical shaft encoder to measure the instantaneous speed. The output of the encoder is a pulse train with width proportional to the speed of the motor shaft. An amplifier is used to strengthen the encoder signal.

Frequency-to-Voltage Converter

The output of the sensor requires conditioning to make it compatible with the data-acquisition card (presented in the next section). One way to accomplish this is to transform the output pulse train to a 0- to 10-V DC voltage signal by the frequency-to-voltage (F/V) converter circuit shown in Fig. 13.84. In this circuit, the output voltage is proportional to the input frequency. The average current from the 40106 Schmitt trigger inverter's ground pin 8 is linearly dependent on the frequency at which the 0.1-μF capacitor discharges into the op-amp summing junction. The op-amp forces this current to flow through the 13.33-kΩ feedback resistor, producing a corresponding voltage drop. This frequency-to-voltage converter yields 0- to -10-V output with 0- to 10-KHz input frequencies.

Driver

The computer should provide the motor with the corrective signal that will keep it running at the desired speed based on the calculation or decisions of the control algorithm. The data-acquisition card will provide a signal output between 0 and 10 V_{DC} with a driving current in the

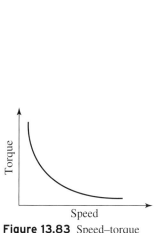

Figure 13.83 Speed–torque curve of a typical series motor.

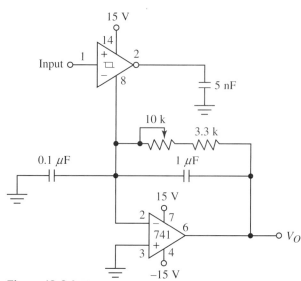

Figure 13.84 Frequency-to-voltage converter.

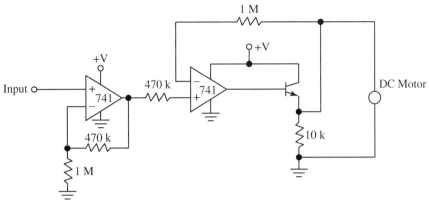

Figure 13.85 The motor driver.

milliwatt range. A special driver is designed to receive the 0- to 10-V_{DC} and produce an amplified output with sufficient driving current to the motor. This circuit is shown in Fig. 13.85. The non-inverting op-amp has a gain of 3.128. The input voltage to the op-amp is variable between 0 and 15 V. The output voltage is therefore variable over an approximate range from 0.5 to 30 V. The transistor is added to the output to boost the output current available to the motor.

Data-Acquisition Card (DAC)

The transducer senses the physical quantity being measured (the measurand) and produces an analog signal (usually voltage) or current loop. The raw output signal from a transducer requires conditioning and must be converted to a digital signal before the computer processing. The data-acquisition system enables the computer to gather, monitor, display, and analyze data. In addition, the DAS provides output capabilities for control purposes. A plug-in card connects directly to the internal bus of a PC through a plug-in slot. The plug-in card is usually connected to an external board where all input/output connections are made. The DAQ mainly converts incoming signals into digital equivalents. The software residing in the PC controls the operation of the DAQ, manipulates the data, and computes the output. Plug-in DAQs are faster because they are connected directly to the computer bus. They are less costly since there is no overhead for packaging or power. As PC prices continue to go down, PC-based DAQs are becoming widely implemented in many areas, such as laboratory automation, industrial monitoring and control, and automatic testing and measurement.

A DAQ typically handles analog/digital inputs, analog/digital outputs, and counter/timer operations. Digital inputs and outputs are important in many applications, such as contact closure and switch monitoring, industrial ON/OFF control, and digital communications. They can also generate waveforms. A counter/timer can be used to perform event counting, flow meter monitoring, frequency counting, and pulsewidth and time-period measurements. Analog inputs require A/D conversion. The criteria for selecting the appropriate A/D converter include the number of input channels, single-ended vs. differential inputs, sampling rate, resolution, input range, noise immunity, and nonlinearity. Although a DAQ card may generate outputs that can directly control equipment in a process, analog outputs require D/A conversion to be useful for this purpose. It is also possible to perform a closed-loop or PID control with a DAQ card.

The data-acquisition card used in this experimental setup is the enhanced Multi-lab PCL 812 PG DAQ card. This multifunction PC-based DAQ offers acceptable performance and cost. It acquires data from external hardware devices and provides analog output signals to the DC motor driver. The key features of this card are:

- 16 single-ended analog input channels
- 12-bit A/D converter with a 30-KHz conversion speed
- Analog input voltage range $(+/-5V, +/-2.5, +/-1.25, +/-0.625, +/-0.3123)$
- Two analog output channels
- 12-bit D/A converter
- Analog output voltage range $(+/-5V, 0–10 V)$.

The PCL-812PG also provides easy-to-use software drivers. A variety of third-party application software packages have been integrated with the system to provide users more application support.

Example 13.18: Experimental comparison of various controllers–Figure 13.86 shows the hardware setup that forms a digital control experimental station where different control algorithms can be implemented, verified, and assessed. The motor will run freely, while the load setup imposes on it a controlled disturbance to compare all control algorithms fairly. The sensor generates a pulse train proportional to the speed of the shaft. The frequency-to-voltage converter converts the train of pulses to $0–10 V_{DC}$ accepted by the DAQ card's A/D converter. A conversion factor $K = 1.45$ was calculated by direct measurement of the shaft speed to convert the voltage value from the F/V converter into the corresponding rpm value. The available rpm value is compared to the reference speed and used to generate the error input to the control algorithm that issues the corrective signal to the D/A converter. The driver receives the signal from the D/A converter and provides the motor with a corrected amplified signal to adjust the motor speed accordingly.

In previous sections of this chapter, the effectiveness of different conventional, modern, fuzzy, and adaptive fuzzy control algorithms was demonstrated via simulation examples. Simulation showed that the PI-like fuzzy controller provides better results, better noise immunity, and better reaction to external disturbances.

The developed controllers are tested and compared with no load and with gradually increasing loads applied for a short duration. The applied loads are 350 g as a small load and 1 kg as a heavy load. Figures 13.87–13.92 show the results obtained under different loading conditions

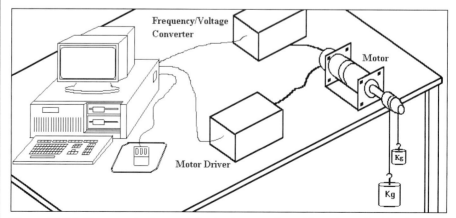

Figure 13.86 The control-scheme-testing hardware station with predetermined loads.

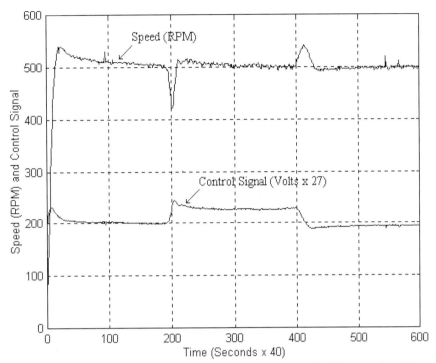

Figure 13.87 Experimental speed output of the station motor with PID controller ($K_P = 0.1$, $K_I = 0.1$, $K_D = 0.005$) and 350-g load at 5 s and off at 10 s (Example 13.18).

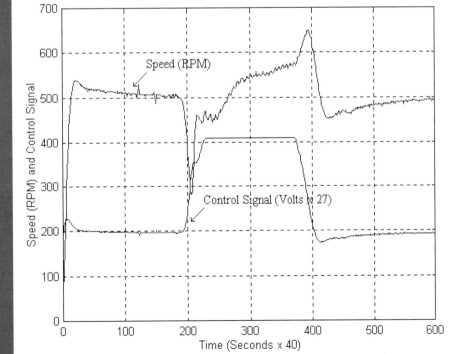

Figure 13.88 Experimental speed output of the station motor with PID controller. 1 kg load (Example 13.18)

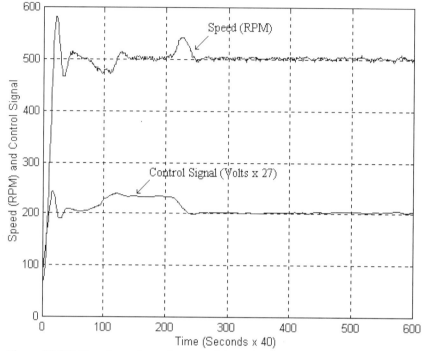

Figure 13.89 Experimental speed output of the station motor with PI-like fuzzy controller. 350 g load (Example 13.18)

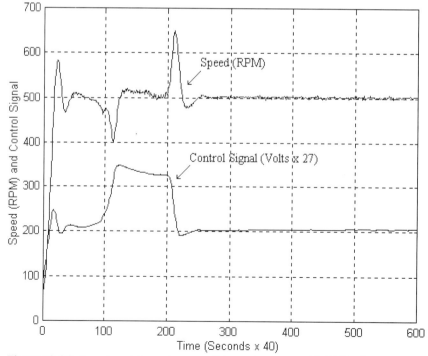

Figure 13.90 Experimental speed output of the station motor with PI-like fuzzy controller. 1 kg load (Example 13.18)

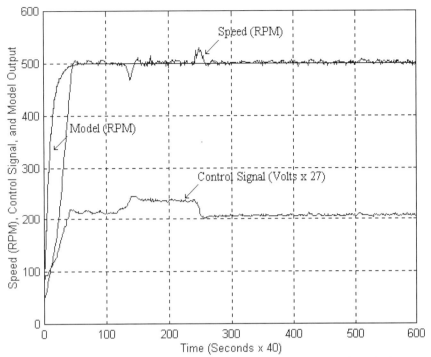

Figure 13.91 Experimental speed output of the station motor with a PI-like adaptive fuzzy MRAC. 350 g load (Example 13.18)

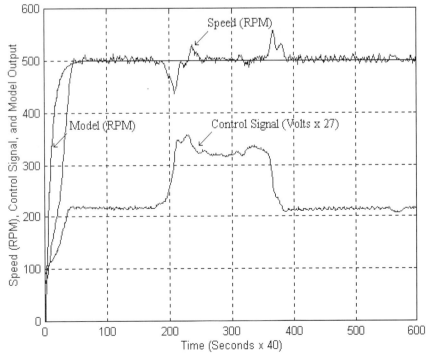

Figure 13.92 Experimental speed output of the station motor with PI-like adaptive fuzzy MRAC. 1 kg load (Example 13.18)

with PID conventional control, PI-like fuzzy control, and PI-like adaptive fuzzy control based on MRAC correction. Tables 13.3–13.5 summarize important features of the responses.

In the no-load case, the fuzzy PI-like controller gave results comparable to those with the PID controller. But the ease of introduction of the integral and derivative action and the simplicity of design of the PID controller make it favorable in similar cases. The adaptive technique improved the response of the fuzzy controller in terms of overshoot and settling time, at the expense of a larger rise time. The fuzzy MRAC signal introduced to the regular fuzzy control signal improvement in the response. The improved performance of the conventional PID controller is due to the fact that the controller was acting on a well-modeled undisturbed

TABLE 13.3 Performance of the PID Controller with Different Loads Applied to the DC Motor Shaft ($K_P = 0.1$, $K_I = 0.1$, $K_D = 0.005$)

Load = 350 g	Fig. 13.87	Note that the speed dropped 75 RPM when the load was applied and increased 40 RPM when the load was removed. The speed was regulated again while loaded, with a settling time of less than 0.5 s.
Load = 1 kg	Fig. 13.88	Note that the speed dropped 200 RPM when the load was applied and increased 200 RPM when the load was removed. The speed could not receover between those two times.

TABLE 13.4 Performance of the PI-like Fuzzy Controller with Different Loads Applied to the Shaft

Load = 350 g	Fig. 13.89	Note that the speed dropped 30 RPM when the load was applied and increased 30 RPM when the load was removed. The speed was regulated again while loaded, with a settling time of less than 1 s.
Load = 1 kg	Fig. 13.90	Note that the speed dropped 100 RPM when the load was applied and increased 120 RPM when the load was removed. The speed was regulated again while loaded, with a settling time of 1 s.

TABLE 13.5 Performance of the Adaptive PI-like Fuzzy MRAC with Different Loads Applied

Load = 350 g	Fig. 13.91	Note that the speed dropped 20 RPM when the load was applied and increased 20 RPM when the load was removed. The speed was regulated again while loaded, with a settling time of less than 0.2 s.
Load = 1 kg	Fig. 13.92	Note that the speed dropped 50 RPM when the load was applied and increased 50 RPM when the load was removed. The speed was regulated again while loaded, with a settling time of 1 s.

system, whereas the FLC counterpart was working with an unmodeled system. In such a case, conventional controllers are expected to perform better and are more suitable.

When a small unmodeled load (350 g) was applied for a short duration (~5 s), all controllers seemed to recover from the imposed disturbance, with acceptable response error. However, fuzzy control algorithms exhibited better response in terms of smaller overshoot and shorter settling time. Nevertheless, conventional PID control showed comparable results.

When a medium load (1 kg) was imposed (for ~5 s), the fuzzy controller handled the load with a permissible control signal and gave acceptable results relative to those of the PID conventional controller. This is manifested in the smaller overshoot for the same load and the faster settling time. The FLC showed better robustness than the conventional scheme. The adaptive techniques showed even better response.

When the heavy load (1.75 kg) was applied over a short duration, the conventional PID techniques failed to respond properly, while the response of the fuzzy techniques was faster and better. It is obvious that the overshoot in the response of the conventional PID techniques increased dramatically (more than 100%), and the settling time was above 2 s.

13.9 CONCLUSION

This chapter presented an overview of historical control strategies, which are algorithms in a microprocessor for issuing adequate input signals to mechatronics drivers to achieve desired output based on feedback-sensed actual output. Specifically, the following outcomes are expected from students who master the material of this chapter.

- Estimate experimentally the transfer function of unknown dynamical systems.
- Analyze the performance of a system model based on frequency domain and time domain tools.
- Design conventional controllers to deliver desired specifications using known system models.
- Design fixed and adaptive modern control schemes for known and partially known systems.
- Design intelligent controllers for regulating the output of unknown systems that are subject to disturbances without relying on past expert operator knowledge.
- Simulate system performance for analysis and design using MATLAB, Simulink, and LabVIEW.
- Experimentally assess the effectiveness of control schemes on a digital test station.

RELATED READING

K. J. Astrom and B. Wittnmark, *Adaptive Control*. Reading, MA: Addison-Wesley, 1995.

A. R. DeCarlo, *Linear Systems*. Englewood Cliffs, NJ: Prentice Hall, 1989.

C. De Silvia, *Control Sensors and Actuators*. Englewood Cliffs, NJ: Prentice Hall, 1989.

R. C. Dorf and R. H. Bishop, *Modern Control Systems*. Reading, MA: Addison-Wesley, 1995.

R. Dorf and R. Bishop, *Modern Control Systems*, 10th ed. New York: Pearson-Prentice Hall, 2005.

Gene Franklin and J. David Powell, *Digital Control of Dynamic Systems*, 2nd ed. Reading, MA: Addison-Wesley, 1990.

M. Ghandakly, M. Shields, and M. Brihoum, "Design of an Adaptive Controller for a DC Motor Within an Existing PLC Framework," *Proceedings of the Annual Meeting of the IEEE Industry Applications Society*, San Diego, CA, 1996.

M. Heidar, L. Huaidong, and C. Guanrong, "New Design and Stability Analysis of Fuzzy Proportional-Derivative Control Systems," *IEEE Transactions on Fuzzy Systems*, vol. 2, No. 4, November 1994.

S. Hwang, "A Real-Time Digital Adaptive Tracking Controller for a DC Motor," *Proceedings of the Annual Meeting of the IEEE Industry Applications Society*, Orlando, FL, 1995.

Y. S. Kung and C. Liaw, "A Fuzzy Controller Improving a Linear Model Following Controller for Motor Drives," *IEEE Transactions on Fuzzy Systems*, vol. 2, no. 3, August 1994.

J. Lee, "On Methods for Improving Performance of PI-Type Fuzzy Logic Controllers," *Letters to IEEE Transaction on Fuzzy Systems*, vol. 1, no. 4, November 1993.

N. E. Leonard and W. Levine, *Using MATLAB to Analyze and Design Control Systems*, 2 ed. Reading, MA: Addison-Wesley, 1995.

A. Lotfi and A. Tsoi, "Learning Fuzzy Inference Systems Using an Adaptive Membership Function Scheme," *IEEE Transactions on Systems, Man, and Cybernetics*, vol. 26, no. 2, April 1996.

T. C. Minh and L. H. Hoang, "Model Reference Adaptive Fuzzy Controller and Fuzzy Estimator for High-Performance Induction Motor Drives," *Proceedings of the Annual Meeting of the IEEE Industry Applications Society*, San Diego, CA, 1996.

D. Misir, M. Heidar, and C. Guanrong, "Design and Analysis of a Fuzzy PID Controller," *Fuzzy Sets and Systems*, vol. 79, no. 3, 1996.

F. Mrad and G. Deeb, "Experimental Comparative Analysis of Adaptive Fuzzy Logic Controllers," *IEEE Transactions on Control Systems Technology*, vol. 10, no. 2, March 2002, pp. 250–255.

F. Mrad, Z. Gao, and N. Dhayagude, "Fuzzy Logic Control of Automated Screw Fastening," *Proceedings of the Annual Meeting of the IEEE Industry Applications Society*. Orlando, FL,1995.

K. Ogata, *Modern Control Engineering*, 4th ed. New York: Prentice Hall, 2002.

W. Z. Qiao and M. Mizumoto, "PID-Type Fuzzy Controller and Parameters Adaptive Method," *Fuzzy Sets and Systems*, vol. 78, no. 11, 1996.

V. Raviraj and P. Sen, "Comparative Study of PI, Sliding-Mode, and Fuzzy Logic Controllers for Power Converters," *Proceedings of the Annual Meeting of the IEEE Industry Applications Society*. Orlando, FL, 1995.

C. von Altrock, *Fuzzy Logic and Neurofuzzy Applications Explained*. New York: Prentice Hall, 1995.

L. Wang, "Stable Adaptive Fuzzy Control of Nonlinear Systems," *IEEE Transactions on Fuzzy Systems*, vol. 1, no. 2, May 1993.

D. Wobschall, *Circuit Design for Electronic Instrumentation*. New York: McGraw Hill, 1979.

T. K. Yin, "Fuzzy Model-Reference Adaptive Control," *IEEE Transactions on Systems, Man and Cybernetics*, vol. 25, no. 12, December 1995.

QUESTIONS

13.1 What are the advantages (if any) of closed-loop control over open-loop systems?

13.2 While meeting controlled system desired specifications, what are the trade-offs?

13.3 In designing a PID controller, what are the effects of the integral and derivative gain contributions?

13.4 Justify the introduction of state space in controlling dynamical systems.

13.5 What issues are addressed in adaptive control?

13.6 What are the advantages of digital control over its analog counterpart?

13.7 What is intelligent control?

13.8 What is the main advantage of fuzzy logic control? Name one main drawback.

13.9 State the main reason behind introducing adaptation to fuzzy logic control.

PROBLEMS

13.1 Consider the following system:

$$\begin{bmatrix} \dot{x}_1 \\ \dot{x}_2 \\ \dot{x}_3 \end{bmatrix} = \begin{bmatrix} 0 & 1 & 0 \\ 0 & 0 & 1 \\ -1 & -5 & -6 \end{bmatrix} \begin{bmatrix} x_1 \\ x_2 \\ x_3 \end{bmatrix} + \begin{bmatrix} 0 \\ 0 \\ 1 \end{bmatrix} u$$

which has the general form

$$\dot{x}(t) = Ax(t) + Bu(t)$$

By using the state feedback control $u(t) = -Kx(t)$, it is desired to have the closed-loop poles at $s = -2 \pm j4$, $s = -10$. Determine the state feedback gain matrix K.

13.2 A boring machine's orientation angle of drilling is critical, so an automatic controller is designed to regulate the guidance angle with a desired reference angle. Determine the values of K, T_1, and T_2 of the system shown here so that the dominant closed-loop poles have $\zeta = 0.5$ and $\omega_n = 3$ rad/sec.

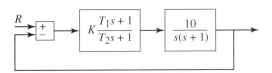

13.3 a. State the basic building blocks constituting a fuzzy control system, and draw them in a block diagram.
b. Assuming the inputs to the fuzzy system are the error and the derivative of the error and that each can be negative, zero, or positive, draw the membership of the error and the derivative of the error.
c. If the output also has three memberships (negative, zero, and positive), construct a valid rule base, and write it in a table form.

13.4 Consider the following unity-feedback transfer function describing a DC motor:

$$G(s) = \frac{672}{s(s + 6)}$$

a. Design a P controller so that that the closed-loop system has an overshoot of less than 20% and a settling time of less than 1 second.
b. Plot the step response of the obtained closed-loop system using MATLAB or LabVIEW, and verify the overshoot and the settling time.
c. Introduce an integral action to the system using the Ziegler–Nichols method, and plot the unit-step response using MATLAB or LabVIEW. What are your conclusions?

13.5 For the unity-feedback control system whose open-loop transfer function models the pointing system of a telescope:

$$G(s) = \frac{5}{s(s + 3)}$$

a. Design PID gains (three of them) for a closed-loop controlled system step-response overshoot of less than 15% and a settling time of less than 2 seconds, (if possible).
b. What is the closed-loop system order?
c. Is the design of part (a) unique? Explain briefly.

13.6 An unknown control system was tested experimentally with sinusoidal inputs of various frequencies, and the following table was recorded:

W	377	314	251	219	157	126	100
Gain (dB)	−7.75	−4.3	−0.2	0.75	5.16	7.97	10.5
W	63	44	16	8	1.38	1.0	
Gain (dB)	15.0	16.9	20.4	21.6	24.0	24.1	

a. Approximate (if possible) an estimate of a transfer function $F(s)$ that is governing the dynamical behavior of the system.
b. Is the obtained $F(s)$ a closed-loop or an open-loop transfer function?

13.7 A voice coil motor is used to position a read/write element suspended at the end of an arm over a rotating media disk in a hard disk drive. A unity-feedback control system with forward transfer function $G(s)$ models the relationship between the input motor voltage and the R/W element position:

$$G(s) = \frac{8000}{s(s + 1)(s + 80)}$$

Design a series type (justify your choice) compensator (if needed!) based on frequency response techniques so that the closed-loop system has at least a 10-dB gain margin and a 30-deg phase margin. Specifically:

a. Approximate the Bode plots of the uncompensated open-loop transfer function.
b. Based on part (a), approximate the P.M. and G.M.
c. Suggest the type of compensator needed, and justify your choice.
d. Design a valid controller based on part (c), and supply it.
e. Confirm the effectiveness of the compensated system by verifying the P.M. and G.M. of the controlled system.

13.8 The discrete equivalent (T is 0.1 s) linear time-invariant state model of a continuous second-order system is given by the following:

$$X(k + 1) = AX(k) + BU(k)$$
$$Y(k) = CX(k)$$

The matrices are:

$$A = \begin{bmatrix} 1 & -1 \\ 0 & 1 \end{bmatrix}; \qquad B^T = [1 \quad 1]; \qquad C = [1 \quad 2]$$

a. Is the given system fully controllable? Justify your answer.

b. Solve for the poles of the open-loop system.

c. Supply the natural transient step-response specifications of the given open loop in the continuous domain.

d. Design the state feedback matrix **G** such that $U(k) = -GX(k)$ will force the closed-loop system to have desired poles in the Z-plane located at 0.4 and 0.6.

e. What are the obtained transient step-response specifications of the closed-loop system in the continuous domain?

13.9 An autonomous rover moves in a straight trajectory, approximated by the model $G(s)$. The desired specifications are given in terms of an ideal second-order system step-response behavior: maximum overshoot not to exceed 5%, and settling time (2% criterion) not to exceed 2 s. A microprocessor-based controller will be used to control the process, hoping to deliver the desired performance.

$$G(s) = \frac{10}{(s + 1)(s + 2)}$$

a. Transform the required specs into a set of desired poles in the s-plane. Is that set unique? Why?

b. Assume the sampling period is $T = 0.1$ s, get the equivalent desired pole location in the Z-plane using zero pole mapping.

c. Using ZOH circuitry, get the equivalent discrete transfer function of $G(s)$.

d. Design a digital PI controller (difference equation) to accomplish the described goals using emulation.

e. Design a direct digital PI controller (difference equation) to accomplish the described goals.

13.10 A DC servo motor is used to rotate a robotic link from its horizontal position to the desired final position of $\pi/2$ rad CCW. The time variation of the gravitational torque is expressed by

$$T_{gr}(t) = 21 \cos\theta(t)$$

The motor parameters are $L_a = 0.1$ H, $R_a = 1.62$ Ω, $J_T = 0.0067$ oz-in.-sec^2, and $B = 0.00955$ oz-in.-sec/rad. The amplifier gain is $A = 10.0$ V/V, and the tachometer constant is $K_{tach} = 0.065$ V/rad/sec. Find the response of the motor to a step input for the following cases.

a. Use a proportional (P) compensator only for $K_P = 10, 20,$ and 40.

b. Use a proportional-derivative (PD) compensator for $K_P = 20$ and $K_D = 0.0, 0.01, 0.02,$ and 0.04.

c. Use a proportional-integral (PI) compensator for $K_P = 20$ and $K_I = 0$ and 75.

d. Use a proportional-derivative-integral (PID) for $K_P = 20, K_D = 0.02,$ and $K_I = 75$.

13.11 A DC motor powered by linear amplifier drives a spindle to position a mechanism at 5° intervals, ±1°. The motor (S6M4HI from PMI motors, www.pmi.com) has the following characteristics: peak torque $T_{PEAK} = 153$ N-cm, continuous stall torque $T_S = 14$ N-cm, peak current $I_{PEAK} = 51$ A, continuous stall current $I_{CONT} = 4.8$ A, peak acceleration (no load) $a_{PEAK} = 256$ krad/s^2, torque constant $K_T = 3.01$ N-cm/A, back-emf constant $K_E = 3.15$ V/krpm, $R_a = 0.94$ Ω, $R_T = 1.207$ Ω, average friction torque $T_f = 0.6$ N-cm, viscous damping constant $B = 0.11$ N-cm/krpm, rotor inertia $J_M = 0.06$ kg-cm^2, and $L_a < 100$ µH. The mechanical time constant $\tau_M = 10$ ms, and its tach gain $K_{tach} = 1$ V/krpm. Assume constant load at $T_L = 10$ oz-in. with negligible inertia.

a. Select an appropriate encoder pulse count per second.

b. Calculate appropriate loop gains.

c. Write an assembly code for its operation.

14

Case Studies

OBJECTIVES

This chapter is an integration of various technological components that will enable the student to:

1 Understand the overall design approach to a mechatronics project

2 Learn from practical examples the integration and interfaces among various components covered in earlier chapters

3 Develop skills needed in realizing mechatronics products, especially in overall design strategy, alternative choices, and software planning and programming

4 Transform project requirements into design characteristics and specifications

5 Interface various blocks of a product

6 Realize a complete project to meet design specifications

7 Adapt parts of presented programs and products for utilization in other mechatronics projects

14.1 INTRODUCTION

In concluding this textbook, it is critical that we integrate all the major content and analysis tools that were introduced earlier in applied cases. This chapter presents a few examples of students' group senior-year graduation projects supervised by the authors at the American University of Beirut in the mechanical engineering and electrical engineering programs. The material in this book was popular and useful reference for the students in realizing their final-year projects.

Some mechatronics course projects and hands-on examples have been presented throughout this book. This chapter highlights larger, complete mechatronics cases. In selecting the projects, the following aims were considered.

• Highlight microcontrollers other than the Motorola-Freescale family, which appears throughout the textbook, in order to familiarize students with other architectures and products in addition to meeting the project design objectives.

- Showcase wireless communication technology in one example because this technology is becoming more popular in mechatronics products.
- Cover the practical issues and realistic constraints that influence motor selection.
- Design alternatives to mechanisms that are an integral part of building mechatronics projects.
- Showcase as part of an applied case unconventional actuators, such as piezoelectric, since such material is becoming more affordable and more practically packaged.

The following sections present report summaries for the following selected cases.

1 *Autonomous Mobile Robot*: This case presents a summary of the report by a group of students of their mechatronics project, which was the design and implementation of a mobile robot equipped with temperature sensors to remotely report critical collected data from hazardous environments. Highlighted technologies include PIC microcontrollers, mobile robot mechanical design, motor sizing and selection, and PIC interfacing circuits.

2 *Wireless Surveillance Blimp*: This case presents a summary of the report by a group of students for their mechatronics project, which was the design and implementation of an unmanned balloon for surveillance purposes that can be controlled remotely with a wireless modem to convey back critical photos or specific measurements from distant targeted areas. Highlighted technologies include wireless modem and communications, flying balloon and UAV, compass, and accelerometers.

3 *Firefighting Robot*: This case presents a summary of the report by a group of students for their mechatronics project, which was the design and implementation of an autonomous mobile robot that can travel through a specified arena in order to detect the presence of a flame (fire) and extinguish it. Highlighted technologies include design process and alternative selection, motor type selection, UV flame sensor, white-line detector, mercury switch (inclination sensor), IR sensors, and fuzzy logic planning.

4 *Active Vibration Control*: This case presents the experimental station that was developed by a graduate student for the verification of active vibration control. This station is a valid platform for testing the effectiveness of the advanced active vibration control of flexible mechanisms. The highlighted technologies include piezo actuators and sensors, LabVIEW software, real-time data acquisition and programming, power electronics, and control schemes.

The selected projects include many design details, programming codes, and circuits details, which are available for downloading by users of this textbook from the associated website. Although there are redundant efforts among these cases, we stress only the distinctive features that contribute to the value added by each individual project.

14.2   CASE STUDY 1: AUTONOMOUS MOBILE ROBOT

Based on the final-year project prepared by Z. Bizri, A. Jabbour, and F. Wakil.

14.2.1   Introduction

This project handles data collection in hazardous situations, such as environments with unknown carbon dioxide concentrations and high/low temperatures that human bodies cannot tolerate. In these environments human intervention is risky. The objective of this project is to design and build a system consisting of an autonomous robot controlled by a master computer via a remote link. The robot's task is to collect data from an unknown environment and to transmit it to the computer at regular intervals.

14.2.2 Mechanical Design Alternatives

The mobile robot has motors for motion and steering. Several motor types were candidates for the drive train. The first requirement that needed to be met was the separation of the two driving wheels in order to provide accurate steering. It also appeared essential to include a gear reduction system; the gears will change the high speed and low torque generated by the electric motor to the low speed and high torque that are required to move the robot. Following are the available options.

Gearhead DC Motors: This type of assembly includes the gear train prepackaged inside the motor and thus eliminates the challenging task of adapting gears to the motor.

Modified Servo Motors: Servo motors are small devices that have built-in control circuitry that are extremely powerful for their size. The control circuits and a potentiometer connected to the output shaft allow for control of motor position. A normal servo is mechanically incapable of turning farther than 180° due to a mechanical stop built on the main output gear. Giving the motor the signal for 0° will cause it to turn at full speed in one direction. The signal for 180° will cause the motor to go in the other direction. Since the feedback from the output shaft is disconnected, the servo will continue in the appropriate direction as long as the signal remains.

As in any moving vehicle, several steering mechanisms could be implemented. The following were the considered options.

Skid Steering: This steering format requires a differential drive (where actuating wheels are independently driven), as illustrated in Fig.14.1. Skid steering is accomplished by creating a differential velocity between the inner and outer wheels. This causes the vehicle to turn, in proportion to the difference in speed, as demonstrated in Fig. 14.2.

Explicit Steering: This is accomplished by the specific rotation of the vehicle's wheels. This mode can be performed by two types of actuators: stepper and servo. Stepper motor steering

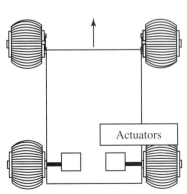

Figure 14.1 Schematic representation of a differential drive.

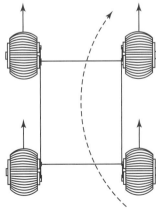

Figure 14.2 Schematic representation of skid steering.

Figure 14.3 Rear-wheel-drive system.

can be rotated by energizing the winding of the motor in a sequential manner. The consecutive energizing pulses enable the motor to turn by the required number of step angles until it reaches the desired steering angle. A regular servo motor can be used, and its operation depends on the received pulse to turn to the required angle.

Since mobile platforms are widely available, the mobile platform of a toy car was adopted (Fig. 14.3). The tricycle wheel drive type was the choice, since three-wheeled vehicles provide continuous wheel-to-ground contact without the need of a suspension system.

A propulsion system is chosen in order to reduce the loss of traction in rapid acceleration: When the robot is accelerating, weight transfers to the rear wheels, which lifts weight off the front wheels; a rear-wheel-drive system is thus preferred in order to minimize traction loss.

Servos are not powerful enough and could not meet basic requirements such as speed. A DC gear-head motor was chosen, with an integrated differential gear (placed midway between the two wheels), allowing the wheels to rotate at independent speeds. This is used to control the traction available to the robot as it attempts to divide the torque equally between the two wheels while cornering.

When driving straight, the differential does not offer additional gear reduction. The gear reduction (which provides a 12.5-to-1 train value) could have been optimized for more reduction in order to provide more torque at low speeds. However this would have entailed mechanical complications; speed regulation was achieved using pulsewidth modulation (PWM).

Servo motors in direct steering have proven to be easier to control than stepper motors since they are pulse coded; i.e., the high time determines the angle to which the motor should move. Another major advantage of servo motors is that every pulse corresponds to a given position; the neutral position could thus easily be initialized by sending the required pulse to the motor, with no need for absolute-position sensing. The direct steering mechanism was adapted to the sole front wheel, simplifying and minimizing the number of linkages. In addition, separate motors used for translation and rotation also make for easier control.

14.2.3 Design Specifications

This section details the various computations performed for the implementation of the specific adopted mechanical design.

Drive Motor Sizing

Assume flat surface operation, properly inflated tires (frictional coefficient $\mu = 0.12$), indoor operation, and, hence, neglecting air drag.

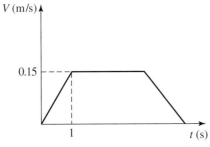

Figure 14.4 Trapezoidal velocity profile.

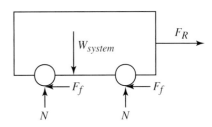

Figure 14.5 System free-body diagram.

Fig. 14.4 presents an estimate of the required trapezoidal velocity profile. The system should reach the maximum velocity in 1 s, so an acceleration of 0.15 m/s² is desired. The wheel radius is estimated to be 3 cm, hence wheel angular velocity is estimated as:

$$v = \omega r = 0.15 \text{ m/s}$$

$$\omega = 5 \text{ rad/s}$$

$$N = 47.746 \text{ rpm}$$

The system weight is estimated as follows:

$$M_{system} = \sum M_{components}$$
$$= M_{body\&wheels} + M_{electronics} + M_{battery} + M_{sensors} + M_{motor} + M_{gears} + M_{misc}$$
$$= 1000 + 200 + 200 + 200 + 500 + 200 + 100 + 600 \tag{14.1}$$
$$M_{system} = 3 \text{ Kg}$$

Figure 14.5 shows the free-body diagram of the system, where W_{system} = system weight, N = ground reaction, F_R = required input force, and F_f = friction force:

$$F_{friction} = N\mu = \mu W_{system} \tag{14.2}$$

From Newton's second law, the resultant required input force F_R is thus:

$$\sum \vec{F}_x = m\vec{a}$$
$$F_R - F_f = ma \tag{14.3}$$
$$F_R = ma + F_f = ma + \mu W_{system}$$
$$F_R = (3 \times 0.15) + (0.12 \times 3 \times 9.81) = 4 \text{ N}$$

The required maximum torque is

$$T_{max} = F_R \times r = 4 \times 0.03 = 0.12 \text{ N-m} \tag{14.4}$$

TABLE 14.1 DC Motor Specs

Model	Op Voltage (V)	No-Load Rpm	Stall Torque (N-mm)	Stall Current (A)
RS-540SH-7520	$3.6 \sim 7.2$	15,400	167	59.0

And thus the required power is

$$P_{requ} = \omega T = 5 \times 0.12 = 0.6 \text{ W}$$

The next step is to determine the available power in DC motors. The maximum output power is attained at half the stall torque, and the corresponding speed at this operating point is equal half the maximum speed:

$$\omega = \frac{1}{2}\omega_{max} \tag{14.5}$$

$$P_{Out} = \frac{1}{4}\omega_{max}T_m \tag{14.6}$$

In order to oversize the motor, we include a factor of safety of 2. The required power is then

$$P_{requ} = 1.2 \text{ W}$$

A Mabuchi RS 540SH (Table 14.1) was selected, with the following characteristics:

$$T_{max} = 167 \times 10^{-3} \text{ N-m}$$

$$W_{max} = \frac{15,400}{60} \times 2\pi = 1612.68 \text{ rad/s}$$

$$P_{max} = \frac{1}{4}W_m T_m = 67.301 \text{ W}$$

The power ranges from 2 W up to the maximum of approximately 65 W. It was chosen as the system main actuator.

Steering Motor Sizing

Assume a steering wheel thickness of 4 cm, a circular area of contact, a maximum steering angle of $60°$, and a response time of 1 s (time to reach maximum angle).

Figure 14.6 presents an illustration of the model used, where t = wheel thickness (m), F_f = frictional force (N), and $r = t/2$.

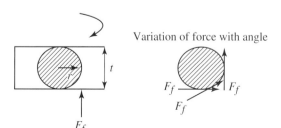

Figure 14.6 Assumed model for steering wheel.

The first step in the design process is to estimate the required power:

$$P = F_f v = F_f \omega r \tag{14.7}$$

where P = required servo motor power (W), ω = angular velocity (assumed to be $\pi/3$ rad/s), and v = wheel velocity. Assuming that the steering wheel carries half of the system weight, we have

$$F_f = \frac{w_{system}}{2} \times \mu \tag{14.8}$$

where w_{system} = system weight, $w_{system} = M_{system} \times g = 3 \times 9.81 = 29.43$ N, and μ = static frictional coefficient, (in order to oversize the system, this coefficient is assumed constant ($\mu = \mu_{static} = 0.3$)). Thus

$$F_f = \frac{0.3 \times w_{system}}{2} = 4.414 \text{ N}$$

$$P_{Required} = 8.829 \times \frac{\pi}{3} \times 0.02 = 0.1 \text{ W}$$

The Futaba S48 was selected, with the following characteristics:

$$T_{out} = 30 \text{ oz-in.}$$

$$\text{Transit time} = \frac{0.22 \text{ s}}{60°}$$

$$T_{out} = 30 \text{ oz-in.} \frac{\text{lb}}{16 \text{ oz}} \frac{1 \text{ N}}{0.2251 \text{ lb}} \frac{2.54 \text{ cm}}{1 \text{ in.}} \frac{1 \text{ m}}{100 \text{ cm}} = 0.2167 \text{ N-m}$$

$$\omega_{max} = \frac{60°}{0.22s} \times \frac{2\pi \text{ rad}}{360°} = 4.76 \text{ rps}$$

The maximum output power is attained at half the stall torque, and the corresponding speed at this operating point is equal half the maximum speed:

$$\omega = \frac{1}{2} \omega_{max}$$

$$P_{out} = \frac{1}{4} \omega_{max} P_{Max}$$

$$P_{out} = 4.6 \times 0.2167 \times \frac{1}{4} = 0.257 \text{ W}$$

The selected motor is acceptable, and the parameters included safety factors.

Gear System

The gearing system serves two main purposes: transmission and transformation of the mechanical energy. For the purposes of a drive train, the gears will change the high speed and low torque of a DC generated by the electric motor to the low speed and high torque that are required to move the robot.

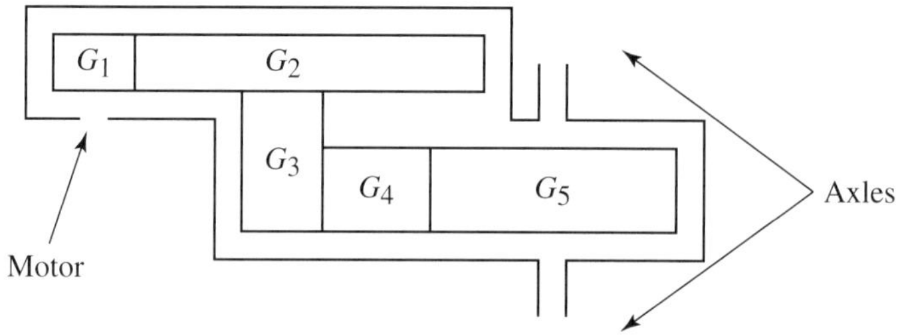

Figure 14.7 Gear reduction schematic drawing.

The chosen differential is a two-stage gear reduction (Fig. 14.7) with the characteristics listed in Table 14.2. The train value of the gear system can be derived as follows:

$$\frac{\omega_{output}}{\omega_{motor}} = \frac{\omega_{G_2}}{\omega_{G_2}} \times \frac{\omega_{G_3}}{\omega_{G_5}} = \frac{N_{G_2}}{N_{G_1}} \times \frac{N_{G_5}}{N_{G_3}}$$

$$\frac{\omega_{output}}{\omega_{motor}} = \frac{80}{16} \times \frac{40}{16} = 12.5 \tag{14.9}$$

$$\frac{\omega_{output}}{\omega_{motor}} = 12.5$$

Thus there is a 12.5-to-1 gear reduction.

The required linear velocity of the robot (after calibration) is 20 cm/s. The diameter of the driving wheels is 6.5 cm. The gear reduction is 12.5 to 1. Thus

$$\omega_{Wheel} = \frac{V_{wheel}}{r_{Wheel}} = \frac{20}{3.25} = 6.15 \text{ rad/s}$$

Or

$$\omega_{Wheel} = 4.6 \left[\frac{\text{rad}}{\text{s}}\right] \times 60 \left[\frac{\text{s}}{\text{min}}\right] \times \frac{1}{2\pi} \left[\frac{\text{rounds}}{\text{rad}}\right] = 58.76 \text{ rpm}$$

The required angular velocity is then

$$\omega_{Motor} = 12.5 \times 58.76 = 734.5 \text{ rpm}$$

TABLE 14.2 Gear Data

Gear Number	Diameter (mm)	Number of Teeth	Function
1	10	16	Fixed to motor shaft
2	50	80	1st-stage reduction
3	15	16	2nd-stage reduction
4	15	16	Transmission
5	37.5	40	Part of differential

The motor is rated at 17,500 rpm (no load). The required modulation signal should therefore have the following duty cycle:

$$\text{Duty cycle} = \frac{734.5}{17,500} = 0.041 \approx 4.1\%$$

Kinematic Analysis

Figure 14.8 presents a drawing of the model used for the kinematic analysis of the tricycle robot, where *ICC* is the instantaneous center of curvature, *a* is the steering angle, *d* is the distance from front to rear axle, θ is the robot heading, *R* is the radius of curvature, ω is the angular velocity, and *V* is the ground linear velocity.

Assumptions
- The instantaneous center of curvature must coincide with the axis of the wheel in contact with the ground; circular motion is exhibited by each wheel around that point.
- No slippage: the traveled distance is assumed to equal $\theta \times r$.
- Lateral slippage is neglected.

The robot has three degrees of freedom, (x, y) is the position, θ is the heading or orientation, and the triplet (x, y, θ) is the pose of the robot in the plane:

$$R = d \tan\left(\frac{\pi}{2} - \alpha\right) \tag{14.10}$$

$$\omega = \frac{V}{B} = \frac{V}{(d^2 + R^2)^{1/2}} = \frac{V}{(d^2 + [d \tan(\pi/2 - a)]^2)^{1/2}} \tag{14.11}$$

$$\theta = \int_0^t \omega + \theta_0 = \int_0^t \frac{V}{(d^2 + [d \tan(\pi/2 - a)]^2)^{1/2}} d\tau + \theta_0 \tag{14.12}$$

$$V_x = V \cos\theta$$

$$V_y = V \sin\theta$$

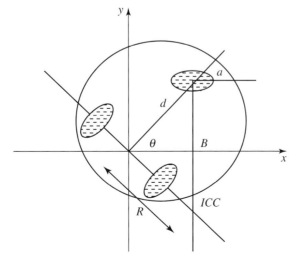

Figure 14.8 Kinematic model.

$$x = \int_0^t V_x = \int_0^r V \cos \theta \, d\tau + x_0$$

$$y = \int_0^t Vy = \int_0^r V \sin \theta \, d\tau + y_0$$

$$(14.13)$$

In this system,

$$a = \pm 45°$$

$$d = 16 \text{ cm}$$

$$V = 50 \text{ cm/s, with 90\% duty cycle measurement}$$

Each steering operation occurs when the robot reaches an obstacle. The robot is required to reach an orientation of $\theta = 0°$. Thus, we replace and get

$$\theta = \int_0^t \omega + \theta_0 = -\frac{0.5}{\left(0.16^2 + \left[0.16 \tan\left(\frac{\pi}{4} \right) \right]^2 \right)^{1/2}} t + \frac{\pi}{2}$$

Required steering time: 0.7109 **s**

Note that the angle θ varies linearly with time: $\theta = \frac{\pi}{2} - 2.2096t$

Replacing: $x = y = 0.22628$ m, or 22.628 cm

Consequently, the robot has to move backward by 22.628 cm before steering. The actual values that were reported and used were: Steer right 0.87 s, left = 0.63 s, backward by 33 cm.

Mechanical Construction

The main aluminum frame constitutes the chassis of the robot; the rest of the components are attached to it (Fig. 14.9). The rest of the components (such as the wheel linkage) were made out of steel. Heavy components such as the battery and the motor were placed as close to the center as possible in order to increase stability. For the same reason, these heavy components were attached from the bottom of the robot, which helps to lower the center of gravity. Sensors were placed on a detection plate positioned over the front steering wheel.

14.2.4 Electronic Circuits and Interfacing

Several circuits were designed and built for the different subsystems of this robot. This section presents the various circuits and selected IC interfacing and communications.

Sensors

For obstacle detection, the choice was proximity detectors using infrared LEDs and detectors or Sharp GP2D02 range detectors. The range detectors have been chosen since they are more accurate and return the distance using an 8-bit value, which can be used inside a program to specify the desired threshold. Sonar ranging can only provide distances between 10 and 80 cm in ideal conditions (100% reflection), decreasing to 10–50 cm in general conditions, and therefore cannot be efficiently used in this application.

Figure 14.9 General view of the robot.

The range detectors return an 8-bit value measuring the distance to the nearest detected object in a range of 10–80 cm. The PIC 16F84A microcontroller will constantly read the values obtained from the range detectors and will activate one of the interrupt lines (through its RB5 line) of the main microcontroller when an object is detected at a distance less than the programmed threshold distance. The range detectors have four lines: Vin, Vout, Gnd, and Vcc. The Vcc and Gnd lines are connected to the main circuit, while the output line, Vout, is connected to an input line of the microcontroller. The Vin line is open drain; therefore a diode is used to interface it with TTL logic. It is used by the microcontroller to notify the range detector to latch the next bit on its Vout line.

Serial Communication Circuit (Fig. 14.10)

The serial port of the station computer is connected to a MAX232 integrated circuit that interfaced RS232 to TTL compatible signals. PIC16F628 microcontroller has one serial communication port and has to be connected to two devices, the computer and the robot main controller; thus communication lines have to be multiplexed while making sure that no signal is misinterpreted. One (I/O) port line was used for selecting the communicating party. The receiving port of the microcontroller is connected to a quad 2:1 multiplexer (74LS157). One input of the multiplexer is connected to the transmission line of the MAX232 chip, and the other input is connected to transmission line coming from the robot main controller.

Robot Circuitry (Fig. 14.11)

The robot main circuit is divided into two separately powered modules for logic circuitry and the motor driving. In this design, two microcontrollers were used: one Microchip PIC 16F877, as the main controller for the robot, and one PIC 16F84A, used solely for obstacle detection. The main controller is connected to the modular sensor through one of its A/D channel input lines (namely, RA0). Also, a power and a ground line are provided for the sensor module. For demonstration, a 15-kΩ thermistor in series with a 10-kΩ resistor were used.

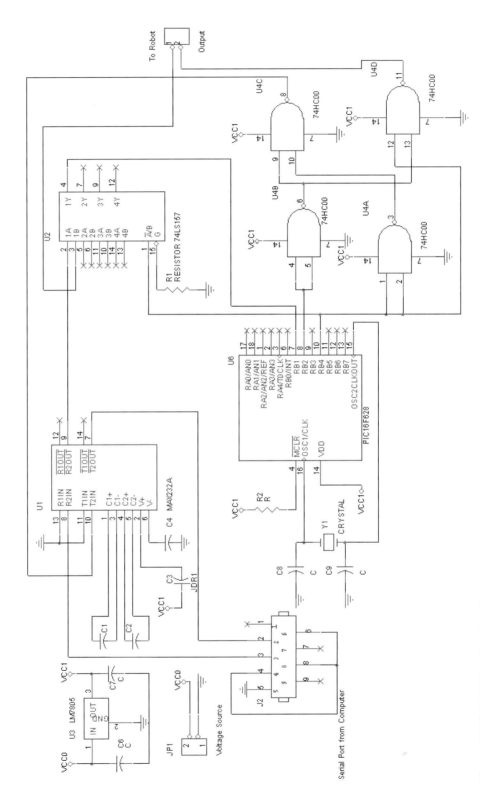

Figure 14.10 Serial interface circuitry.

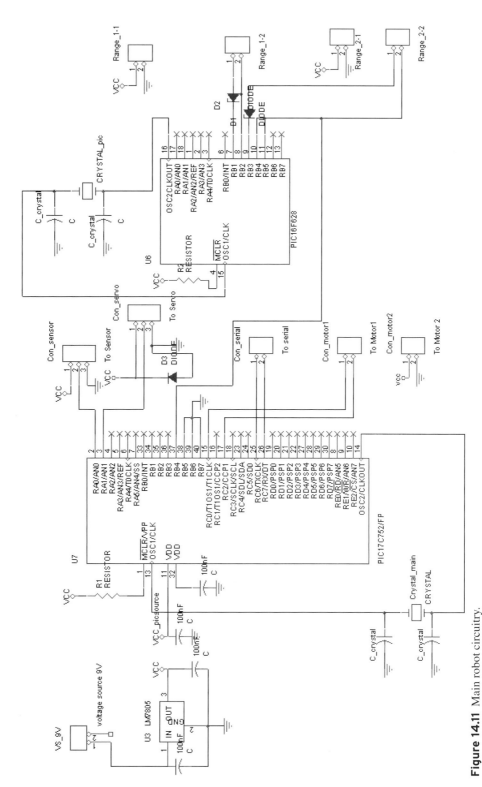

Figure 14.11 Main robot circuitry.

The servo motor used for steering is connected to the RA1 line of the controller. A reverse-biased diode is connected between its power and ground lines, in order to minimize the effects of back emf on the main circuitry when starting the motor.

Also, the RC0 and RC2 lines of the microcontroller are used to control the DC driving motor. RC0 is used for direction (zero for forward motion). The RC2 line is configured as a pulsewidth-modulated line, whose frequency and duty cycle can be programmed, and it is used to control the speed of the DC motor. These lines are input to the motor driving board, together with a power and a ground line. Serial communication is achieved through the RC6 and RC7 lines of the microcontroller, which are in turn connected to the serial communication board. Line RB4 is programmed as interrupt on change and is used for obstacle detection interrupt by being connected to the RB5 pin of the 16F84A controller (used for interfacing the range detectors).

Finally, a voltage regulator is used to enable the powering through batteries, and damping capacitors are placed near the microcontroller for better electric stability.

Motor Driving Circuitry

The motor driving circuitry (Fig. 14.12) is based on the use of an L298N motor driver. Schottky diodes were used in order to have faster switching when pulse-width modulation is used to control speed.

The motor driver has two inputs to determine the direction of motion. When both are at the same logic level, the driver brings the motor to a fast stop; this is therefore equivalent to a brake. When the inputs are of different logic levels, the L298 chip will drive the motor into one direction. Therefore, to control the direction and the speed of the motor, initially it is left as always enabled by connecting the enable pin to the logic high source. One input is connected to the RC0 pin of the main controller (used for direction) and the other pin to the RC2 pin of the main controller (which is the PWM output). By changing the duty cycle of the PWM signal, control of the speed of rotation of the motor is achieved, and alternate motion and

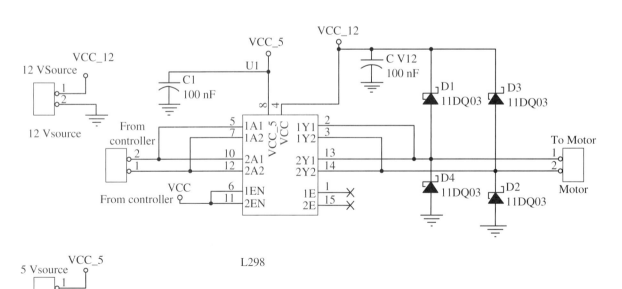

Figure 14.12 Motor driving circuitry.

braking at a high frequency are possible (e.g., 1 kHz). The motor must first start at a high torque to overcome initial friction, then it must decrease its output to a constant lower speed. This is done by generating a 90% duty cycle pulse for starting up and then by generating a 50% duty cycle pulse for constant speed thereafter.

Communication Strategy

The user computer is given priority and is denoted as the master. The adopted strategy is comparable to roll-call polling (a common medium-access control strategy in computer networks). In this scheme, the master initiates the communication with one slave (robot), waits for a reply, and then repeats this process for the next one.

The frames that can be sent by the computer are of two types: active and regular commands. Active commands denote the commands that will instruct the robot to move in a certain direction, stop, steer, or decelerate. On the other hand, regular commands only request the robot to start or stop the A/D conversion process or to send a simple status reply.

If bit 7 is set, then there is an active command. The rest of the bits are interpreted as follows.

Bits 6–4	000	stop
	001	decelerate
	010	forward motion (fast)
	011	backward motion (fast)
	100	rotate servo in center position
	110	rotate servo right
	111	rotate servo left
Bits 3–0	These are unused but can be extended later for ID specification.	

If bit 7 is clear, then there is a regular command. The rest of the bits are interpreted as follows.

Bits 6–5	0x	regular poll
	10	A/D cycle start
	11	A/D cycle stop
Bit 4	This is unused.	
Bits 3–0	These are unused but can be extended later for ID specification.	

Now, the reply from the robot can be either 1 or 2 bytes long, depending on whether a value is to be sent after A/D conversion (since the result is 10 bits long). The general format is as follows.

Bit 7	0	1-byte reply
	1	2-byte reply (i.e., a second byte will be sent after this one)
Bit 6	0	No obstacle has been detected yet
	1	An obstacle has been detected
Bits 5–2	These are unused but can be extended later for further notifications.	
Bits 1–0	xx	(the last two bits in the A/D conversion result).

If a second byte is sent, it will contain only the eight most significant bits of the latest A/D conversion result.

Microcontroller Choice

Between the Motorola-Freescale and the Microchip PIC, the PIC MCUs were chosen for this project since they contain flash memory and thus can be placed directly on test boards, while the HC11 MCUs could not be removed from their evaluation board, for they would lose their program memory. In addition, PIC is a range of MCU family in many different package types, which can be chosen for specific tasks.

Three microcontrollers were integrated. All programs were written in assembly language, in order to have fast and efficient operations. The programs developed are available on the website associated with this textbook.

Interfacing the Servo Steering Motor

The servo motor steering position is achieved by generating a pulse of a certain duration on its control line every 20 ms (approximately). After calibration, a pulse of 1.32-ms duration makes the servo motor rotate to the center position, a pulse of 2.4-ms duration to an approximate 90° position to the right, and a pulse of 0.4-ms duration to an approximate 90° position to the left. Each pulse has to be repeated approximately 16 times to achieve a complete rotation to the desired position.

14.2.5 Software Design

The lengthy and rather complicated involved code, coupled with the difficulty of generating user-friendly graphics eliminated C++ as an option for the programming environment. While Visual Basic was better, since it requires less code and generates better interfaces, concern over the reliability of the serial communication made this option less attractive. MATLAB encapsulates all features needed in the design; it integrates computation, visualization, and programming in an easy-to-use environment.

Serial Communication Algorithm

In order to achieve this task, MATLAB was used to:

- Create a serial port object, which is tied into the serial COM 1 port.
- Connect to the device, which opens the path between the serial port and the robot.
- Establish the correct port behavior. This is achieved by setting the desired baud rate, the parity type, and the number of data bits and stop bits. The baud rate was set to 2400 bps, in order to guarantee perfect compatibility with the programmable interrupt controller (PIC).
- Write and read data.
- Disconnect and clean up; this detaches the device from the serial port.

In this process, asynchronous communication was used. This prevents blocking of the MATLAB input line while communication is taking place. Effectively, data is read asynchronously from the robot and stored in an input buffer. It is then automatically transferred from the input buffer to a MATLAB variable by using the synchronous read function "fread". This means that although the communication is asynchronous, the data stored in the input buffer is retrieved at a certain clock rate.

Data Collection

In normal mode, the computer waits for a data measurement from the robot. This is achieved by keeping the A/D flag set and resetting it in case an obstacle is found. In order to read the data value sent by the robot, the computer waits for two bytes and manipulates them via multiplication and modulo operations to generate the result of the analog-to-digital conversion.

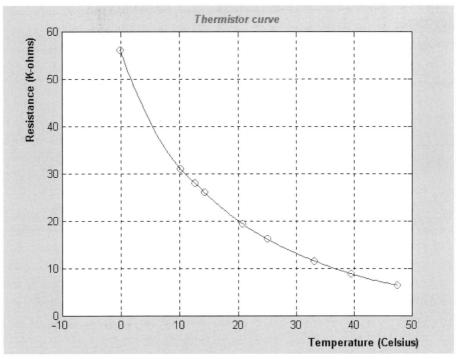

Figure 14.13 Thermistor curve generated from remotely measured and sent samples.

The robot was tested using a temperature sensor—the thermistor. After calibration, the curve in Fig. 14.13 was obtained; the circled values are actual measurements, and the curve is a cubic spline interpolation. The resistance is a decreasing function of temperature, and the temperature range of interest is between 0 and 50° Celsius.

Position determination, robot movement, and data collection all lead to the following discussion of a suitable motion algorithm for the robot.

Motion Algorithm

- The robot starts moving up from the bottom middle of the room.
- The robot is trying to follow a source of temperature; this means that its objective is to find the highest temperature in the room.
- The robot will stop as soon as it finds an obstacle.
- The computer orders the robot to move backward to the point of highest temperature.
- From that point, the robot steers to the right.
- Once the wheels are aligned again, the robot progresses forward.
- It stops again at the next obstacle.
- The computer compares the temperatures; at higher temperature points it moves left.
- The algorithm is then repeated.
- If the temperature at the new position is lower, the robot moves backward and goes right.
- The algorithm is then repeated.

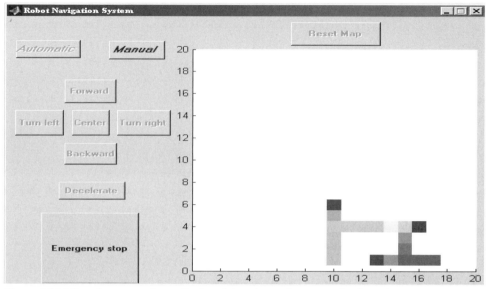

Figure 14.14 Sample generated map display of the robot path, ranges of collected temperatures,

When the system is in automatic mode, the robot will follow the foregoing motion algorithm. After the computer has calculated a new robot position and collected the temperature reading at this position, it should display the results on the map in real time.

Map Generation

The map displays in real time the path traced by the robot, the temperature measurements, and the location of the obstacles encountered—if any. The obstacles are displayed in black; temperatures should be displayed according to a range of colors, going darker with higher temperatures and lighter with lower temperatures. MATLAB stores images as two-dimensional arrays (i.e., matrices), each element of the matrix corresponds to a single pixel in the displayed image; moreover, the decimal value stored in each location represents a different color. A sample map display is given in Fig. 14.14.

14.2.6 Case Outcomes

This case is a typical mechatronics project, integrating the design and the interface of many technology blocks and subsystems. The following are direct outcomes.

1. Selection of mobile robot driving actuators
2. Mobile robot steering choices and actuators
3. Motor sizing based on application dimensions and performance needs
4. Kinematic and dynamic analyses of mobile robots
5. Design and implementation of sensor and motor interfacing circuits
6. PIC microcontroller selection, interfacing, and programming
7. Obstacle-avoidance sensors and strategy
8. MATLAB programming

REFERENCES

D. Apostolopoulos, "Analytical Configuration of Wheeled Robotic Locomotion," Ph.D. thesis, Carnegie Mellon University, May 1998.

J. E. Cooling, *Real-time Interfacing*. New York: Van Nostrand Reinhold (UK), 1986.

G. Dudek and M. Jenkin, *Computational Principles of Mobile Robotics*. New York: Cambridge University Press, 2000.

James E. Duffy, *Modern Automotive Technology*. Tinley Park: Goodheart-Willcox, 2000.

Joseph L. Jones and Anita M. Flynn, *Mobile Robots*. Wellesey: A. K. Peters, 1993.

K. M. Marchek and R. C. Juvinall, *Fundamentals of Machine Component Design*. New York: Wiley, 2000.

The Mathworks, MATLAB 6.1 built-in documentation, 1999.

Microchip Technology, PIC16F84 documentation, 2001.

Microchip Technology, PIC16F628 documentation, 2001.

Microchip Technology, PIC16F877 data sheet, 2001.

John B. Peatman, *Design with Microcontrollers*. New York: McGraw-Hill, 1998.

Benjamin Shamah, "Experimental Comparison of Skid Steering vs. Explicit Steering for a Wheeled Mobile Robot," M.S. Thesis, Carnegie Mellon University, March 1999.

14.3 CASE STUDY 2: WIRELESS SURVEILLANCE BALLOON

Based on the final year project prepared by M. Elhajjar, S. Nakhal, K. Salamy, and A. Majdalani.

14.3.1 Problem Definition

The majority of aerostatic balloons available on the market are small to medium-sized indoor blimps, or extremely large outdoor blimps, which are very expensive. When attempting small outdoor blimps, many challenges are raised, the most important of which are their short flight time and short blimp-support group distance.

Airships and balloons carry passengers and/or cargo, and they obtain their buoyancy from the presence of a lighter-than-air gas, such as hydrogen, helium, or hot air, based on Archimedes principles. An airship can carry a variety of sophisticated surveillance equipment, including data links to keep the commanders on the ground fully in the picture.

This balloon project is an UAV platform for wireless surveillance missions, such as maritime patrol, key installation security, geographic aerial imaging (map generation), and intelligence. It is an assisting tool for critical missions in general and a very good alternative to surveillance helicopters and planes.

14.3.2 Design

The shape selected (Fig. 14.15) combines the positive aspects of a balloon and a blimp. The balloon should be as small as possible while satisfying long flight time without the need for close-by support. It should be able to reach around 5 km from its support crew and should contain various mechanisms for handling loss of communications, power, and buoyancy.

Actuation System

Define an orthonormal reference frame fixed to the body of the balloon, with its origin coincident with the center of buoyancy. Together with a thruster (Fig. 14.16), two motors driving

Figure 14.15 Project balloon.

a pair of propellers are employed to move the balloon up (Fig. 14.17). The pair of propellers is used to move the balloon forward or backward and also to rotate by accelerating one propeller more than the other:

The x-axis pair: F_x, T_z
The y-axis pair: F_y, T_x
The z-axis pair: F_z, T_y

where F_x and T_z, for example, denote thrust and along torque around the x-axis and the z-axis, respectively, and the remaining terms are defined similarly. Other force–torque combinations are also possible.

The balloon needs the thruster to move upward. After the balloon is up in the air to a desired elevation, it needs two propellers to track a predetermined trajectory. The two propellers are used to move the balloon forward or backward by rotating at various individual speeds. Hence, to rotate the balloon left or right, one of the propellers has to move faster than the other.

Figure 14.16 Ducted fan for upthrust.

Figure 14.17 Brushless DC motor with propeller.

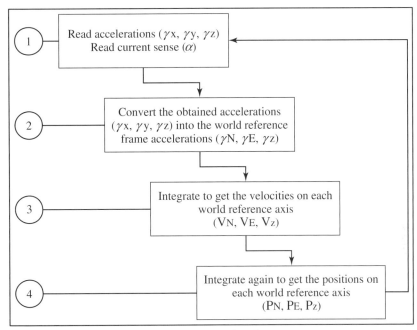

Figure 14.18 General flow of the positioning function.

Navigation System and Kinematics (Fig. 14.18)

The ground station should be able to trace the UAV by knowing its position. However, the balloon and the ground cannot have the same reference frame, since the flying system is moving. Therefore, we have a moving body frame (X1, Y1, Z1), referred to as (B-frame), and an inertial fixed world reference frame (X0, Y0, Z0), referred to as (I-frame). The B-frame has its origin coincident with the center of gravity of the system; its x-axis is parallel to the direction of the forward motion; its z-axis is parallel to the direction of the upward motion; and its y-axis completes the right-hand rule. On the other hand, the I-frame has its origin at a set point on the ground (most probably the position of the ground station). The x-axis of the I-frame is toward magnetic north, because the compass installed on the system gives the direction of the x-axis of the B-frame relative to magnetic north. Thus, to know the position of the system, the x-position of the B-frame is used along with the angle with respect to the magnetic north from the compass. The z-axis of the I-frame will be pointing in the direction parallel and opposite to the direction of gravitational acceleration. Moreover, the y-axis will complete the right-hand rule. α is the angle obtained from the compass, and it represents the angle between X1 and X0.

Using simple trigonometry, the transformation from the B-frame to the I-frame is given by the following equations:

$$X0 = X1 \times \cos(\alpha) - Y1 \times \sin(\alpha)$$

$$Y0 = X1 \times \sin(\alpha) + Y1 \times \cos(\alpha) \tag{14.14}$$

$$Z0 = Z1$$

This will produce the location (X0, Y0, Z0) of the balloon center with respect to the world reference frame. The orientation of the balloon is represented by the roll angle (α) about the z-axis (Fig. 14.19).

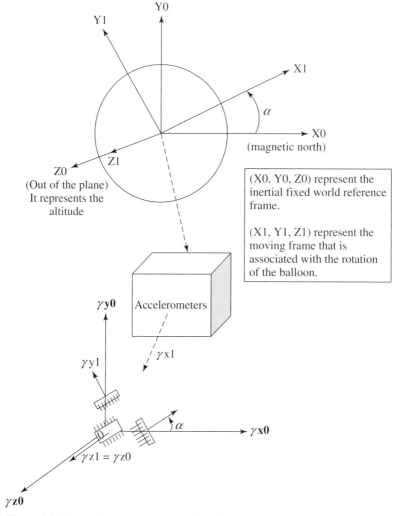

Figure 14.19 Accelerometers in three dimensions.

Control Procedure

The blimp can work in two modes. The first one is the closed-loop control mode, where the user specifies a planned desired trajectory in the ground station before the balloon is operated. In this mode, the balloon will adjust its motion according to the feedback from the accelerometers and compass. The other mode of operation is the open-loop control mode, where the operator on the ground station controls continuously the motion of the balloon by sending throttle signal commands to the onboard microcontrollers through the wireless modems.

In the closed-loop automatic control mode, giving feedback on the position, orientation, and speed of the balloon, a scheme is designed to control the propellers' motors and the ducted fan so that the UAV will be able to track a desired heading, at a specified altitude, and move forward in a prespecified speed. Such closed-loop system requires extensive modeling and controller design, which are included in the complete report on the textbook website. In this section, only open-loop strategy is presented.

The operator can communicate with the balloon via a wireless channel. The ground station reads the acceleration, speed, position, and orientation of the balloon, while the balloon accepts the throttle level from the ground station. The ground station can specify the speed of the ducted fan and each of the propeller motors. It sends these parameters to the balloon, where they are stored in the "ground station interface" (GSI) microcontroller. The GSI microcontroller sends the received values on three of its output ports, which are connected to the throttle signal microcontroller of each motor. The throttle signal is a PWM with a period of 20 ms, where the ON period ranges from 1 to 2 ms. Depending on the ON period, the motor speed will vary from zero to full speed. After receiving the throttle command from the GSI microcontroller, the throttle microcontroller translates the command into the corresponding PWM signal. The GSI microcontroller receives the throttle command via the wireless channel, which is prone to noise, so there is a probability of receiving a wrong throttle command. This would be a major problem in case the wrong throttle signal imposes a sudden change in the speed of the motor that causes a lot of vibrations that might damage the structure. To avoid this problem, a digital filter is embedded in the throttle microcontroller code. The basic idea behind this filter is that the controller passes the average of the last 10 throttle levels, instead of passing directly the current throttle level.

Communication System (Fig. 14.20)

The UAV receives commands from a ground station in order to move according to a prespecified set of desired parameters (position, speed, and orientation). However, the ground station needs to be aware of the current status of the balloon in order to issue the appropriate commands. Therefore, a full-duplex communication system is needed for reliable operation of the system.

The navigation microcontroller will be sampling the orientation as well as the accelerations in the x-, y-, z-directions. Based on these values it will calculate the current velocity and position, as discussed previously. At this stage, it will pass these values to the ground station microcontroller.

Initially this communication scheme was implemented using the computer's serial port. The rule of thumb for this protocol is as follows: If device 1 wants to send data to device 2, it

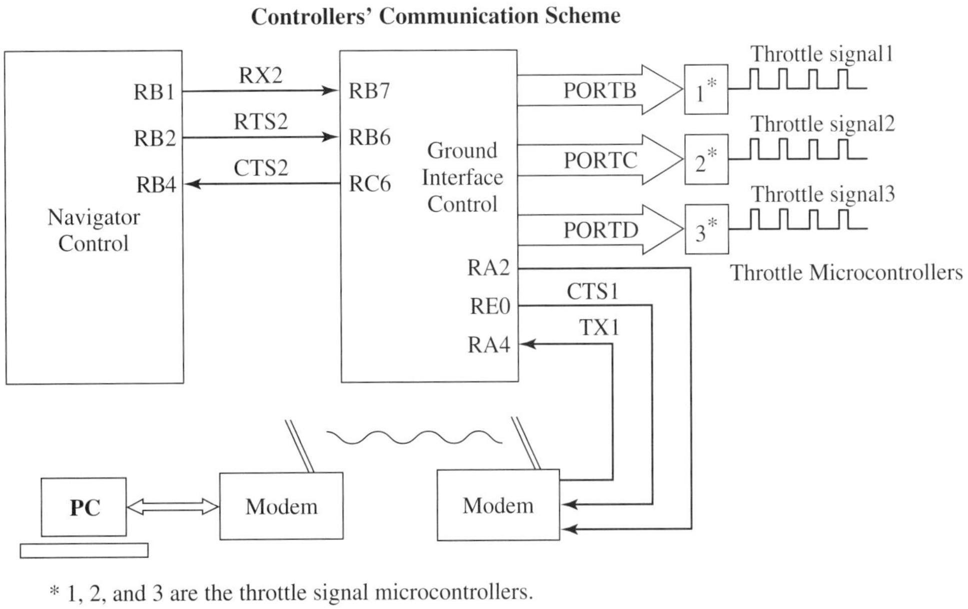

Controllers' Communication Scheme

* 1, 2, and 3 are the throttle signal microcontrollers.

Figure 14.20 Communication flowchart.

would have to issue a request to send (RTS) and then wait until it receives a clear-to-send (CTS) signal from device 2. Receiving the CTS signal indicates that device 2 is ready and waiting for input data. Using this protocol, it was possible to establish a full-duplex reliable communication network between the ground station, the GSI microcontroller, the navigation microcontroller, and the three throttle microcontrollers. The computer sends an RTS to the modem connected to it. When the modem is ready, it sends back a CTS signal to enable the transmission of data. The modem stores the data in its buffer and then sends the data to the other modem, which in its turn stores the data temporarily in its output buffer. The GSI microcontroller, when ready to receive data, will trigger the CTS pin of the modem, causing it to release the data stored in its buffer. The GSI microcontroller communicates with the navigation microcontroller and the throttle signal microcontrollers in the same way.

The GSI microcontroller, after receiving data from the navigation microcontroller, directly sends it to the modem without any RTS/CTS handshaking. This kind of communication is valid because the modem has an input buffer where data can be stored temporarily to be sent to the other modem when possible. The modem on the side of the computer sends an RTS signal to the computer, which will issue a CTS whenever it is ready and then read the data from the modem.

Base Station Interface System

The base station interface (Fig. 14.21) enables the user to control the balloon in two modes: open loop and closed loop. The interface also gives the user all the necessary data for normal operation of the balloon. The interface gives the following telemetry data:

- Motors 1, 2, and ducted-fan rpm
- Batteries 1, 2, and 3 voltage level
- Motors 1, 2, and ducted-fan temperature
- Direction
- Speed
- X,Y, and Z

Drag Calculations and the Balloon Envelope (Table 14.3)

Several lighter-than-air gases could be used inside balloons to achieve lift: hydrogen (H_2), helium (He), and methane (CH_4). Hydrogen and methane are both explosive and highly flammable, so they will not be considered further, for obvious safety reasons, which leaves helium as the gas of choice. The appropriate balloon choice is a 2.6-m-diameter sphere made of polyurethane. Balloon volume is then 9.203 m^3, which lifts 8743 g. Subtracting the weight of the balloon envelope, we get a lift of 6300 g.

The balloon cross-sectional area is 5.31 m^2. The balloon will be subjected to a drag proportional to the square of its forward velocity at zero wind or the square of wind speed when the balloon is stationary.

14.3.3 Parts

The following parts are used in this project.

Brushless DC Motor (Fig. 14.22)

Supplier: Aveox
Quantity: 2
Model: 36/38/3 (Table 14.4)

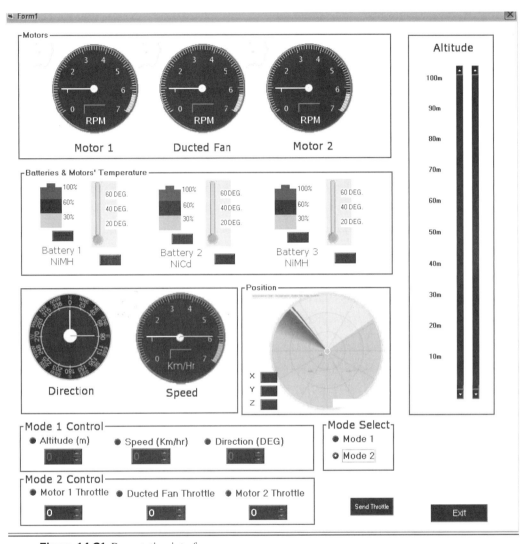

Figure 14.21 Base station interface.

TABLE 14.3 Balloon Estimated Load

Component	Weight (g)	Count	Total Weight (g)
Ducted fan	200	1	200
Brushless motors	560	2	1120
Accelerometer box	100	1	100
NiMH battery pack	225	6	1450
NiCd battery pack	525	1	525
PCB board	310	1	310
Aluminum box	75	2	150
Propeller	50	2	100
Receiver	220	1	220
PVC loop	90 (per m)	10	900
			5075

Figure 14.22 Brushless DC motor.

TABLE 14.4 Brushless DC Motor Specs

Motor Type	Kv (rpm/V)	I_{cont} (A)	I_{peak} (A)	w_{max} (rpm)	Weight (g)
36/38/3	1172	28.0	65	50k	323

Advantages of Brushless DC Motors

- There is no brush; the overall efficiency of the motor is high (greater than 85%).
- Far less electrical noise is generated to interfere with the remote control.
- No maintenance is required on the brushless motor, and there is no deterioration of performance over the life of the motor.
- At lower power levels, the iron losses $= V_{in} \times I_{no\text{-}load}$, and brush drag tends to be the dominant loss in the motor. Since this is much lower in brushless motors, they are efficient at light loads.
- At higher powers, the copper or I^2R losses tend to dominate. Since brushless motors usually have lower resistance in the windings, they are more efficient at higher loads.
- There is no mechanical component in the commutation of a brushless motor; they can reliably operate at higher rpm's.

Motor Drive (Fig. 14.23)

Supplier: Aveox
Quantity: 2
Model: SH-48 (Table 14.5)

Figure 14.23 Aveox motor drive.

TABLE 14.5 Aveox Motor Drive Specs

Model	Cells	Peak Amps	Cont. Amps	Optocoupled	Brake	Beep-on Power	Fwd/Rev	OverTemp. Cutoff
SH-48	10–30	50	40	Yes	Yes	Yes	No	Yes

Since brushless motors are essentially three-phase AC motors that are commutated electronically, a brushed motor speed controller cannot be used. Each brushless motor has three power wires, instead of the two found in a brushed motor, along with five sensor wires. These sensor wires take the place and function of the commutator in a brushed motor. The advantage of a brushless controller is that there must be a minimum of six MOSFETS (Aveox controllers use 12) to commutate the motor. This makes the effective ON resistance lower than a typical brushed controller, and therefore it will stay cooler at high power levels. This motor drive has four DIP switches. The DIP switches have the functions shown in Table 14.6.

The following combination was set in this project:

Switch 1 = OFF

Switch 2 = Not used

Switch 3 = ON

Switch 4 = OFF

Propeller (Fig. 14.24)

Supplier:	JP Zinger
Quantity:	2
Model:	Pusher 14 × 8 (Table 14.7)

Ducted Fan (Fig. 14.25)

Supplier:	AstroFlight
Quantity:	1
Model:	Astro 020 803F

Ducted-Fan Drive (Fig. 14.26)

Supplier:	Castle Creations
Quantity:	1
Model:	Phoenix (Tables 14.8 and 14.9)

TABLE 14.6 Aveox Motor Drive Dip Switch Configuration

Switch Number	Switch ON	Switch OFF
1	Brake	No brake
2	Not used	Not used
3	Soft start	Hard start
4	Competition	Normal

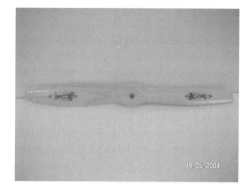

Figure 14.24 Propeller.

Figure 14.25 Ducted fan.

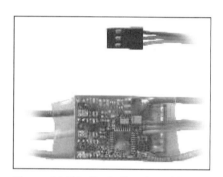

Figure 14.26 Castle Creations drive.

TABLE 14.7 JP Zinger Propeller Specs

Model	Dimensions	Description
JP Zinger	14×8	One-piece propeller

TABLE 14.8 Castle Creations Drive Specs

Motor Type	Cells	Voltage Range (V)	Kv (rpm/V)	I_{cont} (A)	I_{peak} (A)	w_{max} (rpm)	Max. Power (W)
p/n 803F	7–10	8.4–12	3300	21	25	30k	240

TABLE 14.9 Castle Creations Drive I/O Characteristics

Battery	Amps	Watts	RPM
7	13	105	23.1K
8	16	145	25.8K
9	18	180	27.8K
10	21	240	30K

Figure 14.27 Wireless modem.

The sensorless control has a decoder for the radio receiver signal and a microprocessor for controlling the motor. The motor and control can safely handle 25 A.

Wireless Modem (Fig. 14.27)

Supplier: Microdaq
Quantity: 1
Model: X09-019PKT-RA (Table 14.10)

TABLE 14.10 Max Stream Wireless Modem Specs

General

Frequency	902–928 MHz
Spreading spectrum type	Frequency hopping, direct FM
Network topology	Point-to-multipoint, point-to-point multidrop transparent
Channel capacity	Seven hop sequences share 25 frequencies
Serial data interface	Switch selectable RS-232/422/485
I/O data rate	Software selectable 1,200–57,600 bps

Power Requirements

Supply voltage	7–18 V_{DC}
Transmit current	200 mA
Receive current	70 mA

Physical Properties

Enclosure size	2.75 in. $\times$ 5.50″ $\times$ 1.124 in. (7.90 cm $\times$ 13.90 cm $\times$ 3.80 cm)
Weight	7.1 oz (200 g)

Antenna

Type	Half-wave dipole whip
Gain	2.1 dBi
Length	6.75 in. (17.1 cm)
Connector	Reverse-polarity SMA
Impedance	50 ohms unbalanced

Performance

Outdoor LOS range	Up to 7 mi (11 km) with dipole
	Up to 20 mi (32 km) with high gain

TABLE 14.11 Max Stream Modem Pin Configuration

Pin	Signal	Direction of Transmission
1	Data carrier detect	
2	Received data	PIC to computer
3	Transmitted data	Computer to PIC
4	Data terminal ready	
5	Signal ground	
6	Data set ready	
7	Request to send	Computer to PIC
8	Clear to send	PIC to computer
9	Ring indicator	

This modem is a wireless serial port. The pins have the configuration shown in Table 14.11 and Fig. 14.28. One wireless modem communicates with the PC, and the other modem communicates with the PIC micro controller (Fig. 14.29). The handshaking between the modem and its host, PC or microcontroller, is necessary for the smooth flow of data. The handshaking protocol used with these modems is as follows:

• The host device can send data to the modem at any time as long at the CTS line is set low. The modem sets the CTS line high only when the DI buffer is full; this case happens only when the host device is transmitting data at a baud rate higher than that of the modem.

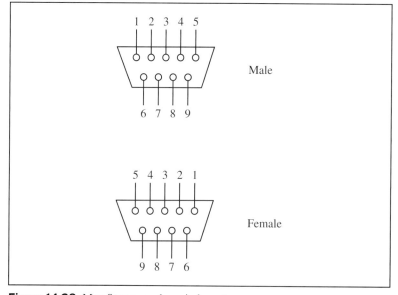

Figure 14.28 Max Stream modem pin layout.

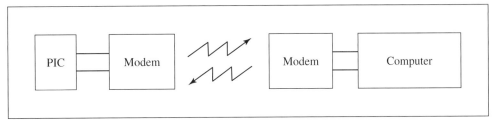

Figure 14.29 Max Stream modem and its host devices.

- The modem can send data to the host device when the RTS line is low. If the RTS line is high, then the modem stores any incoming data from the other modem into its DO buffer until the host device sets the RTS line to a low state.
- The default settings of the modem must be modified to enable the RTS/CTS data flow control. The modem parameters can be modified using the XTU software supplied with the modem. The modem parameters presented in Table 14.12 were used.

Lift Element (Fig. 14.30, Table 14.13)

Supplier:	Mobile Airships
Quantity:	1
Model:	04-B25

Note: Helium used should be 99% pure, otherwise the net lift will not be 6.3 kg; for example, "party balloon helium" is 70% pure and thus can not be used.

Accelerometer (Figs. 14.31 and 14.32, Table 14.14)

Supplier:	Newark InOne
Quantity:	3
Model:	MMA1260D

This accelerometer can be used to measure accelerations in the x-, y-, or z-direction. Yet care must be taken as to the positioning.

TABLE 14.12 Max Stream Modem Interfacing Options

Serial Interfacing Options	
BD—Baud rate	4—19200 (default)
RT—DI2 configuration	2—RTS flow control
FL—Software flow control	0—No software flow control (default)
FT—Flow control threshold	(Default)
CS—DO2 Configuration	0—Normal (Default)
NB—Parity	0–8-bit, no parity (Default)

Figure 14.30 Polyurethane balloon.

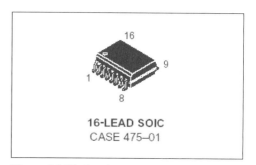

16-LEAD SOIC
CASE 475–01

Figure 14.31 Accelerometer.

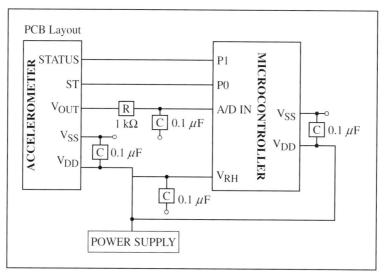

Figure 14.32 Accelerometer connection diagram.

TABLE 14.13 Balloon Specs

Diameter (m)	Volume (m³)	Loss Per Day (m³)	Net Lift (Kg)
2.59	9.118	0.09	6.3

TABLE 14.14 Accelerometer Specs

Characteristic	Typical Value	Unit
Supply Voltage	5	V
Supply current	2.2	mA
Acceleration range	1.5	g
Zero g	0	V
Sensitivity	1200	mV/g

Figure 14.33 Analog compass.

Analog Compass (Fig. 14.33)

Supplier:	The Robson Company
Quantity:	1
Model:	R1655

NiMH and NiCd Batteries

Supplier:	Hobby-Lobby
Quantity:	10 packs
Model:	B2150N6

The chosen batteries are NiMH capable of providing 2.15 A-h at 1.2 V (Fig. 14.34). Nominal 100-W operation for 30 minutes is 50 W-h or 41.6 A-h at 1.2 V. Therefore, 10 batteries will be needed. A battery of this type can be discharged at a maximum rate of 5 A; therefore, 10 batteries will supply only 60 W.

A NiCd battery pack (Fig. 14.35) is used to supply power to the thruster. The rated power needed for the thruster is 75 W, and the battery is rated 17 W-h. Therefore, the battery pack can supply the fan with power for 13 minutes, which is enough time, since the thruster will not be operating continuously.

14.3.4 Case Outcomes

This case is a typical mechatronics project, integrating the design and interface of many technology blocks and subsystems. The following are direct outcomes:

1. Kinematic analysis of UAV
2. Motors interfacing circuit design and implementation
3. PIC microcontroller selection, interfacing, and programming
4. Wireless communications (modems)
5. Accelerometers and compass sensors and interface

Figure 14.34 NiMH battery pack.

Figure 14.35 NiCd battery pack.

REFERENCES

Accelerometer: http://www.motorola.com

The Airship Simon: http://mypage.bluewin.ch/airshipsimon/history.html#introduction

American Blimp Corporation: http://www.americanblimp.com/civil.htm

Analog Compass: http://www.rosonco.com

AstroFlight Brushless DC Motors: http://www.astroflight.com/Brushless.html

AstroFlight Cobalt 90 Airplane motor: http://www.astroflight.com/fai1020.html

Blimps Make Comeback in Aerial Security, WNBC exclusive: http://www.wnbc.com/anniversary/1471149/detail.html

Mobile Airships: http://www.blimpguys.com

The New Military Industrial Complex: http://www.business2.com/articles/mag/0,1640,47023%7C2,00.html

Sanyo NiMH Batteries: http://www.sanyo.com/batteries/specs.cfm

14.4 CASE STUDY 3: FIREFIGHTING ROBOT

Based on the final year project prepared by M. Raydan, L. Elazar, and B. Ghaddar

14.4.1 Problem Statement

Since firefighting is a risky job, the task in this project is to produce a robot that can enter a fire scene and detect and put out a fire without human intervention. F^2R is a robot prototyping a mobile autonomous firefighting robot. It would be able to search for a candle in a predefined arena and put it out. The arena simulates the real world, while the candle simulates the fire. The constraints, limitations, and arena have been set by the Trinity College Firefighting Robot Contest in order to model real-life situations as accurately as possible. The firefighting robot has been the senior project for many universities. There are four levels of competition:

• Junior division (for students below 9th grade)
• High school division (below 12th grade)
• Senior division (university students, graduate engineers, and scientists)
• Expert division

The competition demands the development of a small autonomous robot able to navigate through a model house maze in order to find a candle in minimal time and then to extinguish it. The maze characteristics are known in advance by the contestants, and the candle is randomly placed in this maze. Further, there are obstacles randomly placed in the maze.

Arena Description (Fig. 14.36)

• A staircase in the Standard Arena can be circumvented by taking a longer route.
• Rugs are placed in some or all of the rooms and hallways.
• Small household objects may be placed in the arena.
• All hallways and doorways to room will be approximately 46 cm wide. There will not be a door in the doorways, just a 46 cm opening. There will be a white tape, approximately 2.5 cm wide, fixed to the floor across each doorway to indicate room entrances.

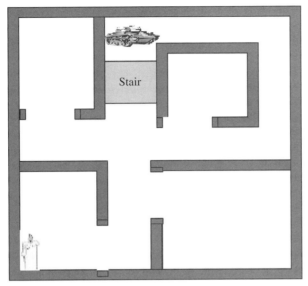

Figure 14.36 Standard-level known layout arena.

- Some robots may use foam, powder or other substances to attack the candle flame and hence there is no guarantee that the floor will remain clean.
- The robot will start at the known Home Circle location.

The contest hall contains ambient lighting, including sunlight, and there would be inconsistent lighting, shadows, glares, etc. This lighting configuration is intentional in order to achieve maximum accuracy in modeling the real world.

Extinguishing Mechanism

The robot must not use any destructive or dangerous methods to put out the candle. It may use such substances as water, air, CO_2, or others. Materials that are harmful to the environment are not allowed. It is allowed to put out the candle by blowing air or other oxygen-bearing gas. There will be a white 30-cm-radius solid circle or circle segment on the floor around the candle, which is placed in the center.

Robot Dimensions

Robot must be able to fit in a box 31 cm long by 31 cm wide by 27 cm high. If the robot has sensors or antennas, these will be counted as part of the robot's total dimensions. There are no restrictions on the weight of the robot or on the types of materials used in the construction of the robot.

The Candle

The candle flame will be from 15 to 20 cm above the nominal floor level. The candle thickness normally will be between 2 and 3 cm. The robot is required to find the candle no matter what size the flame is at that particular moment. The candle will be placed at random in one of the rooms in the arena. The candle has an equal chance of being in any of the four rooms. The candle will not be placed in a hallway, but it might be placed just inside a doorway of a room. The candle circle will not touch the doorway line, which means that the front of the robot will be able to move at least 33 cm into the room before it encounters the candle. The candle will be mounted on a small wooden base painted semigloss yellow.

Sensors

Robots that use laser-based devices must take measures to prevent eye damage to team members and to observers. Besides this, there is no restriction on the type of sensors that can be used in the contest, as long as they do not violate any of the other rules or regulations.

Time Limits

In order to achieve the contest objective of building a robot that can find and extinguish a fire in a house, finding the fire within a reasonable period of time is very important. The maximum time limit for a robot to find the candle will be 5 minutes. After 5 minutes, the trial will be stopped. The maximum time for the robot to return to the Home Circle in the Return Trip mode will be 2 minutes.

14.4.2 Design Alternatives

The method used to select among the different alternatives is the Pugh matrix. The alternative for each subsystem with the highest score is chosen. However, in designing interconnected subsystems, attention should be given to choosing alternatives that fit the already-chosen matching components.

The website associated with this textbook includes all detailed selection matrices for all components. Only one sample selection is presented here, for space considerations. The following are the selected choices: a wheel configuration of a two-wheel differential drive with a castor wheel for stability; a DC motor; a circular chassis configuration; infrared and white-line sensors; an air-blowing fan for candle extinguishing.

Table 14.15 presents the sample Pugh matrix used to evaluate the alternatives for flame sensing. As can be seen from the Pugh matrix, the best design to detect the flame uses a UV sensor coupled with pyroelectric sensors. The pyroelectric sensors are used to determine the direction of the flame.

The evaluation of the design alternatives allowed to reach design. A model of the building blocks of the robotic system is presented in Fig. 14.37.

TABLE 14.15 Pugh Matrix for Selecting a Flame-Sensing Strategy

Weight	Parameters	Constrained UV Sensor Coupled with Motion	Vision-Based Sensing	Pyroelectric Sensor Coupled with Motion	UV Sensor Coupled with Infrared Sensors	UV Sensor Coupled with Pyroelectric Sensors
5	External perception	+	+ +	+	+ +	+ + +
4	Speed	−	− −	−	+	+
3	Weight	+	−	+	+	+
2	Cost	−	− −	+	−	−
3	Response time	+	−	+	+	+
4	Robustness against noise	+ +	+	+ +	+	+ + +
3	Simplicity	+	− −	+	−	−
TOTAL		16	−10	20	19	32

Legend: + = strength (+1), − = weakness (−1).

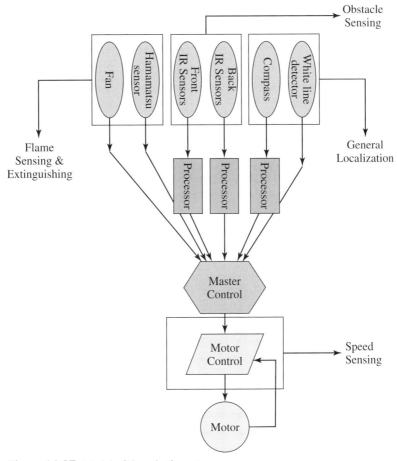

Figure 14.37 Model of the robotic system.

In order to analyze the system, the different subsystems that have been chosen through the decision process are analyzed.

14.4.3 Implementation

Motors

The frequency of the PWM is chosen such that the motor behaves as if a constant DC voltage is across its terminals. This PWM signal generates a varying output DC voltage to drive the DC motor through an H-bridge, which transforms the PWM signal into an output voltage proportional to the duty cycle of the PWM signal. The importance of the H-bridge circuit (L298N) lies in the fact that it can drive the motor forward or backward at any speed using an independent power source, which provides the circuit with increased protection and isolation.

The motor control is made up of two PIC MCUs, one for each motor (Fig. 14.38). The controllers are implemented using a PIC 16F818, since it provides a direct PWM signal; however, the PIC provides an amplitude of only 5 V, with very small current. In order to compensate for this and to isolate the digital component from the motors, the dual H-bridge is used. In

Figure 14.38 Motor controller.

addition to providing isolation between the microcontroller and the motor, the L298 provides the motors with the PWM, though it is adjusted to have a higher-voltage amplitude and higher current. Each PIC receives a 4-bit word from the master control that determines the required speed of the specific motor.

Interfacing Sensors

The sensor control is made up of two PIC 16F818 microcontrollers, which come with built-in analog-to-digital converters (ADCs). One PIC is dedicated to the back while the other is dedicated to the front infrared sensors. The PIC dedicated to the back sensors receives the analog voltage values from the two sensors and converts them into digital values. The PIC then compares the two digital values and transmits the result as a 3-bit word to the master control. Similarly, the PIC dedicated to the three front sensors receives the analog voltages and converts them to digital. Afterwards the PIC processes the three digital values and transmits the result as a 4-bit word to the master controller.

The master controller is made up of a single PIC 18F877A (Fig. 14.39), which makes the decision regarding the next action to be fulfilled by the robot. The master controller receives

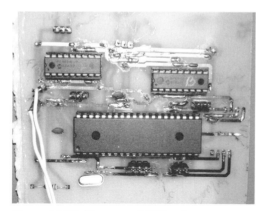

Figure 14.39 The master MCU and sensor control PCB.

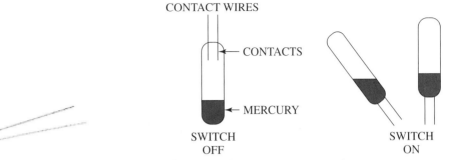

CONTACT WIRES

← CONTACTS

← MERCURY

SWITCH
OFF

SWITCH
ON

Figure 14.40 Mercury switch.　　　**Figure 14.41** Mercury switch behavior.

data from the sensor controller through the 4-bit word and the 3-bit word that are transmitted. The master controller also receives interrupts from the white-line detector and the UVTron in order to confirm a white line and a flame, respectively. The master controller processes all the aforementioned data that it receives and determines the required speed of each motor.

Mercury Switch

The mercury switch (Fig. 14.40) is formed of a tube of glass. Two wires go into the glass tube. When the two wires are perpendicular to the ground and in the downward direction, the mercury inside will touch the two wires, and subsequently the switch is turned ON (Fig. 14.41). When the tube is moved, the mercury moves away from the wires and breaks contact with them, and consequently the mercury disconnects the circuit.

The main purpose of the mercury switch is to detect the existence of the staircase at the beginning of the arena. Climbing stairs would require higher input voltage from the robot. The mercury switch is placed in parallel with the ground, initially the switch is on. Its pins face the front side (the other end faces the rear side) (Fig.14.42).

Hamamatsu Flame Detector

This sensor is formed of a glass bulb coupled with a drive circuit to measure photons in the UV spectrum associated with open flames and fire; the circuit uses little energy (~3 mA) and contains all the necessary high-voltage electronics, drive circuitry, noise cancelling, and signal processing needed (Fig. 14.43).

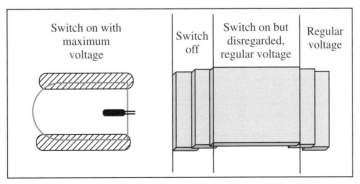

Figure 14.42 Changes in mercury switch over stair path.

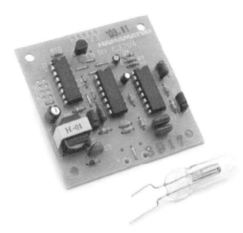

Figure 14.43 Hamamatsu flame sensor with its driving circuit.

Hamamatsu UVtron sensors are used in many different applications: flame detectors for gas/oil lighters and matches, fire alarms, combustion monitors for burners, inspection of ultraviolet leakage, detection of discharge, and ultraviolet switching. It is completely insensitive to visible light because it has a narrow spectral sensitivity of 185–260 nm (Fig. 14.44), and thus it offers excellent performance when it is looking to sources of UV radiation in this range: flames and fire.

Although it has a small size, it ensures wide angular sensitivity (Fig. 14.45) and great reliability in the quick detection of even the weak ultraviolet radiations emitted from a flame. Hence, it can detect the flame of a cigarette lighter at a distance of more than 5 m.

The task of the flame detector begins each time the robot crosses a white line at the door of a room. Since the distance between the door and any corner of the room is less than 5 m, we

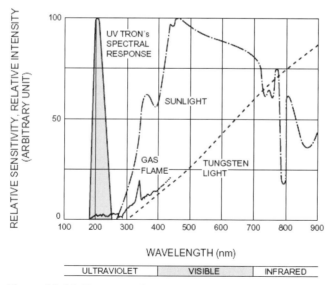

Figure 14.44 Hamamatsu flame sensor's spectral response with different light sources.

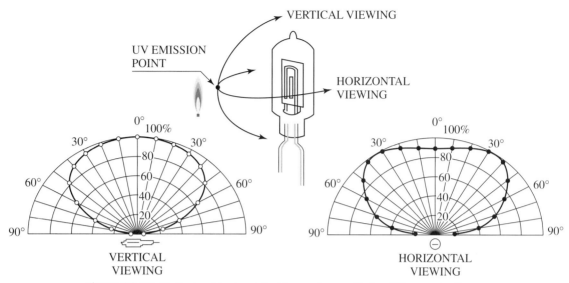

Figure 14.45 Vertical and horizontal wide angular sensitivity of Hamamatsu flame sensor.

can see that the Hamamatsu is able to detect the flame with the robot standing at the door. The robot needs to make a turn at the door in order to cover all the corners because the horizontal sensitivity might have blind spots if the robot just looked straight and since the horizontal sensitivity is not reliable at angles of 90° and −90°. If the Hamamatsu does not detect the flame in the room, the robot has to search in the other rooms until it detects the flame. Since the flame is about 15 cm in height, the sensor was installed at the top of the robot.

Infrared Sensor

For obstacle avoidance and to navigate through the arena, infrared sensors provide the control system with distance between the robot and the obstacle. The specific infrared sensor type that was purchased was Sharp GP2D120 (Fig. 14.46). These detectors offer a small package, minimal current consumption, low cost, and a variety of output options (Table 14.16).

One advantage of infrared sensors is that they are immune from ambient light interference. However, the accuracy is affected by the target area's color and texture. Infrared sensors

Figure 14.46 Sharp GP2D120 IR sensor.

TABLE 14.16 Infrared Sensor Specs

Sensors	Operational Range (cm)	Voltage Signal	Quantity
GP2D120—short range	3–30	Analog	5

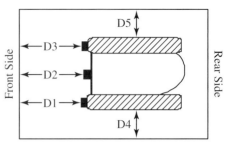

Figure 14.47 Infrared sensor arrangement.

estimate distance in analog form. Three sensors were positioned on the front of the vehicle. In addition, two sensors were placed at the middle of the vehicle sides. D1, D2, and D3 are the distances between the front sensors of the car and the wall or obstacle, D4 and D5 are the distances between the side sensors of the car and the wall (Figs. 14.47 and 14.48). Taking the sensor value at each distance, measured data was plotted vs. actual data for sensor D4 (Fig. 14.49).

White-Line Sensor

A Lynxmotion single-line detector (Fig. 14.50) was used to spot the presence of a white line. The LED is on and the output is LOW when the sensor is positioned over a black or dark surface or if nothing is there to reflect off. The LED goes out and the output is HIGH when the sensor is positioned over a white or light surface. The sensor has an IR-transparent plastic cover for dust protection. The cover looks dark, to filter out much of the non-IR light. Performance in ambient light is of good quality.

The angular geometry also means that even if the white surface is brought too close to the sensor surface, the reflectance will be lost and the sensor will think it is seeing the black tape. The minimum range is 0.32 cm and the maximum is 1.27 cm from the floor. The white-line sensor was placed on the midperimeter of the car, and it will not touch the ground when the robot climbs the stairs. Furthermore, the white line was used in the general localization algorithm so that the robot would have a general idea of global position in the arena.

Analog Compass

The 1525 analog compass sensor (Fig. 14.51) outputs a continuous analog sine/cosine signal capable of being decoded to any degree of accuracy. This compass gives directional information

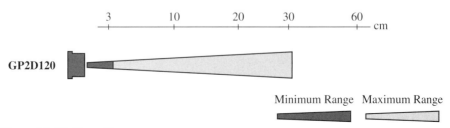

Figure 14.48 Sharp IR sensor range.

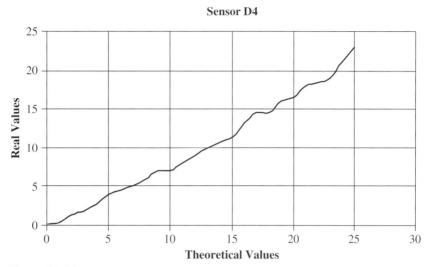

Figure 14.49 IR sensor (D4) collected data vs. actual values.

and enables the robot to leave the room. Once the robot enters the room, it checks if there is a flame; if a flame isn't detected, the robot leaves the room by using the compass. The sensor is constructed to operate in a vertical position with the leads down. The sensor is designed to measure the direction of the horizontal component of the earth's flux field. If it is tilted off vertical, it begins to sense some of the vertical component of the earth's field, introducing some error. For practical purposes, up to about 12° tilt is considered acceptable. The compass is used to permit the robot to make a right angle to enter the room. The analog compass is connected to an ADC for digital values.

Algorithm

The infrared sensors, the flame detector, the white-line detector, the mercury switch, and the compass form the sensory organs of the robot. The motors moving the robot and the fan blowing the candle flame to extinguish it form the movement and functional core of the robot.

Figure 14.50 Lynxmotion single-line detector.

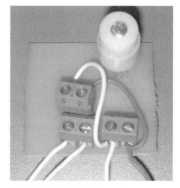

Figure 14.51 1525 analog compass.

The master PIC is the brain that receives the data from the sensors, interprets the signals, recognizes the environment in which the robot exists, and accordingly takes the decision and communicates it for action.

The mercury switch is connected directly to the master PIC. It senses the variation of the ground slope and sends the data to recognize whether the robot is facing stairs, and to order the motors. The front and side infrared sensors sense the existence of an obstacle and its place and then send that information. The white-line detector is connected directly to the master PIC. The task of the MCU is to interpret the data and identify the localization of the robot. The compass has its own PIC. The flame detector is connected directly to the master PIC. It senses the existence of the flame in the room and its direction. When the master PIC determines that the robot is in the room where the flame exists, it gives the motors the order to go straight to the flame and activates the fan.

The decision corresponding to each sensor might conflict with another decision of a different sensor. The combination of all the decisions together is analyzed by the master PIC in order to make the best decision corresponding to the robot's situation. The master PIC solves this problem by giving different priorities to the sensors according to the situation of the robot and consequently makes the best decision that fulfills the task of the robot.

At the beginning, the mercury switch has priority 1. After going down the stairs, the mercury switch is neglected all over the arena, and the side sensors will have priority 1, to detect hallways, walls, and doors, and priority 2 is for the front sensors, to avoid obstacles, and all other sensors are neglected. When there is a door, priority 1 is for the compass, to turn to the room, and priority 2 is for the white-line detector, to know whether the room is entered or not. Priority 3 is for the front sensors, to avoid obstacles, and priority 4 is for the side sensors. When the room is entered, priority 1 is for the flame detector, and priority 2 is for the front sensors, to avoid obstacles in the room.

The three front sensors would detect any obstacle in front of the robot and provide a voltage for the sensor control. Depending on the values indicated by the sensors, the output would be specified as the speed of the motors. For example if the obstacle is near the right sensor (D1) and far from the other two sensors, the motor speed would be specified so that the robot would go left. On the other hand, if all sensors indicate that an obstacle is near, then this obstacle is the wall, since the maximum obstacle diameter is 11 cm and the sensors are 10 cm apart. Hence, in this case the decision of going to the left or right is indicated by the back sensors. The output is the speed of the wheels, which is indicated by providing different voltages to the motors through different PWM duty cycles and hence different robot directions. Fuzzy logic, which was introduced in Chapter 13, is used to estimate the nature and location of the obstacle. Figures 14.52–14.54 illustrate the fuzzyfication membership function for the three inputs (the data sent by the IR front sensors).

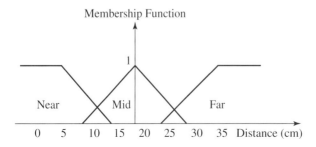

Figure 14.52 IR front sensor data as fuzzy input membership function.

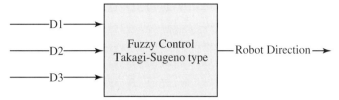

Figure 14.53 Fuzzy system with three IR sensor inputs and robot direction as one decision output.

Robot localization is a key problem when making autonomous robots. The robot can't have an accurate knowledge of its position in the arena due to the uncertainty of the environment. The rear IR sensors were used for general localization, and they used crisp analysis in determining the general robot position in the arena. With this information the robot can navigate the arena and avoid the obstacles to reach the flame, put it out, and then return quickly to its home position. If both back sensors indicate that there is a gap, then the robot would know it is in a hallway (Fig. 14.55). On the other hand, if one gap was detected, then there is a room at that side, and the robot would enter the room to detect if there is a flame. Therefore the back sensors would detect whether the robot is in the hallway or at the beginning of the room, based on the sensor values. Once the robot crosses the white line of the room, the Hamamatsu would indicate the presence of the flame (Fig. 14.56).

Several testings were done to make sure the robot was performing each step in this algorithm. Sensors readings were tested, and the white-line sensing was examined. The algorithm performed as expected.

14.4.4 Case Outcomes

This case is a typical mechatronics project that integrates the design and interface of many technology blocks and subsystems. The following are direct outcomes.

1. Emulating real-life constraints on design and participating in international contests.
2. Applying a systematic process for selecting design alternatives.

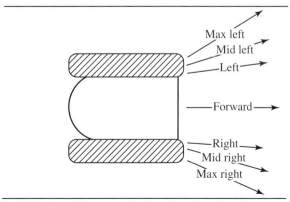

Figure 14.54 Robots different directions from the three IR front sensors.

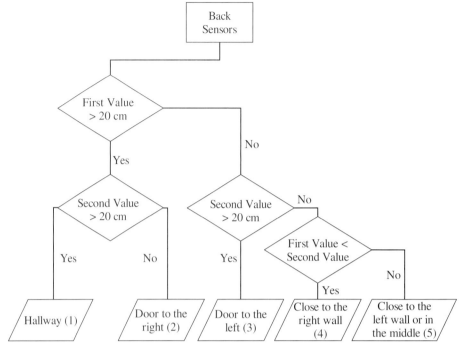

Figure 14.55 Back sensor decision-making flowchart.

3. Specifying sensor fusion and priority settings.
4. Creating a flame detector, a white-line detector, IR sensors, and a mercury-level switch, and selecting and interfacing a compass.
5. Utilizing fuzzy logic for determining direction based on multiple sensor feedback.
6. Devising systems for autonomous robot navigation and localization.

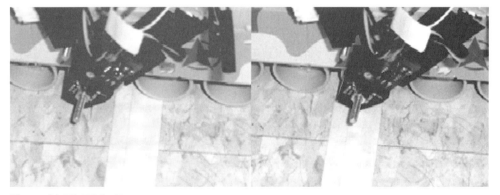

Figure 14.56 White-line sensor.

REFERENCES

D. Clark and M. Owings, *Building Robot Drive Trains*. New York: McGraw-Hill, 2003.

S. Dearie, K. Fisher, B. Rajala, and S. Wasson, *Design and Construction of a Fully Autonomous Fire-Fighting Robot*. New Mexico Institute of Mining and Technology, 2001.

H. R. Everett, *Sensors for Mobile Robots: Theory and Application*. Wellesley: A. K. Peters, 1995.

D. L. Heiserman, *How to Design and Build Your Own Custom Robot*. Blue Ridge Summit: Tab Books, 1981.

J. L. Jones, A. M. Flynn, and B. A. Seiger, *Mobile Robots: Inspiration to Implementations*. Peters, 1999.

L. Miller and others, *Firebot: Design of an Autonomous Fire-Fighting Robot*. Machine Intelligence Laboratory, University of Florida, 2003.

H. H. Poole, *Fundamentals of Robotics Engineering*. New York: Van Nostrand Reinhold, 1989.

R. Siegwart and I. Nourbakhsh, *Introduction to Autonomous Mobile Robots*. Cambridge, MA: MIT Press, 2004.

I. Verner and D. Ahlgren, "Fire-Fighting Robot Contest: Interdisciplinary Design Curricula in College and High School," *Journal of Engineering Education*, vol. 91, no. 3, 2002.

http://www.usfa.fema.gov

http://www.trincoll.edu/events/robot/Rules/default.asp

http://www.oopic.com/thesav.htm

http://www.andrew.cmu.edu/user/apenmets/firebot.html

http://www.trincoll.edu/events/robot

http://www.hpk.co.jp/Eng/products/ETD/uvtrone/uvtrone.htm

14.5 CASE STUDY 4: PIEZO SENSORS AND ACTUATORS IN CANTILEVER BEAM VIBRATION CONTROL

Based on the experimental setup for a thesis on vibration control by K. Joujou

14.5.1 Introduction

The increased demand for faster, quieter, and more efficient lightweight mechanisms has resulted in unacceptable vibration levels in flexible mechanisms and structures. Vibrations increase noise level and reduce operating life, due to fatigue wear of components. Design strategies to suppress the elastodynamic response of high-speed flexible mechanisms and structures have been a main concern in the field of vibration control. Vibration control strategies fall into one of three main categories: passive damping, active damping, and hybrid damping.

With recent advances in materials technology, smart materials, such as electrorheological (ER) fluids, shape memory alloys (SMA), and piezoelectric materials, are playing an increasingly significant role in the active vibration control of flexible mechanisms. Active damping involves sensing the levels of vibrations of the structure and applying appropriate control strategy to provide strategically located actuators with compensated commands to drive the vibration levels to within desired limits. This project aims at verifying a vibration control strategy experimentally. The cantilever beam setup is described, with all its components.

Figure 14.57 Photo of the smart beam under consideration.

14.5.2 Modeling of the Cantilever Beam and PZT Actuator

Simulation of the beam can be carried out by one of two techniques, either by solving the continuous equation of motion or by using finite-element analysis. The finite-element analysis based on thin-beam theory is herein employed. Figure 14.57 shows a cantilever beam.

Modeling of the Beam

The beam is divided into 10 finite beam elements, resulting in 22 degrees of freedom, as shown in Figure 14.58. θ_i are the rotational freedoms and u_i are the translational freedoms. u_{21} and θ_{22} are restrained because they correspond to the node where the beam is connected to ground.

Using proper assembly of the mass and stiffness matrices of all finite elements, the equations of motion that describe the behavior of the beam are formed as follows:

$$[M]\{\ddot{Q}\} + [C]\{\dot{Q}\} + [K]\{Q\} = \{F\} \tag{14.15}$$

where $[M]$, $[K]$, and $[C]$ are, respectively, the system mass, stiffness, and hysteretic damping matrices; $\{\ddot{Q}\}$, $\{\dot{Q}\}$, and $\{Q\}$ are, respectively, the acceleration, velocity, and displacement vectors; and $\{F\}$ is the applied load vector. The hysteretic damping matrix is the equivalent viscous damping, formed according to Rayleigh proportional damping, as a linear combination of the mass and stiffness:

$$[C] = \alpha[M] + \beta[K] \tag{14.16}$$

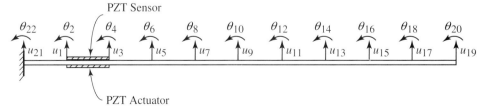

Figure 14.58 Finite-element model for the beam.

where α and β are constants. They are determined by solving the following two algebraic equations involving the natural frequencies of the first two vibration modes of the system:

$$\alpha + \beta\omega_i^2 = 2\omega_i\xi_i; \qquad i = 1, 2 \tag{14.17}$$

A value of $\xi = 0.003$ is used in all calculations to follow. Note that Rayleigh damping tends to overestimate damping in the high-frequency modes. The response of the system is obtained by solving Eq. (14.15) using the Newmark numerical integration method. The second-order, coupled differential equations in physical coordinates given in Eq. (14.15) are transformed into a set of n-uncoupled equations in modal coordinates using r_i modal superposition. Accordingly,

$$u(x, t) = \sum_{i=1}^{n} \phi_i(x)r_i(t) \tag{14.18}$$

or, in matrix form,

$$\{x\} = \left[\Phi\right]\{r\} \tag{14.19}$$

where, $[\Phi]$ is the mass normalized modal matrix formed with the mode shapes $\Phi_i(x)$, $i = 1$, $2, \ldots, n$, as its columns. $\{r\}$ is the vector of modal coordinates u_i. Substituting Eq. (14.18) into Eq. (14.15) yields

$$\{\ddot{r}\} + \left[c_r\right]\{\dot{r}\} + \left[k_r\right]\{r\} = \{F\} \tag{14.20}$$

$\left[c_r\right]$ and $\left[k_r\right]$ are the diagonal modal damping and modal stiffness matrices, respectively. The corresponding rth diagonal elements are

$$\boldsymbol{c_r} = 2\xi_i\omega_i \qquad \boldsymbol{k_r} = \omega_i^2$$

Modeling of the PZT Actuator

The actuator is a lead-zirconate-titenate (PZT) plate bonded to the lower surface of the beam, 1 in. from the fixed end, making the sensor and actuator PZTs a collocated pair (see Fig. 14.57). The actuator is assumed to be perfectly bonded to the surface; that is, the thickness of the bonding layer is infinitely small and perfectly rigid. The shear strain is transferred between the actuator and the beam over an infinitesimal distance near the ends of the actuator. Therefore, the action resulting from the converse piezoelectric effect is represented by a concentrated moment $M_a(x, t)$ applied at the ends of the actuator. This moment is proportional to the control voltage $v_a(t)$ and is expressed by

$$M(x, t) = K_a v_a(t) \tag{14.21}$$

where K_a is an actuator constant derived form the PZT properties, expressed by

$$K_a = Y_p \times d_{31} \times b_a \times (t_b + t_a) \tag{14.22}$$

where Y_p is Young's modulus for the piezoelectric material (N/m^2); d_{31} is the piezoelectric constant (coul/m), in which subscript 3 indicates the direction of applied electric field and subscript 1 indicates the direction of generated shear strains; b_a and t_a are the width and the thickness of the piezoelectric element, respectively; and t_b is the thickness of the beam. The actuator's moments contribute to the formation of the force vector:

$$\{F\} = K_a v_a \{W\} \tag{14.23}$$

where the nonzero elements in the $\{W\}$ vector involve the constants

$$W_i = \left[\phi'_i(x_{2i}) - \phi'_i(x_{1i}) \right] \tag{14.24}$$

W_i couples the moment produced by the ith actuator located between x_{1i} and x_{2i} on the beam to the vibration modes.

Modeling of the Sensor

To measure the vibration levels of the cantilever beam, a PZT plate is bonded to the top surface of the beam, 1 in. from its fixed end (see Figure 14.57). This sensor generates a voltage proportional to the strains incurred in the beam during vibrations. The voltage generated by the PZT is expressed as

$$V_s = K_s \left[w'(x_2) - w'(x_1) \right] \tag{14.25}$$

K_s is the sensor gain, given by

$$K_s = \frac{Y_p d_{31} Z_c b_s}{C_s} \tag{14.26}$$

where Y_p, C_s and d_{31} are Young's modulus, the capacitance, and the piezoelectric constant of the PZT sensor; Z_c and b_s are, respectively, the width of the sensor and the distance from the centroidal axis of the beam. Differentiating the modal transformation equation gives

$$w' = \sum_{i=1}^{n} \phi'_i(x) r_i(t) \tag{14.27}$$

Substituting this into Eq. (14.25) gives

$$V_s = \frac{Q}{C_s} = k_s \left[r_1(t) \{ \phi'_1(x_2) - \phi'_1(x_1) \} + r_2(t) \{ \phi'_2(x_2) - \phi'_2(x_1) \} + \cdots \right]$$

$$= k_s \sum_{i=1}^{n} r_i(t) W_i \tag{14.28}$$

where W_i is the same constant as given by Eq. (14.27) if the sensor and actuator are collacated. It couples the generated voltage with the mode shapes.

14.5.3 Beam Experimental Setup

In order to verify the effectiveness of vibration control strategies, the experimental setup shown in Figure 14.59 was built. The setup consists of the following three main parts: (1) the beam under test, the fixture, and the PZT elements bonded to its surface; (2) the instrumentation

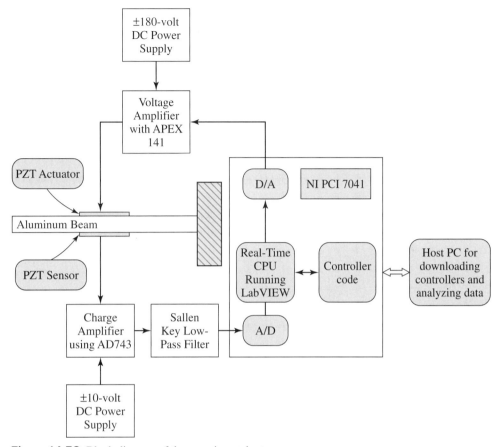

Figure 14.59 Block diagram of the experimental setup.

setup—a charge amplifier, a voltage amplifier, and the data-acquisition board; (3) the software interface—the visuals and the control algorithm to process the measured signal and issue the appropriate control signal.

Properties and Dimensions of the Beam (Table 14.17)

The cantilever beam under test is rigidly fixed to a steel table. The beam is made of aluminum cut out by a sheer machine from a uniform aluminum sheet. The beam is excited by deflecting its free end upward or downward.

TABLE 14.17 Summary of Beam Dimensions and Properties

	Value	Units
Length	0.285	m
Width	0.0254	m
Thickness	0.003	m
Modulus	70.3×10^9	N/m^2
Density	2712	kg/m^3

TABLE 14.18 Summary of PZT Dimensions and Properties

	Value	Units
Length	0.0285	m
Width	0.0127	m
Thickness	0.508×10^{-3}	m
Charge constant (d_{31})	-50×10^{-9}	m/V
Capacitance	12.6	nF

PZT Properties, Dimensions, and Bonding Techniques

The PZT elements used in the experiment are T220-A4-303Y batches from Piezosystems. Their characteristics are presented in Table 14.18. As indicated in the actuator pair, these PZT elements have two layers and act the same way as a single-layer PZT. The lead wires were soldered to the PZT elements before the patches were bonded to the beam's surface. The area was then cleaned with a mild acid solution and then neutralized with a base. Finally, the PZT was bonded to the aluminum surface using a high-strength epoxy from Micro Measurements.

A "two-layer" PZT element eventually includes nine layers: four electrode layers, two piezo ceramic layers, two adhesive layers, and a center shim. Two-layer elements can be made to elongate, bend, or twist, depending on the polarization and wiring configuration of the layers. A center shim laminated between the two piezo layers adds mechanical strength and stiffness but reduces motion. "Two-layer" refers to the number of piezo layers. The two layers reduce drive voltage by half when configured for parallel operation. A two-layer element behaves like a single layer when both layers expand (or contract) together. If an electric field is applied, the element becomes thinner, but extension along the length and width will result, hence the role of an actuator. Typically, only the motion along one axis is utilized. Extender motion on the order of microns to tens of microns and a force from tens to hundreds of newtons are typical.

On the other hand, applying a mechanical stress to a laminated two-layer element will result in electrical generation, which depends on the direction of the force, the direction of polarization, and the wiring of the individual layers. When a mechanical stress causes both layers of a suitably polarized two-layer element to stretch (or compress), a voltage is generated that tries to return the piece to its original dimensions, and the PZT acts like a sensor. Essentially, the element acts like a single sheet of piezo. The metal shim sandwiched between the two piezo layers provides mechanical strength and stiffness while shunting a small portion of the back electromotive force.

14.5.4 Instrumentation Setup

The instrumentation setup consists of three main parts: the charge amplifier used to condition the PZT sensor signal, the high-voltage amplifier used to drive the PZT actuator, and the data-acquisition board used to interface the controller to the physical setup.

Charge Amplifier

The piezoelectric transducer is basically a dielectric with a high leakage resistance; i.e., it has very high output impedance. This requires that the interface electronics used, such as charge-to-voltage and current-to-voltage converters or voltage amplifiers, have high input impedance to prevent loading the sensor. The AC charge amplifier shown in Figure 14.60 was developed

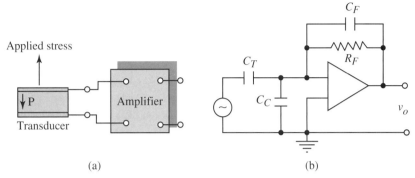

Figure 14.60 PZT and charge amplifier system.

specifically for the high-impedance, piezoelectric crystal-type transducers used in the dynamic measurement of shock, vibration, pressure, force, and sound. The amplifier converts to voltage the very low charges or current signals generated by devices such as capacitive sensors, quantum detectors, and piezoelectric sensors. The gain of the AC amplifier is relatively independent of the shunt capacitances normally associated with the use of piezoelectric transducers. This is an advantage over voltage amplifiers. The transfer function of the converter is

$$v_o = -\Delta q/C_F \tag{14.29}$$

The overall charge conversion gain is $-1/C_F$. The negative sign indicates that the output is 180° out of phase with the input. If C_F is measured in nanofarads, the charge conversion gain is in units of mV/pC. The effective input capacitance of the amplifier is $C_i = C_F \times (1 + A_V) \approx C_F \times A_V$, where A_V is the open-loop gain of the voltage amplifier. A large feedback resistance R_F is used to prevent the nonzero op-amp bias current from developing a significant steady charge on C_F. However, it causes the op-amp to behave as a high-pass filter with a time constant $R_F \times C_F$. A high amplifier input resistance R_i is needed, which is usually attained by using a FET in the input stage. Consequently, the low corner frequency of the amplifier, $F_c = 1/2\pi R_F C_F$, precludes DC operation. Since the high-input-impedance FET amplifier is a virtual ground, the voltage across the equivalent resistance and capacitance of the cable and transducer is essentially zero. Thus, the length of cable has no effect on the sensitivity or the frequency response of the system. The leakage resistance of the transducer must be substantially larger than the impedance of the capacitor C_F at the lowest operating frequency to provide for good low-frequency response. When used with quartz-crystal transducers, the value of C_F is from 10 to 100,000 pF and R_F is from 10^{10} to $10^{14}\Omega$.

The charge amplifier circuitry used for this experiment is shown in Figure 14.61, with a photo shown in Figure 14.62. It consists of two main parts. The first is the traditional charge amplifier, and the second part is a unity-gain low-pass filter with corner frequency equal to 500 Hz and unity gain. The LPF is essential to the system since only the first mode of vibration is being controlled. The PZT width is not modified to render the higher frequencies unobservable.

Voltage Amplifier

A high-voltage amplifier is built to provide the PZT actuator with the necessary drive voltage. This part of the experimental setup consists of two parts: the voltage supply and the voltage amplifier. The voltage supply is a simple rectifier circuit. It consists of a transformer to step

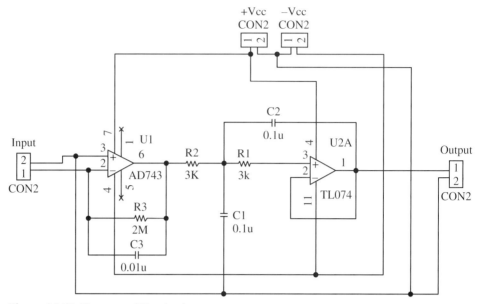

Figure 14.61 Charge amplifier circuit.

down the voltage, a bridge rectifier, and a large capacitor. The capacitor minimizes the ripple voltage after the bridge. The large capacitor used and the low current needed make this circuit adequate to supply ripple-free voltage to the voltage amplifier. The voltage amplifier needs ±175 volts DC; for this reason the circuit is supplied with two rectifier circuits, one for the +175 volts and the other for the −175 volts. The main part of the voltage amplifier is the APEX amplifier PA-141. The controller produces a signal in the range of ±10 volts, whereas

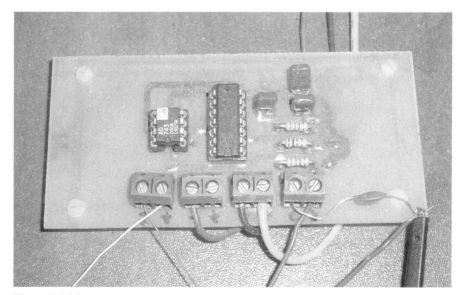

Figure 14.62 A photo of the built charge amplifier circuit.

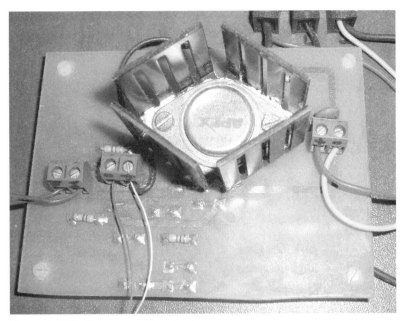

Figure 14.63 Voltage amplifier circuit.

the voltage required by the PZT elements is ± 175 volts. To match the out of the controller to the PZT requirements, a noninverting voltage amplifier based on the Apex PA-141 is employed to give a gain of 18. All connections made to the amplifier are based on the recommendations of the manufacturer. Figures 14.63 and 14.64 show the voltage amplifier and the voltage supply circuits, respectively. The voltage amplifier receives its signal from the data-acquisition board through the analog output channel AO (0).

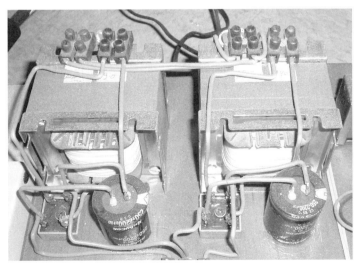

Figure 14.64 DC power supply for the voltage amplifier.

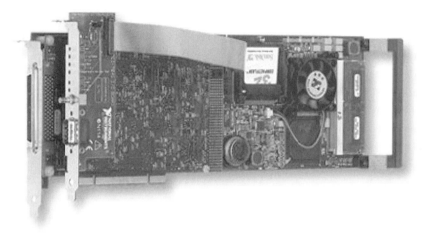

Figure 14.65 The NI-PCI-7041 real-time DAQ.

Data Acquisition

To acquire the amplified sensor signal and to generate the appropriate control signal to the voltage amplifier, the NI-PCI-7041 real-time DAQ shown in Figure 14.65 is used. The board has the following specifications:

- 16 analog inputs; 12-bit resolution; 250-kS/s sampling rate
- Two 12-bit analog outputs; eight digital I/O lines; two 24-bit counter/timers
- Real-time performance using a 700-MHz Pentium III onboard processor
- 32 MB of DRAM for programs; 32 MB onboard compact flash
- 26-kHz maximum single PID loop rate

   The application code is compiled and downloaded to the board's memory. When the system is set in action, the onboard CPU processes the code, collects the data, and stores it in an onboard memory. When the application is stopped, the data files are sent to the host computer to process the data, display the measured response, and calculate the damping ratio.

14.5.5   Controller and Software

A MATLAB/Simulink software code is developed to implement the control, and a LabVIEW source code is developed for experimental implementation. The code was restrained by the real-time data-acquisition board. The P controller was completely implemented in LabVIEW, whereas implementing the fuzzy logic controller required some development in MATLAB before using it on the real-time data-acquisition board. In what follows a more detailed explanation of the code development is presented.

Development of the PID VI

The PID controller applied to the experimental setup is based on the PID toolset provided with LabVIEW. The PID controller compares the set point (SP) to the process variable (PV) to obtain the error (e) defined as:

$$e = SP - PV \tag{14.30}$$

The following relations represent the proportional action, integral action and derivative action, respectively:

$$u(t) = K_c\left(e + \frac{1}{T_i}\int_0^t e\,dt + T_d\frac{de}{dt}\right) \tag{14.31}$$

The PID controller contains the proportional, integral, and derivative actions to calculate the controller action, $u(t)$, as if the error and the controller output have the same range, -100% to 100%; controller gain is the reciprocal of proportional band:

$$u_p(t) = K_c e \tag{14.32}$$

$$u_i(t) = \frac{K_c}{T_i}\int_0^t e\,dt \tag{14.33}$$

$$u_D(t) = K_c T_d\frac{de}{dt} \tag{14.34}$$

where K_c is the proportional gain, T_i is the integral time in minutes, also called the *reset time*, and T_d is the derivative time in minutes, also called the *rate time*. To implement PID control on a digital system, the discrete version of the relation is needed. The discrete form of the current error is

$$e(k) = (SP - PV) \tag{14.35}$$

The discrete form of the proportional action is

$$u_p(k) = (K_c \times e(k)) \tag{14.36}$$

Trapezoidal integration is used to avoid sharp changes in integral action when there is a sudden change in PV or SP. Nonlinear adjustment of integral action is used to counteract overshoot. The larger the error, the smaller the integral action, as indicated by

$$u_i(k) = \frac{K_c}{T_i}\sum_{i=1}^k\left[\frac{e(i) + e(i-1)}{2}\right]\Delta t \tag{14.37}$$

Because of abrupt changes in SP, the derivative action is only applied to the PV, not to the error e, to avoid derivative kick. The partial derivative action is accomplished by

$$u_D(k) = -K_c\frac{T_d}{\Delta t}(PV(k) - PV(k-1)) \tag{14.38}$$

Finally, the discrete form of the PID action is given by

$$u(k) = u_p(k) + u_i(k) + u_D(k) \tag{14.39}$$

The PID algorithm is placed within a While loop that executes 500 times per second. It acquires its input signal from input channel 0 AI (0) and issues the result to the output channel 0 AO (0) on the DAQ board. No further signal conditioning, other than the charge

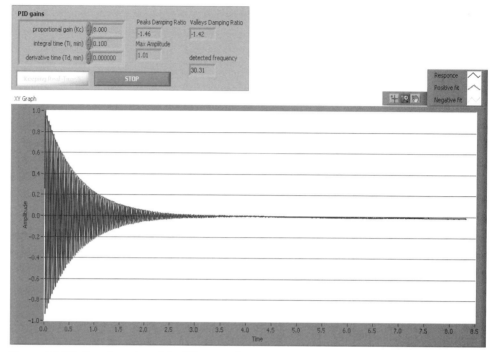

Figure 14.66 Sample setting of the PID controller gains and display of the results.

amplifier and the voltage amplifier discussed earlier, is needed to run the system. The code implemented various signal analyses to compare the results. This part of the code takes the maximum signal acquired by the sensor and uses it to normalize the results to 1. This is done so that a user can visually ascertain the difference between different results. A curve-fitting technique is also applied to the upper and lower peaks to determine the damping ratio. At the same time, the frequency of the system is measured via a built-in LabVIEW VI that uses an FFT technique.

Figure 14.66 shows the front panel of the software developed in LabVIEW. The menu on the top left corner is where the gains of the PID are entered. It also displays the damping ratio resulting from curve-fitting the positive and negative peaks. The positive and negative peaks were curve-fitted to make sure the response is symmetrical. The top left panel also displays the first natural frequency detected. It also includes an alarm signal that turns red if the system is not running deterministically.

Development of the Fuzzy VI

The fuzzy controller is implemented in LabVIEW using a variety of tools. The initial fuzzy set is written in MATLAB using the fuzzy toolbox converted into a Simulink code, where inputs and outputs are added to the fuzzy engine. The Simulink model is then loaded into the MATLAB real-time workshop and then compiled into a DLL file using the Simulation Interface Toolkit (SIT) provided by LabVIEW. The real-time workshop uses a National Instruments engine and Visual C++ to compile the code into the DLL file for later use.

During this process the time step of the system and the solver parameters are set. The time step is chosen to be 0.001 seconds, to give a sampling rate of 1000 samples/second. The rest of the job is implemented in LabVIEW. A driver VI is downloaded into the RT target, and then the DLL file is downloaded onto the flash drive on the board using an FTP server.

The preceding programs cannot run directly on any machine because of the time-critical processes involved. For the controllers to run properly and for the system to be deterministic, the controller has to be implemented in real time. Many platforms can be used to achieve determinism. One way is to run the code on a separate machine using a normal data-acquisition board, on condition that the machine is configured as a remote RT target. In this case data is transmitted by means of a conventional network (TCP/IP). In another way, the same host machine can be used with any data-acquisition board as long as the machine has a dual CPU. In this case one of the CPUs will be reserved for real-time data acquisition. Alternatively, one can use a real-time data-acquisition board that comes with its own built-in CPU and memory. The current investigation used a real-time data-acquisition board (RT). Accordingly, the code is downloaded onto the RT DAQ and the board's CPU will run its control process independent of the computer processes.

14.5.6   Simulation and Experimental PID Results

In order to verify the proposed model and check the effectiveness of the PID controller, many gains were implemented experimentally. Their results are presented next. A measure of improvement is the increase in the damping ratio provided by the control action.

Simulation and Experimental Results: No-Control Case

Figure 14.67 shows the experimental normalized response of the beam over a period of 5.6 s with no control action. Figure 14.68 shows the actual sensor voltage, which ranges between

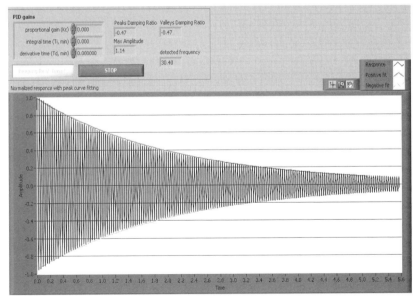

Figure 14.67 Normalized experimental response of the deflections with no control action.

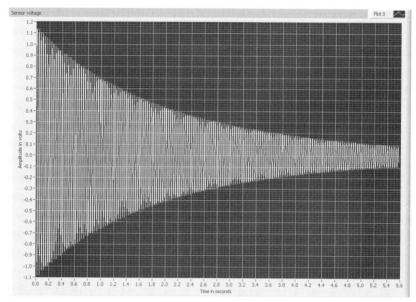

Figure 14.68 Recorded sensor voltage due to deflections in the no-control case.

−1.1 and +1.1 volts. The experimental damping ratio is 0.47 on both the positive and negative curve-fitted peaks, and the experimental natural frequency is 30.4 Hz.

The normalized simulation for a time period of 5-s response is shown in Fig. 14.69. The corresponding damping ratio and natural frequency are, respectively, 0.475 and 30 Hz, which are very close to the experimentally obtained values.

To investigate further the response of the system and the effectiveness of our PID control strategy, more experiments were conducted, with different gains.

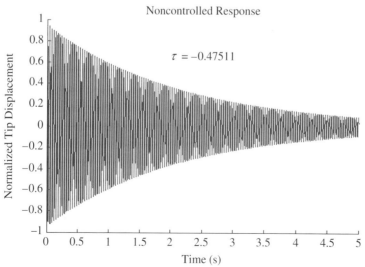

Figure 14.69 Normalized simulation, no control.

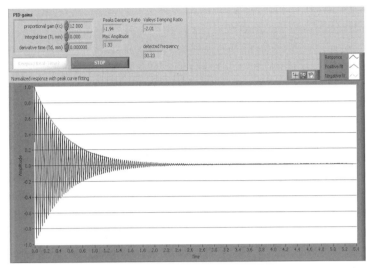

Figure 14.70 Normalized experimental response of the system with PID control.

Simulation and Experimental Results: Case Where *P* = 12

Figures 14.70 through 14.73 show the simulated and experimental response, control voltage, and sensor voltage for a proportional gain of 12. One can clearly note the large increase in the damping ratio, from 0.47 with no control to 2 with P-control. What is also important is that the natural frequency of the system is still the same, about 30 Hz.

Figure 14.70 presents the normalized beam response with the controller action invoked. Similar to the no-control case, the positive and the negative peaks are detected and then curve-fitted to determine the exponential decay rate.

Figure 14.71 plots the control voltage generated by the controller. It is interesting to point out the high control voltage of ±110 volts generated at the beginning of the response. One can

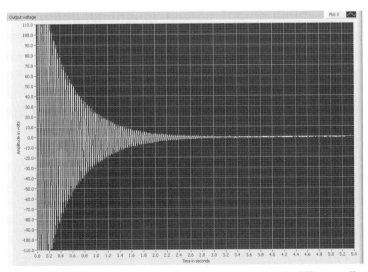

Figure 14.71 Experimental control voltage generated by the PID controller.

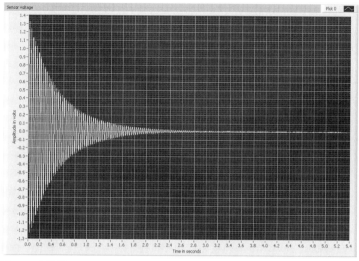

Figure 14.72 Experimental sensor voltage with PID control action.

easily infer the value of the proportional gain by comparing the plots of Figures 14.71 and 14.72.

Figures 14.73 through 14.75 show the simulated normalized response, the simulated controller output voltage, and the simulated sensor voltage, respectively. The controller provided a damping ratio of 1.92, and the maximum control output voltage is ± 110 volts. Comparing experimental and simulation results indicates that they are almost identical. This leads to the conclusion that the beam model used in the simulation is accurate. A P-like fuzzy controller is investigated in the following section.

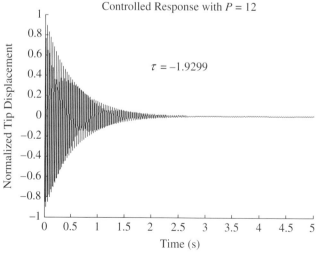

Figure 14.73 Normalized simulation response of the system with PID control.

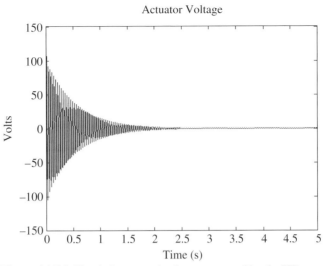

Figure 14.74 Simulation control voltage generated by the PID controller.

14.5.7 Simulation and Experimental Fuzzy Results

Similar to the experiments carried out for the PID controller, a fuzzy controller is herein designed and implemented. Figure 14.76 shows the fuzzy set used. Figures 14.77 through 14.79 present the experimental normalized response of the system, the fuzzy control voltage, and the sensor voltage, respectively. The results indicate an increase in the damping ratio, from 0.47 for the uncontrolled case to 0.58 when the fuzzy controller was engaged.

Figures 14.80 through 14.82 present the corresponding simulation results. The plots and the damping ratios obtained from simulations results are again very close to the experimental ones.

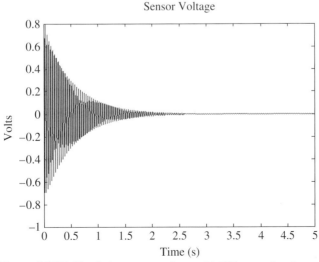

Figure 14.75 Simulation sensor voltage with PID control action.

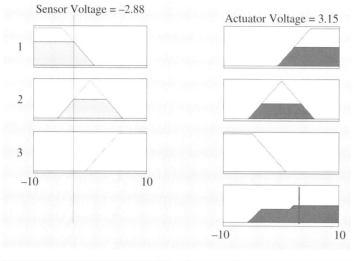

Figure 14.76 Fuzzy set used for both experiment and simulations.

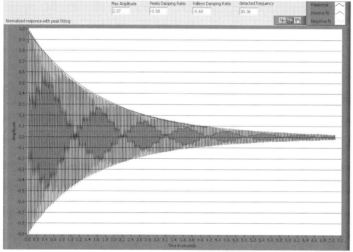

Figure 14.77 Normalized experimental response with fuzzy control.

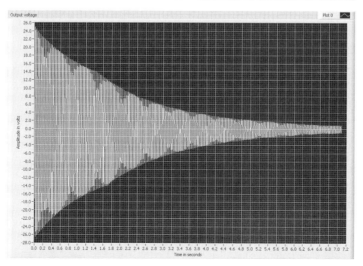

Figure 14.78 Experimental control voltage with fuzzy control.

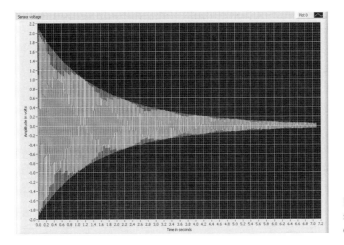

Figure 14.79 Experimental sensor voltage with fuzzy control.

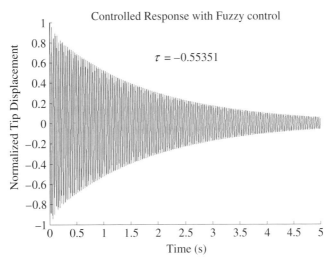

Figure 14.80 Normalized simulation response with fuzzy control.

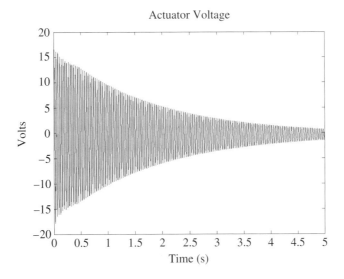

Figure 14.81 Simulation control voltage with fuzzy control.

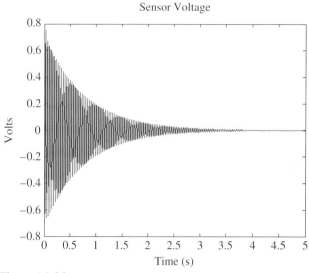

Figure 14.82 Simulation sensor voltage with fuzzy control.

14.5.8 Conclusions

Table 14.19 summarizes the simulation and experimental results with and without control. Comparing simulation and experimental results, one concludes that the model suggested in simulating the response of the cantilever beam system is very accurate. Close scrutiny of the corresponding responses shows that the error between the two is almost negligible. Furthermore, the PID controller performed much better than the fuzzy controller. This is expected because an accurate model of the beam is available. In cases where the model is very well defined, tuning the PID controller is much easier and yields better results than any heuristic technique. Of course one can tune the fuzzy controller to yield results that can compete with those obtained from the simple PID controller.

14.5.9 Case Outcomes

This case study showcased important mechatronics technology integration skills. The students can achieve the following outcomes:

1. Understand the basic operation of piezo electric material as an actuator or a sensor and how to physically bond it on an element.

TABLE 14.19 Summary of Results

	No Control	PID $P = 12$	FLC
Decay rate (simulation)	0.47511	1.9299	0.55351
Decay rate (experimental)	0.47	1.94	0.58
% Attenuation (simulation and experimental)	—	76	19
% Error (between simulation and experimental)	1	0.5%	4.5%

2. Design and build the conditioning and interface circuit for a PZT sensor and actuator.
3. Model a cantilever beam and piezo element.
4. Design an active vibration control strategy using conventional PID and intelligent fuzzy logic controllers.
5. Use MATLAB and LabVIEW for simulation and user interface.
6. Program and use real-time programming with RT DAC.

DC Power Supply

A.1 COMPONENTS OF A POWER SUPPLY

Figure 3.6 shows the essential components of a DC power with the effect each stage has on the generated output. Figure A.1 shows a schematic of a specific triple-output DC power supply.

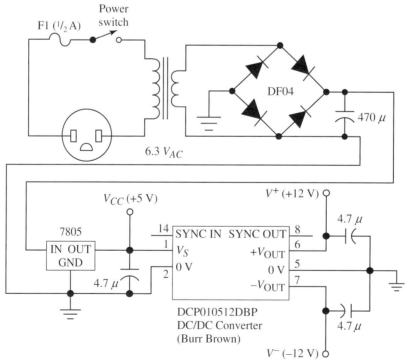

Figure A.1 A triple-output DC power supply.

The power supply provides a +5-V output from the voltage regulator (U1) and +12-V and −12 V output from the DC/DC converter (U2).

Parts List

F1	$\frac{1}{2}$ A fuse
C1	470 μF
C2, C3, C$_4$	4.7 μF
D	DF04 full bridge rectifier
T1	6.3-V$_{AC}$ transformer
U1	7805 voltage regulator IC
U2	DCP010512DBP DC/DC Converter (Burr Brown)

REFERENCE

B. Millier, "High-Resolution Data Acquisition Made Easy," *Circuit Cellar*, no. 140, March 2002.

Pinout of Selected ICs

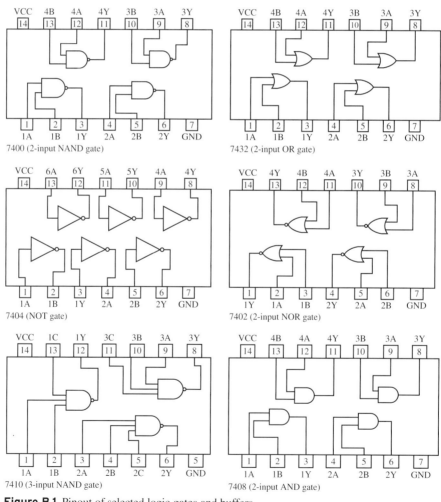

Figure B.1 Pinout of selected logic gates and buffers.

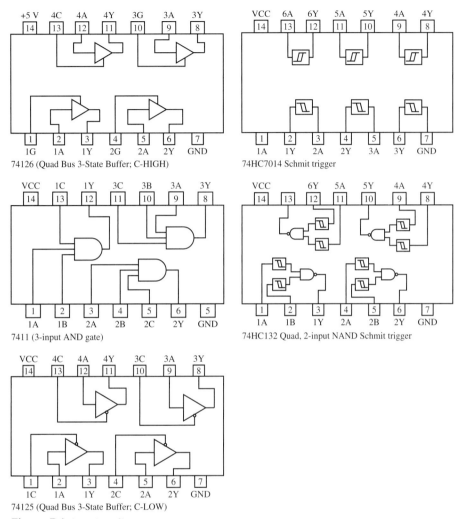

74126 (Quad Bus 3-State Buffer; C-HIGH)

74HC7014 Schmit trigger

7411 (3-input AND gate)

74HC132 Quad, 2-input NAND Schmit trigger

74125 (Quad Bus 3-State Buffer; C-LOW)

Figure B.1 (*continued*)

Instruction Set, Addressing Modes, and Execution Times for the MC9S12C

The following is a sample page of the instruction set for the 9S12C MCU. It is one of 14 pages that list all instructions. The page is taken from the following reference: *M68HC12 & HCS12 Microcontrollers: CPU12 Reference Manual - Rev 3.0*, pp. 418–431. This document is available as "INSTRUCTIONS_CPU12RM.pdf" from the Freescale website at www.freescale.com.

TABLE A.1 Instruction Set Summary (Sheet 1 of 14)

Source Form	Operation	Addr. Mode	Machine Coding (Hex)	Access Detail		S X H I	N Z V C
				HCS12	M68HC12		
ABA	(A) + (B) $\Rightarrow$ A Add Accumulators A and B	INH	18 06	OO	OO	– – Δ –	Δ Δ Δ Δ
ABX	(B) + (X) $\Rightarrow$ X *Translates to* LEAX B,X	IDX	1A E5	Pf	PP[1]	– – – –	– – – –
ABY	(B) + (Y) $\Rightarrow$ Y *Translates to* LEAY B,Y	IDX	19 ED	Pf	PP[1]	– – – –	– – – –
ADCA #*opr8i*	(A) + (M) + C $\Rightarrow$ A	IMM	89 ii	P	P	– – Δ –	Δ Δ Δ Δ
ADCA *opr8a*	Add with Carry to A	DIR	99 dd	rPf	rfP		
ADCA *opr16a*		EXT	B9 hh ll	rPO	rOP		
ADCA *oprx0_xysp*		IDX	A9 xb	rPf	rfP		
ADCA *oprx9,xysp*		IDX1	A9 xb ff	rPO	rPO		
ADCA *oprx16,xysp*		IDX2	A9 xb ee ff	frPP	frPP		
ADCA [D,*xysp*]		[D,IDX]	A9 xb	fIfrPf	fIPrfP		

TABLE A.1 (*Continued*)

Source Form	Operation	Addr. Mode	Machine Coding (Hex)	Access Detail HCS12	Access Detail M68HC12	S X H I	N Z V C
ADCA [*oprx16,xysp*]		[IDX2]	A9 xb ee ff	fIPrPf	fIPrfP		
ADCB #*opr8i*	(B) + (M) + C ⇒ B	IMM	C9 ii	P	P	– – Δ –	Δ Δ Δ Δ
ADCB *opr8a*	Add with Carry to B	DIR	D9 dd	rPf	rfP		
ADCB *opr16a*		EXT	F9 hh ll	rPO	rOP		
ADCB *oprx0_xysp*		IDX	E9 xb	rPf	rfP		
ADCB *oprx9,xysp*		IDX1	E9 xb ff	rPO	rPO		
ADCB *oprx16,xysp*		IDX2	E9 xb ee ff	frPP	frPP		
ADCB [D,*xysp*]		[D,IDX]	E9 xb	fIfrPf	fIfrfP		
ADCB [*oprx16,xysp*]		[IDX2]	E9 xb ee ff	fIPrPf	fIPrfP		
ADDA #*opr8i*	(A) + (M) ⇒ A	IMM	8B ii	P	P	– – Δ –	Δ Δ Δ Δ
ADDA *opr8a*	Add without Carry to A	DIR	9B dd	rPf	rfP		
ADDA *opr16a*		EXT	BB hh ll	rPO	rOP		
ADDA *oprx0_xysp*		IDX	AB xb	rPf	rfP		
ADDA *oprx9,xysp*		IDX1	AB xb ff	rPO	rPO		
ADDA *oprx16,xysp*		IDX2	AB xb ee ff	frPP	frPP		
ADDA [D,*xysp*]		[D,IDX]	AB xb	fIfrPf	fIfrfP		
ADDA [*oprx16,xysp*]		[IDX2]	AB xb ee ff	fIPrPf	fIPrfP		
ADDB #*opr8i*	(B) + (M) ⇒ B	IMM	CB ii	P	P	– – Δ –	Δ Δ Δ Δ
ADDB *opr8a*	Add without Carry to B	DIR	DB dd	rPf	rfP		
ADDB *opr16a*		EXT	FB hh ll	rPO	rOP		
ADDB *oprx0_xysp*		IDX	EB xb	rPf	rfP		
ADDB *oprx9,xysp*		IDX1	EB xb ff	rPO	rPO		
ADDB *oprx16,xysp*		IDX2	EB xb ee ff	frPP	frPP		
ADDB [D,*xysp*]		[D,IDX]	EB xb	fIfrPf	fIfrfP		
ADDB [*oprx16,xysp*]		[IDX2]	EB xb ee ff	fIPrPf	fIPrfP		
ADDD #*opr16i*	(A:B) + (M:M+1) ⇒ A:B	IMM	C3 jj kk	PO	OP	– – – –	Δ Δ Δ Δ
ADDD *opr8a*	Add 16-Bit to D (A:B)	DIR	D3 dd	RPf	RfP		
ADDD *opr16a*		EXT	F3 hh ll	RPO	ROP		
ADDD *oprx0_xysp*		IDX	E3 xb	RPf	RfP		
ADDD *oprx9,xysp*		IDX1	E3 xb ff	RPO	RPO		
ADDD *oprx16,xysp*		IDX2	E3 xb ee ff	fRPP	fRPP		
ADDD [D,*xysp*]		[D,IDX]	E3 xb	fIfRPf	fIfRfP		
ADDD [*oprx16,xysp*]		[IDX2]	E3 xb ee ff	fIPRPf	fIPRfP		
ANDA #*opr8i*	(A) · (M) ⇒ A	IMM	84 ii	P	P	– – – –	Δ Δ 0 –
ANDA *opr8a*	Logical AND A with Memory	DIR	94 dd	rPf	rfP		
ANDA *opr16a*		EXT	B4 hh ll	rPO	rOP		
ANDA *oprx0_xysp*		IDX	A4 xb	rPf	rfP		
ANDA *oprx9,xysp*		IDX1	A4 xb ff	rPO	rPO		
ANDA *oprx16,xysp*		IDX2	A4 xb ee ff	frPP	frPP		
ANDA [D,*xysp*]		[D,IDX]	A4 xb	fIfrPf	fIfrfP		
ANDA [*oprx16,xysp*]		[IDX2]	A4 xb ee ff	fIPrPf	fIPrfP		
ANDB #*opr8i*	(B) • (M) ⇒ B	IMM	C4 ii	P	P	– – – –	Δ Δ 0 –
ANDB *opr8a*	Logical AND B with Memory	DIR	D4 dd	rPf	rfP		

(*Continued*)

TABLE A.1 Instruction Set Summary (Sheet 1 of 14) (*Continued*)

Source Form	Operation	Addr. Mode	Machine Coding (Hex)	Access Detail HCS12	M68HC12	S X H I	N Z V C
ANDB *opr16a*		EXT	F4 hh ll	rPO	rOP		
ANDB *oprx0_xysp*		IDX	E4 xb	rPf	rfP		
ANDB *oprx9,xysp*		IDX1	E4 xb ff	rPO	rPO		
ANDB *oprx16,xysp*		IDX2	E4 xb ee ff	frPP	frPP		
ANDB [D,*xysp*]		[D,IDX]	E4 xb	fIfrPf	fIfrfP		
ANDB [*oprx16,xysp*]		[IDX2]	E4 xb ee ff	fIPrPf	fIPrfP		
ANDCC #*opr8i*	(CCR) • (M) ⇒ CCR Logical AND CCR with Memory	IMM	10 ii	P	P	⇓⇓⇓⇓	⇓⇓⇓⇓

[1] Due to internal CPU requirements, the program word *fetch* is performed twice to the same address during this instruction.

MC9S12C Registers and Control Bit Assignments

The following is a sample page of the register map for the 9S12C MCUs. It is one of 17 pages that list all registers and associated bits. The page is taken from the following reference: *MC9S12C128 Data Sheet for HCS12 Microcontrollers*, pp. 29–46. This document is available as "MC9S12C128_V1.pdf" from the Freescale Semiconductor website at www.freescale.com.

0X0000-0X000F MEBI Map 1 of 3 (HCS12 Multiplexed External Bus Interface)

Address	Name		Bit 7	Bit 6	Bit 5	Bit 4	Bit 3	Bit 2	Bit 1	Bit 0
0x0000	PORTA	Read: Write:	Bit 7	6	5	4	3	2	1	Bit 0
0x0001	PORTB	Read: Write:	Bit 7	6	5	4	3	2	1	Bit 0
0x0002	DDRA	Read: Write:	Bit 7	6	5	4	3	2	1	Bit 0
0x0003	DDRB	Read: Write:	Bit 7	6	5	4	3	2	1	Bit 0
0x0004	Reserved	Read: Write:	0	0	0	0	0	0	0	0
0x0005	Reserved	Read: Write:	0	0	0	0	0	0	0	0
0x0006	Reserved	Read: Write:	0	0	0	0	0	0	0	0
0x0007	Reserved	Read: Write:	0	0	0	0	0	0	0	0

(Continued)

0X0000-0X000F MEBI Map 1 of 3 (HCS12 Multiplexed External Bus Interface) *(Continued)*

Address	Name		Bit 7	Bit 6	Bit 5	Bit 4	Bit 3	Bit 2	Bit 1	Bit 0
0x0008	PORTE	Read: Write:	Bit 7	6	5	4	3	2	Bit 1	Bit 0
0x0009	DDRE	Read: Write:	Bit 7	6	5	4	3	Bit 2	0	0
0x000A	PEAR	Read: Write:	NOACCE	0	PIPOE	NECLK	LSTRE	RDWE	0	0
0x000B	MODE	Read: Write:	MODC	MODB	MODA	0	IVIS	0	EMK	EME
0x000C	PUCR	Read: Write:	PUPKE	0	0	PUPEE	0	0	PUPBE	PUPAE
0x000D	RDRIV	Read: Write:	RDPK	0	0	RDPE	0	0	RDPB	RDPA
0x000E	EBICTL	Read: Write:	0	0	0	0	0	0	0	ESTR
0x000F	Reserved	Read: Write:	0	0	0	0	0	0	0	0

Using the CodeWarrior Integrated Development Environment (IDE)

E.1 DEVELOPING A CODE FOR THE 9S12C

The Integrated Development Environment (IDE) is part of the CodeWarrior Development Studio from Metrowerks for the 68HC(S)12 microcontrollers. Detailed instructions are available in the QUICK_START.pdf file that accompanies the MCUSLK kit available from Freescale Semiconductor. The following is a brief list of steps needed to start the development of a new 9S12C project.

1. Launch the CodeWarrior IDE.
2. From the IDE Main Menu Bar select the **File > New** Option.
3. Create a new project through selecting the **HC(S)12 New Project Wizard**. Several pop up windows will appear that require selecting several options.
4. Select Build Target to connect, for example, to the Simulator.
5. Edit source code.
6. Add files if appropriate.
7. Build the project and select Make from the menu to assemble, compile, and link files.

ASCII Code Table

Char	Hex	Dec	Char	Hex	Dec	Char	Hex	Dec
(space)	20	32	@	40	64	`	60	96
!	21	33	A	41	65	a	61	97
"	22	34	B	42	66	b	62	98
#	23	35	C	43	67	c	63	99
$	24	36	D	44	68	d	64	100
%	25	37	E	45	69	e	65	101
&	26	38	F	46	70	f	66	102
'	27	39	G	47	71	g	67	103
(	28	40	H	48	72	h	68	104
)	29	41	I	49	73	i	69	105
*	2A	42	J	4A	74	j	6A	106
+	2B	43	K	4B	75	k	6B	107
,	2C	44	L	4C	76	l	6C	108
–	2D	45	M	4D	77	m	6D	109
.	2E	46	N	4E	78	n	6E	110
/	2F	47	O	4F	79	o	6F	111
0	30	48	P	50	80	p	70	112
1	31	49	Q	51	81	q	71	113
2	32	50	R	52	82	r	72	114
3	33	51	S	53	83	s	73	115
4	34	52	T	54	84	t	74	116
5	35	53	U	55	85	u	75	117
6	36	54	V	56	86	v	76	118
7	37	55	W	57	87	w	77	119
8	38	56	X	58	88	x	78	120
9	39	57	Y	59	89	y	79	121
:	3A	58	Z	5A	90	z	7A	122
;	3B	59	[	5B	91	{	7B	123
<	3C	60	\	5C	92			7C	124
=	3D	61	]	5D	93	}	7D	125
>	3E	62	^	5E	94	~	7E	126
?	3F	63	_	5F	95	ã	7F	127

Number Systems

G.1 NUMBER REPRESENTATION

All digital circuitry works with binary values. Depending on the application, the binary numbers can be used to represent different numbers. There are four main number systems: decimal, binary, octal, and hexadecimal. A number system is distinguished by its *base*, or *radix R*. The base indicates how many digits n a system uses to represent quantities.

In a decimal system, $R = 10$ because it uses 10 digits (0, 1, 2, . . . , 9). In a binary system $R = 2$ because it uses two *binary digits*, or *bits* (0, 1). The set of eight bits is called a *byte*. The base of the octal system is $R = 8$ because it uses eight digits (0, 1, 2, . . . , 7). The base of the hexadecimal system is $R = 16$ because it uses 16 digits (0, 1, 2, . . . , 9, A, B, . . . , F). Letters have no significance other than the need for single symbols to represent the higher digits: $10 = A$, $11 = B$, $12 = C$, $13 = D$, $14 = E$, and $15 = F$.

The *value* of a number N in any system can be expressed as $N = (d_n \ldots d_1 d_0)_R$. Depending on its position in a number, the ith digit carries a particular weight, which is simply R^i. The weight of the *least significant digit* is $R^0 = 1$ and that of the *most significant digit* is R^n. The value of a number N is found from

$$N = (d_n \ldots d_1 d_0)_R = \sum_{i=1}^{n} d_i R^i \tag{G.1}$$

where d_i is the multiplier. The maximum number that can be represented in any number system is $R^n - 1$. The minus 1 is because all number systems start at 0.

Note: Binary arithmetic involving numbers with decimals (e.g., 11000111.101_2) are treated in the same way as in decimal calculations.

Examples

Binary System (Base 2)

$N = (110100.101)_2$

$= 1 \times 2^{-3} + 0 \times 2^{-2} + 1 \times 2^{-1} + 0 \times 2^0 + 0 \times 2^1 + 1 \times 2^2 + 0 \times 2^3 + 1 \times 2^4 + 1 \times 2^5$

$= 0.125 + 0 + 0.5 + 0 + 0 + 4 + 0 + 16 + 32$

$= (52.625)_{10}$

Octal System (Base 8)

$N = (753)_8$

$= 3 \times 8^0 + 5 \times 8^1 + 7 \times 8^2$

$= 3 + 40 + 448$

$= (491)_{10}$

Hexadecimal System (Base 16)

$N = (B3A)_{16}$

$= A \times 16^0 + 3 \times 16^1 + B \times 16^2$

$= 10 + 48 + 2816$

$= (2874)_{10}$

Two other number systems are important: the *binary-coded decimal* and the *gray code*. The binary-coded decimal is used to decode decimal numbers, four bits for each digit on the basis of 8421 weighting for each group of four bits. It facilitates hardware decoding and generating decimal digits 0–9 for visual display on instrument panels. The gray code has the advantage of only one bit change from one number to the next. It is used in realizing absolute optical encoder devices used as position sensors.

The relationship between the various number systems is shown in Table G.1. Each symbol in a octal number represents three bits in the binary numbers. For example, $(491)_{10} = (753)_8 = (111\ 101\ 011)_2$. Each symbol in a hexadecimal number represents four bits in the binary numbers. For example, $(2876)_{10} = (B3A)_{16} = (1011\ 0011\ 1010)_2$. Hexadecimal is a shorter, more convenient method for writing binary numbers. It simplifies data entry and display to a greater degree than octal. Thus, hexadecimal numbers are widely used to represent data and addresses in microprocessors. For example,

8-bit data: $(1011\ 0001)_2 = (B1)_{16}$

16-bit address: $(0011\ 1111\ 0110\ 1010)_2 = (3F6A)_{16}$

Note that 1010–1111 are illegal BCD numbers.

TABLE G.1 Number Systems Compared

Decimal, B = 10	Binary, B = 2	Octal, B = 8	Hexadecimal, B = 16	Gray	BCD, (8421)
0	0000	0	0	0000	0000
1	0001	1	1	0001	0001
2	0010	2	2	0011	0010
3	0011	3	3	0010	0011
4	0100	4	4	0110	0100
5	0101	5	5	0111	0101
6	0110	6	6	0101	0110
7	0111	7	7	0100	0111
8	1000	10	8	1100	1000
9	1001	11	9	1101	1001
10	1010	12	A	1111	I
11	1011	13	B	1110	L
12	1100	14	C	1010	L
13	1101	15	D	1011	E
14	1110	16	E	1001	G
15	1111	17	F	1000	A
16	10000	20	10		L

G.2 DECIMAL CONVERSION

Conversion of a decimal number to any other number system in base R is done by successive division of the decimal number by the base R. The remainder after each division is a digit in the result, and all the remainders form the number in base R. The MSB digit is the remainder of the last division. For binary, the divisor is 2 and the remainder is 1 or 0. For hexadecimal, the divisor is 16 and the remainder is any hexadecimal digit between 0 and F.

Examples

Converting $(19)_{10}$ to Binary

$$19 \div 2 = 9 \quad \text{rem} \quad 1 \quad \text{(LSB)}$$
$$9 \div 2 = 4 \quad \text{rem } 1$$
$$4 \div 2 = 2 \quad \text{rem } 0$$
$$2 \div 2 = 1 \quad \text{rem } 0$$
$$1 \div 2 = 0 \quad \text{rem } 1 \quad \text{(MSB)}$$
$$(19)_{10} = (10011)_2$$

Converting $(5285)_{10}$ to Hexadecimal

$$21145 \div 16 = 1321 \quad \text{rem } 9 \quad \text{(LSD)}$$
$$1321 \div 16 = 82 \quad \text{rem } 9$$
$$82 \div 16 = 5 \quad \text{rem } 2$$
$$5 \div 16 = 0 \quad \text{rem } 5 \quad \text{(LSD)}$$
$$(21145)_{10} = (5299)_{16}$$

Note: An n-bit binary number can be scaled to represent a decimal fraction X by using the following relation:

$$X = (b_1 \times 2^{-1} + b_2 \times 2^{-2} + \cdots + b_n 2^{-n}) \tag{G.2}$$

G.3 SIGNED AND UNSIGNED NUMBERS

Unsigned numbers carry a sign symbol ($+$ or $-$) separate from themselves to indicate positive and negative numbers. In signed numbers, a sign is inherent within the number itself. Since computers do not carry a sign, the MSB of a number denotes the sign. If n is the number of digits representing a number in base R, the range is divided in half, with positive numbers from 0 to $R^{n-1} - 1$ and negative numbers from -1 to $-R^{n-1}$. Each negative number in the range is the *base complement* of a positive number N; i.e., $N_C = R^n - N$. If n is the number of bits in a word, the range of binary numbers that can be stored is:

$$\text{Unsigned numbers:} \quad 0 \le N \le 2^n - 1$$
$$\text{Signed numbers:} \quad -2^{n-1} \le N \le 2^{n-1} - 1$$

The 68HC11 handles 8-bit word size. The range of numbers it can store is therefore is $2^8 = 256$, from $(00000000)_2$ to $(11111111)_2$. If bit 7 (MSB) of a number is 1, it is considered negative; if it is 0, the number is considered positive. The range of positive numbers is 0 to $2^{8-1} - 1 = 127$, or $(00000000)_2$ to $(01111111)_2$, and that of negative numbers (2s-*complement*) is -1 to $-2^{8-1} = -128$, or $(10000000)_2$ to $(11111111)_2$. This is very similar to a car's odometer. If we construct an 8-bit odometer and set it to show zero initially, then driving the car forward and backward, the odometer would show numbers such as:

$$
\begin{array}{ll}
00000010 & 2 \\
00000001 & 1 \\
00000000 & 0 \\
11111111 & -1 \\
11111110 & -2
\end{array}
$$

The 2s-complement of a binary number is used to handle negative numbers. To find the 2s-complement of a number without subtraction, start with the LSB and locate the *first* 1 in the number. Then take the logical complement of all bits thereafter; e.g, the 2s-complement of (11111010) is (00000110).

G.4 BINARY ADDITION (UNSIGNED NUMBERS)

Binary addition is accomplished in the same way decimal numbers are added. The following rules apply:

$$
\begin{array}{l}
0 + 0 = 0 \\
0 + 1 = 1 + 0 = 1 \\
1 + 1 = 10 \ (0 + \text{carry}) \\
1 + 1 + 1 = 11 \ (1 + \text{carry})
\end{array}
$$

Eight-bit unsigned numbers are limited to $(256)_{10}$. If a carry results in a ninth bit, flag C in the CCR register will set, indicating that the answer is not correct. Here are some examples of 68HC11 instructions used for addition: ABA, ADCB, ADDB, ADDA.

Examples

Adding Two Bytes

```
LDAA    #$30
ADDA    #$2C
```

$$
\begin{array}{ll}
00110000 & (48)_{10} \\
+\ 00100010 & (34)_{10} \\
\hline
01010010 & (82)_{10} \quad [C = 0] \quad \text{correct}
\end{array}
$$

```
LDAB    #$38
ADDB    #$69
```

$$
\begin{array}{ll}
00111000 & (56)_{10} \\
+\ 01101001 & (105)_{10} \\
\hline
10100001 & (161)_{10} \quad [C = 0] \text{ a negative number (incorrect)}
\end{array}
$$

```
LDAB    #$93
ADDB    #$8B
```

$$
\begin{array}{ll}
10010011 & (147)_{10} \\
+\ 10001011 & (139)_{10} \\
\hline
00011110\ (286)_{10} & \quad [C = 1] \text{ a carry to ninth bit (incorrect)}
\end{array}
$$

G.5 BINARY SUBTRACTION (UNSIGNED NUMBERS)

Binary subtraction is similar to decimal subtraction. When the subtrahend (bottom) is larger than the minuend (top), a borrow is extracted from the next digit. For 8-bit numbers the carry flag C in the CCR will set if a borrow is extracted from the ninth bit, indicating the answer incorrect. Here are some examples of instructions used for subtraction: SBCA, SUBA, SUBB, SUBD.

$$
\begin{array}{ll}
11001001 & (201) \\
-\ 10110011 & (179) \\
\hline
00010010 & (22)_{10} \quad [C = 0]
\end{array}
$$

$$00110110 \quad (54)$$
$$- \; 10110011 \quad (179)$$

$$10000011 \; (131)_{10} \; [C = 1], \text{ borrow from C flag.}$$

G.6 BINARY MULTIPLICATION

Binary multiplication follows the same procedure used in decimal multiplication, with the following rules:

$$0 \times 0 = 0$$
$$0 \times 1 = 1 \times 0 = 0$$
$$1 \times 1 = 1$$

The product usually needs as many digits as the sum of the digits of the multiplicand and multiplier. The partial products are always the multiplicand or zeros shifted by the number of places in the multiplier digit. The 68HC11 instruction set includes MUL to carry out multiplication.

G.7 BINARY DIVISION

Binary division is similar to decimal division, with the following rules:

$$0 \div 0 \text{ or } 1 \div 0 \text{ are meaningless}$$
$$0 \div 1 = 0$$
$$1 \div 1 = 1$$

Only whole numbers are used in control applications; the remainder therefore is ignored. 6HC11 supports two instructions to carry out division: IDIV for integer division and FDIV for fraction division.

G.8 BINARY-CODED DECIMAL (BCD) ARITHMETIC

Addition with BCD numbers is accomplished by one of two alternatives: (1) Convert BCD to binary, perform addition, and convert back to BCD; (2) microprocessor circuitry is built to accommodate BCD rules as follows:

1. Add the two BCD numbers as though they are binary.
2. If the least significant nibble (four bits) of the result is a number greater than 9 or if there is a carry from the third bit to the fourth bit (called half-carry, H), add 6 to the four bits; otherwise no change is made.
3. After step 2 is completed, if the four most significant bits of the result is a number greater than 9 or if the normal carry (C) is set, add 6 to these bits; otherwise no change is made.

Example

$$
\begin{array}{llll}
& 0101 & 1001 & (59) \\
& 0100 & 0001 & (41) \\
\end{array}
$$

$$
\begin{array}{lll}
H = 0 \quad 0] & 1001 \quad 1010 & \text{sum } 1010 > 9 \\
& 0000 \quad 0110 & \text{add } 6 \\
\end{array}
$$

$$
\begin{array}{lll}
H = 1 \quad 0] & 1010 \quad 0000 & \text{step 2 complete, sum } 1010 > 9 \\
& 0110 & \text{add } 6 \\
\end{array}
$$

$$
H = 0 \quad 1] \quad 0000 \quad 0000 \quad (100) \quad \text{step 3 complete}
$$

The 68HC11 provides the means to handle BCD addition by inserting the digital adjust instruction DAA immediately after one of the following addition instructions involving ACCA: ABA, ADD, ADC. The instruction uses the C and H bits of the CCR as well as ACCA to determine a valid BCD result. The following instructions illustrate the use of DAA.

G.9 2s-COMPLEMENT ARITHMETIC

Digital computers perform addition and subtraction of signed numbers with 2s-complement arithmetic using the same internal adding circuit. Addition is carried out in the usual way, regardless of the sign of the numbers to be added. The answer is a 2s-complement number with the correct sign. Any carry out of the MSB is always ignored. The answer is correct only if it is within the allowed range. The range is exceeded under two conditions: (1) Two positive numbers are added to produce a negative number and (2) two negative numbers are added to produce a positive number. Under these conditions the overflow flag V in the CCR will set, signaling an incorrect answer for signed numbers. When a positive 2s-complement number and a negative 2s-complement number are added, the sum is always correct.

Examples

$$
\begin{array}{lll}
00001000 & (+8) & \\
+ \; 11110100 & (-12) & \\
\hline
11111100 & (-4) & V = 0 \quad C = 0 \quad \text{(ignore) answer correct}
\end{array}
$$

$$
\begin{array}{lll}
00110011 & (+51) & \\
+ \; 01000001 & (+65) & \\
\hline
01110100 & (+116) & V = 0 \quad C = 0 \quad \text{(ignore) answer correct}
\end{array}
$$

$$
\begin{array}{lll}
01111111 & (+127) & \\
+ \; 00000001 & (+1) & \\
\hline
10000000 & (+128) & V = 1 \quad C = 0 \quad \text{(ignore) answer incorrect}
\end{array}
$$

$$
\begin{array}{lll}
11111011 & (-5) & \\
+ \; 11111001 & (-7) & \\
\hline
11110100 & (-12) & V = 0 \quad C = 1 \quad \text{(ignore) answer correct}
\end{array}
$$

$$\begin{array}{ll} 10000011 & (-125) \\ + \underline{11111010} & (-6) \\ 01111101 & (-131) \quad V = 1 \quad C = 1 \quad \text{(ignore) answer incorrect} \end{array}$$

The CPU's arithmetic logic unit adds two binary numbers and yields correct answer regardless of whether they are signed or unsigned.

Examples

Unsigned Numbers

$$\begin{array}{ll} 10001101 & (+141) \\ + \underline{00010110} & (+22) \\ 10100011 & (+163) \quad C = 0 \quad \text{answer correct} \end{array}$$

Signed Numbers

$$\begin{array}{ll} 10001101 & (-115) \\ + \underline{00010110} & (+22) \\ 10100011 & (-93) \quad V = 0 \quad C = 0 \quad \text{answer correct} \end{array}$$

Unsigned Straight Subtraction

$$\begin{array}{ll} 00100000 & (+32) \\ - \underline{10100000} & (-160) \\ 10000000 & (-128) \quad C = 1 \quad \text{answer correct} \end{array}$$

Signed 2s-Complement Subtraction

$$\begin{array}{ll} 00100000 & (+32) \\ + \underline{01100000} & (+96) \\ 10000000 & (128) \qquad C = 1, N = 1, V = 1, Z = 0 \qquad \text{answer correct} \\ & \qquad\qquad \text{2s-complement is} \quad 10000000 \end{array}$$

G.10   MULTIPLE PRECISION ADDITION AND SUBTRACTION

If is necessary to perform arithmetic operations on numbers that are larger than the limits of the microcontroller, then it is necessary to use more than a word to represent them. For example, 14 bits are required to handle $(10,000)_{10}$. For an 8-bit word processor to handle this number, two words are needed to represent it. Two words per number is usually called *double precision*.

Examples

Multiprecission Addition

Write a code to add the hex numbers $24EB and $3234.

```
LDAA    #$EB    ;load low byte of first number in acc A
ADDA    #$34    ;add low byte of second number to acc A
STAA    $01     ;store the low byte of the sum at $01
LDAA    #$24    ;load high byte of first number in acc A
ADCA    #$32    ;add high byte of second number to acc A
STAA    $00     ;store the low byte of the sum at $00
END
```

$$
\begin{array}{l}
\$24EB \\
\$3234
\end{array}
$$

$$
\begin{array}{cccc}
 & 1 \text{ (carry)} & & \\
 & 00101000 & 11101011 & (+9451) \\
+ & 0110010 & 00110100 & (+12852) \\
\hline
 & 01011011 & 00011111 & (+22303)
\end{array}
$$

Multiprecission Subtraction

Write a code to subtract $52D6 from $4B2A.

```
LDAB    #$2A    ;load low byte of the minuend in acc B
SUBB    #$D6    ;subtract low byte of the subtrahend from contents
                 of acc B
STAB    $01     ;store the low byte of the result at $01
LDAB    #$4B    ;load high byte of the minuend in acc B
SUCB    #$32    ;subtract the high byte of the subtrahend and C bit
                 from acc B
STAB    $00     ;store the high byte of the result at $00
END
```

PROBLEMS

G.1 Convert the following numbers from decimal form to binary form:

a. 175.0

b. −2067.0

c. 44.835

d. 0.042

G.2 Convert the numbers obtained in Problem G.1 from binary form to octal form and then to hexadecimal form.

G.3 Think about a number system in base 3. What are the digits? Demonstrate it by adding $(340)_{10}$ to $(64)_{10}$.

G.4 Convert the following numbers from binary form to decimal form. Assume (1) unsigned, (2) signed numbers.

a. 110110111

b. 101000

c. 0.111

d. 1011.0111

G.5 Convert the binary numbers in Problem G.4 to hexadecimal form.

G.6 What are the largest positive and negative integers that can be represented using 2s-complement arithmetic and the following word lengths?

a. 8 bits

b. 12 bits

c. 16 bits

d. 24 bits

e. 32 bits

G.7 Sixteen-bit 2s-complement arithmetic is used in a control computer. Convert each of the following decimal numbers into the binary form used in this computer.

a. 127

b. −346

c. 2242

d. −3465

e. 18543

f. −32015

G.8 The following binary numbers are found in the memory of a computer that uses 16-bit 2s-complement integer arithmetic. What are the corresponding decimal numbers?

a. 0000011111111000

b. 0110001101110011

c. 1111110011110100

d. 1000000000011101

G.9 Add the binary numbers in Problem G.8 using 16-bit 2s-complement integer arithmetic.

G.10 Convert the binary numbers in Problem G.8 into hexadecimal form.

G.11 Convert the binary numbers in Problem G.8 into octal form.

G.12 What is the minimum number of bits to represent the number of days in a year?

G.13 Convert $(536)_{10}$ to BCD.

G.14 Multiply 1001_2 by 101_2 using unsigned binary arithmetic.

G.15 Divide 1101100_2 by 101_2 using unsigned binary arithmetic.

G.16 Multiply −19 by 23 using binary arithmetic.

G.17 Divide −500 by −7 using binary arithmetic.

G.18 The following equations are implemented in software on a control computer:

$$e_n = r_n - c_n$$
$$m_n = m_{n-1} + 2e_n$$

Assume that $m_{n-1} = 00000111_2$, $r_n = 00111111_2$, and $c_n = 00011100_2$ at the nth sample instant. What is the value of m_n after the binary arithmetic represented by the given equations is performed?

G.19 A variable must represent 4000 increments over a range of ±5? How many bits must be included in the fractional part of a 16-bit binary word used to represent the variable?

G.20 Convert the following decimal fractions into 8-bit unsigned binary fractions.

a. 0.2

b. 113

c. 0.01

d. 13116

G.21 Convert the following decimal fractions into 8-bit signed binary fractions.

a. 0.25

b. −0.3125

c. 0.67

d. −0.14

G.22 Convert the 8-bit signed binary fractions obtained in Problem G.21 back to decimal fractions. What is the error between the binary representation and the original decimal representation?

G.23 Repeat Problems G.21 and G.22 using 16-bit signed binary fractions. Compare the accuracy of the results to the accuracy obtained using 8-bit signed binary fractions.

G.24 Convert π into a binary number. To how many decimal digits of accuracy is your result equivalent?

G.25 Add the decimal numbers 59 + 41, and indicate the status of the H, C, Z, and V flags.

G.26 Subtract $A0 from $20 assuming signed and unsigned numbers.

G.27 Write a program to add the 24-bit number $01 A0 14 to $6C AC 8D.

Mechanisms for Mechatronics

H.1   POWER FLOW IN A MACHINE

A machine is an assembly of many components arranged to operate synchronously to drive a load according to a prescribed motion. The machine receives its energy from a power source and delivers it to the load. The source or the load may belong to one of five energy domains: mechanical translation, mechanical rotation, electrical, fluid, and thermal. For example, the power source for a power-generating windmill is fluid power, the wind. The wind rotates the wind turbine, which drives an electric generator through a mechanical gear train to produce electricity. The load in this case may be regarded as the mechanical rotation of the generator. In an CNC milling machine, the drive motors receive electrical energy and convert it to mechanical work associated with the cutting forces at the tip of the mill. The robot is another example where the actuators receive energy from an electric or hydraulic source and move the mechanical arm to generate any desired output at the end effector, which carries the payload or a tool, such as a welding gun, and a spray-painting gun, among others.

Power flow through each component in a machine occurs as a result of two coexisting power variables: the through variable f and the across variable v. The instantaneous power is simply the product of the two variables:

$$P = vf \tag{H.1}$$

The energy accumulated over a time period T is defined by the integral relation

$$U = \int_0^T P\, dt \tag{H.2}$$

Table H.1 gives the power variables and the instantaneous power in each energy domain. Note the exception that applies to a heat system, for which the thermal power is the heat transfer rate q.

TABLE H.1 Power Variables in Various Energy Forms

System	Across Variable, v	Through Variable, f	Power, P
Electrical	Voltage drop, V	Current, I	$P = VI$
Mechanical translation	Velocity difference, v	Force, F	$P = Fv$
Mechanical rotation	Velocity difference, ω	Torque, T	$P = T\omega$
Thermal	Temperature difference, T	Heat flow rate, q	$P = q$
Fluid	Pressure difference, p	Volume flow rate, Q	$P = pQ$

Two mechanical energy systems are identified in Table H.1: translation and rotation. Elements of a translational system undergo linear (straight-line) motion, whereas components of a rotational system experience rotary motion. A velocity difference referred to in the table means that the system moves relative to a fixed (motionless) reference. Figure H.1 shows a hydraulic actuator driving a translational mass m and an electric motor driving a rotary inertia J.

The input and output power associated with a machine must satisfy the principle of energy conservation, simply stated as

$$\eta P_{IN} + P_{OUT} = 0 \tag{H.3}$$

where P_{IN} is the power input to the machine, P_{OUT} is the output power used up by the load, and η is the efficiency of the machine, which is always less than 1 due to power losses incurred in electrical, thermal, and fluid elements and mechanical friction, if present.

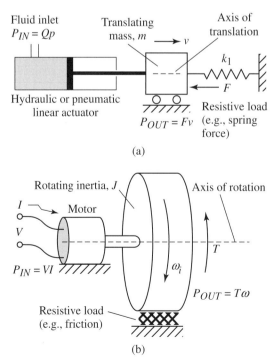

Figure H.1 A linear actuator driving a translational load (a) and a DC motor driving a rotational load (b).

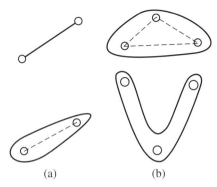

Figure H.2 Skeletal diagram of a binary link (a) and a ternary link (b).

(a)   (b)

H.2 KINEMATIC CHAINS

A kinematic chain is a collection of links connected to each other by joints. A *link* is a word used to designate a machine part or a component of a mechanism. A link may be rigid if it does not experience shape change under applied loads or flexible if it does. The discussion here assumes the links to be rigid bodies. The joint (or *kinematic pair*) connects two links together and consists of two mating elements. The joint allows the two links to rotate relative to one another at the joint's interface. A link is classified as a *binary link* if it contains two different elements of two joints, a *ternary link* if it contains three elements of three different joints, and so on. Skeletal diagrams of these two types of links are shown in Fig. H.2. Other types of links include the gear, the cam, the pulley, and the screw, among others.

Although the links are shown as lines between the corresponding joints, a link can take any shape, as the application requires. However, only the directed length d_i affects the kinematic motion, and the shape of a link has no effect on the kinematics of the mechanism. The shape of the link influences the dynamics of the mechanism where the location of the mass center and mass distribution of the links come to bear. Therefore, for motion or kinematic analysis, only the distances between the joints (link lengths) matter, and the mechanism is represented by a skeleton diagram in which links are represented by straight lines. Such diagrams are referred to as *kinematic diagrams*. A link that can rotate a full cycle (360°) is called a *crank*; if it cannot, it is called a *rocker*.

Figure H.3 shows the joints used in planar mechanisms: the revolute joint (R) and the prismatic joint (P). The revolute joint, such as the hinge of a door, allows the two links it

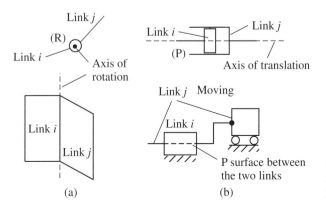

Figure H.3 Main joints for planar mechanisms: revolute (a) and prismatic (b).

connects to rotate only relative to each other. The prismatic joint allows translation between the two connected links. These two joints are classified as lower-order kinematic pairs because the two mating elements of the joint form a contacting surface. Other joints, such as the cam joint between a cam and a follower and the gear joint between two gears, are higher-order joints because the two mating joint elements have either a point or a line contact.

Kinematic chains may be planar if all links are coplanar (move in parallel planes) or spatial if any links move in a three-dimensional manner. Planar kinematic chains are considered here.

H.3   MECHANISMS AS BUILDING BLOCKS

Mechanisms are the fundamental building blocks of which any machine, regardless of its complexity, is built. Depending on the purpose of the machine, several mechanisms can be combined in series or in parallel to transfer energy from the power source to the load.

A mechanism is a single-input, single-output device that receives power through its input link from a single source and delivers it by its output link to the load. Because only the input or the output link could be connected, mechanisms can be combined in one of the following three possible ways:

Parallel combination: All mechanisms receive power from one source.

Serial combination: The mechanisms are assembled sequentially so that the output of one mechanism drives the input of the next mechanism in the series and the output link of the last mechanism drives the load.

Hybrid combination: This includes both the parallel and the serial arrangements.

The input and output links of a mechanism may rotate or translate about axes that are fixed to the ground, allowing the input and output links to experience absolute motion. The axis here refers to the imaginary line that defines the direction of translation or rotation of a link. For a rotational object, the axis is the centerline of the shaft on which the rotational link is mounted and is called the *axis of rotation*. For a translational element, the axis represents the direction along which a mass moves and is called the *axis of translation* (see Fig. H.1).

The input link of a mechanism connects directly to the power source of the machine, or it receives power indirectly through the output link of another mechanism to which it is connected. Similarly, the output link may connect to a load directly if it is the last link in the machine, or it may provide power to the input link of the next mechanism in the series. The connecting link transfers motion from the input link to the output link.

Depending on the type of motion of the input and output links, the mechanism is classified as one of the following motion converters:

- Rotary to rotary (R–R)
- Rotary to translation (R–T)
- Translation to rotary (T–R)
- Translation to translation (T–T)
- Helical to rotary (H–T)
- Helical to translation (H–T)

TABLE H.2 Possible Input-to-Output Motion Relations and the Mechanisms that Generate Them

Motion Conversion	Linear Input/Output Relation	Nonlinear Input/Output Relation
R to R	• Spur gears • Helical gears • Bevel gears • Worm gears • Belt–pulley drive • Chain sprocket drive • Cable (tendon) drives	• Four-bar mechanism • Cam-follower mechanism • Ratchet • Geneva wheel
R to T (or T to R)	• Screw drive • Pinion-rack mechanism	• Slider-crank mechanism • Cam-follower mechanism
T to T		• Double slider mechanism • Cam-follower mechanism
Helical to T	Screw drive	

Furthermore, the relation between the motion of the input and the output links of a mechanism may be either linear or nonlinear. Table H.2 gives a summary of all possible motion relations and the mechanisms that generate them.

The following sections present further details on many useful mechanisms found in mechatronics.

H.4 FOUR-BAR MECHANISM

The mechanism shown in Fig. H.4a is called the four-bar mechanism. The four-bar mechanism is one of the most widely used mechanisms, encompassing a wide range of applications. The mechanism consists of four links labeled d_1–d_4 and connected by four joints. The links are the *ground* (fixed) link, length d_1; the *drive* (*input*) link, length d_2; the *coupler* link, length d_3; and the *follower* (*output*) link, length d_4.

Although the links are shown to lie in one plane, in reality they have offsets between them to allow motion around one another. However, the offsets are considered to be very small and do not affect the motion of the mechanism. Point P in Fig. H.4a is a point on the coupler link and thus is called the *coupler point*. The path it traces during a motion cycle is called the *coupler curve*. Infinite possible coupler curve shapes may be generated and exploited in many useful tasks. For example, the 4R mechanism proportioned as shown in Fig. H.4b generates a straight-line segment by its coupler point P.

Note that only links d_2–d_4 can move, whereas link d_1 is part of the ground to which the mechanism is anchored. Every mechanism must have a ground link, and for all mechanisms in a machine there is only one ground link. d_1 is the length of that part of the ground link that influences the motion of the mechanism. Note also that any of the links may be used as the ground link, while the others are free to move. This results in four different mechanisms with the same link dimensions, called *kinematic inversions*. Although link dimensions are the same for all inversions, the motion characteristics of one inversion is completely different from those of the others. In general, an *n*-link kinematic chain may have *n* different kinematic inversions.

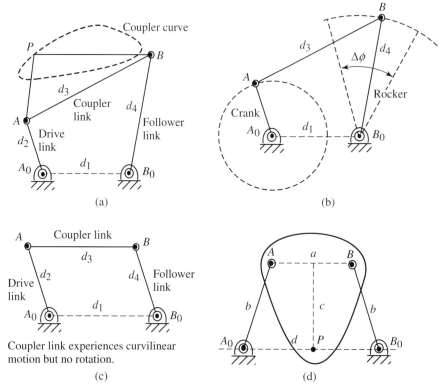

Figure H.4 Four-bar mechanisms: generating a coupler curve (a), crank-rocker (b), parallelogram (c), and straight line motion generator (d).

Types of 4R Mechanisms

Depending on the range of motion of the drive and follower links, a four-bar mechanism may be one of the following types:

- *Crank-crank* (also called *drag link*): The drive and the follower links are both cranks and can rotate 360°.
- *Crank-rocker*: The drive link can make a complete revolution, while the follower link can only oscillate between two limit positions, as shown in Fig. H.4c.
- *Rocker-rocker*: Both the input and the follower links have limited angular rotations.

Grashof's criterion identifies the type of the 4R mechanism from its link dimensions as follows. If s is the length of the shortest link, l the length of the longest link, and p and q the lengths of the other two links, then the class of the 4R mechanism is one of the following:

1. If $s + l < p + q$, the mechanism has at least one crank; that is, at least two links in at least one kinematic inversion can rotate continuously relative to each other. In this case, the following four Grashof 4R possibilities exist.

   a. If s is the crank and any of the adjacent links is the frame, the mechanism is a crank-rocker.

   b. If the shortest link is the ground link, the 4R is a double crank.

   c. If the shortest link is the follower, the 4R is a rocker-crank.

   d. If the link opposite to the shortest is the frame, the 4R is a double rocker.

2. If $s + l > p + q$, continuous relative motion is not possible. All kinematic inversions are non-Grashof triple rocker mechanisms.

3. If $s + l = p + q$, all kinematic inversions belong to case 1, but they suffer from a condition known as the *change point*. At the change point, the centerline of all links become collinear; hence the cranks may change direction of rotation unless given proper guidance.

4. Two special cases of class 3 exist:

   a. *Parallelogram mechanism*: All possible kinematic inversions are of the crank-crank type if controlled at the change point. This is the only 4R mechanism in which the coupler link undergoes curvilinear translation, with all coupler points traversing circular motion (Fig. H.4d).

   b. *Deltoid mechanism*: It is formed by connecting two equal short links to two equal longer links. If a long side is grounded, a crank-rocker results. Meanwhile if a short side is the ground, a crank-crank mechanism is formed. This is the *Gallaway* mechanism, in which the shorter link will turn two revolutions for one turn of the longer link.

H.5 SLIDER-CRANK MECHANISM

The slider-crank mechanism is a special case of the 4R mechanism, in which the length of the follower link d_4 is infinite and the path traced by point B becomes a straight line. Referring to Fig. H.5a, we see that the slider-crank mechanism consists of four binary links, three revolute joints, and one prismatic joint. The links are the *drive (input)* link, length r; the *coupler* link, length l; the *follower (output)* link, which is represented by the block or the slider; and the *ground* link, which includes the pivot where link 2 is connected to the ground and the sliding surface on which the slider moves. The internal combustion engine mechanism is a special

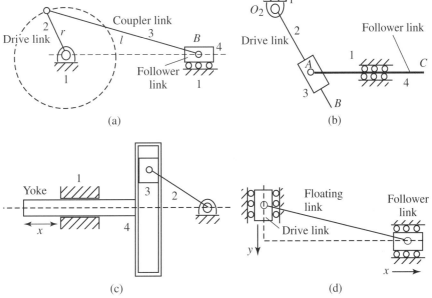

Figure H.5 Basic slider-crank mechanism (a), Rapson's slide (b), Scotch-Yoke (c), and double-slider (d).

case of the slider-crank mechanism, in which link 2 is the crank and is the output link, the coupler link is the connecting rod, the slider is the piston and is used as the input, and the ground is the engine block.

Two other useful mechanisms that belong to the slider-crank family are the Rapson's slide and the Scotch yoke mechanisms. *The Rapson's slide* (Fig. H5b) is used as the marine steering linkage in large ships, where link O_2B represents the tiller and link AC is the actuating rod.

The *Scotch yoke mechanism* (Fig. H.5c) has an infinitely long connecting rod. If link 2 rotates at constant angular velocity, the slider (link 4) generates a *simple harmonic motion*. This mechanism is used in testing machines to simulate simple harmonic vibrations. Figure H.5d is a double slider mechanism that generates a linear output motion from a linear input.

H.6 CAM-FOLLOWER MECHANISMS

A cam–follower mechanism consists of two main components: the *cam*, which is the input, or driving member, and the *follower*, which is the output, or driven member. Cam-follower mechanisms are used to generate nonuniform linear or rotary follower motions that exhibit a special functional relationship to the uniform input motion of the cam. Figure H.6 shows many cam-follower mechanisms. The rotary cam in Figs. H.6a and b is called a *disk cam* because it is a segment of a disk, and the linear cam is called a *wedge cam* (Figs. H.6c and d) because is it a segment of a block or wedge. Other types of cams include the cylindrical (barrel) cam, the conical cam, and the globoidal cam, among others, depending on the shape of the cam.

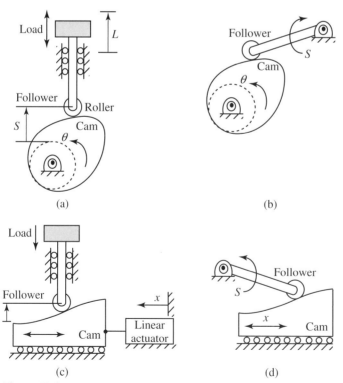

Figure H.6 Radial cam with translating follower (a) and rotary follower (b), and wedge cam with translating follower (c) and rotary follower (d).

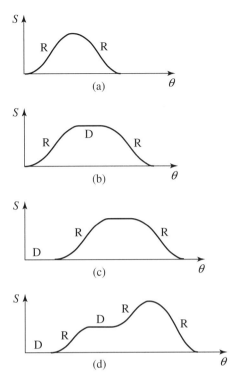

Figure H.7 Various follower motion histories.

Depending on the type of contact with the cam surface, the follower may have a roller or it may have a flat face or a spherical face. The cam and follower must always maintain direct contact. This may be ensured through the force of a spring or by using a groove if the follower is a roller type.

The surface of the cam, or the cam profile, is cut according to the desired motion history of the follower. Limitless follower motion histories are possible. The cam can be fashioned to produce, during one motion cycle (360° in a rotary cam), complex follower motions that combine a series of rise segments, dwells, and return segments. The *rise* is the motion stroke during which the follower moves from its initial to its maximum position, L in Fig. H.6a. A *return* is the opposite of a rise; it is the motion stroke that moves the follower from its maximum position back to its initial position. During a *dwell* zone, the follower is stationary while the cam moves within a specific input stroke. A rise may also be formed to consist of a series of smaller rise segments and possible temporary returns before the maximum position is reached. Similarly the return may be accomplished in smaller return steps with temporary rises in between. Figure H.7 shows a few follower motion regimes. The cam may also be fashioned to produce a specific velocity and acceleration follower profile as well.

Follower Motion Profiles

The follower motion profile may be described by one cam curve or by a combination of many available cam curves. Two basic cam curves are presented here: the simple harmonic motion and the cycloidal motion. The variations of the position $S(\theta)$, velocity $V(\theta)$, and acceleration $A(\theta)$ of the follower during a rise segment β_1 and a return segment β_2 are shown in Figs. H.8 and H.9. The corresponding relations are given next.

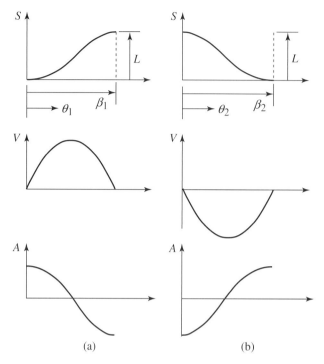

Figure H.8 Simple harmonic motion history during follower rise (a) and return (b).

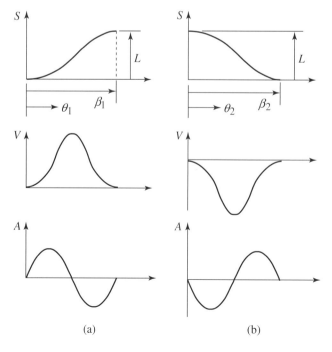

Figure H.9 Cycloidal motion history during follower rise (a) and return (b).

Simple Harmonic Motion Rise

$$S = \frac{L}{2}\left[1 - \cos\frac{\pi\theta_1}{\beta_1}\right]$$

$$V = \frac{\pi L}{2\beta_1}\left[\sin\frac{\pi\theta_1}{\beta_1}\right] \tag{H.4}$$

$$A = \frac{\pi^2 L}{2\beta_1^2}\left[\cos\frac{\pi\theta_1}{\beta_1}\right]$$

Simple Harmonic Motion Return

$$S = \frac{L}{2}\left[1 + \cos\frac{\pi\theta_2}{\beta_2}\right]$$

$$V = \frac{\pi L}{2\beta_2}\left[\sin\frac{\pi\theta_2}{\beta_2}\right] \tag{H.5}$$

$$A = \frac{\pi^2 L}{2\beta_2^2}\left[\cos\frac{\pi\theta_2}{\beta_2}\right]$$

Cycloidal Motion Rise

$$S = L\left[\frac{\theta_1}{\beta_1} - \frac{1}{2\pi}\sin 2\pi\frac{\theta_1}{\beta_1}\right]$$

$$V = \frac{L}{\beta_1}\left[1 - \cos 2\pi\frac{\theta_1}{\beta_1}\right] \tag{H.6}$$

$$A = \frac{2\pi L}{\beta_1^2}\left[\sin 2\pi\frac{\theta_1}{\beta_1}\right]$$

Cycloidal Motion Return

$$S = L\left[1 - \frac{\theta_2}{\beta_2} + \frac{1}{2\pi}\sin 2\pi\frac{\theta_2}{\beta_2}\right]$$

$$V = -\frac{L}{\beta_2}\left[1 - \cos 2\pi\frac{\theta_2}{\beta_2}\right] \tag{H.7}$$

$$A = -\frac{2\pi L}{\beta_2^2}\left[\sin 2\pi\frac{\theta_2}{\beta_2}\right]$$

θ_1 is the cam angle during a rise occurring between 0 and β_1, and θ_2 is the cam angle during a return that varies between 0 and β_2. The same relations apply to translation and rotation alike. S, V, and A in Eqs. (H.4)–(H.7) represent the linear position (m), linear velocity (m/s), and linear acceleration (m/s^2) of a translating follower or the angular position (rad), angular velocity (rad/s), and angular acceleration (rad/s^2) of a rotary follower. L is the follower stroke (m or rad).

H.7 GEAR DRIVES

The fundamental building block of a gear mechanism is the two-gear arrangement, shown in Fig. H.10. The circles shown represent the *pitch circles* of the gears, on which all calculations are based. Each gear is mounted an a separate shaft. The smaller of the two gears is usually the driver, called the *pinion*, and has N_P teeth and a pitch circle radius R_P. The larger of the two gears is usually the driven gear, called the *gear*, and has N_G teeth and a pitch circle radius R_G. The ratio of the angular velocity of the pinion ω_P to the angular velocity of the gear ω_G is called the *velocity ratio VR*, or the *gear ratio*. It is given by

$$VR = \frac{\omega_P}{\omega_G} = \frac{R_G}{R_P} \tag{H.8}$$

Equation (H.8) is applicable to all types of gears. The ratio of the torque on the output shaft T_{OUT} to the torque on the input shaft T_{IN} is *VR*. Thus, a gear drive that reduces the speed multiples the torque, so the power remain the same.

The input and output shafts in Fig H.10 are parallel. However, different types of gears may be used to generate rotary-to-rotary motion between shaft axes that are arranged in different ways. Table H.3 lists the main types of gears and the axis arrangement each type provides.

If two disks (nontoothed) are pressed against each other such that they can only rotate relative to each other without sliding, their motion is similar to that of gears. This type of mechanism is called *traction drive*, and the ratio of the disk radii is the velocity ratio.

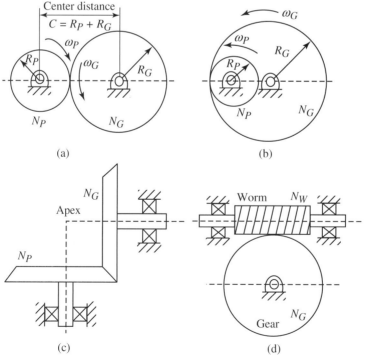

Figure H.10 Pinion-gear drive: external contact (a), internal contact (b), bevel gear set (c), and worm-gear set (d).

TABLE H.3 Gear Types and Input–Output Axis Arrangements They Accommodate

Gear Type	Axis Arrangement
Spur gear	Parallel
Helical gear	Parallel
Crossed helical gear	Nonparallel, nonintersecting
Bevel gear	Intersecting shafts (perpendicular and nonperpendicular
Worm gear	Nonparallel, nonintersecting (shaft angle is 90°)
Pinion-rack	Rack generates linear motion

Gear Trains

Gear trains are gear mechanisms that include more than two gears to generate more than one output (parallel arrangement), increase the speed ratio between the input and output shafts, and bridge the distance between the input and output axes. Two types of gear trains are available: ordinary and planetary. In an ordinary gear train the axes of all gears are fixed to the ground, whereas in a planetary gear train the axes of some gears, called the *planet gears*, are carried by a rotating arm or carrier and rotate with it about the central axis of the gear train. Figure H.11a is a simple ordinary gear train, and Figs. H.11b and 11c are examples of compound ordinary gear trains. The difference between a simple and a compound arrangement is that each shaft

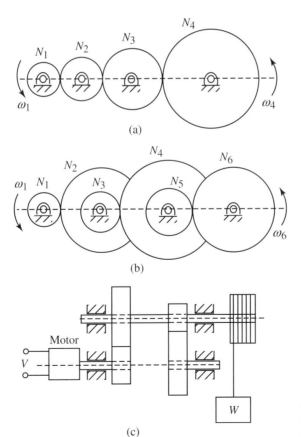

Figure H.11 A simple (a) and a compound (b) ordinary gear train, and a motor hoisting a load via a two-stage gear transmission.

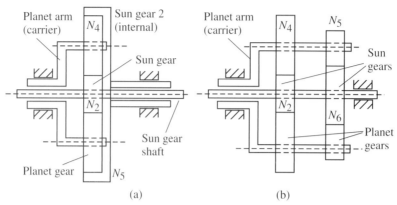

Figure H.12 A simple (a) and a compound (b) planetary gear train.

carries only one gear in the simple gear train, whereas more than one gear can be mounted on the same shaft in a compound gear train. Figure H.12a is a simple planetary gear train, and Fig. H.12b shows a compound planetary gear train.

The kinematic behavior of a gear train is described by its *train value e*. The train value between gear i and gear j in any type of gear train is the same and is given by

$$e_{ij} = (-1)^n \frac{\text{Product of teeth of driving gears from } i \text{ and } j}{\text{Product of teeth of driven gears from } i \text{ to } j} \tag{H.9}$$

where n is the number of externally meshed gear pairs (or number of direction changes) starting from gear i and ending with gear j in the gear train.

The ratio of the speed of gear i to the speed of gear j in an ordinary gear train is the reciprocal of the train value between gear i and gear j; that is, $VR_{ij} = 1/e_{ij}$. However, for a planetary gear train the *VR* depends on the input component of the gear train, the output, and the fixed member.

H.8 BELT-PULLEY DRIVES

Figure H.13a shows an ideal representation of the main components of a belt–pulley or a chain-sprocket drive, which generate a linear input–output motion relation between an input and output axes over a long distance. The smaller pulley is usually the driver and the larger pulley the driven. The velocity ratio generated between the input and output shafts is

$$VR = \frac{\omega_1}{\omega_2} = \frac{R_2}{R_1} \tag{H.10}$$

The ratio of the output torque to the input torque is equal to *VR*. The output motion of a belt–pulley drive is in general rotary. However, a load can be attached to the belt or chain, as shown in Fig. H.13b, to produce a linear output motion from a rotary input motion. The chain-sprocket drive has similar motion and load conversion characteristics as the belt–pulley drive, and the foregoing discussion applies to it equally.

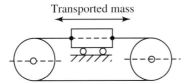

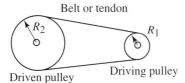

Figure H.13 Basic tendon drive unit.

Cables may also be used as the linking element between the input and output pulleys to form what is referred to as *tendon drives*. Complicated designs may be realized to produce complex motion.

H.9 SCREW DRIVE

Figure H.14 shows a representation of a lead screw driving a load. The load is connected to the nut, which is held against rotation but can slide freely, as shown. The screw is held against translation, but its ends are allowed to rotate freely. As the motor rotates the screw, the load will translate along the screw axis. A screw is characterized by its *lead*, which is the distance traveled by the nut as the screw rotates one complete turn. The relation between the linear displacement of the nut x and the angular rotation of the screw θ is $x = l\theta$. Regular screws suffer from high friction. However, ball screws are available, which reduces friction significantly.

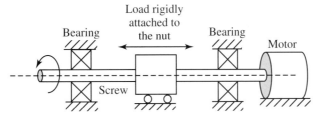

Figure H.14 A motor driving a load via a screw.

Index

2s complement arithmetic, 687
4-to-20 mA transmitters, 453
555 timer, 167, 439
  astable mode, 168
  monostable mode, 169
  sensor interface, 170
9S12C microcontroller, 179
  address bus, 184
  artithmetic logic unit (ALU), 182
  block diagram, 181
  control bus, 184
  CPU12, 182
  data bus, 184
  development tools, 242
  overview, 180

A

absolute encoder, 411
  resolution, 411
absolute stability, 532
absorbtivity, 426
AC induction motors, 461, 515
  single-phase, 515
  three-phase, 516
  speed control, 519
acceleration sensors, 386
accelerometer, 633
accuracy, 392, 435
acoustic sensors, 386
acoustic impedance, 435
acoustic lens, 437
acoustic pressure, 436
acoustic pyrometry, 443
acoustic thermometer, 435
across variable, 692
active filters, 166
  first-order high-pass filter, 167
  first-order low-pass filter, 167
  second- and a third-order, 167
active power, 41
active vibration control, 604, 649
actuators, 460
  electric, 461
  hydraulic, 461
  pneumatic, 461
adaptive control, 528, 571, 556, 559
adaptive fuzzy logic controllers
  fuzzy model-reference adaptive
    controller, 583
    parameter adaptation, 584
    signal adaptation, 584

    membership tuning adaptive
      controller, 586
address register, 192
address register, 184
addressing modes, 198, 250
  direct (DIR) 198, 200
  extended (EXT) 198, 200
  immediate (IMM) 198, 199
  indexed 5-bit offset (IDX) 198, 201
  indexed 9-bit offset (IDX1) 198, 201
  indexed 16-bit offset (IDX2) 198, 201
  indexed accumulator offset (IDX) 202
  indexed auto postdecrement/
    -increment (IDX) 203
  indexed-indirect accumulator D
    ([D,IDX]) 203
  indexed-indirect offset ([IDX2])
    198, 203
  inherent (INH) 198, 199
  relative (REL) 198, 200
admittance 35
aliasing 356
alternating current (AC) 9
amplitude uncertainty 357
analog sensors 387
analog signal 106
analog-to-digital conversion (ADC),
  182, 251, 351, 557
  bipolar, 352
  conversion error, 353
  conversion techniques, 360
  full scale, 352
  I/O mapping, 353
  quanta, 352
  offset, 352
  range, 352
  resolution, 353
  signal-to-noise ratio (SNR), 354
  span, 352
  unipolar, 352
analog-to-digital (ATD) converter,
  9S12C, 365
  channel selection, 366
  channel sampling, 369
  continuous scan control, 369
  conversion complete flag, 368
  conversion time, 366
  external ADC trigger, 371
  input signal range, 373
  registers, 366
  setup, 366
angular frequency, 33
angular velocity, stepper motor, 506

anti-aliasing prefilter, 357
apparent low frequencies, 356
application program, 228
application specific integrated circuit
  (ASIC), 388
arithmetic instructions (see instruction set)
arithmetic logic unit (see 9S12C)
armature, 462
artificial intelligence 529
assembler, 4, 188, 236
assembler directive, 190, 230
  section definition, 231
  constant definitions, 231
  data allocation, 232
  assembly control, 232
  listing file, 232
  conditional assembly, 233
  macro control, 233
assembly language 188, 233
assembly language line 189
  instruction field 189
  label field 190
  comment field 190
  operand field 189
asymptotic stability, 515
asynchronous communication, 291
asynchronous inputs, 162
autonomous mobile robot, 604
average power, 40

B

back diode, 67
back emf, 462, 471
back emf suppression, 456, 500
background debug mode (BDM), 243
band-pass filter, 53
band-reject filter, 53
bandwidth, 53, 148, 513, 532
base, numbers, 681
batteries, NiMH and NiCd 635
baud rate, 297
belt-pulley drive, 475, 705
  optimum pulley radius, 476
Bessel filters, 131
bias, sensors, 391
bifilar winding, 494
bilinear transformation, 574
binary coded decimal arithmetic, 686
binary numbers and arithmetic, 681
binary weighted ladder, 362
Biot-Savart law, 26

bipolar ADC, 352
bipolar stepper motor drive, 494
   driver ICs, 514
bipolar power supply, 479
bipolar converter, 359
bipolar junction transistors (BJTs), 76
   active state, 79
   characteristics, 78
   current amplification factor, 79
   cutoff (or open) state, 79
   DC biasing, 81
   load-line equation, 80
   npn BJT, 76
   pnp BJT , 76
   saturation line, 80
bipolar junction transistor circuits
   transistor switch, 83
   emitter-follower, 83
   current source, 85
   common-emitter amplifier, 85
bistable multivibrators (see latches)
block diagram, 470
Bode plots, 543
Boltzman constant, 66, 388
Boolean algebra, 141
Boolean laws and theorems, 143
bounded-input-bounded-output (BIBO)
   stability, 533
branch instructions (see instruction set)
breakdown voltage, 456
break point, 237
bridge circuit, 18, 431
   operating modes, 432
   null condition, 18
   offset voltage, 18
   sensitivity, 432
   temperature compensation, 432
brushed DC motors (see DC motors)
brushless DC motors (see DC motors)
buffer op-amp (see unity gain amplifier)
buffers, 162, 152, 257, 380
   tri-state buffer, 125
bulk modulus of elasticity, 434
burning the code, 236
bus contention, 165, 311
bus drive termination, 163
Butterworth filters (see filter design)
bypass capacitors, 25, 56, 496

C

C-language, 237
cam-follower mechanisms, 698
candela (see also lumen), 399
controller area network (CAN), 295
capacitance, 20, 451
capacitor, 20
capacitor couplers, 57
capacitive-based, 423
capacitive sensors, 449
   displacement sensor, 450
   liquid-level sensor, 450
carrier signal, 394, 448 (see also
   modulation)
central processing unit (CPU), 180, 182
certainty equivalence principle, 573
charge amplifier, 446, 654
chassis ground, 12
Chebyshev filters, 131
chemical sensors, 386, 432

chopper-drive method, 501
circuit breakers, 10
circuit loading (see loading error)
classical control, 529
clipping, 79
clock, 140, 185, 250
clock monitor reset (see resets)
closed circuit, 10
closed-loop control, 446, 482, 488, 529
CMOS logic family, 155
   Characteristics, 155
CMOS-to-TTL interface, 158
cold junction compensation (CJC)
   (see thermocouples)
combinational circuits, 139
   design of, 147
common-collector amplifier (see bipolar
   junction transistor circuits)
common mode gain (CMG), 115
common mode rejection ratio
   (CMRR), 115
common-emitter amplifier (see bipolar
   junction transistor circuits)
common-mode voltage (CMV), 108
commutator, 463, 465
comparator, 86, 121
compass, sensor, 644
compiler, C-language, 3, 237
complex sensors, 387
computer, 177
computer operating properly (COP)
   (see resets)
condenser microphone, 437
condition code register (CCR), 194
   C carry/borrow bit, 195
   H half-carry bit, 196
   I mask bit, 196
   N negative bit, 196
   S stop disable bit, 195
   V overflow bit, 195
   Z zero bit, 195
condition code register instructions, 221
   (see instruction set)
conduction, 411
conductance, 15
conductivity, 386
conductor, 64
constitutive relations, 429
control computer, 460
controllability, 566
controllable system, 566
convection, 426
convective heat transfer coefficient, 426
conveyor system, 477
corner frequency, 49–53
Coulomb friction torque, 470
Coulomb law, 8
counters, 164
   BCD counter, 164
   decade counters, 164
   up/down counter, 164
coupling ratio, 473, 475, 476
CPU registers, 188
   accumulators A, B, and D, 188
   condition code register (CCR), 190
   index registers X and Y, 189
   program counter (PC), 189
   stack pointer (SP), 190
coupling capacitor, 25, 58, 85, 113
crank-rocker mechanism, 696
critically damped response, 51, 518

crossover frequency, 549, 550, 553, 554
crowbar device, 99
current amplifier, 480
current divider, 13, 16
current limiting resistor, 74
current mirror, 425
current sensing, 429
current source, 10, 82
cutoff frequency, 48, 129
cybernetics, 529
cycloidal motion, 699

D

damped natural frequency, 535
damping coefficient, 492
damping ratio, 50, 375, 535
Darlington, power transistor (see power
   transistors)
data acquisition card (DAQ), 593, 658
data communications equipment,
   296, 306
data direction registers, 258
data handling instructions (see
   instruction set)
data select lines, 349
data selector (see multiplexer)
data terminal equipment, 296, 306
DC motors, brushless, 464, 626
DC motors, brush-type, 461, 463, 592,
   605, 639
   continuous torque, 470, 476
   controller ICs, 489
   electrical time constant, 472
   field coil motors, 461
   generated torque, 466
   heat dissipation, 472
   JDH-2250 motor, 470
   mechanical time constants, 472
   motor constant, 467
   operating principle, 461
   optimum velocity profile, 474
   peak torque, 485
   permanent magnet (PM) motors, 463
   RS 540SH motor, 608
   selection, 476
   stall torque, 468
   state-space model, 470
   torque constant, 466
   transfer function, 471
   voltage constant, 466
DC power supply, 70, 657, 670
DC restorer, 71
DC servos, 490, 605
   Futaba S48 motor, 609
decibel, 43
decimal conversion, 683
decision-making element (see gate)
decoders, 165
deffuzification, 577
delay time, 534
deltoid mechanism, 697
demodulation, 380
demultiplexer, 129
depletion region, 65
derivative causality, 23
detent position, 496
detent torque, 504
DIAC, 101
Dielectric, 20

differential amplifier, 114
differential gain (DG), 109
differential voltage, 109
differentiator, 46, 120
digital control, 574
digital multimeters, 20
digital signals, 138
digital logic circuits, 138
digital signal processing (DSP), 591
digital-to-analog conversion (DAC),
        351, 376, 574
    accuracy, 379
    binary weighted ladder, 377
    bipolar DACs, 380
    ICs, 381
    output voltage, 377
    range, 378
    resolution, 379
diode, 65
    forward-biased, 65
    reverse-biased, 65
    threshold voltage, 67
diode clamp, 70
diode equation, 66
diode limiter, 70
diode thermometer, 72
direct current (DC), 9
direct digital control, 575
direct mode (see addressing modes)
discretization, 574
discretization period, 574
displacement current, 23
displacement sensors, 372
displays, 267
disturbance, 484
dominant pole, 534
DO-WHILE operation, 224, 225
doppler frequency, 442
doppler radar, 442
doppler shift, 442
double buffering, 299
double crank mechanism, 696
double-rocker mechanism, 696
drag link mechanism (see double crank
        mechanism)
drivers, 162, 257
droop, 359
dry circuit, 263
dual slope ADC technique, 361
ducted fan, 629
duty cycle, 331, 468
dynamic response, sensors, 392

E

E-fields, 56
earth ground, 11
efficiency, 692
electret, 439
electric charge, 7
electric current, 8
electric displacement, 428
electric field, 8, 428
electric flux density, 9
electric actuators, 460
electric shielding, 56
electrical axis, piezoelectric, 445
electrical time constant, 472, 499
electrocardiogram (ECG), 117
electrodes, 429
electromagnetic force, 27

electromagnetic interference
        (EMI), 55
electromagnetic spectrum, 399
electromotive force, 8
electron, 7, 402
electron-volt (eV), 399
electrorheological (ER) fluids, 649
electrostatic charge, 156
electrostatic energy, 23
embedded controller, 4
emissivity, 426
emulation, 575
emulator, 237
encoders, 165
energy, 9, 691
energy conservation principle, 692
equates, 189, 231, 236
equation of motion, 650
equivalent inertia, 470, 477
equivalent internal resistance, 37
Euler's identity, 35
extended addressing mode (see
        addressing modes)
external pin reset (see resets)

F

fan out, 149
    current sinking, 149
    current sourcing, 149
Faraday's law, 28
feedback control (see closed-loop
        control)
feedback frequency, 491
field, 8
field coils, 471
field-effect transistors (FETs), 87
    depletion type, 87
    enhancement type, 87
metal oxide semiconductor FETs
        (MOSFETs), 88
    characteristics, 89
filter design
    Bessel filters, 169
    Butterworth filter, 169
    Chebyshev filter, 169
filtering capacitor, 25
filters, 47, 379
    passive filters, 47
    active filters, 130
final value theorem, 484, 485, 536
finite element model, 650
firefighting robot, 604, 636
first-order sensor, 389
flame detector, 641
flash ADC, 364
flash memory, 244
Flicker noise, 55
flip-flops, 139, 264
    clocked D flip-flop, 160
    JK flip-flop, 161
    master-slave flip-flop, 161
    SR flip-flop, 158
    T flip-flop, 159
flow sensors, 386
flux lines, 25
flux linkage, 29
flyback diode, 71, 500
follower, op-amp (see unity gain
        amplifier)
force sensors, 386

forced output compare, 332
four-bar mechanism, 695
framing, 296
free-wheeling diode (see flyback diode)
frequency, 33, 51, 384
frequency, acoustic, 434
frequency, light, 399
frequency compensation, 163
frequency division (see counters)
frequency response, 34, 543
frequency ripple, 433
frequency-to-voltage converter, 592
friction (see Coulomb friction)
full-duplex communications, 296
full-scale input, 377
full-scale output, 377
full step, stepper motor, 496
full-wave rectifier (see rectifiers)
function generator, 43
fuse, 10
fuzzification, 576
fuzzy inference engine, 576
fuzzy logic control (FLC), 576, 560,
        562, 563, 568, 646, 660
    ASIC IC, 591
    PD-like, 578
    PI-like, 580

G

Gain, sensor, 390
gain bandwidth product, 163
gain margin, 545
gain scheduling, 571
gate, 139
    AND gate, 141
    EX-NOR, gates 143
    EX-OR gate, 143
    NAND gate, 143
    NOR gate, 143
    NOT gate, 141
    OR gate, 141
gate-turn-off (GTO), 100
gauge factor, 415
Gauss's law 9
gear drives, 469, 474, 610, 702
gear trains
    ordinary gear trains, 703
    planetary gear trains, 703
gear ratio, 702
    optimum, 475
Grashof's criterion, 696
Ground, 10
ground loops, 56, 381
grounding techniques, 57

H

H-bridge, 485
    L293D 487
    L298 487, 616, 639
    LM18200 487
H-fields, 56
half-duplex communications, 296
half-step, stepper motor, 496
half-wave rectifier (see rectifiers)
Hall coefficient, 428
Hall device, 57, 427
Hall effect, 28, 411
Hall voltage, 427

Hall-effect sensors, 427
  current sensing, 429
  interface to the 9S12C, 428
  position sensing, 429
hardware decoding,
  keyboards, 266
  dedicated ICs, 267
header comments, 226
heat dissipation, 481
heater, 347
heat flux, 426
heat-flux sensor, 425
hexadecimal numbers
  arithmetic, 682
high-level language, 188
high-pass filters
  first-order, 51
  second-order, 53
  third-order, 131
holding torque, 504
hot carrier diode, 67
hybrid stepper motors, 493
hysteresis, Shmitt trigger, 122

I

idle 297
IEEE-P1451 standard, 387
IF-THEN-ELSE operation, 224
IF-THEN fuzzy rules, 560
insulated gate bipolar transistors
  (IGBTs), 96
immediate addressing mode (see
  addressing modes)
impedance, 12, 35
  input impedance, 37
  output impedance, 37, 432
impedance converter, 152
impedance matching, 38
inductive kick, 71
incremental encoder
  resolution, 412
  phase quadrature, 412
  index I channel, 412
indexed addressing mode (see
  addressing modes)
index register X, Y (see CPU
  registers)
inductance, 25
inductive sensors, 451
  motion detection, 451
  LVDT, 451
inductor, 25
inertia matching, 474
infrasound, 419
inherent addressing mode (see
  addressing modes)
input capture, timer 322, 333
  registers, 333
Input/Output (I/O) ports,
  183, 250
  port A, 259
  port AD, 261
  port B, 259
  port E, 259
  port M, 261
  port P, 260
  port S, 261
  port T, 260
input port expansion, SPI, 317
instantaneous power, 9, 39

instruction decoder, 190
instruction queue, 197
instruction set, 207, 249
  accumulator and memory
    instructions, 204
  arithmetic instructions, 213
  addition and subtraction
    instructions, 214
  branch instructions, 222
  condition-code register
    instructions, 221
  data-compare and -testing
    instructions, 220
  data movement instructions, 207
  data transfer and exchange
    instructions, 207
  data modify instruction, 209
  division instructions, 216
  fuzzy logic instructions, 218
  interrupt handling instructions, 229
  jump and subroutine call/return
    instructions, 226
  logic instructions, 220
  multiply instructions, 216
  shift-and rotate instructions, 210
  min/max instructions, 212
  table interpolation instruction, 218
instrumentation amplifier (IA), 116
  common mode rejection ratio
    (CMRR), 117
  differential gain, 117
  voltage gain, 117
insulator, 64
integral causality, 23
integrator, 47, 119
intelligent control, 529, 576
interfering noise, 55
intelligent control, 511, 559
interface, computer, 256
interrupt handling instructions
  (see instruction set)
interrupts, 280
  maskable interrupt, 281, 289
  non-maskable interrupt, 281, 288
  process, 281
  source enable bit, 281, 289
  source flag, 281, 289
  vector, 284
  vector address, 284
  vectored priority, 280, 282
  vector table, 284
interrupt density, 283
interrupt latency, 283
interrupt request, 271
interrupt service routine (ISR), 281
inverting op-amp, 112
IR emitter/detector packages, 407
  optical interruptor, 407
  optical coupler, 408
  optical reflector, 408
IRQ maskable interrupt, 272
isolation, 58
isolator, 58, 408

J

Jitter, 363
Johnson noise (see thermal noise)
junction field effect transistor (JFET),
  87, 415
  characteristics, 88

K

Karnaugh maps, 146
keyboard, 265
  hardware decoding, 266
  rollover, 266
kinematic analysis, tricycle robot, 611
kinematic chains, 693
kinematics, unmanned aerial vehicle, 623
Kirchhoff's laws, 13
knowledge base fuzzy control rules, 577

L

label, 196
label field, 196
LabVIEW, 538
  control design (CD) toolkit, 539
latch, 139, 160
lag compensator, 549
lead compensator, 549
lead screw drive, 476, 705
  optimum pitch, 476
leakage resistance, 398, 432
Lenz's law, 29, 471
level sensors, 386
light emitting diode displays, 271
  hardware decoding 279
  multiplexed display 277
  software decoding 272
light emitting diode (LED), 72, 268
  common anode, 74, 258
  common cathode, 74, 258
light meter, 374
light sensors, 386, 398
  materials, 400
  types, 400
  applications, 400
limiters, 124
line tracing robot, 409
linear amplifiers, 480
linear circuit, 33
linear interpolation, 395
linear steppers, 493
linear time invariant (LTI) control
  system, 532
linearization, 394
linear variable differential transformer
  (LVDT) 32, 435
linker, 236
liquid crystal displays (LCDs), 279
  controllers, 266
  panels, 245
load, 10
loader, 236
loading error, 16, 37, 394, 432
logarithmic amplifier, 137
logic elements, 139
logic equivalence, 143
logic families, 148
  complementary metal oxide
    semiconductor (CMOS), 155
  Emitter-coupled logic (ECL), 148
  Integrated injection logic (IIL), 148
  Transistor-transistor logic (TTL), 151
logic levels, 138, 149
logic signals, 184
logic terms, 171
  complementation, 171
  DeMorgan's theorems, 171
  duality, 139

logic terms (*continued*)
  literal, 171
  maxterm, 173
  minterm, 172
  product of sums, 172
  sum of products, 172
look-up table, 395
loop constructs, 224
Lorentz force, 28
low-pass filters
  first-order, 49
  second-order, 50
  third-order, 131
low-power operating mode, 186
  stop mode, 186
  wait mode, 187
low-voltage reset (LVR), 285
lumen, photometric unit of
      light, 399
Lyapunov stability theory, 573

M

machine language, 3, 188
magnetic compasses, 414
magnetic coupling, 59
magnetic field, 25
magnetic field density, 26
magnetic poles, 471
magnetic reed switch, 426
magnetic sensors, 426
magnetostatic force, 28
magnetomotive force, 27
magnetoresistive effect, 28
manufacturing defects detection, 443
marginal stability, 533
maskable interrupt (see interrupts)
MATLAB, 538, 618
maximum overshoot, 534, 516
maximum power, 478
measurand (*see* stimulus)
measurement system 371
mechanical axis, piezoelectrics, 429
mechanical switches, 262
mechanisms, building blocks, 694
membership functions, fuzzy
      logic, 576
memory map, 187
mercury switch, 641
microelectromechanical systems
      (MEMS), 24, 444
microcontrollers, 179
  Freescale, 179
  Microchip PIC, 179
microfabrication, 403
microsensor, 429
microstepping (see stepper motors)
microstepping, 463
minimum reluctance, 453
MIT rule, 573
mnemonic, 189
model reference adaptive controller
      (MRAC), 572
modern control theory, 565
modulating sensors, 387
modulation (see signal conditioning)
monitor program, 285
motion control processors, 489
  HCTL 1100, 490
  LM628, 489
  LM629, 489

MOSFETs 88, 486, (see transistors)
  threshold voltage, 89
  on resistance, 90
motion-detection sensor, 451
motor commutation, 414
motor constant, 467
multiplexers, 166, 349
multiprecision arithmetic, 688
mutual inductance, 31

N

name, label field, 190
natural frequency, 535
negative charge carrier, 65
negative edge triggered, 108
negative logic, 104
negative temperature coefficient (see
      thermoresistive devices)
neon lamp, 101, 267
NIR receiver/demodulator
      sensor—G1U52X, 410
noise, 53
  random noise, 55
noise immunity, 149
non-inverting op-amp, 113
Norton equivalent circuit, 15
notch filter, 55
null modem, 307
Nyquist frequency, 357
Nyquist plot, 546

O

object code, 236
object detection, 408
observable system, 566
obstacle detection, 409
offset, 352
offset address, relative addressing
      mode, 201
Ohm's law, 15
one shot pulse, 320
op-amps, 107
  differential gain, 109
  gain-bandwidth product, 126
  ideal characteristics, 110
  bias currents, 127
  linear range, 147
  noise model, 128
  non-ideal behavior, 125
  offset voltages, 127
  power supply rejection ratio, 127
  symbol, 108
open circuit, 10
open-loop, 447
open-loop control 412, 487, 529
open-loop transfer function, 537
operand, 186
operand field, (see assembly
      language line)
operating modes, 9S12C, 185
optical encoders, 410
  absolute, 411
  code disk, 411
  incremental, 412
  S5D, 412
  HEDS series, 412
  HCTL 20xx, 412
  PED, 412

optical interrupter, 392
optical isolator, 59
optical reflector, 392
optical switch (see optical interrupter)
optimum gear ratio (see gear ratio)
optimum pitch (see lead screw drive)
optimum pulley radius (see belt-pulley
      drive)
optocoupler, 59, 101
oscillator circuit, 136, 170, 175
output compare function, 9S12C, 324
  registers, 325
  setup, 326
  operation, 326
output compare/input capture/PWM,
      PIC, 251
output data, 257
output impedance, 37, 432
output latch, 257
output port expansion, SPI, 315
overdamped response, 51, 518

P

parabolic function, 536
parallelogram mechanism, 697
parity, 297
passband, filters, 48, 128
passive circuits, 11
passive filters, 47
peak current, 490
peak inverse voltage (PIV), 67
peak time, 534
peak torque, 485
peripherals, 256
period, 32
period measurement, 330
permanent magnet stepper motors, 493
permanent magnets, 471
permeability, 26
permittivity, 21, 429
phase, 449
phase locked loop (PLL), 170
phase margin, 545
phase quadrature, 412 (see also
      optical encoders)
phase shift, 129
phasor, 35
photocell, 400
  interfacing to the 9S12C, 401
  materials, 401
photoconductive effect, 400
photocurrent, 75, 389
photodiode, 75, 402
  applications, 404
  characteristics, 403
  operating modes, 403
photodiode, types, 402
  avalanche, 403
  p-i-n, 403
  p-n, 403
  Schottky, 403
photometric units of light, 399
photon, 398
phototransistor, 86, 405
  characteristics, 405
  applications, 406
photovoltaic effect, 75, 385
PIC microcontrollers, 179, 248
  development suite, 252
  16F84A, 613

16F818, 639
18F877A, 640
PID control, 557, 575, 658
Piezoelectric materials, 444
    constitutive relations, 444
    poling, 444
piezoelectric effects
piezoelectric actuator, 651
piezoelectric microphone, 439
piezoelectric sensor, 445, 652
    amplification, 446
    mass sensitive chemical sensor, 447
piezoresistive effect, 431
pipelining, 197, 247, 250
pitch angle, stepper motor, 452
Plank's constant, 399
p-n junction (see diode)
polarization, 429
pole-placement technique, 559
polling, programming (see programming
        strategies)
poling, (see piezoelectric materials)
port integration module (PIM), 257
port input register, 258
position sensors, 386, 417
position quantization, 483
positive charge carrier, 65
positive logic, 104
positive temperature coefficient
        (see thermoresistive devices)
potential difference (*see* voltage)
potentiometer, 17, 396, 412
    gain, 396
    interface to 9S12C, 398
    resolution, 397
power, 9, 691
power diode, 67
power dissipation, 150
power factor, 42
power op-amps, 133
power supply, 69
power supply rejection ratio (PSRR), 165
power transistors, 89
    bipolar, 89, 487
    Darlingtons, 91, 93, 94, 510
    MOSFETs, 89, 487, 510
power variables, 691
power-on reset (see resets)
pressure sensors, 386
principle of virtual work, 24
probability, 388
program counter (see CPU registers)
program hierarchy, 235
programming basics, 188
programming strategies 234
    interrupt driven, 234
    polling, 234
    state machines, 235
propagation delay, 150
propagation frequency, 388
propeller, 629
proximity sensors, 386, 440
proportional control (see PID control)
proportional-derivative control (see PID
        control)
proportional-integral control (see PID
        control)
proportional-to-absolute temperature
        (PTAT), 66, 423, 425
pull device enable register, 258
pull-in torque, stepper motors, 508
pull-out torque, stepper motors, 508

pull-up resistor, 115, 153, 258, 263, 393
pulse accumulator (PA), 322, 337
    enable, 338
    registers, 337
    event counting mode, 338
    gated time accumulation mode, 339
pulse code modulated (PCM) signals, 490
pulse generator, 43
pulse width measurement, 329
pulse width modulation (PWM), 331, 342
    clock setting, 345
    duty cycle, 343
    emergency shutdown, 346
    enable, 342
    operation, 347
    output form, 345
    period, 343
    polarity, 343
PWM signal, 331, 468
PWM switching amplifiers, 487
pyroelectric sensor, 412
    package, 413
    signal conditioning, 414

Q

quadrature count (see incremental
        encoders)
quality factor, 50
quantum detectors, 400
quartz crystal microbalance, 447
quiescent point, transistors, 81

R

R—2R ladder network, DAC, 364
radiation, heat transfer, 426
radiation thermopile, passive IR
        detector, 418
radio frequency, 381
radiometric units of light, 399
ramp function, 536
ramping, stepper motor, 506
random noise (see noise)
range, 352, 364
range detectors
    GP2D02, 612
    GP2D120, 643
Rapson's slide mechanism, 698
reactance, 35
reactive power, 42
real power (see active power)
real-time control, 229
real-time clock, 341
receive data register, SPI, 292
receive operation, 298
rectification, 67
rectifiers
    half-wave 67, 124
    full wave 69, 124
reduced drive register, 258
reference ground, 11
registers, 164
relative addressing mode (see addressing
        modes)
relative offset, relative addressing
        mode, 203
relative permeability, 26
relative stability, 515
relays, 29

relocatable code, 236
reluctance, 27
repeatability, 377
reset interrupt, 282
reset sequence, 287
resets
    clock monitor reset, 287
    computer operating properly
        (COP), 286
    external pin reset, 285
    power-on reset (see resets), 286
resistance, 15
resistance temperature detectors
        (RTD's), 419
    fabrication, 420
    resistance-temperature relation, 420
resistivity, 15, 431
resistance-based sensors, 20
resistors, 15
resolution, sensors (see sensor
        characteristics)
resolution, ADC (see analog-to-digital
        conversion)
resolver, 32, 448
response (see sensors characteristics)
response time (see sensors characteristics)
return from interrupt, 281
reverse breakdown voltage, 67
reverse saturation current, 388
ringing, 38, 164
ripple band, filters, 129
rise time, 534
root locus method, 537
root mean square (RMS), 34
rotary encoders, 397
rotor (*see* armature)
Routh stability criteria, 533
RS232 (EIA 232), 296, 306
    MAX232 307, 613

S

samarium cobalt, permanent magnets, 472
sample-and-hold (S/H) circuit, 358
sampling frequency, 341, 356
scan code, 254
Schmitt trigger, 122, 264, 393
Scotck-yoke mechanism, 698
second-order system, 389, 375, 534
Seebeck effect, thermocouples, 415
self-generating sensors, 387
self-inductance, 29
self-tuning regulators, control, 573
semiconductor material, 64, 400
semiconductor thermocouples, 416
sensing junction, thermocouples, 400
sensitivity (see sensor characteristics)
sensitivity derivative, 573
sensitivity drift, 390
sensors
    classification, 386
    models, 388
sensor characteristics
    accuracy, 392
    dynamic response, 392
    error, 392
    full-scale input, 391
    gain, 390
    linearity, 391
    repeatability, 391
    resolution, 391

sensor characteristics (*continued*)
  response, 391
  response time, 391
  precision, 392
  sensitivity, 390
  span, 391
  threshold, 391
  uncertainty, 392
sequential circuit, 140
serial communication, COM 1 port, 618
serial communications interface (SCI),
    9S12C, 296
  configuration, 302
  registers, 298
  receive operation, 304
  transmit operation, 303
serial data, 290
serial interface, 295
serial peripheral interface (SPI),
    9S12C, 307
  operation, 313
  registers, 309
  topologies, 311
series resistance (*L/nR*) drive, stepper
    motors, 457
servo amplifiers
  "T"-type, 479
  "H"-type, 479
  selection, 481
servo drive, 482
servo motors (see DC servos)
settling time, stepper motors, 505
settling time, 534
seven segment displays, 245
shape memory alloy (SMA), 649
shift register, 315, 514
short circuit, 10
short pulse, 328
shot noise, 55
sigma-delta (Σ-Δ) ADC, 362
signal common (see reference ground)
signal diode, 67
signal generator, 43
signal ground 12, 395
signal conditioning, 106, 385, 392
  amplification, 393
  conversion, 393
  filtering, 393
  grounding, 395
  impedance buffering, 394
  isolation, 395
  linearization, 394
  modulation/demodulation, 394
signal termination, 164
signed numbers arithmetic, 684
silicon bilateral switch (SBS), 99
silicon controlled rectifier (SCR), 97
  gate threshold voltage, 97
  threshold current, 97
  gate controlled turn-on time, 97
  latching current, 98
silicon unilateral switch (SUS), 101
simple harmonic motion, 699
simple sensors, 387
simulator, 237
Simulink, 581
simplex communication, 296
single input channel, 356
skirt, filters, 129
slew range, stepper motors, 508
slew rate, stepper motors, 506
slew rate, op-amps, 109

slewing error, 360
slewing motion, stepper motors, 505
slider-crank mechanism, 697
slip speed, AC induction motors, 516
small-signal resistance, 18
smart sensors, 387
snubber circuit, 99
solar cell, 75, 390
solenoid, 30, 290
sonar, 440
sound, 434
sound intensity, 435
sound pressure level, 436
source, 10
source clock, 355
source code, 196
  structure, 235
space, 300
span, 352, 377
speed measurement, 341
speed of light,
speed of sound, 434
SPI subsystem, 9S12C, 301
square wave, 328
stack overflow, 190, 213
stack pointer (SP) (see CPU registers)
stall torque, 468
standard, 257
state variable feedback, 550
state space model for the DC motor
    470, 565
state space control, 565
static position error, stepper motor, 505
static torque, 460
stator, 462
steady state accuracy, 535
steady state response 34, 511
steady state error, 536
steering
  skid, 605
  explicit, 605
step angle (see stepper motors)
step function, 536
step rates (see stepper motors)
stepper motor control, 467
  2559, 467
  A3966SA, 468
stepper motors, 491
  back emf suppression, 500
  characteristics, 491
  classification, 491
    variable reluctance, 492
    permanent magnet, 493
    hybrid, 493
    linear steppers, 493
  drive methods
    L/nR drive method, 501, 509
    chopper-drive, 501
  driver ICs, 513
  microstepping, 507
  MS23C motor, 509
  operating principle, 494
  performance, 503
  selection, 508
  step angle, 498
stimulus, 385
stopband, filters, 48, 129
strain, 386, 428
strain gauges, 430
  dimensional change, 431
  gauge factor, 431
strain sensors, 386

stress, 428
subroutine call and return instructions
    (see instructions set)
successive-approximation ADC, 363
summing amplifier, 156
susceptance, 35
switch bounce, 264, 253
switch debounce, 159, 264
switches, 245
switching transistors, selection, 486
synchronous sequential circuits, 140
system stability, 532
system type, 537

T

tach gain, 449
tach feedback, 484
tachometer, 449
target system, 3, 236
temperature sensors, 386, 481
temperature coefficient of bridge arm
    resistance, 418
temperature coefficient of resistance
    (TCR), 418
temperature coefficient of sensitivity
    (TCS), 417
tendon drives, 705
thermal conductivity, 411
thermal detectors, 400, 414
thermal gradient, 411
thermal noise, 55
thermal resistance, 481
thermal runaway, BJT transistors, 96
thermistor, 191, 219, 404
thermocouples, 415
  types, 416
  junction types, 417
  cold-junction compensation (CJC), 417
  interface, 417
thermodiode, 423
thermopile, 418
thermoresistive devices, 418
  resistive temperature detectors, 419
  thermistors, 421
  negative temperature coefficient
    (NTC), 418
  positive temperature coefficient
    (PTC), 418
thermotransistor, 424
  AD590, 425
Thevenin's equivalent circuit, 14, 18
three-state output, TTL, 153
  electrical characteristics, 154
  high-impedance state, 154
through variable, 692
thyristors, 97
time constant, 44
time delay, 322
time-delay circuit, 45
timeout period, 286
time response, 533
timer, 182, 251, 321
timer module (TIM), 9S12C, 322
  enable, 322
  overflow, 324
  prescaler select bits, 316
  timer-counter register, 323
time of flight measurement, 440
toroid, 31
torque constant, 466

torque ripple, 467
tracks, 396
traction drive, 702
traffic-light control, 228, 241
Transducer to Electronic Data Sheet
    (TEDS), 387
transfer function, 389, 470, 483,
    528, 532
transformer, 32, 59
    step-down, 32
    step-up, 32
    power rating, 33
    regulation factor, 33
transient response, 533
trans-impedance amplifier, 403
transistors, 76
    bipolar junction transistors (BJTs), 76
    field effect transistors (FETs), 87
transistor switch (see bipolar junction
    transistors)
transit-time ultrasonic flow meter, 441
transition region, filters, 48, 129
transmission lines, 164
trap filter, 55
TRIACs, 100
triboelectric effect, 8
tricycle robot, 611
trim pot, 17
trimming capacitor, 25
truth table, 141
TTL logic family, 151
    designations, 151
    characteristics, 154
    open-collector, 153
    output configurations, 151
    three-state outputs, 153
    totem-pole, 151
    versions, 151
TTL-to-CMOS interface, 157
tunnel diode, 67
two-phase motors, (see AC induction
    motors)

U

ultrasonic tof ranging system, 425
ultrasound, 434
uncertainty (see sensor characteristics)
undamped natural frequency, 51
underdamped response, 51, 518
unifilar winding, 494
unijunction transistor (UJT), 101
unipolar ADC, 352
unipolar power supply, 479
unipolar step motor, 494
    driver ICs, 513
unipolar signals, 359
unit ramp response, 539
unit step response, 523
unity-gain amplifier, 152, 380
universal asynchronous receiver
    transmitter (UART) device, 295
unsigned numbers arithmetic, 684

V

varactor, 72
variable reluctance stepper motor, 492
velocity sensors, 386
velocity profile, 473
    parabolic, 473
    trapezoidal 473, 607
    triangular, 473
velocity ratio, 702, 704
virtual ground, 150
viscous damping, 470
voltage, 8
voltage amplifier, 431, 481, 657
voltage coefficient, 20
voltage constant, 466, 499
voltage divider, 13, 16, 36
voltage references, 354
voltage regulator, 70, 73
voltage source, 10

voltage-threshold circuit, 279
    MAX817, 279
voltage-to-current converter, 454

W

waiting-time limit, 283
watch dog timer (WDT), 285
water level sensor, 434
wattles power (see reactive power)
wavelength, 384, 419
waves, electromagnetic, 398
waves, acoustic, 434
wheatstone bridge, 406, 416
white noise, 42
    band-limited, 42
    Gaussian, 42
Wien's network, 54, 137
windings, 463, 493
wire ANDing, 153
wireless modem, 631
wireless surveillance balloon, 604, 621

X

XIRQ nonmaskable interrupt, 283, 288

Z

zener diode, 19, 72, 124, 500, 502
zener current, 73
zener voltage, 72
zero drift, 391
zero offset (see bias)
zero-order sensor, 388
zero-order hold, 558
zero-pole mapping, 575
Ziegler-Nichols rules (see PID
    Controllers design), 557